Earth Observation (EO), Remote Sensing (RS), and Geoinformation (GI) Applications in Svalbard

Earth Observation (EO), Remote Sensing (RS), and Geoinformation (GI) Applications in Svalbard

Editors

Shridhar Jawak
Andreas Kääb
Veijo Pohjola
Hiroyuki Enomoto
Geir Moholdt
Kjell Arild Høgda
Małgorzata Błaszczyk
Bo N. Andersen
Ann Mari Fjæraa
Bartłomiej Luks
Roberto Salzano
Frode Dinessen

MDPI • Basel • Beijing • Wuhan • Barcelona • Belgrade • Manchester • Tokyo • Cluj • Tianjin

Editors

Shridhar Jawak
Svalbard Integrated Arctic Earth Observing System
Norway

Andreas Kääb
University of Oslo
Norway

Veijo Pohjola
Uppsala University
Sweden

Hiroyuki Enomoto
National Institute of Polar Research
Japan

Geir Moholdt
Norwegian Polar Institute
Norway

Kjell Arild Høgda
NORCE Technology
Norway

Małgorzata Błaszczyk
University of Silesia in Katowice
Poland

Bo N. Andersen
Norwegian Space Agency and Svalbard Integrated Arctic Earth Observing System—Knowledge Centre
Norway

Ann Mari Fjæraa
Norwegian Institute for Air Research
Norway

Bartłomiej Luks
Institute of Geophysics Polish Academy of Sciences
Poland

Roberto Salzano
National Research Council of Italy
Italy

Frode Dinessen
Norwegian Meteorological Institute
Norway

Editorial Office
MDPI
St. Alban-Anlage 66
4052 Basel, Switzerland

This is a reprint of articles from the Special Issue published online in the open access journal *Remote Sensing* (ISSN 2072-4292) (available at: https://www.mdpi.com/journal/remotesensing/special_issues/SIOS).

For citation purposes, cite each article independently as indicated on the article page online and as indicated below:

LastName, A.A.; LastName, B.B.; LastName, C.C. Article Title. *Journal Name* **Year**, *Volume Number*, Page Range.

ISBN 978-3-0365-7418-9 (Hbk)
ISBN 978-3-0365-7419-6 (PDF)
Cover image courtesy of Anirudh Somadas.

Contents

Preface to "Earth Observation (EO), Remote Sensing (RS), and Geoinformation (GI) Applications in Svalbard"

The Arctic region has become a focus of international attention in recent years due to its rapidly changing environment and its crucial role in global climate change. Svalbard, an archipelago situated in the Arctic Ocean, has been particularly affected by these changes, with rising temperatures, melting ice, and changing ecosystems. As a result, the need for accurate and reliable monitoring of the region has become increasingly important.

Earth observation and remote sensing technologies have been critical in providing scientists with the data needed to understand and monitor the changing environment of Svalbard. This book provides an overview of the current status of Earth observation and remote sensing applications in Svalbard, highlighting their importance and potential for future research. The book covers a range of topics, including the use of satellite imagery for monitoring glaciers and changes in vegetation, the use of spaceborne and ground-based sensors for measuring permafrost and snow cover, and the integration of various data sources to provide a comprehensive understanding of the region's environment.

The book also includes contributions from experts in the field, providing insights into the challenges and opportunities associated with Earth observation and remote sensing in Svalbard. Their perspectives on the current state of the field and the potential for future research will be valuable for students, researchers, and policymakers alike.

Overall, this book provides a comprehensive overview of the current state of Earth observation and remote sensing applications in Svalbard, highlighting the importance of these technologies in understanding and monitoring the rapidly changing Arctic environment.

Shridhar Jawak, Andreas Kääb, Veijo Pohjola, Hiroyuki Enomoto, Geir Moholdt, Kjell Arild Høgda, Małgorzata Błaszczy, Bo N. Andersen, Ann Mari Fjæraa, Bartłomiej Luks, Roberto Salzano, and Frode Dinessen
Editors

remote sensing

Editorial

Status of Earth Observation and Remote Sensing Applications in Svalbard

Shridhar D. Jawak [1,*], Veijo Pohjola [2], Andreas Kääb [3], Bo N. Andersen [4], Małgorzata Błaszczyk [5], Roberto Salzano [6], Bartłomiej Luks [7], Hiroyuki Enomoto [8], Kjell Arild Høgda [9], Geir Moholdt [10], Frode Dinessen [11] and Ann Mari Fjæraa [12]

1 Svalbard Integrated Arctic Earth Observing System (SIOS), SIOS Knowledge Centre, Svalbard Science Centre, Longyearbyen, P.O. Box 156, N-9171 Svalbard, Norway
2 Department of Earth Sciences, Uppsala University, Geocentrum, Villavägen 16, 752 36 Uppsala, Sweden
3 Department of Geosciences, University of Oslo, Blindern, P.O. Box 1047, N-0316 Oslo, Norway
4 Institute of Theoretical Astrophysics, University of Oslo, Blindern, P.O. Box 1047, N-0316 Oslo, Norway
5 Institute of Earth Sciences, University of Silesia in Katowice, Bedzinska 60, 41-200 Sosnowiec, Poland
6 Institute for Atmospheric Pollution Research, National Research Council (CNR-IIA), Sesto Fiorentino, 50019 Florence, Italy
7 Institute of Geophysics Polish Academy of Sciences, Ksiecia Janusza 64, 01-452 Warsaw, Poland
8 National Institute of Polar Research, 10-3, Midori-cho, Tachikawa-shi, Tokyo 190-8518, Japan
9 NORCE Norwegian Research Center AS, Sykehusvn 21, N-9019 Tromsø, Norway
10 Norwegian Polar Institute, Fram Centre, Langnes, P.O. Box 6606, N-9296 Tromso, Norway
11 Norwegian Meteorological Institute, Blindern, P.O. Box 43, N-0371 Oslo, Norway
12 Atmosphere and Climate Department, NILU-Norwegian Institute for Air Research, P.O. Box 100, N-2027 Kjeller, Norway
* Correspondence: shridhar.jawak@sios-svalbard.org

Citation: Jawak, S.D.; Pohjola, V.; Kääb, A.; Andersen, B.N.; Błaszczyk, M.; Salzano, R.; Luks, B.; Enomoto, H.; Høgda, K.A.; Moholdt, G.; et al. Status of Earth Observation and Remote Sensing Applications in Svalbard. *Remote Sens.* **2023**, *15*, 513. https://doi.org/10.3390/rs15020513

Received: 17 December 2022
Revised: 6 January 2023
Accepted: 10 January 2023
Published: 15 January 2023

1. Introduction

Remarkable developments in the fields of earth observation (EO) satellites and remote sensing (RS) technology over the past four decades have substantially contributed to spatial, spectral, and temporal sampling. This has enabled the subsequent retrieval of geographical information (GI) to better understand the current state of the Arctic. Massive data gaps exist in Svalbard, even if it is probably one of the most in situ data-rich regions in the Arctic. These gaps can be filled by installing new research infrastructure (RI), acquiring regular observations, generating new geospatial products using EO and RS, and integrating in situ data with EO-based information. The Svalbard Integrated Arctic Earth Observing System (SIOS) is an international consortium of 28 member institutions from 10 countries, tasked with developing an efficient observation system for long-term in situ and RS measurements in Svalbard and associated waters to address key Earth system science (ESS) questions. RIs, associated with or funded by SIOS, are scattered all over Svalbard to acquire long-term in situ observations. In situ measurements, collected as a part of the observing system, provide value for a variety of existing (e.g., Sentinel and Landsat series) and upcoming (e.g., CRISTAL, CIMR and ROSE-L) satellite missions. The function of these data in such situations are calibration and validation (Cal/Val) efforts and use in ESS modelling community. Such integration of in situ and satellite-based measurements is beneficial for Svalbard science community in addressing broader scientific questions in Arctic science. Our Special Issue (SI) was designed to provide a comprehensive platform for researchers to present regional and Svalbard-wide cross-disciplinary studies, conducted using EO and RS. The SI was focused on attracting ground/field-, underwater-, space-, and airborne platform-based RS studies of Svalbard.

The SI offered broad possibilities to potential authors and encouraged contributions in all areas of Svalbard science, including the following: (1) EO and RS studies pertaining to field studies/campaigns, ESS modelling, and long-term monitoring programs in terrestrial, marine, atmospheric, and cryospheric environments of Svalbard; (2) RS-based studies on cross-sphere interaction of cryosphere with ocean, terrestrial, and atmosphere; (3) The derivation of geophysical and biophysical parameters e.g., chlorophyll concentration, eddies information, snow cover dynamics, vegetation growth, sea ice drift and type, phytoplankton blooms using satellites; (4) Svalbard-wide GI retrieval including geospatial product generation and operationalization using optical (e.g., Sentinel and Landsat series), microwave (e.g., SAR) and Lidar (e.g., ICESat-1/2) applications in Svalbard; (5) Cal/Val activities for various satellite missions, e.g., Pandora (https://www.pandonia-global-network.org/, accessed on 12 January 2023) installation, validation of snow parameters derived using satellite, Cal/Val activities using ocean moorings; (6) Integration of RS, in situ, modelling and previously available GI to advance new knowledge about Svalbard; (7) Artificial intelligence (AI) including deep learning (DL), machine learning (ML), neural networks (NN) and cloud computing-based applications in Arctic areas; (8) RS applications in glaciology including snow cover and its properties, geodetic glacier mass balance, mapping glacier facies, deriving glacier surface elevation changes, etc.; (9) RS of terrestrial and marine cryosphere including snow/firn/ice, sea ice, snow on sea ice, icebergs, ground ice, avalanche studies, and permafrost subsidence studies; (10) RS of terrestrial vegetation, estimating vegetation growing season and primary productivity, mapping vegetation abundance and extent, and time series analysis of terrestrial vegetation; (11) Applications of very high resolution (VHR) satellite and airborne platforms in Arctic areas, including use of airborne imagery and hyperspectral data acquired by Dornier research aircraft, VHR satellites (e.g., WorldView, Planet) and uncrewed aerial vehicles (UAVs); (12) Applications of new technologies, such as robots, autonomous underwater vehicles (AUVs), drone-based mapping using surface from motion (SfM), and terrestrial laser scanners (TLS).

2. Challenges Related to Svalbard Studies and Possible Solutions from SIOS Remote Sensing Activities and Special Issue

2.1. Challenges Related to In Situ Data Collection and Integration with RS Data

The regular collection of in situ measurements by field scientists using field campaigns and ground-based RI is useful for space agencies, Arctic scientists and RS experts for validating geospatial products in Svalbard. Conversely, Arctic field scientists benefit from the geospatial products derived using RS data to widen the scope of their study beyond limited field observations. Therefore, a dedicated platform for facilitating communication between field scientists and RS experts in Svalbard is necessary to make the best use of in situ measurements and RS resources. Presently, SIOS facilitate such dialogue between field scientists and RS experts through its online conferences (https://sios-svalbard.org/RS_OnlineConference, accessed on 12 January 2023) and regular webinars (https://sios-svalbard.org/WebinarSeries, accessed on 12 January 2023). In future, we anticipate launching a dedicated platform or discussion forum on the SIOS website to facilitate discussion between field scientists and RS experts, hopefully widening the scope and usage of RS data in the Svalbard science community.

Even if Svalbard is one of the most in situ data-rich regions in the Arctic, there are gaps that are limiting modelling-based studies. For example, most of the in situ data in Svalbard are clustered in coastal regions and are very sparse in the interiors of Svalbard. More datasets are required from the more remote eastern part of Svalbard, as there are huge differences compared to western parts of Svalbard. Most of the field campaigns are focused on summer seasons, causing an imbalance in intra-seasonal measurements. Due to changing weather conditions and logistical constraints, conducting fieldwork matching with satellite overpasses remains a challenging exercise. Most of the observations are clustered around field bases such as Longyearbyen, Ny Ålesund and Hornsund, causing data gaps in other areas of Svalbard. Filling spatial and temporal gaps in terms of in situ measurements will facilitate further modelling-based studies in the future. Regular maintenance of ground-based RIs is one of the challenges facing researchers, specifically during wintertime and periods of travel restrictions, such as the pandemic. For example, repair and maintenance of several sensors may take months to reinstate the equipment because of logistics, causing gaps in measurements during unexpected failure and maintenance periods. The cost of logistics in Svalbard is one of the challenges for Svalbard researchers. However, generous field grants and access to RI by Svalbard Science Forum (https://www.forskningsradet.no/en/svalbard-science-forum/, accessed on 12 January 2023), SIOS access programme (https://sios-svalbard.org/AccessCall_2023, accessed on 12 January 2023), and INTERACT trans-national access (https://eu-interact.org/, accessed on 12 January 2023) have been useful for supporting field activities in Svalbard.

2.2. Technological Challenges and Innovation

Some of the practical solutions to overcome the challenges, outlined above, for conducting field-based measurements include (1) use of remote or autonomous platforms; (2) reduction in power consumption for remote stations using technological innovation; (3) technical solutions for communication with RIs/sensors in remote locations; (4) development of new measurement methods or infrastructure that can replace manual fieldwork that has a high environmental impact; (5) pilot efforts for community-based observation to support ESS monitoring; (6) cooperation with a manufacturer to further develop existing technology and protocols. To implement these solutions and overcome practical challenges in data collection, SIOS and SSF launched the call for proposals in 2021 (https://sios-svalbard.org/InnovationAward, accessed on 12 January 2023) to promote the development of an innovative technology, solution or method to advance observation capability or reduce the environmental footprint of research and monitoring in Svalbard. The funding awarded supports the development of new technologies, methodologies or means of using products and data within Svalbard science and infrastructure operations.

2.3. Challenges Related to Sunlight Dependent Optical Sensing

Varying resolutions of optical data are useful for many applications in Svalbard. However, traditional issues such as frequent cloud cover and light conditions over Svalbard, limit the usage of sunlight-dependent optical sensing. Time-lapse cameras, UAVs and aircraft-based data can be utilized to complement satellite-based observations; however, such observations can cover only limited areas in Svalbard in one field season. Besides, aircraft-based measurements are also weather-dependent, and low-lying clouds can seriously hamper the quality of the data. Multi-polarization, multi-frequency, and multi-temporal radar observations can be useful to fill data gaps during both cloudy days in summer and limited light conditions in winter. SAR guarantees ground-covering (cloud-free) images, both during summer and dark winter in Svalbard. The failure of Sentinel-1B in December 2021 has impacted the frequency of SAR observations in Svalbard, a situation which is expected to be improved with the launch of Sentinel 1C in 2023.

2.4. Challenges Related to VHR Observations in Svalbard

Very high resolution EO satellite data are mainly available through commercial providers, an arrangement which necessitates dedicated funding for research projects. Most VHR satellite observations are carried out on a request-by-request basis, rather than via continuous monitoring, causing large data gaps in Svalbard. This affects the studies involving temporal change detection, such as surface changes of glaciers, calving events, vegetation dynamics, and many more topics. To address this, SIOS has recently launched the project proposal call with Planet (https://sios-svalbard.org/SIOS_PlanetCall2023, accessed on 12 January 2023) to provide free access to high-resolution optical satellite data, acquired by Planet's satellite constellation (https://www.planet.com/, accessed on 12 January 2023), to stimulate geospatial product generation in Svalbard. Similar to the SIOS–Planet call, SIOS also looks forward to cooperating with other commercial satellite data providers to facilitate the usage of VHR data in Svalbard. SIOS also promotes the usage of UAVs and airborne campaigns using Dornier aircraft (https://sios-svalbard.org/AirborneRemoteSensing, accessed on 12 January 2023) to collect high-resolution data. Since 2020, SIOS has supported around 20 scientific projects by coordinating 50 flight hours to acquire airborne data using the Dornier aircraft and UAVs in Svalbard. Coordinated flight campaigns by SIOS have enabled the optimized usage of aircraft to reduce the environmental footprint of observations, facilitate international collaboration within the Svalbard science community, train the next generation of polar scientists using the collected data, and fill gaps in observations that occurred as a result of pandemic-related travel restrictions.

2.5. Challenges Related to Methods and Applications

Most of the studies published in this Special Issue focus on regional aspects of Svalbard. Methods, applications and data used in these studies should be scaled up to cover the entire archipelago in order to broaden the scope of the study. However, it remains a challenge to address scaling issues because of the variety of issues involved, including varying spatial/spectral/temporal resolutions of satellite and in situ data. Ideally, methods adapted to regional studies such as mapping vegetation or glaciers using VHR data should be scaled up for the entire archipelago using the available coarse resolution or medium resolution satellite data. However, only limited attempts have been demonstrated in this Special Issue. Several manuscripts have shown the potential of having their methods upscaled for Svalbard-wide studies. Such challenges can be circumvented by the colocation of RIs, filling in situ data gaps, improving coverage of in situ data, frequent satellite observations, and the integration of data. To some extent, these challenges can also be addressed using better data management and making data available with FAIR principles. The SIOS Data Management Service (https://sios-svalbard.org/Data, accessed on 12 January 2023) and observation facility catalogue (https://sios-svalbard.org/sios-ri-catalogue, accessed on 12 January 2023) provide effective platforms for making data available to the scientific community in Svalbard.

2.6. Challenges Related to Access Restrictions Such as the Pandemic

Situations such as the pandemic may recur, causing temporary gaps in field measurements due to travel restrictions. To address this, an overview of the maturity of EO technology and its ability to fill in situ gaps using space-borne observations is necessary. In this direction, SIOS started supporting researchers in filling gaps in data using RS observations during the pandemic through a dedicated service (https://sios-svalbard.org/PatchUpData, accessed on 12 January 2023). Besides, SIOS is currently setting the stage for developing citizen science-/community-based observation efforts to involve residents. These efforts will not only involve residents in scientific observation but also complement science studies and long-term monitoring in Svalbard. Such community-based observations will be useful, particularly in future pandemic situations.

3. Overview of Manuscripts Published in Special Issue

Twelve experts from five countries representing nine SIOS member institutions and SIOS-Knowledge Centre (SIOS-KC) have edited this SI. This SI attracted 24 manuscripts (more than 100 co-authors across the globe) over the span of two years (March 2020—June 2022) as an outcome of its broader scope. SIOS also hosted two online conferences on EO- and RS-based studies in Svalbard to attract scientific presentations and publications from the Svalbard science community. After careful review and subsequent revisions, the editorial expert group accepted 19 manuscripts in this volume. All the accepted papers in our SI have the potential to stimulate our knowledge about Svalbard. An accidental result of the unexpected timing of launching the SI at the beginning of the pandemic ultimately presented a stage for Arctic researchers to publish their RS-based studies, of particular value at a time when many researchers were constrained to their home offices. The present volume of the SI covers a wide range of applications of EO and RS in Svalbard, especially with regard to the usage of data, methods, infrastructure, and spheres of the ESS. The available topics included EO and RS applications in snow, glaciers, Arctic vegetation, ocean eddies, permafrost, glacier facies, aerosol, and lake phenology studies in Svalbard. These were investigated using a variety of methods and infrastructure, including interferometric synthetic aperture radar (InSAR), UAVs, TLS, photogrammetry and many more techniques. In brief, the SI consisted of various contributions focused on the usage of various remote sensing platforms and satellite sensors to develop applications in a wide span of environments (atmosphere, terrestrial, marine, and cryosphere) on Svalbard. The contributions are briefly described in the following sections.

3.1. Atmospheric Remote Sensing Applications

Anthropogenic climate change is leading to major changes in the Arctic environment. Arctic aerosol and its related feedback mechanisms are inseparably impacted, whereas the influence of the spatio–temporal variability of aerosols on the radiation budget of the Arctic has not yet been fully understood. To address this gap, Nakoudi et al. [1] exploited synergistic RS observations of a long-range aerosol transport episode in April 2018 using lidar and sun-photometer measurements over two parts of the European Arctic. The aim of the study was to investigate intra-Arctic aerosol alteration and assess its impact on the local radiative budget. The research showed that aged aerosol was transformed, even in neighbouring Arctic upper tropospheric levels, because of changes in the aerosol source regions (north Europe and north-east Asia) and the interaction of removal processes such as nucleation scavenging and dry deposition.

The near-surface warming rate in the Arctic has at least doubled (*Arctic amplification*) in comparison to elsewhere on our planet over the past decades [2] and is reported to be as much as four times the global average [3]. Clouds are an important modulator of the Arctic surface energy budget [4]. Cirrus formations are known to be the only cloud type to cause daytime cooling or heating at the top of the atmosphere. However, the present research into its geometrical and optical properties over the Arctic is inadequate. To address this gap, Nakoudi et al. [5] analyzed the long-term characteristics of cirrus clouds over Ny-Ålesund, Svalbard for the first time. The authors used lidar and radiosonde observations from 2011 to 2020, showing that cirrus layers were geometrically thicker in winter and spring. The authors suggest that for more Arctic cirrus studies in the American and Russian regions of the Arctic are necessary to evaluate potential intra-Arctic variations in cirrus properties.

3.2. Remote Sensing in Arctic Limnology

Arctic lakes are vulnerable to climate change. Ice phonological properties, such as timing and length of ice presence or absence on the surface of lakes, can be used as effective climate indicators. Tuttle et al. [6] compared ice phenology on the Linnévatnet lake surface using satellite-based data, photographs from automatic cameras, and constant observations of lake water temperature. Linnévatnet is one of the largest lakes on Svalbard. Moderate-resolution imaging spectroradiometer (MODIS)-derived surface reflectance data from 2003–2019 were utilized to identify the change in the mean surface reflectance of Linnévatnet. Besides, smoothing splines were used to determine the date of summer ice-off ("break-up end"—BUE). Subsequently, fall ice-on ("freeze-up end"—FUE) was determined using the mean time series (2014–2019) of Sentinel-1 microwave backscatter over the lake. In general, the ice timing dates determined from the satellite-based observations agreed with the in situ observations (RMSE values of $\approx$2–7 days for BUE and FUE). In brief, this regional study indicates that optical and SAR satellite-based measurements can be successfully used to observe changes in the lake ice in the High Arctic, and its results can be extended to study changes in lake ice in regional scales across the Arctic using different optical and SAR satellites and sensor-specific considerations.

3.3. Ocean Remote Sensing Applications

It is well documented that the Arctic coasts are affected by climate change, including the melting of glaciers and variations in precipitation and runoff. These changes are expected to impact productivity in fjordic estuaries. The study by Walch et al. [7] demonstrates the spatio-temporal variability of suspended particulate matter (SPM) in the Adventfjorden, Svalbard. The authors used in situ field measurements, taken in 2019 and 2020, along with Sentinel-2 imagery. This study highlighted the importance of RS in studying fluxes in light-attenuating particles, particularly in the coastal Arctic Ocean. This study also indicates that the Adventfjorden can serve as a representative study area for regions dominated by marine-terminating glaciers and largely glaciated catchments. The authors found that the fine-tuning of the ocean color RS algorithms using field-based measurements resulted in satisfactory estimates of spatiotemporal variability of SPM in the Adventfjorden.

Eddies that form at the ice edge and within marginal ice zones (MIZ) are common geophysical features of ice edge progression in polar oceans. Earlier studies indicate that MIZ eddies are widespread in polar ocean regions, such as the Fram Strait. These studies discuss the wide variety of methods for generating mechanisms eddies using available in situ, airborne, and satellite-borne data. Kozlov and Atadzhanova [8] used Envisat ASAR and Sentinel-1 to investigate the generation of eddies and their properties in the MIZ regions of the Fram Strait and around Svalbard in the winters of 2007 and 2018. The authors analyzed 2039 SAR images to identify 4619 eddy signatures. The authors found that the number of eddies detected in MIZ is similar for both years. Besides they found that the sub-mesoscale and small mesoscale eddies dominate, with cyclones identified twice as repeatedly as anticyclones. The range of Eddy diameters varied from 1 to 68 km, with mean values of 6 km for shallow and 12 km for deep water.

3.4. Remote Sensing for Permafrost Studies

The degradation of carbon-rich permafrost can contribute to global warming through the processing and emission of greenhouse gases (GHGs) that were confined to the frozen ground in the past. Changes in permafrost can also have instant impacts on infrastructure and ecosystems in the Arctic. Rouyet et al. [9] used Sentinel-1 SAR Interferometry (InSAR) to study the seasonal vertical displacement progression on Svalbard. Their study yielded Sentinel-1 InSAR data showing the spatial variability and timing of the seasonal thaw subsidence maxima in three regions of Svalbard. The study indicates the necessity for the development of regional InSAR products in Svalbard to study cyclic subsidence in permafrost areas. The authors examined the possibilities and shortcomings of using the timing of the maximal subsidence as an alternative for the end of the thawing season.

3.5. Glaciological Remote Sensing Applications

In order to understand the contribution of a surging glacier system to sea-level change it is necessary to study the evolution of ice dynamics, mass transfer within the glacier, surface elevation changes, glacial hydrology, and calving resulting in mass transfer from the glaciers to the ocean. A study by Herzfeld et al. [10] evaluated the capabilities of airborne ICESat-2 advanced topographic laser altimeter system (ATLAS) measurements in surface height estimations over crevassed glacial terrain of the surging Negribreen glacier in Svalbard. The authors collected airborne geophysical data in two field campaigns in the summers of 2018 and 2019. Their results of the validation indicated that ICESat-2 ATLAS data, processed with the density–dimension algorithm for ice surfaces (DDA-ice), facilitate surface–height estimations over crevassed terrain.

RS observations of dynamic glacial systems are effective tools for observing the overall changes that are constantly occurring in the cryosphere [11]. Glacier surface facies are among the especially important signatures of variations experienced by a glacial system. These surface facies are shaped by the natural ageing and redeposition of snow and its seasonal melting, and refreezing with dust and debris. Surface facies are easily recognized based on certain variations in surface reflectance values [11]. In a series of three papers, Jawak et al. [12–14] demonstrated the usage of VHR WorldView-2 and WorldView-3 satellites in mapping glacier surface facies on Svalbard and the Himalayas by standardizing image processing routines associated with pixel-based and object-based machine learning methods. These papers are expected to stimulate the operational mapping of glacier facies in the cryosphere using a variety of satellite sensors.

Almost 60% of Svalbard is glaciated and thus glaciers form most of the local ecosystem. Kavan et al. [15] studied tidewater glaciers on the south-east coast of Svalbard to deduce surface elevation changes and retreat rates using satellite images and historic maps. The authors used a 1970 digital elevation model (DEM) and an ArcticDEM mosaic generated from aerial images by the Norwegian Polar Institute. The authors showed that the tidewater glaciers under consideration have retreated at a substantial rate ranging between 10 and 150 m year^{-1}, with a mean retreat rate of 48 m year^{-1} between 1970 and 2019. The maximum retreat rate was revealed at Hambergbreen. The substantial retreat reported by this study revealed that four out of the eleven glaciers completely lost their physical link with the fjords/ocean, resulting in terrestrial glacier systems. Authors showed that the massive retreat of the glaciers has exposed many newly emerged coastlines that are presently being reshaped by coastal processes.

As stated previously, the Arctic is experiencing incessant and substantial variations in land relief, resulting from various geomorphological, glaciological, and hydrogeological processes. DEMs and precise high spatial resolution maps are of crucial importance to studying these changes. Błaszczyk et al. [16] assessed the accuracy of orthomosaics and high-resolution DEMs derived using photogrammetric methods on aerial images captured by Dornier aircraft in 2020 over Hornsund, Svalbard. The authors evaluated and combined the two methods based on terrestrial laser scanning and photogrammetry to produce a continuous and gap-free DEM for the diverse Arctic scientific community. This approach enabled authors to study geomorphological activity over a year, including alterations along the shoreline, the snow thickness in gullies, and the melting of ice cores in the periglacial zone. The study underlines the possibility of combining other RS methods to study the active processes in this region.

3.6. Arctic Studies Using UAVs

Uncrewed aerial vehicles (UAVs) are being progressively utilized as a tool to study rapidly changing arctic environments. Eischeid et al. [17] provided a comprehensive method for producing ground-cover maps using UAVs for effective ecological monitoring in a High Arctic tundra environment. Authors used random forest (RF) classifiers to map up to 15 different ground-cover classes, including two disturbance classes. These were goose grubbing and winter damage which were assessed based on lower NDVI values in comparison to their undisturbed counterparts. The authors showed that UAV image analysis can be an effective tool for studying rapid changes in vegetation at local scales in Arctic tundra ecosystems. The authors also advise ecologists to integrate UAV mapping into long-term ecosystem monitoring programs. The authors suggest stronger interdisciplinary cooperation between EO experts and ecologists to effectively blend the data, knowledge and information to further enhance the quality of ecological monitoring in the Arctic.

3.7. Arctic Vegetation Studies Using Remote Sensing

Variations in the timing of phenological phases in terms of the RS-based onset of vegetation growth are sensitive bio-indicators of climate change. To address this issue, Karlsen et al. [18] focused on the central part of Spitsbergen to study the onset of vegetation growth. The authors prepared, analyzed and presented a time series of daily NDVI products using cloud-free Sentinel-2. The NDVI 2016-2019 time series datasets were merged with ground-based phenocam observations for mapping the onset of growth. The Sentinel-2 NDVI-based mapping and analysis of the onset of growth revealed significant variations between the years. For example, except at higher altitudes, the onset of growth was more than 10 days earlier in 2018 in comparison with 2017. The data presented in this paper were not adequate to clarify these variations. Therefore, authors formulated a future study to examine the connection between early growing season temperature and the timing of snowmelt.

3.8. Snow Remote Sensing in Svalbard

The first day free of snow and the timing of snowmelt are considered reliable markers of the Arctic ecosystem's status in response to global warming. This is mainly because the last day of snow cover and snowmelt timing have a substantial impact on the phenological characteristics of vegetation in Svalbard. In an attempt to assess the seasonal changes in snow, Stendardi et al. [19] used a multi-sensor approach and analyzed the sensitivity of the SAR backscatter to vegetation growth and soil moisture. The authors showed that the C-band SAR data, combined with optical data, can be useful in monitoring the seasonal changes in the timing of snowmelt in Adventdalen. The results also verified that vegetation can be satisfactorily detected by the HV polarization channel. The late melting and resulting disappearance of snow impart a further challenge in the monitoring of vegetation and soil dynamics.

Changes in seasonal snow cover can be considered a sensitive marker of climate change. In response to the changing climate in Svalbard, seasonal snow cover is expected to undergo spatio-temporal changes owing to its sensitivity to temperature and precipitation changes. As vegetation is closely linked to snow cover, the timing of terrestrial productivity is expected to change. Vickers et al. [20] used MODIS Terra data for the period 2000–2019 to produce a 20-year snow cover fraction time series data for Svalbard to study changes in the timing of the first snow-free day (FSFD). The authors have investigated the impact of changes in sea ice concentration (SIC) on FSFD and the effect of FSFD on the phenological growing season. Their results showed obvious patterns of earlier FSFD in the southern and central parts of the archipelago. On the other hand, the northernmost regions exhibited minor changes toward later FSFD. Authors noticed that FSFD led to the onset of the phenological growing season, with an average variation of 12.4 days for the entire archipelago.

Snow models are complementary tools to EO- and RS-based observations for snow cover monitoring, and may be used to fill in spatio-temporal data gaps in field-based observations. In a study involving overlapping periods of data coverage, Vickers et al. [21] used three optical RS datasets and two snow models to study the similarities and inconsistencies in snow cover estimates over Nordenskiöld Land, Svalbard. A 20-year MODIS snow cover dataset was calibrated using Sentinel-2 data. Snow cover fraction estimates, derived using lower-resolution AVHRR and snow model datasets, were corrected using MODIS-based snow cover data. This study practically demonstrated the effectiveness of using current-day high-resolution datasets to improve the consistency of past low spatial resolution snow cover datasets.

3.9. Hyperspectral Remote Sensing in Svalbard

Field-based hyperspectral measurements of ice and snow cover are important for the calibration and validation activities of airborne and spaceborne hyperspectral imageries. However, such measurements are strongly affected by the limited accessibility of data services integrating spectral libraries with comprehensive descriptions of the observed surface cover. To fill this knowledge gap for snow and ice cover, Salvatori et al. [22] presented an updated version of the Snow/Ice Spectral Archive (SISpec 2.0). This version is integrated into a web portal (https://www.cnr.it/en/institutes-databases/database/1076/sispec-snow-ice-spectral-library-project, accessed on 12 January 2023), as exemplified by various functionalities such as accessibility and interoperability.

3.10. Remote Sensing in Response to the COVID-19 Pandemic

The coronavirus (SARS-CoV-2) disease 2019 (COVID-19) pandemic impacted research on Svalbard in a number of ways. In March 2020, Norway announced a nationwide lockdown, including stringent measures to protect the Svalbard community from the possible spread of the infection. Because of the lockdown, associated travel restrictions, and quarantine regulations announced by various countries, most of the field campaigns in Svalbard were cancelled by the first week of March 2020. Jawak et al. [23] highlighted the activities undertaken in response to the exceptional situation enforced by the global pandemic of COVID-19. This study provides a first crucial point of view on general response, potential wider impacts, consequences to other observing systems, and advice for future directions related to field-based research in Svalbard.

4. Vision and Direction for Future Remote Sensing Research in Svalbard

Nearly all polar-orbiting EO satellites pass over or at a marginal distance from Svalbard, enabling high satellite track coverage. Its strategic location has facilitated Svalbard to become the largest downlink site for polar-orbiting EO satellites. The high-latitude geographic location of Svalbard has had a tremendous impact on scientific and monitoring communities in three ways: (1) Regions of Svalbard and associated waters are frequently well covered by EO satellite data, (2) RI in Svalbard offers numerous possibilities for ground-based validation of satellite data for cross-disciplinary research in the Arctic, and (3) effective use of EO satellite data has no adverse impact on the environment. However, the growing volumes of in situ and RS data clearly show the need of using advanced methods in artificial intelligence to make the best use of the ground, airborne, underwater and space-borne research infrastructure in Svalbard. Future research projects should focus on aligning field campaigns and in situ data collection with upcoming satellite missions (e.g., NISAR, CRISTAL, CIMR and ROSE-L) to facilitate cal/val activities and the construction of a sustained EO-based observatory for Svalbard science.

Glacier surface types and zones can be mapped using SAR and optical sensors by applying either pixel- or object-oriented methods. In future, a fusion of various sensor (optical/SAR) data and methods to generate high-resolution spatio-temporal datasets should be considered. For summer seasons, high-resolution optical (multispectral and hyperspectral) data has shown great potential to map surface change. For winter seasons, frequent SAR studies are anticipated to complement optical sensor-based studies. A new generation of VHR optical sensors such as WorldView Legion, Planet, and Pléiades NEO with high temporal, spatial and spectral resolutions are expected to complement field data and medium-resolution optical data such as Landsat series, Sentinel series and MODIS.

Scaling issues across spatial/spectral/temporal scales in Svalbard studies will be essential going forward in stimulating knowledge acquisition from regional studies. This, in turn, will allow researchers to derive Svalbard-wide information using a variety of ground, airborne, underwater and space-borne sensors. Accordingly, training the new generation of cryosphere scientists and field scientists in EO and RS skills is essential. In this vein, SIOS hosts an annual RS training course (https://sios-svalbard.org/Training, accessed on 12 January 2023) on specific topics such as terrestrial/marine RS, hyperspectral RS, and artificial intelligence methods in EO.

Future studies on Svalbard may also consider filling gaps highlighted by several chapters in the State of Environmental Science in Svalbard (SESS) reports (https://sios-svalbard.org/SESSreport, accessed on 12 January 2023) and the recent SESS synthesis report (https://sios-svalbard.org/SESSRecommendationsSynthesis, accessed on 12 January 2023). Additional gaps in snow research in Svalbard have been highlighted in the snow agenda paper (https://sios-svalbard.org/SnowResearch, accessed on 12 January 2023). Future attempts are expected to fill these knowledge and data gaps using in situ and RS data. Several attempts have been already started such as the Snow Pilot project (https://sios-svalbard.org/SnowPilot2022, accessed on 12 January 2023). In situ observations will be enriched by the recently started Cryosphere Integrated Observatory Network on Svalbard (CRIOS) project (https://crios.pl/, accessed on 12 January 2023). Finally, cross-disciplinary studies across various spheres of Svalbard using in situ, space-borne, airborne and underwater platforms are expected to be conducted in future to address broad scientific questions in Earth system science.

Author Contributions: S.D.J. writing—original draft preparation; V.P., A.K., B.N.A., M.B., R.S., B.L., H.E., K.A.H., G.M., F.D., A.M.F.; writing—review and editing. All authors have read and agreed to the published version of the manuscript.

Funding: S.J. would like to acknowledge the support from the Research Council of Norway (RCN) through projects No. 291644 & No. 269927 and Norwegian Space Agency (NoSA) through contract No. 74CO2116. Authors from the Institute of Geophysics Polish Academy of Sciences would like to acknowledge the support from the Polish Ministry of Education and Science DIR/WK/2018/05-1.

Acknowledgments: We acknowledge the time and efforts of academic editors, assistant editors, and reviewers involved in reviewing manuscripts of the Special Issue.

Conflicts of Interest: The authors declare no conflict of interest.

References

1. Nakoudi, K.; Ritter, C.; Böckmann, C.; Kunkel, D.; Eppers, O.; Rozanov, V.; Mei, L.; Pefanis, V.; Jäkel, E.; Herber, A.; et al. Does the Intra-Arctic Modification of Long-Range Transported Aerosol Affect the Local Radiative Budget? *(A Case Study)*. *Remote Sens.* **2020**, *12*, 2112. [CrossRef]
2. Serreze, M.C.; Barry, R.G. Processes and impacts of Arctic amplification: A research synthesis. *Glob. Planet. Chang.* **2011**, *77*, 85–96. [CrossRef]
3. Isaksen, K.; Nordli, Ø.; Ivanov, B.; Køltzow, M.A.; Aaboe, S.; Gjelten, H.M.; Mezghani, A.; Eastwood, S.; Førland, E.; Benestad, R.E.; et al. Exceptional warming over the Barents area. *Sci. Rep.* **2022**, *12*, 9371. [CrossRef] [PubMed]
4. Wendisch, M.; Macke, A.; Ehrlich, A.; Lüpkes, C.; Mech, M.; Chechin, D.; Dethloff, K.; Velasco, C.B.; Bozem, H.; Brückner, M.; et al. The Arctic cloud puzzle: Using ACLOUD/PASCAL multiplatform observations to unravel the role of clouds and aerosol particles in Arctic amplification. *Bull. Am. Meteorol. Soc.* **2019**, *100*, 841–871. [CrossRef]
5. Nakoudi, K.; Ritter, C.; Stachlewska, I.S. Properties of Cirrus Clouds over the European Arctic (Ny-Ålesund, Svalbard). *Remote Sens.* **2021**, *13*, 4555. [CrossRef]
6. Tuttle, S.E.; Roof, S.R.; Retelle, M.J.; Werner, A.; Gunn, G.E.; Bunting, E.L. Evaluation of Satellite-Derived Estimates of Lake Ice Cover Timing on Linnévatnet, Kapp Linné, Svalbard Using In-Situ Data. *Remote Sens.* **2022**, *14*, 1311. [CrossRef]
7. Walch, D.M.R.; Singh, R.K.; Søreide, J.E.; Lantuit, H.; Poste, A. Spatio-Temporal Variability of Suspended Particulate Matter in a High-Arctic Estuary (Adventfjorden, Svalbard) Using Sentinel-2 Time-Series. *Remote Sens.* **2022**, *14*, 3123. [CrossRef]
8. Kozlov, I.E.; Atadzhanova, O.A. Eddies in the Marginal Ice Zone of Fram Strait and Svalbard from Spaceborne SAR Observations in Winter. *Remote Sens.* **2022**, *14*, 134. [CrossRef]
9. Rouyet, L.; Liu, L.; Strand, S.M.; Christiansen, H.H.; Lauknes, T.R.; Larsen, Y. Seasonal InSAR Displacements Documenting the Active Layer Freeze and Thaw Progression in Central-Western Spitsbergen, Svalbard. *Remote Sens.* **2021**, *13*, 2977. [CrossRef]
10. Herzfeld, U.C.; Lawson, M.; Trantow, T.; Nylen, T. Airborne Validation of ICESat-2 ATLAS Data over Crevassed Surfaces and Other Complex Glacial Environments: Results from Experiments of Laser Altimeter and Kinematic GPS Data Collection from a Helicopter over a Surging Arctic Glacier (Negribreen, Svalbard). *Remote Sens.* **2022**, *14*, 1185. [CrossRef]
11. Jawak, S.D.; Wankhede, S.F.; Luis, A.J. Explorative Study on Mapping Surface Facies of Selected Glaciers from Chandra Basin, Himalaya Using WorldView-2 Data. *Remote Sens.* **2019**, *11*, 1207. [CrossRef]
12. Jawak, S.D.; Wankhede, S.F.; Luis, A.J.; Balakrishna, K. Impact of Image-Processing Routines on Mapping Glacier Surface Facies from Svalbard and the Himalayas Using Pixel-Based Methods. *Remote Sens.* **2022**, *14*, 1414. [CrossRef]
13. Jawak, S.D.; Wankhede, S.F.; Luis, A.J.; Balakrishna, K. Effect of Image-Processing Routines on Geographic Object-Based Image Analysis for Mapping Glacier Surface Facies from Svalbard and the Himalayas. *Remote Sens.* **2022**, *14*, 4403. [CrossRef]
14. Jawak, S.D.; Wankhede, S.F.; Luis, A.J.; Balakrishna, K. Multispectral Characteristics of Glacier Surface Facies (Chandra-Bhaga Basin, Himalaya, and Ny-Ålesund, Svalbard) through Investigations of Pixel and Object-Based Mapping Using Variable Processing Routines. *Remote Sens.* **2022**, *14*, 6311. [CrossRef]
15. Kavan, J.; Tallentire, G.D.; Demidionov, M.; Dudek, J.; Strzelecki, M.C. Fifty Years of Tidewater Glacier Surface Elevation and Retreat Dynamics along the South-East Coast of Spitsbergen (Svalbard Archipelago). *Remote Sens.* **2022**, *14*, 354. [CrossRef]
16. Błaszczyk, M.; Laska, M.; Sivertsen, A.; Jawak, S.D. Combined Use of Aerial Photogrammetry and Terrestrial Laser Scanning for Detecting Geomorphological Changes in Hornsund, Svalbard. *Remote Sens.* **2022**, *14*, 601. [CrossRef]
17. Eischeid, I.; Soininen, E.M.; Assmann, J.J.; Ims, R.A.; Madsen, J.; Pedersen, Å.Ø.; Pirotti, F.; Yoccoz, N.G.; Ravolainen, V.T. Disturbance Mapping in Arctic Tundra Improved by a Planning Workflow for Drone Studies: Advancing Tools for Future Ecosystem Monitoring. *Remote Sens.* **2021**, *13*, 4466. [CrossRef]
18. Karlsen, S.R.; Stendardi, L.; Tømmervik, H.; Nilsen, L.; Arntzen, I.; Cooper, E.J. Time-Series of Cloud-Free Sentinel-2 NDVI Data Used in Mapping the Onset of Growth of Central Spitsbergen, Svalbard. *Remote Sens.* **2021**, *13*, 3031. [CrossRef]
19. Stendardi, L.; Karlsen, S.R.; Malnes, E.; Nilsen, L.; Tømmervik, H.; Cooper, E.J.; Notarnicola, C. Multi-Sensor Analysis of Snow Seasonality and a Preliminary Assessment of SAR Backscatter Sensitivity to Arctic Vegetation: Limits and Capabilities. *Remote Sens.* **2022**, *14*, 1866. [CrossRef]

20. Vickers, H.; Karlsen, S.R.; Malnes, E. A 20-Year MODIS-Based Snow Cover Dataset for Svalbard and Its Link to Phenological Timing and Sea Ice Variability. *Remote Sens.* **2020**, *12*, 1123. [CrossRef]
21. Vickers, H.; Malnes, E.; van Pelt, W.J.J.; Pohjola, V.A.; Killie, M.A.; Saloranta, T.; Karlsen, S.R. A Compilation of Snow Cover Datasets for Svalbard: A Multi-Sensor, Multi-Model Study. *Remote Sens.* **2021**, *13*, 2002. [CrossRef]
22. Salvatori, R.; Salzano, R.; Valt, M.; Cerrato, R.; Ghergo, S. The Collection of Hyperspectral Measurements on Snow and Ice Covers in Polar Regions (SISpec 2.0). *Remote Sens.* **2022**, *14*, 2213. [CrossRef]
23. Jawak, S.D.; Andersen, B.N.; Pohjola, V.A.; Godøy, Ø.; Hübner, C.; Jennings, I.; Ignatiuk, D.; Holmén, K.; Sivertsen, A.; Hann, R.; et al. SIOS's Earth Observation (EO), Remote Sensing (RS), and Operational Activities in Response to COVID-19. *Remote Sens.* **2021**, *13*, 712. [CrossRef]

remote sensing MDPI

Perspective

SIOS's Earth Observation (EO), Remote Sensing (RS), and Operational Activities in Response to COVID-19

Shridhar D. Jawak [1,*], Bo N. Andersen [2], Veijo A. Pohjola [3], Øystein Godøy [1,4], Christiane Hübner [1], Inger Jennings [1], Dariusz Ignatiuk [1,5], Kim Holmén [6], Agnar Sivertsen [7], Richard Hann [8,9], Hans Tømmervik [10], Andreas Kääb [2], Małgorzata Błaszczyk [5], Roberto Salzano [11], Bartłomiej Luks [12], Kjell Arild Høgda [7], Rune Storvold [7], Lennart Nilsen [13], Rosamaria Salvatori [14], Kottekkatu Padinchati Krishnan [15], Sourav Chatterjee [15], Dag A. Lorentzen [9], Rasmus Erlandsson [10], Tom Rune Lauknes [7,13], Eirik Malnes [7], Stein Rune Karlsen [7], Hiroyuki Enomoto [16], Ann Mari Fjæraa [17], Jie Zhang [3], Sabine Marty [18], Knut Ove Nygård [4] and Heikki Lihavainen [1]

1 Svalbard Integrated Arctic Earth Observing System (SIOS), SIOS Knowledge Centre, Svalbard Science Centre, P.O. Box 156, N-9171 Longyearbyen, Norway; steingod@met.no (Ø.G.); information@sios-svalbard.org (C.H.); logistics@sios-svalbard.org (I.J.); dariusz.ignatiuk@sios-svalbard.org (D.I.); director@sios-svalbard.org (H.L.)
2 University of Oslo, P.O. Box 1047, Blindern, 0316 Oslo, Norway; b.n.andersen@astro.uio.no (B.N.A.); a.m.kaab@geo.uio.no (A.K.)
3 Department of Earth Sciences, Uppsala University, Geocentrum, Villavägen 16, 752 36 Uppsala, Sweden; Veijo.Pohjola@geo.uu.se (V.A.P.); jie.zhang@geo.uu.se (J.Z.)
4 Norwegian Meteorological Institute, P.O. Box 43, Blindern, 0371 Oslo, Norway; knuton@met.no
5 Institute of Earth Sciences, University of Silesia in Katowice, Bedzinska 60, 41-200 Sosnowiec, Poland; malgorzata.blaszczyk@us.edu.pl
6 Norwegian Polar Institute, Fram Centre, P.O. Box 6606 Langnes, N-9296 Tromsø, Norway; kim.holmen@npolar.no
7 NORCE Norwegian Research Center AS, Sykehusvn 21, 9019 Tromsø, Norway; agsi@norceresearch.no (A.S.); kjho@norceresearch.no (K.A.H.); rust@norceresearch.no (R.S.); tlau@norceresearch.no or tom.r.lauknes@uit.no (T.R.L.); eima@norceresearch.no (E.M.); skar@norceresearch.no (S.R.K.)
8 Centre for Autonomous Marine Operations and Systems, Department of Engineering Cybernetics, Norwegian University of Science and Technology (NTNU), NO-7491 Trondheim, Norway; richard.hann@ntnu.no
9 Department of Arctic Geology, The University Centre in Svalbard (UNIS), P.O. Box 156, 9171 Longyearbyen, Norway; DagL@UNIS.no
10 Norwegian Institute for Nature Research (NINA), Fram Centre, 9296 Tromsø, Norway; hans.tommervik@nina.no (H.T.); rasmus.erlandsson@nina.no (R.E.)
11 Florence Division, Institute for Atmospheric Pollution Research, National Research Council (CNR-IIA), 50019 Sesto Fiorentino, Italy; roberto.salzano@cnr.it
12 Institute of Geophysics, Polish Academy of Sciences, 01-452 Warsaw, Poland; luks@igf.edu.pl
13 Department of Arctic and Marine Biology, UiT The Arctic University of Norway, Hansine Hansens veg 18, 9019 Tromsø, Norway; lennart.nilsen@uit.no
14 Institute of Polar Sciences, National Research Council of Italy (CNR-ISP), 00010 Montelibretti, Italy; rosamaria.salvatori@cnr.it
15 National Centre for Polar and Ocean Research (ESSO-NCPOR), Ministry of Earth Sciences, Government of India, Headland Sada, Vasco-da-Gama, Goa 403 804, India; krishnan@ncpor.res.in (K.P.K.); sourav@ncpor.res.in (S.C.)
16 Arctic Environment Research Center, National Institute of Polar Research, Tokyo 190-8518, Japan; enomoto.hiroyuki@nipr.ac.jp
17 Atmosphere and Climate Department, NILU—Norwegian Institute for Air Research, P.O. Box 100, 2027 Kjeller, Norway; amf@nilu.no
18 Norwegian Institute for Water Research (NIVA), Gaustadalléen 21, 0349 Oslo, Norway; sabine.marty@niva.no
* Correspondence: shridhar.jawak@sios-svalbard.org or shridhar.jawak@gmail.com

Citation: Jawak, S.D.; Andersen, B.N.; Pohjola, V.A.; Godøy, Ø.; Hübner, C.; Jennings, I.; Ignatiuk, D.; Holmén, K.; Sivertsen, A.; Hann, R.; et al. SIOS's Earth Observation (EO), Remote Sensing (RS), and Operational Activities in Response to COVID-19. *Remote Sens.* **2021**, *13*, 712. https://doi.org/10.3390/rs13040712

Academic Editor: Francesco Nex

Received: 15 January 2021
Accepted: 11 February 2021
Published: 15 February 2021

Publisher's Note: MDPI stays neutral with regard to jurisdictional claims in published maps and institutional affiliations.

Abstract: Svalbard Integrated Arctic Earth Observing System (SIOS) is an international partnership of research institutions studying the environment and climate in and around Svalbard. SIOS is developing an efficient observing system, where researchers share technology, experience, and data, work together to close knowledge gaps, and decrease the environmental footprint of science. SIOS maintains and facilitates various scientific activities such as the State of the Environmental Science in Svalbard (SESS) report, international access to research infrastructure in Svalbard, Earth observation and remote sensing services, training courses for the Arctic science community, and open access to data. This perspective paper highlights the activities of SIOS Knowledge Centre, the central hub of SIOS, and the SIOS Remote Sensing Working Group (RSWG) in response to the unprecedented situation imposed by the global pandemic coronavirus (SARS-CoV-2) disease 2019 (COVID-19). The pandemic has affected Svalbard research in several ways. When Norway declared a nationwide lockdown to decrease the rate of spread of the COVID-19 in the community, even more strict measures were taken to protect the Svalbard community from the potential spread of the disease. Due to the lockdown, travel restrictions, and quarantine regulations declared by many nations, most physical meetings, training courses, conferences, and workshops worldwide were cancelled by the first week of March 2020. The resumption of physical scientific meetings is still uncertain in the foreseeable future. Additionally, field campaigns to polar regions, including Svalbard, were and remain severely affected. In response to this changing situation, SIOS initiated several operational activities suitable to mitigate the new challenges resulting from the pandemic. This article provides an extensive overview of SIOS's Earth observation (EO), remote sensing (RS) and other operational activities strengthened and developed in response to COVID-19 to support the Svalbard scientific community in times of cancelled/postponed field campaigns in Svalbard. These include (1) an initiative to patch up field data (in situ) with RS observations, (2) a logistics sharing notice board for effective coordinating field activities in the pandemic times, (3) a monthly webinar series and panel discussion on EO talks, (4) an online conference on EO and RS, (5) the SIOS's special issue in the Remote Sensing (MDPI) journal, (6) the conversion of a terrestrial remote sensing training course into an online edition, and (7) the announcement of opportunity (AO) in airborne remote sensing for filling the data gaps using aerial imagery and hyperspectral data. As SIOS is a consortium of 24 research institutions from 9 nations, this paper also presents an extensive overview of the activities from a few research institutes in pandemic times and highlights our upcoming activities for the next year 2021. Finally, we provide a critical perspective on our overall response, possible broader impacts, relevance to other observing systems, and future directions. We hope that our practical services, experiences, and activities implemented in these difficult times will motivate other similar monitoring programs and observing systems when responding to future challenging situations. With a broad scientific audience in mind, we present our perspective paper on activities in Svalbard as a case study.

Keywords: earth observation; remote sensing; COVID-19; Svalbard; earth system science; SIOS

1. Introduction

Svalbard Integrated Arctic Earth Observing System (SIOS) [1] is a Norwegian initiated international consortium of 24 research institutes from 9 nations building and optimising a sustained regional multidomain distributed Arctic observing system of long-term measurements in and around the High-Arctic archipelago of Svalbard. The observing system consists of long-term in situ and remotely sensed measurements in and around Svalbard addressing Earth System Science (ESS) questions. SIOS research infrastructures (RI) can be used for various current and future satellite missions for calibration and validation (Cal/Val) activities. Eventually, integration of in situ and satellite-based measurements will benefit the entire ESS community to address broad scientific questions. Within SIOS, researchers can cooperate to access instruments, acquire remotely sensed and in situ data, and address questions that would not be practical or cost effective for a single institution or nation alone. SIOS focuses on cross-disciplinary processes and their interactions between the different spheres, i.e., biosphere, geosphere, atmosphere, cryosphere, and hydrosphere. The observing system strives to provide the members with systematic high-quality observations in a cost-efficient and environment friendly way. Other networks such as the Sustaining Arctic Observing Networks (SAON)/Arctic GEOSS, stakeholders, and users outside the SIOS consortium also benefit from these data series. The SIOS Knowledge Centre (SIOS-KC), located in the Svalbard Science Centre in Longyearbyen, is the central hub of SIOS. SIOS-KC coordinates the services provided by SIOS for the international research community and ensures sustainability and maintenance of the network. The services offered include: (1) integration of the distributed research infrastructure and data to optimise the observing system, (2) access to the research infrastructure (SIOS ACCESS programme), (3) data management including storing and curating of scientific data (ground-based, airborne and spaceborne), (4) tools and support for scientists to make use of the extensive Earth observation (EO) and remote sensing (RS) resources available, (5) coordination of logistical services, (6) training and education programmes, and (7) information and outreach. EO and RS activities are important parts of the observing system as certain key data can be measured with satellite-based observations. Over the past three decades, tremendous developments in EO satellite missions have made significant contributions to spatial-spectral-temporal sampling and subsequent extraction of geoinformation (GI) from the Arctic. Due to the harsh and remote Arctic environment, there is a lack of basic in situ observations that can support scientific understanding of key processes and support operational services. Most of the existing data are collected via time limited research projects, and many of these via manual and on-site work [2]. Svalbard is probably one of those regions in the Arctic with the most in situ measurements spanning across various spheres; still, there are massive gaps spatially distributed in various regions. Such data gaps can be filled using frequent satellite-based acquisitions, new product generation using EO/RS/GI, and integration of ground-based, airborne, and satellite-based measurements.

The novel coronavirus SARS-CoV-2 caused the current global pandemic of Coronavirus disease 2019 (COVID-19), resulting in massive infection and mortality around the world [3]. Besides, situations like the pandemic have the potential to contribute to issues related to exacerbating personal mental health and socio-economic inequalities. The virus is still surging across the world in multiple waves of outbreaks in different countries, causing considerable societal and global economic impacts. Many national governments implemented lockdowns of various forms, including prohibition on large gatherings, travel restrictions, and implementing or encouraging social distancing, to decrease virus spread and reduce pressure on healthcare systems [4]. International researchers, including those from SIOS member institutions, carry out field campaigns annually in different parts of Svalbard and associated waters, especially focussing on Ny-Ålesund, Longyearbyen, and Hornsund. However, due to the sudden and rapid outbreak of the COVID-19, the World Health Organization (WHO) declared a global pandemic on 11th March 2020 and most nations closed their borders. Norway was locked down from the 13th of March 2020. Because of the small community and limited health services in Longyearbyen, even stricter measures were taken in Svalbard in comparison to mainland Norway. The worldwide lockdown in the beginning of the Arctic spring season has affected many field campaigns to Svalbard. Glaciologists and snow scientists (including those associated with biosciences) were initially worst affected as spring is a crucial period to monitor changes in glaciers and amount of snow stored in the catchments. Most of the scientists in the countries with active Svalbard research activities, including Norway, were grounded. Some of RIs (e.g., Polish Polar research facility at Hornsund) were temporarily closed for visitors. By the time Norway initiated the lockdown, also most of the planned conferences were already cancelled (Figure 1), postponed, or converted to online mode. This was a completely new situation and needed special attention.

COVID-19 has impacted the polar research in many ways since the beginning of the year and it continues to affect future programmes. These include cancellation of field campaigns, cancellation and/or postponement of important conferences, workshops, and training courses, delays in delivery of scientific outputs because of shutdown of campuses, cancellations and/or delay in funding and many more. Several transnational access projects in the Arctic supported by EU-INTERACT (https://eu-interact.org/ (accessed on 15 January 2021)) have been affected during the field season in 2020 with around 120 scientists had to cancel their field campaigns. Frame and Hemmings [5] reviewed the potential impact of COVID-19 in Antarctica through tourism and scientific research over three (short-term, mid-term, and long-term) time periods. According to these authors, Antarctic tourism and field-based research will be severely reduced in the short term. We believe that this conclusion is also applicable to the impact on Svalbard field-based research in a short term. In response to this situation, a few activities have been started by leading networks such as Scientific Committee on Antarctic Research (SCAR) and International Centre for Integrated Mountain Development (ICIMOD) on studying social aspects and impact on communities due to COVID-19. ICIMOD has published a comprehensive report on COVID-19 impact and policy responses in the Hindu Kush Himalaya [6]. COVID-19 has also heavily impacted scientific and logistical activities in Antarctica [7]. The Antarctic COVID-19 Research Group of SCAR has conducted a survey [8] to learn more about the impact of pandemic on the Antarctic research community. Outcomes of this survey are expected to understand effects of the COVID-19 situation on the Antarctic science community to inform strategic decisions to mitigate impacts. On the other hand, SIOS focused more on mitigating short term impacts of COVID-19 during the core field season in Svalbard. The Rapid Action on Coronavirus and EO (RACE) [9] platform launched by joint cooperation between European Space Agency (ESA) and the European Commission demonstrates the use of wide range of EO data from Copernicus Sentinels and third-party missions to track societal, economic, and environmental changes in times of COVID-19 pandemic. Recent publications and services in literature highlight the contribution of EO and RS in tracking the spread of virus and monitoring the pandemic events or the effect of the lockdown on the environment [10–12]. Our study, however, is probably the first attempt of highlighting the role of EO and RS for mitigating the damage in terms of possible data gaps in long time data series of scientific observations in one of the most remote places on planet Earth.

Figure 1. Most conferences were either cancelled, postponed, or converted to online versions due to the global COVID-19 pandemic.

SIOS-KC provides services including logistics, remote sensing, communication, online events, training courses, data access, observation facility catalogue, and satellite data visualisation on the website and provide support via online tools. In the beginning of the pandemic (January–February 2020) in mainland China and Italy, regular activities were not yet affected but SIOS-KC was alert and developed a pandemic plan of action. When the Norwegian lockdown began on 13th March 2020, SIOS-KC quickly initiated several activities to adapt to the new situation. The Remote Sensing Working Group (RSWG) of SIOS took an active part in developing new activities to keep the scientific community engaged in these difficult times. SIOS responded with several initiatives: to keep international research community up-to-date about the fast-changing regulations concerning travel restrictions to Svalbard and to the different research facilities, a compilation of all relevant information resources was made available on the SIOS web page; a logistics sharing notice board was launched as a platform to offer and request help with issues related to cancelled field activities during the lockdown; the SIOS remote sensing service published an offer to patch up field data gaps with RS techniques, and a monthly webinar mini-series was started in March 2020 to provide a social experience and keep the Svalbard research community engaged. SIOS also approached the situation with a general attitude of flexibility and service mindedness, for example being generous with extensions to deadlines and trying to find solutions to logistical problems in field campaigns. Detailed descriptions of these activities are presented in this article.

2. SIOS's Response to the COVID-19 Pandemic

In the following sections, we present our key activities conducted in the period March–December 2020 in response to changes in travel restrictions in Svalbard in chronological order (Figure 2).

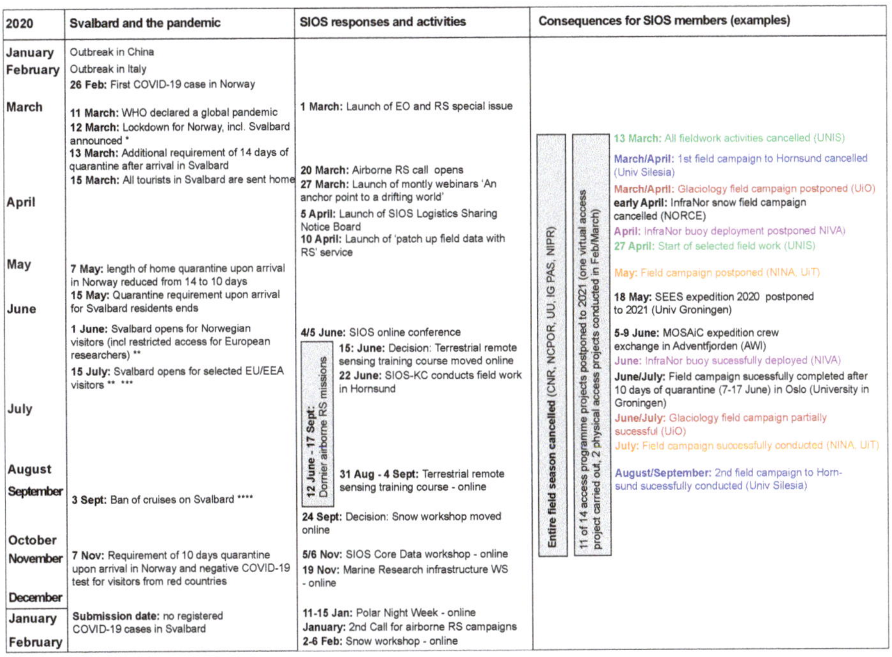

Figure 2. Timeline of events since outbreak of the pandemic, including travel restrictions related to Svalbard, Svalbard Integrated Arctic Earth Observing System (SIOS) responses and activities and some examples of consequences for SIOS member institutions. CNR: National Research Council of Italy; NCPOR: National Centre for Polar and Ocean Research, India; UU: Uppsala University, Sweden; IG PAS: Institute for Geophysics Polish Academy of Sciences, Poland; NIPR: National Institute of Polar Research, Japan; UNIS: University Centre in Svalbard, Norway; UiO: University of Oslo, Norway; NORCE: Norwegian Research Institute, Norway; NINA: Norwegian Institute for Nature Research, Norway; UiT: UiT The Arctic University of Norway; AWI: Alfred Wegener Institute, Germany. [* 14 days of quarantine after arrival in Norway required, ** except EU/EEA countries with high infection risk ('red countries'), *** no longer 10 days quarantine after arrival in Svalbard required, **** For cruises with more than 30 people (crew and passenger), except day cruises].

2.1. *Special Issue in an International Journal to Facilitate Publications from the International Scientific Community*

An unintended consequence of fortuitous timing of an already planned activity was the launch of a special issue to facilitate publications from the international scientific community in Svalbard. Since the beginning of March 2020, SIOS has hosted the special issue "Earth Observation (EO), Remote Sensing (RS), and Geoinformation (GI) Applications in Svalbard" [13] in the Remote Sensing (https://www.mdpi.com/journal/remotesensing (accessed on 15 January 2021)) journal. This special issue is being edited by 12 experts from

SIOS member institutions and SIOS-KC with a strong RS, GI and EO background. This activity was not started in response to COVID-19 but was already planned at the beginning of the year. However, this activity has since provided a platform for researchers to publish their EO and RS based studies in Svalbard during a time when many are confined to home offices. It also helps scientists to expedite their pending publications during lockdown time.

2.2. SIOS Webinar Series: An Anchor Point to a Drifting World

This activity was SIOS's first response specifically to the new situation, when the RSWG launched the first of the SIOS webinar series on EO and RS. We aimed to provide a social experience to the Svalbard research community to remedy the lack of a social environment because of cancelled field campaigns and conferences by launching a webinar series. The SIOS webinar series "an anchor point to a drifting world!" includes talks on EO and RS [14] with expert scientists working on different environments of Svalbard, followed by panel discussions (Figure 3). In our experience, keeping the community together in times of chaos was a wise decision. All the talks of this webinar series are available online as PDF and video files, which are open for the entire scientific community. This constitutes an excellent learning resource for the research community, especially early career researchers. Other online seminar series and events were announced subsequent to the launch of the SIOS webinar series. The International glaciological society launched an IGS global seminar series [15] on 15th April 2020. Polar Geospatial Centre (PGC) webinars [16] also gained attention during the lockdown period. All these online events were a means to bring the scientific community together and share their scientific results with the community in these challenging times.

Figure 3. In March 2020, SIOS launched a monthly webinar series to keep the SIOS community engaged with each other and to provide a social experience. This webinar series acts as a symbolic anchor point to a drifting Svalbard science community.

2.3. SIOS's Online Conference on Earth Observation (EO), Remote Sensing (RS), and Geoinformation (GI) Applications in Svalbard

Encouraged by the success of the SIOS webinar series and lessons learned from the online version of the International Arctic Science Committee's (IASC's) Arctic Science Summit Week (ASSW) in the beginning of lockdown (March 2020), SIOS decided to host its own online conference in June 2020 [17]. The conference aimed to (1) promote the opportunity for PhD students, postdocs, researchers, scientists, and academicians to contribute actively to SIOS's special issue in Remote Sensing journal (MDPI), (2) review the state-of-the-art EO/RS/GI applications in Svalbard and (3) provide a social experience to the Svalbard scientific community. The conference was hosted on 4–5 June 2020. A total of 53 talks were delivered, including 3 keynote talks and 3 invited talks, by presenters from 24 institutions from 12 countries. Out of these talks, 14 talks were delivered by early career researchers. A special session was hosted on the airborne remote sensing applications in Svalbard, in which project investigators from the "SIOS announcement of opportunity in airborne remote sensing" delivered short presentations on the planned activities. Additionally, a special session on EO and RS activities supported by SIOS was conducted. We also conducted open remote sensing dialogue for the first time to receive inputs from the EO and RS community on further developments on COVID-19 related activities. The participation in the conference was wide, with more than 370 registered participants (around 140 attendees on the first day and around 90 attendees on the second day). A very positive response was observed on feedback polls with an average score of 9 out of 10 for all the presentations (Table 1). Most of the presentations and an abstract book are available on the SIOS website (https://sios-svalbard.org/RS_OnlineConference2020 (accessed on 15 January 2021)) and is an important knowledge resource in the pandemic times.

Table 1. Overview of virtual activities conducted by SIOS to bring the community together.

No.	Webinar Theme	Date	No. of Talks	No. of Registrations	No. of Attendees	Feedback (Out of 10)
1	An anchor point to a drifting world: EO and RS applications in Svalbard and panel discussion on COVID-19 driven damage mitigation in Svalbard	27 March 2020	5	Registration was not required	55	≈8
2	An anchor point to a drifting world: EO and RS applications in Svalbard and panel discussion on data management	24 April 2020	3	Registration was not required	48	≈8
3	SIOS Online Conference on EO, RS, and GI	4–5 June 2020	53	372	≈90–140	≈9
4	SIOS Terrestrial Remote Sensing Training Course-Lecture series	31 August–4 September 2020	12	163	≈40–50	≈8
5	An anchor point to a drifting world: Grand Challenge Initiative (GCI) cusp rocket missions in Svalbard and their relevance to atmospheric studies	23 October 2020	4	55	30	≈9
6	SIOS Core data workshop	5–6 November 2020	4	58	43	≈7
7	SIOS Marine research infrastructure workshop	19 November 2020	16	67	60	≈9
8	An anchor point to a drifting world: Marine science in Svalbard—global to local perspective and panel on the outcomes of the SIOS marine infrastructure network workshop	27 November 2020	4	124	65	≈9

2.4. Shifting from Physical to the Virtual Mode of Terrestrial Remote Sensing Training Course

After the successful and interactive online conference organised in June 2020, we were still not confident that travel restrictions would be lifted in September 2020 when our annual remote sensing training was planned to take place in Longyearbyen. Therefore, on 15th June 2020, we decided to convert the training course to a fully online mode. With this decision, SIOS wanted to ensure that applicants from all countries can participate in the training course, independent of the current travel restrictions. SIOS held this online training course [18] during 31st August–4th September 2020. The course covered topics on how to effectively use EO and RS data acquired from satellites, from the air or from the ground, and their associated tools and software in the context of terrestrial research in Svalbard. The course was intended for field scientists, PhD students, and technicians with no or little experience with EO and RS techniques. The training was delivered by ten EO and RS experts and international academicians. A virtual field excursion activity and hands-on sessions were developed to make this training course to include practical exercises. This was our first experience of hosting a fully online training course and the feedback from this activity provided us with valuable insight into how to deliver future online courses. Apart from providing training to selected participants, we also decided to open the course lectures for anyone interested to join, as a special edition of our SIOS webinar series. This essentially provided a free opportunity for many budding scientists to learn new EO and RS skills necessary for Svalbard research. Around 150 registered participants attended different lectures during the whole training course.

2.5. *Patch Up Your Field Data with Remote Sensing Observations*

Continuous long-term data is essential for modellers and other scientists for interpreting and tracking long term changes and is an essential component of the SIOS observing system. However, the pandemic situation posed a new challenge in measuring in situ parameters due to travel restrictions and quarantine regulations in Svalbard. EO and RS are thus more relevant in the current times than ever. While in the beginning we focused more on online events and carrying out discussions with scientists, in the next step, we started working on practical suggestions from the community to respond to their needs. The first step was mapping the needs of researchers and evaluating the possibilities to provide RS observations in the absence of field campaigns in Svalbard. Since we had no generic solution, in the first stage, we wanted to understand the anticipated damage to data series within the Svalbard scientific community. With the launch of the service "patch up your field data with remote sensing observations" [19], we tried to help those scientists who were unable to conduct field campaigns in Svalbard this year to fill the gaps in observations using RS (Figure 4). The data collected from this tool would provide a summary of data and logistics requirements from researchers, role of EO and RS observations in absence of field campaigns, and mitigation measures that can be considered in the future similar situations to save or fill the long-term time series of data and the continuous operation of instruments in Svalbard. This strategy would probably be applicable to many cryospheric regions including Antarctica and Himalayas, where field campaigns often need to be cancelled or postponed. We believe that our initiative can be expanded to other regions to map requirements from scientists to avoid gaps in essential data in the future.

2.6. *Logistics Sharing Notice Board*

One problem that arose during the onset of COVID-19 was by scientists that were scheduled to maintain and download data from various installations, both as permanent infrastructures in research facilities, but also temporary installations in the field. In addition, field parties usually collect various data during the maintenance of the installations, essential to complement data collected by various stationary instruments. In response to this tricky situation, SIOS came forward to provide a helping hand to look after essential instruments in the field when scientists were not able to travel to Svalbard. Although we had started providing EO and RS based solutions, we realised that there were many scientific observations for which it was not possible to fill the gap using only RS. As such, our next level of response to the changing situation was more hands-on. In response to cancelled fieldwork in Svalbard, Svalbard Science Forum (SSF) (https://www.forskningsradet.no/en/svalbard-science-forum/ (accessed on 15 January 2021)) stepped forward at the very beginning of lockdown to help connect researchers in need of support for field logistics by facilitating dialogues on a social media platform [20]. Since SIOS had already started developing the "Logistics Sharing Notice Board" for regular sharing of information on logistics and resources, we decided to expand this activity to meet the new challenge. The most significant issue at the beginning of the lockdown was that most of the glaciologists who are visiting Svalbard every year at the beginning of

March 2020 were now unable to travel and carry out fieldwork. In these special times, SIOS offered to assist where possible to avoid gaps in long-term data series. SIOS introduced the Logistics Sharing Notice Board [21] in April 2020, which gives an overview of the planned fieldwork in Svalbard with the possibility to offer or request spare capacity (Table 2).

Figure 4. Overview of SIOS's Earth observation (EO), Remote Sensing (RS) and operational activities in response to COVID-19 pandemic including (**a**) Platform for publishing studies: SIOS's special issue in an international journal motivate scientists to consider submitting manuscripts in indoor times. (**b**) EO and RS activities: Patch up missing data with remote sensing observations in times of lockdown and travel restrictions. (**c**) Social experience and networking platform: SIOS online conference as a platform to connect the Svalbard research community when physical meetings are not feasible, and (**d**) Logistics Sharing Notice Board: a tool to help scientists in times of lockdown to maintain their field instruments for the season.

Table 2. Fieldwork assistance requested via SIOS's Logistics Sharing Notice Board (More details: https://sios-svalbard.org/logistic-notice-board# (accessed on 15 January 2021)). Those highlighted with grey coloured background were not fulfilled because of logistics and technical constraints. * Research in Svalbard, RiS Id (https://www.researchinsvalbard.no/ (accessed on 15 January 2021)).

No.	Project Name/Brief Description	RiS ID *	Location in Svalbard
1	Lomonosovfonna Firn Aquifers and mass balance	3395	Lomonosovfonna / Nordenskiöldbreen
2	Hydro-condition of Werenskioldbreen 2020	11198	Werenskioldbreen, Hornsund
3	Plant stable isotopes	N.A	Longyearbyen
4	Antibiotic resistance genes and integrons as indicators of biotic pollution and resistance load of Arctic ecosystems	11563	Longyearbyen
5	Bogerbreen mass balance	N.A	Bogerbreen in Endalen (Longyearbyen)
6	Hydro-meteorological condition of the Hornsund area 2020	11198	Catchments of the Hornsund fjord region including Werenskioldbreen, Brateggbreen, Ariebreen, and Fuglebekken

SIOS-KC is staffed by professionals with fieldwork experience in a range of disciplines including biology, environmental science, atmospheric sciences, glaciology, and EO/RS. Therefore, we decided to make use of the SIOS-KC's expertise by offering to conduct campaigns to carry out field measurements, collect and/or deploy equipment in the field and carry out basic maintenance. This was also an opportunity for us to support local businesses by cooperating with local logistics providers to solve such tasks alongside the expertise of SIOS partners and SIOS-KC. This is significant for the current situation because the tourist industry, including local tourist guides with extensive field experience, is suffering heavily with loss of income due to dwindling numbers of tourists.

The first request entered on the notice board was from a glaciologist from Uppsala University, who has been monitoring glaciological and meteorological parameters on the Lomonosovfonna icefield since 2006. The team was hopeful in mid-March that they could make it to Svalbard, but the travel restrictions to Svalbard made it impossible for them to carry out their field campaign. The research group wanted to recover data and remove instruments from the glacier before the sensors installed last year would be buried by snow and eventually lost, and likely hard to recover after next winter. This would be a huge loss in terms of funding and resources and create more than a year-long gap in the data that has been monitored since 2006. They requested a rescue mission to save these instruments, which SIOS-KC planned and was almost ready to execute. Unfortunately, when a different scientific field party had an accident on another glacier, all activities on glaciers were stopped to minimise the pressure on the limited search and rescue resources in Svalbard and the plan to rescue the instruments was cancelled. This example highlights the uncertainty in carrying out field observations in polar regions. More successful uses of the notice board are described in Section 3.

2.7. SIOS's Announcement of Opportunity (AO) in Airborne Remote Sensing—2020

In parallel to the efforts based on satellite data gap filling and ground-based logistics support to scientists, we channelised dedicated funding for scientists to use aircraft and unmanned aerial vehicles (UAVs) to collect data in Svalbard (https://sios-svalbard.org/AirborneRS (accessed on 15 January 2021)). This is one of the best ways to fill the data gap in the current situation as it is practically possible to conduct field campaigns using airborne platforms despite travel restrictions with minimal resources. SIOS allocated 1 million Norwegian kroner to cover 25 h of flight time in the field season of 2020 in collaboration with its member institutions NORCE Norwegian Research Centre and University Centre in Svalbard (UNIS). The available platforms offered aerial images, UAV images, and hyperspectral data which will be freely available to scientists [22]. In total, 10 scientific research projects (Table 3 and Figure 5) were able to acquire aerial images and hyperspectral data from various locations in Svalbard through this initiative. The aircraft Dornier DO228 aircraft operated by the local company Lufttransport is permanently stationed on Svalbard and used regularly for transporting personnel and cargo to the science community in Ny-Ålesund and mining community in Svea. The aircraft has recently been fitted with a camera (Phasone IXU-150, Schneider LS 55 mm f/2.8) and a hyperspectral imager (VNIR-1800, Norsk elektrooptikk) (Figure 6). The hyperspectral sensor images the ground in 186 spectral bands covering the range 400–1000 nm. Hyperspectral data can be used to map minerals, vegetation, and the presence of animals. In addition, aerial photos have many uses in Svalbard, such as mapping snow, sea ice, and glaciers, counting seals, and making 3D models of the terrain. The cameras installed on the Dornier aircraft can acquire images with a ground resolution of 10 cm from flight altitude of 1000 m. A big advantage for this platform is that we can acquire images when it is cloudy above the airplane, unlike satellites that depend on cloudless conditions—something which can be a rare sight on a typical Svalbard summer. This is a well-timed activity which coincided with scientists who usually travel to Svalbard every year to continue their scientific measurements being grounded in their home countries. Data collected from the SIOS funded airborne missions will not only help to fill a few of the data gaps resulting from the lockdown, and will be used by glaciologists, biologists, hydrologists, and other Earth system scientists to understand the state of the environment of Svalbard during these times.

Table 3. Airborne remote sensing projects from SIOS's announcement of opportunity (More details: https://sios-svalbard.org/AirborneRS (accessed on 15 January 2021)). Those highlighted with grey coloured background were not implemented because of logistics and technical constraints.

No.	Project Name/Brief Description	RiS ID	Location in Svalbard
1	Mapping surface properties on Lomonosovfonna (SurfPro)	3395; 3231	Lomonosovfonna
2	Kongsvegen surge: Digital Elevation Model 2020 (KING_SURGE_DEM2020)	11431	Kongsvegen
3	Airborne Remote Sensing in South of Spitsbergen (current evolution of polar environment) (AirborneSOS)	10511, 11411, 10218, 6823, 11500	Wedel-Jarlsberg Land (Southwestern Spitsbergen)
4	Hindcasting and projections of hydro-climatic conditions of Southern Spitsbergen (HyMote)	11198	Southern Spitsbergen
5	The Vanishing White (VANWHITE)-Airborne Remote Sensing campaign Svalbard 2020.	11411	Coraholmen and Flintholmen
6	Automatic system for monitoring vegetation and environmental seasonal changes on Svalbard using hyperspectral data (ASMoVEn).	11063	Adventdalen
7	Long term changes in vegetation and permafrost in Rosenbergdalen (Rosenbergdalen)	11497	Rosenbergdalen
8	Barnacle Goose Ecology: interactions with a changing environment (GOOSE)	6359	Ny-Ålesund and surroundings
9	De-icing of Arctic Coasts: Critical or new opportunities for marine biodiversity and Ecosystem Services?	N/A	Adventdalen and Agardhfjorden
10	Icebergs study by Centre for Integrated Remote sensing and Forecasting for Arctic Operations (CIRFA)	10373	Icebergs around Nordaustlandet
11	Mapping of perennial firn aquifers and firn characteristics on Svalbard ice fields (PFA)	3395; 3231	Holtedahlfonna to Lomonosovfonna
12	Sea ice observations using aerial imagery from the Dornier during the Coordinated Arctic Acoustic Thermometry Experiment (CAATEX) cruise to 84 N with KV Svalbard (CAATEX-DORNIER)	N/A	Sea ice areas north of Svalbard

The call for access to aerial photos opened at about the same time as Norway closed down in March 2020, scientific review and logistics feasibility was done during May, and we began aircraft missions at the beginning of June 2020. In September 2019, NORCE had already conducted pilot missions to test sensor performance, and while planning these flight campaigns researchers were invited to request data from their area of interest. These test flights helped us to rapidly plan and execute this first operational activity in the time of the pandemic when it is most relevant for the scientific community. Researchers will be able to analyse the data from the aerial photographs by comparing them with satellite data and their own measurement series from previous fieldwork. Such comparisons are important both for researchers working on Svalbard and for owners of Earth observation satellites for validation of satellite data.

One of the major goals of SIOS is to work towards reducing the environmental footprint of scientific observations. Therefore, SIOS strongly supports and coordinates the usage of UAVs and aircraft platforms to acquire aerial imageries and hyperspectral data for the Svalbard research community. The most significant advantage of this activity is that the Dornier is always stationed in Longyearbyen. With its regular flights to Ny-Ålesund and

Svea, it makes it even more economical to collect data from the regions that are covered by the flight route. Flights to Villum Research Station at Station Nord, Greenland expands the applicability of these missions outside Svalbard. All the data collected during these missions will follow the FAIR principles [23]. The huge volume of airborne data being collected during these missions will make Svalbard one of the richest regions on the Earth in terms of the amount of hyperspectral data available. This years' flight missions continued until mid-September and currently we are planning our next call of opportunity to use the aircraft and UAVs for flight operations in 2021. This call would add more value to fill the gaps for consecutively two years as we are still not sure whether scientists would be able to carry out field activities next year yet.

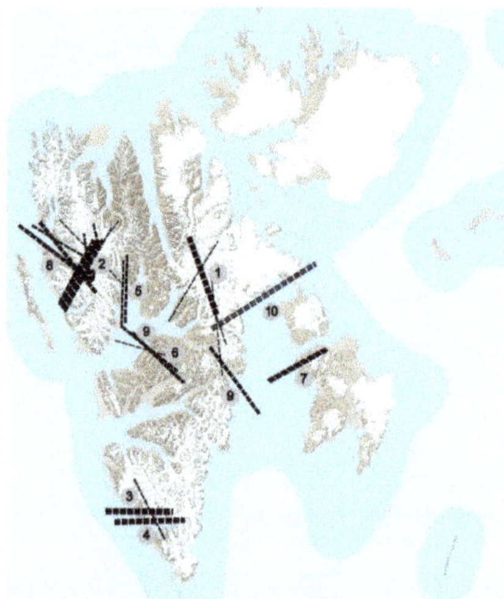

Figure 5. All flight lines covering an area of approximately 700 sq. km in the period 09 June 2020–20 September 2020. Ground resolution varies between 10–30 cm for the RGB images and 20–50 cm for the hyperspectral data.

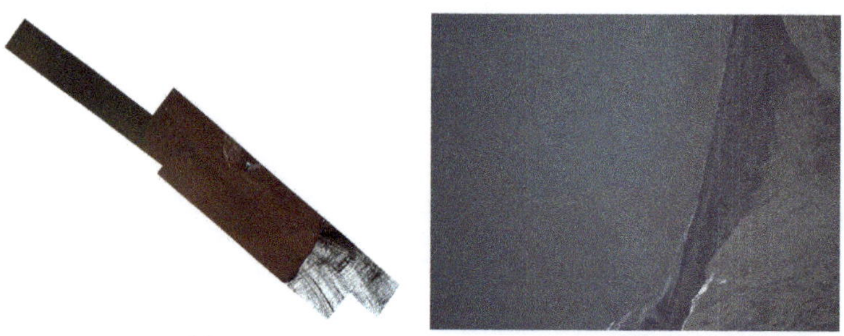

(a) (b)

Figure 6. SIOS-NORCE aircraft-based sample imagery and hyperspectral data over Svalbard. (**a**) A RGB (460 nm, 550 nm and 640 nm) composition of three hyperspectral lines at the bottom of Kongsvegen glacier and (**b**) an image of the coastline at Agardhbukta.

2.7.1. Possible Use of Dornier Based Image Acquisition to Fill the Data Gap in Various Fields of Science in Svalbard during Pandemic Times

The successfully completed flight campaigns in 2020 have been indicated as a flight map (Figure 5). SIOS hopes to attract more researchers and anticipates an increase in the number of requests in 2021. This effort can be continued even after the pandemic to reduce the environmental footprint of observations. The flights over Lomonosovfonna will be used to reconstruct mass changes for the mass balance years 2019–20 and 2020–21 by using high resolution RGB photography from the aerial missions in September 2020 to create a digital elevation model and compare the airborne digital elevation model (DEM) with the DEMs created from mobile differential global positioning system (dGPS) campaigns in spring 2019 and planned in 2021. The high resolution DEMs will also be used to identify newly opened crevasses and utilise safer routes along the glaciers for the spring 2021 to minimise the risk of accidents. The flights over Flintholmen and Coraholmen will be used for mapping of lichen vegetation extent and health in the region using different combinations of spectral bands and indices from hyperspectral data. Acquisition of hyperspectral data is critical to accurately detect and map lichen and biological soil crust dominated vegetation communities. Complementary on-site fieldwork was conducted (carbon flux, carbon economy, albedo, species abundance, and other relevant vegetation features) by NINA/UiT in the same week as the acquisition of hyperspectral data in July 2020.

2.7.2. Applications of Unmanned Aerial Vehicles (UAVs) in Svalbard during the Pandemic

During recent years, the use of UAVs has become an important element for earth observation activities—especially in Svalbard [24]. Typically, UAV-based fieldwork is conducted during spring and summer in Svalbard. In spring, cryosphere studies in remote locations can be reached with snowmobiles. In summer, most sites are accessed by boats or by foot. UAV fieldwork in the summer is conducted in a wide range of disciplines, most of which require snow-free ground conditions (e.g., biology, geology). The impact of the pandemic on UAV applications in Svalbard is hard to assess because no complete overview of planned field activities exists. However, we have a good overview of UAV-based field applications that were conducted in cooperation with SIOS and UNIS—although only on an anecdotal basis. For example, campaigns have been conducted by various groups in UNIS, NTNU, NIVA, the Nicolaus Copernicus University in Torun. Other campaigns have been cancelled by NORCE, Czech Academy of Sciences, University of Silesia in Katowice, and UNIS.

During the early phase of the pandemic in the spring, all fieldwork activities including UAV operations were stopped due to the lockdown. The local situation relaxed after Easter and field excursions with low risk were permitted. In practice, this meant that day trips around Longyearbyen were made possible again. During this time, strict international travel restrictions were in place in Europe, which meant that only researchers located in Svalbard had the opportunity to conduct fieldwork. This left a window of less than a month for snowmobile fieldwork until the snow cover disappeared in mid-May. This was a severe limitation, especially to cryospheric studies.

During the summer, international travel restrictions were relaxed which allowed researchers to travel to Svalbard for fieldwork. A 10-day quarantine had to be fulfilled on the Norwegian mainland before onward travel to Svalbard, resulting in it being mostly Norwegian scientists making the journey. However, some other nationalities, especially German, Polish, and Czech are also known to have come to Svalbard. To our knowledge, several UAV-based field activities have been conducted based in Longyearbyen, Ny-Ålesund, and Hornsund. UAVs were used for fieldwork in wide range of disciplines, including glaciology (glacier mapping), biology (vegetation index mapping), geology (outcrops mapping), cultural heritage preservation (mapping), and meteorology (atmospheric measurements). Most of these activities were conducted with simple, off-the-shelf UAVs and to our knowledge no advanced UAV operations occurred. It should be noted that field activities usually require a good amount of planning and preparation. As a result, several summer campaigns were cancelled in March and April because the confusing and unpredictable pandemic situation prevented proper planning.

In summary, the assessment of the impact of the pandemic on UAV-based fieldwork is difficult and must remain incomplete. To our knowledge, most UAV fieldworks during the springtime (March-June) were cancelled or postponed. Fieldwork in the summer and thereafter seemed to be conducted more successfully. This means that the largest knowledge gaps due to the pandemic are occurred in long-term studies in the cryospheric sciences. We know of at least one case where long-term monitoring of crevassed glaciers with UAVs was substantially affected by the lockdown, severely limiting their scope [25].

2.8. Ongoing and Upcoming Activities in Times of Pandemic

Our regular activities such as the ACCESS programme [26], that provides an opportunity to scientists from around the globe to utilise research infrastructure in Svalbard, have been affected badly by the current crisis. Due to COVID-19, 11 of the 14 projects that were funded in the 2019 call were postponed until 2021. The SIOS ACCESS programme supported the lunar aerosol optical depth (AOD) intercomparison campaign carried out in Ny-Ålesund in February 2020 (https://sios-svalbard.org/AOD_intercomparison_2020 (accessed on 15 January 2021)). The NY-alesund TurbulencE Fiber-Optic eXperiment (NYTEFOX, https://sios-svalbard.org/NYTEFOX_2020 (accessed on 15 January 2021)) is a project led by the University of Bayreuth and supported by SIOS. This project was also fortunate enough to complete fieldwork before the lockdown in Norway started. These are the only planned projects supported by SIOS that successfully completed their field campaigns in Svalbard in 2020. In addition, a virtual access project was completed (https://sios-svalbard.org/SWESOS_2020 (accessed on 15 January 2021)). It was anticipated that SIOS partners would have reduced capacity to accept new projects in the 2020 call because their research facilities/infrastructure were underutilised in 2020 and are expected to be under pressure in 2021 by national programmes. It was also considered likely that fewer applications would be received due to the uncertainty for researchers when planning field campaigns. We did not anticipate that virtual (access to data) or remote (sample collection by local operators) access will be affected by COVID-19 and saw the current situation as an opportunity to promote these options and increase remote support to scientists. When the call for access closed on 30th September 2020, 17 applications had been submitted, down from 29 in the 2019 call. For the first time we received applications for remote access to the research infrastructure, 3 in total. This shows how researchers are changing their behaviour in response to the current crisis.

Another regular activity of SIOS is the annual State of Environmental Science in Svalbard (SESS) [27] report. The report involves collaborative writing of a chapter by a team of authors but planned meetings and writing workshops have had to quickly move online in response to the pandemic. Despite this, the writing teams have worked hard and managed to deliver their contributions to the report in time. The report will be released in January 2021 and contains 11 chapters and a joint statement with topics ranging from the upper and lower atmosphere, pollution, coastal ecosystems, hydrology to snow cover and permafrost (Figure 7). Several chapters are focussing on remote sensing methodologies and almost all have remote sensing sections included.

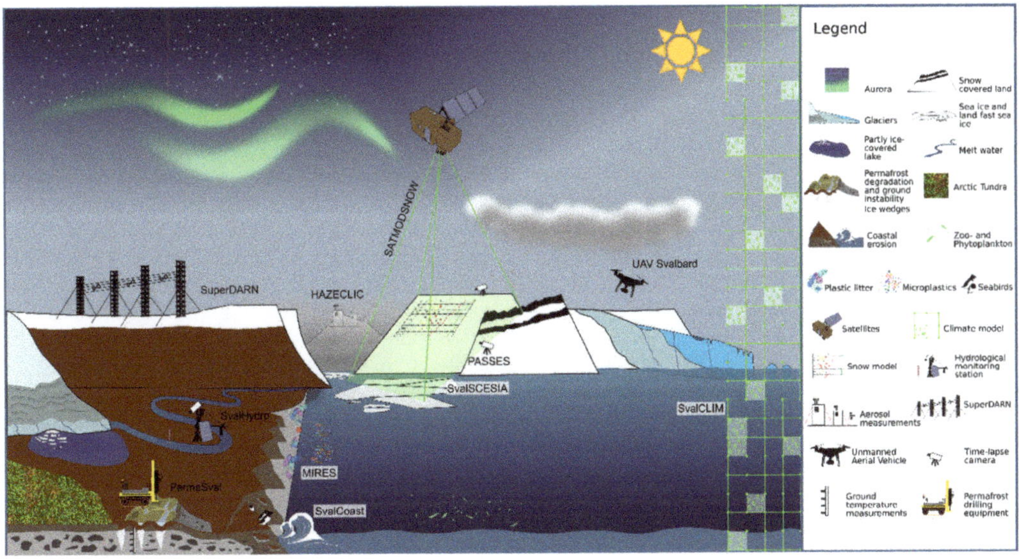

Figure 7. The State of Environmental Science in Svalbard (SESS) report 2020 includes chapters on the use of satellite images and in situ cameras to estimate snow cover, a review on unmanned aerial vehicle (UAV) activities and many more. The report can be downloaded from the SIOS website: https://sios-svalbard.org/SESS_Issue3 (accessed on 15 January 2021) (Illustration: Floor van den Heuvel).

3. Success Stories from SIOS Initiatives

The academic response of COVID-19 [28] shows that writing papers for scientific publication has been the most common task during the pandemic as this task can readily be performed remotely. The SIOS special issue which was launched in the beginning of the March-2020 proved to be an effective platform for researchers to publish their research. At the time of writing, we have 3 submissions [29,30] in this special issue. The top-five presentations by early career researchers from the SIOS online conference have been selected and the authors have been invited to submit full manuscripts to the special issue. In addition, SIOS's regular activity of SESS report provided an additional platform for scientists to write a collaborative report on Svalbard science.

We have received a few inputs on our mapping of the requirements from researchers and how SIOS can help fill data gaps using EO under "patch up your field data with RS observation" service. A group of researchers are trying to investigate the aerodynamic roughness length of crevassed glaciers, which influences the rate at which the glacier is exchanging heat with the atmosphere. With globally rising air temperatures, this mechanism is likely to lead to an accelerated warming and melting of highly crevassed glaciers. During this project they aimed to conduct UAV-based RS on crevassed glaciers on Svalbard. UAV flights were planned to map the glaciers and then the aerodynamic roughness length would be estimated based on the high-resolution digital elevation models (DEMs). However, the researchers were unable to conduct most of the spring and summer fieldwork. This means that they lack data for several glaciers (e.g., Nordenskiöldbreen, Wahlenbergbreen, Tunabreen, Mohnbukta), as well as a time-series that they planned to build (e.g., on Tunabreen). Researchers hoped to conduct as much fieldwork as possible in the summer, but it was not possible to fulfil all missions during the spring and early summer because of restrictions on field activities and lack of field personnel. Researchers are now investigating the possibility to use some satellite-based RS to estimate aerodynamic surface roughness on glaciers. UAV based data collected using limited flights in summer can be used for validation of the results retrieved from satellites.

A PhD student had planned her field campaigns in April and August 2020 to undertake ground penetrating radar (GPR) measurements to study the bedrock of glaciers and subglacial landforms, e.g., overpassed moraines. The spring campaign and GPR measurements were cancelled because of travel restrictions. Due to the lack of aerial images from the southern part of Wedel Jarlsberg Land, she would instead have to work on an older DEM which might affect the results of the study. Without her field campaigns she would not be able to understand the morphological changes in the glacier and estimate the relative age of glacial landforms. Further, without this data it will be difficult for her in working with her doctoral research without prolongation of her schedule. However, the team conducted geomorphological analysis over one glacial moraine with the use of UAV during August/September 2020. With the acquired aerial images and hyperspectral data she is able generate a DEM for geomorphological analysis and vegetation analysis for estimating the relative age of glacial landforms. Photos from UAV collected during summer will be used to validate data generated from photos collected by Dornier aircraft. In this example, it is evident that the SIOS support to acquire airborne data was very important to fill the data gap.

Another researcher had planned to conduct continuous velocity measurements of Hansbreen using dGPS over the period August–September 2020, operating mainly from on the Polish Polar Station Hornsund. This project was designed to expand current glacial monitoring and integrate it with oceanographic monitoring. The research group collected dGPS measurements, atmospheric and marine observations on the Hansbreen and in the Hornsund fjord during summer 2020. Additionally, terrestrial laser scanning was used to collect data over the Hansbreen front. The ground control points (GCP) measured with dGPS technique will serve to generate DEM from aerial imagery captured by Dornier aircraft in June 2020. Field data together with the aerial imageries will allow them to estimate the velocity of Hansbreen in the period July–August 2020.

A team of researchers had planned a series of activities including acquiring aerophotogrammetry data by UAV in order to generate a digital elevation model, radio-echo soundings on Werenskioldbreen and Hansbreen, digging snowpits, and collecting snow cores. However, because the field campaign was cancelled, they were left with a gap in the perennial continuity of geophysical data collection (radio-echo soundings), and lack of data from manual snowpits for data validation, causing a lot of damage to this scientific study to understand glacier behaviour. In order to fill the data gaps and mitigate the damage this season, the team has requested to use the Dornier to acquire airborne remote sensing data to process and obtain a DEM of Werenskioldbreen and Hansbreen, southwestern Spitsbergen. The DEM will be used to determine the internal structure of the glacier as a response to glacial processes and complex environmental conditions. The DEM will be combined and compared with other data from this region (e.g., geophysical).

The SIOS team received a few requests on the Logistics Sharing Notice Board. One of these requests was from the University of Silesia in Poland, who needed help to install instruments and collect samples in Wedel Jarlsberg Land. We were able to organise a short fieldwork campaign in the vicinity of the Polish Polar Station Hornsund in close cooperation with the University Centre in Svalbard (UNIS). Our team spent three days at the Baranowski Station (University of Wroclaw) next to the Werenskioldbreen maintaining automatic weather stations, time-lapse cameras, water gauge station, collecting water samples and measuring the water runoff in two catchments. More information on the fieldwork conducted in June 2020 as available online: https://sios-svalbard.org/News_20200703 (accessed on 15 January 2021). At this point we have fulfilled a few supporting campaigns to help scientists collect essential data and save instruments in these difficult times. The number of tourists visiting Svalbard is at a record low and this is accompanied by record high unemployment in Longyearbyen, especially among tourist guides. In this difficult situation, SIOS has tried to involve local skilled guides in scientific measurements which is a need of the hour. These activities will surely make a difference in paying back to our society in these challenging times.

SIOS has also taken the opportunity when many of our collaborators are confined to their offices and home offices to discuss new initiatives. One examples of such is the developing marine infrastructure network, which will be a forum for discussing all aspects of marine observational infrastructure (technical solutions, data sharing, measurement gaps, logistical cooperation). The inaugural workshop of this network was held online in November 2020. Networks such as these are extremely important to SIOS, as it helps the scientific community to be more robust in response to future disruptions such as pandemics. It is envisioned that should a similar situation occur, it may be possible for partners to cooperate to ensure the most essential mooring maintenance is carried out. By creating an integrated mooring network within which data are shared we can also ensure greater spatial coverage of essential variables, which makes the loss of data from one mooring less critical than it would otherwise be.

4. Response of SIOS Member Institutions to Mitigate Damage Caused by COVID-19

In this section, we focus on examples of activities conducted by SIOS member institutions in response to the COVID-19 pandemic.

Cancellation of the spring expedition of University of Silesia to Hornsund affected many researchers and students. Students of the International Environmental Doctoral School [31] could not conduct field work of their projects. The service and maintenance of automatic weather stations (AWS) was also suspended. As the Dornier aircraft collected data over Hornsund only once, scientists were not able to realise plans such as: analysis of geometry changes of Hansbreen glacier, dynamics of the glacier, evolution of glacial zones of Hansbreen and Werenskiolbreen, producing vegetation health maps, analysis of snow cover extent in the coastal zone of Hornsund fjord and continuation of a sedimentological study of a glacio-marginal landform and glacio-fluvial deposit at Werenskioldbreen. However, after relaxation of lockdown in July 2020, the team was able to conduct a summer field campaign in August/September 2020 to partly fill the gaps with the fieldworks such as measurements of Hansbreen surface by terrestrial laser scanning, dGPS measurements of mass balance stakes on Hansbreen and collecting imagery with UAV over the glacial moraine. This limited field data will be used as validation data for Dornier based aerial imagery. Aerial images from Dornier aircraft in connection with field data will enable the team to fill certain gaps, e.g., orthophotos together with laser scanning data will be used to estimate glacier velocity, DEMs from aerial photos will be used in geomorphological and glaciological analysis and hyperspectral data will be used for assessment of age of glacial landforms based on succession of vegetation.

Major parts of the annual field campaign of the Institute of Geophysics Polish Academy of Sciences (IG PAS) in Hornsund during the spring–autumn 2020 were affected by the pandemic. In Spring, the research facility in Hornsund closed to visitors of any origin, while in the late Autumn the disruption was mainly due to travel restrictions to Svalbard. A longstanding Hansbreen glacier mass balance monitoring programme was heavily affected, as glaciologists failed to reach the research facility in Hornsund. For safety reasons, crew from the research facility only managed to conduct limited snow depth measurements at the accessible mass balance stakes installed on the glacier. The lack of a specialised team in the field lead to gaps in measurements of glacier velocity and snow stratigraphy. In the autumn, ablation was measured at only 3 stakes. Most of the stakes melted out, as maintenance of the stake network planned for Spring 2020 failed. IG PAS plans an extensive field campaign in Spring 2021, to rebuild the stake network and estimate mass balance of Hansbreen glacier. It is planned to use EO and RS products provided within the SIOS framework to establish new mass balance monitoring network.

The Norwegian Institute for Nature Research (NINA) and UiT The Arctic University of Norway field campaigns (ASMoVEn) (https://www.researchinsvalbard.no/project/8922 (accessed on 15 January 2021)) were delayed due to travel restrictions from late April / beginning of May to beginning of June 2020. The plan was to set up near-surface remote sensing racks and a spectrometer in Adventdalen as a part of the SIOS-InfraNor project. Apart from the early setbacks, fieldwork was conducted smoothly with a new visit to Adventdalen carried out during 19–26 July 2020. Description of vegetation (species cover/frequency) was conducted (by UiT) on the sites where the instruments are situated. The racks and instruments were taken down for maintenance and retrieval of data in the last week of September 2020. Data gaps that exist during spring and some of dates in summer can be replaced by available cloud-free satellite data from Moderate Resolution Imaging Spectroradiometer (MODIS Terra and Aqua) and Sentinel-2A/2B if they exist. The collection of data from the rack-sites and the FLoX-spectrometer was successfully accomplished and the team will begin analysis of the spectrometer data in order to assess them against carbon-flux data from the eddy covariance tower situated nearby. SIOS airborne campaigns in Adventdalen were carried in three periods, and the instruments on the racks and spectrometer act as controls/calibration on the ground. Especially, the early acquisition airborne campaign during May–June may fill a gap during spring when the team was not able to monitor the vegetation properly due to travel restrictions.

A team of scientists from Uppsala University cancelled their whole field campaign in 2020, but a few essential tasks of maintenance and replacement of instruments were taken care of by engagement of residents in Svalbard. The research team is hopeful to make best use of aerial images and hyperspectral data acquired by Dornier aircraft-based sensors. Even after relaxation of lockdown in July 2020, it was too late to organize helicopter-based operations in the field site. The team worked indoors during the lockdown period to fill missing observations using EO and RS and also organise maintenance of instruments with the help of local residents in Svalbard. Around 75% of field data has been lost during the field season and we spent only a fraction of the budgeted costs. Organising a field campaign by involving residents has been emerged as an effective alternative for field activities, even after pandemic times.

Researchers from NORCE Norwegian Research Centre cancelled a 10-day field campaign involving UAV, GPR, and snow science satellite validation in March 2020, the week after lockdown was announced, while equipment was already shipped to Longyearbyen. To fill the data gap, the team ordered satellite acquisitions by TerraSAR-X but in situ observations could not be carried out. Parts of the scientific objectives related to measuring snow water equivalent with a UAV borne ground penetrating radar was performed in Tromsø in late April. Accordingly, some of the scientific objectives of time series of snow product development were met, but without Cal/Val activities with in situ measurements as this needed to be done in the spring season before the snowmelt. Further, NORCE chose to cancel a field trip to the Longyearbyen area in the last half of May. The purpose of the trip was to place time-lapse cameras (trail cameras) for phenological observations. These cameras have been used since 2014 and capture photos each hour between 10 a.m. and 3 p.m. every day from late May to mid-September, and the images then show the plant development. These images are then used as validation data for MODIS and Sentinel-2 data in mapping onset, peak, and end of the growth season.

The Japanese biological field campaign of National Institute of Polar Research (NIPR) was postponed, and maintenance of field equipment was not completed in 2020. The team is looking into the availability of hyperspectral cameras for considering possible future monitoring remotely. Japanese researchers left Svalbard in the critical period of middle of March when the lockdown period started. In the middle of October, they could revisit the field, however the ground was already frozen and no maintenance of ground sensors was possible.

The University of Oslo (UiO) had to cancel or postpone many of the planned field campaigns but managed to conduct three on short notice: installation and replacement of autonomous instruments on Kongsvegen glacier related to glacier hydrology and glacier dynamics; test and application of novel miniature loggers to explore subglacial drainage systems; exploration of ice-filled permafrost cases around Ny-Ålesund to assess soil dynamics and sorting in them. Whereas the two last ones cannot be supported much or even replaced by RS, the team will use high-resolution satellite radar and optical images to supplement the glaciological ground measurements on Kongsvegen.

The Pandora spectrometer instrument was installed by the Norwegian Institute for Air Research (NILU) at Sverdrup research facility in Ny-Ålesund in 2019 and calibrated in March 2020. This acts as a fiducial reference measurement for the Sentinel-5P mission. Being permanently mounted at the Sverdrup station and self-calibrated and remote-controlled, the Pandora spectrometer is relatively invulnerable to general travel cancellations and postponed campaigns during the Covid-19 outbreak. However, manual, and physical inspection of the instrument is required on a regular basis, especially in the winter season when there is a need to remove snow and ice on the sensor head and cables. The manual inspection is performed by local operators in Ny-Ålesund, and there is thus a risk for less maintenance, or worst-case scenario critical damage, of the instrument linked to a COVID-19 disease outbreak affecting the daily operations at the station.

The characterisation of the snow optical behaviour is usually performed during field surveys that require efforts in terms of preparation. Accordingly, one field campaign scheduled by National Research Council of Italy (CNR) during April–May 2020 in the framework of the iCUPE H2020 project [32] was cancelled due to the COVID-19 related restrictions and the relaxation of travel restrictions announced right at the end of the snow-melting season. The need for ground observations during the pandemic times has been attempted to be met by using in situ facilities such as time-lapse cameras and the Continuous Reflectance Monitor (CReM) that are continuously operating in the Ny-Ålesund area. While time-lapse cameras are facilities already operating in the area for decades, the continuous detection of the snow spectral reflectance is an innovative tool based on different setups [33,34]. The CReM setup and time-lapse cameras are operating at the Amundsen-Nobile Climate Change Tower in Ny-Ålesund and provide detailed information about the spectral behaviour of snow in the visible and in the short-wave infrared wavelength domains. The use of the Zeppelin webcam image time-series [35] processed using the automatic classification algorithm described by [36], provided information about the snow cover at different spatial scales. The ground-truth data collected by these instruments provides a quick solution for studying the melting season even if fieldwork is limited by the pandemic situation. The heavy impact of travel restrictions forced the team to use time-lapse cameras and the development of terrestrial photography applications in order to substitute field campaign. All the collected images, even if they do not replace direct snow reflectance measurements, can be integrated with the continuous data taken from the CReM and with those remotely sensed. The combination of different data sources could represent a valid approach for reconstructing the seasonality of the snowpack and the vegetation phenology. In spring 2021, the team is planning to install updated instruments and to evaluate drifts in old devices. Possibilities of making easy-to-install devices are being considered so that in the unfortunate case of cancelled field campaigns in future, they can be substituted by involving local residents in installing and maintaining instruments. This example shows the need of the development of innovative technology to make future field instruments in the Arctic which can be handled easily by indigenous people and residents. For broader relevance in the future, the team created strong synergies with international partners aimed at creating a network and harmonising the different procedures related to terrestrial camera infrastructure operations. Furthermore, the creation of a network focused on "terrestrial Photography ApplicationS on Snow covEr in Svalbard" (PASSES) [37] could be a seed for the growth of a camera network useful to the research community for compensating, at least partially, the lack of field observations in future.

As is the case with most of the in situ observational programmes around Svalbard from different organisations, the Indian Arctic Programme (IAP) of the National Centre for Polar and Ocean Research (ESSO-NCPOR) has been significantly affected by the restrictions imposed due to the COVID-19 pandemic resulting in cancelling all the field activities in Svalbard for the year 2020. The field visits which start around the same time as the national lockdown (March) by glaciologists, atmospheric scientists, and oceanographers were completely hampered by international travel restrictions. All such field campaigns for the current year, and probably for next year as well, had to be cancelled. This creates significant discontinuity for the in situ observations that are systematically recorded every year around the same time. Unlike the neighbouring countries in Europe, international travel restrictions imposed in India did not allow for field campaigns even after the first phase of lockdown in Svalbard was relaxed around mid-July. This has particularly affected the projects dealing with mass balance studies of targeted glaciers around Svalbard (e.g., Vester Broggerbreen, Feiringbreen), permafrost health monitoring, terrestrial, and marine biodiversity projects. However, the automated instruments used for observation of different multi-disciplinary parameters of the fjord and atmosphere have produced valuable information uninterrupted. Furthermore, with the help of the local company Kings Bay AS, a few planned CTD castings could also be conducted in the fjords. The major effect on these automated instruments was in terms of their routine physical check-up and calibrations, particularly when they remain unattended during a long period of time in winter. However, the observations from these instruments could be very crucial to decipher the effect of the pandemic and associated restrictions on different environmental parameters in Svalbard.

At the beginning of lockdown, the Norwegian Institute for Water Research (NIVA) had to postpone the annual deployment of the SIOS-InfraNor buoy in Adventfjorden until June 2020 when the travel restrictions were relaxed. Additionally, NIVA cancelled a few cruises with the cargo ship M/S Norbjørn. All the field sampling activities and laboratory experiments of the FreshFrate project (funded by the FRAM centre), addressing effects of freshwater runoff on Svalbard's coastal ecosystems, have been postponed to 2021 due to travel restrictions. The relaxation of lockdown did not help NIVA researchers as the sampling strategy was focused on seasonal variations which cannot be captured when there are gaps in sampling. In response, NIVA tried to mitigate data gaps with the help of local scientists from UNIS to collect water samples, in addition to the SIOS-InfraNor buoy deployment for example. Data collected from the buoy was intended for the Cal/Val activities of satellites but the delay in deployment affected this activity. NIVA will actively participate in training local researchers in buoy deployment experiments and to collaborate with UNIS to prepare for similar situations in the future. Besides that, the M/S Norbjørn cargo ship sailing plan has not been impacted by the COVID-19 situation, so the NIVA ferrybox installation on board has been measuring continuously and picked up discrete water samples.

The operational instrumentation that the Norwegian Meteorological Institute (MET Norway) maintains in Svalbard is designed to operate on its own for long periods. Due to weather and ice conditions, yearly maintenance is not always possible, implying that all automatic weather stations are designed for sustained standalone operation. A consequence of this thinking is that extension of the observation programme at the automatic weather stations is done in a manner not affecting the operational capability of the stations, meaning that the importance of sensors and their geophysical output is prioritised. As a consequence, the power supply is split between critical and non-critical sensors which may be shut down if power supply issues arise. The primary consequence of the COVID-19 lockdown on the institute's activities in Svalbard was on the planned upgrade of several stations. Originally the institute planned to establish a borehole for permafrost measurements co-located with the automatic weather station at Edgeøya–Kapp Heuglin, and to extend the sensor suite at Verlegenhuken with measurements of permafrost, snow, and surface irradiance. These activities were planned as part of the yearly maintenance for the automatic weather stations with transportation support from the Norwegian Coast Guard. Installations at Verlegenhuken were successfully completed and the flow of data established, but preparation of the borehole at Edgeøya–Kapp Heuglin had to be postponed due to the number of personnel involved. Furthermore, construction of the foundation for two inland automatic weather stations, in Nedre Sassendalen and Klauva, was cancelled due to COVID-19 and work is postponed until 2021. Operation of the manned stations of the institute has continued as planned, but with stronger restrictions on personnel exchange.

The Norwegian Polar Institute (NPI) is responsible for a large set of long-term observational programmes in Svalbard. These are related to climate, ecotoxicology, and biodiversity. NPI also has a geological mapping programme in Svalbard. Many of these activities overlap and are merged into the SIOS observations and priorities. The COVID-19 lockdown hampered some programmes severely. Notably the polar bear monitoring programme was unable to perform any activities during the essential spring field season. This was partially compensated for with a fortified activity in the autumn. Oceanographic observations around Svalbard have been hampered with several cruises planned for 2020 being cancelled. Other important programmes (ptarmigan and reindeer population studies as well as botanical inventories) were able to adapt and perform full programmes with field groups where key scientists had trained local personnel (mainly laid-off field adept individuals from the tourist industry). COVID-19 has thus opened a new perspective on how to manage some programmes within NPI which probably will both benefit local employment, lower total costs and decrease the environmental footprint (e.g., less travel). In Ny-Ålesund NPI has had a modified rotation of personnel (some individuals have spent longer periods in Ny-Ålesund than previously) and been able to run all long-term observational programmes. NPI has also supported several of the international institutions such that some of their programmes have been able to be maintained despite the institutions' inability to send their own personnel to Ny-Ålesund. This may also lead to new approaches to collaboration well beyond COVID-19 that are viewed as a positive outcome of the 2020 ordeal.

The Arctic Geophysics department (AGF) of the University Centre in Svalbard (UNIS) has two research groups—one in space physics and one in air–ice–sea interaction. For space physics, all fieldwork takes place in the dark season. Thus, the space physics group has not had cancellations of fieldwork due to COVID-19 related issues, as at the start of the outbreak, all fieldwork had already been done. For the dark season 2020/21 there has been cancellations of sounding rocket launches which the group participates in, and these have been postponed until the 2021/22 dark season. Apart from this, the space physics group will be able to maintain its standard observational programme at the Kjell Henriksen Observatory. The oceanographers usually use academic course related fieldwork cruises to collect research data from buoys. Due to COVID-19 the planned cruise was cancelled; thus the oceanographers were not able to collect data in April 2020 (winter) as has been done every year since 1999. However, they were able to obtain two days on a research vessel in order to secure the September data (summer) of the hydrography in Isfjorden and the shelf outside, thus maintaining the long time series for the Isfjorden transect. The meteorologists also use course fieldwork to collect data for ongoing research projects. Due to COVID-19, fieldwork was cancelled in autumn 2020, and data could not be collected. The data is used for e.g., validation of weather models in the Arctic. This has not been mitigated.

It is essential to mention that other institutions (including those not listed in this article) also planned to conduct measurement campaigns during the pandemic as part of the already developed procedures and due precautions. Unfortunately, the biggest obstacle that caused the campaigns to be cancelled was rapidly changing regulations on the border control to Norway and quarantine rules. In the absence of stable international traffic regulation in times of a pandemic, research activities in the following months may be subject to significant risk, despite appropriate measures being taken.

5. Discussion

Arctic regions are particularly vulnerable to situations like the pandemic [38]. Therefore, Arctic science has been significantly influenced by COVID-19. Even during the absence of pandemic, field research in the Arctic regions including Svalbard is relatively limited, with scarce scientific observations for studying Earth System Science [39]. Therefore, the effects of the persistent gap in field activities will reverberate for decades across scientific disciplines of Svalbard research.

At the time of writing this manuscript, Svalbard is still officially one of those regions on the planet that has escaped the first few waves of the COVID-19 pandemic. This was aided by the island's relative isolation, relatively low population (approx. 2800 across an area of 61,022 km^2, population density: 0.044/km^2), highly restrictive travel and quarantine measures adopted during the beginning of the pandemic, and relatively lower number of visits of tourists and tourist ships during 2020. However, based on our current knowledge of the variety of known and unknown modes of spread of the virus and looking at the speed of its spread, it is difficult to prevent the entry of virus to Svalbard even if strict travel and quarantine measures are still in place since the beginning of the pandemic.

The global pandemic has resulted in significantly decreased field activity in Svalbard leading to reduction in environmental impacts for a brief period at the expense of gaps in long-term data series and monitoring programmes. International cooperation through networks/consortia is necessary to save the essential scientific data and infrastructure in Svalbard. Norway started relaxing the lockdown slowly and Svalbard was open for Norwegian researchers and tourists on 1st June 2020, while it reopened to a few EU/EEA countries on 15th July. Currently travel from most countries is banned with restrictions and exemptions. With the temporary reopening of Svalbard during summer 2020, a few researchers could conduct their field campaigns during the rest of the period of summer and autumn, but the reopening came too late for those exclusively interested in the spring season. Even after the reopening of Svalbard, there was limited field activity. Since the beginning, the COVID-19 pandemic forced us to develop alternative activities for the utterly new, unforeseen, and unique situation. In the beginning, the trickiest part was none of us were prepared for such a working style and nobody was sure about the success of these activities. However, we wasted no time in starting these activities in response to the continuously changing situation. Our resources and activities have been listed in the GEO repository (https://www.earthobservations.org/covid19.php (accessed on 15 January 2021)) of COVID-19 response activities and tools developed by various organisations in the world. Here you can find many more relevant activities in response to COVID-19, but SIOS's activities are the only ones highlighting practical services to field scientists in the Arctic.

5.1. Perspectives on Impact of Data Loss in Long Term Monitoring and Science

Our knowledge of gradually more dramatic impacts of continuously changing climate will be held back if critical long-term monitoring work is interrupted [40,41]. It is foreseen that the pandemic will cause delays of up to or exceeding one year in planned field activities. The loss of data for the Svalbard related science for a year or more may have very differing implications on different projects. The impact is clearly highest for the experiments where the lack of regular maintenance can lead to loss of the instruments as well as the yearly data (e.g., glaciers mass balance and burial of AWSs). Cancelled fieldwork could affect not only a single campaign but also the possibility to maintain instruments and observation sites including drifts in field instruments during the whole season. This can cause large data gaps spanning over long periods during years with rapid environmental changes. For some measurement series, especially long-term monitoring, data gaps are immensely problematic in terms of missing important data in particular seasons or years and may complicate the statistical analysis.

5.2. Suggested Mitigation Measures

The pandemic lockdown has shown that it is beneficial to expand the engagement of local residents, research infrastructure providers, logistics operators, and collaborators to carry out joint research activities on mutual terms. This will require some local capacity building, with particular focus on training, equipment, and funding to cover fair pay. In a funding perspective, a coordinated and flexible response to delays and changes in field activities is required by funding agencies and responsible institutions. It is necessary to reserve funding for emergency mitigation to ensure sustained support for researchers and covering losses of local logistics and/or infrastructure providers.

It is also clear that the negative impact could partly have been mitigated if the scientific and logistics groups had prepared for this situation beforehand. However, such prior preparations or backup plans are not common for all kinds of scientific activities. For instance, AWSs are generally installed to work for years without major maintenance but glacier stakes and other instruments may need entirely different approach to prepare for similar situations in future.

In a situation where scientists suddenly are cut off from their field area—as the Svalbard research community experiences now—it would be favourable to have well planned or established citizen science projects that can be conducted by the local community. This could include collection of both ground truth data for validation of satellite-based products and collecting samples for scientific experiments. Current citizen science projects are often closely related to tourism activities, e.g., cruises, which also were halted during 2020. However, the lockdown left most tourist guides without work and a Svalbard specific citizen science model could be developed to involve these and others from the local community. The field expertise of local tourist guides, students, and parts of the general public would facilitate their involvement in scientific data collection activities. By developing easy-to-understand standard field data protocols and documents on regular maintenance of instruments, in combination with regular courses, a pool of potential citizen science contributors could be built. This would be a strong resource in general and especially in a situation like now to back-up scientific research in remote places like Svalbard.

It is imperative to establish a holistic perspective of the observing system with prioritisation of the importance of various observing efforts and their sensitivity to data gaps. This can result in specialised efforts to avoid gaps in top-priority data series in future situations. Some observing efforts already have mitigation plans for such circumstances (e.g., automatic weather stations that can operate for years without maintenance) with sufficiently scaled power supply (large batteries with additional solar panels and wind turbines) and communication capabilities (online access with delayed mode data dump locally) to compensate for weather and ice conditions preventing the scheduled maintenance. Such approaches have proven handy in the current situation, although not originally designed for a pandemic situation. For other activities alternative approaches relying on remote sensing applications (satellites, airplanes, UAVs) and to increase the cooperation between scientists and organisations in the areas of logistics, maintenance support and data sharing could be more appropriate in filling data gaps.

5.3. Earth Observation and Remote Sensing Perspective

Satellite coverage from polar orbiting satellites is good in the Arctic and Svalbard is probably one of those areas in the High Arctic that has the best potential coverage for satellite data. Utilisation of optical sensors is challenged by the Arctic night and extensive cloud cover. The latter is partly compensated by the high repeat cycle at high latitudes, but the Arctic night leaves imagers in the visible part of the electromagnetic spectrum useless parts of the year. Microwave sensors on the other hand do not have issues with the Arctic night or clouds, but have challenges related to spatial resolution (passive microwave), surface emissivity and/or weather (atmospheric column, wind effects etc.). Challenges vary depending on the application area and whether the study area is terrestrial or marine. Applications requiring high temporal resolution depend on the repeat cycle (specified by the inclination and footprint of the satellite/sensor) of the satellite platform and instrument considered. For studying glacial movements and spring thaw of permafrost shifts, Synthetic Aperture Radar (SAR) measurements can cover most of the requirements using proven algorithms, albeit, with the lack of ground verification or limited verification. For these applications, the impact will clearly be the lowest while better preparation for the pandemic situation could have reduced the impact even further. However, for other application areas, remote sensing algorithms are still under development and development of these will stall due to a lack of in situ validation data. In the future, it is necessary to critically assess the requirements of scientific projects in Svalbard aligned with preparation to reduce potential impacts of pandemic situations.

Space agencies have major roles to play in such circumstances where EO is highly relevant to fill data gaps. Proprietary or commercial satellite owners should take an active role in collecting frequent data in Svalbard and Arctic during pandemic times to help scientists filling the data gaps. One of such initiatives include Maxar's Open Data Program that has released an initial set of high-resolution satellite imagery in support of the COVID-19 response efforts. However, such efforts are limited to those systems where data collection and mode specification for satellite acquisitions are actively controlled.

The major non-commercial space agencies already collect the maximum amount of data possible over Svalbard and surrounding waters. Integration of remote sensing data into the field scientist workflow is covered in many disciplines, however, it is still pending in some disciplines. For the commercially available systems, dominantly in the optical region, increased collection of data depends on the availability of funds to get specific images at given times. For historical data, several of the commercial providers have low-cost to free access for scientific use which can readily be used. In order to increase the utilisation of satellite remote sensing products by field scientists, training is required to increase the knowledge and closer collaboration between communities is required to identify best practises and gaps in the current product portfolio. SIOS is annually arranging courses for field scientists to increase their practical knowledge on how to utilise satellite remote sensing.

Airplane and UAV observations are promoted in SIOS to make scientists aware of the capabilities that airborne remote sensing offer to acquire the data for their studies. This will hopefully encourage scientists to explore the potential of remote sensing observations. Use of UAVs enables minimised environmental footprint research in and on sensitive Arctic ecosystems [42] and may be the only approach to near ground sensing. There are also some opportunities created by the pandemic in the fields of airborne remote sensing. For example, the substantial cut in air travel opens the airspace for increased UAV usage—an example for this can be found in the Nepal where UAVs were used for mapping during the pandemic [43]. Additionally, UAV application may replace or complement ground-based fieldwork in environments that are limited due to social distancing requirements (e.g., during transportation) or due to shortened campaign durations (e.g., due to quarantine). In such cases, advanced UAV operations (e.g., with fixed-wing UAVs) could be used to access a field sites that are not possible to access physically due to social distancing or time limitations. Simpler UAV operations (i.e., with off-the-shelf UAVs) may be able to accelerate field activities, for example by replacing ground-measurements.

Remotely sensed observations in combination with autonomous surface based remote sensing equipment like normalised difference vegetation index (NDVI)-sensors, cameras and spectrometers [44] can supplement missing field data for terrestrial projects (e.g., vegetation mapping) for scientists who lost access to Svalbard during the summer of 2020. Very high spatial and temporal resolutions RS data is one of the important characteristics needed by field scientists, especially terrestrial scientists and glaciologists. Data acquisition by very high resolution (VHR) satellites (WorldView-2/3, GeoEye, Pleiades) may also fill in specific gaps for a few fields of studies during pandemic situations as well as gaps due to environmental restrictions or other events. This is particularly dependent on application area, existence of robust and validated methods/algorithms for deriving information, and availability of specific modes of data at a given time. Frequent VHR satellite acquisitions would be valuable in case of lack of field data for glaciologists and geomorphologists. For instance, temporal acquisitions of such datasets can be used to derive glacier velocities, changes in frontal ablations, generating DEMs, and many more. However, the optical satellites exhibit large repeat cycles complicating the use in most of the applications. Frequent observation should especially cover dedicated areas where long term monitoring is being conducted. Determination of such areas of interest would need a contact between researchers and provider of RS data. Here, the role of SIOS as a platform for such communication is necessary. Critical assessment of satellite products derived from VHR data, proved in situ validation, and the usefulness of these products for various scientific communities working in Svalbard is necessary. Continuous availability of high-resolution satellite data with repetitive acquisitions are not feasible if not planned in advanced. Most of the high-resolution satellite datasets are acquired by commercial satellite owners making the acquisition expensive for projects with limited funding. On the other hand, most of the funds on travelling and field campaigns have been saved due to the pandemic and can be substituted for acquiring high resolution satellite images.

Concerning measurements that normally rely dominantly on satellite measurements, a single year without ground truthing will only have minor impact on the long-term measurements. The lack of ground reference measurements may increase the uncertainty of the satellite measurements. This uncertainty will increase if there are several consecutive years without ground measurements. This year, the capability of using VHR measurements to replace some of the in situ measurements has not been materialized, because of lack of planning and the unavailability of funds. Even if there may be larger possibilities, due to less air traffic, for local aircraft and UAV measurements the lack of locally available personnel has reduced the amount of these measurements during the pandemic.

Filling data gaps using EO observations in various applications is limited by the limited quantity of satellite and airborne (UAV and aircraft) observations and limited resources including funds for skilled personnel and data processing infrastructure. Therefore, quantitative assessment on the number of data gaps that can be filled using EO is not possible until geospatial products that can replace in situ measurements are generated using EO data.

Based on our case studies and examples provided in previous sections, a few of the long-term data monitoring programmes that have been affected and can be partially mitigated using EO include; (1) Glacier front monitoring measurements, which can be partially mitigated using satellite-based and UAV-based measurements. (2) Missing GNSS based glacier surface velocity measurements can be replaced by offset tracking repeat optical and radar satellite data. (3) Vegetation mapping projects are affected due to limited ground-truthing. (4) Snow cover extent data is being partially mitigated using EO observations.

There are of course serious limitations in filling data gaps using EO and RS (see for example UiO subglacial and cave measurements). Even if a process is observable from space at all, spatial and temporal resolution of the RS data often set clear limitations. The temporal resolution of optical data interferes with cloud cover which often renders data useless even if acquired successfully. A related question is if one can design field measurements in a different way, so that EO and RS can fill gaps. For instance, instead of relying on ground instruments directly design an observing-modelling framework where ground and satellite measurements are assimilated. Unavailability of ground measurements will then still hurt, but perhaps less than in a purely ground-based scheme. However, modelling and assimilation maturity for different fields of science in Svalbard are at different levels, which makes such efforts challenging. For example, modelling in oceanography and meteorological applications are well proved and practiced, while it is difficult to implement in biodiversity research.

It is also realised during this period that more efforts are required on accurately downscaling the resolution of remote sensing products to match with or substitute a few in situ datasets. Providing these downscaled products particularly over Svalbard can be useful for the studies affected due to in situ data gaps during the pandemic. In summary, there is a need to critically assess quantitatively how much data gaps can be filled with purely EO and RS activities in the absence of field data in future situations in Svalbard. Financial analysis is also needed to determine the break-even point to conduct costly long-term observations to fill the ground-based measurements gaps. Together with scientists, funding agencies should develop a realistic financing plan for substitute research for the coming years.

5.4. Data Management and Its Importance in Pandemic Times

One of the missions of SIOS is to provide free and open data sharing. The pandemic has again demonstrated the need for the scientific community to address sharing of scientific data and publications to facilitate open science, collaboration and cooperation, and rapid dissemination of scientific results to the community [45]. The SIOS Data Management System (SDMS) federates information on in-situ and remote sensing data from a number of data repositories. This is achieved through harvesting information about datasets into a unified catalogue. When scientists cannot perform fieldwork, such integration of information becomes increasingly important. In the beginning of November 2020, SIOS hosted the first online workshop on SIOS Core Data (SCD), which are Earth System Science data for Svalbard that fulfil the defined criteria of scientific requirements, data availability and collecting commitment. SIOS core data are connected to Essential Climate Variables (ECVs) as defined by The Global Climate Observing System (GCOS), the State of Environmental Science in Svalbard Report (SESS) and the needs of stakeholders and users. The SCD workshop engaged scientists in defining, standardising and harmonising observation protocols as well the related issues in data management when fieldwork was not possible. In a way, this activity brought the scientific community of SIOS together to work on various aspects of making scientific data in Svalbard easier to integrate and reuse. There are a lot of data, and this situation gives some time to discuss, learn, and explore best practises of metadata at discovery and use level, encoding formats to identify data management gaps that have to be addressed.

5.5. Importance of Bringing the Community Together in Pandemic Times

SIOS's activities to bring the community together such as webinars, online conference, and a training course have been proved to be good initiatives not only for the Svalbard community but the broader area of the Arctic. If not beforehand, participants from online activities and the different aid packages delivered by SIOS now understand the strengths of a regional observing system in Svalbard. SIOS played an important role on welding the Svalbard scientific community into a more united organisation by bringing scientists together via annual SESS report, ACCESS programme, workshops, and airborne remote sensing opportunities during the pandemic situation.

Nevertheless, due to the lack of personal conversation and cooperation during the fieldworks or exchanging of the crew in the research stations, the discussions on joint international projects and joint activities are challenging compared to the time before the pandemic. The fieldwork is an essential part to provide a learning experience for early career researchers (ECRs) on communicating with senior researchers and fieldwork management. The lack of fieldwork may also impact the level of interest of students in polar science-related disciplines because of training course without a practical part in the field. We must remember that all of this cannot be replaced solely by online meetings.

5.6. Perspectives on How Sudden Perturbations Can Kindle Reforms in Our Approaches

Our case studies and practical examples have potential to enhance preparedness for future pandemic or similar situations. Examples and lessons learned from our experiences are probably transferrable to many fields of science and technology beyond their applications in polar regions. Many solutions were to some extent ad hoc choices on short notice but are here offered as a permanent resource for responding to future situations for international organizations, observing systems, and scientists in a more weighed manner.

For various reasons, the COVID-19 outbreak may have triggered a paradigm shift in the way we are conducting science studies in polar regions. This situation triggered a reduction of carbon footprint by reducing travels associated with meetings and fieldwork.

This means that there is probably smarter way to maintain scientific installations using the assistance and expertise from local technicians and cut travel costs. All the same, scientists will of course prefer to manage the maintenance themselves, since this is a way to ensure best performance of data and resulting science questions. Our activities and examples show that there is a need of re-assessing our ways of conducting field campaigns in Svalbard.

5.7. Importance of International Networks Like SIOS in Pandemic Times

The pandemic yet again pointed out the need of international coordination to optimise the observational networks, particularly in key scientific regions like Svalbard. Certainly, many organisations involved in observation of different aspects in Svalbard have faced data gaps to some extent. The only way to fill this is through the use of remote sensing and sharing of data within organisations. SIOS has been instrumental in this, by not only promoting and making remote sensing products easily accessible, but also bringing a large number of international organisations under a common platform which makes researchers aware of the developments and provide required information to fill the data gaps due to the restrictions during the pandemic. This is also a great opportunity to boost interdisciplinary studies in Svalbard given that the restrictions implemented during the pandemic may allow studying more clear signals on changes in different environmental parameters. For example, the changes in the aerosol emissions in lower latitudes can be expected to be significantly altered due to the restrictions. These relatively large signals in a pristine environment can make their influence on the ecosystem more evident helping researches to understand and quantify their role more accurately. Overall, while the pandemic has halted progress in enhancing the observation networks around Svalbard, it has opened up new dimensions which help the scientific communities to realise the requirement of international cooperation, coordination, and usefulness of common platforms like SIOS in optimum use of observations around Svalbard. We stress that at this time the scientific community should connect more and open up for increased data and infrastructure sharing. Integrated observing systems, like SIOS, should play a key role in this setting. We have time to think about our results and our scientific needs: we can discuss inside the community and find areas for cooperation and interaction. It is important to learn from the past and to plan for the future. In future proposals and planning of field campaigns should include a contingency plan that discusses backups in case of pandemics, lockdowns, or travel restrictions. SIOS could become an important element for supporting such backup plans in and around Svalbard.

6. Conclusions

COVID-19 has posed a distinctive challenge to the scientific community working in Svalbard in terms of lockdown, remote working, cancelled social and networking events, cancelled fieldwork, changing working practices, and isolation and anxiety. Even though the pandemic has changed our regular activities for the unforeseeable time, the Svalbard research community continues in its resilience under the extraordinary and unparalleled current circumstances. In this study, we highlighted a bunch of EO, RS, and operational activities of SIOS and its member institutions to respond to COVID-19 related travel restrictions and resulting limited field activities in Svalbard. These activities were developed for the Svalbard research community, but we believe that similar activities can be expanded to the whole Arctic, Himalayas, or Antarctica or other remote places where field activities were affected during the COVID-19 situation. This set of activities will form a base for all such activities being implemented in other parts of the world and could significantly contribute to developing a long-term sustainable plan for responding to future similar situations. A few of these activities were developed for the new normal situation created by the pandemic, but we believe that these activities are transferable for the regular activities in future. Our efforts show that even after the lockdown of Svalbard due to the pandemic, scientific observations can be done using various means including EO, RS, and planning logistical activities using local skilled personnel. Laid off workers can be better utilised in conducting field campaigns in a future pandemic or similar situations. Times of crisis often result in innovation, new technology and new ways of working and thinking. Our experience and lessons learned during the pandemic would help to accelerate the development of remote sensing methods and solutions in order to reduce the environmental footprint of research in the fragile Arctic region. Our activities were successful in bringing the whole Svalbard science community together in pandemic times by facilitating collaboration via the SESS report, ACCESS programme, airborne measurements, webinars, online conference, training course, and finally documenting our response to COVID-19 in the form of a manuscript. The pandemic is not over yet, but these positive activities could stimulate more activities using EO and RS. Today, usage of satellites and other new technologies in RS such as UAVs, and aircraft-based data collection are more relevant than ever. It also gives the opportunity to carry out monitoring and research that minimises the environmental footprint of such activities in sensitive High Arctic ecosystems. It is not possible to observe all essential Svalbard variables (SIOS core data) using EO and RS, but we can secure a few of these variables when in situ measurement is not possible. This article attempts to share our experience, expertise, and activities for a broader audience in observing systems in Earth System Science. Overall, the results of these activities will be realised in the coming years when we move forward into a post-pandemic world. The pandemic situation is changing rapidly, and the relevance of this work would be realised at later stages. We have summarised all the activities at the time of writing this paper while we continue preparing for the next year. A more detailed analysis of the short term (next field season), mid-term (next 3 years), and long-term (7–10 years) effects of the pandemic on Svalbard research have not been analysed yet. We anticipate that our activities will provide a foundation to respond to future similar situations.

Author Contributions: Conceptualization and development, S.D.J., V.A.P., H.L., C.H., I.J., D.I., B.N.A., Ø.G., K.H.; Figures and illustrations, C.H., S.D.J.; Writing—Original Draft Preparation, S.D.J., B.N.A.; Writing—Review & Editing, V.A.P., H.L., C.H., I.J., D.I., B.N.A., Ø.G., A.S., R.H., H.T., A.K., M.B., R.S. (Roberto Salzano), B.L., K.A.H., R.S. (Rune Storvold), L.N., R.S. (Rosamaria Salvatori), K.P.K., S.C., D.A.L., R.E., T.R.L., E.M., S.R.K., H.E., A.M.F., J.Z., S.M., K.O.N.; Project Administration, S.D.J., V.A.P., H.L., C.H., I.J., D.I., B.N.A., Ø.G., K.H. All authors have read and agreed to the published version of the manuscript.

Funding: The SIOS KC would like to acknowledge the support from the Research Council of Norway (RCN) through projects No. 291644 & No. 269927 and Norwegian Space Agency (NoSA) through contract No. NIT 12.20.5. The research team of UiT and NINA was funded by RCN project VanWhite, No. 287402 and SIOS-InfraNor, project No. 269927. Authors from the Institute of Geophysics Polish Academy of Sciences would like to acknowledge the support from the Polish Ministry of Education and Science Project No. DIR/WK/201805.

Data Availability Statement: No new data were created or analyzed in this study. Data sharing is not applicable to this article.

Acknowledgments: We thank all the participants of the webinar series, online conference and all our activities conducted during the summer of 2020. We thank editors of the special issue, RSWG members and all participants for sharing their experience during webinar series and online conference.

Conflicts of Interest: The authors declare no conflict of interest.

References

1. Svalbard Integrated Arctic Earth Observing System (SIOS). Available online: https://sios-svalbard.org/ (accessed on 23 July 2020).
2. Buch, E.; Madsen, M.S.; She, J.; Stendel, M.; Leth, O.K.; Fjæraa, A.M.; Rattenborg, M. Arctic In Situ Data. Availability, Issue: 2.1 Date: 09/12/2019 EEA/IDM/15/026/LOT1. Available online: https://insitu.copernicus.eu/library/reports/CopernicusArcticDataReportFinalVersion2.1.pdf (accessed on 2 November 2020).
3. World Health Organisation (WHO). Coronavirus Disease (COVID-19) Situation Report—174. 2020. Available online: https://www.who.int/docs/default-source/coronaviruse/situation-reports/20200712-covid-19-sitrep-174.pdf?sfvrsn=5d1c1b2c_2 (accessed on 21 December 2020).

4. Anderson, R.M.; Heesterbeek, H.; Klinkenberg, D.; Hollingsworth, T.D. How will country-based mitigation measures influence the course of the COVID-19 epidemic? *Lancet* **2020**, *395*, 931–934. [CrossRef]
5. Frame, B.; Hemmings, A.D. Coronavirus at the end of the world: Antarctica matters. *Soc. Sci. Humanit. Open* **2020**, *2*, 100054. [CrossRef]
6. ICIMOD. *COVID-19 Impact and Policy Responses in the Hindu Kush Himalaya*; International Centre for Integrated Mountain Development: Kathmandu, Nepal, 2020. Available online: https://lib.icimod.org/record/34863 (accessed on 2 November 2020).
7. Hughes, K.; Convey, P. Implications of the COVID-19 pandemic for Antarctica. *Antarct. Sci.* **2020**, *32*, 426–439. [CrossRef]
8. SCAR. Survey: The Impact of COVID-19 on the Antarctic Research Community. 2020. Available online: https://www.scar.org/scar-news/humanities-and-social-science-news/covid19-survey/ (accessed on 2 November 2020).
9. European Space Agency and European Commission. Rapid action on Coronavirus and EO (RACE). 2020. Available online: https://race.esa.int/ (accessed on 2 November 2020).
10. Nichol, J.E.; Bilal, M.; Ali, M.A.; Qiu, Z. Air Pollution Scenario over China during COVID-19. *Remote Sens.* **2020**, *12*, 2100. [CrossRef]
11. Fan, C.; Li, Y.; Guang, J.; Li, Z.; Elnashar, A.; Allam, M.; de Leeuw, G. The Impact of the Control Measures during the COVID-19 Outbreak on Air Pollution in China. *Remote Sens.* **2020**, *12*, 1613. [CrossRef]
12. Liu, Q.; Sha, D.; Liu, W.; Houser, P.; Zhang, L.; Hou, R.; Lan, H.; Flynn, C.; Lu, M.; Hu, T.; et al. Spatiotemporal Patterns of COVID-19 Impact on Human Activities and Environment in Mainland China Using Nighttime Light and Air Quality Data. *Remote Sens.* **2020**, *12*, 1576. [CrossRef]
13. SIOS Special Issue on "Earth Observation (EO), Remote Sensing (RS), and Geoinformation (GI) Applications in Svalbard". Available online: https://sios-svalbard.org/SpecialIssueRemoteSensing (accessed on 23 July 2020).
14. SIOS Monthly Webinar Series on Earth Observation (EO) and Remote Sensing (RS) Talks. Available online: https://sios-svalbard.org/RSWebinarSeries (accessed on 23 July 2020).
15. International Glaciological Society Global Seminar Series. 2020. Available online: https://www.igsoc.org/igswebinar/ (accessed on 2 November 2020).
16. Polar Geospatial Centre-Webinar Series. Available online: https://www.pgc.umn.edu/tag/webinar/ (accessed on 2 November 2020).
17. SIOS's Online Conference on EO/RS/GI in Svalbard—4–5 June 2020. Available online: https://sios-svalbard.org/RS_OnlineConference2020 (accessed on 23 July 2020).
18. Training Course on Terrestrial Remote Sensing in Svalbard. Available online: https://sios-svalbard.org/TRST_2020 (accessed on 20 July 2020).
19. Patch up Your Field Data Gaps with Remote Sensing. Available online: https://sios-svalbard.org/News_20200324a (accessed on 23 July 2020).
20. SSF Connecting Svalbard Researchers for Field Logistics. Available online: https://www.facebook.com/groups/139776787422274/ (accessed on 23 July 2020).
21. SIOS Logistics Sharing Notice Board. Available online: https://sios-svalbard.org/logistic-notice-board (accessed on 23 July 2020).
22. SIOS's Announcement of Opportunity (AO) in Airborne Remote Sensing-2020. Available online: https://sios-svalbard.org/AirborneRS (accessed on 23 July 2020).
23. Wilkinson, M.; Dumontier, M.; Aalbersberg, I.; Appleton, G.; Axton, M.; Baak, A.; Blomberg, N.; Boiten, J.-W.; da Silva Santos, L.B.; Bourne, P.E.; et al. The FAIR Guiding Principles for scientific data management and stewardship. *Sci. Data* **2016**, *3*, 160018. Available online: https://www.nature.com/articles/sdata201618 (accessed on 7 January 2021). [CrossRef] [PubMed]
24. Hann, R.; Altstädter, B.; Betlem, P.; Deja, K.; Dragańska-Deja, K.; Ewertowski, M.; Hartvic, F.; Jonassen, M.; Lampert, A.; Laska, M.; et al. Scientific Applications of Unmanned Vehicles in Svalbard (UAV Svalbard). In *State of Environmental Science in Svalbard (SESS) Report 2020*; Svalbard Integrated Arctic Earth Observing System (SIOS): Longyearbyen, Norway, 2021. [CrossRef]
25. Hodson, A.; Jonassen, M.O.; Hann, R.; Hess, C.; Garreau, A.; Dachauer, A. Using UAVs to Investigate Effects of Crevasses upon Glacier Surface Melting (Crevasse UAV), 2019–2020, RIS-ID 11148. Available online: https://www.researchinsvalbard.no/project/9044 (accessed on 23 November 2020).
26. SIOS ACCESS Programme. Available online: https://sios-svalbard.org/RIAccess (accessed on 24 July 2020).
27. SIOS State of Environmental Science in Svalbard (SESS) Report. Available online: https://sios-svalbard.org/SESSreport (accessed on 20 July 2020).
28. Chantelle, R.; Frederick, F. The academic response to COVID-19. *Front. Public Health* **2020**, *8*, 621563. [CrossRef]
29. Nakoudi, K.; Ritter, C.; Böckmann, C.; Kunkel, D.; Eppers, O.; Rozanov, V.; Mei, L.; Pefanis, V.; Jäkel, E.; Herber, A.; et al. Does the Intra-Arctic Modification of Long-Range Transported Aerosol Affect the Local Radiative Budget? (A Case Study). *Remote Sens.* **2020**, *12*, 2112. [CrossRef]
30. Vickers, H.; Karlsen, S.R.; Malnes, E. A 20-Year MODIS-Based Snow Cover Dataset for Svalbard and Its Link to Phenological Timing and Sea Ice Variability. *Remote Sens.* **2020**, *12*, 1123. [CrossRef]
31. International Environmental Doctoral School. Available online: https://www.mssd.us.edu.pl/en/ (accessed on 23 July 2020).
32. Petäjä, T.; Duplissy, E.M.; Tabakova, K.; Schmale, J.; Altstädter, B.; Ancellet, G.; Arshinov, M.; Balin, Y.; Baltensperger, U.; Bange, J.; et al. Overview—Integrative and Comprehensive Understanding on Polar Environments (iCUPE): The concept and initial results, Environments (iCUPE). *Atmos. Chem. Phys.* **2020**, *20*, 8551–8592. [CrossRef]

33. Picard, G.; Libois, Q.; Arnaud, L.; Verin, G.; Dumont, M. Development and calibration of an automatic spectral albedometer to estimate near-surface snow SSA time series. *Cryosphere* **2016**, *10*, 1297–1316. [CrossRef]
34. Salzano, R.; Lanconelli, C.; Salvatori, R.; Esposito, G.; Vitale, V. Continuous monitoring of spectral albedo of snowed surfaces in Ny-Ålesund. *Rend. Lincei. Sci. Fis.* **2016**, *27*, 137–146. [CrossRef]
35. Pedersen, C. *Zeppelin Web Camera Time-Series, Dataset*; Norwegian Polar Institute: Tromsø, Norway, 2013. [CrossRef]
36. Salzano, R.; Salvatori, R.; Valt, M.; Giuliani, G.; Chatenoux, B.; Ioppi, L. Automated Classification of Terrestrial Images: The Contribution to the Remote Sensing of Snow Cover. *Geosciences* **2019**, *9*, 97. [CrossRef]
37. Salzano, R.; Aalstad, K.; Boldrini, E.; Gallet, J.C.; Kępski, D.; Luks, B.; Nilsen, L.; Salvatori, R.; Westermann, S. Terrestrial Photography Applications on Snow Cover in Svalbard (PASSES). In *State of Environmental Science in Svalbard (SESS) Report 2020*; Svalbard Integrated Arctic Earth Observing System (SIOS): Longyearbyen, Norway, 2021. [CrossRef]
38. Petrov, A.N.; Hinzman, L.D.; Kullerud, L.; Degai, T.S.; Holmberg, L.; Pope, A.; Yefimenko, A. Building resilient Arctic science amid the COVID-19 pandemic. *Nat. Commun.* **2020**, *11*, 6278. [CrossRef] [PubMed]
39. Lee, C.; Eicken, H.; Jakobsson, M. Introduction: The Arctic Observing Summit 2013. *Arctic* **2015**, *68*, iii. [CrossRef]
40. IPCC. IPCC Special Report on the Ocean and Cryosphere in a Changing Climate. 2019. Available online: https://www.ipcc.ch/srocc/ (accessed on 20 December 2020).
41. Laing, A.; Garrison, C. 'Isolated within Isolation': Keeping out Coronavirus in the Frozen Antarctic. 2020. Available online: https://uk.reuters.com/article/uk-health-coronavirus-antarctica-feature/isolated-within-isolation-keeping-out-coronavirus-in-the-frozen-antarctic-idUKKCN21W2OE (accessed on 21 December 2020).
42. Tømmervik, H.; Karlsen, S.R.; Nilsen, L.; Johansen, B.; Storvold, R.; Zmarz, A.; Beck, P.S.; Johansen, K.S.; Høgda, K.A.; Goetz, S.; et al. Use of unmanned aircraft systems (UAS) in a multiscale vegetation index study of Arctic plant communities in Adventdalen on Svalbard. *EARSeL eProceedings* **2014**, *13*, 47–52.
43. Wight, A.J. COVID clears the skies for Earth-observing drones in Nepal. *EOS* **2020**, *101*. [CrossRef]
44. Anderson, H.B.; Nilsen, L.; Tømmervik, H.; Karlsen, S.R.; Nagai, S.; Cooper, E.J. Using Ordinary Digital Cameras in Place of Near-Infrared Sensors to Derive Vegetation Indices for Phenology Studies of High Arctic Vegetation. *Remote Sens.* **2016**, *8*, 847. [CrossRef]
45. Data Together COVID-19 Appeal and Actions. Available online: https://codata.org/data-together-covid-19-appeal-and-actions/ (accessed on 21 December 2020).

remote sensing

MDPI

Article

A 20-Year MODIS-Based Snow Cover Dataset for Svalbard and Its Link to Phenological Timing and Sea Ice Variability

Hannah Vickers *, Stein Rune Karlsen and Eirik Malnes

NORCE Norwegian Research Centre AS, P.O. Box 6434, 9294 Tromsø, Norway; skar@norceresearch.no (S.R.K.); eima@norceresearch.no (E.M.)

* Correspondence: havi@norceresearch.no

Received: 4 March 2020; Accepted: 31 March 2020; Published: 1 April 2020

Abstract: The climate in Svalbard has been warming dramatically compared with the global average for the last few decades. Seasonal snow cover, which is sensitive to temperature and precipitation changes, is therefore expected to undergo both spatial and temporal changes in response to the changing climate in Svalbard. This will in turn have implications for timing of terrestrial productivity, which is closely linked to the disappearance of seasonal snow. We have produced a 20-year snow cover fraction time series for the Svalbard archipelago, derived from MODIS (Moderate Resolution Imaging Spectroradiometer) Terra data to map and identify changes in the timing of the first snow-free day (FSFD) for the period 2000–2019. Moreover, we investigate the influence of sea ice concentration (SIC) variations on FSFD and how FSFD is related to the start of the phenological growing season in Svalbard. Our results revealed clear patterns of earlier FSFD in the southern and central parts of the archipelago, while the northernmost parts exhibit little change or trend toward later FSFD, resulting in weaker trends in summer and winter duration. We found that FSFD preceded the onset of the phenological growing season with an average difference of 12.4 days for the entire archipelago, but with large regional variations that are indicative of temperature dependence. Lastly, we found a significant correlation between variations of time-integrated SIC and variations in FSFD, which maximizes when correlating SIC northeast of Svalbard with FSFD averaged over Nordaustlandet. Prolonged sea ice cover in the spring was correlated with late snow disappearance, while lower-than-average sea ice cover correlated with early snow disappearance, indicating that proximity to sea ice plays an important role in regulating the timing of snow disappearance on land through influencing the regional air temperature and therefore rate of spring snowmelt.

Keywords: snow cover; remote sensing; sea ice variability; vegetation growth; arctic climate change

1. Introduction

Seasonal snow cover evolves continually due to meteorological factors affecting the physical properties of the seasonal snowpack. As such, seasonal snow cover may be considered a sensitive indicator of climate change. During the last few decades, the Arctic in particular has experienced twice the rate of warming compared with the global average, also referred to as the Arctic amplification [1,2]. Warming has been linked to loss of Arctic sea ice cover [3–6], as larger areas of open water absorb greater amounts of solar radiation, which is re-emitted as long-wave radiation and warms the lower troposphere. Sea ice loss is also accelerating [7], resulting in an overall reduction in albedo, which serves to amplify snow-temperature feedback processes [8]. In particular, sea ice loss north of the Svalbard archipelago has been found to be more pronounced during winter than summer [9], and the air temperature is expected to increase on average by 3–4 °C on the west coast, while even greater temperature increases of up to 6–8 °C are expected to occur in the northeastern regions by the end of the 21st century. Wintertime warming is expected to be stronger than in other seasons [10–12], accompanied by increased precipitation as sea ice loss enhances surface evaporation and moisture transport. Winter warming in combination with greater precipitation has meant that rain-on-snow events during winter have also become more frequent in the most recent decades [13–15] in Svalbard. Such events cause the formation of basal ice layers at the ground surface, which negatively affects the food availability for grazing animals [16] since their potential food supply is essentially locked off.

Trends of increasing temperature and precipitation on Svalbard have been associated with significant changes in seasonal snow cover during the last few decades. These changes include an upward trend in the equilibrium line altitude (ELA) of perennial snow, later onset of snow in response to autumn warming [17], earlier disappearance of seasonal snow, shorter duration of snow cover and both a decrease in and earlier timing of maximum snow depth [18]. Traditional methods for snow cover change monitoring include in-situ point measurements of snow depth as well as combining snowpack modeling with regional climate models [17]. While models have been used to provide estimates of snow cover at large spatial scales, in-situ measurements are time consuming and limited in spatial extent where long-term and large-scale monitoring of snow cover changes are desirable. On the other hand, remote sensing methods are an effective tool for providing measurements from which snow cover fraction can be derived [19] at spatial resolutions on the order of 1 km and over large areas. For example, imaging spectrometers such as the Advanced Very High Resolution Radiometer (AVHRR), Moderate Resolution Imaging Spectroradiometer (MODIS) and the Visible Infrared Imaging Radiometer Suite (VIIRS) have been utilized for this purpose since as far back as 1982, providing a large archive of measurements that cover periods long enough for identifying climate-related trends in snow cover characteristics. These sensors provide observations at daily time scales as long as the surface is not obscured by cloud cover and is illuminated by sunlight. Methods also exist for remotely sensing snow cover that are not dependent on cloud-free conditions or solar illumination; one such example is the use of Synthetic Aperture Radar (SAR), which has been used to detect and map the occurrence of wet snow cover [20]. However, a weakness of using SAR is that it cannot distinguish dry snow from bare ground, which limits its uses during winter when dry snow cover largely dominates. Passive microwave radiometers are also utilized to obtain daily information on snow cover but typically suffer from substantially poorer resolution (~20 km), and measurements are therefore subject to a greater degree of uncertainty in complex terrain such as along coastlines and in mountainous areas, where the scale size of features is typically smaller than the spatial resolution. Instruments with low spatial resolution are thus not optimal for studies of snow cover over Svalbard, which is predominantly made up of mountainous terrain and fjords. Active microwave systems such as the QuikSCAT SeaWinds scatterometer instrument are also sensitive to liquid water content in snow and as such have been successfully exploited in earlier studies to map the timing of total melt days and summer melt onset over glaciated areas of Svalbard [21,22]. However, while the onset of snowmelt can give an indication of when snow volume begins to decrease systemically, it does not provide information on when snow has completely disappeared, which is important for phenological processes.

A remote sensing technique that exploits MODIS Terra data together with temporal interpolation methods that correct for presence of cloud cover has been developed for mapping snow cover characteristics over Northern Scandinavia [23], demonstrating an approach that can address challenges related to Arctic weather and the absence of solar illumination for several months of the year. It is therefore a primary goal of this study to use the MODIS Terra dataset for the period 2000–2019 and adapt this method for application to the Svalbard archipelago, in order to study changes and trends in patterns of snow cover disappearance, duration and onset.

Since snowmelt is driven by temperature changes, it follows that in a warming climate, ecosystems with seasonal snow cover may experience earlier plant development of some of the species as snow disappears earlier [24]. The onset of the growing season on Svalbard has been studied and mapped using a method based on MODIS Terra surface reflectance products [25] for the period 2000–2013. Based on this relatively short period, the authors noted only a very weak delayed onset of 1.3 days for the whole of Svalbard. Other studies have alluded to similarities between spatial patterns of snow season duration and growing season duration for Northern Scandinavia, but they did not present a quantitative examination of the relationship between those two parameters [23]. Hence, the secondary goal of this study is to evaluate how the timing of snow disappearance affects the timing of the growing season in Svalbard at MODIS pixel resolution scale. Moreover, we wish to establish whether the relationship between snow disappearance and growing season onset is changing as a result of the ongoing warming trends in the Arctic. This knowledge is important since it is already well known that earlier snow disappearance may expose vulnerable plants to low temperatures, leading to frost damage and negative impacts on plant growth and productivity [26–28]. Indeed, at Arctic latitudes it has been shown that earlier snowmelt due to warming leads to earlier onset of plant growth but has varied effects on overall plant production [29].

Proximity to sea ice is known to affect air temperature on adjacent land masses [30], thereby constituting an indirect factor that also influences the timing of snowmelt and disappearance and hence onset of the phenological growing season. Indeed, an anticorrelation between September sea ice extent and total number of melt days was found under a study of two glaciated archipelagos in the Russian High Arctic [31] using multiple passive and microwave sensor data. Other studies have discussed the apparent link between variability in the sea ice edge and onset of snowmelt in Svalbard [21], though this relationship was not quantitatively established. On the other hand, a decline in sea ice may lead to mass growth in certain glaciated regions on Svalbard due to enhanced wintertime precipitation resulting from increased evaporation over open water [32,33]. Outside of Svalbard, a link between sea ice decline in the Barents Sea and more severe Eurasian winter snowfall has been reported, which was attributed to increased frequency of atmospheric blocking patterns and cold air advection [34]. There is therefore a complex relationship between variations and trends in Arctic sea ice cover that affects regional temperature and precipitation trends and consequently wintertime snow cover and timing of snow disappearance in spring, which in turn controls onset of growth. Our final goal of this study is therefore to quantify how sea ice concentration around Svalbard is related to the timing of snow disappearance in Svalbard, which is derived by optical remote sensing. In summary, the main objectives of this paper are as follows: firstly, to map snow cover fraction over Svalbard for the period 2000–2019 by applying earlier developed algorithms for estimating and mapping snow cover fraction using MODIS Terra data (MOD10A1 at 500 m resolution); secondly, to calculate the timing of snow disappearance and onset for each year to further derive trends of snow cover change over 20 years; thirdly, to study how the timing of snow disappearance was related to both timing of plant phenology processes and sea ice variability around Svalbard. Moreover, we aim to quantify the observed relationships at both local and regional scale and for the entire Svalbard archipelago.

2. Materials and Methods

2.1. Study Area

The Svalbard archipelago is located in the Arctic ocean, approximately halfway between the northernmost point of the Norwegian mainland and the North Pole. It consists of a group of nine islands, of which the largest is Spitsbergen. The Svalbard archipelago covers a total area of 61,000 km^2 of which 60% is glaciated, 30% is covered by barren rock and the remaining 10% is covered by vegetation. The largest glacier, Austfonna, is situated on the island Nordaustlandet. The archipelago spans latitudes between 74–81°N and longitudes ranging from 10–35°E, which means both midnight sun and polar night can dominate for large parts of the year. In the administrative center Longyearbyen, polar night conditions are present from 26 October until 15 February, with midnight sun from 20 April to 23 August. To the west of the archipelago, the West Spitsbergen Current brings warm salty water from the Atlantic Ocean northwards, resulting in milder climate than experienced at similar latitudes elsewhere. Meteorological data recorded at Svalbard Airport, close to Longyearbyen, show that the mean annual temperature in the period of study (2000–2019) ranges from −6.1 °C (2003) to 0 °C (2016), while annual precipitation ranges from a minimum of 142.1 mm (2005) up to 310 mm (2016).

Figure 1 presents an overview of the Svalbard archipelago showing both the vegetated part of the land and the locations of six selected meteorological stations from where data were collected, while in Figure 2 we present an illustration of the research scheme, datasets and procedures that are used and carried out in this study.

Figure 1. Location of the Svalbard archipelago showing the six meteorological sites where snow depth measurements are made. Areas with mean July (2000–2013) Normalized Difference Vegetation Index (NDVI) above 0.1 are labeled "vegetated areas", extracted from Karlsen et al. [25].

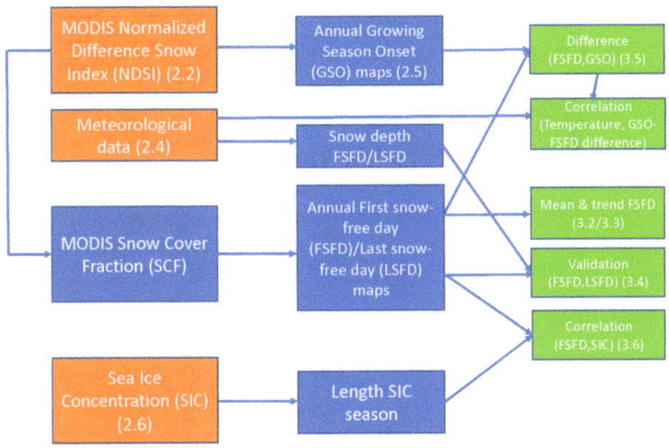

Figure 2. Overview of the research scheme and datasets and methods used in this study. Orange components refer to datasets, blue boxes indicate derived products and green boxes are end results.

2.2. MODIS Terra Dataset

2.2.1. Snow Cover Fraction Algorithm

In this study we used the 2000–2019 NASA MODIS/Terra Snow Cover Daily L3 Global 500 m Grid, Version 6 (MOD10A1). This product provides Normalized Difference Snow Index (NDSI) as a daily product [35]. Snow cover fraction (hereafter, SCF) can be derived from NDSI using a universal approach [36] defined by Equation (1):

$$SCF = (0.06 + 1.21NDSI) \times 100\% \quad (1)$$

Equation (1) was originally developed for use with version 5 of the MODIS dataset but is nevertheless compatible with the upgraded version 6. We masked out areas that corresponded to cloud cover, ocean and winter darkness, and we converted projection to WGS 1984 UTM Zone 33N. Then, we used a multitemporal interpolation technique (c.f. Section 2.2.2) to fill the gaps due to missing observations. We processed daily images beginning from 1 March, 2000 (DOY 60) to 1 November, 2019 (DOY 306). With the exception of a few days, in particular DOY 166–183 in 2001, daily datasets have been available for the whole period. Due to the absence of solar illumination during the polar night period, the MOD10A1 is only produced between 1 March to 1 November for the entire Svalbard archipelago. In this study we are primarily interested in studying the temporal characteristics of the seasonal snow cover in the periglacial landscape. We therefore used a vegetation map produced from classification of Landsat TM/ETM+ data [37] to mask out inland water bodies and glaciated areas.

2.2.2. Temporal Interpolation Methods to Obtain a Cloud-Free Climatology

Due to the high cloud cover fraction over Svalbard, which is on average 85% for all areas over the 20-year period, there is a need to interpolate between cloud-free pixels to obtain useful estimates of SCF. Several methods have been discussed in detail in literature [20,21], which includes the use of both pure temporal interpolation between cloud-free pixels, a combination of spatial filtering and temporal interpolation and also combinations of the two aforementioned methods joined with multisensor fusion [38]. In the current approach, we processed MODIS Terra data using temporal interpolation as outlined in the technique developed by Malnes et al. [23] for Northern Scandinavia using a small modification to account for the higher latitude of the Svalbard archipelago. We assumed that the entire land area is snow covered on DOY = 365 (366) each year as a boundary condition. Due to interpolation, this assumption will force all land pixels to have 100% snow cover fraction late in the autumn, irrespective of the last observation. By applying the interpolation operator to the MODIS time series, we obtained an entirely cloud-free time series that can be used to estimate total SCF for the entire Svalbard archipelago. On average, the number of cloud-free samples per year per pixel is 52 ± 16 days for the entire time series, based on the period of 240 available days where the polar night mask does not apply. Some regions however, predominantly in the northern parts of Spitsbergen and over Austfonna, had very few cloud-free samples per year (c.f. Figure A1). Since the absolute minimum number of pixels required to obtain interpolated estimates is three, there are ultimately some pixels in these regions with missing SCF estimates for certain years.

2.2.3. First and Last Day of Snow

The timing of snow disappearance and onset, which we will refer to using the snow metrics first snow-free day (FSFD) and the last snow-free day (LSFD), are defined as the days when SCF falls below or rises above 50%, respectively. The algorithm of Malnes et al. [23] additionally constrains the window for FSFD and LSFD detection to the periods between DOY 90–230 and DOY 230–306, respectively, and the corresponding SCF must remain below (above) 50% for a minimum of 10 days after the estimated date in order for the result to be retained. The requirement on SCF to remain below (above) the 50% threshold for at least 10 days was imposed to avoid detecting short-lived fluctuations in SCF around the threshold (e.g., due to late season snowfall or snowmelt), while the constraints on date for detection of FSFD and LSFD were applied to avoid problems resulting from low solar angle. The latest date for detection of LSFD was set to DOY 306 (1 November), which marks the approximate start of the polar night in Svalbard when optical data cannot be used to infer information on the snow cover.

The estimates of FSFD and LSFD were further used to estimate the length of summer and the length of winter, for which trends will be studied over the 20-year period. For the purpose of this study, the length of summer *LS* is defined as the difference between LSFD and FSFD from the same year, while the length of winter *LW* is the difference between LSFD from the previous year and FSFD from the current year.

2.2.4. Accuracy of SCF and Derived Snow Metrics

The MODIS NDSI single-day accuracy has been assessed by numerous authors on a global scale [35,36,39]. In the majority of studies, comparisons with in-situ snow depth data recorded at meteorological stations and with high resolution optical data such as those obtained by the Landsat and Sentinel-2 satellites are accepted as the best way to assess the accuracy. To assess the accuracy of interpolated gap-filled time series, the average number of days between cloud-free observations was used as a proxy for the accuracy of the interpolated snow product [19,23]. We adopted this approach to assess the uncertainty incurred in the estimates of FSFD and LSFD by calculating the number of days since the last cloud-free observation. The average number of cloud-free days for our estimates of the FSFD and LSFD are mapped for the entire Svalbard archipelago in Figure A2. Using all 20 years of the dataset, the average number of days since the last cloud-free day is 8.3 ± 3.8 days for FSFD and 7.7 ± 4.2 days for LSFD. The difference is minimal, but the number of pixels where we cannot estimate LSFD (26.0% of all land pixels for all years) is marginally higher than for FSFD (25.3%) and concentrated primarily in the northern parts of the archipelago. Hence the overall estimate of LSFD is marginally less certain over these areas than for FSFD. However, there are both large spatial and interannual variations, with some years and regions experiencing long periods of the order 1–2 months without cloud-free observations. For the meteorological sites and available years of data we used to validate MODIS FSFD (c.f. Section 2.4), the mean number of days since the last cloud-free observation for FSFD is 4.4 days, while for estimates of LSFD, the average number of days since the preceding cloud-free observation is 2.8 days.

2.3. *SIOS Database and Open Science*

SIOS (Svalbard Integrated Arctic Earth Observing System) was first initiated in 2007 and has been fully operational since 2018. SIOS is a regional observing system that aims to establish a database of long-term measurements in Svalbard, bringing together both field, remote sensing and model observations to address Earth system science questions. As such, the MODIS 20-year snow cover fraction time series will also contribute to the growing volume of data being gathered by the SIOS project and has been stored in the SIOS database [40]. The dataset can be easily accessed through FTP (File Transfer Protocol) or other software frameworks such as OPeNDAP (Open-source Project for a Network Data Access Protocol) [41] that facilitates the retrieval and exchange of remotely stored, structured datasets, streaming of subregions and effective use of the data. In the newly started ESA project Cryosphere Virtual Laboratory (CVL) [42], the MODIS snow cover dataset and its relations to other climate time series will be developed as one of five use cases. The project will use Open Science standards, specifically OPeNDAP, to demonstrate that the results of this work can be reproduced and correlated with other datasets using the CVL system that operate on virtual machines. This study aims to demonstrate how the current dataset can be used with complementary satellite-derived datasets, also available through the SIOS database, to study the factors that influence the temporal characteristics of snow cover over Svalbard and the impact of snow cover on terrestrial productivity.

2.4. Meteorological Data

Meteorological data managed by the Norwegian Meteorological Institute have been made available for download via OPeNDAP; however, this is currently limited to the years from 2016 onwards. In order to cover the full time interval of MODIS data, we downloaded data from six meteorological stations in Svalbard from eklima.no, starting from the year 2000. In particular, we obtained measurements of precipitation, snow depth, mean, minimum and maximum air temperature. However, only a few of the stations had a complete time series of snow depth observations that could be utilized to validate our snow metrics.

A threshold of 2 cm was applied to snow depth measurements to determine the meteorological FSFD, and this had to be maintained for at least 5 consecutive days to qualify for the condition "snow-free". The meteorological FSFD was then compared with MODIS FSFD by performing linear regression and deriving the correlation between the two estimates. This was done firstly to validate the satellite-derived product and secondly to obtain the general relationship between the meteorological and MODIS FSFD. Several stations had incomplete time series of FSFD either due to missing in-situ measurements or missing data in the MODIS dataset. Therefore, measurements from all sites were combined in the regression analysis. Table 1 summarizes data availability at each site and the total number of measurements used in validation of both FSFD and LSFD.

Table 1. Meteorological stations, observation periods and years with first/last day of snow validations. The total number of validations (i.e., sum of first snow-free day (FSFD) and last snow-free day (LSFD)) is given in the third column. Note that many years were lost due to unstable snow settlement.

Station	Years	Validations (FSFD + LSFD)
Longyearbyen	2009–2019	24
Adventdalen	2018–2019	3
Ny-Ålesund	2009–2019	24
Platåberget	2018–2019	4
Sveagruva	2017–2019	7
Hornsund	2002–2019	35

2.5. Growing Season Onset Mapped with MODIS-Normalized Difference Vegetation Index (NDVI) Data

This study used estimates of growing season onset (GSO) for 2000–2016 based on MODIS-NDVI data, which have been correlated with field phenology observations. The method for estimating GSO from MODIS data was previously developed by Karlsen et al. [25] for 2000–2013, but the dataset has since been extended for plant biomass studies [43] and for studies of the coupling between tundra productivity and sea ice distribution [44]. The NDVI dataset is based on the 8-day reflectance MODIS products MOD09Q1 and MOD09A1. A combination of three different cloud removal methods was applied to obtain a clear-sky time series of NDVI data. From this NDVI time series, GSO estimates were extracted for each year using a pixel-by-pixel threshold method. The NDVI threshold level was reached after several iterations, and the level that gave the highest correlation with general flowering of Arctic willow (*Salix polaris*) across several different vegetation types was used. Arctic willow is among the most common plant species on Svalbard, and the timing of flowering of Arctic willow occurs on average six days before the mean flowering date of twelve other vascular plants in Longyearbyen areas and should indicate well the general time of onset of growth [25]. This study aims to quantify the relationship between the timing of snow disappearance, as denoted by FSFD and the onset of growth. We therefore mapped the difference of the dates in order to identify spatial patterns and endeavor to establish the dependence of this length of time on the mean daily temperature using data from the three meteorological stations Hornsund, Longyearbyen and Ny-Ålesund, since temperature is an important factor for vegetation growth. In addition to mapping the spatial patterns of the difference, the relationship between timing of snow disappearance and start of the growing season was further investigated by reducing the dataset through averaging. This was done by selecting the pixels corresponding to each DOY in the range of FSFD. For each of FSFD DOY, we averaged GSO, excluding pixels where the GSO date occurred before that DOY for FSFD, since it is unlikely that growing season can begin before snow disappearance. In addition, we only accepted the average GSO date where there was a minimum of 50 pixels to average. The averaging was performed individually for each year of data in order to identify interannual variations in the relationship between timing of FSFD and GSO.

2.6. *Sea Ice Concentration*

Sea ice concentration maps (hereafter, SIC) in the Arctic sea are available via OPeNDAP for the period 1967–present and are produced by the Norwegian Meteorological Institute. Ice charting has been carried out manually by GIS (Geographic Information Systems) experts [45]. Currently, the data are available on a daily basis and cover a latitude range from 53.8–88.8°N and longitudes from −84.0 to 87.2°E. During the ice charting procedure, ice types are classified into different classes depending on the ice concentrations. The SIC limits and ice type definitions used in the classification can be found in Table 2 of Dahlke et al. [46]. We accessed the SIC map catalogue [47] using OPeNDAP to extract SIC maps of the sea region around Svalbard from 2000 until 2019 in order to investigate whether there are observable correlations between the timing of snow disappearance and sea ice concentration. An example of the data product extracted from the catalogue is shown in Figure 3 for 5 March 2001. We used the SIC classifications of Dahlke et al. [46] to define the color scale used in Figure 3. This subregion of data was further divided to calculate the mean daily ice concentration over four different sea areas surrounding Svalbard in order to investigate whether the region where sea ice is present influences the timing of snow disappearance and, if so, which area of sea has strongest influence. These four areas comprise the sea to the northeast, east and west of Svalbard, plus averaging over the entire sea region surrounding Svalbard. The boundaries of the areas used for averaging SIC are indicated in Figure 3. For each sea region, we smoothed each year of ice concentration measurements with a median filter of 15-day width to remove short-lived fluctuations. The smoothed time series were then temporally averaged over all years in the dataset, for each day of the year.

Table 2. Summary of the decadal trends in first snow-free day, last snow-free day, length of summer and length of winter, for the three regions Nordaustlandet, Edgeøya and Nordenskiöld Land as well as the average for the whole of Svalbard.

Region	FSFD	LSFD	LS	LW
Nordaustlandet	+2.9	+2.9	+1.4	+0.6
Edgeøya	−2.8	+2.8	+5.6	−7.1
Nordenskiöld Land	−1.5	+1.6	+3.3	−4.0
All Svalbard	−1.1	+2.5	+4.2	−4.3

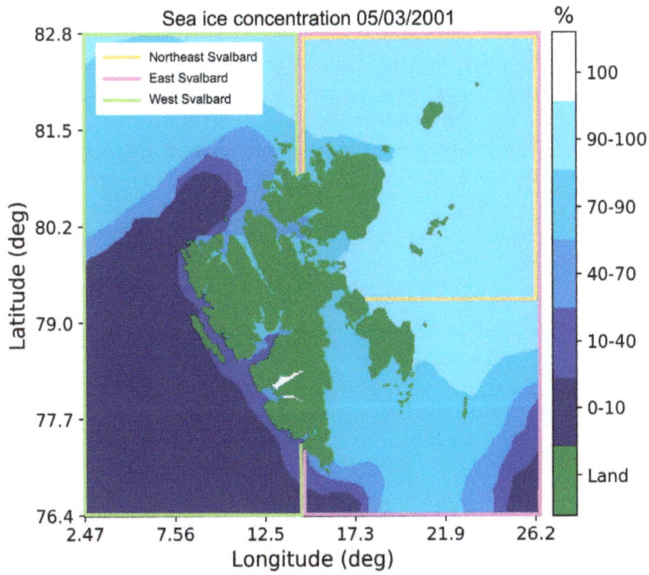

Figure 3. Example of the sea ice concentration around Svalbard on 5 March 2001. The yellow, pink and green colored boxes indicate the subregions of data that are averaged and correlated with the snow cover fraction dataset.

This study aims to determine whether ta correlation exists between SIC in different regions around Svalbard and the timing of snow disappearance, for which we use the MODIS-derived estimates of FSFD as a proxy. Furthermore, we investigated if proximity to sea ice is important for snow disappearance by averaging the FSFD estimates over four different areas of Svalbard: Nordaustlandet, in the northeastern corner of the archipelago; Edgeøya, located in the southeast of the archipelago; Nordenskiöldland, in the central part of the archipelago; and lastly, in the case of the land-averaged values of FSFD. We integrated the area under the annual ice concentration curves, starting from 1 September of the year prior to the first snow-free day estimates (SIC minimum) until 30 June of the year after. We did not integrate to dates after 30 June, since it was found that the average date of the FSFD was 29 June; thus, integrating the ice concentration to a later date will not reveal if the timing of FSFD was the consequence of the amount of sea ice at a date after snow disappearance has occurred. The integration approach provided a better measure for distinguishing between years with similar maximum ice concentrations but different rates of melting. That is to say, the approach better separated years when SIC declined rapidly after the April maximum from years when SIC remained high for a longer duration after the April maximum. The integration was carried out by first converting SIC from percentages to a fraction and applying a trapezoid approximation on the daily mean ice fraction values, averaged over the four sea regions (Northeast Svalbard, East Svalbard, West Svalbard and the entire sea region around Svalbard).

The time series of the time-integrated sea ice concentration (hereafter, TI-SIC) and time series of FSFD were both linearly detrended before performing linear regression. Detrending both time series allows us to establish whether a correlation exists between the variations in each parameter, rather than using the absolute values that may include a trend due to the effect of temperature, which likely affects both parameters. We therefore obtained a set of 16 maximum correlation coefficients and linear fits corresponding to the maximum correlation for each time series of SIC and FSFD that have been averaged over four different areas. We also noted the date defining the end of the integration interval for which the detrended TI-SIC time series was best correlated with the FSFD time series.

3. Results

In this section, we start by presenting the results of the snow cover fraction mapping, which forms the basis for estimation of the first and last snow-free days. In Sections 3.2 and 3.3 we present results from mapping FSFD and the decadal trends in the snow cover parameters. A comparison of FSFD and LSFD estimated from in-situ snow depth measurements against the MODIS-derived estimates is shown in Section 3.4. Finally, we use the estimates of FSFD to examine firstly the relationship with onset of the growing season in Section 3.5 and thereafter the relationship between snow disappearance and sea ice concentration in Section 3.6.

3.1. *Snow Cover Fraction*

The total (land-averaged) SCF for Svalbard for all 20 years in the time series is shown in Figure 4. It can be seen that the variability in total SCF is greatest during the summer period from July to the end of September. The SCF converges to a minimum value of close to 20% for the majority of years in the time series, but SCF did not fall below approximately 40% during a few years (most notably 2000 and 2008) due to late season snowfall in August over the northernmost areas of Svalbard.

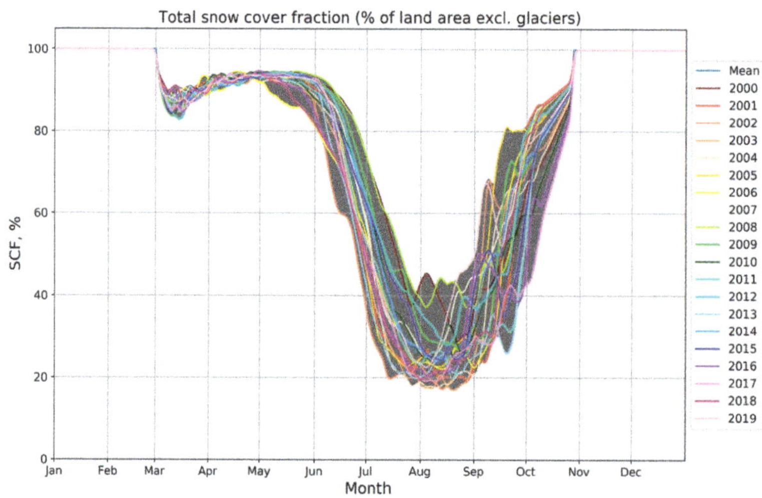

Figure 4. Total snow cover fraction over Svalbard as a function of months for all 20 snow seasons.

The average date of the minimum SCF occurs on 8 August (DOY 220), with an average snow cover fraction of 27.4%. The number of years where the SCF exceeded 50% on 8 August is illustrated in Figure 5a, while the snow cover fraction on 8 August is mapped in Figure 5b. Figure 5b therefore allows the areas that contribute most to the land-averaged minimum SCF remaining above 20% during the period studied (Figure 4) to be identified. While Figure 5b shows that there are large parts of the archipelago that are close to snow-free (<20%), there is a clear tendency towards higher minimum SCF in the northernmost areas on Spitsbergen and Nordaustlandet. These areas are fully or close to fully snow covered in August for many of the years during the study period. In particular, the years 2000 and 2008 experienced late snowfalls in July/August, which resulted in large areas that never became snow-free (SCF < 50%). The same areas retain high minimum SCF for more than 5 years of the 20-year time series. Hence, when SCF is averaged over the entire archipelago, the land-averaged minimum SCF never falls to zero due to these areas with persistent snow cover.

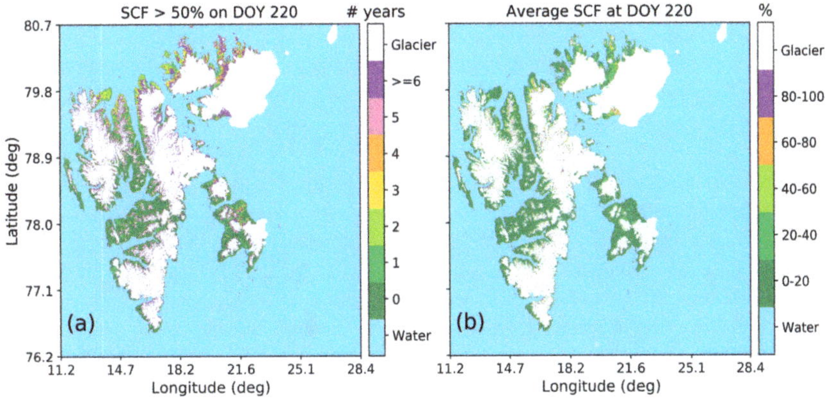

Figure 5. (**a**) Number of years when snow cover fraction (SCF) >50% on 8 August and (**b**) average snow cover fraction at minimum snow cover 8 August (DOY 220) for 2000–2019.

3.2. Patterns in First Snow-Free Day

Since we wish to study the relationship between snow disappearance and timing of growing season onset, and between timing of snow disappearance and sea ice concentration, we will mainly focus on the results of the spatial patterns in FSFD in Sections 3.2 and 3.3. The mean date of the first snow-free day, averaged over the entire period 2000–2019, is shown in Figure 6. The land-averaged date for FSFD is DOY 180 (29 June) ± 17.8 days. It can be seen from this figure that there are clear north–south variations as well as variations with elevation. In the northern- and easternmost areas of the archipelago for example, the mean date for FSFD is typically after DOY 190, or mid-July, while over Nordenskiöld Land it can be seen that the low-lying valley areas exhibit a mean FSFD of DOY 130–170, which is one to one and a half months earlier than compared with the northern parts of the archipelago. In the mountainous regions over Nordenskiöld Land, FSFD occurs as late as the lower lying areas of the northern parts of the archipelago.

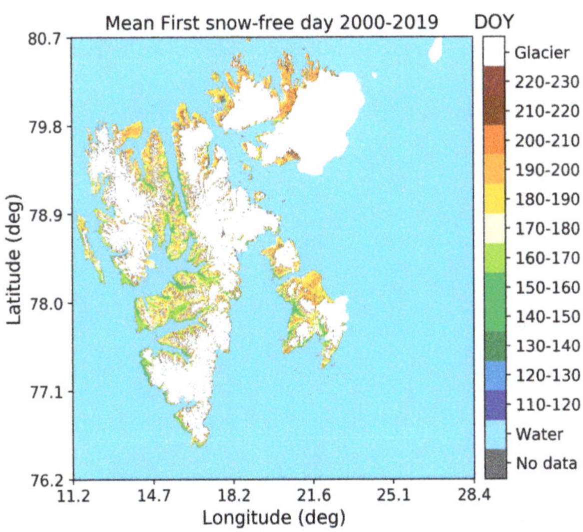

Figure 6. Mean FSFD for 2000–2019.

Anomalies in FSFD for each year are shown in Figure 7. These have been calculated by subtracting the individual pixel values of FSFD from the 2000–2019 average FSFD. The anomalies, or differences from the 2000–2019 average have been categorized into either >1 week earlier, ±1 week change or >1 week later in order to identify years where the timing of snow disappearance occurred much earlier or much later than the 20-year average. Figure 7 shows that 2000, 2008 and 2009 stand out as years where the FSFD occurred much later than average over the whole archipelago, while in 2002 and 2013 the timing of FSFD was much earlier than the 2000–2019 average. During other years in the time series, the timing of FSFD was either later or earlier than the average for different parts of the archipelago. For example, in 2007, FSFD occurred >1 week earlier than average on Nordaustlandet in the northeast part of the archipelago, while on Edgeøya FSFD occurred >1 week later than average, even though it can be seen from Figure 6 that the mean FSFDs for these two areas are actually very similar and quite late. In 2014, the opposite situation is observed, with Nordaustlandet and western coastal parts of Spitsbergen exhibiting FSFD which was >1 week later than average, while the timing of snow disappearance on Edgeøya occurred >1 week earlier than average.

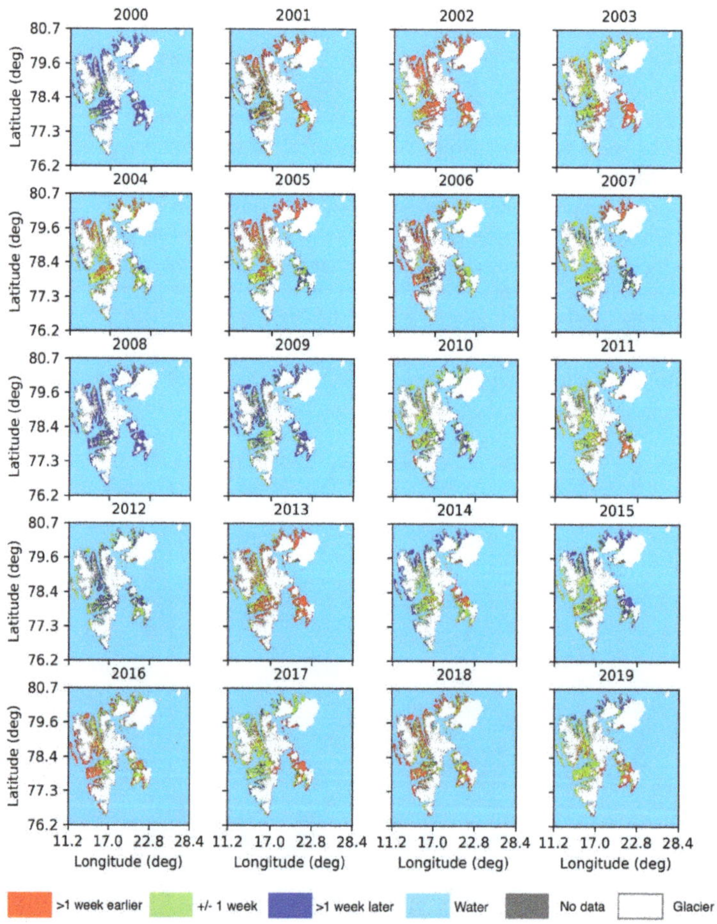

Figure 7. Figure showing the anomalies in first snow-free day compared with the mean first snow-free day over the period 2000–2019.

3.3. Decadal Trends in Snow Cover Characteristics

The four snow cover parameters FSFD, LSFD, LS and LW are mapped for each year in the time series. Pixel-wise linear regression was performed on the time series of each parameter in order to obtain the linear trend over the 20-year period. Pixels with more than 1 year of missing data were excluded from the fitting in order to obtain as reliable fits as possible. The results of the trend mapping are illustrated in Figure 8a–d for each of the four parameters.

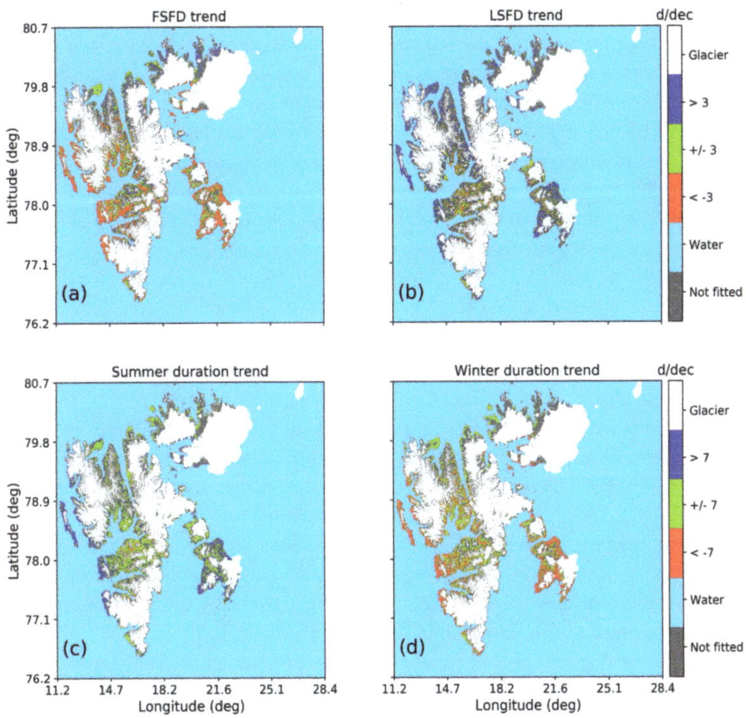

Figure 8. Trends in (**a**) first snow-free day (**b**) last snow-free day (**c**) length of summer and (**d**) length of winter between 2000 and 2019. Trends are given in days decade^{-1}.

Figure 8a shows that there are both north–south and east–west differences, with a shift towards earlier FSFD of >3 days decade^{-1} in the western and central parts of Spitsbergen and across the island of Edgeøya in the east, while in the northern parts of the archipelago, there is a trend toward delayed FSFD of >3 days decade^{-1}. Some areas within the central parts of the archipelago exhibit smaller trends of ±3 days decade^{-1}. The parameters have been averaged over three specific regions to quantitatively determine the regional differences in FSFD and LSFD: Nordaustlandet in the northeast of the archipelago, Edgeøya in the east and Nordenskiöld Land in the central part of Spitsbergen. We found that on Nordaustlandet there is a trend of +2.9 days decade^{-1} for FSFD and +2.9 days decade^{-1} for LSFD. In contrast, the trend in FSFD averaged over Edgeøya, in the east of the archipelago, was found to be −2.8 days decade^{-1} with corresponding delay in LSFD of +2.8 days decade^{-1}. Similar but weaker trends are obtained for Nordenskiold Land, with a trend in FSFD of −1.5 days decade^{-1} and a trend toward later LSFD of +1.6 days decade^{-1}. Thus, while the trend in onset of autumnal snow (Figure 8b), toward later dates is quite consistent across the archipelago, the timing of snow disappearance is becoming later in the north but earlier in the southern parts of the archipelago.

These changes in FSFD and LSFD in turn affect the length of summer and winter at different parts of Svalbard, as shown in Figure 8c,d respectively. For the entire archipelago on average, the length of summer is increasing by +4.2 days decade^{-1}, though Figure 8c indicates that there are distinct regional variations. The greatest increases in the duration occur at the western edges of Spitsbergen, while the remainder of the archipelago exhibits an increase of up to 7 days. On Nordaustlandet the trend in length of summer is only +1.4 days decade^{-1}, while on Edgeøya and Nordenskiöld Land the trends are somewhat stronger, with increases of +5.6 and +3.3 days decade^{-1}, respectively. With regard to trends in winter duration, the westernmost and easternmost areas exhibit the greatest reductions, of >7 days decade^{-1}. On Edgeøya the average trend in winter duration is −7.1 days decade^{-1}, while the central parts of the archipelago exhibit somewhat weaker reductions of up to 7 days decade^{-1}. On Nordenskiöld Land, for example, the reduction was on average −4 days decade^{-1}. This contrasts with a few isolated areas in the north which exhibit an increase in the length of winter of >7 days decade^{-1}. Averaged over Nordaustlandet, the mean trend in winter duration is a negligible +0.6 days decade^{-1}. It should be noted, however, that a large proportion of pixels on Nordaustlandet did not possess sufficient data points in the time series for a linear fit to be made for LSFD, summer and winter duration trends, as illustrated by the grey colored pixels.

In Table 2 below, a summary of the decadal trends in each snow parameter (FSFD, LSFD, LS, LW) is provided for each of the regions Nordaustlandet, Edgeøya and Nordenskiöld Land as well as the land-averaged decadal trends.

3.4. Correlation with Meteorological Data

Figure 9 (left) shows that there was a significant and positive correlation between the meteorological estimate of FSFD and the MODIS estimate, with correlation coefficient of 0.77. The bias of −12.32 indicates that FSFD estimates obtained from the MODIS snow cover fraction product were on average 12 days later than those observed from the point snow depth measurements. Qualitatively, the dataset exhibits more spread in MODIS estimates when FSFD as observed from snow depth measurements occurs earlier than DOY 150 and better correlation between the two estimates when the snow depth estimate of FSFD occurs after DOY 150. In the case of LSFD, in the right panel of Figure 9, there is significantly greater scatter around the regression line, indicating less agreement between the two estimates and therefore weaker correlation, as noted from the correlation coefficient of 0.48. The regression line intersects with the equality line when the meteorological LSFD is around DOY 270; hence, when the meteorological LSFD occurs before DOY 270, the MODIS LSFD estimate is later, and for meteorological LSFD after DOY 270, the MODIS estimate LSFD is earlier.

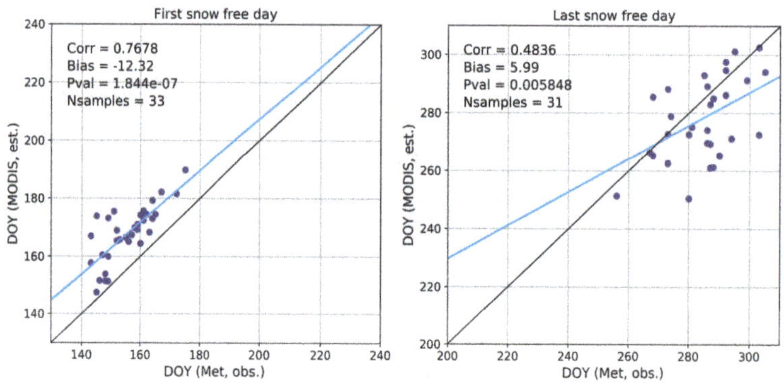

Figure 9. Comparison of the first (left) and last (right) snow-free day estimates derived from snow depth measurements made at the six observational sites listed in Table 1 and those derived from the Moderate Resolution Imaging Spectroradiometer (MODIS) snow cover fraction product. Both datasets have been filtered to exclude estimates during the polar night period (DOY > 306) where MODIS data are unavailable. The black line indicates where the estimates would be equal and the linear regression result is highlighted by the solid blue line.

3.5. Relationship between Snow Disappearance and Growing Season Onset

The mean difference, or length of time from snow disappearance to onset of growth averaged over the 17-year dataset is shown in Figure 10. The difference has been split into four categories corresponding to; <7 days, 7–14 days, 14–21 days and >21 days. In instances where the difference between GSO and first snow-free day was less than zero, i.e., growing season onset date occurred before first snow-free day, these pixels were excluded from the averaging. Figure 10 shows that the shortest differences of <7 days are concentrated on the east side of Edgeøya and on Reinsflya, which is situated at the northern end of Svalbard. These areas have cold summers with short growing seasons. On Nordenskiöld Land there is also much variation in the difference, with the central and southern parts exhibiting a mean difference of typically <7 days, mostly at higher elevations, or 7–14 days, mostly in the warm valley floors. However, in the valley floor of Adventdalen, close to Longyearbyen, the mean difference is typically in the range of 14–21 days, with some parts >21 days. Further north of Nordenskiöld Land there are not clear spatial patterns in the difference, with some coastal areas exhibiting differences of up to 14 days, while other coastal areas show differences between snow disappearance and GSO of >14 days. However, we found that for the whole of the nonglaciated parts of Svalbard, the difference between snow disappearance and the start of the growing season was on average 12.4 ± 1.7 days (minimum = 9.5 days, maximum = 17.7 days) when the entire 17-year period was considered.

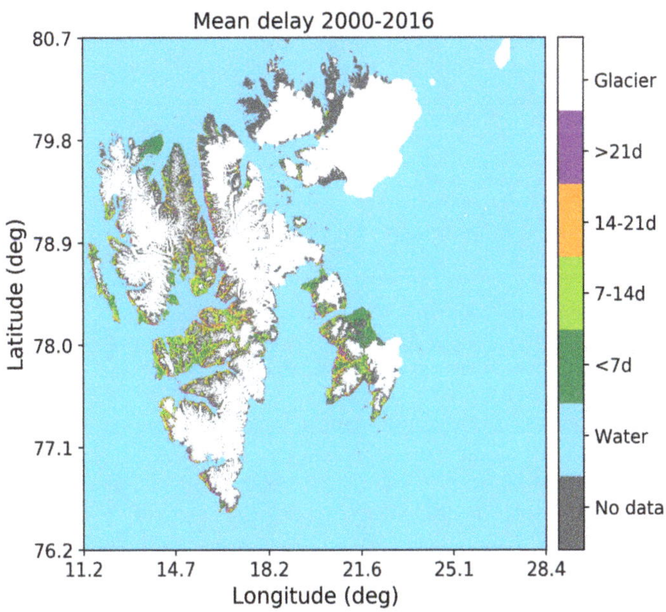

Figure 10. Mean difference between growing season onset and first snow-free day, averaged over 2000–2016.

Figure 11 shows the average relationship between FSFD and onset of the growing season, as obtained through the data reduction procedure described in Section 2.5. Overall there is a linear relationship between GSO date and FSFD for the majority of years, especially over the interval where FSFD occurred between DOY 160 and DOY 210. For earlier FSFD occurrence, there is a greater degree of spread in the timing of GSO, indicating that there was a larger variation in the difference for earlier snow disappearance dates, likely determined by the average air temperatures after snow disappearance. Certain years emerge with greater occurrence of early snow disappearance; for example, in 2006 there was a greater proportion of pixels where the FSFD occurred before DOY 135 compared with all other years in the dataset. However, the corresponding average GSO date remains almost unchanged at approximately DOY 165 over the interval where FSFD occurred between DOY 115–135, which is equivalent to a difference that is on average 30–50 days between FSFD and the start of the growing season. Figure 11 also shows that the date of GSO was consistently early in 2016 compared with other years. Similarly, the earliest occurrence of FSFD in 2016 was rather early compared with other years, with the exception of 2006. In contrast, the average growing season date in 2000 and 2008 occurred late relative to the timing of FSFD, and the high minimum value of FSFD for these two years indicates that snow disappearance also occurred late compared with other years in the dataset.

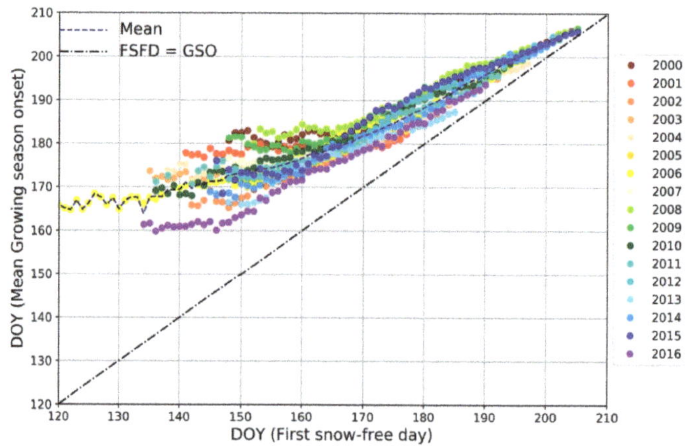

Figure 11. Scatterplot showing the relationship between first snow-free day and the average growing season onset (GSO) date. The mean growing season date for all years is indicated by the dashed line and the dashed-dot line indicates where first snow-free day would be equal to growing season onset.

In general, Figure 11 shows that when snow disappearance occurred late (DOY 190, 9 July), there is a difference of 5 to 10 days until onset of the growing season, while for earlier FSFD (DOY 160, 9 June) the difference ranges from 10 to 25 days. Hence, in years with early FSFD, there is a greater difference between GSO and FSFD compared with years when FSFD occurred later.

The relationship between the difference between FSFD and GSO and mean daily temperature in this period for the sites Hornsund, Longyearbyen and Ny-Ålesund is illustrated in Figure 12. Only Longyearbyen and Ny-Ålesund had complete time series for the mean daily temperature, while for Hornsund there were some years with missing temperature data. For all three sites there were complete time series of the difference between GSO and FSFD. Figure 12 indicates that occurrences of greater (smaller) differences are due to cooler (warmer) average temperatures, indicated by the negative slope of the regression line. At Ny-Ålesund for example, where the difference between snow disappearance and GSO was generally shorter (minimum = 5.2 days, maximum = 17.5 days), the mean daily temperatures are also higher (mean = 4.4 °C, variance = 1.3 °C) than at Longyearbyen (mean = 2.3 °C, variance = 1.0 °C) where the difference between FSFD and GSO was longer (minimum = 13.7 days, maximum = 36.7 days). At Hornsund, the spread in difference is quite large (minimum = 1.5 days, maximum = 22.5 days), while the corresponding mean daily temperature during the period between snow disappearance and GSO was confined to a smaller range of about 2 °C. It should be noted though, that the mean daily temperatures being compared in Figure 12 correspond to different times of year. This is because the mean FSFD at Longyearbyen typically occurs up to a month earlier than at, for example, Ny-Ålesund (c.f. Figure 6) which explains why the data in Figure 12 shows lower average temperatures for Longyearbyen despite its more southerly location in comparison with Ny-Ålesund. Despite the data points being spread about the regression line, the correlation is quite strong and negative (r = −0.75) and significant ($p < 0.05$). The range and mean values of the difference and mean daily temperature for the three sites are summarized in Table 3.

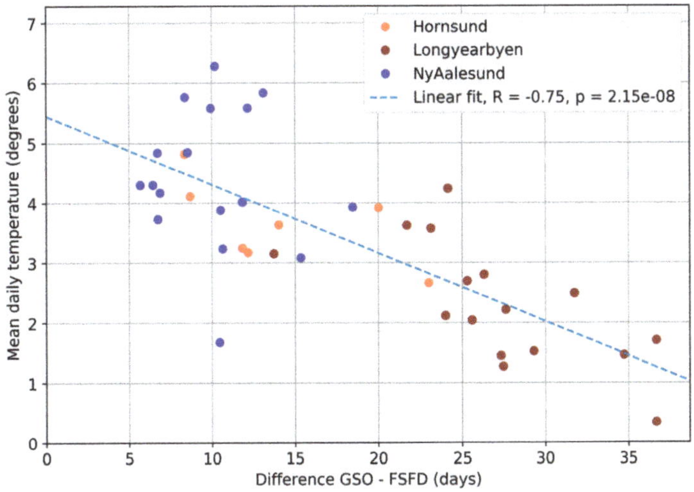

Figure 12. Relationship between mean daily temperature (in the period between snow disappearance and GSO), and the length of the time between snow disappearance and GSO calculated for Hornsund, Svea, Longyearbyen and Ny-Ålesund.

Table 3. Summary of the minimum, maximum, mean and variance in the difference between first snow-free day and growing season onset, and the mean daily temperature in the period between FSFD and GSO.

Station	Years	Difference Range (Days)	Mean Difference (Days)	Temperature Range (°C)	Mean Temperature (°C)
Hornsund	7	1.5–22.5	10.9	2.7–4.8	3.7
Longyearbyen	16	13.7–36.7	27.5	0.3–4.2	2.3
Ny-Ålesund	17	5.2–17.5	9.6	1.7–6.3	4.4

3.6. Correlation with Sea Ice Concentration

Time series of the spatially and temporally averaged SIC are illustrated for each of the four sea regions (c.f. Figure 3) in Figure 13a–d together with the minimum and maximum SIC indicated by the dashed lines. For all regions, the maximum mean SIC is reached in the beginning of April and falls to a minimum in the start of September. Similarly, for all regions, the variation is smallest between lowest and highest SIC at April maximum, while there is greatest variation between the lowest and highest values of the mean SIC at the start and end of year. The mean SIC is typically highest in the sector northeast of Svalbard, ranging from a mean minimum concentration of just over 40% to a mean maximum concentration of approximately 85%. This contrasts with SIC in the sector west of Svalbard, where warm Atlantic water is driven northwards by the West Spitsbergen Current and inhibits the formation of sea ice. As a result, SIC is overall lower compared with SIC in the east of Svalbard, with much smaller range between the mean minimum and maximum SIC, ranging from 20% to 40%.

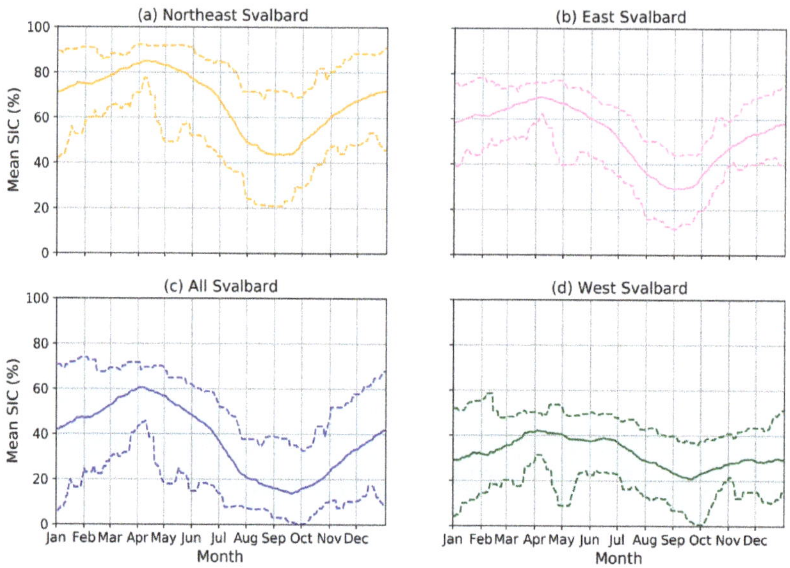

Figure 13. Sea ice concentration per DOY averaged over 19 years for subregions of the sea area around Svalbard as shown in Figure 4: (**a**) sea region northeast of Svalbard; (**b**) sea region east of Svalbard; (**c**) entire sea region around Svalbard; (**d**) sea region west of Svalbard. Solid lines indicate the average sea ice concentration (SIC), and the minimum and maximum SIC are shown by dashed lines.

As outlined in Section 2.6, linear regression was performed on detrended time series of TI-SIC, and FSFD and correlation coefficients were obtained for each pair of TI-SIC and FSFD linear fits, using the four different sea ice regions and four land regions for which FSFD has been averaged. The set of 16 maximum correlation coefficients, together with the corresponding p-values, are summarized in Table 4.

Table 4. Summary of the maximum correlation between the detrended time series of FSFD for four averaging areas and time-integrated sea ice concentration (TI-SIC) for four subareas of the sea around Svalbard. Asterisks indicate correlations that were statistically significant at the 95% level.

Sea Ice Area	Land Average	Nordaustlandet	Edgeøya	Nordenskiöldland
NE Svalbard	0.59, p = 0.010 *	0.69, p = 0.002 *	0.54, p = 0.020 *	0.43, p = 0.073
East Svalbard	0.54, p = 0.021 *	0.64, p = 0.004 *	0.56, p = 0.016 *	0.42, p = 0.081
Entire Svalbard	0.44, p = 0.066	0.60, p = 0.008 *	0.48, p = 0.044 *	0.30, p = 0.220
West Svalbard	0.48, p = 0.042 *	0.63, p = 0.005 *	0.39, p = 0.110	0.38, p = 0.116

Table 4 shows that the strength of the correlation between TI-SIC and FSFD variations varied according to both the region the sea ice that data was averaged over and the area over which the FSFD was averaged. Strongest correlations were obtained when TI-SIC variations averaged over the sea northeast of Svalbard were correlated with FSFD variations averaged over Nordaustlandet. Furthermore, this maximum correlation (r = 0.69, p = 0.002) was achieved when SIC had been integrated over the entire period investigated (1 September to 30 June). The correlation between FSFD variations over Nordaustlandet and TI-SIC variations was significant for all four sea ice regions investigated. In contrast, using the time series of FSFD variations that had been averaged over Nordenskiöldland, representing the central parts of Svalbard, resulted in the weakest correlations with SIC variations regardless of the region of sea ice investigated. For Edgeøya, in the southeast of the archipelago, we note that the correlation between FSFD and TI-SIC was weaker but was greatest when using SIC averaged over the sea region east Svalbard. It is interesting to note that the maximum correlation between FSFD variations on Edgeøya with three of the sea ice regions (northeast, east, all of Svalbard) was obtained when SIC had been integrated from 1 September to mid-October (not shown) and not over the entire period (which was the case for Nordaustlandet). Correlation coefficients obtained for the land-averaged FSFD variations were somewhat weaker compared with those obtained for FSFD variations on Nordaustlandet only but were nevertheless greater than the correlation using FSFD variations averaged only over Edgeøya and Nordenskiöld Land.

Figure 14a shows the time series of the correlation coefficients that were obtained when correlating the detrended time series of TI-SIC for the four sea regions with FSFD variations for Nordaustlandet, where the strongest correlations with TI-SIC variations were obtained. For reference, the detrended time series of TI-SIC at DOY 180 for Northeast Svalbard, which produced maximum correlation, is shown together with the detrended time series of FSFD in Figure 14b. What is common for all sea regions is that the correlation with the FSFD variations continued to increase when the integration period was extended beyond April, which coincides with the declining phase of SIC. Equivalent figures for FSFD on Edgeøya, Nordenskiöld Land and the land average are included in the Appendix A. Figure 15 shows the TIC-SIC variations for the sea region Northeast Svalbard, which resulted in the greatest correlation together with the FSFD variations averaged over Nordaustlandet. Linear fits are shown for correlations that were significant at the 95% level.

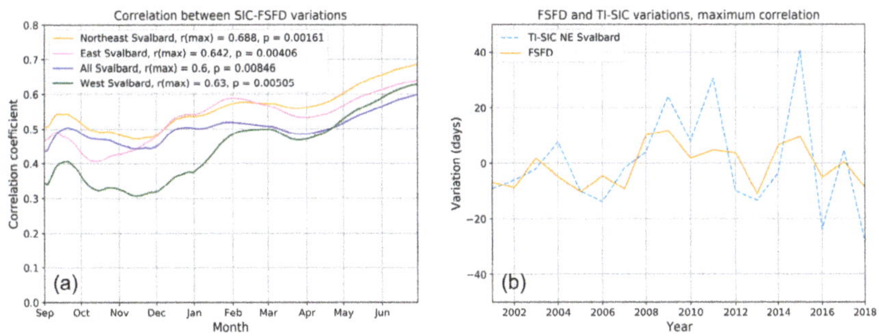

Figure 14. (**a**) Correlation between TI-SIC and FSFD variations averaged over Nordaustlandet, calculated for SIC integrated from 1 September of the previous year to each consecutive day of year until 30 June of the following year, for the four sea ice regions. The maximum correlation coefficients for each sea region and the associated p-values are stated in the legend. (**b**) Time series of the TI-SIC variations for the sea region northeast of Svalbard that produced maximum correlation with the time series of FSFD variations over Nordaustlandet.

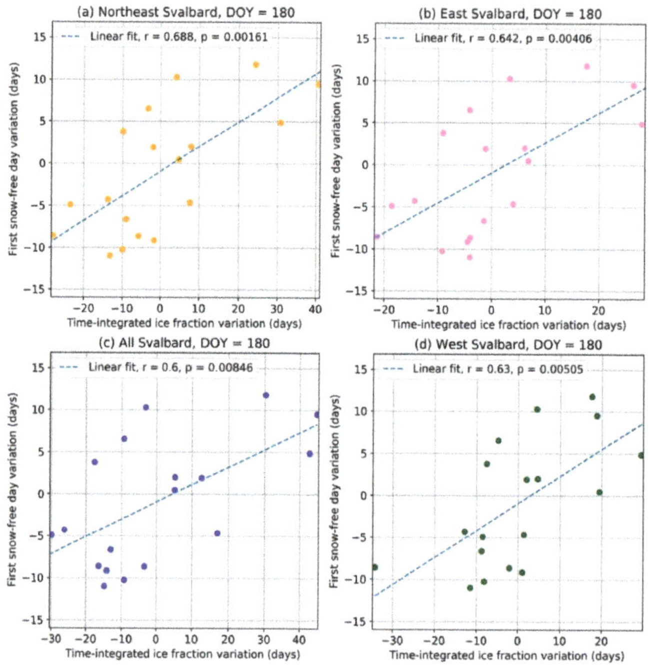

Figure 15. Scatterplots showing the relationship between detrended TI-SIC and FSFD variations averaged over Nordaustlandet, using the TI-SIC time series that gave the highest correlation with the detrended FSFD time series for the four sea ice regions (**a**) Northeast Svalbard (**b**) East Svalbard (**c**) All Svalbard and (**d**) West Svalbard. Values of maximum correlation (r) and p-values are stated in the legend. The day of year associated with maximum correlation between TI-SIC and FSFD is also stated.

4. Discussion

This study has utilized a 20-year MODIS Terra snow cover dataset to derive the snow cover fraction for the periglacial landscape of Svalbard. The time series of snow cover fraction has been used together with an algorithm that was previously developed for Northern Scandinavia [23] to estimate both the first snow-free day (FSFD) and last snow-free day (LSFD) for the nonglaciated areas on Svalbard. The FSFD estimates have been used as an indication of timing of snow disappearance for seasonal snow cover on Svalbard, from which we have studied its link to the timing of phenological processes, namely the start of the growing season, and its relationship to variations in sea ice concentration surrounding the archipelago. This section will address the main findings of each aspect of this work and discuss how they fit in with current knowledge.

4.1. Geographic Patterns of First and Last Snow-Free Days

For Svalbard as a whole, the average date of FSFD was found to be DOY 180 or 29 June, while the average date of minimum snow cover fraction for Svalbard as a whole was 8 August, with the average snow cover fraction on this date being 27.4%. Our results indicate that while the majority of the central and southern areas of Svalbard are completely snow-free on this date, there are nevertheless areas in the northernmost part of Spitsbergen and Nordaustlandet that remain partially snow covered, with >50% snow cover fraction for many years of the 20-year dataset, explaining why the average minimum snow cover fraction for the whole of Svalbard remains above zero. However, the time series of the land-averaged FSFD did not reveal any significant trend, which is in agreement with earlier modeling studies of seasonal snow disappearance on Svalbard [17,48]. On the other hand, studies of snow disappearance date over glaciated regions of Svalbard using QuikSCAT datasets have earlier reported on a general trend toward earlier snow disappearance [21], despite their time period of study being less than half that which was studied in this work.

The spatial patterns in FSFD trends indicated that predominantly western, southern, eastern and some central parts of Spitsbergen experience a trend toward earlier FSFD, while isolated areas furthest north on Spitsbergen and Nordaustlandet exhibit a trend of later snow disappearance over the two decades studied. These regional trends are in partial agreement with the results of Rotschky et al. [21], for scatterometer detection of snow melt onset over the glaciated parts of Svalbard, and Van Pelt et al. [17], where trends in snow disappearance date were mapped for both the periglacial and glaciated landscape of Svalbard. In the modeling results of Van Pelt et al. [17] for the period 1961–2012, a trend toward earlier snow disappearance of around −1 day per decade was obtained over western and central parts of Svalbard, while a trend toward later snow disappearance of up to 4 days decade^{-1} was found over northern and eastern parts of Svalbard. Thus, our results differ primarily over Edgeøya, in the east of the archipelago and over southern Spitsbergen. It may be noted that the spatial patterns of FSFD trends coincide qualitatively rather well with the spatial patterns of precipitation trends from modeling rather than the trends in warming [17], which may provide some insight into the factors that have greatest influence on the trends in snow disappearance. However, on a land-averaged basis, the date of snow disappearance derived from the MODIS dataset leads the average date of snow disappearance obtained from snowpack modeling [17] by approximately 10 days. This is most likely explained by the threshold used to define snow disappearance in this study, where snow cover fraction at the individual pixel level may still be up to 50% while being defined as snow-free.

Across the majority of Svalbard, the timing of snow onset in the autumn, or LSFD, is becoming later, with the land-averaged trend being +2.5 days decade^{-1}, which was not significant ($p > 0.05$). These patterns in snow onset are also consistent with the earlier snowpack modeling results, where positive trends in snow onset date (i.e., later LSFD) were present over the entire archipelago. Earlier snow disappearance and later autumn snow onset resulted in an increase in the duration of the summer season for these areas and a corresponding reduction in the duration of winter. We found that the greatest changes in the duration of summer and winter occur on Edgeøya, where the increase in duration of summer was on average 5.6 days decade^{-1}, while winter has become shorter by on average 7.1 days decade^{-1}. In contrast, the smallest changes in duration of summer and winter are found in the furthest north of the archipelago, where the trend in summer duration averaged over Nordaustlandet was found to be only +1.4 days decade^{-1}, while the duration of winter changed by only +0.6 days decade^{-1}.

From the spatial patterns of trends in precipitation and temperature over Svalbard, Van Pelt et al. [17] concluded that the later disappearance of snow in the northern regions was to a greater extent controlled by trends of increased winter snowfall rather than spring snowmelt. Increased winter precipitation over the northeastern parts of Svalbard has been projected from regional climate model simulations [49] and has previously been suggested to be a likely driver of observed ice cap growth on Austfonna, an ice cap located on Nordaustlandet [32], whereby winter snow accumulation exceeds spring snow loss due to melting. Here, the increased winter precipitation was linked to sea ice decline in the adjacent Barents Sea and resulting increased moisture transport over Nordaustlandet. A decline in autumn Arctic sea ice extent has in many studies been linked to increased snowfall in Northern Europe during late autumn/early winter, as it provides an enhanced moisture source that increases the humidity of the Arctic air mass [50,51]. Whether increased winter precipitation over the northern parts of the archipelago can explain the trend toward later FSFD has not been ascertained in this study since we have not used meteorological data from these parts of the Svalbard. However, we have explored the topic of how the timing of FSFD is related to variations in sea ice concentration, which may provide deeper insight into whether this is the mechanism behind the observed spatial variations in FSFD trends across Svalbard, and this is dealt with in Section 4.2.

4.2. Link between FSFD and Sea Ice Concentration Variability

We have found a significant positive correlation between variations in the TI-SIC and FSFD, which are strongest when considering SIC over the sea to the northeast and east of Svalbard (c.f. Figure 13) and when correlating SIC that has been integrated over the entire winter/spring. Moreover, the strongest correlation was obtained when correlating TI-SIC variations with FSFD variations over Nordaustlandet, located in the northeastern corner of the archipelago. This result would indicate that proximity to the sea ice region may be important for sea ice variations to have an influence on the timing of snow disappearance on land, possibly by influencing the local/regional temperatures, since surface air temperatures in Svalbard are known to be sensitive to sea ice edge proximity [30]. Rotschky et al. [21], in their study of snowmelt over glaciated regions of Svalbard, also discussed that melt onset and duration were sensitive to interannual variations in the sea ice edge location, though no quantitative analysis of the link between sea ice and melt onset was presented. Our results also show that the strength of the correlation between TI-SIC variations and FSFD increase when SIC is integrated to later intervals following the start of April, which is typically the timing of maximum SIC. This suggests that SIC, during its declining phase after maximum (which is controlled by the rate of melting), becomes increasingly stronger linked to the timing of snow disappearance on Svalbard. Here, years with lower-than-average SIC were increasingly distinguished from years with above average SIC as the melting period progresses toward the end of spring, when snow disappears on Svalbard.

Terrestrial productivity on Svalbard is coupled to sea ice at the subregional/local scale through cold air advection from ice-covered ocean [44], and correlation between productivity and sea ice concentration has been observed to decrease with distance from the coast [6]. Since the onset of terrestrial productivity follows the timing of snow disappearance, it could be expected that the processes responsible for sea ice–terrestrial productivity coupling may also explain the observed relationship between sea ice and timing of snow disappearance. A recent study on the link between surface temperature, atmospheric circulation (AC) types and SIC in Svalbard found strongest correlations between SIC and anticyclonic AC types originating from the north and northeast during winter [12] which are associated with lowest air temperatures regardless of season. Coupled with earlier knowledge that air temperatures are influenced by proximity to sea ice [30], we therefore suggest that sea ice presence, allowing cold air masses of northerly origin to be advected onto surrounding land, would also result in lower air temperature and thereby preserve snow cover on land for longer than average. Snow cover itself also possesses high albedo, thereby maintaining lower air temperatures in regions close to ice through feedback effects and therefore hindering ice melt. This may explain why the correlation is also strongest between TI-SIC and FSFD on Nordaustlandet, which is situated closest to the region of highest SIC in the north and northeast of Svalbard. However, while the strong correlation between surface air temperature (SAT) and SIC for air masses originating from the north and east of Svalbard has earlier been reported [12], a very recent study on trends in SAT and sea ice extent (SIE) across Svalbard for 1980–2016 reports that correlation between SAT and SIE is greater when SIE time series lag SAT by one month compared with the opposite case when time series of SAT lag SIE by one month. This led the authors to conclude that sea ice is more prone to SAT forcing rather than SIE driving variations in SAT [46]. Hence, while it appears that sea ice presence can directly affect air temperatures on adjacent land masses and therefore influence when snow melts during the same season, the relationship between sea ice and air temperature is rather complex, with feedbacks going in both directions. One additional aspect of the effect of sea ice on timing of snow disappearance that we have not addressed in this current study is whether sea ice loss in the preceding spring season may have a link to the observed timing of snow disappearance the following year. This is a topic deserving of further study, since it is well known that sea ice loss during the warm months allows larger areas of open water to absorb more solar radiation, thereby creating a heat sink that later acts as a source of upward long-wave radiation which warms the lower troposphere during the following winter [3].

Earlier studies have linked September sea ice decline to increased moisture transport due to a greater area of open water and hence increased precipitation as snow during late autumn and winter [51–53]. An increase in autumn/winter snowfall due to this mechanism would produce a deeper snowpack, which would possibly require a longer interval to melt and hence a later-than-average FSFD. We note from our dataset that even though the correlation between FSFD and TI-SIC variations in September was not as great as that obtained from integrating SIC over the entire winter and spring, the correlation between FSFD and TI-SIC variations in the region northeast and east of Svalbard was both positive (~0.5) and significant (p-values of ~0.02–0.04), indicating that lower-than-average SIC is correlated with earlier-than-average FSFD and not later-than-average FSFD. Furthermore, we did not find a significant declining trend in the September-averaged sea ice concentrations for any of the four regions during this period (2000–2019) studied. Even though our results seem to suggest that there is, on average, a trend toward longer winters over the northern parts of Nordaustlandet, which is also the part of the archipelago most strongly correlated with TI-SIC variations, our results do not point to increased winter precipitation resulting from declining sea ice as an important contribution, though overall increases in precipitation over this part of the archipelago may well be explained by large-scale warming trends. However, since we have not utilized gridded precipitation data, and since the trends in winter duration are not significant, these conclusions remain nevertheless rather speculative.

4.3. Timing of Snow Disappearance and Its Effect on Phenology

The timing of snow disappearance is an important factor that determines the timing of greening and vegetation growth on Svalbard. We have combined our dataset for first snow-free day with a MODIS NDVI-based dataset for growing season onset [25] in order to quantify the relationship between timing of snow disappearance and the start of the growing season. Our results indicate that the mean difference, averaged over the period 2000–2016 for the vegetated part of Svalbard, is 12.4 ± 1.7 days, but there was a lack of trend in the land-averaged difference. In Adventdalen valley, close to Longyearbyen, the difference is mostly between three and four weeks. Cooper et al. [24] studied the average time from snow disappearance to green-up and flowering of 13 vascular species in Adventdalen and found large individual differences between the species, however with averages of 3 weeks to green-up and 5 weeks to flowering. This study works on an aggregate level, with 500 m pixel resolution, but our results are still in accordance with the earlier findings [24]. Furthermore, our results show that for areas with early snow disappearance, the length of time to GSO is longer than for areas with late snow disappearance (for example, in the north of the archipelago). This relationship has also been found for the timing of flowering [26] with respect to snow disappearance in North America. In addition to mapping the difference between snow disappearance and GSO for the entire archipelago, we have examined the influence of temperature on the observed variations in the length of time between snow disappearance and GSO. The northernmost site, Ny-Ålesund, with mean daily temperature of 4.4 °C, typically exhibited the smallest range in difference, varying from 5 days to 2.5 weeks, while the length of time at Longyearbyen was of the order 2 to 5 weeks, with a mean difference of approximately 4 weeks between snow disappearance and the date of GSO and lower mean daily temperature of 2.3 °C. The higher mean temperature at the more northerly latitude of Ny-Ålesund is explained by the later timing of snow disappearance, which is typically one month later than at Longyearbyen and is therefore associated with warmer daily temperatures during the period between snow disappearance and onset of the growth season. We also found a significant negative and linear relationship ($r = -0.75$, $p < 0.05$) between the length of time between snow disappearance and GSO and the mean daily temperature during this period, indicating the dependence of the GSO on the mean air temperatures following snow disappearance.

Potential implications of the trend toward earlier snow disappearance over large parts of the archipelago may include reduced productivity, since both timing of snowmelt as well as snowpack accumulation is strongly linked to the levels of soil water storage and hence water availability during the growing season [54–56]. In addition, earlier snow disappearance can expose plants at a time of year when the air temperature is still low, which is known to lead to frost damage [26] and may result in slower rates of growth [29]. Moreover, a study of maximum NDVI (Normalized Difference Vegetation

Index), which is an indicator of vegetation productivity, over central parts of Svalbard from 1986–2015 reported on a general greening trend which was linked to trends of increasing temperature over the study period [57]. However, it was shown that the rate of greening slowed during the second half of the study period, which has subsequently been attributed to more frequent occurrences of extreme winter warming events [58,59], but we do not rule out the possibility that the slower rates of greening since the start of the millennium may also have links to timing of spring snow disappearance, which can be affected by winter warming events through changes in the snowpack accumulation.

4.4. Comparison with Meteorological Data

One of the primary limitations of the method for estimating FSFD and LSFD implemented in this study was cloud cover presence in MODIS images, which introduced uncertainties on the order of 7 days based on the average number of days between cloud-free images for Svalbard as a whole. Using in-situ ground-based measurements of snow depth recorded at six meteorological stations, we found that there was a strong and positive correlation between the two estimates, but MODIS FSFD estimates were on average 12 days later than those obtained from snow depth measurements. A weaker but nevertheless significant positive correlation was also found for estimates of LSFD but with a bias indicating that MODIS estimates were on average 6 days earlier than those inferred from snow depth measurements. Due to the large differences in spatial area associated with the two types of estimate of FSFD and LSFD (point measurements vs. 500 m grid), it is unsurprising that there are discrepancies between the estimates. Further validation of the MODIS dataset is planned to follow this study, and we endeavor to combine multiple and complementary remote sensing snow cover datasets in order to carry out a more in-depth accuracy assessment and thereby identify potential improvements to the methodology outlined in this study.

5. Conclusions

This work has exploited a 20-year MODIS Terra dataset to derive snow cover fraction maps for the Svalbard archipelago from 2000–2019. From this product, we have estimated and mapped the first and last snow-free days as a proxy of snow disappearance and onset timing. For large parts of the archipelago, we found trends toward earlier FSFD and later LSFD, resulting in longer summer duration along with a corresponding decrease in the duration of winter. The exception to this trend was found at the northernmost end of the archipelago, where both snow disappearance and snow onset are trending toward later dates. Consequently, there were marginal increases of the order +1 day decade^{-1} in both the duration of summer and following winter for these areas. For the archipelago as a whole, we obtained decadal trends in FSFD and LSFD of −1.1 and +2.5 days decade^{-1}, respectively, but in both cases the trends were not significant.

Furthermore, we have investigated the difference in date between snow disappearance and the timing of the phenological growing season. For the Svalbard archipelago as a whole, we found that the average difference between FSFD and GSO was 12.4 days; however, there were both significant regional and interannual variations. In areas with very early snow disappearance, the average time to onset of growth was much larger (>20 days) compared with areas with very late snow disappearance (<10 days). Using meteorological data recorded at Hornsund, Longyearbyen and Ny-Ålesund, we found a significant negative correlation between the length of the time to GSO and mean daily temperature recorded in the interval between snow disappearance and start of the GSO.

In the final part of this study, we conducted an analysis of the relationship between timing of snow disappearance and time-integrated sea ice concentration by correlating detrended time series of FSFD and TI-SIC for four different regions of sea ice and four land regions. A significant positive correlation between TI-SIC variations and FSFD variations was obtained when FSFD averaged over Nordaustlandet was correlated with TI-SIC variations that had been averaged over the sector to the northeast of the archipelago, indicating that proximity to sea ice allows the variations in TI-SIC to play an important role in regulating the regional climate and hence the timing of snow disappearance.

Moreover, the correlation reached a maximum when SIC had been integrated over the whole period and exhibited an increasing trend during the melting phase after April. Our interpretation of these results is that the correlation becomes stronger when integrating to intervals after maximum SIC because years of lower-than-average SIC become more strongly distinguished from years of greater-than-average SIC and hence variations in the time-integrated product follow more closely the interannual variations exhibited by timing of snow disappearance. Since we have not made use of gridded temperature and precipitation data in this study, it has not been possible to establish whether there is a connection between sea ice concentration, winter precipitation and timing of spring snow disappearance; however, this will be one of the future goals of a follow-up study.

The results of this paper will be expanded upon in the ESA Cryosphere Virtual Laboratory project using Open Scientific standards, with the main objective being to compare the MODIS snow cover dataset with a wide range of openly available datasets for Svalbard. As such, we aim to improve the temporal and spatial resolution of the current snow cover fraction dataset by investigating the possibilities of data fusion, especially for the MODIS snow cover product with other higher resolution datasets.

Author Contributions: Conceptualization, E.M. and H.V.; methodology, E.M. and H.V.; software, E.M. and H.V.; validation, E.M., H.V. and S.R.K.; formal analysis, H.V.; investigation, H.V; resources, E.M.; data curation, E.M., H.V. and S.R.K.; writing—original draft preparation, H.V and E.M.; writing - review and editing, H.V.; project administration, E.M.; funding acquisition, E.M. All authors have read and agreed to the published version of the manuscript.

Funding: This work was supported by the Research Council of Norway under the project Svalbard Integrated Arctic Earth Observing System—Infrastructure development of the Norwegian node (SIOS-InfraNor Project No. 269927). This project is partly funded by the Norwegian Space Agency (NoSA). The European Space Agency has also funded parts of the work through Cryosphere Virtual Laboratory (ESRIN/Contract 4000128808/19/I-NS).

Acknowledgments: MODIS Terra snow cover data were retrieved from NSCDC NASA DAAC: National Snow and Ice Data Center; meteorological data and sea ice maps were retrieved from the Norwegian Meteorological institute.

Conflicts of Interest: The authors declare no conflict of interest.

Appendix A

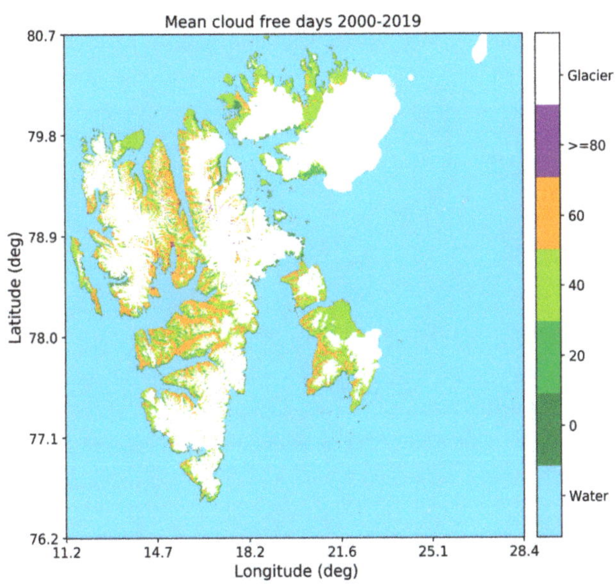

Figure A1. Average number of cloud-free days per year.

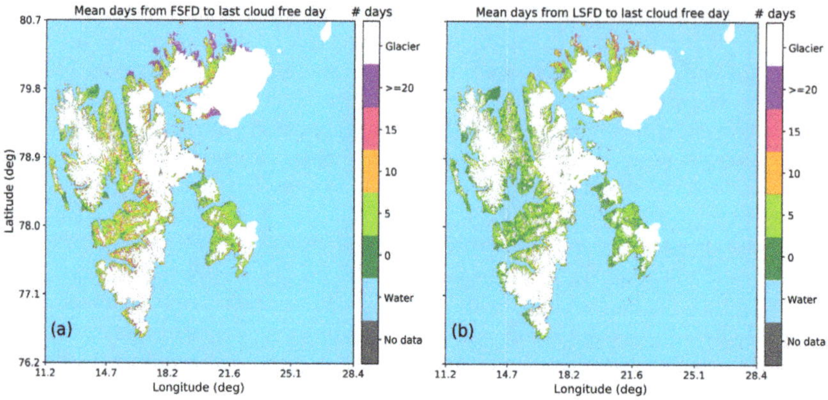

Figure A2. Mean number of days since the last cloud-free observation for (**a**) FSFD and (**b**) LSFD.

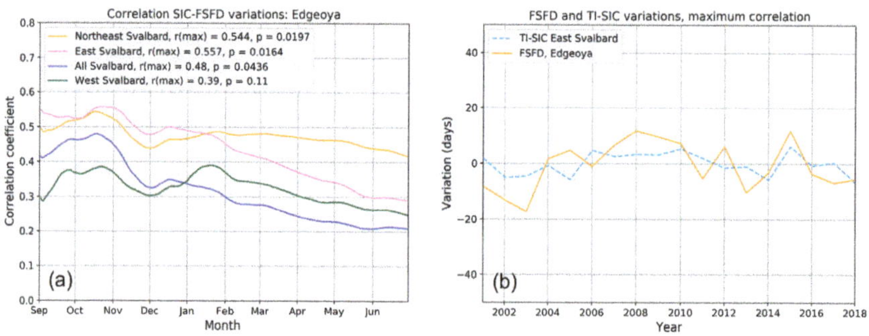

Figure A3. (**a**) Correlation between TI-SIC and FSFD variations averaged over Edgeøya, calculated for SIC integrated from 1 September of the previous year, to each consecutive day of year until 30 June the following year, for the four sea ice regions (**b**). Time series of FSFD variations averaged over Edgeøya together with the time series of TI-SIC variations for the sea region East Svalbard that produced maximum correlation.

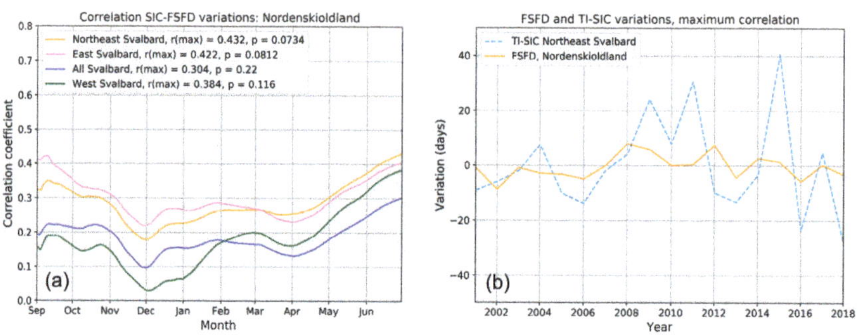

Figure A4. As for Figure A3 but for FSFD variations over Nordenskiöld Land and TI-SIC for Northeast Svalbard.

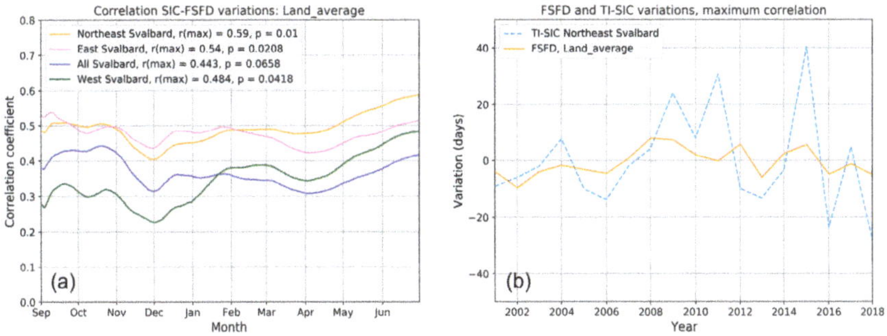

Figure A5. As for Figure A3 but for FSFD variations averaged over the whole of Svalbard and TI-SIC for Northeast Svalbard.

References

1. Chylek, P.; Folland, C.K.; Lesins, G.; Dubey, M.K.; Wang, M. Arctic air temperature change amplification and the Atlantic multidecadal oscillation. *Geophys. Res. Lett.* **2009**, *36*, L14801. [CrossRef]
2. Serreze, M.C.; Francis, J.A. The Arctic amplification debate. *Clim. Chang.* **2006**, *76*, 241–264. [CrossRef]
3. Dai, A.; Luo, D.; Song, M.; Liu, J. Arctic amplification is caused by sea-ice loss under increasing CO_2. *Nat. Commun.* **2019**, *10*, 121. [CrossRef] [PubMed]
4. Screen, J.A.; Simmonds, I. The central role of diminishing sea ice in recent Arctic temperature amplification. *Nature* **2010**, *464*, 1334–1337. [CrossRef] [PubMed]
5. Serreze, M.C.; Barrett, A.P.; Stroeve, J.C.; Kindig, D.M.; Holland, M.M. The emergence of surface-based Arctic amplification. *Cryosphere* **2009**, *3*, 11–19. [CrossRef]
6. Dutrieux, L.P.; Bartholomeus, H.; Herold, M.; Verbesselt, J. Relationships between declining summer sea ice, increasing temperatures and changing vegetation in the Siberian Arctic tundra from MODIS time series (2000–11). *Environ. Res. Lett.* **2012**, *7*, 044028. [CrossRef]
7. Comiso, J.C.; Parkinson, C.L.; Gersten, R.; Stock, L. Accelerated decline in the Arctic sea ice cover. *Geophys. Res. Lett.* **2008**, *35*, L01703. [CrossRef]
8. Pithan, F.; Mauritsen, T. Arctic amplification dominated by temperature feedbacks in contemporary climate models. *Nat. Geosci.* **2014**, *7*, 181–184. [CrossRef]
9. Onarheim, I.H.; Smedsrud, L.H.; Ingvaldsen, R.B.; Nilsen, F. Loss of sea ice during winter north of Svalbard. *Tellus A Dyn. Meteorol. Oceanogr.* **2014**, *66*, 23933. [CrossRef]
10. Førland, E.J.; Benestad, R.; Hanssen-Bauer, I.; Haugen, J.E.; Skaugen, T.E. Temperature and Precipitation Development at Svalbard 1900–2100. *Adv. Meteorol.* **2011**, *2011*, 14. [CrossRef]
11. Bintanja, R.; van der Linden, E. The changing seasonal climate in the Arctic. *Sci. Rep.* **2013**, *3*, 1556. [CrossRef]
12. Isaksen, K.; Nordli, Ø.; Førland, E.J.; Łupikasza, E.; Eastwood, S.; Niedźwiedź, T. Recent warming on Spitsbergen - Influence of atmospheric circulation and sea ice cover. *J. Geophys. Res. Atmos.* **2016**, *121*, 11913–11931. [CrossRef]
13. Hansen, B.B.; Aanes, R.; Herfindal, I.; Kohler, J.; Sæther, B.-E. Climate, icing, and wild arctic reindeer: Past relationships and future prospects. *Ecology* **2011**, *92*, 1917–1923. [CrossRef]
14. Bintanja, R.; Andry, O. Towards a rain-dominated Arctic. *Nat. Clim. Chang.* **2017**, *7*, 263–267. [CrossRef]
15. Peeters, B.; Pedersen, A.O.; Loe, L.E.; Isaksen, K.; Veiberg, V.; Stien, A.; Kohler, J.; Gallet, J.-C.; Aanes, R.; Hansen, B.B. Spatiotemporal patterns of rain-on-snow and basal ice in high Arctic Svalbard: Detection of a climate-cryosphere regime shift. *Environ. Res. Lett.* **2019**, *14*, 015002. [CrossRef]
16. Hansen, B.B.; Isaksen, K.; Benestad, R.E.; Kohler, J.; Åshild, Ø.P.; Loe, L.E.; Coulson, S.J.; Larsen, J.O.; Varpe, Ø. Warmer and wetter winters: Characteristics and implications of an extreme weather event in the High Arctic. *Environ. Res. Lett.* **2014**, *9*, 114021. [CrossRef]
17. Van Pelt, W.J.; Kohler, J.; Liston, G.E.; Hagen, J.O.; Luks, B.; Reijmer, C.H.; Pohjola, V.A. Multidecadal climate and seasonal snow conditions in Svalbard. *J. Geophys. Res. Earth Surface* **2016**, *121*, 2100–2117. [CrossRef]

18. Osuch, M.; Wawrzyniak, T. Variations and changes in snow depth at meteorological stations Barentsburg and Hornsund (Spitsbergen). *Ann. Glaciol.* **2017**, *58*, 11–20. [CrossRef]
19. Dietz, A.; Kuenzer, C.; Gessner, U.; Dech, S. Remote sensing of snow—A review of available methods. *Int. J. Remote. Sens.* **2011**, *33*, 4094–4134. [CrossRef]
20. Nagler, T.; Rott, H. Retrieval of wet snow by means of multitemporal SAR data. *IEEE Trans. Geosci. Remote. Sens.* **2000**, *38*, 754–765. [CrossRef]
21. Rotschky, G.; Schuler, T.V.; Haarpaintner, J.; Kohler, J.; Isaksson, E. Spatio-temporal variability of snowmelt across Svalbard during the period 2000—08 derived from QuikSCAT/SeaWinds scatterometry. *Polar Res.* **2011**, *30*, 5963. [CrossRef]
22. Sharp, M.; Wang, L. A five-year record of summer melt on Eurasian Arctic Ice Caps. *J. Clim.* **2009**, *22*, 133–145. [CrossRef]
23. Malnes, E.; Karlsen, S.R.; Johansen, B.; Bjerke, J.W.; Tømmervik, H. Snow season variability in a boreal-Arctic transition area monitored by MODIS data. *Environ. Res. Lett.* **2016**, *11*, 125005. [CrossRef]
24. Cooper, E.J.; Dullinger, S.; Semenchuk, P. Late snowmelt delays plant development and results in lower reproductive success in the High Arctic. *Plant Sci.* **2011**, *180*, 157–167. [CrossRef]
25. Karlsen, S.R.; Elvebakk, A.; Høgda, K.A.; Grydeland, T. Spatial and Temporal Variability in the Onset of the Growing Season on Svalbard, Arctic Norway—Measured by MODIS-NDVI Satellite Data. *Remote Sens.* **2014**, *6*, 8088–8106. [CrossRef]
26. Inouye, D.W. Effects of climate change on phenology, frost damage, and floral abundance of montane wildflowers. *Ecology* **2008**, *89*, 353–362. [CrossRef]
27. Wipf, S.; Stoeckli, V.; Bebi, P. Winter climate change in alpine tundra: Plant responses to changes in snow depth and snowmelt timing. *Clim. Chang.* **2009**, *94*, 105–121. [CrossRef]
28. Bjerke, J.W.; Tømmervik, H.; Zielke, M.; Jørgensen, M. Impacts of snow season on ground-ice accumulation, soil frost and primary productivity in a grassland of sub-Arctic Norway. *Environ. Res. Lett.* **2015**, *10*, 095007. [CrossRef]
29. Livensperger, C.; Steltzer, H.; Darrouzet-Nardi, A.; Sullivan, P.F.; Wallenstein, M.; Weintraub, M.N. Earlier snowmelt and warming lead to earlier but not necessarily more plant growth. *AoB Plants* **2016**, *8*, plw021. [CrossRef]
30. Benestad, R.E.; Førland, E.J.; Hanssen-Bauer, I. Empirically downscaled temperature scenarios for Svalbard. *Atmos. Sci. Lett.* **2002**, *3*, 71–93. [CrossRef]
31. Zhao, M.; Ramage, J.; Semmens, K.; Obleitner, F. Recent ice cap snowmelt in Russian High Arctic and anti-correlation with late summer sea ice extent. *Environ. Res. Lett.* **2014**, *9*, 045009. [CrossRef]
32. Bamber, J.; Krabill, W.; Raper, V.; Dowdeswell, J. Anomalous recent growth of part of a large Arctic ice cap: Austfonna, Svalbard. *Geophys. Res. Lett.* **2004**, *31*, L12402. [CrossRef]
33. Day, J.J.; Bamber, J.L.; Valdes, P.J.; Kohler, J. The impact of a seasonally ice free Arctic Ocean on the temperature, precipitation and surface mass balance of Svalbard. *Cryosphere* **2012**, *6*, 35–50. [CrossRef]
34. Mori, M.; Watanabe, M.; Shiogama, H.; Inoue, J.; Kimoto, M. Robust Arctic sea-ice influence on the frequent Eurasian cold winters in past decades. *Nat. Geosci.* **2014**, *7*, 869–873. [CrossRef]
35. Hall, D.K.; Riggs, G.A.; Salomonson, V.V.; DiGirolamo, N.E.; Bayr, K.J. MODIS snow-cover products. *Remote Sens. Environ.* **2002**, *83*, 181–194. [CrossRef]
36. Salomonson, V.V.; Appel, I. Estimating fractional snow cover from MODIS using the normalized difference snow index. *Remote Sens. Environ.* **2004**, *89*, 351–360. [CrossRef]
37. Johansen, B.; Karlsen, S.; Tømmervik, H. Vegetation mapping of Svalbard utilising Landsat TM/ETM data. *Polar Rec.* **2012**, *48*, 47–63. [CrossRef]
38. Dietz, A.J.; Wohner, C.; & Kuenzer, C. European snow cover characteristics between 2000 and 2011 derived from improved MODIS daily snow cover products. *Remote Sens.* **2012**, *4*, 2432–2454. [CrossRef]
39. Metsämäki, S.; Mattila, O.P.; Pulliainen, J.; Niemi, K.; Luojus, K.; Böttcher, K. An optical reflectance model-based method for fractional snow cover mapping applicable to continental scale. *Remote. Sens. Environ.* **2012**, *123*, 508–521. [CrossRef]
40. Svalbard Integrated Arctic Earth Observing System. Available online: https://sios-svalbard.org (accessed on 20 February 2020).
41. Cornillon, P. OPeNDAP: Accessing data in a distributed, heterogeneous environment. *Data Sci. J.* **2003**, *2*, 164–174. [CrossRef]

42. Cryosphere Virtual Lab. Available online: https://cvl.eo.esa.int (accessed on 20 February 2020).
43. Karlsen, S.R.; Anderson, H.; van der Wal, R.; Hansen, B. A new NDVI measure that overcomes data sparsity in cloud-covered regions predicts annual variation in ground-based estimates of high arctic plant productivity. *Environ. Res. Lett.* **2018**, *13*, 025011. [CrossRef]
44. Macias-Fauria, M.; Karlsen, S.R.; Forbes, B.C. Disentangling the coupling between sea ice and tundra productivity in Svalbard. *Sci. Rep.* **2017**, *7*, 8586. [CrossRef] [PubMed]
45. Zakhvatkina, N.; Smirnov, V.; Bychkova, I. Satellite SAR Data-based Sea Ice Classification: An Overview. *Geosciences* **2019**, *9*, 152. [CrossRef]
46. Dahlke, S.; Hughes, N.E.; Wagner, P.M.; Gerland, S.; Wawrzyniak, T.; Ivanov, B.; Maturilli, M. The observed recent surface air temperature development across Svalbard and concurring footprints in local sea ice cover. *Int. J. Climatol.* **2020**, 1–20. [CrossRef]
47. MET Norway Thredds Service Ice Concentration [Data set]. Available online: http://thredds.met.no/thredds/catalog/arcticdata/met.no/iceChartSat/catalog.html (accessed on 20 February 2020).
48. Van Pelt, W.; Pohjola, V.; Pettersson, R.; Marchenko, S.; Kohler, J.; Luks, B.; Hagen, J.O.; Schuler, T.V.; Dunse, T.; Noel, B.; et al. A long-term dataset of climatic mass balance, snow conditions, and runoff in Svalbard (1957–2018). *Cryosphere* **2019**, *13*, 2259–2280. [CrossRef]
49. Hanssen-Bauer, I.; Førland, E.J.; Hisdal, H.; Mayer, S.; Sandø, A.B.; Sorteberg, A. *Climate in Svalbard 2100—A Knowledge Base for Climate Adaptation*; NCCS Report no. 1/2019; Norwegian Meteorological Institute: Oslo, Norway, 2018.
50. Liu, J.P.; Curry, J.A.; Wang, H.; Song, M.; Horton, R.M. Impact of declining Arctic sea ice on winter snowfall. *Proc. Natl. Acad. Sci. USA* **2012**, *109*, 4074–4079. [CrossRef] [PubMed]
51. Wegmann, M.; Orsolini, Y.; Vázquez, M.; Gimeno, L.; Nieto, R.; Bulygina, O.; Jaiser, R.; Handorf, D.; Rinke, A.; Dethloff, K.; et al. Arctic moisture source for Eurasian snow cover variations in autumn. *Environ. Res. Lett.* **2015**, *10*, 054015. [CrossRef]
52. Cohen, J.L.; Furtado, J.C.; Barlow, M.A.; Alexeev, V.A.; Cherry, J.E. Arctic warming, increasing snow cover and widespread boreal winter cooling. *Environ. Res. Lett.* **2012**, *7*, 014007. [CrossRef]
53. Park, H.; Walsh, J.; Kim, Y.; Nakai, T.; Ohata, T. The role of declining Arctic sea ice in recent decreasing Arctic snow depths. *Polar Sci.* **2012**, *7*, 174–187. [CrossRef]
54. Potopová, V.; Boroneanţ, C.; Možný, M.; Soukup, J. Driving role of snow cover on soil moisture and drought development during the growing season in the Czech Republic. *Int. J. Climatol.* **2016**, *36*, 3741–3758. [CrossRef]
55. Parida, B.R.; Buermann, W. Increasing summer drying in North American ecosystems in response to longer nonfrozen periods. *Geophys. Res. Lett.* **2014**, *41*, 5476–5483. [CrossRef]
56. Barichivich, J.; Briffa, K.R.; Myneni, R.; Van Der Schrier, G.; Dorigo, W.; Tucker, C.J.; Osborn, T.J.; Melvin, T.M. Temperature and snow-mediated moisture controls of summer photosynthetic activity in Northern terrestrial ecosystems between 1982 and 2011. *Remote Sens.* **2014**, *6*, 1390–1431. [CrossRef]
57. Vickers, H.; Høgda, K.A.; Solbø, S.; Karlsen, S.R.; Tømmervik, H.; Aanes, R.; Hansen, B.B. Changes in greening in the high Arctic: Insights from a 30 year AVHRR max NDVI dataset for Svalbard. *Environ. Res. Lett.* **2016**, *11*, 105004. [CrossRef]
58. Treharne, R.; Bjerke, J.W.; Tømmervik, H.; Stendardi, L.; Phoenix, G.K. Arctic browning: Impacts of extreme climatic events on heathland ecosystem CO_2 fluxes. *Glob. Chang. Biol.* **2019**, *25*, 489–503. [CrossRef]
59. Phoenix, G.K.; Bjerke, J.W. Arctic browning: Extreme events and trends reversing arctic greening. *Glob. Chang. Biol.* **2016**, *22*, 2960–2962. [CrossRef]

remote sensing

MDPI

Article

A Compilation of Snow Cover Datasets for Svalbard: A Multi-Sensor, Multi-Model Study

Hannah Vickers [1,*], Eirik Malnes [1], Ward J. J. van Pelt [2], Veijo A. Pohjola [2], Mari Anne Killie [3], Tuomo Saloranta [4] and Stein Rune Karlsen [1]

1 NORCE Norwegian Research Centre AS, P.O. Box 6434, NO-9294 Tromsø, Norway; eima@norceresearch.no (E.M.); skar@norceresearch.no (S.R.K.)
2 Department of Earth Sciences, Uppsala University, 75105 Uppsala, Sweden; ward.van.pelt@geo.uu.se (W.J.J.v.P.); Veijo.Pohjola@geo.uu.se (V.A.P.)
3 Norwegian Meteorological Institute, P.O. Box 43, NO-0313 Oslo, Norway; mariak@met.no
4 Hydrology Department, Norwegian Water Resources and Energy Directorate, P.O. Box 5091, NO-0301 Oslo, Norway; tus@nve.no
* Correspondence: havi@norceresearch.no

Abstract: Reliable and accurate mapping of snow cover are essential in applications such as water resource management, hazard forecasting, calibration and validation of hydrological models and climate impact assessments. Optical remote sensing has been utilized as a tool for snow cover monitoring over the last several decades. However, consistent long-term monitoring of snow cover can be challenging due to differences in spatial resolution and retrieval algorithms of the different generations of satellite-based sensors. Snow models represent a complementary tool to remote sensing for snow cover monitoring, being able to fill in temporal and spatial data gaps where a lack of observations exist. This study utilized three optical remote sensing datasets and two snow models with overlapping periods of data coverage to investigate the similarities and discrepancies in snow cover estimates over Nordenskiöld Land in central Svalbard. High-resolution Sentinel-2 observations were utilized to calibrate a 20-year MODIS snow cover dataset that was subsequently used to correct snow cover fraction estimates made by the lower resolution AVHRR instrument and snow model datasets. A consistent overestimation of snow cover fraction by the lower resolution datasets was found, as well as estimates of the first snow-free day (FSFD) that were, on average, 10–15 days later when compared with the baseline MODIS estimates. Correction of the AVHRR time series produced a significantly slower decadal change in the land-averaged FSFD, indicating that caution should be exercised when interpreting climate-related trends from earlier lower resolution observations. Substantial differences in the dynamic characteristics of snow cover in early autumn were also present between the remote sensing and snow model datasets, which need to be investigated separately. This work demonstrates that the consistency of earlier low spatial resolution snow cover datasets can be improved by using current-day higher resolution datasets.

Keywords: polar regions; snow cover; remote sensing; snow modelling; MODIS; Sentinel-2

Citation: Vickers, H.; Malnes, E.; van Pelt, W.J.J.; Pohjola, V.A.; Killie, M.A.; Saloranta, T.; Karlsen, S.R. A Compilation of Snow Cover Datasets for Svalbard: A Multi-Sensor, Multi-Model Study. *Remote Sens.* **2021**, *13*, 2002. https://doi.org/10.3390/rs13102002

Academic Editor: Gareth Rees

Received: 7 April 2021
Accepted: 11 May 2021
Published: 20 May 2021

1. Introduction

Snow cover is a crucial component of the climate system, with its high albedo allowing up to 90% of incoming solar radiation to be reflected. Snow is also an important insulator, and in cold climates such as those found in the high latitude regions, it protects underlying soil and vegetation from frost damage. However, past and present changes in the global climate have been producing pronounced effects in the polar regions, as increasing temperatures lead to loss of snow, glacier and sea ice cover which in turn reduce the surface albedo and increase absorption of solar radiation, producing even greater warming [1]. The Svalbard archipelago, located in the High Arctic, is heavily glaciated and glaciers alone make up 57% of the total land area of Svalbard [2]. However, as a result of a warming climate, the spatiotemporal characteristics of seasonal snow cover on Svalbard have undergone significant changes in the past two decades, with large parts of the archipelago exhibiting trends of earlier spring snowmelt and disappearance [3,4]. Projected changes in climate, as outlined in a recent report on the future climate of Svalbard, indicate that by 2100, increases of 3–4 °C and 6–8 °C in the mean annual temperature can be expected for the west coast and northeastern regions respectively, compared with the 1961–1990 average [5]. Such marked warming will inevitably lead to significant impacts across the cryosphere, hydrosphere and biosphere, as changes in snow and ice cover affect timing and intensity of surface runoff and water storage and availability. On average, predictions for Svalbard indicate more than doubling of the snow-free season length and total runoff from glaciers and seasonal snow between 1957–2018 and 2019–60 for a RCP4.5 emission scenario [6]. Snowpack water content is an important component of the hydrological cycle; therefore, snow cover mapping is useful in both assessing water resources and for calibration of hydrological models [7–9] through data assimilation [10]. Up-to-date, detailed information on snow cover and conditions is also an important element for forecasting of natural hazards such as avalanches, slush flows and snowmelt floods, all of which may occur more frequently in a warming climate. Knowledge of snow water equivalent (SWE) in mountain catchments is also crucial to the hydropower industry, especially for management of seasonal water resources. Operational daily maps of simulated snow conditions have already existed for 15 years for mainland Norway. However, there is an absence of detailed, spatiotemporal information of snow conditions on Svalbard, despite the obvious relevance and need for such information in, for example, natural hazard forecasting on Svalbard and planning of outdoor and tourism activities.

The evolution of snow parameters can be simulated continuously in space and time through utilization of snow models. These require a surface meteorological forcing, which is either obtained from output of regional climate/numerical weather prediction models or reanalysis datasets for large-scale modelling. The evolution of the seasonal snowpack over land in Svalbard is dominated by snow accumulation during autumn and winter and subsequent melting during late spring and summer. Snow accumulation and spring maximum snow depth is mostly determined by cumulative precipitation in the form of snow during autumn and winter, while snowmelt depends on land-atmosphere interactions that can be estimated using simple melt-air temperature relationships such as the positive-degree day model, or more sophisticated models that solve the surface energy balance. Snow models therefore represent a valuable tool for filling spatial and temporal gaps in observational datasets. Moreover, they can simulate snow over longer time-periods and larger spatial domains than observational datasets. Essential to snow model calibration and validation is the use of in situ and/or remote sensing snow products, estimates of SWE, snow depth, density, temperature and water content. However, measuring snow parameters traditionally by means of in situ observations provide only point measurements and is limited in spatial coverage. Furthermore, the installation and maintenance of networks of meteorological instruments is often challenging in high mountain and remote terrain environments.

Remote sensing of snow cover provides a means of observing snow cover over large spatial areas that cannot be fulfilled by in situ observations alone and has been well-reviewed in recent years [11,12]. Optical sensors make detection of snow possible by utilizing the reflectance characteristics of snow at different wavelengths. Snow is distinguishable from other types of surface cover due to its high reflectance properties at visible wavelengths, low reflectance in the near infrared band and shortwave infrared wavelengths [13]. Several generations of optical sensors have now been acquiring data globally for several decades; the Moderate Resolution Imaging Spectroradiometer (MODIS) onboard the Terra and Aqua satellites has been acquiring optical images since 2000, from which the Normalized Difference Snow Index (NDSI) can be derived [14] as well as fractional snow cover [15]. Recently, a 20-year MODIS snow cover fraction (SCF) dataset for Svalbard based on the NASA MOD10A1-product [16] at 500 m spatial resolution has been produced and investigated [4]. Other spaceborne optical sensors include the Advanced Very High-Resolution Radiometer (AVHRR) instrument, which has flown onboard polar orbiting satellites since the late 1970s and provides observations for monitoring snow cover extent (SCE). The instrument has approximately 1 km spatial resolution, but only data at a reduced effective resolution of approximately 4 km is permanently archived and available with global area coverage (GAC). Meanwhile, newer more sophisticated optical sensors on board the Sentinel-2 A and B satellites have been delivering data over Svalbard since 2016 at a nominal 10 m pixel spacing. Since the launch of the Sentinel-2B satellite in 2017, daily coverage of Svalbard has also been possible, therefore providing unrivalled opportunities to study snow cover changes in Svalbard at high temporal and spatial resolution. A recent study has begun to address the similarities and differences in fractional snow cover retrievals using three optical sensors at different resolution and extracted using retrieval algorithms for a study site in northwest Svalbard [17]. The remote sensing observations were further validated by very high-resolution terrestrial photography. Even though the study area and time period were limited in extent, the results nevertheless indicate that there exist discrepancies when comparing observations from lower and higher resolution sensors, as well as the methods used to retrieve them. Furthermore, compiling long term climate records and linking observations to climatic variations by combining data from different sensors that cover different time periods inevitably becomes challenging due to mixed-pixel problems creating biases in fractional snow cover estimates when aggregated over large areas [18].

Despite the existence of multiple optical satellite datasets and snow model datasets, there is an obvious lack of continuity and consistency with respect to spatial resolution and periods of data coverage. Moreover, few attempts have been made that demonstrate how current-day, high resolution remote sensing datasets can be used to reconstruct and upscale snow cover observations of the earlier, low resolution datasets that often provide long time periods of data coverage. This is especially true for the high latitude Svalbard archipelago, where changes in seasonal snow are occurring faster than snow-covered areas at lower latitudes. Furthermore, studies of snow cover to date have often only utilized either remote sensing or modelling and there is therefore a need for large scale comparisons between different resolution sensors and models as well as evaluating the relative strengths and weaknesses of each dataset. Therefore, the objective of this study is to demonstrate the similarities and differences of snow cover observations made using remote sensing datasets and snow models, and how these differences can affect the extraction of derived parameters linked to the dynamical processes such as snow melt and disappearance. We examine snow cover fraction derived using the AVHRR dataset at 4 km resolution and Sentinel-2 observations produced at 20 m resolution and how they compare with the recently published MODIS SCF dataset for Svalbard at 500 m resolution [4]. In addition, two independent snow models developed by the University of Uppsala (Energy balance—snow and firn model; EBFM) and the Norwegian Water Resources and Energy Directorate (seNorge) that provide estimates of SWE and fractional snow-covered area, are used to derive snow cover extent maps. These are compared with the snow cover extent derived from binarization of the MODIS SCF maps to examine the temporal and spatial differences in snow cover. The area of study for this work is defined by the overlapping area common to all datasets available. For this work, we have therefore carried out the data analysis for the Nordenskiöld Land region in the central part of Svalbard.

An overview of the study area will be presented in Section 2 together with a description of the remote sensing and snow model datasets and an outline of the data processing and analysis methods used to perform the comparisons between the datasets. The results of the comparisons as well as a quantitative evaluation of the consistency between the datasets are presented in Section 3. In Section 4, a discussion of the main results is made in the context of current knowledge and earlier studies of relevance. A summary of the primary findings of this study, as well as suggestions for further work is given in the conclusion, in Section 5.

2. Materials and Methods

2.1. Study Area

The Svalbard archipelago is a group consisting of nine islands, located approximately halfway between the northernmost point of the Norwegian mainland and the North Pole. The archipelago covers a total area of 61,000 km^2 of which 60% is glaciated and the remaining area is covered by barren rock and vegetation. The archipelago is spread over latitudes in the range 74–81°N and longitudes ranging from 10–35°E. Both midnight sun and polar night are therefore present for a large proportion of the year. Nordenskiöld Land is situated approximately in the center of the archipelago in terms of latitude and it is in this region the administrative center Longyearbyen is located. The climate of this area is affected by the West Spitsbergen Current, which brings warm salty water from the Atlantic Ocean northwards, producing a milder climate than experienced at similar latitudes elsewhere. Meteorological data recorded at Svalbard Airport, close to Longyearbyen, show that the mean annual temperature in the period of study (2000–2019) ranges from −6.1 °C (2003) to 0 °C (2016), while annual precipitation ranges from a minimum of 142.1 mm (2005) up to 310 mm (2016).

Figure 1 illustrates the area of study chosen for this work. This study area corresponds to about one and a half Sentinel-2 tiles for which NDSI data have been processed and is also the area that overlaps with all datasets that have been compared.

Figure 1. Study area of central Svalbard. Sentinel-2 based map from 1 August (mean 2016–2019) with colors extracted from NDSI and NDVI (Normalized Difference Vegetation Index) values.

2.2. MODIS

The MODIS instrument on board the Terra and Aqua satellites provides multispectral imagery at visible-shortwave infrared (VSWIR) wavelengths, with a daily revisit period and spatial resolution of approximately 500 m for all VSWIR bands. This study uses the NASA MODIS/Terra Snow Cover Daily L3 Global 500 m Grid, Version 6 (MOD10A1) dataset to derive NDSI as a daily product [16], which is calculated using the spectral band 4 (visible light) and band 6 (short wave infrared) through the relation,

$$NDSI = ((band4 - band6)/(band4 + band6)) \quad (1)$$

Snow cover fraction, expressed as a percentage, is then estimated using a universal approach [15] defined by the regression formula,

$$SCF = (0.06 + 1.21\ NDSI) \times 100 \quad (2)$$

During the polar night when there is no optical data coverage, SCF is set to 100 %. As such, the MODIS SCF estimates are produced only for the period 1 March–1 November for the entire periglacial landscape in Svalbard as a temporally interpolated product at daily intervals and 500 m resolution. In addition, cloud cover can often contaminate MODIS data, which is detected and masked out. There are a number of reconstruction approaches that have been developed to produce daily cloud-free MODIS snow products [19,20]. However, in this study we have implemented a temporal interpolation technique to obtain cloud-free maps for the Svalbard archipelago, which has been outlined in earlier work [4]. The reader is referred to this work for a more detailed description of the method. Since MODIS has moderate spatial resolution and excellent temporal overlap with both older and newer available satellite and modelled data products being used in this study, the MODIS dataset is used throughout this study as the baseline for comparisons.

2.3. AVHRR

The AVHRR Global Area Coverage (GAC) data are used to produce a fundamental climate data record (FCDR) for radiances and brightness temperatures. This dataset has been made available by the EUMETSAT Climate Monitoring Satellite Application Facility (CM SAF). The current release, CLARA-A2, covers 1982–2015 [21]. A time series of daily snow cover maps covering the Svalbard archipelago at 4 km grid spacing was derived from the CLARA-A2 FCDR using the probabilistic snow cover algorithm provided by MET Norway. This snow cover algorithm uses a set of instrument channel combinations and statistical coefficients, the latter of which are derived from prior knowledge of the typical behavior of the surface classes across the spectrum. Cloud-free pixels from the AVHRR GAC swath products are averaged and gridded to produce daily maps of average snow probability, to which a threshold of 50% is applied to derive a binary snow cover extent product. Data are provided only for the period between March 1 and September 30 each year due to no data availability during the polar night. This results in one less month with data to compare with MODIS for the overlapping period. In addition, temporal gap filling was applied to achieve daily cloud-free mosaics using information from cloud-free pixels up to 9 days forward or backward in time. The product also indicates the age of the reference image used to make the cloud cover corrections and is the product used for comparisons with MODIS data. Furthermore, a vegetation map for Svalbard produced using Landsat data [22] is used to mask out glaciers and water bodies in the AVHRR dataset, as was done to produce the MODIS SCF dataset.

2.4. Sentinel-2 (S2)

In this study we use all the available Sentinel-2 data for the period 15th April to 15th September, each year from 2016 to 2019. From 1 July 2017 data from Sentinel-2B is acquired in addition to Sentinel-2A. From this point onwards, data are available from both twin satellites on most of the days for the study area. Due to the low solar elevation angle, both early and late in the snow season (solar zenith angles higher than 70°) the Sentinel-2 Level 2A (bottom-of-atmosphere) data are not reliable, and we only used scene classification data from Level 2A as reference data and rely on Level 1C data (top-of-atmosphere) in the snow mapping.

For cloud detection we examined the cloud probability in Level 2A (Sen2Cor processor) and "s2cloudless" machine-learning based algorithm [23]. In addition, we developed our own cloud detection algorithms from multi-spectral values and multi-temporal tests. However, none of the methods work well for sparsely vegetated areas (bright surfaces), which is very common in the study area. For thin semi-transparent clouds, and for cloud shadows the algorithms did not show sufficient accuracies. To detect clouds, we performed a visual inspection in the visual and SWIR bands, and masked out cloud free areas, only using the cloud masks as a reference. One exception was the cirrus clouds, which could be accurately detected with band 10. However, cirrus clouds only appear in a few of the images. This time-consuming method for cloud detection based on visual inspection, ensures fewer errors, but not all cloud free data are included. Some of the images had many small cumulus clouds which were too time-consuming to mask out and so were not used.

NDSI values were calculated (Equation (1)) for the cloud-free pixels and interpolated to produce daily NDSI maps at 20 m resolution. This was achieved by performing linear interpolation and smoothing with a Savitzky–Golay filter. In this study we have primarily focused on retrieving estimates of the land-averaged snow cover fraction by applying a fixed threshold of 0.4 [16,24] to the Sentinel-2 NDSI which produces a binary (snow/no snow) snow cover extent map. We have chosen to utilize a thresholding method since regression coefficients have not been thoroughly tested and validated on Sentinel-2 data yet, as has previously been done for the MODIS datasets.

2.5. SeNorge

The seNorge snow model [25] requires 3-hourly or daily mean air temperature and a sum of precipitation as input forcing. Solid precipitation is defined as precipitation occurring at an air temperature ≤0.5 °C. Snow and ice melt are calculated using the extended degree-day model including air temperature and solar radiation terms. The two parameters of the melt algorithm have subsequently been estimated based on 3356 quality controlled daily melt rates observed by the Norwegian snow pillow network [26]. The sub- grid snow distribution algorithm in the model [25] assumes a uniform probability distribution of snow amounts within the grid cells. In addition, an even layer of new snow can form on top of the uniformly distributed "old" snowpack and snow-covered fraction is then set to 1. The main effect of the sub-grid snow distribution is to reduce the grid cell average melting rates towards the late melt season rates when significant areas of bare ground are present in the grid. The 3-hourly input data are aggregated from the hourly meteorological forcing data obtained and downscaled from the AROME Arctic numerical weather prediction model (NWP). Input precipitation in the current model application is scaled by a factor 0.75, based on initial evaluation of the first model results. Model parameter values are set to the same values as those in the application for mainland Norway, except the spatial snow distribution parameter CF is increased from the default value of 0.5 to 0.85, giving larger variance for sub-grid snow distribution. The model application for Svalbard starts at bare ground initial conditions in September 2012. Following this, snow/firn older than 1 year is removed from the model's snow store on September 1 each year. The two first snow seasons may therefore be considered as a model spin-up period at higher elevation areas with perennial snow. SCF estimates for this study are provided at 3-h intervals daily for the years 2013–2019. This study has utilized the seNorge SCF product corresponding to 1200 UTC.

2.6. EBFM

The coupled energy balance—snow and firn model (EBFM) [27] has been used to study the long-term climatic mass balance of glaciers [28], as well as seasonal snow conditions and runoff on glaciers and land since 1957. EBFM solves the surface energy balance to calculate surface melt and temperature, which provides upper boundary conditions for a subsurface model, simulating the multi-layer evolution of snow density, temperature and water content [27]. In Van Pelt et al. (2019) [28], the model was forced by downscaled meteorological fields of precipitation, air temperature, relative humidity, wind speed and air pressure from the High-Resolution Limited Area Model (HIRLAM) [29]. For calibration and validation of the model and meteorological downscaling, in situ on-glacier measurements of weather conditions, stake mass balance and subsurface density were utilized; no calibration or validation was performed for snow conditions in non-glacier terrain, potentially deteriorating performance in these areas. For more details about the methods and dataset, the reader may refer to [28] and references therein. From this large dataset, SWE is extracted across Svalbard at 1 × 1 km spatial resolution and daily temporal resolution for 2000–2019. The dataset overlaps for a longer part of the MODIS period, as well as with the seNorge snow model dataset.

2.7. Processing of the Datasets

For all datasets being compared only the period for which MODIS data are both available and overlap with the other datasets are used. That is to say, the period where SCF is assumed to be 100% (November–February) is excluded from the study. Table 1 summarizes the spatial resolution, period of data availability and actual time period that has been compared with MODIS.

Table 1. Summary of the datasets used in the study, together with the spatial resolution, time period used and type of snow cover product provided.

Dataset	Data Cover	Years	Resolution	Snow Product
MODIS	1 March–31 October	2000–2019	500 m	Snow cover fraction
AVHRR	1 March–30 September	2000–2015	4 km	Snow cover extent
Sentinel-2	15 April–15 September	2016–2019	20 m	Normalized Difference Snow Index
seNorge	1 January–31 December	2013–2019	1 km	Snow cover fraction
EBFM	1 January–31 December	2000–2019	1 km	Snow water equivalent

In Figure 2, a simple flowchart illustrates the workflow and extraction of the parameters of interest for the area studied. For all datasets being compared with MODIS, a resampling was performed in order to reproduce the datasets with a common grid and identical coverage area. The land-averaged snow cover fraction (SCF) was calculated for the study area using either the binary or fractional snow cover products. For all snow products, glaciers and water bodies were excluded from the spatial averaging. SCF was computed for only the periods with overlapping data coverage.

As described in Section 2.4, a fixed threshold of 0.4 was applied to the Sentinel-2 NDSI maps to extract binary snow cover extent maps. For the EBFM SWE datasets, an optimum threshold was determined using the MODIS SCF as a reference dataset. Ten different thresholds on SWE ranging from 0.001 to 0.01 m were used to first obtain a binary snow cover maps and subsequently to calculate the corresponding land-averaged snow cover fraction time series for each threshold. The squared difference of the EBFM and MODIS-derived SCF time series was calculated and summed over the full year, for each year in the dataset. The threshold producing land-averaged SCF time series that gave the smallest squared-sum was identified as the best threshold for that year. Since the optimum threshold varied from year to year, the median value of 0.003 m for all 20 years was taken as a fixed threshold to be applied to the SWE data.

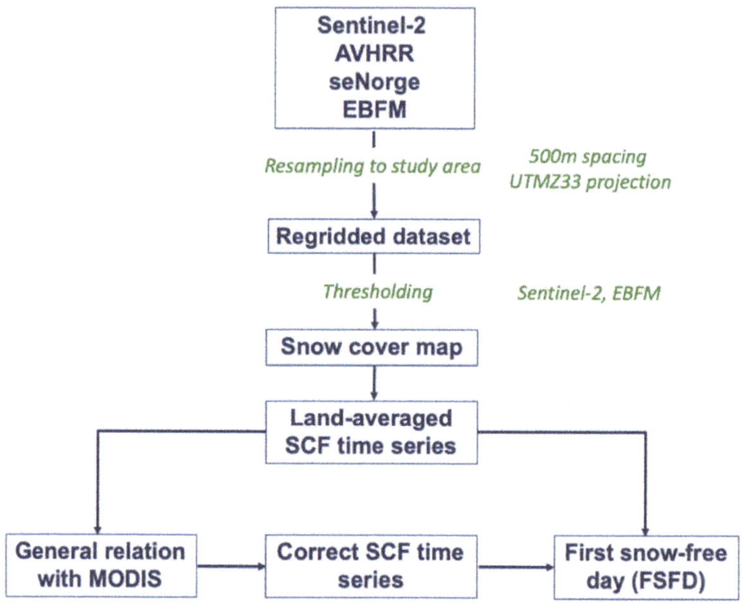

Figure 2. Flow diagram illustrating the data processing and parameters extracted from the datasets.

For all datasets, a general relationship was obtained with the MODIS time series by making a cubic spline fit to each pair of datasets (MODIS-AVHRR, MODIS-S2, MODIS-seNorge, MODIS-EBFM). These relationships were further utilized to demonstrate how simple corrections to the datasets can be used to obtain better consistency with the baseline MODIS dataset. This procedure is described in greater detail in Section 3. Lastly, an estimate of the first snow-free day was made using the uncorrected and corrected land-averaged SCF time series in order to study the effect of the corrections on the timing of snow disappearance deduced from the datasets.

3. Results

In this section, we present the results of the data processing and analysis outlined in Section 2. This section is divided into five subsections, which describe the different aspects of the data comparisons made. In Section 3.1, a description of the general relationship between the datasets is made, while in Section 3.2, we present a more specific comparisons of the geographical and altitudinal differences between the datasets. Section 3.3 is dedicated to the results of the normalization, or correction of the snow cover time series using the results of Section 3.1, and in Section 3.4, we quantify the effect of the corrections on derived estimates of first snow-free day, compared with the original time series. A quantitative evaluation of the differences between the datasets, before and after corrections is made in Section 3.5.

3.1. General Relationship between the Datasets

In this section, a comparison of the snow cover fraction time series is made for each dataset, with respect to the MODIS SCF time series. These comparisons are made using the SCF which is obtained by averaging the snow cover products over the study area. Figure 3 shows the time series of the land-averaged snow cover fraction for all datasets used in this study. There is some overlap between datasets: for example, between MODIS, AVHRR and the EBFM dataset, from 2000–2019, and from 2013–2015, there is overlap between MODIS, AVHRR, EBFM and the seNorge datasets. In the final four years of the period (2016–2019), there is overlap between MODIS, seNorge, EBFM and Sentinel-2. There is good agreement in the SCF minimum values for MODIS, AVHRR and EBFM for the first five years of the period, though with AVHRR exhibiting greater and more frequent fluctuations in SCF during the summer minimum compared with MODIS and EBFM. From 2006 onwards the fluctuations in SCF derived from AVHRR during the minimum period become more pronounced; moreover, the SCF values during this period also tend to be some tens of percent greater compared with MODIS. It may also be noticed that in spring the MODIS SCF begins to fall slightly earlier compared with AVHRR and EBFM, while the increase in SCF at the end of the summer is somewhat misleading due to the different periods of coverage of the dataset, with the two remote sensing datasets ending at either September 30 (AVHRR) or October 31 (MODIS). The alternative snow model dataset, seNorge provides SCF from 2013 until 2019 inclusive. Here it can be seen that like AVHRR, there are large fluctuations in the land-averaged SCF during the period where SCF is at a minimum. These fluctuations can also be several tens of percent in magnitude. The curves in 2014 and 2015 display these large variations in the seNorge SCF quite clearly. Moreover, the lowest SCF reached in the seNorge dataset is some 20% lower than those exhibited by the MODIS and Sentinel-2 datasets. In general, the Sentinel-2 snow cover fraction follows closely the temporal variations of the MODIS estimates, though in 2016 and 2018 the Sentinel-2 SCF appears to fall marginally earlier than MODIS in the spring.

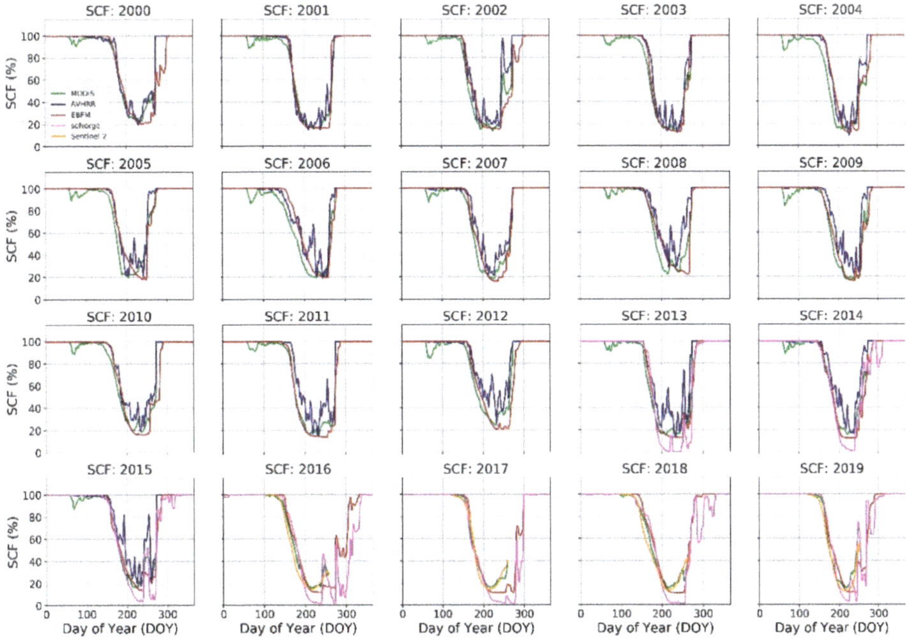

Figure 3. Time series comparing the land-averaged SCF derived using the MODIS dataset (green, 2000–2019), AVHRR (blue, 2000–2015), EBFM (dark red, 2000–2019), seNorge (purple, 2013–2019) and Sentinel-2 (orange, 2016–2019).

Figure 4 is a representation of the data shown in Figure 3, as a scatter plot. Here the SCF values from each dataset were plotted against the MODIS SCF, with the time series separated into two periods corresponding to data from 1 March–31 August (blue) and from 1 September–31 October (coral), or until the end of the dataset being compared. This was done to separate snow cover estimates made during the spring melt period and those from autumn snow onset. In all but the case of the Sentinel-2 dataset, the lower resolution data from AVHRR and the snow models were plotted on the x-axis, with MODIS along the y-axis. Since the Sentinel-2 data are at higher resolution than the MODIS dataset, these were plotted along the y-axis to derive the relation that transforms the lower resolution dataset to the SCF estimates of the higher resolution sensor. Since this study endeavors to examine the differences in timing of snow disappearance between the datasets, the most critical period for dataset correction is the snow melt period. As such, the function used to transform the datasets is obtained by fitting a cubic spline function to the pairs of data obtained only in the period from 1 March–31 August (blue datapoints) to derive the general relationship with the MODIS values; this is displayed by the light blue curves. The number of scatter points in each plot reflects the size of the dataset, with the AVHRR and EBFM datasets being the largest with respectively 16 and 20 full years of data that overlap with MODIS. Of all four datasets being compared against MODIS, there is poorest agreement with the AVHRR dataset, in terms of magnitude. This is especially true when the land-averaged MODIS SCF lies the range 30–60%, with the corresponding AVHRR SCF being on average 20% greater than MODIS. On the other hand, the land-averaged SCF obtained from Sentinel-2 generally agrees well with MODIS at low (<25%) and high (>90%) snow cover fractions. At all other snow cover fractions, the Sentinel-2 snow cover fraction is in general lower than MODIS and can reach up to nearly 10% lower than MODIS. For the seNorge dataset there is generally a good but non-linear correlation with the MODIS values, when considering only the SCF data from between 1 March–31 August. There is clearly a large spread in values for SCF obtained in the period 1 September–31October, but for the melt period of interest the land-averaged SCF derived from seNorge is of the order 5–10% greater than MODIS when MODIS SCF is >30%. For the EBFM dataset, shown in Figure 4d the relationship between the MODIS SCF and the model-derived SCF is similar to that of seNorge snow model, but the EBFM estimates can be on average up to 15% greater than MODIS for the period between 1 March–31 August, as indicated by the largest offset between the fitted spline (light blue curve) and the equality line (dark blue, dashed) while EBFM estimates of SCF obtained from after September 1 which corresponds to the start of the hydrological year, are consistently lower than MODIS.

3.2. Geographical and Elevation Differences

The difference in annual number of days with snow cover derived from each of the data products and that obtained from MODIS was mapped. To make this geographical comparison, a binary snow map was first obtained. For the fractional snow cover products SCF was thresholded at 50%, where SCF below the threshold is defined as "no snow" and SCF greater than the threshold, "snow". Hence, for each pixel in the grid, the number of days the pixel was classified as snow covered/not-snow covered during each year of data coverage using the datasets was calculated. The difference in number of days with snow cover between AVHRR/Sentinel-2 and MODIS, and the two snow model datasets and MODIS was then calculated. For the remote sensing datasets, the difference in number of days with snow applies only to the part of the year outside of the polar night period. For the AVHRR dataset, this is the difference in number of days with snow between 1 March and 30 September, while for Sentinel 2 it is restricted to only the period 15 April–15 September. Since the two snow models produce SWE and SCF for the entire year, the full period of MODIS coverage from 1 March–31 October is used in the comparisons.

In Figure 5, the mean difference in number of days with snow derived for each of the datasets is shown. The mean difference was calculated per pixel by averaging the difference in number of days with snow over all the years in each dataset. Since the difference is calculated by subtracting the MODIS number of days with snow from the AVHRR number of days with snow, positive numbers indicate a greater number of days with snow per year on average with respect to MODIS and negative numbers indicate fewer days with snow per year on average, compared with MODIS. In the case of the AVHRR dataset, it has already been shown in Figure 4a that snow cover fraction is on average for the study area, always greater than that estimated using the MODIS instrument, for the period of data cover 2000–2015. Figure 5a illustrates this pattern geographically, where there are large areas of blue (positive difference in number of days with snow) that correspond to low-lying valleys. Areas with light yellow tone indicate where the number of days with snow each year estimated by AVHRR and MODIS are roughly the same (zero difference). Since the AVHRR data were georeferenced to the MODIS grid, the downscaling from 4 km to 500 m pixel spacing is also clear from the square-like edges of blue areas. Areas where the difference in number of days with snow cover estimated by AVHRR was less than MODIS, indicated by the dark red areas, can also very likely be attributed to the resolution differences.

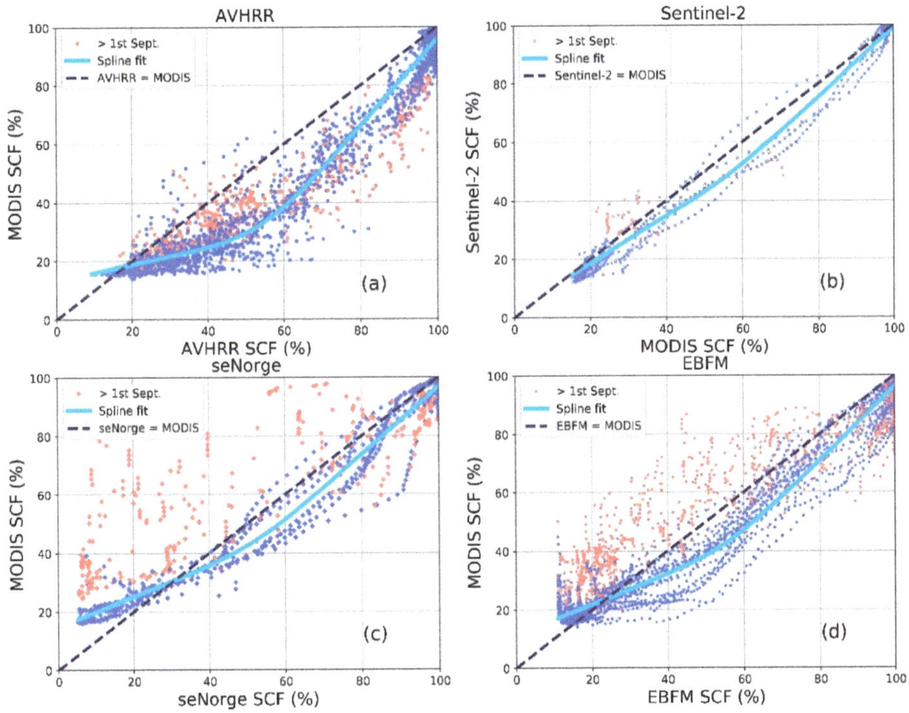

Figure 4. Scatter plots illustrating the relationship between the land-averaged SCF derived from MODIS and (**a**) AVHRR (2000–2015) (**b**) Sentinel-2 (2016–2019) (**c**) seNorge snow model (2013–2019) (**d**) University of Uppsala EBFM (2000–2019). In each case the datasets have been georeferenced to the MODIS 500 m grid and all days with overlapping data coverage have been used to produce the scatter plots. The time series of SCF for each dataset were split into two time periods, with blue datapoints corresponding to 1st March–31st August and coral-colored datapoints corresponding to data produced for the period 1st September–31st October.

For Sentinel-2 data on the other hand, which have much higher spatial resolution than MODIS, the difference in mean number of days with snow for the period 2016–2019, shown in Figure 5b is close to zero or below zero across the area of study, as exhibited by the prevalence of yellow and orange. This implies that MODIS always estimates a greater number of days with snow per year than Sentinel-2, with greatest differences on mountain slopes. Figure 5b would suggest that MODIS estimates greater than 60 days more with snow in these areas, compared to Sentinel-2.

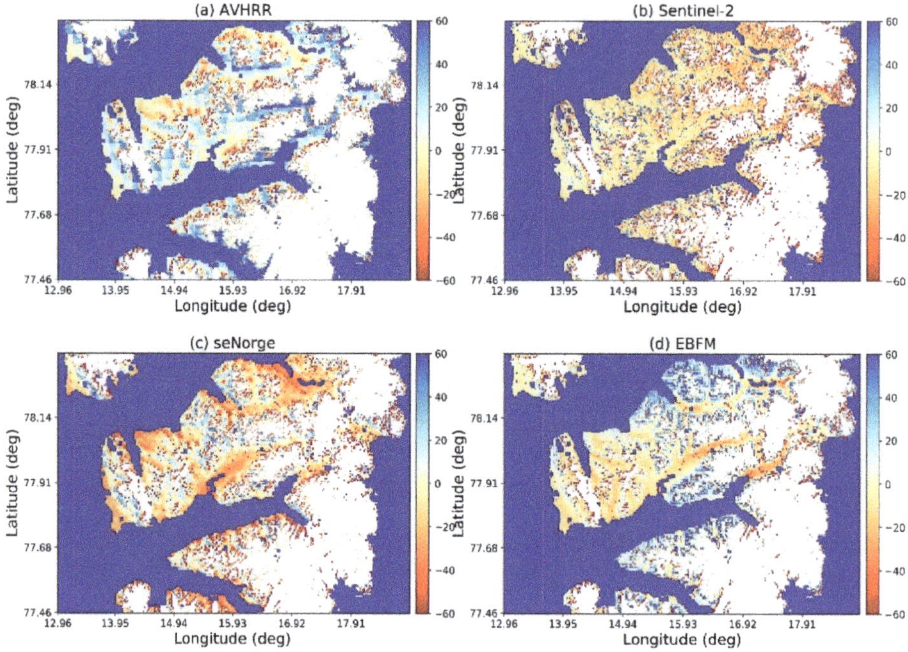

Figure 5. The geographical differences in the average number of days with snow per year, comparing the average number of days of snow per pixel derived using the MODIS dataset and (**a**) AVHRR (2000–2015) (**b**) Sentinel-2 (2016–2019) (**c**) seNorge snow model (2013–2019) and (**d**) University of Uppsala EBFM (2000–2019). The average number of days with snow is calculated using different number of years due to the time period of data coverage/availability as stated in the parentheses.

The comparison of mean days with snow cover between MODIS and the two models, seNorge and EBFM is shown in Figure 5c,d respectively. The geographical distribution and magnitude of the differences are quite different; for the seNorge snow cover area dataset, the snow model estimates on average fewer days with snow cover per year in the low-lying valleys and around all coastal areas, when compared with MODIS. On the other hand, blue areas corresponding to the highest elevation mountain zones indicate that the seNorge estimates on average a greater number of days with snow in these areas compared to the MODIS dataset. For the EBFM dataset shown in Figure 5d, there is also a tendency toward moderate to large underestimation in mean number of days with snow cover per year for the valley areas when compared with MODIS, as shown by the light yellow, orange and red regions. However, different to the seNorge dataset, EBFM tends to produce larger underestimation in number of days with snow in the inland parts of the valleys, whereas the largest underestimations for the valley areas in the seNorge dataset tends to be situated closer to the coastal areas. In addition, the EBFM dataset also transitions to greater number of days with snow cover compared to MODIS in the mountainous regions in the northern and eastern part of the study area but begins at much lower elevations than for the seNorge dataset. This is especially noticeable for the mountain slopes that are also located along coastal areas.

These elevation-dependent differences in the mean number of days with snow per year, between the different remote sensing and snow model datasets, and MODIS, is illustrated in Figure 6. This figure was produced by averaging the mean differences in number of days with snow (Figure 5) over all elevation zones at intervals of 200 m between 0 and 1200 m.a.s.l. using a high resolution (20 m) Digital Elevation Model (DEM). Here, the x-axis of Figure 6 represents the middle point of each elevation interval. Figure 6 therefore demonstrates that the degree of overestimation in mean number of days with snow by AVHRR in the low altitude areas (0–200 m.a.s.l.) is in fact only of the order of 10 days on average for the whole area of interest. The magnitude of underestimation in mean number of days with snow estimated by AVHRR with respect to MODIS increases with elevation, with the data suggesting underestimations of approximately 40 days or more for all altitudes intervals above 600 m.a.s.l. As shown in Figure 5b, Sentinel-2 data exhibit of the order 10–15 fewer days with snow on average compared to MODIS at all elevations, though at the lowest altitudes the difference is only 5 days.

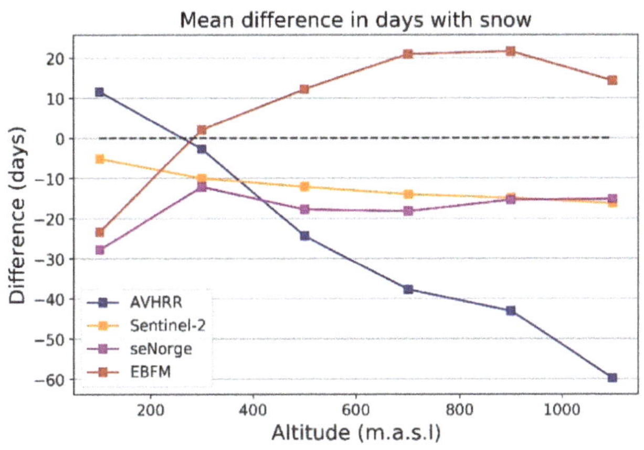

Figure 6. The altitude distribution of the average difference in number of days with snow for the four datasets being compared with MODIS. This figure is the equivalent altitude distribution of the data shown in Figure 5a–d.

Common to both snow models, there is a relatively large underestimation in mean number of days with snow at elevations between 0–200 m.a.s.l. when compared with the MODIS dataset. The mean number of days with snow is on average of the order of 25 days less than MODIS in this elevation zone for the EBFM and seNorge datasets. However, the elevation distributions are noticeably different for the two snow models at elevations above 400 m.a.s.l. While the seNorge dataset exhibits a negative difference in mean number of days with snow compared with MODIS on average (15–20 days) for the whole region at all elevations >200 m.a.s.l., EBFM estimates on average 10–20 days greater snow cover per year compared with MODIS when averaged over all the elevation intervals above 400 m.a.s.l., with largest differences present at elevations between 600–1000 m.a.s.l. This pattern was also described earlier for the geographical distribution of the differences shown in Figure 5d.

3.3. Correction of the Datasets

The Sentinel-2 dataset, with the highest spatial resolution of all the datasets but relatively small temporal period of coverage, was used to adjust the lower-resolution MODIS SCF time series by applying the obtained spline fit (Figure 4b) to the MODIS time series. With the adjusted MODIS time series, spline fits to the three remaining datasets were subsequently updated and applied to the respective time series to obtain a final corrected SCF for the three remaining lower resolution datasets (AVHRR, seNorge and EBFM). The objective of this procedure was to normalize the snow cover observations from each dataset to a baseline in order to achieve better consistency between the products. The final corrected SCF time series and the corresponding spline fits associated with these are presented in Figure 7a–d. While there remains a degree of spread in the land-averaged SCF values, the corrected time series are on average much closer to the MODIS values, as exhibited by the final spline fits which lie close to the equality line (dark blue, dashed). In particular, the relationship between the corrected datapoints and the MODIS datasets for the melt period between 1 March–31 August is more linear compared with their original time series. This is true for all datasets.

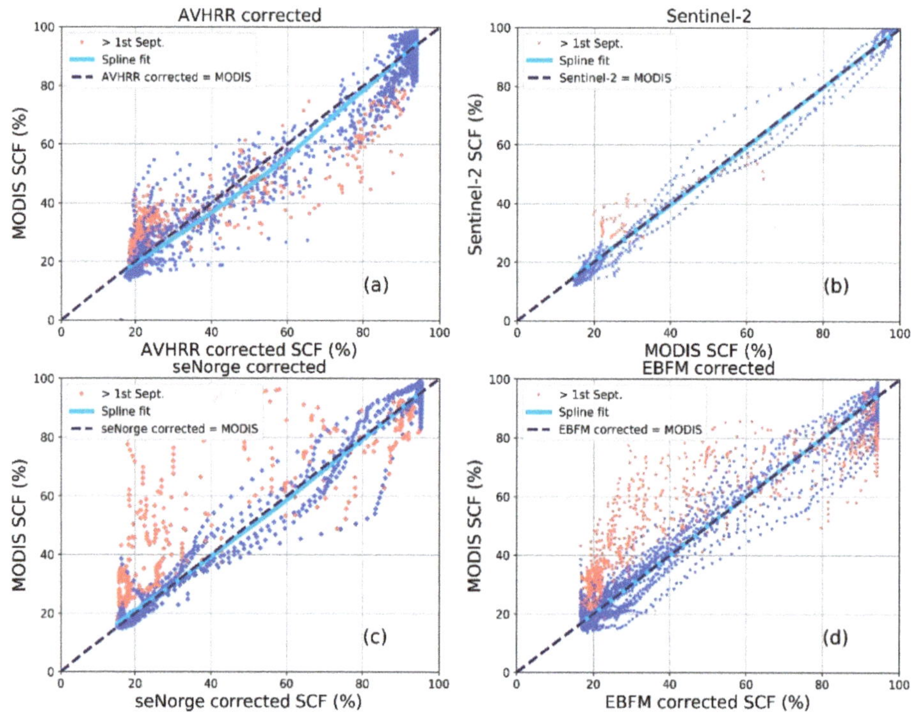

Figure 7. Scatter plots showing the relationship between the corrected MODIS dataset and the adjusted lower resolution datasets. The initial correction of the MODIS dataset was performed using the spline fit shown in Figure 4b. The corrected MODIS dataset was then used to obtain updated spline fits with the AVHRR, seNorge and EBFM datasets (not shown) which were subsequently applied to produce the final adjusted datasets shown in (**a**,**c**,**d**).

3.4. Estimation of First Snow-Free Day (FSFD)

The SCF time series produced using the different datasets were used to extract estimates of the first snow-free day, which is taken to be the point at which the SCF first falls below 50%. FSFD was estimated using the uncorrected and corrected land-averaged SCF time series for each dataset and compared with the FSFD values produced using the corresponding MODIS time series. These are shown in Figure 8a,b. Qualitatively speaking, the FSFD estimates made using the uncorrected datasets lie much further from the MODIS estimates (Figure 8a) compared with those made using the corrected SCF time series, as shown in Figure 8b. Significant improvements in the estimates of FSFD occur following correction of the SCF time series, which before the dataset was corrected, were up to 15 days greater than the corresponding MODIS FSFD estimates. Following correction of all datasets, FSFD estimates obtained from AVHRR, seNorge and EBFM all lie close to the MODIS FSFD estimates, with the corrected FSFD occurring within approximately 5 days later or earlier than MODIS. This is as expected since the spline fit made to the corrected datasets, shown in Figure 6, all closely follow the line which indicates where the two datasets would be equal.

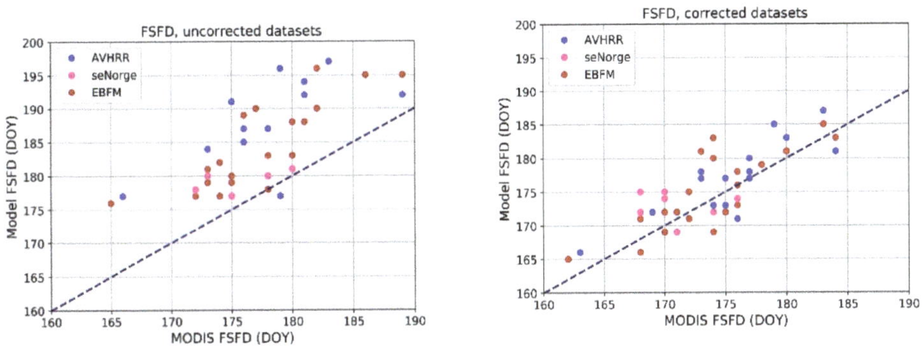

(a) (b)

Figure 8. (**a**) First snow-free day estimates shown for the three uncorrected lower resolution AVHRR, seNorge and EBFM datasets, plotted against the uncorrected MODIS land-averaged FSFD estimates. A dark blue dashed line indicates where the FSFD estimates would be equal and (**b**) FSFD estimates obtained from the corrected MODIS, AVHRR, seNorge and EBFM datasets.

In the case of the AVHRR dataset, the entire 34-year time series of snow cover extent maps for 1982–2015 are used to first extract the land-averaged SCF which is then corrected using the updated spline fits. By estimating FSFD for the entire 34–year period, decadal trends in FSFD have also been calculated using both the original uncorrected AVHRR SCF time series. This has allowed us to examine differences in FSFD trends when the data had not been upscaled using the fitted spline function. For the two corrected SCF time series corresponding to the snow models, only estimates of FSFD were made and compared with MODIS, since the temporal period of coverage of each dataset was not long enough to obtain meaningful estimates of decadal trends.

For the FSFD estimates extracted using the original AVHRR time series from 1982–2015, shown in Figure 9, FSFD occurs later compared with those extracted from the corrected time series. The offset between the two linear trend lines is approximately 15 days, indicating that using the original AVHRR SCF time series results in FSFD estimates that are later by on average 2 weeks when compared with the FSFD estimates extracted from the corrected time series. Moreover, the slope of the linear trend line is also smaller, with a decadal trend in FSFD of −3.36 days/decade (p = 0.08) for the uncorrected time series, while the decadal trend in FSFD estimated with the corrected AVHRR SCF time series is −2.81 days/decade (p = 0.01). Hence, not only does the decadal trend in FSFD derived from the corrected time series become significant at the 95% confidence level, but the advance in FSFD is >0.5 days/decade slower than that suggested by the original AVHRR dataset.

3.5. Evaluation Metrics

To evaluate the effect of the corrections on each dataset, four metrics were selected for evaluation using both the uncorrected and corrected datasets, following the approach of recent similar studies that compare snow cover retrievals from lower and higher resolution datasets [17,30,31]. We have first calculated the mean error, which indicates the bias of the dataset being evaluated. This was done firstly using Sentinel-2 as a baseline for the MODIS snow cover dataset and subsequently for the AVHRR, seNorge and EBFM datasets with MODIS as the baseline dataset. We have therefore implicitly assumed that Sentinel-2 is more accurate than MODIS and that MODIS is more accurate than AVHRR and the two snow models for the purpose of this evaluation. The root mean-squared error (RMSE) was also calculated, as well as the Spearman rank correlation coefficient. The Spearman rank correlation coefficient was chosen over the Pearson correlation coefficient since it is known to be more appropriate for non-linear relationships between datasets. For both the uncorrected and corrected datasets, each of the four metrics was calculated for two cases: firstly, using the period from 1 March–31 August and second, the period from 1 September–31 October as a basis for the calculations.

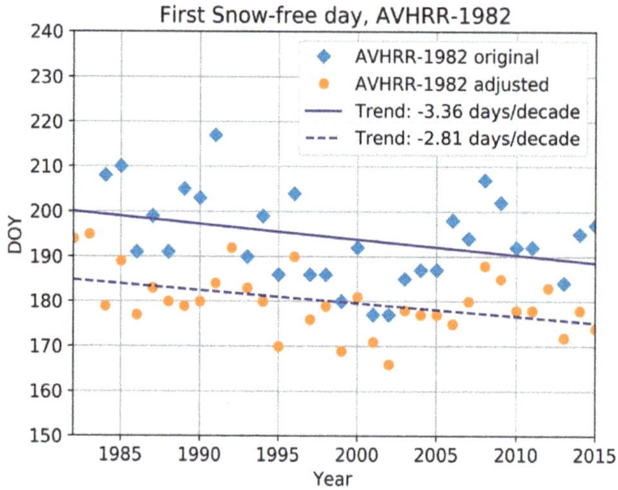

Figure 9. Estimates of FSFD obtained using the original (blue diamonds) 34-year AVHRR time series of the land-averaged SCF for Nordenskiöld Land, from 1982–2015 and the adjusted (orange circles) AVHRR SCF time series. Linear fits were made to each time series, shown by the dark blue solid line for the original AVHRR datasets and the dashed line for FSFD obtained using the adjusted dataset. Decadal trends derived from the linear fits are indicated in the legend.

Table 2 summarizes the RMSE, mean error (bias) and Spearman correlation coefficient for the AVHRR, seNorge and EBFM datasets for the main period of interest (1 March –31 August) as well as the autumn data acquired after 1 September, which are stated in italic. The upper section presents the metrics calculated using the uncorrected SCF time series, while the lower section of Table 2 displays the same metrics calculated following correction of the datasets. Sentinel-2 was used as a baseline for obtaining the spline model with which corrections to the MODIS time series was made and we show only the metrics that were calculated for the datasets sharing the same (MODIS) baseline.

Table 2. Evaluation metrics for the uncorrected and corrected (land-averaged) SCF time series in the period 1 March–31 August. Values in parentheses correspond to the metrics calculated only for data acquired in the period 1 September–31 October (or 30 September for AVHRR).

Metric (*uncorrected*)	AVHRR	seNorge	EBFM
RMSE (%)	11.22 (*17.07*))	6.82 (*23.84*)	7.45 (*13.86*)
Mean error (%)	7.85 (*13.56*)	2.11 (*−12.05*)	3.65 (*−3.66*)
R	0.85 (*0.89*)	0.85 (*0.76*)	0.91 (*0.94*)
(*corrected*)			
RMSE (%)	5.95	4.64	4.80
Mean error (%)	0.39	−0.19	0.0
R	0.87	0.90	0.91

Firstly, comparison of the metrics calculated for the spring (1 March–31 August) and autumn (1 September–31 October) data reinforces the patterns described by the scatter plots shown in Figure 3. For all three datasets, there is greater spread in the data acquired after 1 September, as indicated by the larger RMSE values; the mean error is also greater and in the case of the snow models, of the opposite polarity compared with the spring data. Comparison of the metrics calculated before and after corrections were made to the time series shows an obvious improvement and reduction in the RMSE and mean error. Greatest changes in RMSE occur in the correction of the AVHRR time series, resulting in RMSE being nearly halved, from 11.22 to 5.95%. The mean error is also reduced from 7.85% to 0.39%, indicating that the positive bias, or overestimate becomes almost minimal following correction of the dataset. This is indeed reflected by the scatter plot in Figure 6a. The Spearman correlation coefficient was only marginally increased by correcting the time series and this is not surprising since there was qualitatively a good and non-linear correlation between AVHRR and MODIS before corrections were made. For the MODIS dataset itself, the corrections made to the time series using the relationship with Sentinel-2 also resulted in a small reduction in RMSE (not shown), as well as removal of the positive bias of 2.8% which was present in the uncorrected time series. The Spearman correlation coefficient of 0.97 remained unchanged. For the two snow models, the corrections made using the spline fits also resulted in a reduction in the RMSE by between 2–2.5% while the slight positive biases in both uncorrected datasets were reduced to almost zero. For the seNorge snow model, the Spearman correlation coefficient increased from 0.85 to 0.90 while for the EBFM dataset there was no change.

4. Discussion

This study has utilized snow cover observations from three optical remote sensing satellites and two snow models and made comparisons of the land-averaged snow cover fraction and the derived FSFD over Nordenskiöld Land in central Svalbard in the period 2000–2019. In this section we review and discuss the results presented in Section 3.

4.1. Comparison of the Datasets

A consistent pattern was identified, whereby lower resolution datasets are found to overestimate SCF when compared with a higher spatial resolution dataset. SCF estimated using the AVHRR dataset were found to be up to some 20% greater than the corresponding MODIS estimates, especially for intermediate snow cover fractions in the range 50–60%. Similar findings have earlier been reported using AVHRR observations over the Canadian Arctic, where an evaluation of the NOAA AVHRR snow cover dataset was shown to consistently overestimate snow cover extent during spring snowmelt period [32], ultimately resulting in apparent delays of up to 4 weeks in estimates of melt onset. We also find that SCF overestimations in the AVHRR dataset are predominantly found at lower elevation areas, but with a tendency to underestimate the number of days with snow in higher elevation mountainous areas when compared with the MODIS dataset. Since the topography in mountainous areas can change over spatial distances smaller than the 4 km resolution of the AVHRR data, the observations would indicate that the AVHRR instrument is unable to capture the spatial variations in snow cover occurring over distances smaller than the AVHRR resolution in these areas of high relief, resulting in an overall underestimation with respect to higher resolution datasets. However, since the land-averaged snow fraction in the AVHRR is systemically greater than MODIS, the findings would suggest that the underestimation in snow cover at higher elevations does not contribute greatly to the overall differences between MODIS and AVHRR.

Similarly, it was also found that the MODIS SCF was on average 5–10% greater than SCF estimates made by the Sentinel-2 sensor, dependent on the magnitude of the snow cover fraction. It has recently been demonstrated that for high resolution sensors such as Sentinel-2, both thresholding and regression-based algorithms can systematically overestimate fractional snow cover when the data are aggregated to lower resolution grids [17], such as that of the MODIS dataset which was done in this work. This suggests that the Sentinel-2 estimates of the land-averaged SCF could in fact be lower than was illustrated, had more superior retrieval methods such as spectral unmixing been implemented to estimate SCF. As a result, the difference between the Sentinel-2 and MODIS SCF estimates would be greater than what has been shown in this study. An additional factor which has not been accounted for in this work is the effect of terrain shadow on Sentinel-2 NDSI. In autumn, the low sun elevation on Svalbard causes topographic shadow, which could be misinterpreted as snow cover in the NDSI values, which would also cause the derived Sentinel-2 snow fraction to be greater than if these effects were corrected for. A systematic study of the impact of such shadow for all the remote sensing datasets is needed to quantify this problem. Other recent studies that have investigated methods to retrieve fractional snow cover from Sentinel-2 have modelled the NDSI-snow cover fraction relation as a non-linear sigmoid-shaped function [31], which may present a better approach for calibrating the MODIS data against Sentinel-2 in future. However, since this study was based on a test site in alpine mountainous terrain, the calibration coefficient may not necessarily be universal and may need to be revised for polar regions such as Svalbard and is the main reason this approach was not adopted in this study.

For the comparison between MODIS and snow cover fraction derived from the two snow model datasets that were produced at 1 km resolution, there was also an offset of 5-15% between the MODIS dataset and SCF provided by the seNorge snow cover product or by thresholding the EBFM SWE dataset, with slightly closer agreement for the seNorge dataset when only data from the spring melt period, between 1 March–31 August were considered. In contrast to the AVHRR and Sentinel-2 comparisons, both snow models exhibited distinct differences in the relationship with MODIS SCF when data from the spring melt period and autumn snow onset (September 1 onwards) were compared. For the seNorge model, there was a consistent underestimation of SCF compared with MODIS for the post-September 1 data, while SCF estimated by the snow models during the spring melt period from 1 March–31 August, were almost always greater than the corresponding MODIS SCF. This pattern was not observed in the AVHRR and Sentinel-2 datasets, where SCF estimated in the period after 1 September tended to follow the same relationship with MODIS as those which had been obtained from the spring melt period. Furthermore, the snow models tended to underestimate SCF for low SCF (<20%), which is true of the data from both the spring melt period (1 March–31 August) as well as from after 1 September, with the geographical variations showing that the underestimations were greatest in lower elevation zones, as indicated in Figures 5 and 6. During the spring snowmelt period, this may for example be attributed to snow accumulation in terrain formations such as couloirs and lee slopes which melt later. These may also contribute to perennial/multi-year snow patches that are present during the summer months and can be captured by MODIS but not the snow models. On the other hand, since the calculation of the average number of days with snow includes data from part of the autumn period, the underestimation of snow is due to a combination of the models missing snow during spring melt as well as underestimating new snow in the autumn, which may have other underlying reasons.

There were also elevation-dependent differences in snow cover obtained from the two snow models. For SCF derived from the EBFM SWE dataset, there was on average more days with snow compared with MODIS at altitudes above 400 m.a.s.l. In contrast, the seNorge snow model exhibited consistently fewer days with snow on average compared with MODIS, at all altitudes. However, there are differences between the two snow model data products as well as the method by which they are derived. The EBFM dataset covers a period of 20 years between 2000–2019, while the seNorge product only covers 7 full years in the latter part of the same period, from 2013–2019. On the other hand, we have also analyzed the EBFM data only for the 2013–2019 interval, but only minor differences were found, and these are shown in Figure 10; there is still the same pattern of overestimation in the land-averaged SCF compared with MODIS in the 1 March–31 August period and these were predominantly found at higher elevations. The underestimation in average number of days with snow was still confined to the lower elevation valley areas but were in fact more pronounced than when averaged over the 20-year period. On the other hand, both the RMSE and mean error in the EBFM SCF estimates were reduced to 5.63% and 1.91% respectively, which are slightly lower than for the 2000–2019 period, given in Table 2.

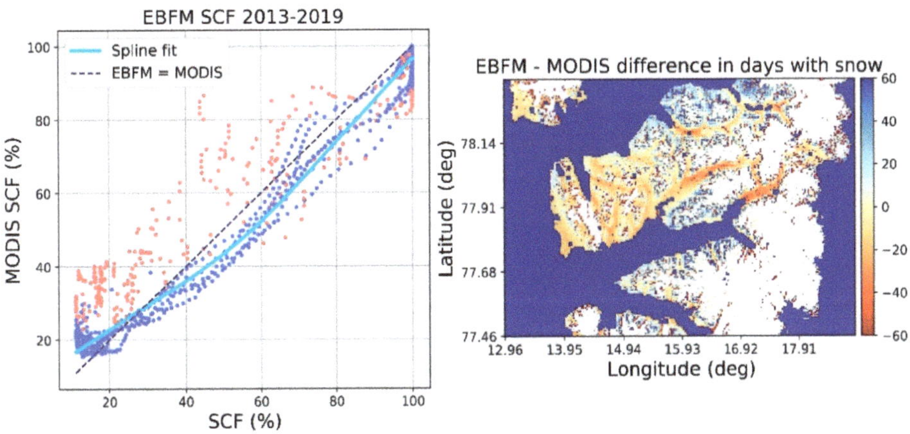

Figure 10. (**left**) Scatter plot showing the relationship between MODIS and the EBFM land-averaged snow cover fraction using data corresponding to the seNorge period (2013–2019) and (**right**) the average difference in number of days with snow comparing EBFM and MODIS for 2013–2019.

Moreover, the seNorge dataset is produced using the AROME NWP model forecast data and model archives for the historical period as inputs, while for EBFM the input is the HIRLAM reanalysis data. Thus, the snow models are not driven by the same inputs and cannot be expected to produce outputs that behave identically. Since there was little change between the geographical differences in average number of days with snow for the EBFM dataset shown in Figures 5d and 10, the dissimilarities between the seNorge and EBFM models are most likely attributed to differences in the way the models distribute new or melting snow, as well as due to different meteorological input data. In addition, the land-averaged SCF has been estimated from the EBFM SWE product by first converting the SWE maps into a binary snow cover map using a fixed threshold of 0.003 m. The choice of SWE threshold may therefore have contributed to the apparent greater number of days with snow in higher elevation zones in the EBFM snow cover dataset for certain years where the optimum threshold was found to be greater. However, we have chosen to use a fixed value based on the threshold which was found to produce best agreement with MODIS for the majority of the EBFM dataset, thereby maintaining a consistent method of snow cover extent extraction from the SWE dataset.

4.2. Correction of the Datasets

Using the spline fit to the MODIS SCF time series and the higher resolution Sentinel-2 dataset, the MODIS dataset was adjusted and subsequently utilized to obtain updated spline fits with the three remaining datasets. In doing so, a correction to the lower resolution SCF time series from the AVHRR instrument and the two snow models was also made. Evaluation metrics calculated for the MODIS time series in this study, using Sentinel-2 as a baseline before correction indicated a mean error, or bias of approximately 5.7%. A similar study that used high-resolution (0.5 m) terrestrial photography as a reference for the evaluation of Sentinel-2, Landsat 8 and MODIS datasets for a case study area in north-western Svalbard [17], reported comparable values, with the MODIS MOD10A1-v6 dataset exhibiting a mean error of 5% with respect to the high-resolution reference dataset. The RMSE of 8.6% in the uncorrected dataset is somewhat smaller than the RMSE obtained in their study, which was of the order of 14–15%, but here we have used Sentinel-2 data rather than very high-resolution optical images as a reference. In several earlier studies, the RMSE of the dataset being evaluated is often reduced when the reference data are aggregated to coarser resolution [17,31] as was done with the Sentinel-2 data in this study. Moreover, at the resolution of the MODIS data, the correction using the spline relationship resulted in just 1% reduction in RMSE and no change in correlation, while the mean error was reduced to 0%. Similarly, we found that largest decreases in the metrics being evaluated, occurred in the mean error following correction of the AVHRR, seNorge and EBFM datasets, while more modest reductions were observed in RMSE. The exception to this is for the AVHRR dataset, where RMSE was almost halved as a result of the corrections. The Spearman correlation coefficient remained virtually unchanged for AVHRR and EBFM while an increase in correlation from 0.85 to 0.90 occurred from the correction to the seNorge dataset. Hence it may be deduced that the main effect of the SCF time series using a spline model fit, was to almost remove the bias in the datasets.

A pattern of earlier FSFD following correction of the AVHRR SCF time series was also found. In this case FSFD obtained with the uncorrected AVHRR SCF time series was on average 15 days later than FSFD estimated using the corrected AVHRR SCF time series. In both cases, decadal trends revealed that FSFD became earlier over the 34-year period, but the advance in FSFD estimated using the uncorrected SCF dataset, of –3.36 days/decade ($p > 0.05$) was greater than the advance observed from the corrected dataset, which was found to be −2.81 days/decade ($p < 0.05$), by 0.5 days per decade. Hence, the results obtained in this study would imply that not only does the lower resolution AVHRR snow cover data tend to overestimate snow cover with respect to higher resolution datasets, but it also results in estimates of timing of snow disappearance that can be too late by up to two weeks, as well as a small overestimation in the decadal advance in FSFD, or the rate at which snow cover is declining. This result can have significant impacts, for example in applications where remote sensing observations of snow cover are used as inputs to calibrate climate and hydrological models. In such cases the accuracy of predictions for both present and future climate changes will be dependent on the representation of snow processes, and there is therefore a need to ensure that these parameters are reliable.

On the other hand, the calibration of the AVHRR data are based only on the 15 years of overlap with MODIS and may not necessarily be valid for the earlier part of the AVHRR dataset (1982–1999) which was not used in the comparisons. However, the decadal trends in FSFD found in this work are comparable to those published in a recent where the same 34-year AVHRR snow cover dataset was used to identify the relationship between snow cover variability and sea ice variability [33]. In this work, the authors found a decadal trend of −2.6 days/decade change in melt onset, defined as the point where SCF crosses a threshold of 95%, rather than examining the timing of snow disappearance which uses an SCF threshold of 50% in this work. Nevertheless, it should be noted that the melt onset trend found by their study was calculated using the snow fraction averaged over the entire Svalbard archipelago, whereas this study focuses only on the Nordenskiöld Land region. Therefore, the similar trends found in both this study and that of [33] support each other, despite the differences in study area as well as the snow melt parameter analyzed.

Both uncorrected snow model datasets produced FSFD estimates that were consistently later than MODIS by up to 15 days, with largest differences exhibited by the EBFM (SWE) dataset. The seNorge dataset resulted in FSFD estimates that were later than MODIS by around 5 days, both before and after correction with the spline fit. Since it was earlier also found that the seNorge dataset estimated on average fewer days with snow per year compared with MODIS, the result that the seNorge FSFD estimates are consistently later than MODIS would indicate that the discrepancy is most likely attributed to a much later onset of snow in the autumn, producing fewer days with snow compared with MODIS. This is to some extent verified in the comparison of the yearly SCF time series shown in Figure 3, where it was observed that the seNorge SCF rises later in autumn in relation to the other datasets, as well as exhibiting a lower minimum. Datapoints corresponding to SCF estimates obtained after September 1 for the seNorge dataset also point to much lower SCF than MODIS (Figure 4c) which further verifies this conclusion. Possible reasons for this difference may lie in the NWP input, if for example the model forecasts have a warm bias or too dry conditions, leading to too little early autumn snowfall in the 1 September–31 October period. For the EBFM dataset, the uncorrected land-averaged SCF time series also produced much later FSFD compared with MODIS, but geographically the low elevation areas exhibited on average fewer days with snow than MODIS, while higher elevation mountain slopes were found to have a greater number of days with snow on average compared with MODIS and would therefore be expected to contribute to later snow disappearance. Therefore, the temporal variations in SCF produced by EBFM may be in better agreement with MODIS at higher elevations, while being poorer at lower elevations. Moreover, there exist differences between the snow models and remotely sensed SCF for the period when snow cover is decreasing in the spring and when snow cover begins to increase again after the summer minimum snow cover extent, which suggests the need for a separate treatment in order to produce accurate corrections to the time series. The results found in this study therefore show that there is potential to integrate remote sensing observations into the calibration of snow models in order to improve the description of snow cover produced by the models.

5. Conclusions

Accurate maps of snow cover and characterization of the dynamic processes are critical in applications such as calibration of hydrological models and climate predictions, and especially so in regions where seasonal snow cover is responding rapidly to ongoing changes in climate. This study has investigated the similarities and differences between snow cover observations over Nordenskiöld Land in Svalbard, obtained by three optical remote sensing datasets and two snow models. The purpose of this work was to attempt to use high spatial resolution snow cover observations to make corrections to earlier, lower spatial resolution snow cover products to reconstruct long term snow cover datasets at both high spatial and temporal resolution. To achieve this, relationships between the higher and lower resolution datasets were obtained for the land-averaged snow cover fraction over the study area. Sentinel-2, with its high spatial resolution, was first utilised to adjust the moderate resolution MODIS dataset, which had excellent temporal overlap with the AVHRR dataset as well as the two snow models. This adjusted MODIS dataset was subsequently used to correct the lower resolution datasets and estimates of the timing of snow disappearance were made. For all the uncorrected datasets, estimates of FSFD were found to be later than MODIS FSFD by 10–15 days. Following correction of these datasets to the higher resolution of the MODIS dataset, FSFD estimates were significantly improved and varied by up to ± 5 days from the MODIS estimates. Furthermore, the decadal advance in FSFD estimated from the uncorrected 34-year AVHRR time series were found to be 0.5 days/decade greater than the decadal trend in FSFD following correction of the AVHRR time series, indicating that interpretation of lower resolution datasets requires some care when put in the context of climate-related change.

This work has demonstrated that there is potential to improve the consistency in snow cover observations made using earlier generation remote sensing instruments which have lower resolution than current day sensors that have comparatively high temporal and spatial resolution. Specifically, we have presented a method to implement relatively simple corrections to land-averaged snow cover fraction estimated by different remote sensing datasets and snow models. This approach has been shown to produce significant updates in estimates of snow disappearance timing and decadal trends in this parameter. However, since this study has focused on improving the consistency between snow cover observations at a land-averaged scale, further work is required to upscale lower resolution snow cover datasets at the pixel level and reconcile the associated geographical and elevation dependent differences in snow cover made by remote sensing and snow model datasets, as was highlighted in this study. Ultimately, it would be desirable to reproduce older snow cover maps at the high resolution of the newest sensors. This could for example be achieved by establishing a statistical average of the snow cover distribution at high spatial resolution, given a land-averaged snow cover fraction obtained from a lower resolution dataset.

Author Contributions: Conceptualization, E.M.; methodology, E.M. and H.V.; software, H.V.; formal analysis, H.V.; investigation, H.V.; resources, E.M., W.J.J.v.P., V.A.P., T.S., M.A.K., S.R.K.; writing—original draft preparation, H.V.; writing—review and editing, H.V., E.M., W.J.J.v.P., V.A.P., T.S., M.A.K., S.R.K.; project administration, E.M.; funding acquisition, E.M. All authors have read and agreed to the published version of the manuscript.

Funding: The work was partly based on the results of the project SATMODSNOW: https://zenodo.org/record/4294072#.YKO46bUzbIU (accessed on 18 May 2021) funded by the Research Council of Norway under the project Svalbard Integrated Arctic Earth Observing System—Infrastructure development of the Norwegian node (SIOS-InfraNor Project. 801 No. 269927. WvP acknowledges funding from the Swedish National Space Agency (project 189/18).

Data Availability Statement: Several of the datasets used in this study are available via the SIOS data access portal: https://www.sios-svalbard.org/metadata_search (accessed on 18 May 2021). The MODIS dataset is described in Vickers et al. (2020) https://www.mdpi.com/2072-4292/12/7/1123 (accessed on 18 May 2021). For the Uppsala Snow model: https://bit.ly/3nkfu18 (accessed on 18 May 2021) and AVHRR snow cover extent maps can be accessed at https://thredds.met.no/thredds/dodsC/arcticdata/sios/SvalSCE-agg.html (accessed on 18 May 2021). SeNorge for Svalbard is not yet currently publicly available but will be published later on www.senorge.no.

Acknowledgments: MODIS Terra Snow cover data were retrieved from NSCDC NASA DAAC: National Snow and Ice data Center; Copernicus Sentinel-2 data were retrieved from ESA SciHub.

Conflicts of Interest: The authors declare no conflict of interest.

References

1. Matsumura, S.; Zhang, X.; Yamazaki, K. Summer arctic atmospheric circulation response to spring eurasian snow cover and its possible linkage to accelerated sea ice decrease. *J. Clim.* **2014**, *27*, 6551–6558. [CrossRef]
2. Nuth, C.; Schuler, T.V.; Kohler, J.; Altena, B.; Hagen, J.O. Estimating the long-term calving flux of Kronebreen, Svalbard, from geodetic elevation changes and mass-balance modelling. *J. Glaciol.* **2012**, *58*, 119–133. [CrossRef]
3. Van Pelt, W.J.; Kohler, J.; Liston, G.E.; Hagen, J.O.; Luks, B.; Reijmer, C.H.; Pohjola, V.A. Multidecadal climate and seasonal snow conditions in Svalbard. *J. Geophys. Res. Earth Surf.* **2016**, *121*, 2100–2117. [CrossRef]
4. Vickers, H.; Karlsen, S.R.; Malnes, E. A 20-Year MODIS-based snow cover dataset for svalbard and its link to phenological timing and sea ice variability. *Remote Sens.* **2020**, *12*, 1123. [CrossRef]
5. Førland, E.J.; Benestad, R.; Hanssen-Bauer, I.; Haugen, J.E.; Skaugen, T.E. Temperature and precipitation development at svalbard 1900–2100. *Adv. Meteorol.* **2011**, *2011*, 14. [CrossRef]
6. Van Pelt, W.J.J.; Schuler, T.V.; Pohjola, V.A.; Pettersson, R. Accelerating future mass loss of Svalbard glaciers from a multi-model ensemble. *J. Glaciol.* **2021**, *67*, 485–499. [CrossRef]
7. Udnaes, H.C.; Alfnes, E.; Andreassen, L.M. Improving runoff modeling using satellite-derived snow cover area? *Nordic Hydrol.* **2007**, *38*, 21–32. [CrossRef]
8. Parajka, J.; Merz, R.; Blöschl, G. Uncertainty and multiple objective calibration in regional water balance modelling—Case study in 320 Austrian catchments. *Hydrol. Process.* **2007**, *21*, 435–446. [CrossRef]
9. Parajka, J.; Blöschl, G. The value of MODIS snow cover data in validating and calibrating conceptual hydrologic models. *J. Hydrol.* **2008**, *358*, 240–258. [CrossRef]
10. Clark, M.P.; Slater, A.G.; Barrett, A.P.; Hay, L.E.; McCabe, G.J.; Rajagopalan, B.; Leavesley, G.H. Assimilation of snow covered area information into hydrologic and land-surface models. *Adv. Water Resour.* **2006**, *29*, 1209–1221. [CrossRef]
11. Dietz, A.J.; Kuenzer, C.; Gessner, U.; Dech, S. Remote sensing of snow—A review of available methods. *Int. J. Remote Sens.* **2012**, *33*, 4094–4134. [CrossRef]
12. Nolin, A. Recent advances in remote sensing of seasonal snow. *J. Glaciol.* **2010**, *56*, 1141–1150. [CrossRef]
13. Warren, S.G. Optical properties of snow. *Rev. Geophys.* **1982**, *20*, 67–89. [CrossRef]
14. Dozier, J. Spectral signature of alpine snow cover from the landsat thematic mapper. *Remote Sens. Environ.* **1989**, *28*, 9–22. [CrossRef]
15. Salomonson, V.V.; Appel, I. Estimating fractional snow cover from MODIS using the normalized difference snow index. *Remote Sens. Environ.* **2004**, *89*, 351–360. [CrossRef]
16. Hall, D.K.; Riggs, G.A.; Salomonson, V.V.; DiGirolamo, N.E.; Bayr, K.J. MODIS snow-cover products. *Remote Sens. Environ.* **2002**, *83*, 181–194. [CrossRef]
17. Aalstad, K.; Westermann, S.; Bertino, L. Evaluating satellite retrieved fractional snow-covered area at a high-Arctic site using terrestrial photography. *Remote Sens. Environ.* **2020**, *239*, 111618. [CrossRef]

18. Selkowitz, D.J.; Forster, R.R.; Caldwell, M.K. Prevalence of pure versus mixed snow cover pixels across spatial resolutions in alpine environments. *Remote Sens.* **2014**, *6*, 12478–12508. [CrossRef]
19. Li, X.; Jing, Y.; Shen, H.; Zhang, L. The recent developments in cloud removal approaches of MODIS snow cover product. *Hydrol. Earth Syst. Sci.* **2019**, *23*, 2401–2416. [CrossRef]
20. Da Ronco, P.; De Michele, C. Cloud obstruction and snow cover in Alpine areas from MODIS products. *Hydrol. Earth Syst. Sci.* **2014**, *18*, 4579–4600. [CrossRef]
21. Karlsson, K.G.; Anttila, K.; Trentmann, J.; Stengel, M.; Fokke Meirink, J.; Devasthale, A.; Hanschmann, T.; Kothe, S.; Jääskeläinen, E.; Sedlar, J.; et al. CLARA-A2: The second edition of the CM SAF cloud and radiation data record from 34 years of global AVHRR data. *Atmos. Chem. Phys.* **2017**, *17*, 5809–5828. [CrossRef]
22. Johansen, B.; Karlsen, S.; Tømmervik, H. Vegetation mapping of Svalbard utilising Landsat TM/ETM data. *Polar Rec.* **2012**, *48*, 47–63. [CrossRef]
23. Zupanc, A. Improving Cloud Detection with Machine Learning. 2019. Available online: https://medium.com/sentinel-hub/improving-cloud-detection-with-machine-learning-c09dc5d7cf13 (accessed on 29 November 2019).
24. Gascoin, S.; Hagolle, O.; Huc, M.; Jarlan, L.; Dejoux, J.F.; Szczypta, C.; Marti, R.; Sánchez, R. A snow cover climatology for the pyrenees from modis snow products. *Hydrol. Earth Syst. Sci.* **2015**, *19*, 2337–2351. [CrossRef]
25. Saloranta, T.M. Operational snow mapping with simplified data assimilation using the seNorge snow model. *J. Hydrol.* **2016**, *538*, 314–325. [CrossRef]
26. Saloranta, T.M. *New Version (V.1.1.1) of The Senorge Snow Model and Snow Maps for Norway, Rapport 6-2014*; Norwegian Water Resources and Energy Directorate: Oslo, Norway, 2014. Available online: http://publikasjoner.nve.no/rapport/2014/rapport2014_06.pdf (accessed on 18 May 2021).
27. Van Pelt, W.J.J.; Oerlemans, J.; Reijmer, C.H.; Pohjola, V.A.; Pettersson, R.; van Angelen, J.H. Simulating melt, runoff and refreezing on Nordenskiöldbreen, Svalbard, using a coupled snow and energy balance model. *Cryosphere* **2012**, *6*, 641–659. [CrossRef]
28. Van Pelt, W.J.J.; Pohjola, V.A.; Pettersson, R.; Marchenko, S.; Kohler, J.; Luks, B.; Hagen, J.O.; Schuler, T.V.; Dunse, T.; Noël, B.; et al. A long-term dataset of climatic mass balance, snow conditions and runoff in Svalbard (1957–2018). *Cryosphere* **2019**, *13*, 2259–2280. [CrossRef]
29. Reistad, M.; Breivik, Ø.; Haakenstad, H.; Aarnes, O.J.; Furevik, B.R.; Bidlot, J.-R. A high-resolution hindcast of wind and waves for the North Sea, the Norwegian Sea, and the Barents Sea. *J. Geophys. Res.* **2011**, *116*, 18. [CrossRef]
30. Rittger, K.; Painter, T.H.; Dozier, J. Assessment of methods for mapping snow cover from MODIS. *Adv. Water Resour.* **2013**, *51*, 367–380. [CrossRef]
31. Gascoin, S.; Barrou Dumont, Z.; Deschamps-Berger, C.; Marti, F.; Salgues, G.; López-Moreno, J.I.; Revuelto, J.; Michon, T.; Schattan, P.; Hagolle, O. Estimating fractional snow cover in open terrain from sentinel-2 using the normalized difference snow index. *Remote Sens.* **2020**, *12*, 2904. [CrossRef]
32. Wang, L.; Sharp, M.; Brown, R.; Derksen, C.; Rivard, B. Evaluation of spring snow covered area depletion in the Canadian Arctic from NOAA snow charts. *Remote Sens. Environ.* **2005**, *95*, 453–463. [CrossRef]
33. Killie, M.A.; Aaboe, S.; Isaksen, K.; Van Pelt, W.; Pedersen, A.Ø.; Luks, B. Svalbard Snow and Sea-Ice Cover: Comparing Satellite Data, On-Site Measurements, and Modelling Results (Svalscesia). In *The State of Environmental Science in Svalbard (SESS) Report 2020*; SIOS: Longyearbyen, Norway, 2021.

remote sensing

MDPI

Article

Multi-Sensor Analysis of Snow Seasonality and a Preliminary Assessment of SAR Backscatter Sensitivity to Arctic Vegetation: Limits and Capabilities

Laura Stendardi [1,2,*], Stein Rune Karlsen [3], Eirik Malnes [3], Lennart Nilsen [4], Hans Tømmervik [5], Elisabeth J. Cooper [4] and Claudia Notarnicola [2]

1 Department of Agriculture, Food, Environment, and Forestry (DAGRI), University of Florence, Piazzale Delle Cascine 18, 50144 Florence, Italy
2 Institute for Earth Observation, Eurac Research, 39100 Bolzano, Italy; claudia.notarnicola@eurac.edu
3 Climatic Department, NORCE, Norwegian Research Centre AS, P.O. Box 6434, N-9294 Tromsø, Norway; skar@norceresearch.no (S.R.K.); eima@norceresearch.no (E.M.)
4 Department of Arctic and Marine Biology, University of Tromsø, PO Box 6050, N-9037 Tromsø, Norway; lennart.nilsen@uit.no (L.N.); elisabeth.cooper@uit.no (E.J.C.)
5 Norwegian Institute for Nature Research (NINA), FRAM—High North Research Centre for Climate and the Environment, P.O. Box 6606, N-9296 Tromsø, Norway; hans.tommervik@nina.no
* Correspondence: laura.stendardi@unifi.it

Abstract: Snow melt timing and the last day of snow cover have a significant impact on vegetation phenology in the Svalbard archipelago. The aim of this study is to assess the seasonal variations of the snow using a multi-sensor approach and to analyze the sensitivity of the Synthetic Aperture Radar (SAR) backscatter to vegetation growth and soil moisture in an arctic environment. A combined approach using time series data from active remote sensing sensors such as SAR and passive optical sensors is a known technique in snow monitoring, while there is little knowledge of the radar C-band's response pattern to vegetation dynamics in the arctic. First, we created multi-sensor masks using the HV backscatter coefficients from Sentinel-1 and the Normalized Difference Snow Index (NDSI) time series from Sentinel-2, monitoring the snow dynamics in Adventdalen (Svalbard) for the season from 2017 to 2018. Second, radar sensitivity analysis was performed using the HV polarized channel responses to vegetation growth and soil moisture dynamics. (1) Our results showed that the C-band radar data are capable of monitoring the seasonal variability in timing of snow melting in Adventdalen, revealing an earlier start by approximately 20 days in 2018 compared to 2017. (2) From the sensitivity analyses, the HV channel showed a major response to the vegetation component in areas with drier graminoid dominated vegetation without water-saturated soil ($R = 0.69$). However, the temperature was strongly correlated with the HV channel ($R = 0.74$) during the years with delayed snow melting. Areas of frozen tundra with drier vegetation dominated by graminoids had delayed soil thawing processes and therefore this may limit the ability of the radar to follow the vegetation growth pattern and soil moisture.

Keywords: remote sensing; Sentinel-1 and Sentinel-2; time series analysis; snow melt; Svalbard; tundra; plant phenology

Citation: Stendardi, L.; Karlsen, S.R.; Malnes, E.; Nilsen, L.; Tømmervik, H.; Cooper, E.J.; Notarnicola, C. Multi-Sensor Analysis of Snow Seasonality and a Preliminary Assessment of SAR Backscatter Sensitivity to Arctic Vegetation: Limits and Capabilities. *Remote Sens.* **2022**, *14*, 1866. https://doi.org/10.3390/rs14081866

Academic Editor: Xinghua Li

Received: 26 February 2022
Accepted: 9 April 2022
Published: 13 April 2022

1. Introduction

The timing of snow melt and the first day free of snow are considered indicators of Arctic climate and ecosystem status in response to global warming [1]. Moreover, the depth of snow and the period of snow melt contribute significantly to defining the phenological phases of vegetation [2,3] and plant biomass [4]. Due to an increase in winter precipitation [5], as well as in the frequency of extreme weather events in the Svalbard archipelago [6,7], snow cover monitoring is highly important.

Along with the ability to detect snow seasonality, there is a growing interest in studying the relationship between the timing of snow melt and vegetation phenology (late snow, onset of the growing season, and advanced snow melting). Malnes et al. (2016) [8] used satellite remote sensing (MODIS) over northern Norway in order to detect the snow covered season with significant accuracy due to start (r = 0.51, $p > 0.05$) and end (r = 0.79, $p > 0.05$) of the snow covered season for most meteorological stations for the 2000–2010 period. However, in some areas and some years, the snow covered season could not be detected due to long overcast periods.

Since cloud cover often dominates the Arctic environment and especially in Svalbard, it would be useful to monitor phenology with sensors that are unaffected by clouds. Satellite remote sensing imagery can be used in the analysis and interpretation of the snow-pack variations of Svalbard [9]. Multi-sensor analysis has proven to be effective to monitor snow coverage and its relation to vegetation phenology [10,11]. In the optical domain, the Normalized Difference Snow Index (NDSI) has been successfully used to detect snow cover [12,13]. This index takes advantage of the high reflectance of snow in the visible wavelengths and low reflectance in the short wave infrared wavelengths. While optical sensors are based on surface reflectance, SAR sensors allow evaluating change that is occurring within the surface. This is because the strong dielectric contrast between the solid and liquid phases of the water generates changes in the backscatter coefficients. Indeed, even a small amount of liquid water reduces the depth of penetration of the radar signal, allowing identifying the start of the snow melting process [14].

The most common approach to snow mapping with Synthetic Aperture Radar (SAR) is based on multi temporal comparison of images of the same area in snow-free/dry and wet snow conditions [15–18]. Attenuation due to snow in the frequency range 1–12 GHz is very low, and dry snow covers cannot be discerned from the bare ground. Conversely, there is a signal attenuation in wet snow conditions, due to a change in the dielectric properties of the surfaces [16,19,20].

Due to polar night, SAR data are essential in the arctic during the winter period. However, because of the high cloud coverage that limits optical images, SAR data are also important during the summer period. Once the snow has disappeared, the radar beam can simultaneously penetrate both the vegetation and the soil to a depth that is difficult to determine [21,22]. Innovative approaches are based on the use of SAR and optical sensors to follow the phenological phases of vegetation [23]. These approaches may be useful in areas where cloud cover is widespread during the growing season. Moreover, the sensitivity of backscattering coefficients to the vegetation of high arctic not yet been investigated to our knowledge.

The Copernicus program allowed free access to time series of different sensors, which can be used in synergy. Sentinel-1A and 1B SAR sensors (centre frequency of 5.405 GHz) provide medium and high-resolution time series of C-band data, while Sentinel-2A and 2B optical sensors acquire 13 spectral bands in the optical domain [24,25]. The two sensors used simultaneously can improve the characterization of the snow season and vegetation growth, allowing us to derive multi-sensor products with a high temporal and spatial resolution. The central aim of this study was to derive wet snow maps and snow maps from the time series of S-1 and S-2, in order to assess the seasonal variations of the snow. Then, a sensitivity analysis of the backscattering coefficient σ^0 to vegetation and soil moisture was made and related to snow dynamics. The final phase enabled an overall understanding of the sensitivity of the SAR signal to vegetation in relation to snow dynamics. The novelty of our study is to assess the impact of vegetation growth and soil moisture on the SAR signal in the Arctic, taking into account the variability of the snow seasonality.

Compared to the studies presented above, our main objectives of the study are:

1. to apply a combination of radar and optical satellite data (Sentinel-1 and Sentinel-2) to map the spatial and temporal pattern of wet and dry snow conditions, and its relation to the vegetation growth season;

2. to understand how polarized radar data (HV- horizontal transmit and vertical received) can contribute to detect the pattern of arctic vegetation growth and soil moisture.

2. Materials and Methods

2.1. Study Area

The study area is the Adventdalen valley and the surrounding plateaux, close to the town of Longyearbyen (78°13′N 15°33′E), in the Svalbard archipelago (Figure 1).

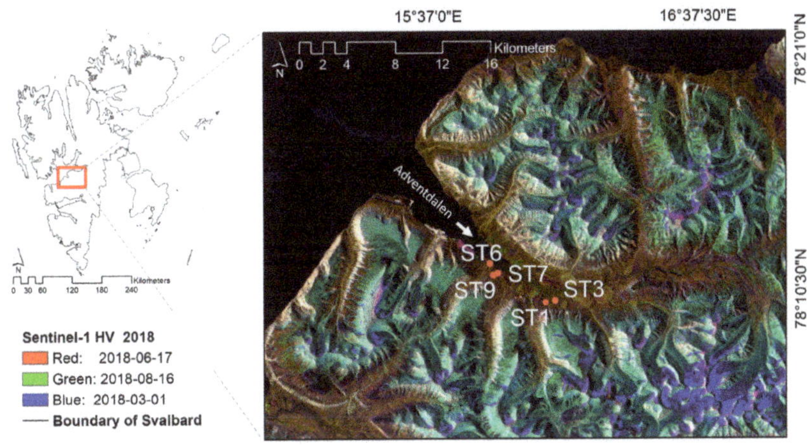

Figure 1. Adventdalen valley and the study area as shown by the Sentinel-1 HV channel in a RGB composition: 17 June 2018 Red band; 16 August 2018 Green band; 1 March 2018 Blue band. The location of the study area on Svalbard (the white arrow points to the Adventdalen valley) and the five selected ground stations are shown.

The study area lies approximately between latitudes 78°20′ and 78°07′N, and longitudes 15°10′ and 17°10′E. The periglacial landscape is characterized by vast plateaux intersected by wide glacial valleys and alluvial plains. Precipitation is low and the maritime environment strongly influences the snowpack characteristics [26]. Adventdalen is characterized by a polar-tundra climate [27] and is located in the Middle-Arctic and Northern-Arctic tundra zone. The Middle-Arctic zone is characterized by dwarf-shrub heaths, where *Cassiope tetragona* often dominates, and in the study area small patches of *Betula nana* are found, whereas in the Northern Arctic Tundra zone, the genus *Luzula* is characteristic with *Salix polaris*, *Saxifraga oppositifolia* and *Dryas octopetala* [28,29] also common. The meteorological station located in Adventdalen (Station number SN99870, Norwegian Meteorological Institute) recorded an average air temperature in July of 6.8 °C and 7.2 °C for the years 2017 and 2018, respectively. Within the projects SnoEco (NRC ref. 230970), Sentinels Synergy Framework (EC FP7 collaborative project), and SIOS (www.svalbard-sios.org, accessed on 18 September 2019), ten ground stations were set up in Adventdalen. Each station [30] was equipped with data loggers, environmental sensors for soil temperature and moisture at 10 cm depth and time-lapse ordinary RGB cameras and NDVI sensors positioned at 2 m above the ground. Since the location of some stations changed between 2017 and 2018, we selected only five stations; their characteristics are listed in Table 1, while an example of images is shown in Figure 2. Camera images covered an area of approx. 1.4 square meters.

Table 1. Description of ground stations location (UTM zone 33 north, datum WGS84), equipment and vegetation in Adventdalen during season 2017 and 2018.

Station	UTM X	UTM Y	Vegetation and Site Description	Sensors
ST1	523620	8677555	Moist moss tundra with *Alopecurus ovatus*, *Bistorta vivipara* and *Salix polaris*. Depressions with *Equisetum arvense*, patches of *Saxifraga hirculus*, and scattered *Dupontia fisheri* and *Eriophorum scheuchzeri*. Vegetation cover: 100%	NDVI sensor, soil temperature/moisture, Infrared radiometer, PhenoCams
ST3	524461	8677707	Mosaic of the shrub *Dryas octopetala*, and graminoids, e.g., *Luzula confusa*, *Poa pratensis alpigena* and *Alopecurus ovatus*. Lots of *Salix polaris* and *Bistorta vivipara* on moist to wet moss tundra dominated by silty sand. Small landscape feature dominated by soil frost polygon with little vegetation in the center. Vegetation cover: 90–100%	NDVI sensor, soil temperature/moisture, Infrared radiometer, PhenoCams
ST6	519008	8680756	Grass dominated sandy sediment plain. *Festuca rubra*, *Poa pratensis ssp alpigena*, and *Alopecurus ovatus*.Thin organic layer, with lots of *Salix polaris* in between the grasses. Vegetation cover: 80–100%	NDVI sensor, soil temperature/moisture, PhenoCams
ST7	519655	8679964	Wetland vegetation on flat silty and sandy substrate, dominated by large polygon soil patterns. *Puccinellia phryganodes*, *Dupontia fisheri* and *Eriophorum scheuchzeri* in the interior part of polygons, while *Ranunculus pygmeaus* and bryophytes such as *Scorpidium cossonii* and *Scorpidium revolvens* dominate the wettest part in polygon cracks. Vegetation cover: 100%	NDVI sensor, soil temperature/moisture, PhenoCams
ST9	519280	8679794	Heath dominated by *Luzula confusa*. Other species present are *Salix polaris*, *Poa pratensis alpigena*, *Cerastium arcticum* and bryophytes such as *Sanionia uncinata* and *Tomentypnum nitens*. Some cryoturbation and silty soil. Vegetation cover: 70–100%	NDVI sensor, soil temperature/moisture, PhenoCams

Figure 2. Images recorded on 27 July 2018 from stations (**a**) ST1, (**b**) ST3, (**c**) ST6, (**d**) ST7, and (**e**) ST9.

2.2. Datasets

The analyzed data sets are composed of time series from Sentinel-1, Sentinel-2, and ground station data. 115 S-1A and S-1B images acquired from February 2017 (4 February 2017) to December 2018 (26 December 2018) were processed. With a revisit frequency of six days, the C-band (wavelength, λ = 5.5 cm) images were acquired in Interferometric Wide swath mode (IW), with one relative orbit (track 014), and ascending pass. The images were available in cross polarized HV ('Horizontal transmit' and 'Vertical receive') channel, with a spatial resolution of 10 m. The S-2 Multi Spectral Instrument (MSI) acquires 13 spectral bands in the Visible, Near-Infra-Red (NIR) and Short Wave Infra-Red (SWIR) domains, with a spatial resolution of 10 to 60 m. Time series of S-2A and S-2B were processed starting from Level-1C Top-Of-Atmosphere reflectance. The high latitude results indeed in low solar elevation angles, which generate an under-correction of the atmospheric signal for the Bottom-Of-Atmosphere (BOA) processing level 2A [31] Due to polar night, images were acquired from April to September of each year. The polar orbit of S-2 enables daily data of Adventdalen to be obtained. However, only 43 images were processed in this study because of fog and frequent clouds, which reduced the number of images available. The ground stations were equipped with Meter Group SRS-NDVI Sensor [32] and soil temperature/moisture sensors (5 TM [33]). Water content and temperature of soil were measured at 10 cm depth. Station ST1 and ST3 additionally included infrared thermometers (SI-421 [34]) to monitor surface temperature. Data were recorded every four hours from mid-April to the end of September. The 5TM soil moisture and temperature sensors were not installed before the beginning of June, as the active layer was frozen until then. Time lapse cameras (WingScapes, model WCT-00122), with 8 MP of resolution, recorded images three times a day (9 am, 12 am and 3 pm). April to the beginning of October for three of the stations (6, 7 and 9), while station 1 recorded from mid of August to October, and station 3 from end of July to October)

3. Methodology

In this study, we primarily applied known techniques to define the melting of snow and the last day of snow. Following snowmelt, a study on the radar response to vegetation growth and soil moisture was carried out. The scheme of the suggested methodology is shown in Figure 3 and briefly described in the following sections.

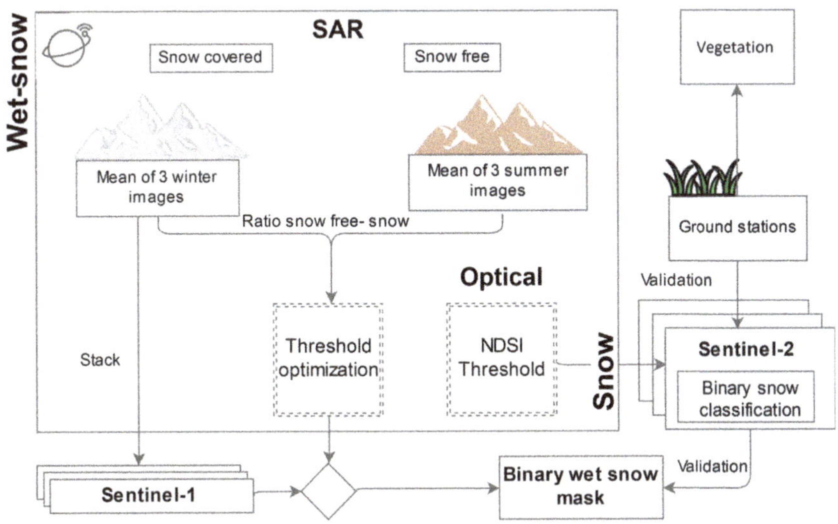

Figure 3. Flow chart of the proposed approach to detect the main features of snow season and phenology.

3.1. SAR, Optical and Ground Data Processing

The pre-processing of S-1 data includes several standard steps to derive geocoded intensity images from Level-1 GRD (Ground Range Detected) data. Each scene was geocoded in the Norut software package GSAR [35] and stored as geotiff files. To reduce speckle-affecting backscattering values and obtain a more homogenous snow cover pattern, a Frost filter with a 7×7 window was applied to the images [36]. Top-of-atmospheric Sentinel-2 products (L1C) were processed to obtain clear-sky time-series. The cloud detection provided with the Sentinel-2 (L2A) product was often not sufficiently robust, and so clouds were detected by a visual analysis of the images combined with different cloud-detection algorithms for cloud removal. Specifically, we used algorithms from the literature [37,38], along with our own developed algorithms based on our experiences with time series processing in Svalbard [39,40]. After this step, in the period from late April to late September, 16–22 days were cloud-free in 2017, and 15–21 days in 2018. Then, to reduce the time discrepancy with SAR data, the cloud-free pixels were interpolated to daily data, by using a Kernel Ridge Regression machine learning method [41]. Finally, time series from ground sensors were filtered, selecting only the acquisition between 12 and 4 pm. The values within this interval of time were averaged in order to reduce the noise.

3.2. Snow-Melting Detection

To detect temporal changes in snow structural and dielectric properties due to the presence of water with S-1 C band backscatter, we used the ratio method [42]. Starting from the climate database of the Norwegian Meteorological Institute (https://klimaservicesenter.no/ Norwegian Centre for Climate Services (NCCS), accessed on 5 August 2019,) we selected winter intervals with specific characteristics of snow depth, air temperature and time between extreme events in days (snow depth ≥ 5 cm, mean and max of temperature ≤ 0 °C, ±5 days before and after). The same procedure was applied to the summer period (snow depth = 0 cm, mean and min of temperature ≥ 5 °C, ±5 days before and after, no precipitation). The meteorological station is located at 15 m a.s.l in the central part of Adventdalen (latitude 78.2022°–longitude 15.831°). Using this information, it was possible to select images during the winter as a reference for dry snow/snow covered, as well during the summer period as snow free references. The specification of the selected images are shown in Table 2.

Table 2. Sentinel-1 master images and meteorological data years 2017–2018.

Image Date	Snow Depth (cm)	Air Temperature Mean (°C)	Air Temperature Max (°C)	Air Temperature Min (°C)	Precipitation (mm)
6 March 2017	9	−11.8	−7.2	−13.3	-
12March 2017	9	−16.9	−14.9	−21.5	0.4
24 March 2017	12	−11.2	−8.9	−12.5	1.6
10 July 2017	-	7.4	8.7	5.7	-
22 July 2017	-	8.3	10.9	5.9	-
3 August 2017	-	8.1	10.6	7.4	-
17 February 2018	21	−16.3	−10.4	−22.6	-
7 March 2018	13	−16	−13.9	−16.7	-
13 March 2018	14	−18.3	−14.8	−20.4	-

As a first step, the 2017 time series was used to optimise a wet-snow threshold (Th). Three images acquired during the winter period were averaged, representing the snow-covered conditions ($\sigma_{\text{sc}\,2017}$).

The same procedure was applied to three summer images, in order to obtain a reference image for the snow-free surface ($\sigma_{\text{sf}\,2017}$). The threshold for defining wet-snow surfaces was quantified by calculating the ratio (R_{Th}) of these two images (Equation (1)):

$$R_{\text{Th}} = \frac{\sigma_{\text{sc}\,2017}}{\sigma_{\text{sf}\,2017}} \quad (1)$$

Then, a sample of 30,000 snow-covered and snow-free pixels was extracted from the ratio, respectively. RGB S-2 images from the corresponding period were used as a guide to define snow-covered and snow-free surfaces. From the intersection of the frequency distribution of pixels values, a threshold was obtained. As shown in Figure 4, the intersection between the distribution of snow-free and snow-covered S-1 pixels was −2.8 dB.

This threshold was applied to discriminate wet-snow in the 2017 and 2018 time series, for a total of 115 wet-snow masks.

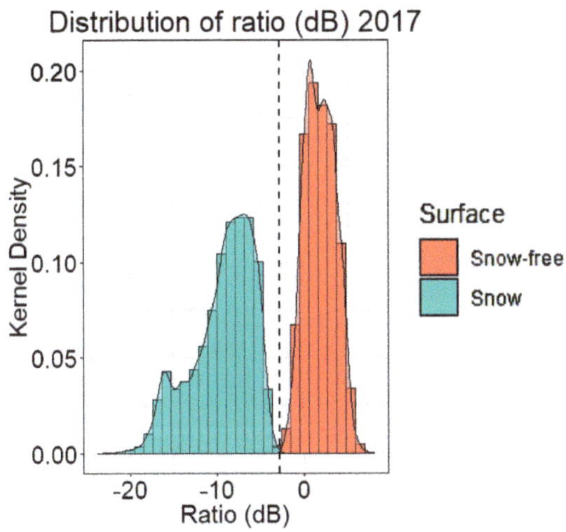

Figure 4. Distribution of Ratio values (dB) in snow covered and snow-free condition.

Simultaneously, a stack of images was created between the winter reference images of the years 2017 and 2018 (σ_{sc2017}, σ_{sc2018}) and each image of the time series, calculating their ratio according to Equation (2):

$$R_{TS} = \frac{\sigma_{\text{time series}}}{\sigma_{\text{sc reference}}} \tag{2}$$

where R_{TS} (Equation (2)) is the ratio of each image of the time series ($\sigma_{\text{time series}}$) and the reference images ($\sigma_{\text{sc reference}}$) of 2017 ($\sigma_{sc2017}$) and 2018 ($\sigma_{sc20178}$), respectively. After rationing each image with the reference one, we applied the previously fixed threshold. Finally, a binary classification of wet-snow ($Mask_{\text{wet-snow}}$) was produced for the years 2017 and 2018 (Equation (3)).

$$Mask_{\text{wet-snow}} = R_{TS} < \text{Th} \tag{3}$$

3.3. Snow-Free Mapping and Accuracy Assessment

The Normalized Difference Snow Index (NDSI) [12] was calculated from S-2 time series for the years 2017 and 2018 according to Equation (4) :

$$\text{NDSI} = \frac{(b03 - b11)}{(b03 + b11)} \tag{4}$$

where b03 (Equation (4)) corresponds to green band (central wavelength 0.560 µm) and b11 (Equation (4)) to shortwave infrared band (central wavelength 1.610 µm). A threshold of 0.6 was then applied to create a binary snow/snow-free map for the season 2017 and 2018 [36]. The threshold was chosen based on our experience from previous work on defining snow-free surfaces in Arctic environments [8].

The comparison between the wet-snow maps derived from Sentinel-1 and the snow maps derived from Sentinel-2 was carried out.

At the same time, the ground stations and the meteorological station n. SN99870 (Norwegian Meteorological Institute) were used to validate the Sentinel-2 masks. The validation of the optical snow cover/snow free mask was carried out using photos from the PhenoCams. When images were not available (ST1, ST3 season 2017), the NDVI from the ground stations was used [43] as:

$$\text{Snow free} = \text{NDVI} > 0.3 \tag{5}$$

To obtain a more robust validation, air temperature at 2 m height and snow depth from the meteorological station (snow depth = 0 cm, mean and min of temperature > 0) were incorporated into the validation process.

3.4. Ground Sensor Data Analysis

The next phase focused on understanding the dominant components in the backscatter signal in relation to vegetation development. S-1 σ^0 time series were extracted in homogeneous areas of around 50 × 50 m corresponding to the ground stations. To remove the noise, a linear interpolation and a moving average filter [44] were applied to the time series.

Correlation analyses were performed between time series of S-1 and ground sensor data to understand the contribution of vegetation, and soil moisture on the SAR signal. Subsequently, in order to measure the importance of the factors in a quantitative basis, a dominance analysis [45] of the variables was carried out.

4. Results

First, a validation of the masks were carried out. Subsequently, the S-1 melting season was calculated for the year 2017 and 2018, together with the snow season from S-2. Finally, a sensitivity analysis between S-1 backscatter and ground information was performed.

4.1. Snow Masks, Inter-Satellite Cross-Comparison and Ground Validation

Applying the −2.8 dB threshold, wet snow masks were created for the 2017 and 2018 seasons. The masks were selected for the May-September period, when optical images were also available. Next we created a binary daily snow cover mask by applying the S-2 NDSI threshold. The S-1 wet-snow masks and S-2 snow masks were overlaid and visually verified together with RGB images. An example of S-1 and S-2 masks is illustrated in Figure 5; using a RGB image as the base (Figure 5a), the mask obtained from the NDSI (Figure 5b) and the mask obtained from the coefficients of backscatter (Figure 5c) are overlaid.

Then, the first snow-free day of the S-2 masks was compared with the data of the ground sensors. From the regression model, the coefficient of determination R^2 between S-2 masks and ground data was 0.73. The results are summarized in Table 3.

Table 3. First snow-free Day Of Year (DOY) detected by ground station and S-2 snow masks. The 2018 ground stations acquired data staring from May 23, already a snow free period. For this reason only the information of the climatic station (SN99870) was available.

Station	Ground Sensors	Snow Mask
SN99870-2017	150 (30 May)	142 (22 May)
ST1-2017	151 (31 May)	154 (3 June)
ST3-2017	150 (30 May)	146 (26 May)
ST6-2017	141 (21 May)	144 (24 May)
ST7-2017	152 (1 June)	147 (27 May)
ST9-2017	154 (3 June)	166 (15 June)
SN99870-2018	128 (8 May)	124 (4 May)

4.2. Snow Seasonality

The total number of pixels affected by both by the presence of snow and melting snow during the two seasons was comparable. However, a significant difference (quantified in terms of km^2 of wet-snow) was observed in the temporal distribution of the process. Figure 6 illustrates late melting of snow in 2017, with a maximum area of wet snow around DOY 150–160 (start of June). On the contrary, the snow melting season in 2018 was earlier, with a large amount of wet snow already in May. The peak of melting in 2018 corresponded to DOY 170–180 (end of June). Therefore, the melting process started early in 2018, whereas in 2017, there was a late but faster snow melt. During the season, after the different start dates of snowmelt in 2017 and 2018, the pattern of melting proceeded similarly from from the middle of June (around DOY 170) until the end of August (around DOY 240). To assess the spatial distribution of the snowmelt, please refer to Appendix A.1.

(a)

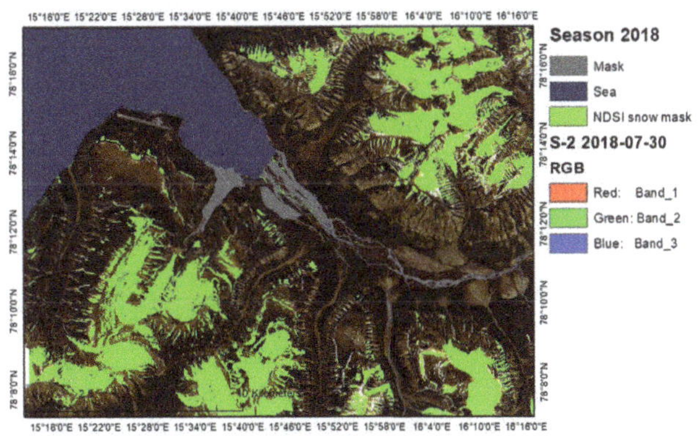

(b)

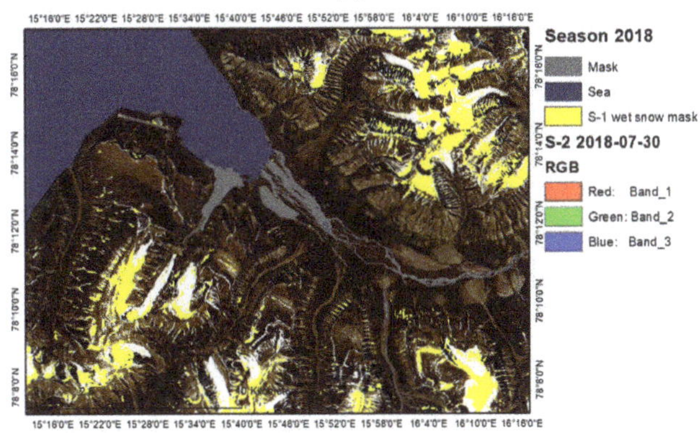

(c)

Figure 5. Sentinel-2 RGB images of 2018-07-30 (**a**) with overlapped in green the snow mask obtained using an NDSI threshold (**b**), and in yellow the wet snow mask obtained with 2.8 threshold on $sigma^0$ (**c**).

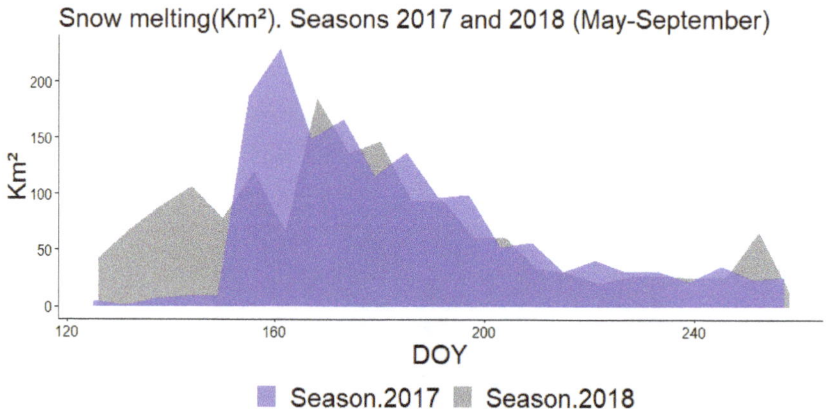

Figure 6. Surface (km^2) covered by wet-snow during the melting season 2017 and 2018. The range of days expressed in DOY (from approximately DOY 130, which corresponds to early May, to DOY 250, which corresponds to early September) represents the window in which Sentinel-2 images are usable due to polar night.

The S-2 results also indicated that the number of snow-free areas in 2018 were greater than in the previous year (Figure 7). In 2017, until the start of June (approximately DOY 150) snow slowly disappeared from the surfaces, and then had a rapid decrease until the end of June (approximately DOY 180). In 2018, on the contrary, since the beginning of the season (DOY 130), the km^2 occupied by snow progressively decreased, forming a bell-shaped trend. Again, after a very different starting phase, the pattern proceeded similarly in both years from the end of June (about DOY 180) until the end of August (about DOY 240).

The average difference in days between the first snow free day of 2017 and 2018 obtained by the validation of the stations positioned in Adventdalen valley was 23 days. The spatial distribution of the snow-free surfaces over the two years is shown in Appendix A.2.

Finally, to obtain an overview of the snowmelt and snow seasons for both years we created maps of 'wet snow', 'dry snow', and 'snow-free' areas. An example is shown in Figure 8, where (a) and (c) are derived from 2017, while (b) and (d) from the same period of 2018 ($\pm$1 day).

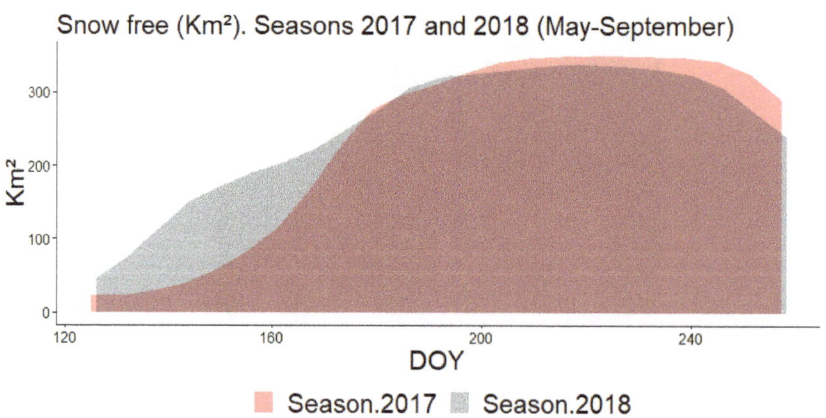

Figure 7. Surface [km^2] free of snow during the season 2017 and 2018.

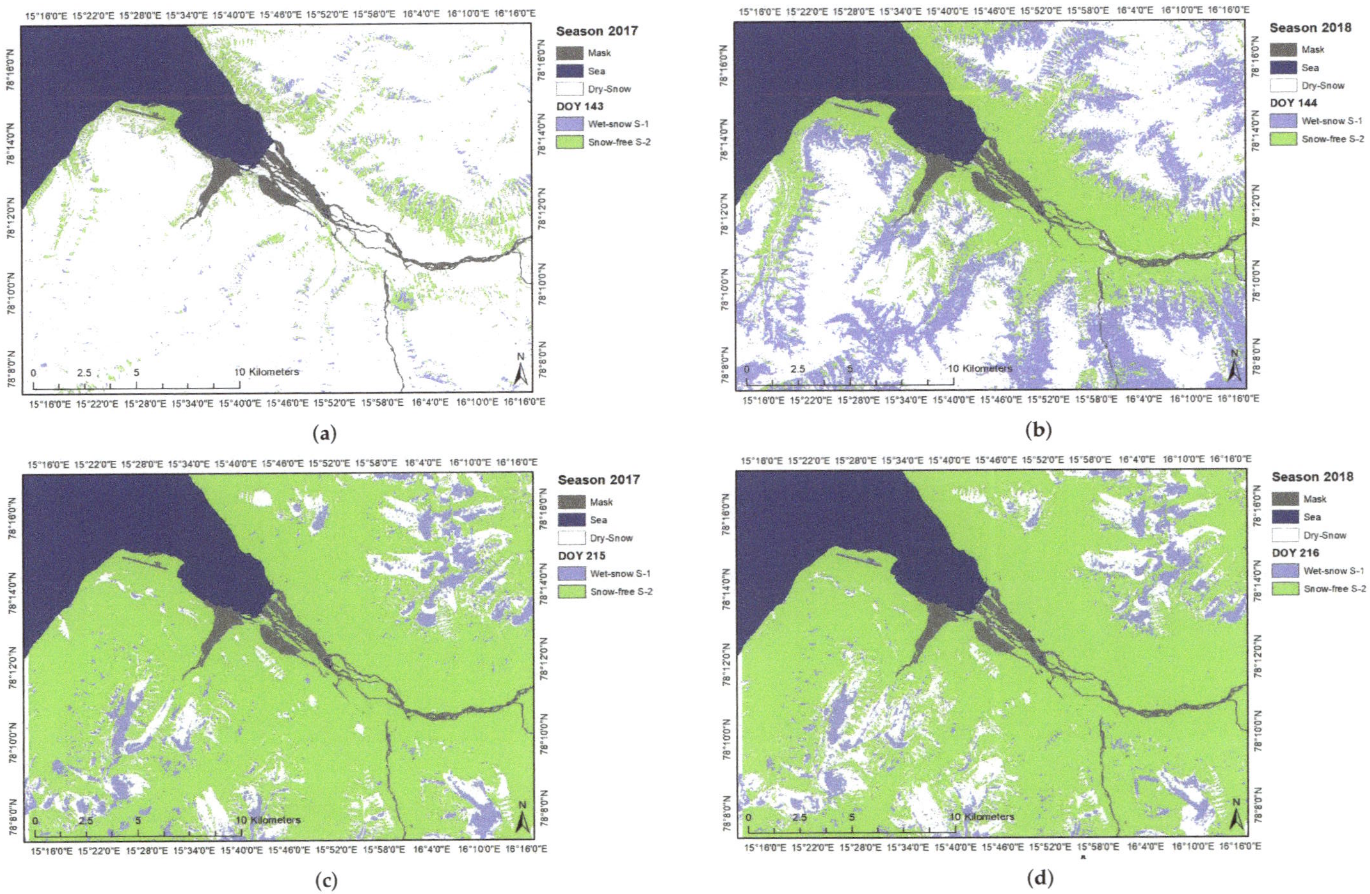

Figure 8. Snow cover classification of Sentinel-1 and Sentinel-2, based on backscatter and NDSI thresholds. The two maps at the top represent the month of May (DOY 143 2017 (**a**) and DOY 144 2018 (**b**)). In the lower part represents the month of August (DOY 215 2017 (**c**) and DOY 216 2018 (**d**)).

4.3. Multi Sensor Analyses and Vegetation

To understand the evolution of the HV signal in respect of the NDVI, the Soil Water Content (SWC) and the temperature of soil, a temporal analysis was computed. Figure 9 shows an example of the S-1 time series related to ground information; the time series are from 2017, station ST3.

The Pearson correlation coefficient (R) and p-value between time series are shown in Table 4.

Table 4. Pearson correlation coefficient (R) and p-value for the five stations.

Time Series	Station	Pearson 2017	p-Value 2017	Pearson 2018	p-Value 2018
S1 HV $\sim$ NDVI	ST1	0.36	0.000	0.52	0.000
	ST3	0.09	0.379	0.23	0.014
	ST6	0.64	0.000	0.69	0.000
	ST7	0.45	0.000	0.56	0.000
	ST9	0.53	0.000	0.14	0.147
S1 HV $\sim$ SWC	ST1	0.17	0.094	0.61	0.000
	ST3	0.81	0.000	0.71	0.000
	ST6	0.42	0.000	0.53	0.000
	ST7	−0.34	0.000	−0.34	0.000
	ST9	0.44	0.000	0.65	0.000
S1 HV $\sim$ Temp	ST1	0.38	0.000	0.19	0.047
	ST3	0.30	0.002	0.19	0.043
	ST6	0.79	0.000	0.32	0.000
	ST7	0.77	0.000	0.50	0.000
	ST9	0.68	0.000	−0.12	0.195

In the 2017 season, except for the ST3 and ST6 areas, the R between HV $\sim$ NDVI and HV $\sim$ SWC was below 0.6, with p-values less than 0.05 (excluding ST3 and ST1). Conversely, the R between HV and temperature in the ST6, ST7, and ST9 stations were significant (R mean of 0.74). Through the Pearson correlation coefficient, the positive correlation between the time series was defined, except for the station ST7 (HV $\sim$ SWC). In 2018, the results of the R were significant for HV $\sim$ NDVI and HV $\sim$ SWC. Instead, the HV $\sim$ temperature correlation showed a low value in all five areas. Again, the ST7 station was negative correlated with HV and SWC, along with the HV and temperature of the ST9 area. To determine the order of factors soil temperature and moisture content, dominance analyses of the linear models were performed. The results in Table 5 illustrate the dominance, expressed as a percentage, of the NDVI and SWC variables in linear regression with the HV channel. However, in 2018, for the stations ST1 and ST3 the SWC influences the HV channel more than the vegetation.

Table 5. Dominance analyses of the linear models between HV $\sim$ NDVI and HV $\sim$ SWC.

Station	2017 NDVI %	2017 SWC%	2018 NDVI%	2018 SWC%
ST1	87	13	42	58
ST3	1	99	8	92
ST6	71	29	64	36
ST7	64	36	74	26
ST9	59	41	2	98

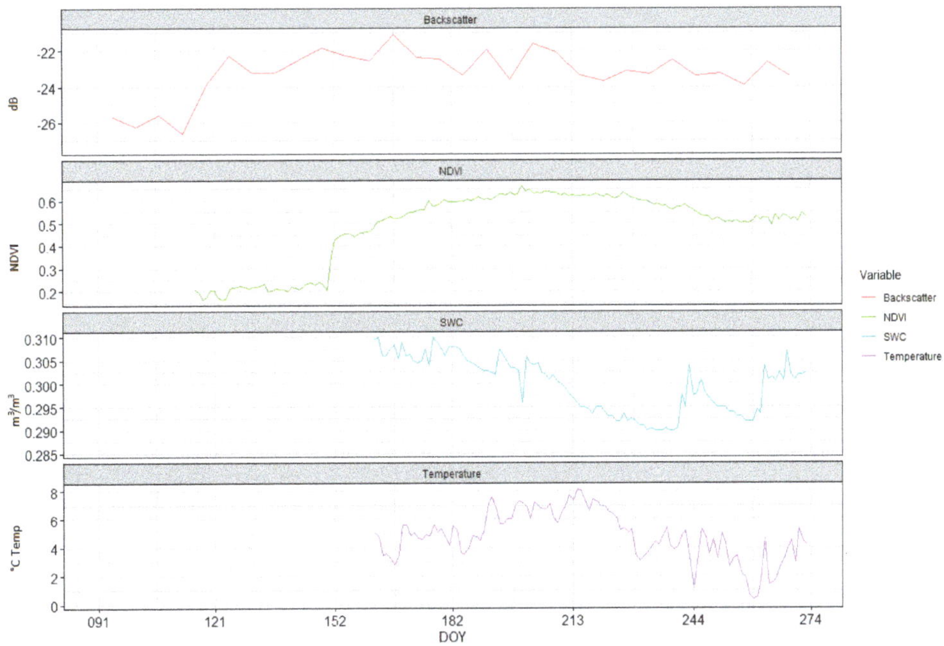

Figure 9. Time series of S-1 HV channel and ground sensors for the area ST3 2017: in the graphs, in red is shown the HV backscatter, in green the NDVI, in blue the Soil Water Content (SWC) and in purple the soil temperature at 10 cm depth. A moving average filter was applied to the time series.

4.4. Discussion

The threshold for detection of wet snow was calculated and applied to 115 S-1 C band images. To minimize the threshold dependency on the reference images, it was essential to average three images during the winter ($\sigma_{sc\,2017}$) and summer ($\sigma_{sf\,2017}$) periods. The threshold identified in this study was in line with the thresholds found in the literature [15,16,46].

However, the time lapse between two acquisitions was decisive in the definition of wet snow, because there were strong changes from one day to the next. For this reason, together with the low amount of snow, it was not possible to detect wet snow in some areas of the valley floor. A specific threshold analysis for vegetated areas, rocks, and sediments could improve our wet snow masks [17].

In terms of optical images, the validation of the NDSI masks by PhenoCams/NDVI was not possible for the 2018 season. Due to the early snow melting, the start of the RGB image acquisition occurred only in a snow-free period. For this reason, it was only possible to use the information from meteorological station. The NDSI snow-mask/field data correlation, considering both years, was found to have an R^2 of 0.73. This is significantly better than the results obtained by the optical sensor MODIS over northern Norway [8].

The two seasons investigated showed a variability in the melting period. The 2017 season experienced late melting, with peaks around the end of June. On the contrary, in 2018 the snow began to melt much earlier. On average, the discrepancy in the valley between the two years was around 20 days.

Once the masks were obtained and validated, it was possible to get an overview of the snow season 2017 and 2018. The SAR and optical time-series were fused into a multi-sensor and multi-temporal snow cover masks. At this latitude, due to low incident angle, after September it was not possible to detect the first day of snow covered with optical data in the valley. The discrepancy between snow melting and the last day of snow corresponded in both cases to about 20 days at the lowest elevations in the two seasons under consideration. Therefore, in 2018, in these areas, snow melting was about 20 days earlier, as well as snow-free surface in comparison with the 2017 season. Considering that the vegetation phenology is strongly influenced by the dynamics of snow [2,3], it would be interesting to estimate the impact of the two seasons on a large scale [10], with a spatial resolution equal to 20 m (the spatial resolution of the S-2 short wave infrared band).

An exploratory investigation was conducted on the sensitivity of backscattering coefficients to snow-free surfaces in the high arctic environment. The structure of the vegetation and its dielectric properties, soil roughness and moisture influence the backscattering coefficients. For dense vegetation and small leaves, the C band has proved useful in detecting the dynamics of vegetation compared to other frequencies [47,48]. Moreover, in the cross polarized channels the contribution of vegetation is predominant [49]. Our results confirmed at most stations a dominance of the vegetation factor, expressed by the NDVI, on soil moisture (SWC). Nevertheless, the R between HV and NDVI showed, especially in 2017, a low significance and a variation depending on the areas under consideration. In the ST3 area, in both years, soil moisture was the prevalent component of σ^0 HV channel. The area has cryoturbated soil and is dominated by polygons of frost patterned ground, with little vegetation in the center. Soil moisture in this area averaged 0.30 m^3/m^3 in both years, with maximum values not exceeding 0.34 m^3/m^3. Furthermore, the dominant vegetation of this area are graminoid that grow in tussocks; an example is the genus *Luzula*. These plants have short leaves (1–6 cm and 3–5 mm wide) and keep withered leaves and sheaths from previous years [50,51]. These factors could explain the dominance of the soil component over the vegetation in this area. On the contrary, station ST6 has a plain of sandy sediments dominated by herbs, such as the genus *Festuca*. In this case, the leaves are up to 10–20 cm long, with 0.7–1.0 mm broad when rolled up, up to 2.5 mm broad when flat [50,51]. For this reason, the sensitivity of the HV channel is greater with respect to the vegetation component than to the soil component. A similar argument applies to station ST1 and ST7, which have graminoid dominated cover. A further element in the analysis of the results of the station ST7 is the saturation of the soil. The SWC reaches maximum values of 0.6–0.8 m^3/m^3 in 2017 and 2018, respectively. This could limit the penetration of the radar signal into the soil, resulting in limited sensitivity to this component. Another important point is the relevance of the correlations with temperature in the two years under consideration. In 2017, the year with delayed melting and disappearance of the snow, the SAR signal is more related to temperature at stations ST6, ST7, and ST9, than to vegetation and soil. On the contrary, in 2018, the relevance of temperature is almost negligible at all the stations. The change in snow cover determines the thawing rates of the soil, and thus controls the temperatures in the soil and soil surface, as well as the thickness of the active layer [52].

As the soil temperature decreases, there is lower liquid water content, which causes the backscattered signal to reduce [53,54]. The amount of frozen soil/tundra (with drier vegetation dominated by graminoids) may therefore limit the radar's capability to follow the vegetation growth pattern and soil moisture. In addition, links were observed between thawing of tundra soils and rainfall, especially from May to June [55]. For a phenological state estimation model, the incorporation of the soil temperature and precipitation could improve the outcomes. A better and deeper understanding of the backscatter dependence on soil and vegetation properties could be achieved by using data from boreholes in Svalbard (e.g., Adventdalen and Kapp Linne') in combination with fieldwork procedures such as those conducted by Bergstedt et al. (2018) [53] in their study covering circumpolar regions including Scandinavia. As a final consideration, the variation in the correlation between the vegetation and the backscatter signal could be determined by the presence of reindeer and migratory birds on the vegetated areas. The grazing of Svalbard reindeer (*Rangifer tarandus platyrhynchus*), whose population is increasing with an estimated mean population size of 22,435 [56], as well as grazing and grubbing by an exponentially increasing population of geese (e.g., *Anser brachyrhynchus*) cause a severe loss of plant biomass [57] that should be taken into account in understanding the response of backscatter coefficients to vegetation dynamics.

4.5. Conclusions

To follow snow dynamics on Svalbard, Sentinel-1 wet snow masks and Sentinel-2 snow masks were created and validated using ground data. An optimized threshold, applied to SAR backscattering coefficients, was used to detect wet snow during the years 2017 and 2018. Using the NDSI index, a daily snow mask was extracted from the Sentinel-2 sensor. Then Sentinel-1 and NDVI were used in synergy to follow vegetation dynamics. Our results have shown a variability in the length of snow seasons of 2017 and 2018. In 2018, the snow melted and the surface was free of snow about 20 days earlier than in 2017, and this was most clearly visible in the valley bottoms. In future studies, it would be interesting to evaluate the impact of these snow seasons on the vegetation phenology. Since high cloud coverage limits optical satellite data in the Arctic environment, a sensitivity study of cross-polarized channels to phenology and soil dynamics was performed.

The results of our study confirmed that vegetation is best detected by the HV channel in the Arctic environment. However, the ability to detect is limited due to the structure of the vegetation and the saturation of the soil. The late melting and disappearance of snow causes a further challenge in the monitoring of vegetation and soil dynamics. Therefore, when studying vegetation and soil dynamics, the amount of frozen soil/tundra should also be considered to understand the response of backscattering coefficients. Furthermore, the grazing of reindeer and arctic geese should be taken into account to fully understand the data.

Author Contributions: All authors contributed extensively to the work presented in this paper. Project conceived and designed by C.N., L.S., S.R.K., E.M. and H.T.; pre-processing of ground station data by L.N.; Fieldwork designed and data acquired by H.T., E.J.C. and L.N.; data analyses carried out by L.S. with support from S.R.K. and C.N.; Manuscript preparation lead by L.S., with support (substantial critical feedback, revisions and additions to text) from all co-authors. Final version of the manuscript read and approved by all co-authors. All authors have read and agreed to the published version of the manuscript.

Funding: This research was funded by Eurac research. The field work was funded by The Research Council of Norway through the 'SnoEco' project (project No. 230970) to E.J.C. and by SenSYF from The Fram Centre Terrestrial Flagship to S.R.K., L.N. and H.T. was funded by RCN project SIOS-InfraNor (No. 269927) and European Commission Research and Innovation Action project CHARTER no. 869471.

Acknowledgments: The authors thank the Department of Innovation, Research, University and Museums of the Autonomous Province of Bozen/Bolzano for covering the Open Access publication costs. The station network and some of the satellite data processing is funded by The Research Council of Norway through the InfraNor project (instruments #42, #43, #44, #49, and #52) managed by the Svalbard Integrated Arctic Earth Observation System (SIOS).

Conflicts of Interest: The authors declare no conflict of interest.

Appendix A

Appendix A.1

A visual representation of the difference between years is shown in Figure A1, where each colour matches the wet snow cumulatively at a specific month. The snow cover for a specific month is coloured blue for May, red for June, orange for July, green for August and purple for September.

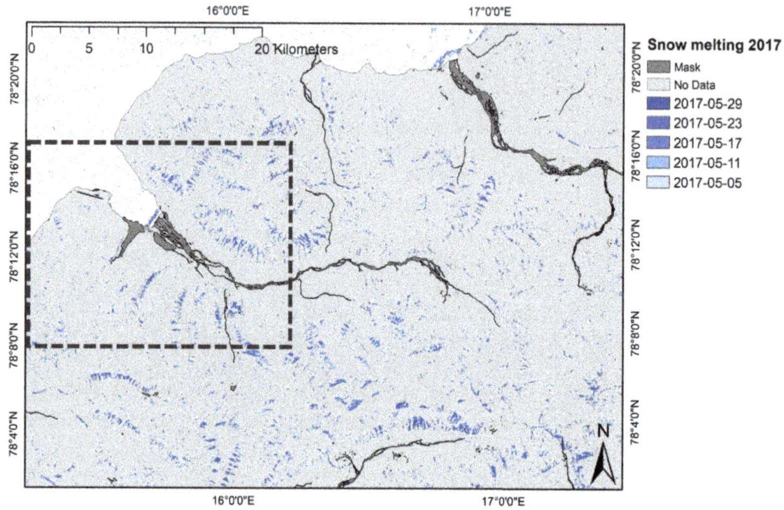

(a)

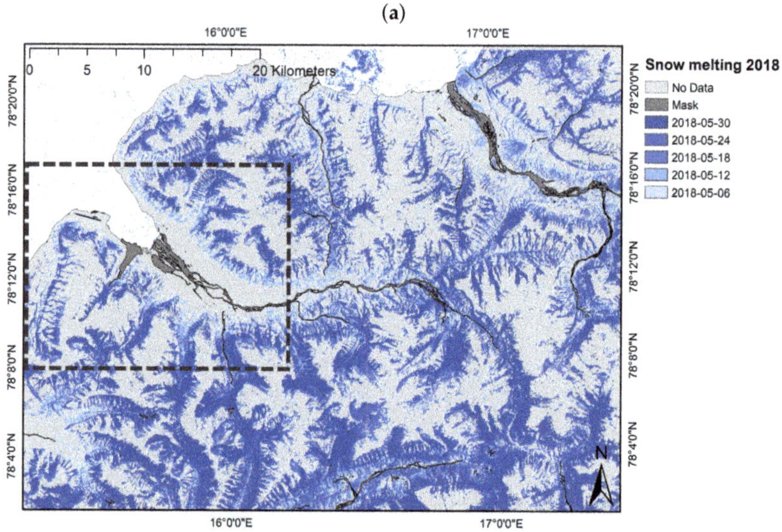

(b)

Figure A1. *Cont.*

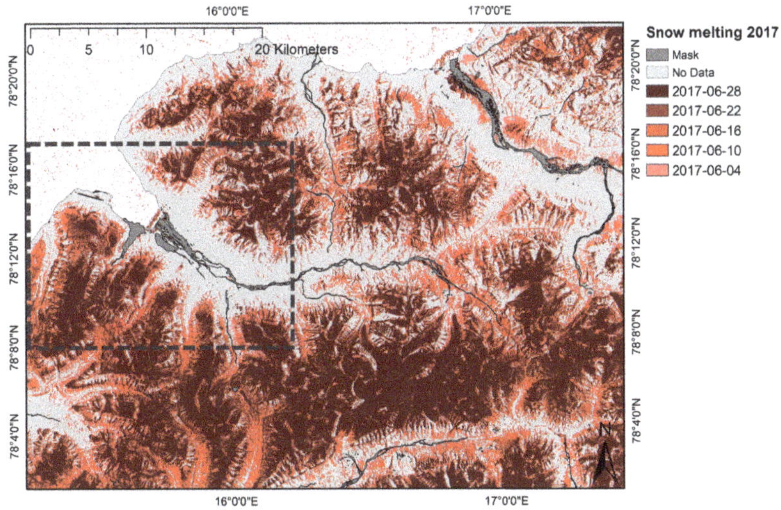

(**c**)

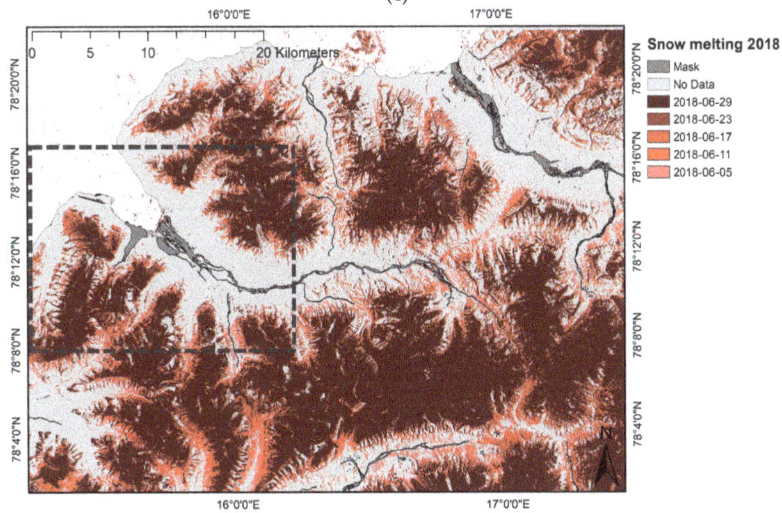

(**d**)

Figure A1. *Cont.*

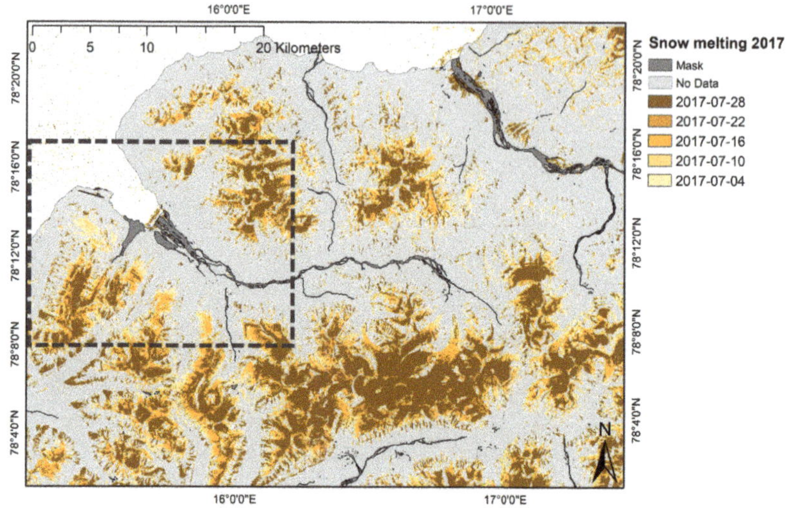

(e)

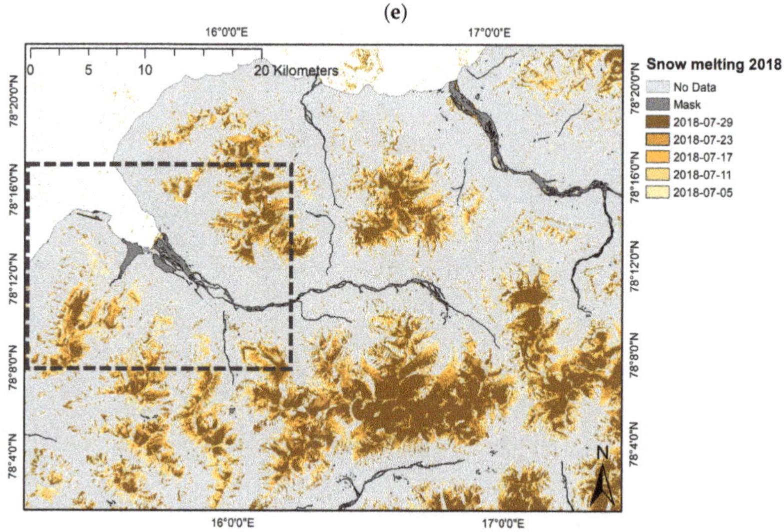

(f)

Figure A1. *Cont.*

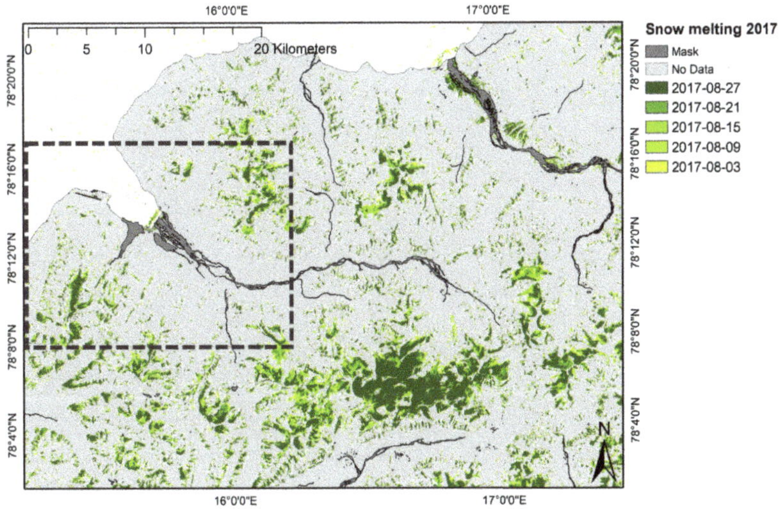

(g)

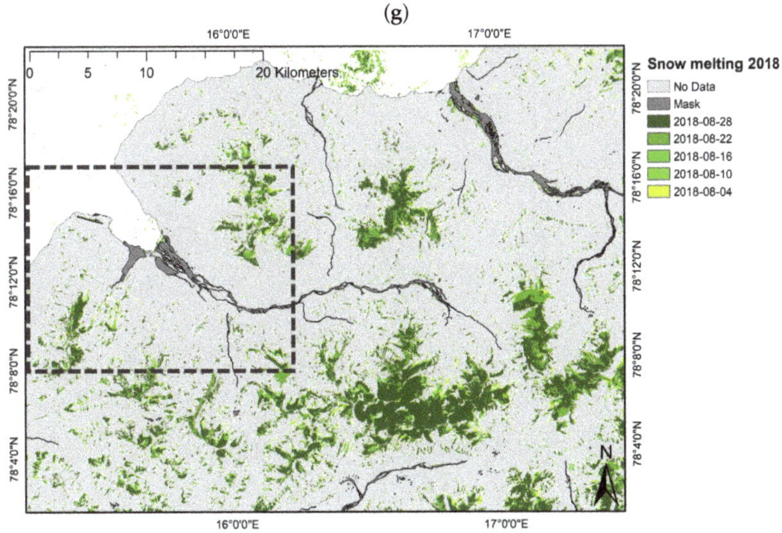

(h)

Figure A1. *Cont.*

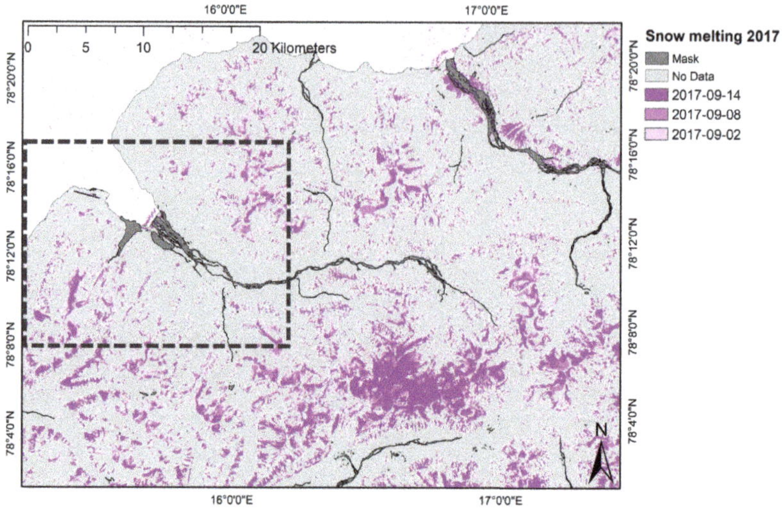

(i)

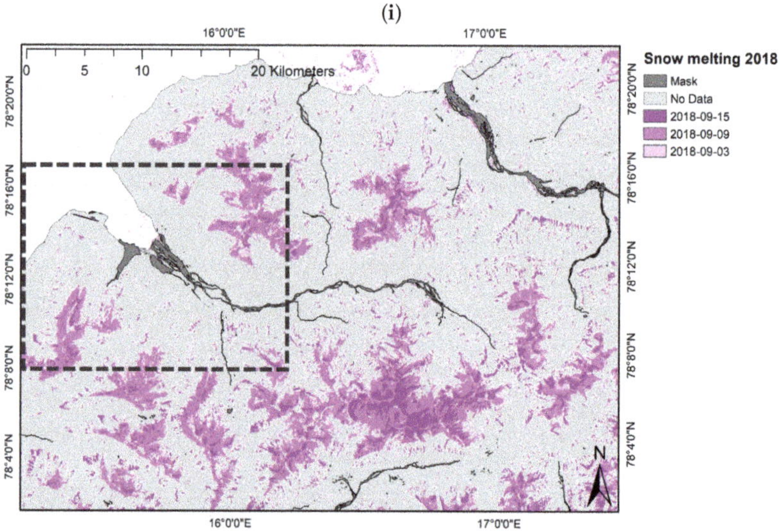

(j)

Figure A1. Wetsnow masks of the season 2017 and 2018: May is shown in blue (**a**,**b**), June in red (**c**,**d**), July in orange (**e**,**f**), August in green (**g**,**h**) and September in purple (**i**,**j**). The masks of the different days of the months are overlapped by a dark to light gradient for each type of color (cumulative of month periods). The picture shows the surrounding area of Adventdalen to provide an overview of the region, with a dotted frame representing the investigated area.

Appendix A.2

The periods without snow (expressed in DOY) are shown in Figure A2 by red to dark purple shading. In both years the first surface free from snow was in the area of the airport situated at the mouth of Adventfjørd, then on roads and on the steepest slopes of the mountains. After that, the snow disappeared from the flat bottom and lower slopes of the main Advent valley, in the secondary vallies and lastly in the areas with the highest altitude. The 2018 snow-free season was earlier than in 2017. The maps show a discrepancy in the red-orange colours of the vallies. The difference between the maps is smaller for areas at higher elevations.

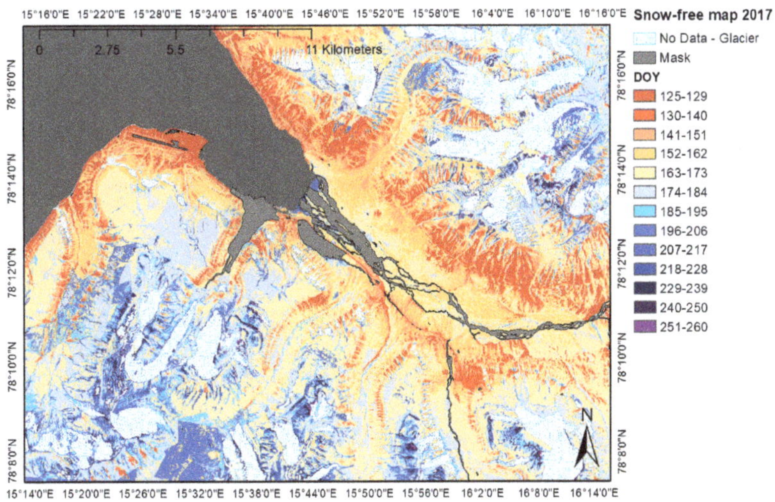

(a)

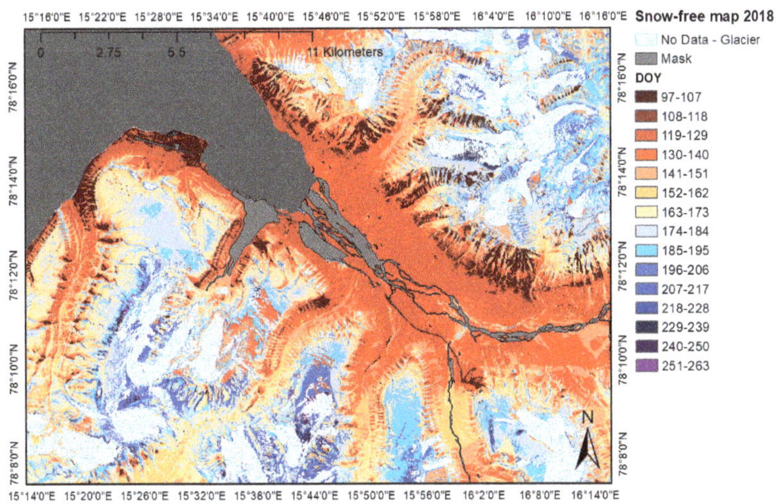

(b)

Figure A2. Snow-free maps for the years 2017–2018 (Period May-September). Each colour is matched by a period of 10 days expressed in DOY. (**a**) 2017 and (**b**) 2018.

References

1. Box, J.E.; Colgan, W.T.; Christensen, T.R.; Schmidt, N.M.; Lund, M.; Parmentier, F.J.W.; Brown, R.; Bhatt, U.S.; Euskirchen, E.S.; Romanovsky, V.E.; et al. Key indicators of Arctic climate change: 1971–2017. *Environ. Res. Lett.* **2019**, *14*. [CrossRef]
2. Gillespie, M.A.K.; Baggesen, N.; Cooper, E.J.H. Arctic flowering phenology and plant–pollinator interactions in response to delayed snow melt and simulated warming. *Environ. Res. Lett.* **2016**, *11*, 115006. [CrossRef]
3. Semenchuk, P.R.; Gillespie, M.A.K.; Rumpf, S.B.; Baggesen, N.; Elberling, B.; Cooper, E.J.H. Arctic plant phenology is determined by snowmelt patterns but duration of phenological periods is fixed: An example of periodicity. *Environ. Res. Lett.* **2016**, *11*, 125006. [CrossRef]

4. Treharne, R.; Bjerke, J.W.; Tømmervik, H.; Stendardi, L.; Phoenix, G.K. Arctic browning: Impacts of extreme climatic events on heathland ecosystem CO2 fluxes. *Glob. Chang. Biol.* **2019**, *25*, 489—503. [CrossRef] [PubMed]
5. Saha, S.K.; Rinke, A.; Dethloff, K.C. Future winter extreme temperature and precipitation events in the Arctic. *Geophys. Res. Lett.* **2006**, *33*. [CrossRef]
6. Hansen, B.B.; Isaksen, K.; Benestad, R.E.; Kohler, J.; Pedersen, Å.; Loe, L.E.; Coulson, S.J.; Larsen, J.O.; Varpe, Ø. Warmer and wetter winters: Characteristics and implications of an extreme weather event in the High Arctic. *Environ. Res. Lett.* **2014**, *9*, 114021. [CrossRef]
7. Vikhamar-Schuler, D.; Isaksen, K.; Haugen, J.E.; Tømmervik, H.; Luks, B.; Schuler, T.V.; Bjerke, J.W. Changes in winter warming events in the Nordic Arctic Region. *J. Clim.* **2016**, *29*, 6223–6244. [CrossRef]
8. Malnes, E.; Karlsen, S.R.; Johansen, B.; Bjerke, J.W.; Tømmervik, H. Snow season variability in a boreal-Arctic transition area monitored by MODIS data. *Environ. Res. Lett.* **2016**, *11*, 125005. [CrossRef]
9. Winther, J.-G.; Bruland, O.; Sand, K.; Gerland, S.; Marechal, D.; Ivanov, B.; Gøowacki, P.; König, M. Snow research in Svalbard—An overview. *Polar Res.* **2003**, *22*,125–144. [CrossRef]
10. Malnes, E.; Karlsen, S.R.; Johansen, B.; Haarpaintner, J.; Hogda, K.A. Monitoring of the snow coverage and its relation to vegetation and growing seasons on Svalbard using ENVISAT ASAR and TERRA MODIS data. In Prcoceedings of the ESA Living Planet Symposium, Bergen, Norway, 28 June–2 July 2010; Volume 686, pp. 28.6–28.7.
11. Vickers, H.; Karlsen, S.R.; Malnes, E. A 20-year MODIS-based snow cover dataset for Svalbard and its link to phenological timing and sea ice variability. *Remote Sens.* **2020**, *12*, 1123. [CrossRef]
12. Hall, D.K.; Riggs, G.A.; Salomonson, V.V. Development of methods for mapping global snow cover using moderate resolution imaging spectroradiometer data. *Remote Sens. Environ.* **1995**, *54*, 127–140. [CrossRef]
13. Salomonson, V.V.; Appel, I. Estimating fractional snow cover from MODIS using the normalized difference snow index. *Remote Sens. Environ.* **2004**, *89*, 351–360. [CrossRef]
14. Shi, J.; Dozier, J. Mapping seasonal snow with SIR-C/X-SAR in mountainous areas. *Remote Sens. Environ.* **1997**, *59*, 294–307. [CrossRef]
15. Nagler, T.; Rott, H. Retrieval of wet snow by means of multitemporal SAR data. *IEEE Trans. Geosci. Remote Sens.* **2000**, *38*, 754–765. [CrossRef]
16. Baghdadi, N.; Gauthier, Y.; Bernier, M. Capability of multitemporal ERS-1 SAR data for wet-snow mapping. *Remote Sens. Environ.* **1997**, *60*,174–186. [CrossRef]
17. Notarnicola, C.; Ratti, R.; Maddalena, V.; Schellenberger, T.; Ventura, B.; Zebisch, M. Seasonal snow cover mapping in alpine areas through time series of COSMO-skymed images. *IEEE Geosci. Remote Sens. Lett.* **2012**, *10*, 716–720. [CrossRef]
18. Buchelt, S.; Skov, K.; Rasmussen, K.K.; Ullmann, T. Sentinel-1 time series for mapping snow cover depletion and timing of snowmelt in Arctic periglacial environments: Case study from Zackenberg and Kobbefjord, Greenland. *Cryosphere* **2022**, *16*, 625–646. [CrossRef]
19. Hallikainen, M.; Ulaby, F.; Abdelrazik, M. Dielectric properties of snow in the 3 to 37 GHz range. *IEEE Trans. Antennas Propag.* **1986**, *34*, 1329–1340. [CrossRef]
20. Mätzler, C.; Schanda, E. Snow mapping with active microwave sensors. *Remote Sens.* **1984**, *5*,409–422. [CrossRef]
21. Ulaby, F. Radar response to vegetation. *IEEE Trans. Antennas Propag.* **1975**, *23*, 36–45. [CrossRef]
22. Attema, E.P.W.; Ulaby, F.T. Vegetation modeled as a water cloud. *Radio Sci.* **1978**, *13*, 357–364. [CrossRef]
23. De Bernardis, C.; Vicente-Guijalba, F.; Martinez-Marin, T.; Lopez-Sanchez, J.M. Contribution to real-time estimation of crop phenological states in a dynamical framework based on NDVI time series: Data fusion with SAR and temperature. *IEEE J. Sel. Top. Appl. Earth Obs. Remote Sens.* **2016**, *9*, 3512–3523. [CrossRef]
24. Torres, R.; Snoeij, P.; Geudtner, D.; Bibby, D.; Davidson, M.; Attema, E.; Potin, P.; Rommen, B.; Floury, N.; Brown, M.; et al. GMES Sentinel-1 mission. *Remote Sens. Environ.* **2012**, *120*, 9–24. [CrossRef]
25. Drusch, M.; Del, B.U.; Carlier, S.; Colin, O.; Fernandez, V.; Gascon, F.; Hoersch, B.; Isola, C.; Laberinti, P.; Martimort, P.; et al. Sentinel-2: ESA's optical high-resolution mission for GMES operational services. *Remote Sens. Environ.* **2012**, *120*, 25–36. [CrossRef]
26. Christiansen, H.H.; Humlum, O.; Eckerstorfer, M. Central Svalbard 2000–2011 meteorological dynamics and periglacial landscape response. *Arct. Antarct. Alp. Res.* **2013**, *45*, 6–18. [CrossRef]
27. Peel, M.C.; Finlayson, B.L.; McMahon, T.A. Updated world map of the Köppen-Geiger climate classification. *Hydrol. Earth Syst. Sci. Discuss.* **2007**, *4*, 439–473. ISSN 1812-2116. [CrossRef]
28. Elvebakk, A. Tundra diversity and ecological characteristics of Svalbard. *Ecosyst. World Polar Alp. Tundra* **1997**, *347*, 347–360.
29. Elvebakk, A. A vegetation map of Svalbard on the scale 1: 3.5 mill. *Phytocoenologia* **2005**, *35*, 951–967. [CrossRef]
30. Anderson, H.B.; Nilsen, L.; Tømmervik, H.; Karlsen, S.R.; Nagai, S.; Cooper, E.J. Using ordinary digital cameras in place of near-infrared sensors to derive vegetation indices for phenology studies of High Arctic vegetation. *Remote Sens.* **2016**, *8*, 847. [CrossRef]
31. Clerc, S.; Devignot, O.; Pessiot, L.; MPC Team. S2 MPC Level 2A Data Quality Report. In *PDGS-MPC-L2ADQR*, 39th ed.; ESA Copernicus: Lima, Peru, 2019; pp. 1–18.
32. Spectral Reflectance Sensor for NDVI. METER Group, Inc. USA. Available online: https://www.ai-nex.co.jp/SRS-N%20Integrators%20Guide.pdf (accessed on 18 September 2019).

33. Soil Temperature and Moisture Sensor (5 TM). METER Group, Inc. USA. Available online: http://publications.metergroup.com/Manuals/20424_5TM_Manual_Web.pdf (accessed on 18 September 2019).
34. Infrared Radiometer (SI-421). Apogee Instruments, INC. | 721 WEST 1800 NORTH, LOGAN, UTAH 84321, USA. Available online: https://www.apogeeinstruments.com/content/SI-400-manual.pdf (accessed on 18 September 2019).
35. Larsen, Y.; Engen, G.; Lauknes, T.R.; Malnes, E.; Høgda, K. Arild A generic differential interferometric SAR processing system, with applications to land subsidence and snow-water equivalent retrieval. In Proceedings of the Fringe 2005 Workshop, ESA ESRIN, (ESA SP-610), Frascati, Rome, 28 November–2 December 2005 .
36. Schellenberger, T.; Ventura, B.; Zebisch, M.; Notarnicola, C. Wet snow cover mapping algorithm based on multitemporal COSMO-SkyMed X-band SAR images. *IEEE J. Sel. Top. Appl. Earth Obs. Remote Sens.* **2012**, *5*, 1045–1053. [CrossRef]
37. Hollstein, A.; Segl, K.; Guanter, L.; Brell, M.; Enesco, M. Ready-to-use methods for the detection of clouds, cirrus, snow, shadow, water and clear sky pixels in Sentinel-2 MSI images. *Remote Sens.* **2016**, *8*, 666. [CrossRef]
38. Zupanc, A. Improving Cloud Detection with Machine Learning. Available online: https://medium.com/sentinel-hub/improving-cloud-detection-with-machine-learning-c09dc5d7cf13 (accessed on 20 September 2019).
39. Karlsen, S.R.; Elvebakk, A.; Høgda, K.A.; Grydeland, T. Spatial and temporal variability in the onset of the growing season on Svalbard, Arctic Norway—measured by MODIS-NDVI satellite data. *Remote Sens.* **2014**, *6*, 8088–8106. [CrossRef]
40. Karlsen, S.R.; Stendardi, L.; Tømmervik, H.; Nilsen, L.; Arntzen, I.; Cooper, E.J. Time-Series of Cloud-Free Sentinel-2 NDVI Data Used in Mapping the Onset of Growth of Central Spitsbergen, Svalbard. *Remote Sens.* **2021**, *13*, 3031. [CrossRef]
41. Matthew, J.M.; Christopher, S.S.; Kyle, J.; Martha, K.R. Landsat-based snow persistence map for northwest Alaska. *Remote Sens. Environ.* **2015**, *163*, 23–31. [CrossRef]
42. Rignot, E.J.M.; Van Z.J.J. Change detection techniques for ERS-1 SAR data. *IEEE Trans. Geosci. Remote Sens.* **1993**, *31*, 896–906. [CrossRef]
43. Metsämäki, S.; Vepsäläinen, J.; Pulliainen, J.; Sucksdorff, Y. Improved linear interpolation method for the estimation of snow-covered area from optical data. *Remote Sens. Environ.* **2002**, *82*, 64–78. [CrossRef]
44. Qiu, D.; Shao, Q.; Yang, L. Efficient inference for autoregressive coefficients in the presence of trends. *J. Multivar. Anal.* **2013**, *114*, 40–53. [CrossRef]
45. Budescu, D.V. Dominance analysis: A new approach to the problem of relative importance of predictors in multiple regression. *Psychol. Bull.* **1993**, *114*, 542–551. [CrossRef]
46. Floricioiu, D.; Rott, H. Seasonal and short-term variability of multifrequency, polarimetric radar backscatter of alpine terrain from SIR-C/X-SAR and AIRSAR data. *IEEE Trans. Geosci. Remote Sens.* **2001**, *39*, 2634–2648. [CrossRef]
47. Ferrazzoli, P.; Paloscia, S.; Pampaloni, P.; Schiavon, G.; Sigismondi, S.; Solimini, D. The potential of multifrequency polarimetric SAR in assessing agricultural and arboreous biomass. *IEEE Trans. Geosci. Remote Sens.* **1997**, *35*, 5–17. [CrossRef]
48. Macelloni, G.; Paloscia, S.; Pampaloni, P.; Marliani, F.; Gai, M. The relationship between the backscattering coefficient and the biomass of narrow and broad leaf crops. *IEEE Trans. Geosci. Remote Sens.* **2001**, *39*, 873–884. [CrossRef]
49. McNairn, H.; Brisco, B. The application of C-band polarimetric SAR for agriculture: A review. *Can. J. Remote Sens.* **2004**, *30*, 525–542. [CrossRef]
50. Scholander, P.F. Vascular Plants from Northern Svalbard: With Remarks on the Vegetation in North-East Land. 1934. Available online: http://hdl.handle.net/11250/173806 (accessed on 23 September 2019).
51. Available online: https://svalbardflora.no/ (accessed on 30 September 2019).
52. Eckerstorfer, M.; Malnes, E.; Christiansen, H.H. Freeze/thaw conditions at periglacial landforms in Kapp Linné, Svalbard, investigated using field observations, in situ, and radar satellite monitoring. *Geomorphology* **2017**, *293*, 433–447. [CrossRef]
53. Bergstedt, H.; Zwieback, S.; Bartsch, A.; Leibman, M. Dependence of C-band backscatter on ground temperature, air temperature and snow depth in arctic permafrost regions. *Remote Sens.* **2018**, *10*, 142. [CrossRef]
54. Baghdadi, N.; Bazzi, H.; El Hajj, M.; Zribi, M. Detection of frozen soil using Sentinel-1 SAR data. *Remote Sens.* **2018**, *10*, 1182. [CrossRef]
55. Rydén, B.E.; Kostov, L. Thawing and freezing in tundra soils. *Ecol. Bull.* **1980**, *30*, 251–281. [CrossRef]
56. Le Moullec, M.; Pedersen, Å.Ø.; Stien, A.; Rosvold, J.; Hansen, B.B. A century of conservation: The ongoing recovery of Svalbard reindeer. *J. Wildl. Manag.* **2019**, *83*, 1676–1686. [CrossRef]
57. Speed, J.D.M.; Woodin, S.J.; Tømmervik, H.; Van der Wal, R. Extrapolating herbivore-induced carbon loss across an arctic landscape. *Polar Biol.* **2010**, *33*, 789–797. [CrossRef]

remote sensing
MDPI

Article

Time-Series of Cloud-Free Sentinel-2 NDVI Data Used in Mapping the Onset of Growth of Central Spitsbergen, Svalbard

Stein Rune Karlsen [1,*], Laura Stendardi [2], Hans Tømmervik [3], Lennart Nilsen [4], Ingar Arntzen [1] and Elisabeth J. Cooper [4]

1 NORCE Norwegian Research Centre AS, P.O. Box 6434, 9294 Tromsø, Norway; inar@norceresearch.no
2 Department of Agriculture, Food, Environment and Forestry (DAGRI), University of Florence, Piazzale delle Cascine, 18-50144 Firenze, Italy; laura.stendardi@unifi.it
3 Norwegian Institute for Nature Research (NINA), FRAM—High North Research Centre for Climate and the Environment, P.O. Box 6606, Langnes, 9296 Tromsø, Norway; Hans.Tommervik@nina.no
4 Department of Arctic and Marine Biology, UiT—The Arctic University of Norway, 9037 Tromsø, Norway; lennart.nilsen@uit.no (L.N.); elisabeth.cooper@uit.no (E.J.C.)
* Correspondence: skar@norceresearch.no; Tel.: +47-934-19904

Abstract: The Arctic is a region that is expected to experience a high increase in temperature. Changes in the timing of phenological phases, such as the onset of growth (as observed by remote sensing), is a sensitive bio-indicator of climate change. In this paper, the study area was the central part of Spitsbergen, Svalbard, located between 77.28°N and 78.44°N. The goals of this study were: (1) to prepare, analyze and present a cloud-free time-series of daily Sentinel-2 NDVI datasets for the 2016 to 2019 seasons, and (2) to demonstrate the use of the dataset in mapping the onset of growth. Due to a short and intense period with greening-up and frequent cloud cover, all the cloud-free Sentinel-2 data were used. The onset of growth was then mapped by a NDVI threshold method, which showed significant correlation ($r^2 = 0.47$, $n = 38$, $p < 0.0001$) with ground-based phenocam observation of the onset of growth in seven vegetation types. However, large bias was found between the Sentinel-2 NDVI-based mapped onset of growth and the phenocam-based onset of growth in a moss tundra, which indicates that the data in these vegetation types must be interpreted with care. In 2018, the onset of growth was about 10 days earlier compared to 2017.

Keywords: Sentinel-2; NDVI; time-series; onset of growth; Svalbard

Citation: Karlsen, S.R.; Stendardi, L.; Tømmervik, H.; Nilsen, L.; Arntzen, I.; Cooper, E.J. Time-Series of Cloud-Free Sentinel-2 NDVI Data Used in Mapping the Onset of Growth of Central Spitsbergen, Svalbard. *Remote Sens.* **2021**, *13*, 3031. https://doi.org/10.3390/rs13153031

Academic Editor: Dino Ienco

Received: 9 June 2021
Accepted: 29 July 2021
Published: 2 August 2021

Publisher's Note: MDPI stays neutral with regard to jurisdictional claims in published maps and institutional affiliations.

1. Introduction

Global temperature is increasing, and particularly so in Svalbard [1–3]. This has strong impacts on terrestrial ecosystems. Whilst the 1982–2015 greening trend for the Arctic is most pronounced at very high latitudes in continental tundra regions, maritime areas including the high Arctic islands (here defined as oceanic to moderately continental regions), show weaker trends [4–6].

In order to monitor climate-induced change in vegetation, time-series of optical satellite data can be used to map a range of different biophysical parameters, and such data have been used in Svalbard to map phenological stages such as the onset and peak of the growing season, as well as plant productivity [6–8]. A strong link between the plant productivity and sea ice distribution was found [9]. Cloud detection is the most crucial step during the pre-processing of time-series of optical satellite images. Failure to mask out (remove) the clouds from the image will have a significant negative impact on any subsequent analyses, such as differentiating the phenological stages [8,10], change detection and land-use classification [11]. Cloudiness in Svalbard, as well as in the other arctic islands, has caused problems when monitoring the different phenological stages using satellite data [7,9]. Snow and ice, frequent events with fog, low solar elevation angles, short growing seasons, and weak vegetation responses characterize Svalbard, and make cloud masking on the archipelago a challenging task. Further, Svalbard is undergoing dramatic climatic changes, with periods of heavy rain instead of clear cold weather with steady snow cover in autumn–early winter, and mild periods in midwinter, creating ice- or even snow-free spots that occurred several times in the last few years [12]. How these extreme climatic events and changes in snow duration, snow properties, and time of green-up affect the plant growth is largely unknown at the plant community and ecosystems levels, and the provision of more accurate proxies of phenology at large spatial scales is necessary [13].

Extensive datasets of ground-based visual observations of phenology have been collected in Adventdalen, Svalbard since 2007 [14] and elsewhere in arctic and alpine areas over the last 30 years through the International Tundra Experiment (e.g., [15–17]). However, these are very time-consuming to collect and require constant presence in the field, and so an automated system of observations would be more efficient. It is a challenge to make the links between sensors at different distances from the vegetation, from near-scale up to satellite-derived data [18]. Clearly, there is an urgent need for methodologies to be developed that combine field-based observations and data using near-ground sensors such as those obtained from phenocams [10,19], in order to validate satellite data, e.g., Sentinel-2 (S2) [13,20–22] and provide more accurate estimates of plant phenology and productivity at different spatial scales [13].

Compared with previous studies of the onset of growth in Svalbard based on 231.65 m pixel resolution in MODIS data [7] S2 data with 10–20 m pixel resolution offers exciting opportunities to monitor and study plant phenology at much greater detail [23]. Due to the northern location of Svalbard, and the polar orbit of S2, the study area has frequent acquisition, with often two images of S2 data per day, and is therefore a suitable area for vegetation phenology studies.

The main aim of this study is thus to (1) prepare, analyze and present a cloud-free time-series of daily S2 NDVI datasets for the 2016 to 2019 seasons covering central parts of Spitsbergen in Svalbard, and (2) to demonstrate the use of the dataset in mapping the onset of vegetation growth.

2. Materials and Methods

2.1. Study Area

The study area was the central part of Spitsbergen (Nordenskiöld Land peninsula, and Nathorst Land) in the archipelago of Svalbard, located between approximately 77.28° N–78.44° N and 12.40° W–19.10° W (Figure 1), which is S2 tile 33XWG and most of tile 33XVG.

Nordenskiöld Land is characterized by large valleys with dense vegetation cover, where the mean NDVI value locally reaches above 0.5 (Figure 1), indicating high plant biomass [8,24]. At higher elevations, the vegetation is sparse, as shown in NDVI values below 0.1 (Figure 1). Nathorst Land is mainly covered by glaciers, and vegetation is only found along the coast and in small valleys. Altogether, 1889 km^2 (20.7%), of the land area has NDVI values above 0.1; the remaining area is made up of glaciers or is very sparsely vegetated.

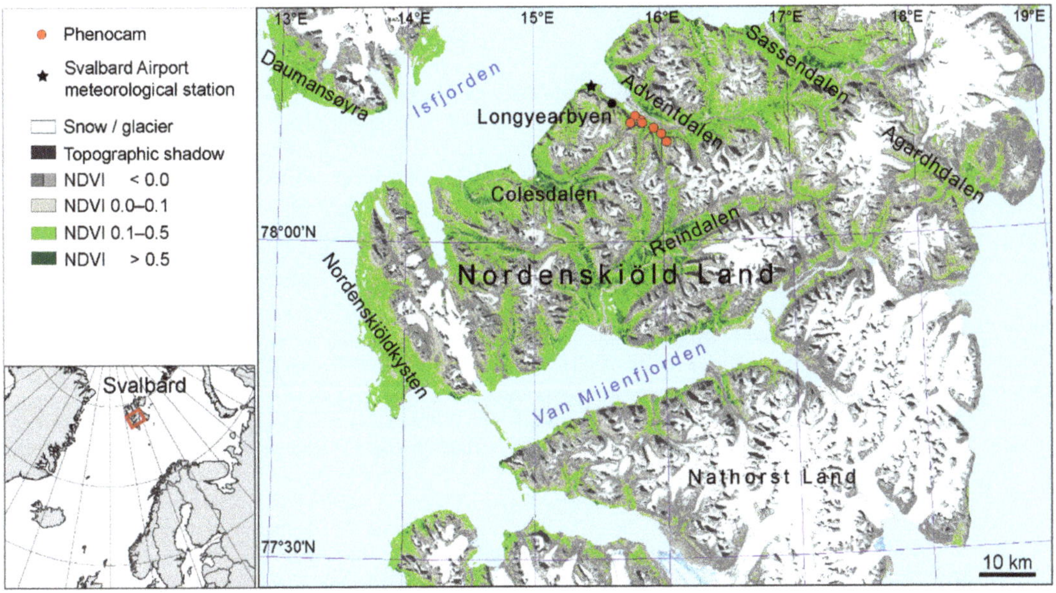

Figure 1. The study area of central Spitsbergen, Svalbard, showing mean NDVI values for the 25 July to 1 August period (4-year mean (2016–2019)). Snow/glacier mean values are shown for the same period, extracted from NDSI values (see text). The topographic shadow is approximate when the S2 passes in late July. The map also shows the location of the phenocams used in this study.

The meteorological station at Svalbard Airport, located close to the archipelago's administrative center in Longyearbyen (Figure 1), recorded a mean July air temperature of 7.0 °C and mean annual temperature of −3.8 °C for the 1991−2020 normal period (Table 1). This study is based on use of S2 data for the years 2016 to 2019. For Nordenskiöld Land, the snow melted early in both 2016 and 2018 compared to the 2000–2019 average [6], reflected by high May temperatures of 1.4 °C in 2016 and 1.8 °C in 2018, compared to the −2.2 °C average from the last 30 years (Table 1). On the other hand, the year 2017 had a May temperature of only −3.9 °C. The onset of growth starts in June in the warmest valleys, and in early July at higher elevations [7]. All the years examined in this study (2016–2019) had mean June temperatures above the 1991–2020 average.

Table 1. Monthly and annual temperature (°C) for the meteorological station at Svalbard Airport (station number 99840), located close to Longyearbyen (Figure 1). Temperatures for the years with S2 data used in this study compared to the 1991–2020 normal period (Norwegian Meteorological Institute 2021).

Month/Year	2016	2017	2018	2019	1991–2020
May	1.4	−3.9	1.8	−2.3	−2.2
June	5.0	4.6	4.0	4.8	3.6
July	9.0	6.9	7.2	8.4	7.0
Annual	−0.1	−2.2	−1.8	−3.4	−3.8

2.2. Phenological In-Situ Data

Phenological observations designed to be up-scaled by S2 data were established in Adventdalen and adjacent Endalen valleys (Figure 1). In the field, large homogeneous areas of the following vegetation types were located: *Dupontia fisheri* marsh, *Dryas octopetala* tundra, mixed *Dryas octopetala-Cassiope tetragona* tundra, moss tundra (*Aulacomnium turgidum-Tomentypnum nitens* type), *Equisetum arvense* ssp. *alpestre* snowbed, and *Luzula confusa* tundra. The mixed *Dryas octopetala-Cassiope tetragona* tundra site was located in Endalen, which has a slightly warmer local climate and represents the warmest/earliest site for growth onset. In addition, we also monitored phenology on a mixed exposed tundra type on a mountain plateau (350 m.a.s.l.). All the selected areas are homogeneous, except the site at the mountain plateau, which is moss rich and where gravel covers about one third, and *Luzula confusa* and *Salix polaris* are the most common vascular plants. Together, these vegetation types make up a large part of the vegetation diversity and thereby the phenological variation in the Adventdalen valley and on a nearby mountain plateau. The smallest site, a very homogeneous *Equisetum* snowbed where the moss *Sanionia uncinata* dominates the ground layer, covers fourteen 10×10 m^2 S2 pixels (1400 m^2) and the other six sites from 73 to 361 pixels (7300–36,100 m^2). The phenological observations were obtained by manual examination of photos captured by time-lapse cameras (phenocams) (trail camera model Acorn LtL-5310WA with 12-Megapixel and 100° wide angle) placed on tripod 40–60 cm above the surface, covering a plot of 0.95–1.43 m^2. The cameras were used in the selected vegetation types at the same site each year, established after snowmelt but before green-up, and removed from the field in September. The phenocams captured images each hour from 10 a.m. to 2 p.m. each day. In the moss tundra and in the mixed *Dryas-Cassiope* tundra we used two cameras, while the other sites had one camera.

In each site, the images capture photos of the dominant and/or the sub-dominant species, except the moss tundra, where both *Dryas octopetala* and *Salix polaris* only occurs scattered on dryer parts (Figure 2). The shrub *Salix polaris* is among the most widespread vascular species in Svalbard, and is very common or even sub-dominant in all the observation sites, except in the *Dupontia fisheri* marsh, and *Salix polaris* was included in some of the images on the other sites as well.

Figure 2. A phenocam on a tripod in Adventdalen valley established to capture images of *Dryas octopetala* in a moss tundra. Photo from 3 June 2016, eight days before the defined onset of growth of *Dryas octopetala* that year.

Within the phenocam images, we counted and observed leaves/steams close to a stick or at the same spot each year, since the camera had the same placement each year. The area of interest examined was about 5×5 cm and had at least 5 leaves of the study species.

From the close-up phenocam images, we followed the phenological growth stages, and we defined the phenophases according to the extended BBCH scale [25]. The abbreviation BBCH derives from the names of the originally participating stakeholders: "Biologische

Bundesanstalt, Bundessortenamt und CHemische Industrie". The BBCH scale is a system for the uniform coding of phenologically similar growth stages of all mono- and dicotyledonous plant species. Similar phenological stages of each plant species are given the same code. In the two-digit code, the first number shows the principal growth stage (0–9) and the second number the secondary stage (0–8). In this study, we manually extracted the date (day of the year) on which the plant reached the BBCH scale 15, which is on the shrubs we observed: '>5 leaves unfolded, but not yet full size', and the corresponding phenophase on the graminoids we observed: '>5 leaves (>3 cm) clearly visible'. For the horsetail *Equisetum arvense* ssp. *alpestre*, no BBCH code exists, so for this species we defined the onset of growth as: '>5 stem (>3 cm) having branches (>0.5 cm)'. The BBCH scale was established primarily for agricultural plants, but BBCH code 15 gives a good definition of the first opening of leaves of our study species. This phenological stage is called 'onset of growth' in this study and corresponds to 'green-up' in many tundra publications (e.g., [15]). It happens rapidly and is easily observed by eye in the field in Svalbard.

2.3. Processing Sentinel-2 Data—Cloud Removal and Interpolation

This study makes use of all available S2 imagery from April/May to mid-September and spans four seasons (2016 to 2019). Svalbard's location close to the North Pole as well as the polar orbit of S2 enables a higher number of captured images. On 1 July 2017, the satellite Sentinel-2A was joined by its twin, satellite Sentinel-2B, and from this point onwards we often have two images per day.

At the same time though, the high latitude implies low solar elevation angles (solar zenith angles higher than 70°) for its measurements, and particularly so late in the growth season. For this reason, processing level 2A (bottom-of-atmosphere (BOA)) is not reliable for this area, as it results in the under-correction of the atmospheric signal [26]. The study is therefore based on processing level 1C (top-of-atmosphere (TOA)) reflectance data. Level 2A data are only used as reference data for scene classification and the estimation of cloud probability.

Overcoming cloud cover is a crucial step during the pre-processing of time-series of optical satellite images. For cloud detection, we examined the cloud probability in Level 2A (Sen2Cor processor), from the 's2cloudless' machine-learning-based algorithm [27], and from previous S2-based algorithms [28]. In addition, we developed our own cloud detection algorithms from multi-spectral values and multi-temporal tests, using experiences from MODIS data time-series processing of Svalbard [7], as, for instance, the algorithm: the value of band 8 (NIR) for a given pixel is compared to its median value for that pixel in the 2016–2019 period, and values less than 30% of the median indicated cloud shadows covering vegetated areas [28]. However, none of the cloud detection methods work well for sparsely vegetated areas (bright surfaces), which is very common in the study area. Additionally, for thin semi-transparent clouds, and for cloud shadows, the algorithms did not show sufficient accuracies. To detect clouds, we performed a visual inspection in the visual and SWIR bands, and visually masked out cloud free areas, only using all the different cloud masks as references. One exception was the cirrus clouds, which could be accurately detected with the S2 band 10. However, cirrus clouds only appear in a few of the images. This time-consuming method for overcoming cloud cover, based on visual inspection, ensured few errors, but not all cloud-free data are included. Some of the images had many small cumulus clouds. Areas scattered with such clouds were too time-consuming to mask out and were therefore not used.

Normalized Difference Vegetation Index (NDVI) values were calculated for the cloud-free pixels and interpolated to daily data. The gap-filling to daily data was achieved by performing a linear interpolation and smoothing with a Savitzky–Golay filter [29]. Areas with NDVI <0.1 (4-year (2016–2019) mean NDVI for the 25 July to 1 August period, Figure 1) were not included in the dataset, as they have no, or at most, very sparse vegetation cover. The interpolation and smoothing method used led to some unrealistically low NDVI values early in the season when it was snow covered, and all NDVI values below −0.1 were reclassified to −0.1. Finally, due to the short and intense growth season in the study area, we believed that data with more than 10 days since last cloud-free observation would be too inaccurate to be used, and were removed from the dataset. The resulting dataset is daily clear-sky NDVI maps with 10×10 m^2 resolutions for the 2016 to 2019 seasons.

In addition, from the clear-sky S2 dataset, the NDSI (Normalized Difference Snow Index) values were calculated and interpolated to daily data for all land areas in the study area. This S2-NDSI dataset is used in a study where different snow mapping methods are compared [30], and the mean snow cover for 25 July to 1 August (mean 2016–2019) was extracted and used on the maps in this study for improved illustrations.)

2.4. Mapping the Onset of Growth

To map the onset of growth each year, we used the time-series of cloud-free daily S2-NDVI data. First, we computed the 4-year (2016–2019) mean NDVI value for every pixel in the study area for the 10 July to 5 August period. This period was chosen as the period where the leaves are fully open and before senescence starts and reduced the "noise" from snow-covered ground. The onset of growth at each pixel each year was defined as the time when the NDVI value each year exceeded 69% of the 10 July to 5 August 4-year mean NDVI value. This NDVI threshold level was reached after several iterations as the level that gives the highest correlation to—and least bias with—the 'onset of growth' observed using phenocam photos. Variations of this method were first used in Fennoscandia [31–35] and were used in Svalbard on MODIS data [7].

3. Results

3.1. A Clear-Sky Daily NDVI Dataset

The availability of cloud-free data from May to September limits the possibility to map the onset of growth. In 2016, we used S2 data for the 30 April to 12 September period. For this 135-day period, S2 images were captured on 111 days. From these 111 days with S2 data, we obtained on average only 8.2 days with cloud-free data (Figure 3), indicating a cloud cover of greater than 90% of the time in which the satellite captured images that season. In total, 17.5% of the pixels had more than 10 days since the last cloud-free day, and so were removed from the dataset, leading to main gaps in the time-series 9–11 May, 2–3 June, 25–28 June and 21–22 July. In particular, Agardhdalen, some parts of Adventdalen, and south-westernmost parts of the study area had limited S2 data (<8 cloud free days) in the 2016 season, creating restrictions to the reliable mapping of the onset of growth in parts of the study area.

In 2017, we used data for the 9 May to 12 August period; for these 95 days, we had on average 10.5 days with cloud-free pixels. In the west (Nordenskiöldkysten) there was little cloud-free data (less than 8 days), but the main valleys (Reindalen, Adventdalen, and Sassendalen) had about 13–16 days with cloud-free pixels (Figure 3). Some main gaps in the data were 19 June and in the 2–12 July period, and altogether, 16.8% of the pixels had more than 10 days since the last cloud-free day and were removed.

In 2018, S2 data from 30 April to 23 August was processed. However, for this 115-day period, on average, only 7.6 days of cloud-free data were found, and these cloud-free pixels were unevenly spread: while Reindalen, Sassendalen and Agardhdalen had mostly less than 7 days with cloud-free pixels, Daumansøyra had more than 13 days (Figure 3). Due to the cloudy conditions in this season, a total of 21.9% of the pixels had more than 10 days since the last cloud-free day, and this led to large gaps in the dataset between 25 May and 7 June, and from 14 to 19 July.

For the year 2019, we used S2 data in the period from 30 April to 15 August. For these 107 days, we obtained cloud-free data on an average of 18.8 days. In parts of Reindalen, up to 29 days had cloud-free data (Figure 3). Only 4.2% of the pixels had more than 10 days since the last cloud-free day, leading to few gaps in the time-series, and these gaps were mainly in the northwestern parts of the study area between 29 June and 16 July.

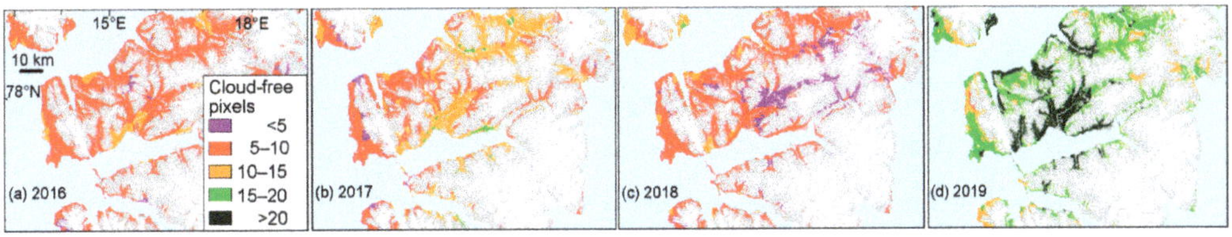

Figure 3. Central Svalbard. Number of cloud-free pixels for the season: (**a**) 2016; (**b**) 2017; (**c**) 2018; (**d**) 2019.

3.2. Timing of the Onset of Growth

The calculation of the S2-NDVI-based onset of growth for each of the seven polygons, representing the seven vegetation types with phenocams, were compared with the phenocam-based records of onset of growth (Table 2). The correlation between these two methods was highly significant (Figure 4), and on average, the NDVI-based onset of growth occurs one day later than the phenocam-based (day of year 171 vs. 170, 19 vs. 20 June).

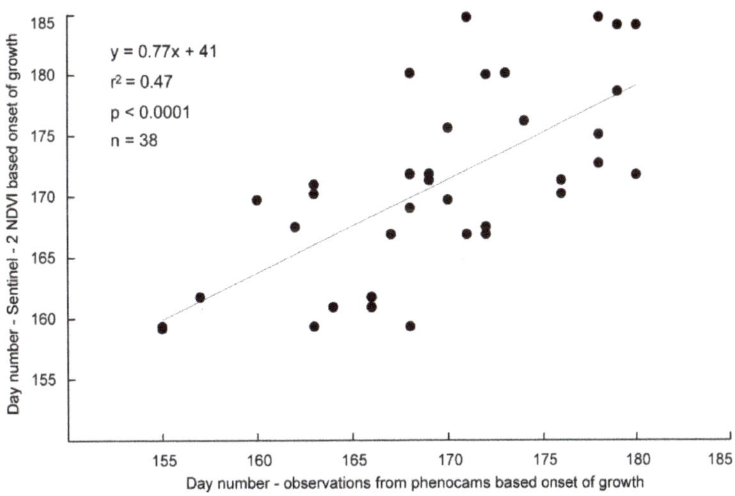

Figure 4. Relationship between the phenocam-based observations of onset of growth and the S2-NDVI-based onset of growth.

Some data are missing in Table 2. *Dupontia fisheri* was monitored in field only in 2016, and *Equisetum arvense* ssp. *alpestre* only in 2018 and 2019. The other missing data is due to technical problems with the cameras, or because the cameras fell down during the season.

The earliest onset of growth recorded by phenocams was on *D. octopetala* in 2018 (doy 155/4 June), in the *D. octopetala–C. tetragona* tundra in Endalen, and this polygon also had the earliest S2 NDVI-based onset (doy 159/8 June). In general, the latest onset of growth measured both by phenocam and S2 NDVI was found in 2017. Among the seven vegetation types, there were some differences between species and years in the onset of growth, as observed by phenocams. In general, *Luzula confusa* in the mixed exposed tundra on the mountain plateau, and *E. arvense* in the snowbed site had a late onset of growth compared with the other species/sites, while *D. octopetala* in the mixed *D. octopetala–C. tetragona* tundra in the relatively warm Endalen had an early onset of growth, followed by *D. octopetala* in the *D. octopetala* tundra (Table 2).

Table 2. Onset of growth (day of year) as measured from phenocams (left value), and from S2 NDVI (right value) for the seven polygons representing seven vegetation types, see location in Figure 5.

Vegetation Types/Species	2016	2017	2018	2019
1. Mixed exposed tundra				
Luzula confusa	178/175	179/179	174/176	
2. *Equsetum arvense* snowbed				
Equsetum arvense ssp. *alpestre*			180/172	178/173
3. *Dupontia fisheri* marsh				
Dupontia fisheri	170/176			
4. *Dryas octopetala* tundra				
Dryas octopetala	160/170		155/159	162/168
Salix polaris	170/170			172/168
5. *Luzula confusa* tundra				
Luzula confusa	168/172	168/180	164/161	168/172
Salix polaris	169/172	173/180	166/161	169/172
6. Moss tundra				
Dryas octopetala	163/170	171/185	157/162	168/169
Salix polaris	176/170	178/185	166/162	
7. Mixed *D. octopetala*–*C. tetragona* tundra				
Dryas octopetala	163/171	172/180	155/159	167/167
Betula nana	169/171	179/184	168/159	171/167
Salix polaris	176/171	180/184	163/159	172/167

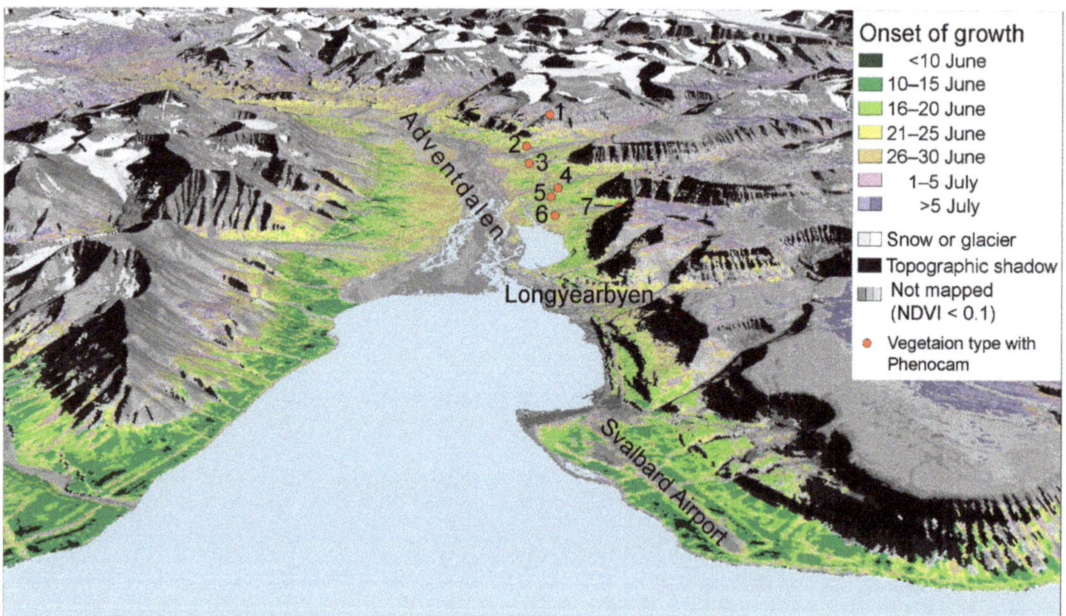

Figure 5. Longyearbyen–Adventdalen valley area seen from northwest towards southeast. Showing mean onset of growth for the 2016 to 2018 period, and the placement of the phenocams.

The largest bias and variability between the phenocam and NDVI-based onset of growth were found in the moss tundra, where the scattered occurrences of vascular plants only contribute very little to the NDVI signal. In the moss tundra, the phenocam-observed onset of growth in vascular plants occurs up to 14 days earlier and up to 5 days later compared with the S2-NDVI-based onset.

A large bias of timing was also seen in the *Equisetum* snowbed, where the moss *Sanionia uncinata* dominate the ground layer, which had phenocams in 2018 and 2019, and where the phenocams showed 8- and 5-day later onset of growth, respectively, compared with the S2 NDVI-based onset (Table 2).

3.3. Onset of Growth

Figure 5 shows a 3D view of the mean (2016–2019) onset of growth of the Longyearbyen–Adventdalen valley area and illustrates the level of detail used for mapping the onset of growth. An early average onset of growth (<15 June) is found at the coast and in areas with favorable slope (e.g., facing S or SW) and exposure at lower altitudes. The map also reveals a clear altitude gradient, with the onset of growth in July at higher altitudes.

In 2019, the cloud-free S2 data availability was high, and the onset of growth could be mapped in most of the study area (Figure 6d), while in other years (2016–2018), extensive areas could not be mapped for growth onset due to gaps in the S2 NDVI time-series during the spring period (Figure 6a–c). The mapping reveals large differences in the onset of growth between the years. The year 2018 had an early onset, mostly before 15 June at lower altitudes. The year 2016 had an early onset of growth in the western parts, but more average in the eastern parts of the study area. The year 2017 had a very late onset, in early July in most of the study area, except along the coast and in parts of Sassendalen and Agardhdalen, which had an onset of growth in mid-June. In 2019, the onset of growth was close to the 2016–2019 average. In Adventdalen valley, the onset of growth was more than 10 days earlier in 2018 compared to 2017, while at higher altitudes, there were much lower differences between these years.

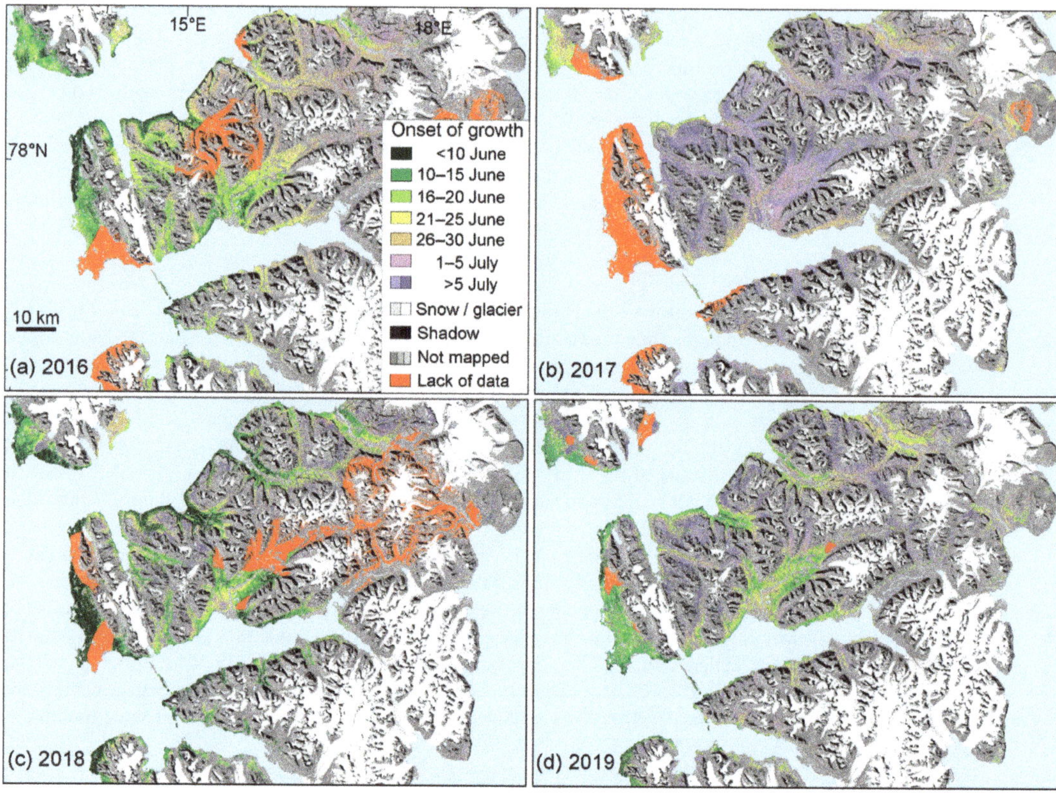

Figure 6. Central Svalbard. Onset of growth: (**a**) 2016; (**b**) 2017; (**c**) 2018; (**d**) 2019.

4. Discussion

Svalbard has a short growing season, and when green-up occurs, it progresses very rapidly. Hence, it is important to use all cloud-free data to gain an accurate date for the onset of growth. The 2018 season had above-normal cloud cover, and on average only 7.2 cloud-free days were found, which is marginal for processing a NDVI curve. In 2018, the dataset had many gaps (>10 days since last cloud-free pixel), and relying on cloud detection algorithms to clean the dataset would lead too to many errors to create a NDVI curve, at least for most parts of the study area. On the other hand, for the 2019 season, more than 20 cloud-free days were found in some parts of the study area, and for these areas in the 2019 season, we believe cloud detection algorithms could be used, since a suitable gap-filling and smoothing method would 'erase' most of the errors in the cloud masking. Further, including Landsat-8 data [36], which has a close to daily revisit time in the study area, would increase the temporal sampling and reduce the need for visual inspections of cloud cover. In particular, in 2018, the year with less cloud-free data, even after removing large gaps in the NDVI time-series (>10 days since last cloud-free pixel), some errors are found, occurring as unrealistic transitions in the onset of growth maps (Figure 6c).

Previously, Karlsen et al. [8] found that the peak NDVI on Svalbard had much less correlation with plant productivity compared to an integrated NDVI value from the onset to peak of growth, due to the cloudy weather with rarely cloud-free pixels during the time of peak NDVI. This is also the case for this dataset.

Several methods for extracting phenophases from NDVI time-series exist [23]. This study used a NDVI threshold method to detect and map the onset of growth. In high Arctic areas with snow cover for most of the year and a long period with polar night, methods with the use of the NDVI data only for the snow-free period were shown to be the most accurate [7]. This study showed that the method performs well for all vegetation types studied except the moss tundra.

The NDVI time-series could be improved by including atmospheric correction (from top-of-atmosphere to bottom-of-atmosphere) dealing with the low solar elevation in the study area [26]. Further, NDVI is sensitive to changes in the solar angle, and correcting the dataset with a bidirectional reflectance distribution function (BRDF) would give more accurate NDVI values. However, by including atmospheric correction and BRDF correctionwould only to some degree influence the NDVI trend [37], but the threshold value for estimating the onset of growth then must be adjusted.

The extraction of phenophases from phenocam images on tripods works well in most cases. A few of the cameras fell down during the season, probably kicked down by reindeer. The exact interpretation of the phenocam photos of unfolded leaves on the shrubs *Salix polaris* was difficult due to its small size. However, due to the rapid growth in spring, the period from first leaf development to full-size leaf is often less than two weeks, and by comparing images between the years, to ensure equal interpretation, we believe the inaccuracy in the timing of the phenophase BBCH code 15 [25] used in this study is less than 4 days, and this is in accordance with Anderson et al. [10], who showed a very rapid and similar development for *Salix polaris* in 2016 using both phenocams and Decagon NDVI sensors (8–20 June). Moreover, visual observations in the field in Adventdalen in 2015 at 60 permanent plots gave dates of green-up within a similar timeframe [14], supporting the findings presented here.

We had two phenocams situated in a moss tundra, with only scattered occurrences of vascular plants. These few vascular plants contribute insignificantly to the NDVI values, and we found large bias between the S2 NDVI-based onset of growth and the phenophase BBCH code 15 observed in field on vascular plant. This is an indication that the moss tundra does not directly follow the phenophases of the vascular plants, and here, Anderson et al. [10] showed that the moss tundra had a later greening than vegetation types dominated or characterized by vascular plants such as *Salix polaris*, *Dryas octopetala*, *Cassiope tetragona* and *Luzula* spp. Hence, the separate definition of growth season stages should be defined for the moss tundra, and is an important theme for a future study.

Large differences between years in the date of onset of growth were found using both satellite and phenocam methods. It is not within the remit of the data presented in this paper to explain the annual differences, but it is highly likely that the date of snowmelt and the cumulative temperatures experienced by the plants play an important role in this [14].

5. Conclusions

1. In Svalbard, cloud, fog, and haze, in combination with a low sun elevation angle during the short growing season, hinders the acquisition of time-series of ground reflectance data. Cloud detection is hence the most crucial step during the pre-processing of time-series of optical satellite images from such areas.
2. Cloud detection algorithms in S2 data have proven to perform poorly in sparsely vegetated areas (bright surfaces) which are widespread in the study area, and when thin semi-transparent clouds or cloud shadows are present. Thus, additional visual inspection of the visible and SWIR bands was applied to mask-out cloud-free data and ensure as few errors as possible.
3. Normalized Difference Vegetation Index (NDVI) values were calculated for the cloud-free pixels, and interpolated to daily data. A close to complete time-series of daily cloud-free S2 NDVI data could be processed for the 2019 season. For the other years studied (2016–2018) there are several gaps in the time-series. The removal of these pixels resulted in spatial gaps in the onset of growth map.
4. Ground based time-lapse cameras (phenocams) were used within seven vegetation types, and the date of a precisely defined phenophase "onset of growth" with an

accurate botanical definition (BBCH code 15) was extracted from the phenocam images.

5. By applying an NDVI threshold method on the clear-sky time-series of S2 data, the mapping of "onset of growth", shows a significant correlation ($r^2 = 0.47$, $n = 38$, $p < 0.0001$) with timing of onset of growth as defined from the phenocam images.
6. However, in moss tundra where vascular plants play an insignificant role for the NDVI value, this correlation showed large bias; hence, a separate definition of growth season stages should be defined for the moss tundra.
7. The S2 NDVI-based mapping of onset of growth reveals large differences between the years. In 2018, the onset of growth was more than 10 days earlier compared with 2017, except at higher altitudes. The data presented in this paper are not sufficient to explain these differences, but a future study will examine the relationship between the timing of snowmelt and early growing season temperature.

Author Contributions: Conceptualization, S.R.K.; methodology, S.R.K. and L.S.; software, S.R.K., L.S. and I.A.; validation, S.R.K., L.S. and L.N.; formal analysis, S.R.K.; investigation, S.R.K.; resources, S.R.K., H.T. and L.N.; writing—original draft preparation, S.R.K., H.T., L.N., I.A. and E.J.C.; writing—review and editing, S.R.K., H.T. and E.J.C.; visualization, S.R.K.; project administration, S.R.K.; funding acquisition, S.R.K., H.T. and L.N. All authors have read and agreed to the published version of the manuscript.

Funding: This work (S.R.K. and L.N.) was supported by the Research Council of Norway under the project Svalbard Integrated Arctic Earth Observing System—Infrastructure development of the Norwegian node (SIOS-InfraNor Project No. 269927). This SIOS project (InfraNord instrument #52) is funded by the Norwegian Space Agency (NoSA). H.T. was supported by European Commission Research and Innovation Action project CHARTER no. 869471.

Institutional Review Board Statement: Not applicable.

Informed Consent Statement: Not applicable.

Acknowledgments: We are grateful to senior advisor John Richard Hansen, at the Norwegian Polar Institute, who in 2010–2016 initiated and supported satellite-based phenological monitoring on Svalbard as a part of the Environmental Monitoring of Svalbard and Jan Mayen (MOSJ). That previous work gave valuable experience for this study. Copernicus Sentinel-2 data were retrieved from ESA SciHub.

Conflicts of Interest: The authors declare no conflict of interest.

References

1. Isaksen, K.; Nordli, Ø.; Førland, E.J.; Lupikasza, E.; Eastwood, S.; Niedźwiedź, T. Recent warming on Spitsbergen—Influence of atmospheric circulation and sea ice cover. *J. Geophys. Res. Atmos.* **2016**, *121*. [CrossRef]
2. Vikhamar-Schuler, D.; Isaksen, K.; Haugen, J.E.; Tømmervik, H.; Luks, B.; Schuler, T.; Bjerke, J. Changes in Winter Warming Events in the Nordic Arctic Region. *J. Clim.* **2016**, *29*, 6223–6244. [CrossRef]
3. Hanssen-Bauer, I.; Førland, E.; Hisdal, H.; Mayer, S.; Sandø, A.; Sorteberg, A. *Climate in Svalbard 2100—A Knowledge Base for Climate Adaptation*; NCCS Report no.1/2019; Norwegian Meteorological Institute: Oslo, Norway, 2019.
4. Park, T.; Ganguly, S.; Tømmervik, H.; Euskirchen, E.S.; Høgda, K.-A.; Karlsen, S.R.; Brovkin, V.; Nemani, R.R.; Myneni, R. Changes in growing season duration and productivity of northern vegetation inferred from long-term remote sensing data. *Environ. Res. Lett.* **2016**, *11*, 084001. [CrossRef]
5. Park, T.; Chen, C.; Macias-Fauria, M.; Tømmervik, H.; Choi, S.; Winkler, A.J.; Bhatt, U.; Walker, D.A.; Piao, S.; Brovkin, V.; et al. Changes in timing of seasonal peak photosynthetic activity in northern ecosystems. *Glob. Chang. Biol.* **2019**, *25*, 2382–2395. [CrossRef]
6. Vickers, H.; Karlsen, S.R.; Malnes, E. A 20-Year MODIS-Based Snow Cover Dataset for Svalbard and Its Link to Phenological Timing and Sea Ice Variability. *Remote. Sens.* **2020**, *12*, 1123. [CrossRef]
7. Karlsen, S.R.; Elvebakk, A.; Høgda, K.A.; Grydeland, T. Spatial and Temporal Variability in the Onset of the Growing Season on Svalbard, Arctic Norway Measured by MODIS-NDVI Satellite Data. *Remote. Sens.* **2014**, *6*, 8088–8106. [CrossRef]
8. Karlsen, S.R.; Anderson, H.B.; Van Der Wal, R.; Hansen, B. A new NDVI measure that overcomes data sparsity in cloud-covered regions predicts annual variation in ground-based estimates of high arctic plant productivity. *Environ. Res. Lett.* **2018**, *13*, 025011. [CrossRef]

9. Macias-Fauria, M.; Karlsen, S.R.; Forbes, B. Disentangling the coupling between sea ice and tundra productivity in Svalbard. *Sci. Rep.* **2017**, *7*, 1–10. [CrossRef]
10. Anderson, H.B.; Nilsen, L.; Tømmervik, H.; Karlsen, S.R.; Nagai, S.; Cooper, E.J. Using Ordinary Digital Cameras in Place of Near-Infrared Sensors to Derive Vegetation Indices for Phenology Studies of High Arctic Vegetation. *Remote. Sens.* **2016**, *8*, 847. [CrossRef]
11. Johansen, B.E.; Karlsen, S.R.; Tømmervik, H. Vegetation mapping of Svalbard utilising Landsat TM/ETM+ data. *Polar Rec.* **2011**, *48*, 47–63. [CrossRef]
12. Bjerke, J.W.; Treharne, R.; Vikhamar-Schuler, D.; Karlsen, S.R.; Ravolainen, V.; Bokhorst, S.; Phoenix, G.K.; Bochenek, Z.; Tømmervik, H. Understanding the drivers of extensive plant damage in boreal and Arctic ecosystems: Insights from field surveys in the aftermath of damage. *Sci. Total. Environ.* **2017**, *599–600*, 1965–1976. [CrossRef]
13. Karlsen, S.R.; Stendardi, L.; Nilsen, L.; Malnes, E.; Eklundh, L.; Julitta, T.; Burkart, A.; Tømmervik, H. Sentinel Satellite-Based Mapping of Plant Productivity in Relation to Snow Duration and Time of Green-up (GROWTH). In *SESS Report 2019 The State of Environmental Science in Svalbard—An Annual Report*; Van den Heuvel, F., Hübner, C., Błaszczyk, M., Heimann, M., Lihavainen, H., Eds.; Svalbard Integrated Arctic Earth Observing System: Longyearbyen, Norway, 2020; pp. 42–54.
14. Semenchuk, P.R.; Gillespie, M.A.K.; Rumpf, S.; Baggesen, N.; Elberling, B.; Cooper, E.J. High Arctic plant phenology is determined by snowmelt patterns but duration of phenological periods is fixed: An example of periodicity. *Environ. Res. Lett.* **2016**, *11*, 125006. [CrossRef]
15. Oberbauer, S.F.; Elmendorf, S.; Troxler, T.G.; Hollister, R.; Rocha, A.V.; Bret-Harte, M.S.; Dawes, M.A.; Fosaa, A.M.; Henry, G.H.R.; Hoye, T.T.; et al. Phenological response of tundra plants to background climate variation tested using the International Tundra Experiment. *Philos. Trans. R Soc. B Biol. Sci.* **2013**, *368*, 20120481. [CrossRef] [PubMed]
16. Prevéy, J.; Vellend, M.; Rüger, N.; Hollister, R.; Bjorkman, A.; Myers-Smith, I.H.; Elmendorf, S.; Clark, K.; Cooper, E.J.; Elberling, B.; et al. Greater temperature sensitivity of plant phenology at colder sites: Implications for convergence across northern latitudes. *Glob. Chang. Biol.* **2017**, *23*, 2660–2671. [CrossRef] [PubMed]
17. Prevéy, J.S.; Rixen, C.; Rüger, N.; Høye, T.T.; Bjorkman, A.D.; Myers-Smith, I.H.; Elmendorf, S.; Ashton, I.W.; Cannone, N.; Chisholm, C.L.; et al. Warming shortens flowering seasons of tundra plant communities. *Nat. Ecol. Evol.* **2018**, *3*, 45–52. [CrossRef]
18. Parmentier, F.-J.; Nilsen, L.; Tømmervik, H.; Cooper, E. A Distributed Time-Lapse Camera Network to Track Vegetation Phenology with High Temporal Detail and at Varying Scales. *Earth Syst. Sci. Data Discus.* **2021**, *13*, 3593–3606. [CrossRef]
19. Westergaard-Nielsen, A.; Lund, M.; Hansen, B.; Tamstorf, M. Camera derived vegetation greenness index as proxy for gross primary production in a low Arctic wetland area. *ISPRS J. Photogramm. Remote. Sens.* **2013**, *86*, 89–99. [CrossRef]
20. Stendardi, L.; Karlsen, S.R.; Niedrist, G.; Gerdol, R.; Zebisch, M.; Rossi, M.; Notarnicola, C. Exploiting Time Series of Sentinel-1 and Sentinel-2 Imagery to Detect Meadow Phenology in Mountain Regions. *Remote. Sens.* **2019**, *11*, 542. [CrossRef]
21. Descals, A.; Verger, A.; Yin, G.; Peñuelas, J. Improved Estimates of Arctic Land Surface Phenology Using Sentinel-2 Time Series. *Remote. Sens.* **2020**, *12*, 3738. [CrossRef]
22. Tian, F.; Cai, Z.; Jin, H.; Hufkens, K.; Scheifinger, H.; Tagesson, T.; Smets, B.; Van Hoolst, R.; Bonte, K.; Ivits, E.; et al. Calibrating vegetation phenology from Sentinel-2 using eddy covariance, PhenoCam, and PEP725 networks across Europe. *Remote. Sens. Environ.* **2021**, *260*, 112456. [CrossRef]
23. Misra, G.; Cawkwell, F.; Wingler, A. Status of Phenological Research Using Sentinel-2 Data: A Review. *Remote. Sens.* **2020**, *12*, 2760. [CrossRef]
24. Johansen, B.; Tømmervik, H. The relationship between phytomass, NDVI and vegetation communities on Svalbard. *Int. J. Appl. Earth Obs. Geoinf.* **2014**, *27*, 20–30. [CrossRef]
25. Meier, U. *Growth Stages of Mono—And Dicotyledonous Plants, BBCH Monograph*; Julius Kühn-Institut: Quedlinburg, Germany, 2018; pp. 1–204.
26. Clerc, S. *S2 MPC Level 2A Data Quality Report, PDGS-MPC-L2ADQR*, 39th ed.; ESA Copernicus: Lima, Peru, 2021; pp. 1–28.
27. Zupanc, A. Improving Cloud Detection with Machine Learning. Available online: https://medium.com/sentinel-hub/improving-cloud-detection-with-machine-learning-c09dc5d7cf13 (accessed on 4 May 2021).
28. Hollstein, A.; Segl, K.; Guanter, L.; Brell, M.; Enesco, M. Ready-to-Use Methods for the Detection of Clouds, Cirrus, Snow, Shadow, Water and Clear Sky Pixels in Sentinel-2 MSI Images. *Remote. Sens.* **2016**, *8*, 666. [CrossRef]
29. Chen, J.; Jönsson, P.; Tamura, M.; Gu, Z.; Matsushita, B.; Eklundh, L. A simple method for reconstructing a high-quality NDVI time-series data set based on the Savitzky–Golay filter. *Remote. Sens. Environ.* **2004**, *91*, 332–344. [CrossRef]
30. Vickers, H.; Malnes, E.; Van Pelt, W.; Pohjola, V.; Killie, M.; Saloranta, T.; Karlsen, S. A Compilation of Snow Cover Datasets for Svalbard: A Multi-Sensor, Multi-Model Study. *Remote. Sens.* **2021**, *13*, 2002. [CrossRef]
31. Karlsen, S.R.; Elvebakk, A.; Høgda, K.A.; Johansen, B. Satellite-based mapping of the growing season and bioclimatic zones in Fennoscandia. *Glob. Ecol. Biogeogr.* **2006**, *15*, 416–430. [CrossRef]
32. Karlsen, S.R.; Solheim, I.; Beck, P.S.A.; Høgda, K.A.; Wielgolaski, F.E.; Tømmervik, H. Variability of the start of the growing season in Fennoscandia, 1982–2002. *Int. J. Biometeorol.* **2007**, *51*, 513–524. [CrossRef]
33. Karlsen, S.R.; Tolvanen, A.; Kubin, E.; Poikolainen, J.; Høgda, K.A.; Johansen, B.; Danks, F.S.; Aspholm, P.; Wielgolaski, F.E.; Makarova, O. MODIS-NDVI-based mapping of the length of the growing season in northern Fennoscandia. *Int. J. Appl. Earth Obs. Geoinf.* **2008**, *10*, 253–266. [CrossRef]

34. Høgda, K.A.; Tømmervik, H.; Karlsen, S.R. Trends in the Start of the Growing Season in Fennoscandia 1982–2011. *Remote. Sens.* **2013**, *5*, 4304–4318. [CrossRef]
35. Hogda, K.A.; Karlsen, S.R.; Solheim, I. Climatic change impact on growing season in Fennoscandia studied by a time series of NOAA AVHRR NDVI data. In *IEEE 2001 International Geoscience and Remote Sensing Symposium (Cat. No. 01CH37217)*; IEEE: Piscataway, NJ, USA, 2002. [CrossRef]
36. Bolton, D.K.; Gray, J.; Melaas, E.K.; Moon, M.; Eklundh, L.; Friedl, M.A. Continental-scale land surface phenology from harmonized Landsat 8 and Sentinel-2 imagery. *Remote. Sens. Environ.* **2020**, *240*, 111685. [CrossRef]
37. Seong, N.-H.; Jung, D.; Kim, J.; Han, K.-S. Evaluation of NDVI Estimation Considering Atmospheric and BRDF Correction through Himawari-8/AHI. *Asia Pac. J. Atmos. Sci.* **2020**, *56*, 265–274. [CrossRef]

remote sensing

MDPI

Article

Impact of Image-Processing Routines on Mapping Glacier Surface Facies from Svalbard and the Himalayas Using Pixel-Based Methods

Shridhar D. Jawak [1], Sagar F. Wankhede [2,*], Alvarinho J. Luis [3] and Keshava Balakrishna [2]

[1] Svalbard Integrated Arctic Earth Observing System (SIOS), SIOS Knowledge Centre, P.O. Box 156, N-9171 Longyearbyen, Svalbard, Norway; shridhar.jawak@sios-svalbard.org

[2] Department of Civil Engineering, Manipal Institute of Technology, Manipal Academy of Higher Education, Manipal 576104, Karnataka, India; k.balakrishna@manipal.edu

[3] Earth System Sciences Organization, National Centre for Polar and Ocean Research (NCPOR), Ministry of Earth Sciences, Government of India, Headland Sada, Vasco-da-Gama 403804, Goa, India; alvluis@ncpor.res.in

* Correspondence: sagar.wankhede@learner.manipal.edu

Citation: Jawak, S.D.; Wankhede, S.F.; Luis, A.J.; Balakrishna, K. Impact of Image-Processing Routines on Mapping Glacier Surface Facies from Svalbard and the Himalayas Using Pixel-Based Methods. *Remote Sens.* **2022**, *14*, 1414. https://doi.org/10.3390/rs14061414

Academic Editor: Alberto Godio

Received: 24 December 2021
Accepted: 9 March 2022
Published: 15 March 2022

Abstract: Glacier surface facies are valuable indicators of changes experienced by a glacial system. The interplay of accumulation and ablation facies, followed by intermixing with dust and debris, as well as the local climate, all induce observable and mappable changes on the supraglacial terrain. In the absence or lag of continuous field monitoring, remote sensing observations become vital for maintaining a constant supply of measurable data. However, remote satellite observations suffer from atmospheric effects, resolution disparity, and use of a multitude of mapping methods. Efficient image-processing routines are, hence, necessary to prepare and test the derivable data for mapping applications. The existing literature provides an application-centric view for selection of image processing schemes. This can create confusion, as it is not clear which method of atmospheric correction would be ideal for retrieving facies spectral reflectance, nor are the effects of pansharpening examined on facies. Moreover, with a variety of supervised classifiers and target detection methods now available, it is prudent to test the impact of variations in processing schemes on the resultant thematic classifications. In this context, the current study set its experimental goals. Using very-high-resolution (VHR) WorldView-2 data, we aimed to test the effects of three common atmospheric correction methods, viz. Dark Object Subtraction (DOS), Quick Atmospheric Correction (QUAC), and Fast Line-of-Sight Atmospheric Analysis of Hypercubes (FLAASH); and two pansharpening methods, viz. Gram–Schmidt (GS) and Hyperspherical Color Sharpening (HCS), on thematic classification of facies using 12 supervised classifiers. The conventional classifiers included: Mahalanobis Distance (MHD), Maximum Likelihood (MXL), Minimum Distance to Mean (MD), Spectral Angle Mapper (SAM), and Winner Takes All (WTA). The advanced/target detection classifiers consisted of: Adaptive Coherence Estimator (ACE), Constrained Energy Minimization (CEM), Matched Filtering (MF), Mixture-Tuned Matched Filtering (MTMF), Mixture-Tuned Target-Constrained Interference-Minimized Filter (MTTCIMF), Orthogonal Space Projection (OSP), and Target-Constrained Interference-Minimized Filter (TCIMF). This experiment was performed on glaciers at two test sites, Ny-Ålesund, Svalbard, Norway; and Chandra–Bhaga basin, Himalaya, India. The overall performance suggested that the FLAASH correction delivered realistic reflectance spectra, while DOS delivered the least realistic. Spectra derived from HCS sharpened subsets seemed to match the average reflectance trends, whereas GS reduced the overall reflectance. WTA classification of the DOS subsets achieved the highest overall accuracy (0.81). MTTCIMF classification of the FLAASH subsets yielded the lowest overall accuracy of 0.01. However, FLAASH consistently provided better performance (less variable and generally accurate) than DOS and QUAC, making it the more reliable and hence recommended algorithm. While HCS-pansharpened classification achieved a lower error rate (0.71) in comparison to GS pansharpening (0.76), neither significantly improved accuracy nor efficiency. The Ny-Ålesund glacier facies were best classified using MXL (error rate = 0.49) and WTA classifiers (error rate = 0.53), whereas the Himalayan glacier facies were best classified using MD (error rate = 0.61) and WTA (error rate = 0.45). The final comparative analysis of classifiers based on the total error rate across all atmospheric corrections and pansharpening methods yielded the following reliability order: MXL > WTA > MHD > ACE > MD > CEM = MF > SAM > MTMF = TCIMF > OSP > MTTCIMF. The findings of the current study suggested that for VHR visible near-infrared (VNIR) mapping of facies, FLAASH was the best atmospheric correction, while MXL may deliver reliable thematic classification. Moreover, an extensive account of the varying exertions of each processing scheme is discussed, and could be transferable when compared against other VHR VNIR mapping methods.

Keywords: glacier facies; atmospheric correction; pansharpening; WorldView-2; Ny-Ålesund; Chandra–Bhaga basin; target detection; supervised classification

1. Introduction

Remotely sensed observations of glaciers are an efficient means of monitoring the overall changes occurring in the cryosphere. Partly or fully inaccessible glacial regions have greatly benefitted from temporal and resolution advancements in remote sensing (RS) technology. Multispectral observations of glaciers have led to the development of a range of methods for mapping glacier extents [1,2], deriving albedo [3,4], monitoring of the equilibrium line altitude (ELA) [5,6], surface temperature [7,8], and identification of glacier zones [9,10]. Zones of a glacier refer to the distinct variations of snow and ice found on a glacier's surface, occurring due to the natural accumulation, melt, refreezing, and flow of precipitated snow. These zones are also called facies. Different facies exhibit different reflectance characteristics, which can be monitored using multispectral sensors [11,12], SAR sensors [13,14], and radiometric observations [15,16], and are visually discernible in very-high-resolution (VHR) data [17]. Hence, mapping facies is usually tasked on a variety of RS products. Reflectance-based RS products are extremely versatile and can be incorporated into calibrating distributed mass balance models [18], spectral libraries, development of indices, and testing of methodologies, and can be compared band by band against existing literature [17]. Prior to identifying facies on a VHR multispectral image, a methodical preprocessing protocol of the image is of paramount importance to minimize signal errors and extract maximum information. Processes such as deriving reflectance and enhancing spatial resolution are necessary when looking for details such as sporadic distribution of snow. The current study aimed to map facies on glaciers of two distinct cryosphere zones while determining the best methods of pre-processing and band selection using VHR data. The following literature review presents an account of glacier facies, data preparation, and mapping methods to ascertain the objectives of this study.

1.1. Glacier Facies

First documented by Benson [19], the concept of glacier facies was thoroughly described by Pope and Rees [10]. Concisely stated, glaciological facies are the variations of snow and ice in the accumulation and ablation zones that stretch within the body of the glacier and can differ across seasons and years. However, the range of facies available on a glacier is most efficiently observed at the end of the ablation season. Surface observations of facies often are the intermix of supraglacial debris, particulate matter, crevasses, and meltwater streams, which represent the individual zones. Therefore, Jawak et al. [17] used the term "surface glacier facies", derived from the surface classes used by Pope and Rees [10]. Glacier facies mapped through Synthetic Aperture Radar (SAR) data are often called radar zones [20] or radar facies [21]. Barzycka et al. [13] derived the changes of glacier facies on Hornsund glaciers in Svalbard using unsupervised classification of multisensor SAR data and compared it with ground-penetrating radar (GPR) observations and excavated ice cores. Like Barzycka et al. [22], they advocated the use of the Internal Reflection Energy (IRE) derived from GPR data for validation of SAR mapping and validation of facies extents. Mapping facies using SAR is usually conducted for identifying snowlines, firn lines, or the equilibrium line altitude (ELA), and/or monitoring glacier extents predominantly during winter months or early spring. However, the full range of ablation facies do not appear on the surface of the glacier until the end of summer, when the temporary snow cover is at its minimum. Cloud-free optical remote sensing, on the other hand, relies on ablation/summer season data to obtain maximum information of the available range of facies.

1.2. Multispectral Mapping of Glacier Facies

Supraglacial terrain classification usually falls into three broad categories: 1. debris and their associated phenomena; 2. identification of generalized facies; and 3. sensor-specific responses and methodologies for mapping facies. A large volume of literature is devoted to mapping glacial debris and understanding its associated complexities [8,23–28]. Usually, debris mapping entails usage of shortwave infrared (SWIR), and thermal infrared (TIR) bands in addition to the visible NIR (VNIR) range of optical data [29]. Elevation data is also necessary to adjust for topographic influence on retrieved debris characteristics [25]. However, Jawak et al. [17] mapped ice mixed debris and debris facies using only the VNIR range of the spectrum. Yousuf et al. [15] described the distinctions between studies in which mapping glacier facies was the main aim and studies in which facies were mapped due to a different primary objective such as outlining glacial extents. We focused on studies wherein mapping facies was the primary objective. After Dozier [30] highlighted the utility of Landsat TM for obtaining reflectance characteristics of snow, Hall et al. [31] and Williams et al. [12] used Landsat TM band ratios to distinguish between the reflectance of glacier facies and the terrain. Several TM band ratios were used to identify bare ice, debris-covered ice, slush, two facies of snow, and shadows in supervised and unsupervised classification schemes [32]. Summer facies derived by an ISODATA algorithm by Braun et al. [18] using TM and ETM+ bands were employed to validate distributed mass balance modelling. The potentials of band ratios of Landsat 8 (operational land imager) OLI, and a thermal infrared sensor (TIRS) were tested for mapping clean glacier ice, dirty glacier ice, slush zone, snow, and supraglacial debris [9]. Using pansharpened and atmospherically corrected imagery, the authors were also able to map crevasses, and observed a reduction in the derived at-satellite brightness temperature. However, Jawak et al. [17,33], who used 2 m resolution WorldView-2 (WV-2) data, were able to map crevasses using customized indices in the VNIR range. Their study focused on comparing pixel- and object-based methods of mapping facies using VHR data. Ali et al. [34] mapped spatiotemporal variation in facies in the Indian Himalayas using a range of optical sensors by creating ancillary layers using band ratios, elevation, and thermal data. Similar multisensor image and ancillary layer-based methods were tested by Shukla and Ali [35] and Yousuf et al. [15,36]. Pope and Rees [10] used Airborne Thematic Mapper (ATM) imagery, in situ spectral reflectance, and Landsat ETM+ imagery over Midtre Lovénbreen to map the facies using linear combinations of principal components derived from the spectral signatures. They highlighted the importance of sensor-specific indices, particularly in the VNIR range, for the most efficient surface classification of facies. Paul et al. [11] mapped facies using Sentinel 2A imagery and compared the results against Landsat 8. Their results suggested that a higher resolution would yield a higher-quality product. Optical remote sensing of glacier facies is dependent on sensor and scene characteristics, resolution of data, mapping technique, and ancillary information. Finer resolution scenes and processing parameters will invariably lead to better-quality facies products.

1.3. Pansharpening

Visual identification of glacier facies requires fine spatial resolution for observing textural differences and geometric characteristics, and good spectral resolution for associating the textural and tonal changes with reflectance characteristics of target facies. Pansharpening, the process of fusing panchromatic (PAN) and multispectral (MS) images, retains the spectral diversity of MS data while integrating the spatial sharpness of PAN data [37]. In addition to being a common image manipulation method prior to information extraction, enhancing spatial resolution is of paramount importance when the features are relatively small, or the terrain is homogenous. Xu et al. [38] compared pansharpened and non-pansharpened soil spectral indices on images from WV-2, Landsat 8, and GeoEye-1 to create soil mineral indices. After testing the Brovey, Gram–Schmidt (GS), and IHS methods, they inferred that the GS method was better at identifying structural and textural details. A recent review [39] suggested that the GS pansharpening method was the most optimal choice among the available methods. In glacial areas, the GS method delivered higher accuracies for Jawak and Luis [40], who tested it against other pansharpening methods by developing land cover mapping indices using WV-2 imagery. GS-sharpened imagery has also enabled minute-scale vegetation mapping in Antarctica [41]. Jawak et al. [17,33] identified glacier facies in the Himalayas using GS-sharpened WV-2 data by devising customized spectral index ratios. Although the GS method has proven to be a reliable method of pansharpening high-resolution images, Hyperspherical Color Space (HCS) sharpening was developed by Padwick et al. [42] specifically for WV-2 imagery. Their tests revealed that the HCS method retained high spectral and spatial performance against the GS, IHS, and PCA methods [42]. Wyzcalek and Wyzcalek [43] tested the efficacy of PCA against weighted HCS pansharpening in an object-oriented domain using NDVI thresholds to classify the segmented objects. Their results suggested that that the weighted HCS performed better than the PCA. However, other studies that compared HCS against other pansharpening methods suggested that it yielded quantitatively inferior results to methods such as Fuse Go and Ehlers, which retained better spatial and spectral details [44]. Snehamani et al. [45] compared 27 pansharpening algorithms, including the HCS, using QuickBird and WV-3 images captured over urban settings. Their findings suggested that the selection of pansharpening methods must be sensor- and scene-specific. This agreed with the observations of Nikolakopoulos and Oikonomidis [46]. Rayegani et al. [47] also arrived at the same conclusion while noting that the HCS method could induce some pepper noise effect, but it closely retained the histogram of the original image. It is worth noting that Wu et al. [48] proposed an enhanced HCS to correct some of the spatial distortion produced in the sharpened imagery. However, the effects of the HCS sharpening were not tested over glacierized landscapes, whereas the GS method was proven to be effective in the same settings. Furthermore, the effects of these methods on identifying glacier facies have not yet been observed.

1.4. Atmospheric Correction

Conversion of digital brightness numbers to at sensor radiance and subsequently to surface/apparent surface reflectance is an important image preprocessing step in any reflectance-based feature identification protocol. Atmospheric correction aims to resolve the influence of scattering and absorption by atmospheric molecules and aerosols occurring in the Field-of-View (FOV) of the acquiring sensor. Several atmospheric correction models have been developed through empirical statistical methods and atmosphere radiative transfer codes [49]. Some of the most popular atmospheric correction models are Dark Object Subtraction (DOS), the Quick Atmospheric Correction (QUAC), and the Fast-Line-of-Sight Atmospheric Analysis of Spectral Hypercubes (FLAASH). DOS rectifies the additive scattering effect [50], QUAC corrects the multiplicative scattering effect [51], and FLAASH [52] is based on the moderate-resolution atmospheric transmission 4 (MODTRAN4) radiative transfer code [53]. Marcello et al. [54] compared the performance of DOS, QUAC, FLAASH, Atmospheric Correction (ATCOR), and Second Simulation of a Satellite Signal in the Solar Spectrum (6S) models to retrieve vegetation and soil sites in semiarid areas. Their analysis was performed on WV-2 imagery and compared to in situ spectral signatures. They recommended the 6S model for information extraction using vegetation indices. However, except for the blue band, all the atmospherically corrected signatures and in-situ signatures closely matched. Shi et al. [55] observed that the FLAASH corrected reflectance was closest to in situ reflectance when compared to DOS, and QUAC for hyperspectral data over bloom water. This opposed the findings made by Dewi and

Trisakti [56], who found that the FLAASH algorithm delivered inferior soil spectral patterns in comparison to DOS and QUAC. However, when assessed by factoring location and time consistency, they observed that FLAASH had the highest absolute value over DOS and QUAC. A comparative study between ATCOR, FLAASH, and DOS1 [57] for geological mapping in arid and semiarid environments using Landsat 8 data suggested that DOS1 provided a simpler alternative to the other two methods. While the FLAASH algorithm performed slightly better than DOS1, they maintained that usage of DOS1 delivered good performance in complex semiarid regions. Cryosphere studies have adopted all the above methods for retrieving surface reflectance. For example, Guo et al. [58] used FLAASH to retrieve albedo for mapping the spatiotemporal variability of the snow line altitude at the end of the melt season across High Mountain Asia (HMA) glaciers. Albert [59] used DOS to correct atmospheric scattering for ice area classification of an ice cap using Landsat TM 5 imagery. Karimi et al. [8] mapped debris-covered glaciers in Iran using QUAC-corrected satellite data. Jawak et al. [17,33] identified glacier facies in the Chandra basin using FLAASH-rectified WV-2 imagery. Luis and Singh [60] attempted to map facies on the Edithbreen glacier in Svalbard using FLAASH-corrected WV-3 data. All the reviewed literature on atmospheric correction invariably pointed toward selection of a method based solely upon the unique requirements of that study [61]. As the effects of DOS, FLAASH, and QUAC are yet to be observed for mapping of glacier facies in one comprehensive study, it would be premature to suggest one optimal method.

1.5. Research Motivation and Aim

SAR data operates mainly for winter assessment of glacier facies (Section 1.1). This is beneficial for accumulation-area estimations and winter facies of glaciers when the snowpack is mostly frozen, and SAR can penetrate well. However, seasonality of facies relies on summer season data. Summer facies on a glacier imply that most of the ablation zone would be wet. The varying degrees of wetness, thickness, and debris would determine the kinds of surface facies visible. Gore et al. [62] stated that these variable and altitudinal properties of melt exposed the full range of supraglacial features. Thakur et al. [63] suggested that the low penetrating depth of SAR into wet snow was a limiting factor [64,65]. Melt-induced reduction in reflectance [66] would, however, be identifiable in the multispectral bands of optical satellite data. Therefore, utilizing cloud-free optical remote sensing data during summer would greatly complement ongoing SAR efforts and provide reflectance-based products, such as spectral profiles and thematic outputs for further testing. While moderate-resolution mapping of supraglacial features using optical data is conducted at a basin-level [28], high-resolution (HR) data is mainly used for facies mapping on selected glaciers [9,11], with VHR data being used for validation purposes [67]. Efforts made by Luis and Singh [60] using VHR WV-3 data were only for a single glacier. Jawak et al. [17] used VHR WV-2 data, but only for selected glaciers of the same area. Hence, multiregion testing of VHR optical data for mapping glacier facies has not yet been comprehensively performed.

Moreover, optical data necessitates image-rectification procedures such as atmospheric correction (Section 1.4) and pansharpening (Section 1.3). The effect of different atmospheric corrections on resultant reflectance spectra was tested for other applications such as soil and vegetation mapping [54,56], bloom water hyperspectral retrieval [55], and geological mapping [57]. Cryosphere studies have relied on single atmospheric corrections before proceeding to extract albedo [58], classifying ice area [59], mapping debris-covered glaciers [8], monitoring seasonal variations [68], and characterizing glacier facies [15]. This suggested that a comparative assessment of the effects of different atmospheric corrections on resultant reflectance spectra of glacier facies and their subsequent classification has not been conducted. Lee and Yum [61] reviewed research-based usage of different atmospheric corrections and recommended selection of any correction method by evaluating requirements of the study. However, the evaluation for selecting a correction method for glacier facies mapping using VHR data itself has not yet been conducted.

Pansharpening supports better visual characterization of glacial features. This effect can be beneficial when mapping glacial extents and boundaries when coupled with relevant DEMs [9]. Xu et al. [38] observed that the GS pansharpening method was better suited to derive structural and textural details in an ensemble analysis of VHR and MR resolution images when applying soil indices. The same method yielded better results among a comparative study that mapped land cover in polar regions using WV-2 VHR data [40]. However, the HCS method [42] was shown to deliver superior results on segmented objects ratioed using NDVI [43]. Snehamani et al. [45] and Nikolakopoulos and Oikonomidis [46] concluded that pansharpening must be application- and image-oriented. To the best of the authors' knowledge, a testing of the effects of pansharpening on the identification and mapping of glacier facies has not been carried out. While Jawak and Luis [17] and Luis and Singh [60] both used GS to sharpen WV-2 MS data, neither evaluated the effects of HCS, which was developed for WV-2 itself [42]. This study presents an efficient test for comparing the GS, which is purportedly the most suited method for land cover classification (GS) [39] against the HCS when using WV-2 imagery.

Based on the literature described, it was evident that there are research gaps in the schemes of image-processing routines for multispectral mapping of glacier facies. The gaps are summarized as follows: (a) optical VHR data has not been comprehensively tested on multiregion glaciers for mapping facies; (b) the effects of atmospheric correction are yet to be observed on the spectral and thematic results of mapped facies; and (c) the compounding effects of pansharpening on characterizing glacier facies has not been clearly studied. In addition to these, an exhaustive test of conventional and advanced pixel-based classification methods would aid in identifying which algorithms are the most efficient for mapping facies. A thorough evaluation of the effects of atmospheric correction, pansharpening, and various pixel-based classification algorithms on thematic outputs of glacier facies would result in robust recommendations for their operational mapping using VHR multispectral data. This summarizes the motivation for the current study. To accomplish this, the following research aims were set: (1) effective characterization of glacier facies using VHR multispectral data using pixel-based methods; (2) testing the effect of atmospheric correction procedures on glacier facies mapping; and (3) testing the effect of pansharpening methods on glacier facies mapping.

The current study evaluated the FLAASH, QUAC, and DOS atmospheric correction algorithms, and found the FLAASH correction to deliver the best reflectance pattern. GS and HCS pansharpening algorithms were tested, and the HCS was found to deliver the least decrement in spectral reflectance. A total of 12 conventional and advanced classification algorithms were employed to test the effects induced in thematic classification by variations in atmospheric corrections and pansharpening. Among the tested methods, the maximum likelihood classifier delivered the most consistent results across each atmospheric correction and pansharpening method. Based on the results of the study, conventional classifiers were more efficient and delivered higher accuracies in comparison to advanced classification algorithms. The results were consistent across two distinct study areas, Ny-Ålesund, Svalbard; and Chandra–Bhaga basin, Himalayas, using VHR WorldView-2 imagery.

2. Study Area and Data Used

2.1. Spatial Extent of the Test Sites

2.1.1. Site A: Ny-Ålesund, Svalbard

The Nordic archipelago of Svalbard is a pristine mass of glacial landscapes in the Arctic Ocean between 75° and 82°N [69]. An interplay of different oceanic currents and variation in atmospheric circulation causes this landscape to experience climates ranging from continental to coastal, with further fluctuations between winter, spring, and summer months [70]. Currently, this system is one of the most rapidly warming areas on the planet. The rate of its increase in temperature is reportedly double the global average [71]. The direct effect of this warming is visible on its glaciers in the form of glacier thinning [72], recession of perennial snow cover to higher elevations [73], and near-surface densification of the accumulation zone [74]. Due to the aforementioned factors, the entire region is of international scientific significance. The research base at Ny-Ålesund is an ode to this significance, and is the primary hub for scientific endeavors in western Svalbard. The glaciers near Ny-Ålesund are polythermal in nature [75,76]. The glaciers selected for this study included Vestre Brøggerbreen (VB), Austre Lovénbreen (AL), Austre Brøggerbreen (AB), Midtre Lovénbreen (ML), Edithbreen (EB), Botnfjellbreen (BB), Pedersbreen (PB), and Uvérsbreen (UB) (Figure 1). ML and AB are perhaps the most well documented of the selected glaciers. One of the earliest accounts of ML is a photographed documentation by Hamberg [77]. The Norwegian Polar Institute set up regular monitoring of mass balance for glaciers AB and ML in 1966 and 1967, respectively [69,78]. Furthermore, glacier surface facies were mapped on ML prior to this attempt by Pope and Rees [10,16], thereby presenting a working knowledge basis for direct comparison. Luis and Singh [60] also attempted to identify facies on the nearby Edithbreen glacier.

2.1.2. Site B: Chandra–Bhaga Basin, Himalaya

Known as the "Water Tower of Asia" [79], the Himalayas are a mountain chain of extreme cultural, sociological, economic, geopolitical, and strategic significance. Their cumulative scientific importance is, hence, phenomenal. In response to changing climates, the Himalayan cryosphere is receding, and has been observed to be losing frozen mass at an alarming rate [80]. The Indian Himalayas are well documented through both Survey of India (SOI) topographic maps and remote observations [81]. The hostile mountain terrain, vast landscape, and harsh weather conditions are often incumbent to field investigations, which result in certain pockets of glacier basins being selected for continuous monitoring. The Chandra–Bhaga basin is one such region. This basin is in the Lahaul and Spiti district of Himachal Pradesh, India. It lies within the monsoon–arid transition zone, and was therefore an optimal choice for studying glacial climatic response [82]. Himansh, the Indian Himalayan research base, is situated here at an altitude of 4080 m above mean sea level [17]. The glaciers selected were Samudra Tapu (ST), CB1, CB2, CB3, CB4, CB5, and CB6 (Figure 1). Samudra Tapu in the Chandra–Bhaga basin is analogous to Midtre Lovénbreen in Ny-Ålesund, as both are well monitored and provided established results for comparison. Alphanumeric identifiers were assigned to glaciers for which, to the best of our knowledge, no name has ever been assigned. Some of the studies over ST consisted of snow cover change analysis over four decades [83], snowline altitude changes for three decades [79], glacier facies mapping using VHR data [17,84], and debris cover variation analysis [85,86].

Table 1 documents all the selected glaciers from Ny-Ålesund and the Chandra–Bhaga basin, their respective areal extents, and their Global Land Ice Measurements from Space (GLIMS) reference ID [87].

2.2. *Geospatial Data*

The core datasets of this study were LV2A-processed images obtained from Digital Globe, Inc., Westminster, CO, USA [88]. The Himalayan image was acquired on 16 October 2014 (imagery © 2014 Maxar). In the Chandra–Bhaga basin, that period is just after the ablation season and the early onset of winter. It had a multispectral (MSL) resolution of 2 m and a panchromatic (PAN) resolution of 0.5 m. The Svalbard image was acquired on 10 August 2016 (imagery © 2016 Maxar). In Ny-Ålesund, this is right at the end of the ablation season. This arctic product had an at-nadir spatial resolution of 1.24 m, whereas the SWIR bands and PAN band had resolutions of 3.7 m and 0.31 m, respectively. The datasets had a radiometric resolution of 16 bits per pixel. The spectral resolution of WV-2 consisted of the bands PAN (0.45–0.80 μm), coastal (0.40–0.45 μm), blue (0.45–0.51 μm), green (0.51–0.58 μm), yellow (0.585–0.625 μm), red (0.63–0.69 μm), red edge (0.705–0.745 μm), near-infrared 1 (NIR1) (0.770–0.895 μm), and near-infrared 2 (NIR2) (0.86–1.04 μm). The projection and datum of the Svalbard image were done with WGS 1984 UTM Zone 33N, whereas the Himalayan image was projected with WGS 1984 UTM Zone 43N.

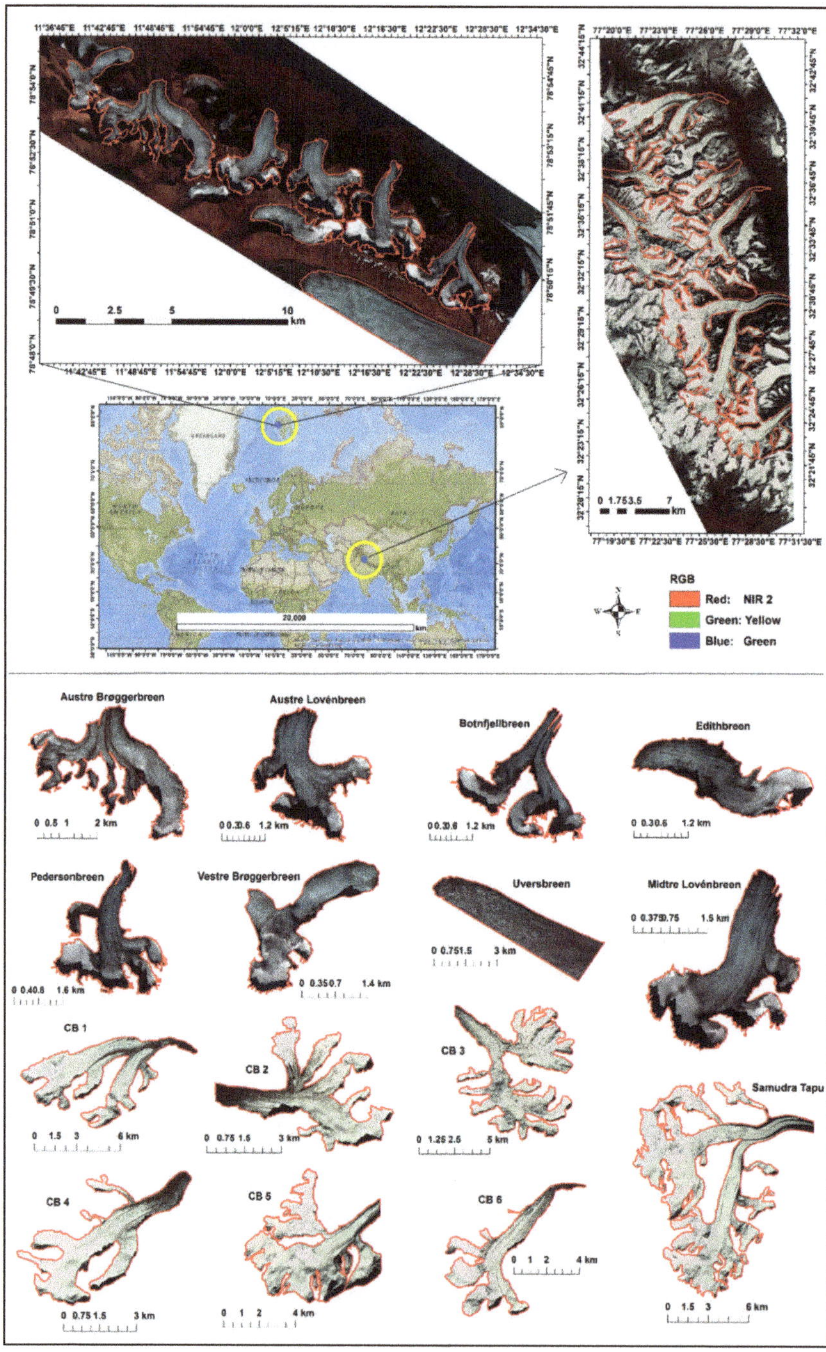

Figure 1. Location of the test sites (**upper** section) and insets of the selected glaciers (**lower** section). The satellite imagery used in the manuscript was obtained from Digital Globe, Inc., Westminster, CO, USA. Chandra–Bhaga basin image: imagery © 2014 Maxar; Ny-Ålesund image: imagery © 2016 Maxar.

Table 1. The selected glaciers of the study, their areal extents, and GLIMS reference IDs. The areal extents were calculated from the delineated shapefiles using the geometry calculator in ArcGIS.

Region	Glacier	Areal Extent in km^2	GLIMS Reference ID
Ny-Ålesund Svalbard	Vestre Brøggerbreen	2.89	G011694E78906N
	Austre Lovénbreen	4.64	G012161E78870N
	Austre Brøggerbreen	8.08	G011895E78886N
	Midtre Lovénbreen	4.75	G012039E78878N
	Edithbreen	3.27	G012119E78852N
	Botnfjellbreen	4.82	G012405E78843N
	Pedersbreen	5.87	G012286E78855N
	Uvérsbreen	13.85	G012520E78787N
Chandra–Bhaga basin Himalayas	Samudra Tapu	76.00	G077426E32511N
	CB 1	27.70	G077376E32671N
	CB 2	12.44	G077368E32619N
	CB 3	37.43	G077369E32564N
	CB 4	12.05	G077421E32604N
	CB 5	24.93	G077485E32394N
	CB 6	16.65	G077438E32563N

Pansharpened scenes were draped on 30 m Advanced Spaceborne Thermal Emission and Reflection Radiometer (ASTER) and Global Digital Elevation Model (GDEM) v2 [89] for the Chandra–Bhaga basin, and 5 m Arctic DEM [90,91] for Ny-Ålesund. The resultant 3D view of the study areas presented reliable surfaces for digitization of glacial boundaries [9,17].

3. Data Processing Methodology

3.1. Experimental Setup

The review of literature suggested utilizing the VNIR range of VHR optical data to exploit sensor-specific methods of mapping facies, specifically band ratioing. Comparison of outputs between ratios is beneficial to outlining the role of each band in mapping procedures. Based on the findings by Paul et al. [11] and Pope and Rees [16], it was observed that higher spatial and spectral differences in VHR data could improve the mapping outputs of glacier facies and deliver potentially superior results. Conventional and advanced supervised pixel-based classifiers (PBC) methods have also been shown to deliver good results in glaciological applications [17,59]. Finally, assessment of the effects of atmospheric corrections and pansharpening on the spatial and spectral differences induced by VNIR VHR data would thoroughly define an ideal processing protocol, as well as mapping mechanism. Figure 2 outlines the broad experimental setup of the current study.

This study aimed to map facies for selected glaciers in Ny-Ålesund, Svalbard; and the Chandra–Bhaga basin, Indian Himalayas, using VNIR VHR WV-2 data. Three different atmospheric corrections; viz., DOS, FLAASH, and QUAC, were used to derive reflectance, followed by pansharpening using GS and HCS. Glacial extents were defined by delineating 3D raised images over ASTER GDEM v2 and Arctic DEM, respectively. The subsets were then classified using conventional and advanced PBC methods, and the results were assessed using error matrices and qualitative assessment against published literature. Thus, the study used three atmospheric corrections, two pansharpening methods, and a host of classification algorithms to test the effects of pansharpening, atmospheric corrections, and classification algorithms on mapping glacier surface facies. Henceforth in the manuscript, processing "levels" of datasets will be referred to as processing schemes to describe the stage of processing. Table 2 displays each processing scheme and its associated nomenclature in the study. This nomenclature will be used to refer to the datasets, the classification, and respective workflows.

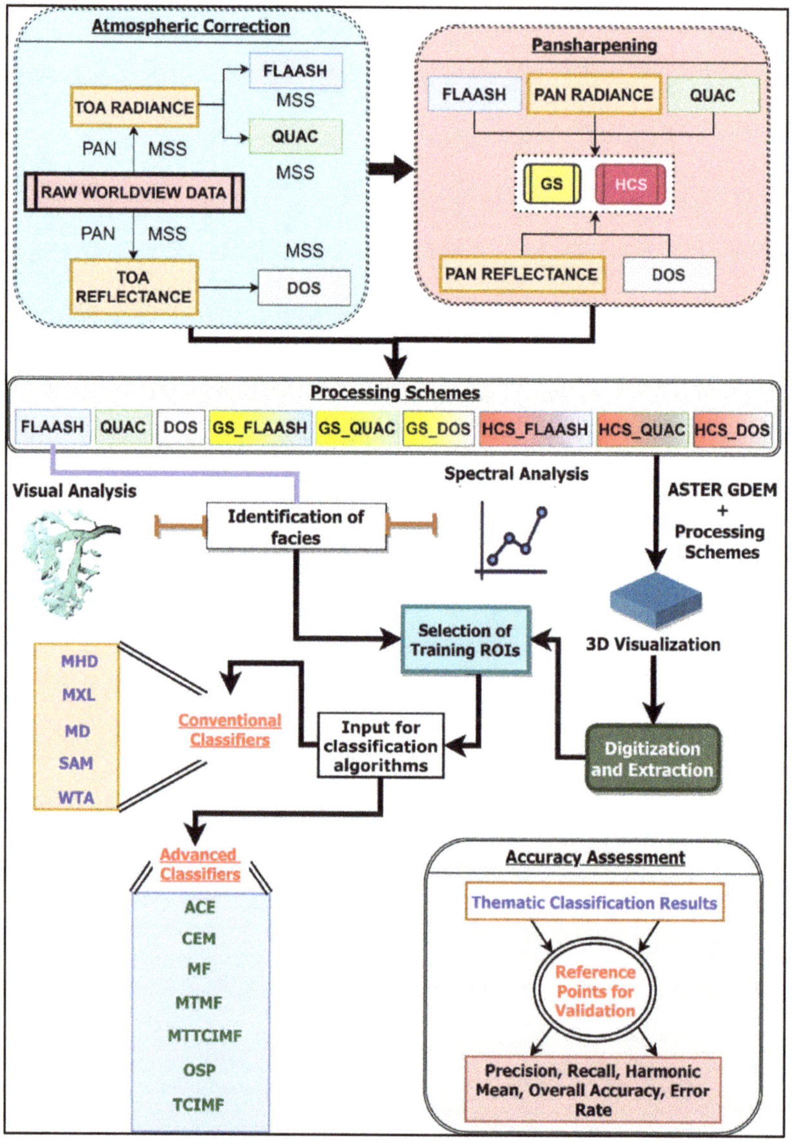

Figure 2. Experimental set up of the study. TOA: Top of Atmosphere; PAN: Panchromatic; MSS: Multispectral; DOS: Dark Object Subtraction; FLAASH: Fast-Line-of-Sight Atmospheric Analysis of Spectral Hypercubes; QUAC: Quick Atmospheric Correction; GS: Gram–Schmidt; HCS: Hyperspherical Color Space; MHD: Mahalanobis Distance; MXL: Maximum Likelihood; MD: Minimum Distance; SAM: Spectral Angle Mapper; WTA: Winner Takes All; ACE: Adaptive Coherence Estimator; CEM: Constrained Energy Minimization; MF: Matched Filtering; MTMF: Mixture-Tuned Matched Filtering; MTTCIMF: Mixture-Tuned Target-Constrained Interference-Minimized Filter; OSP: Orthogonal Space Projection; TCIMF: Target-Constrained Interference-Minimized Filter.

Table 2. Nomenclature of processing schemes used in the current study. AC: Advanced Classifiers; CC: Conventional Classifiers.

Nomenclature/Abbreviation	Description/Definition
DOS	DOS-corrected
FLAASH	FLAASH-corrected
QUAC	QUAC-corrected
GS_DOS	DOS followed by GS sharpening
GS_FLAASH	FLAASH followed by GS sharpening
GS_QUAC	QUAC followed by GS sharpening
HCS_DOS	DOS followed by HCS sharpening
HCS_FLAASH	FLAASH followed by HCS sharpening
HCS_QUAC	QUAC followed by HCS sharpening
DOS_AC/CC	DOS followed by AC or CC classification
FLAASH_AC/CC	FLAASH followed by AC or CC classification
QUAC_AC/CC	QUAC followed by AC or CC classification
GS_DOS_AC/CC	DOS followed by GS followed by AC or CC classification
GS_FLAASH_AC/CC	FLAASH followed by GS followed by AC or CC classification
GS_QUAC_AC/CC	QUAC followed by GS followed by AC or CC classification
HCS_DOS_AC/CC	DOS followed by HCS followed by AC or CC classification
HCS_FLAASH_AC/CC	FLAASH followed by HCS followed by AC or CC classification
HCS_QUAC_AC/CC	QUAC followed by HCS followed by AC or CC classification
AC: ACE/CEM/MF/MTMF/MTTCIMF/OSP/TCIMF	Individual processing schemes are followed by the abbreviations for each advanced classifier
CC: MHD/MXL/MD/SAM/WTA	Individual processing schemes are followed by the abbreviations for each conventional classifier

3.2. Image Processing

3.2.1. Radiometric Calibration and Atmospheric Correction

Conversion of DN to reflectance is a dual-step procedure, which involves: (a) converting digital number/brightness values to at-sensor spectral radiance; and (b) retrieving apparent surface spectral reflectance from at-sensor spectral radiance through atmospheric correction. The first step was carried out using the radiometric calibration module in Environment for Visualizing Images (ENVI) 5.3. This study tested three atmospheric correction models; each procedure is described as follows.

The FLAASH correction is a two-step process that requires: (1) retrieval of atmospheric parameters such as aerosol description and the water column amount; and (2) using the model atmosphere and aerosol description to convert radiance to reflectance using the radiative transfer code [92]. The atmosphere model [93] and aerosol model [94] were defined using the guidelines prescribed by Abreu and Anderson [95]. Other parameters such as initial visibility and GMT were user-defined using the image metadata. Factors including pixel size, aerosol height, CO2 mixing ratio, water column multiplier, zenith angle, sensor altitude, and scene center location were computed automatically upon definition of the sensor type. Table 3 highlights the parameters used for the images of each study area.

Table 3. Input parameters for FLAASH atmospheric correction.

Parameter	Chandra–Bhaga Basin	Ny-Ålesund	Computation
Flight date	16 October 2014	10 August 2018	Imagery metadata
Scene center location	Lat: 32.5324 Long: 77.4175	Lat: 78.8816 Long: 12.0734	Automatic computation
GMT	5.6825	12.7456	User-defined
Sensor altitude (km)	770	770	Automatic computation
View zenith angle (degrees)	180.00	180.00	Automatic computation
Initial visibility (km)	40.00	40.00	User-defined
Atmospheric model	1 (Tropical)	4 (Subarctic Summer)	User-defined [93]
Aerosol model	6 (Tropospheric)	4 (Maritime)	User-defined [93]
Water column multiplier	1.00	1.00	Automatic computation
Pixel size (m)	2.00	0.90	Automatic computation
Aerosol scale height	1.50	1.50	Automatic computation
CO2 mixing ratio (ppm)	390.00	390.00	Automatic computation

Unlike the FLAASH model, the QUAC correction is relatively straightforward. Described in [51] as an in-scene approach, it relies primarily on central wavelengths and the first step of sensor calibration. Due to this, the procedure directly involves input of the image into the QUAC module, which delivers the output scaled to a reflectance factor of 10,000 [92]. A simple band math correction brings the reflectance values into the range of 0–1. Like QUAC, the DOS correction is also an image-based corrective procedure. DOS is based on the principle that contributions of atmospheric scattering cause upwelling of the path radiance in dark pixels of an image in the concerned spectral [96]. Zhang et al. [97] outlined the DOS equation while stating that a single dark value was used to determine path radiance. For analyzing spectral signatures and subsequent classification, the imagery must be converted to reflectance. Therefore, following Rumora et al. [98], TOA reflectance was used an input to the DOS correction. Moreover, DOS correction in ENVI can incorporate user-defined dark pixel values. The procedure simply involves an operator-assisted identification of a few dark pixels and calculation of the average reflectance of each of the dark pixels. These average values can then be manually added into the DOS module window in ENVI 5.3 for each spectral band. Table 4 displays the average user-defined dark pixel reflectance values.

Table 4. Spectral-band-wise at-sensor reflectance values of selected dark pixels for input into DOS correction module in ENVI 5.3.

Spectral Bands	Mean at-Sensor Reflectance of Selected Dark Pixels	
	Ny-Ålesund	**Chandra–Bhaga Basin**
Coastal	0.09	0.17
Blue	0.06	0.14
Green	0.04	0.11
Yellow	0.03	0.09
Red	0.03	0.08
Red Edge	0.02	0.08
NIR1	0.01	0.06
NIR2	0.01	0.06

3.2.2. Pansharpening and Digitization

Pansharpening was performed in this study to test the differences between the GS, which is the most purported algorithm for retaining spectral information [40], and the HCS, which was designed for WV-2 imagery [42] against non-pansharpened imagery. GS estimates the panchromatic data based on the spectral response function of a given sensor [99]. The procedure requires direct input of the images into the GS module in ENVI. HCS sharpens MS imagery by replacing the intensity component of MS data in the hyperspherical color space with the intensity-matched form of the PAN band [44]. The procedure requires input of the PAN and MS images into the HCS fusion module in ERDAS IMAGINE.

Pandey and Venkatraman [79] experienced difficulties while manually digitizing ice divides and the glacier terminus. Bhardwaj et al. [9] resolved this issue by generating a 3D perspective of the area to observe and delineate the glacial boundaries. Jawak et al. [17] followed a similar approach, and highlighted the efficiency by which ice divides can be observed using the same method. Therefore, the current study followed suit by draping the GS-pansharpened imagery over the Arctic DEM for Ny-Ålesund and over the ASTER GDEM v2 for the Chandra–Bhaga basin. The glacial boundaries and ice divides were then digitized and extracted from the complete image using ArcGIS.

3.3. Glacier Facies Mapping Using Advanced Image Processing

Pixel-Based Classification

A wide variety of pixel-based algorithms can be employed for information extraction. Pope and Rees [10] used an unsupervised ISODATA algorithm to map facies. However, their base image was acquired using an Airborne Thematic Mapper (ATM). While this provided a good comparison against the Linear Combinations (LCs) of their study, it cannot be directly applied to satellite data. Supervised algorithms, on the other hand, have been used to map facies using satellite data [17,32,35,60,82] with good accuracies. Moreover, this study intended to improve upon comparisons between supervised classifiers [100] and test the effects of image-processing routines on the classification outputs. Such a test acts on the mathematical and computational differences between each classifier, a discussion that is beyond the scope of this paper. Nevertheless, as end users of classification algorithms, it was necessary to identify and evaluate their thematic performance. A comprehensive test of this scale would necessitate assessment of conventional and advanced pixel-based classifiers. ENVI offers both under its Terrain Categorization (TERCAT) and Target Detection (TD) workflows. Selected algorithms comprised Mahalanobis Distance (MHD), Maximum Likelihood (MXL), Minimum Distance (MD), Spectral Angle Mapper (SAM), Winner Takes All (WTA), Adaptive Coherence Estimator (ACE), Constrained Energy Minimization (CEM), Matched Filtering (MF), Mixture-Tuned Matched Filtering (MTMF), Mixture-Tuned Target-Constrained Interference-Minimized Filter (MTTCIMF), Orthogonal Space Projection (OSP), and Target-Constrained Interference-Minimized Filter (TCIMF). Table 5 describes each algorithm, the workflow under which they were available in ENVI, and reference studies in which the algorithms were used for information-extraction applications.

Table 5. Pixel-based classifiers used in this study, their descriptions, and their reference applications. Assessment: wherein different pixel-based methods are assessed for their comparative performance. Descriptions of the classifiers were modified from [101].

Approach/Workflow	Algorithm	Description	Reference Applications
Conventional Classifiers	Mahalanobis Distance (MHD)	Assumes equal class covariances and assigns pixels to closest training samples based on direction sensitive highest probability.	Landcover Mahmon et al. [102]; Aerosol classification: Hamill et al. [103]; Assessment: Doma et al. [104]; Gao and Mas [105]; Glacier facies: Jawak et al. [17]
	Maximum Likelihood (MXL)	Assigns pixels according to highest probability based on an assumption of normal distribution of the statistics for each training sample in each band.	Landcover: Mahmon et al. [102]; Assessment: Doma et al. [104]; Vegetation area: Gevana et al. [106]; Glacier facies: Shukla and Ali [35]; Jawak et al. [17]
	Minimum Distance (MD)	Calculates the average of training samples and computes the Euclidean distance from each unknown pixel to the average sample for each class.	Face Recognition: ChandraBhensle and Raja [107]; Landcover: Mahmon et al. [102]; Assessment: Doma et al. [104]; Crop area: Ahmed et al. [108]; Glacier facies: Jawak et al. [17]
	Spectral Angle Mapper (SAM)	Uses an n (spectral band numbers)-D angle of spectral similarity to assign pixel spectra to training samples with the smallest angle (hence, closest probable class).	Crop area: Ahmed et al. [108]; Canopy species identification: Cho et al. [109]; Burnt area mapping: Petropoulos et al. [110]; Glacier facies: Jawak et al. [17]
	Winner Takes All (WTA)	A voting method that classifies pixels based on the majority compiled from all other methods in the TERCAT workflow.	Pattern recognition: Chen et al. [111]; Polar land cover mapping: Jawak and Luis [40]; Multisource object extraction: Mancini et al. [112]
Advanced Classifiers	Adaptive Coherence Estimator (ACE)	Derived from the generalized likelihood ratio (GLR). It does not require knowledge of all target classes in an image.	Mineral mapping: Ni et al. [113]; Shoreline mapping: Sukcharoenpong et al. [114]; Tree crown classification: Zou et al. [115]; Sonar systems: Soules and Broadwater [116]
	Constrained Energy Minimization (CEM)	Classifies pixels through a covariance matrix using a constrained finite impulse filter based on the provided training samples.	Assessment: Ren et al. [117]; Du et al. [118]; Mineral mapping: Pour et al. [119]; Glacier facies: Jawak et al. [17]
	Matched Filtering (MF)	Minimizes the unknown background spectra according to the training sample through partial unmixing, assigning classes based on mean pixel spectra abundances.	Surface water pollution: Gursoy and Atun [120]; Lithology: Harris et al. [121]; Gas plumes: Funk et al. [122]; Glacier facies: Jawak et al. [17]
	Mixture-Tuned Matched Filtering (MTMF)	Adds an infeasibility image to the results to reduce the number of false positives that may occur in MF results.	Lithology: Mehr et al. [123]; Hyperspectral leafy spurge cover: Williams and Hunt Jr. [124]; Mineral mapping: Zadeh et al. [125]

Table 5. *Cont.*

Approach/Workflow	Algorithm	Description	Reference Applications
	Orthogonal Space Projection (OSP)	Matches pixels to training samples by using an orthogonal subspace projector to remove nontargets and then applying MF.	Assessment: Du et al. [118]; Face recognition: Singha et al. [126]; Glacier facies: Jawak et al. [17]
	Target-Constrained Interference-Minimized Filter (TCIMF)	Constrained to eliminate the response of nontargets rather than minimize their energy. It can minimize interferences in classification.	Hyperspectral subpixel target detection: Ren and Chang [127]; Assessment: Du and Ren [128]; Flood area mapping: Millan et al. [129]
	Mixture-Tuned Target-Constrained Interference-Minimized Filter (MTTCIMF)	Adds infeasibilty to TCIMF in order to reduce misclassification after using a minimum noise fraction transformation to perform TCIMF	Assessment: Seyedein et al. [130]; Subpixel mineral mapping: Kumar et al. [131]; Oil spill spectral unmixing: Sidike et al. [132]

MHD, MXL, MD, and SAM are some of the most widely used and popular classifiers [133]. WTA is an ensemble of the majority classification from MHD, MXL, MD, and SAM. The CEM, MF, and OSP classifiers have been used for mapping glacier facies (Table 5). However, the other AC algorithms have been tested in applications requiring minute spectral differentiation (Table 5), and theoretically should be capable of discriminating between closely matching spectra of surface facies. Therefore, a comparative assessment of thematic classification among the popular and advanced algorithms using different processing schemes would lead to a greater understanding of their capabilities at identifying facies.

PBC is usually a two-step procedure requiring: (1) selection of training samples (regions of interest); and (2) application of supervised algorithms. Training samples were assigned based on visual and spectral analysis of the available facies. Polygonal regions of interest (ROIs) were outlined for each facies to accommodate their spectral variations (Supplementary Figure S1). Distribution of ROIs in percentage for the Chandra–Bhaga basin were as follows: snow was assigned 42.12%, glacier ice was assigned 20.82%, ice mixed debris was assigned 8.74%, and crevasses were assigned 9.10%; whereas shadowed area and debris were assigned ROIs containing 12.04% and 7.18%, respectively. For ROIs of the Ny-Ålesund glaciers, snow was distributed at 3.42%, streams and crevasses at 4.19%, shadowed snow at 22.36%, saturated snow at 11.57%, melting snow at 7.42%, melting glacier ice at 18.71%, and glacier ice at 20.23%; while dry snow and dirty ice were distributed at 3.34% and 8.76%, respectively. The MF, MTTCIMF, and MTMF algorithms required the imagery to undergo a minimum noise fraction (MNF) transformation prior to classification. Algorithms that did not require this transformation were processed directly after input of the image and ROIs into the respective workflows. Default parameters were retained in the workflows, and postclassification processing was avoided to negate unintentional analyst bias. Therefore, as performed by Jawak et al. [17], the stretch (square root) and rule thresholds (0.4) were common for each facies, for all classifiers, and for all the processing schemes.

3.4. Identification of Surface Facies

In this study, glacier surface facies were identified in Ny-Ålesund and the Chandra–Bhaga basin using visual and spectral characteristics. Jawak et al. [17] described the visual identification of surface facies in the Chandra–Bhaga basin using the smoothness and higher elevation of snow, rougher and flow-induced striated texture of ice, disheveled structure of crevasses, and brightness variations between debris and ice mixed debris (IMD).

The same characterization was incorporated here to identify facies and derive spectral signatures for the Himalayan glaciers. A similar logic of surficial appearance, texture, and tonal variations, along with location on the glacier, was incorporated to identify surface facies on Ny-Ålesund glaciers. Figure 3 highlights the visual characteristics of observable surface facies on the ML glacier. The Chandra–Bhaga basin image was obtained at the start of early winter; hence, it showcased a large distribution of snow and glacier ice. However, the Ny-Ålesund image was obtained at the end of the ablation season, thereby showcasing the full range of surface facies. Facies identified on Ny-Ålesund glaciers consisted of dry snow, wet snow, melting snow, saturated snow, glacier ice, melting glacier ice, dirty ice, and streams and crevasses. Dry snow was characterized by its bright appearance at the highest elevation of the glacier. Wet snow appeared next, having a reduced brightness due to an increased moisture content than dry snow, but still maintaining an overall smoothness. Melting snow had more visible tonal roughness than wet snow, whereas saturated snow was much darker. This could be due to the high moisture and low integrity of the surface cover. Glacier ice was distinguishable through flow-induced striations and was distinct from melting glacier ice by its brighter appearance. Greater moisture and surface dust were the next possible reasons for its visual characteristic. Streams and crevasses were grouped together, as their individual characterization was difficult when much of the glacier's surface appeared crevassed due to flowing supraglacial stream channels. Dirty ice was the darkest ablation facies characterizable, and functionally comprised more dust and debris toward the end of the glacial tongue.

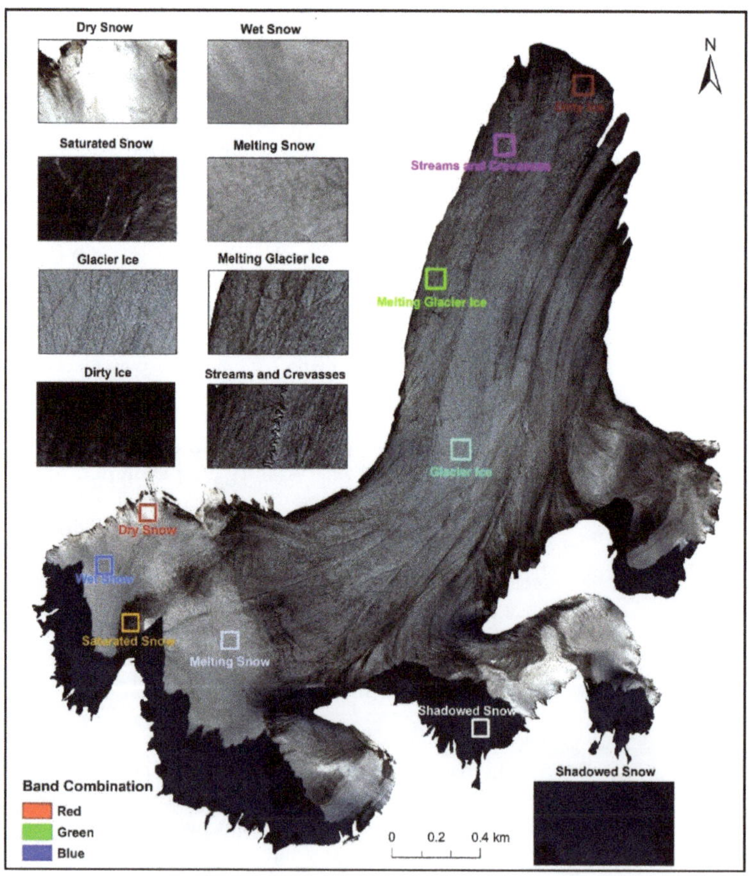

Figure 3. Visual characteristics of identifiable glacier facies displayed on the ML glacier. Boxes on the glacier highlight location of target facies. Zoomed and labelled insets display the visual characteristics of each facies.

Multispectral mapping of any earth feature entails utilization of spectral characteristics of the target feature. This requires good operator/analyst knowledge when field data is not available for reference. In this study, the spatial characteristics of facies were instrumental in identifying target sites. However, it was the spectral signatures that ultimately determined the separability of surface facies. As accuracies of spectral signatures rely upon the image-processing schemes employed, the current study presents the derived reflectance of facies identified on Ny-Ålesund glaciers in Figure 4 and the reflectance of facies identified on the Chandra–Bhaga basin glaciers in Figure 5. A detailed discussion of the variations in spectral signatures using each processing scheme is provided in Section 4.1.

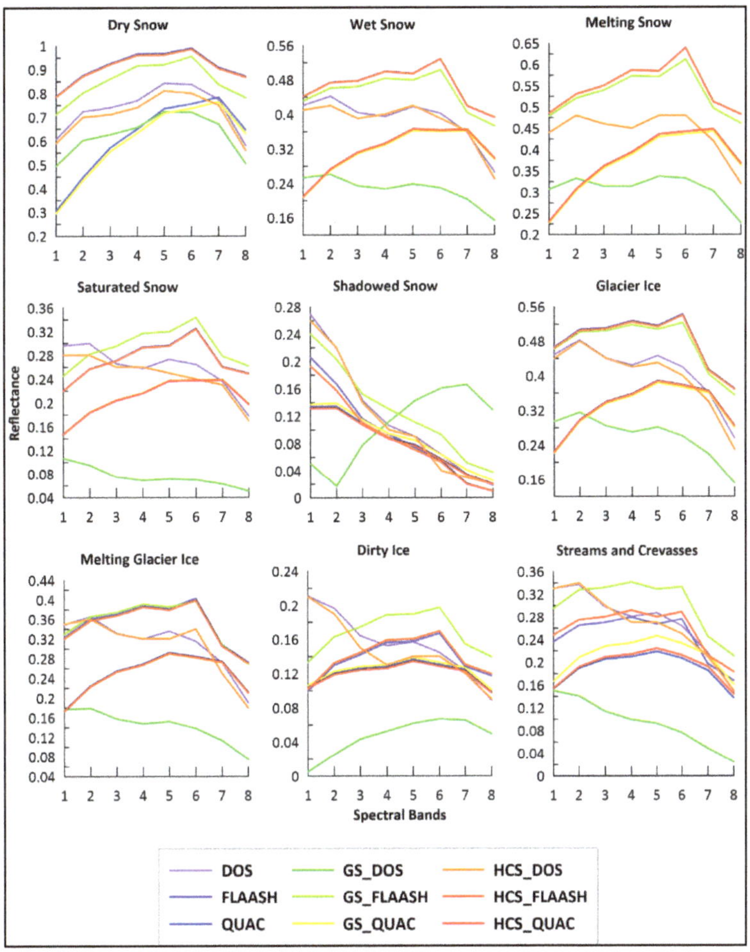

Figure 4. Variations in spectral signatures of identified facies in Ny-Ålesund for each processing scheme. 1: Band 1/Coastal; 2: Band 2/Blue; 3: Band 3/Green; 4: Band 4/Yellow; 5: Band 5/Red; 6: Band 6/Red Edge; 7: Band 7/NIR 1; 8: Band 8/NIR 2.

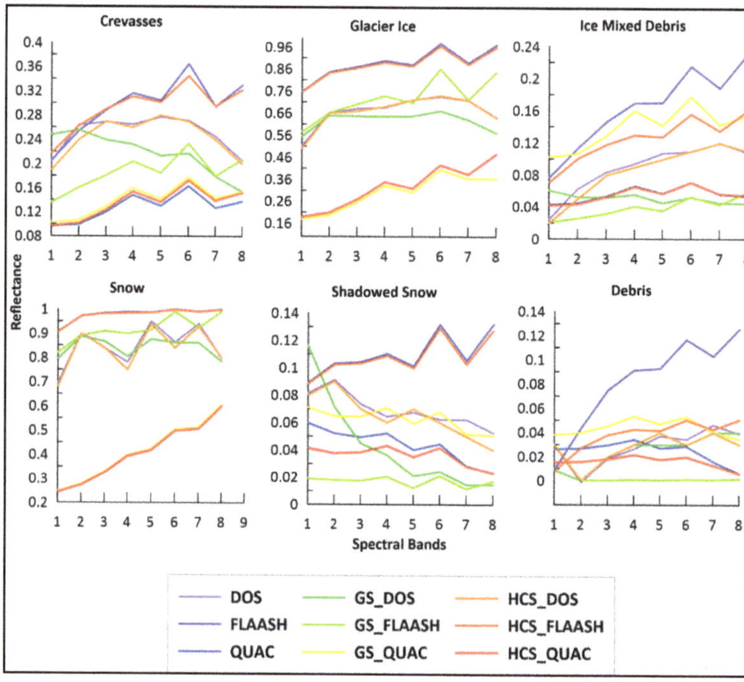

Figure 5. Variations in spectral signatures of identified facies in Chandra–Bhaga basin for each processing scheme. 1: Band 1/Coastal; 2: Band 2/Blue; 3: Band 3/Green; 4: Band 4/Yellow; 5: Band 5/Red; 6: Band 6/Red Edge; 7: Band 7/NIR 1; 8: Band 8/NIR 2.

3.5. Thematic Accuracy Assessment

The acquired satellite data could not be corroborated with field data due to harsh field conditions in the season of acquisition in the Himalayas and logistical difficulties in transit to Svalbard. In order to overcome this, the study followed an equalized random-sampling approach to assign reference points for analysis [134]. This approach provided every thematic class with an equal number of reference points [135]. To accommodate spatially limited facies, every class was allotted 10 reference points. Thus, for Ny-Ålesund glaciers, it resulted in 90 points per glacier, resulting in 720 pixels in total. For the Chandra–Bhaga basin glaciers, it resulted in 420 pixels in total. Bias was avoided in determining reference pixels by developing them independent of each other and by using polygons for ROIs and points for reference data. Confusion matrices were generated to calculate measures such as precision, recall, F1 score, overall accuracy (OA), error rate (ER), and specificity. Maxwell and Warner [136] described each measure, and Supplementary Table S1 displays the equations used to calculate precision, recall, F1 score, specificity and OA. ER was defined simply as "1-OA". The measures were computed for every facies over each glacier individually, and were sequentially averaged to obtain mean values for each measure.

4. Results and Discussion

4.1. Spectral Signatures

The spectral characteristics of glacier surface facies are an account of snow ageing, metamorphosis, dust and debris entrainment, atmospheric depositions and a historical archive of glacial health, if monitored over long temporal scales. A change in reflectance characteristics is not only significant for ascertaining the state of changing snow and ice, but also for the understanding of possible causes for the visible change. This places a large importance on the selection of atmospheric-correction algorithms for deriving reflectance, and subsequently, spectral signatures. Figure 4 highlights the changes in spectral reflectance for each facies from Ny-Ålesund for each processing scheme. Figure 5 displays the variations induced by the respective processing schemes in derived spectral reflectance for facies observed in the Chandra–Bhaga basin. Both figures display the average reflectance for each facies. The reader is referred to Jawak et al. [17] for a detailed analysis of the spectral signature of facies derived in the Chandra–Bhaga basin. As the FLAASH atmospheric correction previously delivered the reflectance spectra most comparable with previous works, the characterizing of facies in the present analysis of Ny-Ålesund glaciers was initially performed using reflectance delivered by the same. Cassachia et al. [137] defined dry snow as not being subjected to melting or infiltration of water, and in principle it should be limited to the highest elevations on the glacier. No presence of water implied that facies with the highest reflectance would correspond to dry snow. Warren [138] described snow reflectance as being influenced by grain size, thickness of snow pack, mixing of dust and impurities, and moisture/liquid water content. Facies having a lower reflectance in the NIR region, but higher than other facies, would directly correspond to snow having greater moisture, and little to no surface impurities. Wet snow in this study was characterized by similar features. Wet snow and melting snow had a higher reflectance than fully saturated snow. The FLAASH-derived spectral signature of saturated snow closely matched that observed by Hinkler et al. [139]. The reflectance curves of dirty ice and glacier ice observed here were similar in trend to the curves depicted by Gao and Liu [140], who adapted their method from Zeng et al. [141]. Melting glacier ice identified here corresponded to that observed by Pope and Rees [10]. Table 6 displays bandwise-calculated variances for atmospheric corrections, GS pansharpening, and HCS pansharpening, according to maximum and minimum variance for each facies, averaged across all glaciers from the two test sites. Among the Chandra–Bhaga basin glaciers, snow class showed the maximum variation (0.31) among the atmospheric correction methods, whereas debris class showed the minimum variation (0.01). Snow and debris classes showed the same trend among the GS-sharpening (max: 0.29, min: 0.01) and HCS-sharpening processing schemes (max: 0.31, min: 0.01). For the Ny-Ålesund glaciers, dry snow exhibited a maximum variation of 0.20, while dirty ice showed a minimum variation of 0.00, for the atmospheric correction methods. For the GS sharpening schemes, dirty ice showed the maximum variation at 0.6; whereas the lowest variation was shown by shadowed snow at 0.01. The HCS sharpening schemes showed a maximum variation of 0.20 for dry snow, and a common minimum of 0.00 for shadowed snow and dirty ice. Section 4.3 discusses these variations in further detail.

Table 6. Maximum and minimum variations in spectral reflectance of facies derived from the average spectral spectra from each processing scheme. Atmospheric corrections: calculated between DOS, FLAASH, and QUAC; GS sharpening: calculated between GS_DOS, GS_FLAASH, and GS_QUAC; HCS sharpening: calculated between HCS_DOS, HCS_FLAASH, and HCS_QUAC.

Test Site	Facies	Variations in Spectral Reflectance					
		Atmospheric Corrections		GS Sharpening		HCS Sharpening	
		Max.	*Min.*	*Max.*	*Min.*	*Max.*	*Min.*
Ny-Ålesund	Dry snow	B1 (0.20)	B7 (0.06)	B2 (0.17)	B7 (0.07)	B1 (0.20)	B7 (0.07)
	Wet snow	B1 (0.10)	B7 (0.03)	B4, B6 (0.11)	B2, B7, B9 (0.09)	B1 (0.10)	B7 (0.03)
	Melting snow	B1 (0.11)	B7 (0.03)	B6 (0.12)	B7 (0.08)	B1 (0.11)	B7 (0.04)
	Saturated snow	B1 (0.06)	B7 (0.01)	B6 (0.11)	B1 (0.06)	B1 (0.05)	B7 (0.01)
	Shadowed snow	B1 (0.06)	B7 (0.01)	B1, B2 (0.08)	B4 (0.01)	B1 (0.05)	B8 (0.00)
	Glacier ice	B1 (0.11)	B7 (0.02)	B6 (0.11)	B7, B8 (0.08)	B1 (0.11)	B7 (0.03)
	Melting glacier ice	B1 (0.08)	B7 (0.02)	B6 (0.11)	B1 (0.07)	B1 (0.08)	B7 (0.02)

Table 6. *Cont.*

Test Site	Facies	Variations in Spectral Reflectance					
		Atmospheric Corrections		GS Sharpening		HCS Sharpening	
		Max.	*Min.*	*Max.*	*Min.*	*Max.*	*Min.*
	Dirty ice	B1 (0.05)	B7 (0.00)	B1, B2, B4 (0.6)	B7, B8 (0.04)	B1 (0.05)	B7 (0.00)
	Streams and crevasses	B1 (0.07)	B7, B8 (0.01)	B6 (0.11)	B1 (0.06)	B1 (0.07)	B7 (0.01)
Chandra–Bhaga basin	**Crevasses**	B2, B3, B5, B6, B8 (0.08)	B1 (0.05)	B1, B2 (0.06)	B6, B7, B8 (0.02)	B2–B6, B8 (0.7)	B1 (0.05)
	Glacier ice	B2 (0.27)	B7, B8 (0.21)	B2 (0.22)	B7 (0.15)	B1 (0.27)	B8 (0.20)
	Ice mixed debris	B8 (0.07)	B1 (0.02)	B6 (0.06)	B1, B2 (0.03)	B8 (0.04)	B1 (0.02)
	Shadowed snow	B8 (0.05)	B1 (0.01)	B1 (0.04)	B2–B8 (0.02)	B8 (0.05)	B1 (0.02)
	Debris	B8 (0.05)	B1 (0.01)	B2–B8 (0.02)	B1 (0.01)	B8 (0.16)	B1–B7 (0.01)
	Snow	B2 (0.31)	B8 (0.16)	B2 (0.29)	B7 (0.18)	B2 (0.31)	B8 (0.16)

4.2. *Quantitative Analysis of Mapped Facies*

This section analyzes classification results generated from the AC and CC workflows as an average of all the processing schemes for areal distribution and accuracy. Results are displayed for the glacier subsets ML and ST, and are presented according to the outputs for each facies.

4.2.1. Area per Facies Produced by Each Classifier

Areas of facies for Ny-Ålesund (reference glacier ML) and for the Chandra–Bhaga basin (reference glacier ST) are provided in Tables 7 and 8.

Table 7. Classified area of each facies as an average of the results for all AC classification algorithms for each processing scheme.

Facies		ACE	CEM	MF	MTMF	MTTCIMF	OSP	TCIMF
Ny-Ålesund	**Unclassified**	0.04	0.10	0.03	0.18	0.05	0.03	0.20
	Dry Snow	0.64	0.28	0.31	0.45	0.14	0.29	0.43
	Wet Snow	0.46	0.45	0.47	0.68	0.47	0.53	0.63
	Melting Snow	0.31	0.36	0.44	0.37	0.63	0.42	0.33
	Saturated Snow	0.63	0.62	0.59	0.46	0.72	0.64	0.50
	Shadowed Snow	0.70	0.78	0.65	0.81	0.69	0.50	0.76
	Glacier Ice	0.38	0.37	0.44	0.32	0.42	0.60	0.29
	Melting Glacier Ice	0.74	0.58	0.70	0.59	0.63	0.62	0.67
	Dirty Ice	0.57	0.93	0.89	0.64	0.58	0.88	0.61
	Streams and Crevasses	0.29	0.28	0.23	0.24	0.42	0.24	0.33
Chandra–Bhaga basin	**Unclassified**	0.59	0.54	0.81	0.55	0.81	0.35	0.78
	Crevasses	4.75	5.58	4.79	4.05	6.22	5.17	4.46
	Glacier Ice	20.62	15.61	21.45	22.68	17.22	22.13	22.55
	Ice Mixed Debris	4.04	6.89	8.42	4.82	8.69	10.44	4.55
	Shadowed Snow	8.44	6.98	1.93	8.20	3.09	3.97	8.27
	Debris	8.25	6.23	8.76	7.81	10.77	8.35	7.59
	Snow	29.30	34.17	29.84	27.88	29.21	25.59	27.81

Table 8. Classified area of each facies as an average of the results for all CC classification algorithms for each processing scheme.

	Facies	MHD	MXL	MD	SAM	WTA
Ny-Ålesund	Unclassified	0.00	0.00	0.00	0.76	0.02
	Dry Snow	0.16	0.15	0.16	0.26	0.16
	Wet Snow	0.43	0.32	0.50	0.49	0.38
	Melting Snow	0.62	0.47	0.76	0.64	0.65
	Saturated Snow	0.56	0.79	0.54	0.40	0.66
	Shadowed Snow	0.69	0.74	0.88	0.65	0.79
	Glacier Ice	0.50	0.62	0.35	0.45	0.54
	Melting Glacier Ice	0.59	0.81	0.73	0.52	0.76
	Dirty Ice	0.78	0.43	0.44	0.32	0.48
	Streams and Crevasses	0.43	0.41	0.39	0.27	0.32
Chandra–Bhaga basin	Unclassified	0.00	0.00	0.00	6.02	0.37
	Crevasses	3.93	8.46	4.58	1.89	4.81
	Glacier Ice	19.20	24.95	24.42	35.85	27.71
	Ice Mixed Debris	2.84	2.39	1.72	0.57	1.84
	Shadowed Snow	3.56	1.47	0.94	1.90	1.65
	Debris	2.40	2.33	2.90	0.64	2.37
	Snow	44.07	36.41	41.45	29.13	37.26

Among the AC classifiers, the largest unclassified area for ML was given by TCIMF (0.20 km^2), whereas OSP and MF produced the lowest unclassified area of 0.03 km^2. Dry snow was given the largest distribution by ACE (0.64 km^2) and the lowest by MTTCIMF (0.14 km^2). Wet snow was classified with maximum areal distribution by MTMF (0.68 km^2). The lowest distribution was reported by CEM at 0.45 km^2. Melting snow achieved maximum distribution through MTTCIMF and the lowest through ACE at 0.63 km^2 and 0.31 km^2, respectively. MTTCIMF classified saturated snow at 0.72 km^2, whereas the lowest distribution of saturated snow was produced by MTMF at 0.46 km^2. The largest distribution of shadowed snow was provided by MTMF at 0.81 km^2, and the lowest was delivered by OSP at 0.50 km^2. Glacier ice was assigned the largest area by OSP at 0.60 km^2, and the lowest at 0.29 km^2 by TCIMF. Melting glacier ice was given the largest area at 0.74 km^2 by ACE, and the lowest at 0.58 km^2 by CEM. Dirty ice was distributed by CEM to a maximum of 0.93 km^2, while ACE provided it with the lowest at 0.57 km^2. Streams and crevasses were assigned the most area by MTTCIMF (0.42 km^2), and the lowest by MF at 0.23 km^2. Among the CC classifiers for ML, WTA assigned 0.02 km^2 to unclassified, 0.16 km^2 to dry snow, 0.38 km^2 to wet snow, and 0.65 km^2 to melting snow. Saturated snow, shadowed snow, and glacier ice were assigned 0.66 km^2, 0.79 km^2, and 0.54 km^2, respectively. Melting glacier ice, dirty ice, and streams and crevasses were assigned 0.76 km^2, 0.48 km^2, and 0.32 km^2, respectively.

For the reference glacier ST, among the AC methods, unclassified areas were provided with the largest area (0.81 km^2) by MTTCIMF and MF, and with the least area by OSP (0.35 km^2). Crevasses were mapped with the largest areal distribution at 6.22 km^2 by MTTCIMF and the lowest at 4.05 km^2 by MTMF. Glacier ice was mapped at 22.68 km^2 by MTMF and 15.61 km^2 by CEM. IMD was given the largest area at 10.44 by OSP km^2, and the lowest at 4.04 km^2 by ACE. Shadowed snow was mapped at 8.44 km^2 by ACE and 1.93 km^2 by MF. Debris cover was given a maximum areal extent of 10.77 km^2 by MTTCIMF, and a lowest at 6.23 km^2 by CEM. Snow was given the largest area at 34.17 km^2 by CEM, and the lowest area for snow was given by OSP at 25.59 km^2. Among the CC classifiers for ST, WTA classified 37.26 km^2 as snow, 2.37 km^2 as debris, and 1.65 km^2 as shadowed snow. IMD, glacier ice, and crevasses were distributed as 1.84 km^2, 27.71 km^2 and 4.81 km^2, respectively. Unclassified area totaled 0.37 km^2. WTA was the ensemble of all CC methods, therefore its areal distribution was cumulatively affected by the area provided for each facies by the preceding CC algorithms.

4.2.2. Accuracy Achieved by Each Classifier

All measures of accuracy for all the classifiers are provided in Supplementary Sheet S1. This section aims to analyze the classification results of the AC and CC classifiers, and presents the F1 score as the harmonic mean of precision and recall [136] for each classifier as an average of all the processing schemes. This was to utilize the F1 score as a measure of reliability of the classification, independent of processing schemes.

(a) F1 score for classification in Ny-Ålesund

Among the AC classifiers, dry snow was classified with a F1 score of 0.80 by ACE and 0.19 by TCIMF, whereas CEM, MF, MTMF, MTTCIMF, and OSP yielded an F1 score of 0.00. Similarly, wet snow was classified with an F1 score of 0.11 by ACE, 0.07 by TCIMF, and 0.00 by the other AC classifiers. Melting snow was classified with an F1 score of 0.36 by ACE, 0.18 by MTMF, 0.05 by OSP, and 0.04 by TCIMF. Saturated snow was mapped by ACE with an F1 score of 0.40, 0.22 by CEM and MF, 0.19 by OSP, 0.17 by MTMF, 0.04 by TCIMF, and 0.00 by TCIMF. Shadowed snow was classified with an F1 score of 0.00 by MTTCIMF, 0.10 by TCIMF, 0.20 by OSP, 0.21 by MTMF, 0.80 by ACE, and 0.33 by both CEM and MF. OSP and MTTCIMF classified glacier ice with an F1 score of 0.00, while TCIMF resulted in 0.04, MTMF in 0.15, and ACE in 0.51. CEM and MF each resulted in an F1 score of 0.21. Melting glacier ice was classified with an F1 score of 0.37 by CEM and MF, 0.11 by MTMF and OSP, 0.29 by ACE, and 0.04 by TCIMF. Dirty ice was mapped with a F1 score of 0.00 by MTTCIMF. CEM and MF yielded scores of 0.46 each, while ACE scored the highest with 0.54. OSP, TCIMF, and MTMF yielded scores of 0.13, 0.18, and 0.05, respectively. Streams and crevasses each were mapped by CEM and MF with an F1 score of 0.32, whereas ACE delivered an F1 score of 0.51, and MTMF delivered 0.07; MTTCIMF, OSP, and TCIMF delivered 0.00. Among the CC classifiers, MXL delivered a full F1 score of 1.00 for dry snow and shadowed snow, followed by 0.81 for streams and crevasses. Glacier ice and dirty ice were mapped by MXL with F1 scores of 0.74 and 0.73, respectively. Wet snow and melting snow were mapped with F1 scores of 0.69 and 0.68, respectively, by MXL. Melting glacier ice and saturated snow were classified by MXL with output F1 scores of 0.57 and 0.53, respectively. MHD classified both dry snow and shadowed snow with an F1 score of 0.93, while glacier ice achieved 0.72, dirty ice yielded 0.67, streams and crevasses resulted in 0.58, and saturated snow and melting snow achieved 0.47 and 0.45, respectively. Melting glacier ice and wet snow were classified with F1 scores of 0.39 and 0.33, respectively. MD classified dry snow with an F1 score of 0.96, while shadowed snow was classified with a score of 0.64. Melting snow and glacier ice scored 0.48 each. Dirty ice was mapped with a result of 0.47; glacier ice, saturated snow, and wet snow were classified with scores of 0.35, 0.30, and 0.24, respectively. Streams and crevasses were given a classified F1 score of 0.13. SAM classified shadowed snow with an F1 score of 0.83 and dry snow with an F1 score of 0.75, whereas melting glacier ice achieved 0.54, followed by wet snow at 0.48 and melting snow at 0.44. Glacier ice and saturated snow were classified with scores of 0.33 and 0.29, respectively; followed by dirty ice and streams and crevasses at 0.22 and 0.17, respectively. The WTA classifier achieved an F1 score of 0.95 for dry snow and shadowed snow, followed by streams and crevasses and glacier ice at 0.78 and 0.75, respectively. Dirty ice and saturated snow were classified with scores of 0.69 and 0.53, respectively. Melting snow and melting glacier ice achieved scores of 0.50 and 0.47 each. Lastly, wet snow was classified with a score of 0.38. Reliability orders for each facies are presented according to the classifiers. The reliability order for dry snow was: MXL > MD > WTA > MHD > ACE > SAM > TCIMF > CEM = MF = MTMF = MTTCIMF = OSP. The reliability for wet snow was: MXL > SAM > WTA > MHD > MD > ACE > TCIMF > CEM = MF = MTMF = MTTCIMF = OSP. The reliability of melting snow was: MXL > WTA > MD > MHD > SAM > ACE >

MTMF > OSP > TCIMF > CEM = MF = MTTCIMF. Saturated snow presented the following reliability order: MXL = WTA > MHD > ACE > MD > SAM > CEM = MF > OSP > MTMF > TCIMF > MTMF. The reliability for shadowed snow was: MXL > WTA > MHD > SAM > ACE > MD > CEM = MF > OSP > MTMF > TCIMF > MTTCIMF. Glacier ice presented the following reliability order: WTA > MXL > MHD > ACE > MD > SAM > CEM = MF > MTMF > TCIMF > MTTCIMF = OSP. The reliability order for melting glacier ice was: MXL > SAM > MD > WTA > MHD > CEM = MF > ACE > MTMF = OSP > TCIMF > MTTCIMF. Dirty ice presented the following reliability order: MXL > WTA > MHD > ACE > MD > CEM = MF > SAM > TCIMF > OSP > MTMF > MTTCIMF. The reliability order for streams and crevasses was: MXL > WTA > MHD > ACE > CEM = MF > SAM > MD > MTMF > MTTCIMF = OSP = TCIMF.

(b) F1 score for classification in the Chandra–Bhaga basin

Among the AC classifiers, crevasses were mapped with an F1 score of 0.69 by ACE, 0.53 for both CEM and MF, 0.17 for both MTMF and OSP, 0.21 for TCIMF, and 0.07 for MTTCIMF. Glacier ice was classified by CEM and MF with a common F1 score of 0.83, ACE delivered a score of 0.77, and MTMF scored 0.32. OSP and TCIMF delivered F1 scores of 0.21 each, whereas the lowest score was delivered by MTTCIMF (0.00). IMD was classified by ACE with an F1 score of 0.40. CEM and MF resulted in scores of 0.30 each. TCIMF resulted in a score of 0.21, whereas OSP produced a score of 0.19. MTMF delivered a score of 0.10, and MTTCIMF produced a score of 0.02. Shadowed snow was mapped by ACE with a score of 0.79, and TCIMF produced a score of 0.39. CEM, OSP, and MF delivered F1 scores of 0.34 each. MTMF and MTTCIMF delivered F1 scores of 0.33 and 0.00, respectively. Debris was mapped with a common F1 score of 0.23 for CEM, MF, and TCIMF. OSP and ACE provided scores of 0.22 and 0.16 each. MTMF and MTTCIMF resulted in scores of 0.07 and 0.00, respectively. ACE, CEM, and MF produced a common F1 score of 0.67 for snow, followed by MTMF with a score of 0.26. OSP and TCIMF mapped snow with a score of 0.23, whereas MTTCIMF delivered 0.00. Among the CC classifiers, MXL mapped crevasses with an F1 score of 0.80, MHD delivered a score of 0.79, whereas MD and SAM delivered scores of 0.67 and 0.28, respectively. Glacier ice was mapped with the highest F1 score of 0.96 by MD, followed by MXL with a score of 0.92. SAM and MHD were mapped with scores of 0.87 and 0.84, respectively. IMD was given the highest F1 score of 0.55, whereas SAM resulted in the lowest score of 0.00. MXL and MHD mapped IMD with scores of 0.47 and 0.41 each. Shadowed snow was classified with high scores of 0.93 and 0.90 by MHD and MXL, respectively; whereas MD and SAM were mapped with scores of 0.20 and 0.17, respectively. Debris was classified by MD with a score of 0.62. SAM and MHD delivered scores of 0.27 and 0.25, respectively; while MXL resulted in a score of 0.47. MD mapped snow with a high F1 score of 0.98, followed by MXL with a score of 0.97, and MHD at 0.95. SAM delivered a score of 0.77. WTA classified snow with an F1 score of 0.96, followed by shadowed snow with an F1 score of 0.95. Crevasses were subsequently mapped with a score of 0.89, followed by glacier ice at 0.88. IMD and debris were mapped with scores of 0.44 and 0.61 each. The order for reliable classification of crevasses was: WTA > MXL > MHD > ACE > MD > CEM = MF > SAM > TCIMF > MTMF = OSP > MTTCIMF. The reliability order for glacier ice was: MD > MXL > WTA > SAM > MHD > CEM = MF > ACE > MTMF > OSP = TCIMF > MTTCIMF. Ice mixed debris presented the following reliability order: MD > MXL > WTA > MHD > ACE > CEM = MF > TCIMF > OSP > MTMF > MTTCIMF > SAM. The reliability order for shadowed snow was: WTA > MXL > MHD > ACE > TCIMF > CEM = MF = OSP > MTMF > MD > SAM > MTTCIMF. Debris presented the following reliability order: MD > WTA > MXL > SAM > MHD > CEM = MF = TCIMF > OSP > ACE > MTMF > MTTCIMF. The reliability order for snow was: MD > MXL > WTA > MHD > SAM > ACE = CEM = MF > MTMF > OSP = TCIMF > MTTCIMF.

Table 9 gives an overall representation of classifier performance averaged over all the processing schemes. Apart from the ACE classifier, all other classifiers showed a greater error rate while classifying facies in Ny-Ålesund in comparison to the Chandra–Bhaga basin. The difference between them, however, was 0.01. The best-performing classifier was the MXL, while the worst was the MTTCIMF. The order of overall classifier performance was thus: MXL > WTA > MHD > ACE > MD > CEM = MF > SAM > MTMF = TCIMF > OSP > MTTCIMF.

Table 9. Cumulative error rate independent of atmospheric corrections and pansharpening methods, calculated by averaging all the error rates over all processing schemes for individual test sites. The classifiers with the lowest error rate are emboldened and italicized.

Algorithm/Classifier	Error Rate	
	Himalayas	Ny-Ålesund
ACE	0.60	0.59
CEM	0.65	0.75
MF	0.64	0.75
MTMF	0.78	0.82
MTTCIMF	0.82	0.91
OSP	0.77	0.88
TCIMF	0.73	0.87
MHD	0.47	0.56
MXL	***0.44***	***0.49***
MD	0.61	0.68
SAM	0.78	0.69
WTA	0.45	0.53

4.2.3. Comparison between Atmospheric Correction Methods

Figure 6 displays the overall accuracy (OA) achieved by each of the classification algorithms across the atmospheric corrections. An analysis of the trends of the OA revealed a similar trend for all three atmospheric corrections. Averages and variances were calculated for both study areas for DOS, FLAASH, and QUAC. Visual analysis of Figure 6 depicts FLAASH as having the least total variance, with DOS and QUAC showing consistent variances. MXL and WTA achieved the highest OA.

The DOS_MXL classification showed no variance across both study areas, FLAASH_MXL classification achieved 0.01 variance, and the QUAC_MXL classification achieved a variance of 0.03. WTA classification varied by 0.05, 0.03, and 0.06 for DOS_WTA, FLAASH_WTA, and QUAC_WTA, respectively. DOS_MHD classification resulted in a variance of 0.07, FLAASH_MHD resulted in 0.03, and QUAC_MHD resulted in 0.04. DOS_MD resulted in a variation of 0.14, FLAASH_MD delivered a variance of 0.15, and QUAC_MD resulted in 0.04. DOS_ACE produced a variance of 0.06, FLAASH_ACE resulted in 0.04, and FLAASH_QUAC resulted in 0.10. DOS_SAM produced a classification variance of 0.03, FLAASH_SAM produced a variance of 0.03, and QUAC_SAM delivered 0.08. DOS_MTTCIMF delivered a variance of 0.02, FLAASH_MTTCIMF produced 0.01, and QUAC_MTTCIMF resulted in 0.16. DOS_MTMF classified with a resultant variance of 0.04, FLAASH_MTMF classified with a variance of 0.02, and QUAC_MTMF produced a variance of 0.07. DOS_CEM resulted in a variance of 0.24, FLAASH_CEM produced a variance of 0.38, and QUAC_CEM delivered 0.07. DOS_MF classified facies with a variance of 0.24, followed by FLAASH_MF at 0.01and QUAC_MF at 0.01. DOS_OSP classified facies with a variance of 0.03, FLAASH_OSP delivered 0.05, and QUAC_OSP produced 0.13. DOS_TCIMF produced a variance of 0.12, FLAASH_TCIMF delivered a variance of 0.00, and QUAC_TCIMF produced a variance of 0.09.

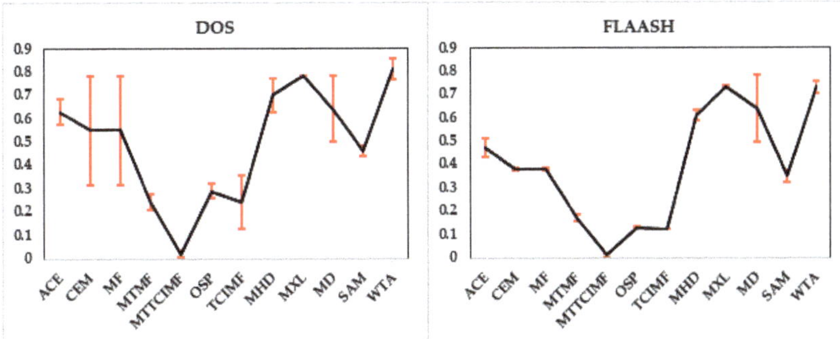

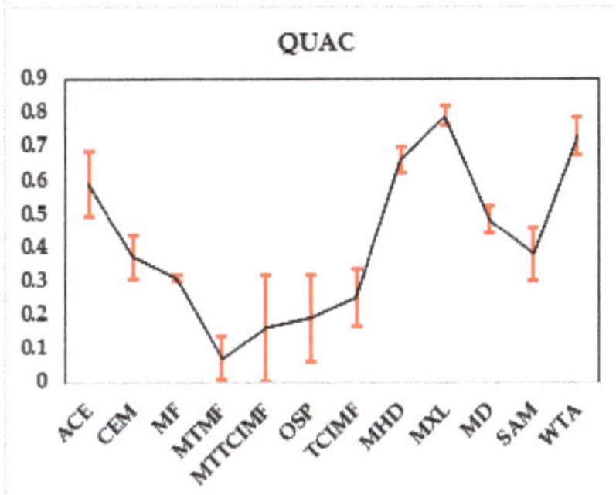

Classifier	Overall Accuracy		
	DOS	FLAASH	QUAC
ACE	0.63	0.47	0.59
CEM	0.55	0.38	0.37
MF	0.55	0.38	0.31
MTMF	0.24	0.17	0.07
MTTCIMF	0.02	0.01	0.16
OSP	0.29	0.13	0.19
TCIMF	0.24	0.12	0.25
MHD	0.70	0.61	0.66
MXL	0.78	0.73	0.79
MD	0.64	0.64	0.48
SAM	0.46	0.35	0.38
WTA	0.81	0.73	0.73

Figure 6. Overall accuracy depicted as a line graph and calculated variances as error bars across all classification algorithms for each atmospheric correction. Bottom-right inset table shows the values of overall accuracy for each of the correction methods.

In summary, the classifier showing the most consistent performance across different atmospheric corrections and test sites was the MXL, followed by WTA. The atmospheric correction showing the least variation across the test sites was FLAASH. The highest OA was achieved by DOS_WTA (0.81), whereas the lowest was achieved by FLAASH_MTTCIMF (0.01). The order of reliability among the atmospheric corrections was: FLAASH > QUAC > DOS. This reliability was based upon the total variance in the OA across all the classifiers and both test sites. The reliability order of classifier performance averaged across both test sites for each atmospheric correction (based on OA) was: DOS_WTA > QUAC_MXL > DOS_MXL > FLAASH_MXL = FLAASH_WTA = QUAC_WTA > DOS_MHD > QUAC_MHD > DOS_MD = FLAASH_MD > DOS_ACE > FLAASH_MHD > QUAC_ACE > DOS_CEM = DOS_MF > QUAC_MD > FLAASH_ACE > DOS_SAM > FLAASH_CEM = FLAASH_MF = QUAC_SAM > QUAC_CEM > FLAASH_SAM > QUAC_MF > DOS_OSP = QUAC_TCIMF > DOS_MTMF = DOS_TCIMF > QUAC_OSP > FLAASH_MTMF > QUAC_MTTCIMF > FLAASH_OSP > FLAASH_TCIMF > QUAC_MTMF > DOS_MTTCIMF > FLAASH_MTTCIMF.

4.2.4. Effect of Pansharpening

Table 10 showcases the average error rate (both study areas) achieved by each classifier for each processing scheme when not pansharpened, GS sharpened, and HCS sharpened. The error rate was a suitable measure to highlight the changes in performance for each scheme. An initial analysis of Table 10 presented a general trend of decrease in classifier performance after pansharpening. GS_DOS_ACE showed an increase in error by 0.18 from DOS_ACE, whereas HCS_DOS_ACE increased by 0.28. The errors in GS_DOS_CEM and GS_DOS_MF increased by 0.32, whereas HCS_DOS_CEM increased by 0.33 and HCS_DOS_MF by 0.34. GS_DOS_MTMF showed a 0.00 increase in error, whereas HCS_DOS_MTMF increased by 0.06. MTTCIMF presents a case of decreasing error by 0.09 and 0.08 for the GS_DOS and HCS_DOS processing schemes, respectively. OSP classification had an increase in error of 0.17 for GS_DOS and 0.14 for HCS_DOS, respectively. TCIMF performance decreased by 0.12 for GS_DOS and 0.08 for HCS_DOS subsets.

Table 10. Average performance of each classifier w.r.t. pansharpening using error rate as the comparative measure. Values of the best-performing classifiers are emboldened and italicized.

Classifier	DOS	FLAASH	QUAC	GS			HCS		
				DOS	FLAASH	QUAC	DOS	FLAASH	QUAC
ACE	0.38	0.53	0.42	0.56	0.80	0.79	0.66	***0.48***	0.77
CEM	0.46	0.63	0.64	0.78	0.79	0.84	0.79	0.62	0.81
MF	0.46	0.63	0.64	0.78	0.79	0.84	0.80	0.52	0.81
MTMF	0.77	0.84	0.78	0.77	0.87	0.84	0.83	0.72	0.81
MTTCI-MF	0.99	0.99	1.00	0.90	0.89	0.88	0.91	0.64	0.59
OSP	0.71	0.88	0.81	0.88	0.85	0.84	0.85	0.72	0.88
TCIMF	0.77	0.88	0.76	0.89	0.80	0.87	0.85	0.54	0.88
MHD	0.30	0.40	0.34	0.42	***0.75***	0.81	0.66	***0.48***	0.52
MXL	0.22	***0.28***	***0.21***	***0.25***	***0.75***	0.77	***0.49***	0.78	***0.45***
MD	0.36	0.37	0.52	0.68	0.80	0.79	0.82	0.81	0.69
SAM	0.55	0.66	0.62	0.67	0.89	0.83	0.79	0.91	0.73
WTA	***0.20***	***0.28***	0.28	0.35	0.76	***0.76***	0.61	0.76	0.46

MHD classification resulted in an increase in error of 0.36 for HCS_DOS and 0.12 for GS_DOS. MXL classification showed an increase in error of 0.27 for HCS_DOS and 0.03 for GS_DOS. MD performance showed an increase in error of 0.46 for HCS_DOS and 0.32 for GS_DOS. SAM showed an increase in the resultant error by 0.24 for HCS_DOS and 0.12 for GS_DOS.WTA classification increased in error by 0.41 for HCS_DOS and 0.15 for GS_DOS. For the FLAASH subsets, GS _FLAASH_ACE decreased in performance by 0.27, whereas the HCS_FLAASH_ACE classification showed an increase in performance by 0.05. HCS_FLAASH_CEM, HCS_FLAASH_MF, HCS_FLAASH_MTMF, HCS_FLAASH_MTTCIMF, HCS_FLAASH_OSP, and HCS_FLAASH_TCIMF showed an increase in classification performance by 0.01, 0.11, 0.12, 0.35, 0.16, and 0.34, respectively. GS_FLAASH_MTTCIMF, GS_FLAASH_OSP, and GS_FLAASH_TCIMF increased in performance by 0.10, 0.03, and 0.08, respectively. GS_FLAASH_CEM and GS_FLAASH_MF each showed a common decrease in performance by 0.16. GS_FLAASH_MTMF increased in error by 0.03. MHD classification decreased in performance by 0.35 and 0.08 for the GS_FLAASH and HCS_FLAASH subsets.

MXL classification increased in error by 0.50 and 0.47 for the HCS_FLAASH and GS_FLAASH subsets. GS_FLAASH_MD, GS_FLAASH_SAM, and GS_FLAASH_WTA increased in resultant error by 0.43, 0.23, and 0.48, respectively. HCS_FLAASH_MD, HCS_FLAASH_SAM, HCS_FLAASH_WTA decreased in performance by 0.44, 0.25, and 0.48, respectively. For the QUAC subsets, only GS_QUAC_MTTCIMF and HCS_QUAC_MTTCIMF showed a decrease in error by 0.12 and 0.41, respectively. GS_QUAC_ACE and HCS_QUAC_ACE showed an increase in error by 0.37 and 0.35, respectively. GS_QUAC_CEM and GS_QUAC_MF each showed a common increase in error of 0.20. Similarly, HCS_QUAC_CEM and HCS_QUAC_MF each resulted in a common increase in error of 0.17. GS_QUAC_MTMF and HCS_QUAC_MTMF resulted in an increase in classification error of 0.06 and 0.03, respectively. GS_QUAC_OSP and HCS_QUAC_OSP delivered an increase in error by 0.03 and 0.07 each. TCIMF classification of GS_QUAC and HCS_QUAC subsets resulted in an increase in classification error by 0.11 and 0.12, respectively. GS_QUAC_MHD and HCS_QUAC_MHD resulted in a decrease in performance by 0.47 and 0.18 each. MXL classification resulted in an increase in error by 0.56 for GS_QUAC and 0.24 for HCS_QUAC. GS_QUAC_MD and HCS_QUAC_MD increased in error by 0.27 and 0.17

each. SAM classification of the GS_QUAC and HCS_QUAC subsets delivered an increase in error by 0.21 and 0.11, respectively. GS_QUAC_WTA and HCS_QUAC_WTA resulted in an increase in classification error by 0.48 and 0.18, respectively. The average variance between DOS, GS_DOS, and HCS_DOS for each classifier produced the following order of increasing variability: MTMF < MTTCIMF < TCIMF < OSP < SAM < ACE = MXL < CEM < MF < SAM < WTA < MD. The average variance between FLAASH, GS_FLAASH, and HCS_FLAASH for each classifier resulted in the following variability order: MTMF < OSP < CEM < MF = SAM < ACE < MTTCIMF = TCIMF = MHD < MD < MXL < WTA. Variability between QUAC, GS_QUAC, and HCS_QUAC resulted in the following order: MTMF < OSP < TCIMF < CEM = MF = SAM < MD < ACE = MTTCIMF < MHD < WTA < MXL. While the MTMF produced the least varying classification between nonsharpened and pansharpened imagery, it possessed a high error for all the subsets of imagery tested in this study.

The cumulative average error rate for the pansharpened subsets revealed the following order of reliability: GS_DOS > HCS_FLAASH > HCS_QUAC > HCS_DOS > GS_FLAASH > GS_QUAC. Upon averaging the effects of atmospheric corrections, the HCS pansharpening was calculated to have a lower error rate (0.71) for all classification algorithms and all atmospheric corrections, when compared to the GS method (0.76). Figures 7–9 display the thematic results of FLAASH_MXL and HCS_FLAASH_MXL.

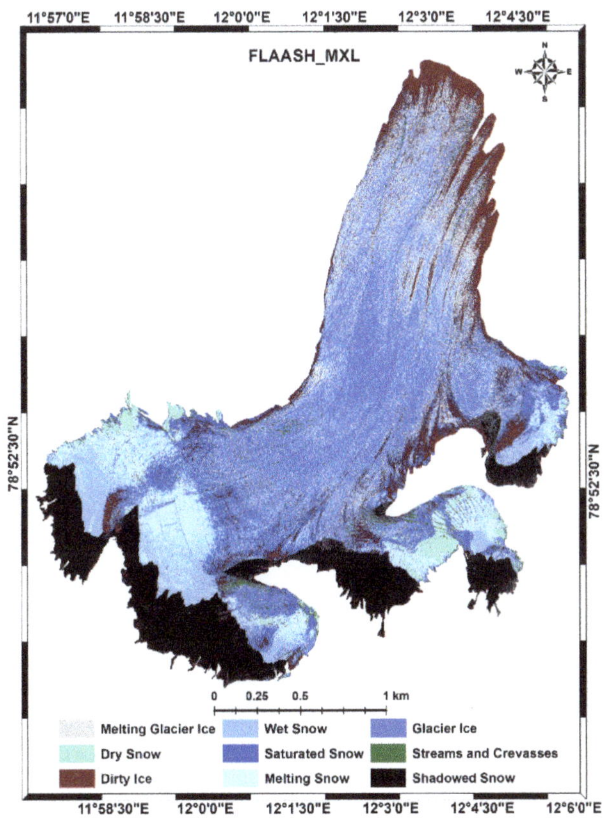

Figure 7. Thematic classification results of the MXL algorithm for the FLAASH atmospheric correction for the ML glacier.

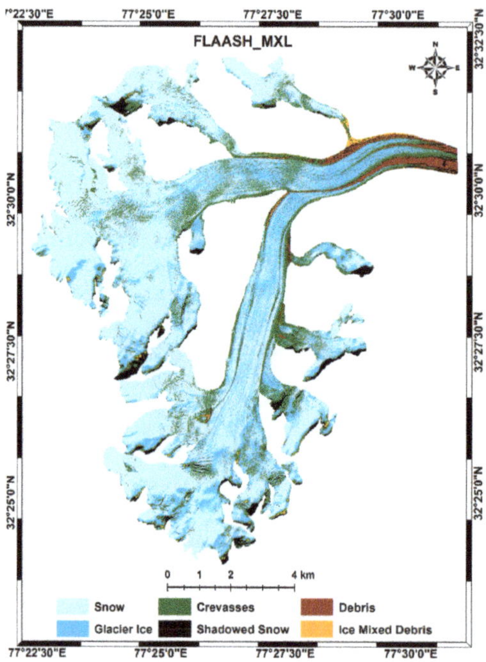

Figure 8. Thematic classification results of the MXL algorithm for the FLAASH atmospheric correction for the ST glacier.

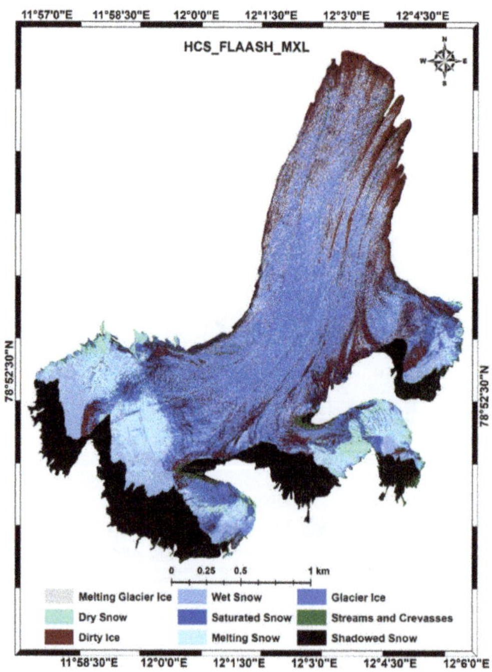

Figure 9. Thematic classification results of the MXL algorithm for the HCS_FLAASH processing scheme for the ML glacier.

4.3. Discussion

Atmospheric correction impacts the quality of observable and derivable spectral reflectance of desired targets. Section 1.4 highlights studies that compared atmospheric corrections and the complexity imbued in the selection process. The overarching consensus states that the choice of atmospheric correction is application-centric [61]. A careful assessment of the impact of atmospheric correction on the spectral signature of target facies in this study was presented in Section 4.1. Previously, the FLAASH correction was used to derive target spectral reflectance and then matched against existing literature for its validity [17]. In the current study, spectral signatures of facies derived from the FLAASH correction were used in the same capacity. The extracted signatures closely matched those observed in previous efforts (Section 4.1), based on the properties of reduced reflectance [62] and mixing of moisture, dust, impurities, and debris [138]. Subsequent extraction of spectral signatures of the same facies from subsets of each processing scheme revealed a variance between each scheme, and thereby, differences in the resultant classifications. The highest variances between DOS, FLAASH, and QUAC were most prominent in B1, B2, B7, and B8. This was potentially because B1 and B2 (Coastal and Blue) were used to provide atmospheric information [142,143], and were therefore affected the most by its effects, whereas B7 and B8 (NIR1 and NIR2) were predominantly affected by water absorption and dispersion of suspended particles [144]. Moreover, Chakouri et al. [57] suggested that the green-to-NIR spectrum is least affected by atmospheric scattering. Analysis of Figures 4 and 5, as well as the maximum and minimum variations of spectra depicted in Table 6, highlighted that the reflectance spectra of facies derived from non-pansharpened imagery showed the least variations between B3 (Green) to B6 (Red Edge), and the most variations were observed at B1 (Coastal) for most of the facies. Reliability based on variance in OA across both study areas for all classifiers suggested that DOS was the worst performer. This was likely because DOS does not emulate atmospheric absorption, and produces a decrement of surface reflectance [145,146]. Moreover, as the minimum value of dark pixels was a combination of atmospheric effects, specular reflection, and skylight scattering from the entire image [147], the DOS configuration was too simplistic for separating overlapping spectral signatures from different classes. FLAASH was the most reliable atmospheric correction method, as the results of the classification from its subsets were the most consistent. This was believed to be due to close matching of FLAASH-derived reflectance with surface reflectance [148]. QUAC performed poorer than FLAASH in the current study, perhaps due to the nonexistence of a minimum of 10 distinct spectral classes [93,149]. Saini et al. [149] further went on to reiterate the realistic reflectance derived through FLAASH.

The literature review of pansharpening (Section 1.3) arrived at a task-specific conclusion, similar to that for atmospheric correction. A visual analysis of Figures 4 and 5 highlighted the decrease in reflectance derived from pansharpened imagery. This implied an overall decrement in spectral signature characteristics. GS_DOS was found to have the most deterioration. Resultant spectra from HCS_DOS/FLAASH/QUAC were found to match closely with the nonsharpened DOS-, FLAASH-, and QUAC-derived spectra. This agreed with Rayegani et al. [47] and Padwick et al. [42]. Spectra of the Chandra–Bhaga basin snow class showed the highest variance (0.29) across GS_DOS, GS_FLAASH, and GS_QUAC. The highest variance across HCS_DOS, HCS_FLAASH, and HCS_QUAC was found for the spectra of the Chandra–Bhaga basin snow class (0.31). Moreover, variance in GS- and HCS-sharpened spectra for the Chandra–Bhaga basin facies was found to be higher than that of the Ny-Ålesund facies. Shadowed snow for both CB and Ny-Ålesund showed a much lower variance across the GS and HCS subsets than other facies. The spectral bands showing the most variance were B1, B2, B7, and B8. This was due to the obvious reason that atmospheric influences (for B1 and B2) were only reduced by appropriate correction algorithms, whereas the effects of moisture and particle mixing (B7 and B8) were target- and scene-specific. Ablation facies were characterized by increasing moisture, saturation, densification to ice and subsequent discharge of melt water. Therefore, the variations in B7 and B8 would most likely be persistent. Although HCS was more stable than GS across all its subsets according to its classification performance (Section 4.2.4), both resulted in a high error rate. One reason could be the compounding effect of the previous atmospheric corrections and subsequent variety of classification algorithms. Previous assertions of the utility of the GS method [38–40] did not hold true in the current study.

4.3.1. Classifiers and Surface Facies: Performance and Comparison

Jawak et al. [41] tested a variety of band ratios and classifiers, such as MTTCIMF, CEM, ACE, OSP, MTMF, MF, MXL, SVM, NNC, and SAM, to map Antarctic vegetation using WV-2 data. Their work showed the prowess of MTMF in mapping sparse vegetation patches. Moreover, the MXL was inefficient in their analysis by creating a maximum number of misclassified pixels. The current study, however, delivered the opposite results. Here, MXL and MD performed better than MTMF in an overall identification of facies. This contrasting result could be because the adjacent classes in their work; namely, landmass/rocks, water bodies, snow/ice on lakes and rocks, shadowed ice, shadowed landmass, melt water, and surface debris on snow/ice, were all highly distinctive in their spectral characteristics. In the case of the surface facies, distinct classes were based on reduction in reflectance properties induced by melt and mixing of particles. This was also noted by Jawak et al. [41] when shadowed ice and melt water on the surface of rocks caused significant misclassification in the AC methods. Kumar et al. [130] attempted to identify minerals at the subpixel level using MTTCIMF in mountainous areas of Rajasthan, India. Their results suggested that MTTCIMF performed poorly when there was high interclass similarity. Portions of the reflectance spectra of facies overlapped each other (Figures 5 and 6), which could have caused poorer performance of MTTCIMF in facies applications. Curiously, MTTCIMF delivered higher accuracy for the GS and HCS subsets for all three atmospheric corrections. This could imply that pansharpening VHR imagery may improve target detection using MTTCIMF. An increase in accuracy was also observed for classification results of ACE, CEM, MF, MTMF, OSP, and TCIMF for the HCS_FLAASH subset. The greatest increase was for the HCS_FLAASH_MTTCIMF classification (0.35 increase, 35% improvement). Although the increase in performance did not improve the overall ranking of the classifiers to a large extent, it was an important factor to note, as an increase in the total number of pixels under the same polygonal unit of the concerned ROI could improve target detection when sharpened by HCS in this instance. Millan et al. [129] tested five AC methods (CEM, ACE, SAM, TCIMF, and MTMF) for estimating reflectance of targets of interest in mine-related flooded areas of Nord-Rhein Westphalia, Germany. They found variable performance of the classifiers for their targets, but recommended SAM, ACE, and MTMF, as targets were better classified using these. They inferred this, as the recommended classifiers showed low sensitivity to Bidirectional Reflectance Distribution Factor (BRDF) effects on target classes. In the current case, each glacier was carefully extracted from the complete imagery, nullifying any valley rock/nonglacier influence. Moreover, any BRDF influence observed over each glacier for the spectral bands would be uniform throughout the classification process, and should not have been a hindrance in the results of the current study [17]. Moreover, Millan et al. [129] also described the effectivity of specific classifiers for individual targets. Here, for the Ny-Ålesund facies, dry snow, wet snow, melting snow, shadowed snow, glacier ice, melting glacier ice, dirty ice, and streams and crevasses were best classified by MXL, whereas saturated snow was equally well classified by MXL and WTA. For the Chandra–Bhaga basin facies, crevasses and shadowed snow were best classified by WTA, whereas glacier ice, IMD, debris, and snow were best classified by MD. Jin et al. [150] tested MF, SAM, CEM, TCIMF, ACE, and OSP for target detection at subpixel and full-pixel scales over Cooke City, Montana, USA. Their findings suggested that classifiers based on matched filters (MF, CEM, MTMF, and TCIMF) had poor generalization, causing misclassification of pixels belonging to the same class but with slightly different spectral signatures. Poor performance of the MF-based classifiers in the current study supported the same inference. Furthermore, they found that the ACE classifier achieved better target visibility and classification than other AC methods. Here, ACE was the best performer of all the AC algorithms across all processing schemes, and therefore validated the same inference. Jawak and Luis [151] assessed the performance of SVM, MXL, NNC, SAM, and an ensemble WTA to map land cover in Larsemann Hills, Antarctica, using HCS-sharpened WV-2 imagery. They found that WTA performed best, while MXL performed worst. However, the accuracies of all methods were quite high. In the current study, MXL and WTA performed well across all processing schemes. The performance of SAM as shown by Jawak and Luis [151] depended upon the abundance and separability of spectral classes. A significant feature of their study was the nonoverlap of land cover classes, thereby resulting in high accuracy. The differences in facies, however, was not essentially a sharp contrast. The variations from accumulation to ablation can cause confusion if spectral signatures are not carefully considered. Moreover, SAM performed poorly here, leading to unclassified pixels, consequently causing unclassified pixels in the WTA classification. The default settings in ENVI were used to enable an unbiased analysis of classifier performance. This may have led to low representation of the maximum angle between the ROI and input pixel spectrum. Luis and Singh [60] attempted to map surface facies on the Edithbreen glacier in Ny-Ålesund, Svalbard, using VHR WV-3 and Landsat 8 OLI data. Their attempt focused on comparing pixel- and object-based methods. While the object-based methods thoroughly achieved the best results, the pixel-based results for MHD, MXL, and MD highlighted the robustness of MXL. However, MD performed poorly in their analysis, which countered the current findings; this could be possibly be due to the larger number of overlapping classes defined by Luis and Singh [60]. Albert [59] compared MXL, MD, parallelepiped (PP), SAM, ISODATA, linear unmixing, MTMF, a range of fuzzy classification methods, and band math indices/techniques to delineate ice cover around the tropical Quelccaya ice cap in Peru. The author used a DOS-corrected Landsat 5 TM image. The final processing steps involved conversion of radiance to reflectance units. However, in the current study, DN was first converted into radiance and then to TOA reflectance before performing user-defined DOS [98]. Albert [59] found SAM to be the most accurate of the supervised classifiers. This occurred due to testing of a variety of angular thresholds. The author noted that that the supervised classifiers performed well because snow and ice were one thematic class, rather than bifurcated into different snow and ice facies. This suggested that additional categories of closely matching spectra, but with distinct thematic features, may reduce the performance of many supervised classifiers. The results of the current study were consistent with this.

Pope and Rees [10] mapped glacier facies on the ML glacier using in situ spectral reflectance on field-observed surface classes of facies through ETM+ imagery classification. The authors used an unsupervised ISODATA algorithm and principle-component-derived linear combinations (LCs) to categorize surface classes. Several classes identified by the authors were based upon field assessment of grain size and visible/flowing water on the surface. The spectral reflectance signatures used for validation were collected a decade after the image acquisition. This was the opposite of the recent recommendation by Yousuf et al. [36], who suggested snow and ice validation data should be closely timed with the image capture. Due to nonavailability of field spectra, the current study relied on image and spectra interpretation and published literature for references and a comparison of spectral signatures (refer to Sections 3.4 and 4.1, and [17]). The comparative pixel- and object-based characterization of glacier facies on the Edithbreen glacier by Luis and Singh [60] presented a curious case. Common classes between their work and the current study included dry and wet snow, melting ice, and shadow. Uncommon classes included percolation snow, dirty snow, debris, off-glacier, water stream, and crevasses. The separation of water stream and crevasses was avoided in the current study due to increased spectral confusion and probability of misclassification. Moreover, absence of in situ/reference spectra enhanced the chances of incorrect training of ROIs. Manual digitization of glacial boundaries negated the need for an off-glacier class. Furthermore, when comparing Landsat 8 (L8) and WV-

3, their L8 classification showed a class called wet semisaturated snow. This class was identified by Pope and Rees [10] after in situ collection, and was separated from dry semisaturated ice by an increased amount of water on the surface. The quantity of water that marked the spectral contrast between wet snow and wet semisaturated snow was not easily identified with medium-resolution satellite data. Luis and Singh [60] touted the effectivity of the object-based indices; however, the thematic results showcased dirty "snow", a snow surface facies class on the ablation area of the glacier, using WV-3, but then labeled the same as dirty ice in classification of L8 data. Nevertheless, their work provided an important reconnaissance for future effective mapping of surface facies in the region. The current study built upon their findings and objectively characterized facies using visual and spectral characteristics. Jawak et al. [17,33] mapped surface facies using a combination of pixel- and object-based classification techniques on FLAASH-corrected WV-2 data. Their goal was to effectively characterize facies for the Chandra–Bhaga basin region using VNIR VHR satellite data. The current study aimed to test the impacts of varying processing schemes on the overall classification of surface facies to determine the most efficient and accurate method for future mapping attempts. Testing of additional supervised classifiers such as TCIMF, MTTCIMF, MTMF, and ACE proved useful in highlighting the utility of matched filtering methods and the performance of ACE. Moreover, improvement in classification accuracy by MTTCIMF after pansharpening was an important clue to the potential implementation of the algorithm using in situ spectra and/or aerial imagery.

4.3.2. Computer Processing Time and Limitations

Successful data-driven remote sensing applications depend upon stable computational infrastructure. In the preceding sections, this study qualitatively and quantitatively assessed classified thematic results of different processing schemes. Therefore, it is now necessary to evaluate the computational requirements and loads of the individual schemes. To this end, the study considered factors such as data acquisitional challenges, system properties, time taken for processing, and storage space needed. The specifications of the primary system used to process data in this study (System A) consisted of 16 GB of RAM (DDR4), an SSD with 256 GB, an HDD with 1 TB, a ninth-generation Intel® Core™ i7-9750H (64-bit) processor, and a NVIDIA® GeForce® GTX 1650 (4GB) graphics card. In addition to this, an additional portable hard drive with more than 4 TB of storage was needed to store all the generated data. All files, beginning from calibration to classification, were saved in default ENVI formats to maintain uniformity. The HCS sharpening performed in ERDAS IMAGINE 15 necessitated input files in its default IMG imagine format. The HCS-sharpened files were then exported back into the default ENVI format (.dat) for further classification. This standardization of file formats was in line with the recommendations by Shcadt et al. [152] for big data management. Table 11 highlights the complete processing time and storage space needed, from calibration to classification, for each of the image-processing schemes for glaciers ML and ST. This presents an account of the practical limitations when processing data for such applications. In terms of the processing schemes, the GS and HCS products occupied the maximum disk space and processing time with no superior enhancements in accuracy. The Himalayan glacial subsets, being bigger in size, took the most space and time. AC classification of the Samudra Tapu GS_FLAASH subset had the largest file size (312 GB), and consequently took the most processing time (593 h) of all the subsets and processing schemes tested in the current study.

Table 11. Stepwise break down of the time taken and storage space occupied at each processing step, which included radiometric calibration, pansharpening, band math conversion (0 to 1 reflectance units), classification, conversion of raster thematic data to vector files, and export of the resultant vector files. The time is displayed in hours, and the space occupied is provided in parenthesis after the time.

Test Site Image Subset	Time in Hours (and Storage Space Occupied)					Total Time in h	Total Storage in GB
	Radiometric Calibration	Pansharpening	Band Math	Classification	Exporting Shapefiles (h)		
Midtre Lovénbreen (ML)	*DOS:* 0.50 h (0.44 GB)	–		*TD:* 1.58 h (5.04 GB)	4.00	6.08	5.48
				TERCAT: 1.08 h (1.94 GB)	2.00	3.58	2.38
		GS: 1.00 h (7.25 GB)	–	*TD:* 56.00 h (96.80 GB)	100.00	157.5	104.49
				TERCAT: 48.00 h (34.50 GB)	35.00	84.50	42.19
		HCS: 0.38 h (8.15 GB)	–	*TD:* 51.00 h (90.80 GB)	86.00	137.88	99.39
				TERCAT: 44.00 h (31.50 GB)	29.00	73.88	40.09
	FLAASH: 0.83 h (0.23 GB)	–	0.33 h (0.45 GB)	*TD:* 2.17 h (5.05 GB)	9.00	12.33	5.73
				TERCAT: 1.68 h (0.14 GB)	1.00	3.84	0.82
		GS: 1.13 h (6.57 GB)	0.57 h (6.60 GB)	*TD:* 60.00 h (81.10 GB)	83.00	145.53	94.5
				TERCAT: 50.00 h (31.30 GB)	32.00	84.53	44.7
		HCS: 0.42 h (9.31 GB)	0.50 h (6.57 GB)	*TD:* 54.00 h (80.70 GB)	82.00	137.75	96.81
				TERCAT: 45.00 h (31.30 GB)	30.00	76.75	47.41
	QUAC: 0.70 h (0.59 GB)	–	0.30 h (0.64 GB)	*TD:* 1.77 h (5.05 GB)	9.00	11.77	6.28
				TERCAT: 1.50 h (1.95 GB)	4.00	6.50	3.18
		GS: 1.00 h (6.57 GB)	0.50 h (6.70 GB)	*TD:* 55.00 h (76.5 GB)	64.00	121.20	90.36
				TERCAT: 46.00 h (29 GB)	25.00	73.20	42.86
		HCS: 0.47 h (9.31 GB)	0.41 h (6.60 GB)	*TD:* 51.00 h (80.7 GB)	74.00	126.58	97.2
				TERCAT: 43.00 h (29.9 GB)	28.00	72.58	46.4
Samudra Tapu (ST)	*DOS:* 1.00 h (2.15 GB)	–		*TD:* 3.28 h (19.80 GB)	24.00	28.28	21.95
				TERCAT: 2.45 h (6.75 GB)	10.00	13.45	8.9
		GS: 1.25 h (35.6 GB)	–	*TD:* 65.00 h (219.00 GB)	374.00	441.25	256.75
				TERCAT: 57.00 h (70.60 GB)	61.00	120.25	108.35
		HCS: 1.56 h (43.50 GB)	–	*TD:* 68.00 h (221.00 GB)	336.00	338.56	266.65
				TERCAT: 58.50 h (75.6 GB)	71.00	132.06	121.25
	FLAASH: 1.56 h (0.81 GB)	–	1.58 h (2.62 GB)	*TD:* 4.12 h (19.80 GB)	24.00	31.26	23.23
				TERCAT: 3.34 h (0.58 GB)	3.00	9.48	4.01
		GS: 2.40 h (16.50 GB)	1.85 h (33.00 GB)	*TD:* 76.00 h (312.00 GB)	512.00	593.81	362.31
				TERCAT: 66.00 h (130.00 GB)	104.00	175.81	180.31
		HCS: 1.00 h (50.00 GB)	1.75 h (102.00 GB)	*TD:* 70.00 h (282.00 GB)	432.00	506.31	434.81
				TERCAT: 59.10 h (109.00 GB)	96.00	159.41	261.81
	QUAC: 1.35 h (1.10 GB)	–	1.50 h (2.06 GB)	*TD:* 3.80 h (19.80 GB)	24.00	30.65	22.96
				TERCAT: 3.10 h (0.683 GB)	3.00	8.95	3.843
		GS: 2.10 h (16.50 GB)	1.80 h (33 GB)	*TD:* 72.6 h (254 GB)	418.00	495.85	304.6
				TERCAT: 64.00 h (108 GB)	96.00	165.25	158.6
		HCS: 0.90 h (20.50 GB)	1.72 h (33 GB)	*TD:* 68.40 h (282 GB)	432.00	504.37	336.6
				TERCAT: 56.10 h (108 GB)	96.00	156.07	162.6
Total time and storage						**5247.05 h**	**3909.80 GB**

The smallest input file was the Midtre Lovénbreen DOS subset (0.44 GB). As an example of the comprehensive approach of the study, the total time for image calibration and classification of the glacial subsets of Samudra Tapu and Midtre Lovénbreen was 5247.05 h, with a combined disk size of 3909.80 GB. The CC methods, being lesser in number, consequently occupied less disk space. Considering the accuracies delivered by AC and CC algorithms (Section 4.2.2), and the respective times for processing and storage space (Table 11), it was evident that the CC methods were much more efficient.

4.3.3. Inherent Challenges and Limitations

In the quest for assessing image-processing impacts on mapping glacier facies, the current study encountered and attempted to resolve several computational challenges, as described in the sections above. However, some challenges for mapping facies and supraglacial terrain were inherent to this application itself. Factors such as cloud cover, seasonal snow, precipitation, and crevassed surfaces pose difficulties for efficient snow and ice delineation [153]. Debris cover is a challenge for glacier terrain mapping due to its spectral confusion with the surrounding valley rocks [154], as most of the debris on the surface of a glacier is deposited either by rockfalls from the surrounding valley or is entrained into the glacial mass during its movement from the bed rock. This can cause erroneous mapping, as spectral signature-based classification methods may misclassify supraglacial debris and valley rock due to the resultant spectral similarity. Shadowed snow, dependent on sun azimuth and solar elevation [155], can create areas of spectral mixing, causing misclassifications [17]. Cumulatively, the topography of the area [156], the time/season of capture, illumination conditions, and local weather [155] all play key roles in determining the features visible on a glacier's surface and the "quality" of the image. Bernardo et al. [142] found that a residue of atmospheric attenuation was left in the image after atmospheric correction. This influenced band-by-band comparison against known spectral libraries and in situ spectra. In such a case, selection of appropriate atmospheric correction algorithms would be of prime importance when mapping methods rely upon reflectance characteristics. For example, DOS correction ignores the effect of atmospheric dispersion on spectral signatures and is often too simplistic, resulting in a decrement in surface reflectance [146]. Algorithms such as FLAASH provide the most realistic reflectance pattern, as they consider variables such as sensor altitudes and atmospheric and aerosol models to reduce the compounding effects of atmospheric attenuation. Therefore, while simplistic methods such as DOS are convenient and time-efficient, they may retain more noise than sophisticated methods such as FLAASH. WV-2 applicability for glacier facies mapping was demonstrated in the past [17,33,60] and in this study as well. However, the WV-2 dataset is expensive to procure; this adds logistic impediments to its usage for long-term temporal monitoring of facies.

Remotely sensed imagery can acquire noise at any moment, beginning from acquisition/image capture, rectification procedures, geometric corrections to enhancements, and even compression from data storage and transmission procedures [157,158]. VHR satellite data consists of noise acquired during the acquisition and transmission [159]. According to Liang et al. [160], this is an impulse noise, also called the salt-and-pepper noise, which presents as white and black pixels in the spectral image. Estimation of this noise is an important part of information-extraction procedures, specifically for hyperspectral data. The minimum noise fraction (MNF) transformation [161] in ENVI is a two-step principal component analysis that enhances the quality of data by reducing computational requirements. This is performed by reducing data dimensionality and segregating the noise to yield higher-order components comprising noise-free, coherent eigen images [17,162,163]. Noise statistics in the form of eigenvalues for each spectral band are generated in the forward MNF transformation [163]. Table 12 highlights the estimated noise in the raw image and for each processing scheme for subsets of Samudra Tapu, Chandra–Bhaga basin, and Ny-Ålesund, Svalbard.

Table 12. Eigenvalues of noise contained in the VHR WorldView-2 data of the test sites. Noise was computed for the raw data and each processing scheme for two selected glacial subsets. Noise was calculated following the MNF operation in ENVI.

Test Subset	Spectral Bands	Noise within the Processing Scheme Subsets									
		Raw DN	*DOS*	*FLAASH*	*QUAC*	*GS*			*HCS*		
						DOS	*FLAASH*	*QUAC*	*DOS*	*FLAASH*	*QUAC*
Samudra Tapu	**B1**	443.81	152.56	290.19	180.59	834.25	647.47	493.75	350.27	179.23	624.17
	B2	16.72	15.68	13.11	15.49	411.80	207.93	412.07	172.58	43.79	312.83
	B3	5.52	3.75	6.65	5.29	108.70	51.59	158.97	36.17	11.86	92.01
	B4	2.57	2.42	2.39	3.07	58.31	20.20	88.51	25.48	6.98	77.26
	B5	1.68	1.21	2.00	1.21	14.59	16.33	25.34	10.11	6.66	23.59
	B6	1.18	1.10	1.16	1.10	12.83	15.44	20.81	9.19	4.18	19.04
	B7	1.08	1.03	1.05	1.03	8.26	12.34	19.54	5.46	3.73	18.21
	B8	0.96	0.94	0.99	0.95	2.15	7.62	14.99	3.01	3.37	14.01
Midtre Lovénbreen	**B1**	144.38	116.77	56.14	50.05	866.56	314.90	203.19	41.28	37.54	51.60
	B2	27.41	26.67	10.13	17.57	57.59	52.55	50.12	13.57	10.72	21.78
	B3	2.25	2.05	2.07	3.04	47.93	35.13	30.15	9.32	8.93	11.77
	B4	1.18	1.17	1.29	1.72	19.88	18.72	17.20	6.41	8.23	11.28
	B5	1.15	1.17	1.18	1.65	18.79	18.14	18.02	8.49	7.69	10.83
	B6	1.11	1.12	1.11	1.51	17.58	15.60	13.19	5.36	7.36	9.95
	B7	1.00	1.00	1.01	1.24	13.78	13.90	14.24	5.21	6.70	9.48
	B8	0.99	0.98	0.99	1.00	13.13	12.83	12.21	3.01	5.06	8.97

Spectral bands with eigenvalues (Table 12) closer to 1 contained noise, and those with values greater than 1 contained data [163]. Hence, when averaged across all processing schemes, B1 contained the maximum data and least noise, followed by B2, B3, B4, B5, B6, B7, and B8, which comprised the least data with maximum noise. While the performance of each band was consistent, the processing schemes had a variable level of noise in each subset. In the case of hyperspectral data, this noise statistic is important when considering which spectral bands can be retained for further classification/processing (bands with high data) and which bands can be rejected (bands with high noise). In the present study, only eight spectral bands comprised the current set of imagery. Moreover, the aim of this experiment was to gauge variations induced by each processing scheme in the resultant spectral reflectance and thematic classification. An inverse MNF transform was suggested to denoise imagery [163], and a comparative test of denoised imagery has promising future potential for assessing the impact on glacier facies extraction. The current study served as a baseline for such a potential test.

In the current study, surrounding valley rocks were not incumbent to classification, as the glacial subsets were manually digitized and extracted. A greater number of atmospheric corrections and pansharpening methods were not tested here. However, the current methods were sufficient to observe the overall effects of different methods. B1, B2, B7, and B8 displayed the maximum variation in spectral reflectance across each processing scheme and for most of the identified facies (Table 6). This can be a limitation for the FLAASH correction, as its execution has been noted to have significant dependence upon the visible bands, especially the blue (here, B2) band [56]. QUAC was limited by less than 10 spectral classes, as the discernible surface facies numbered 9 for the Ny-Ålesund glaciers and 6 for the Chandra–Bhaga basin glaciers. Pansharpened data took the most time for classification and occupied the most disk space, without resulting in any improvements in accuracy. Some classification algorithms, such as MTMF [41] and SAM [59], have shown better results for identifying land cover and snow classes, but were not apt at discriminating the minute variation in facies. Ramezan et al. [164] suggested utilizing larger training data samples for improving overall accuracy across classification methods. This may improve AC methods in future. Lack of field data limited the current study. However, by utilizing an equalized sampling approach described by Keshri et al. [134], the study assigned equal points to each facies class, and gauged them not only for their accuracies and average error rates, but also for the variances between each processing scheme. This ensured that the end goal—understanding the impacts of image-processing schemes on the VHR classification of surface facies—was achieved.

Albedo plays a crucial role in the analysis of surface characteristics of glacial bodies. As glacier surface facies vary in terms of their reflectance characteristics, the albedo of these facies will differ. Moreover, glacier surface facies are completely discernible only at the end of the ablation season. Any precipitation event prior to image acquisition will cover the target facies with snow and hamper effective characterization. However, the influence of precipitation on reflectance and the variations in albedo of surface facies are influential characteristics that require their own independent study. Nevertheless, freely available albedo products such as the Moderate Resolution Imaging Spectroradiometer (MODIS) MCD43A3 [165], the Suomi National Polar-Orbiting Partnership (Suomi NPP), NASA Visible Infrared Imaging Radiometer Suite (VIIRS) VNP43IA3 [166] and VNP43MA3 [167], and the Copernicus Global Land Service (CGLS) VEGETATION sensor on the Project for On-Board Autonomy platform (PROBA-V) Surface Albedo (SA) [168] version 1.5.1., are available for assessment against glacier facies maps. In this context, however, spatial resolution and data gaps play a large role in determining usability of these products. For facies mapped at resolutions of 2 m and less (akin to the current study), direct comparison between albedo data of 500 m resolution or more can be difficult for small glaciers such as Midtre Lovénbreen. For comparison, the current study downloaded the VIIRS VNP43IA3 (500 m resolution) product for Ny-Ålesund, and CGLS PROBA-V v1.5.1 (1 km resolution) for the Chandra–Bhaga basin. Supplementary Figure S2 highlights the gaps in CGLS PROBA-V SA v1.5.1 during the month and year of image acquisition near the Samudra Tapu glacier. Such large gaps rendered the dataset unsuitable for the current study. Similar, albeit smaller, gaps were observed for Ny-Ålesund for the month and year of image acquisition in the VIIRS VNP43IA3 dataset. The VNP43IA3 products are available for download at [169], and the CGLS PROBA-V SA v1.5.1. products can be downloaded at [170]. In studies that involve mountain glaciers and assessment of glacier surface characteristics, coarse resolution albedo is not useful [3]. Hence, Naegeli et al. [3] utilized Sentinel-2 and Landsat 8 data to derive albedo products using narrow-to-broadband conversion formulae described by Knap et al. [171] and Liang [172] and compared them to albedos derived from the Airborne Prism Experiment (APEX) imaging spectrometer. Zhou et al. [173] concluded that albedo derived from moderate-resolution sensors such as Landsat 8 is almost free of the mixed pixel effect, and thus results in greater accuracy than coarse-resolution albedo (500 m). Moderate-resolution albedo can be useful for binary glacial surface characterization (snow and ice albedo). However, when multiple supraglacial features are mapped, finer-resolution products are better suited at capturing the small-scale heterogeneity of glacier surfaces [3].

Local weather conditions and sudden precipitation events or dust storms will impact albedo and spectral reflectance. Dust exerts direct and indirect effects on the earth's radiation absorption, scattering, and energy balance [174]. Mineral or light-absorbing dust on the surface of a glacier can influence the spectral reflectance of facies, decrease surface albedo, and thereby increase melting of snow and ice [175,176]. Global and regional air mass circulation highlight a significant contribution from Iceland in air masses containing submicron dust particles reaching the Arctic [177,178]. The most effective measure of dust and debris mineralogy is performed using in situ analytical techniques such as X-ray diffraction (XRD), as performed by Moroni et al. [175] to differentiate between local and transported dust in Ny-Ålesund. However, satellite image analysis of supraglacial mineralogy as performed by Casey et al. [179] for the Ngozumpa and Khumbu glaciers in the Himalayas required application of mineral indices, inclusion of shortwave and thermal infrared (SWIR and TIR) spectral bands, and hyperspectral reflectance. While the separation of dust and facies spectral reflectance can improve the identification of glacier facies and enhance complex distributed mass balance modeling, the current study was limited by a lack of SWIR/TIR wavelengths, hyperspectral data, and in situ analytical verification. The present experiment focused upon the reflectance variations introduced by changing image-processing parameters in the easily observable surface facies.

Moreover, testing the effects of precipitation would also need corroborative seasonal/multitemporal imagery. At present, this was beyond the scope of this study. Nevertheless, freely available precipitation data for Ny-Ålesund [180] showed zero precipitation on the date of image acquisition. Similar data for the Chandra–Bhaga basin was not found at the time of writing this manuscript.

4.3.4. Significances and a Path Forward

Selection of image-processing schemes is of paramount importance for accurate identification of image targets, subsequent analysis of spectral reflectance, and thematic classification. The literature described in Section 1 highlighted the application-centric notion of selection of image-processing methods. Robust processing routines can prepare satellite data for a variety of information-extraction methods. A standardized processing routine, if defined for glacier surface facies mapping, would go a long way in enabling temporal monitoring. In the current study, the FLAASH correction retrieved the most reliable reflectance in comparison to DOS and QUAC. Pansharpening did not necessarily improve classification accuracy. GS produced the worst spectral reflectance when coupled with the atmospheric corrections, whereas HCS showed detrimental performance with QUAC and DOS, but an improved performance for AC methods with FLAASH. The MTTCIMF classifier showed improved performance for GS sharpening as well, but only for FLAASH correction. Matched-filtering-based classifiers are poor at generalization [150], and as such can misclassify pixels of the same facies with a small variation in spectral reflectance. Therefore, the MF, CEM, MTMF, TCIMF, and MTTCIMF classifiers are disadvantaged for mapping facies, as often elevation and illumination differences on the glacier can cause the same facies to show a small deviation from its average spectral signature. Moreover, Jin et al. [150] concluded that the ACE classifier delivered better performance than the OSP. The findings of the present study corroborated the performance of the matched-filtering-based methods, as well as the better performance of ACE. Therefore, while ACE was not as accurate as the CC methods, it was the best performer of the AC methods. The CC methods delivered the best performance with a more efficient computer processing time. The biggest disadvantage of the AC methods was the time needed for processing. This also was true for the GS and HCS sharpening. Pansharpening of VHR imagery did not improve glacier facies mapping; rather, it added an excessive computational load. MXL is a robust and efficient information-extraction method and provides the most consistent results across a range of VHR processing schemes. The MXL classifier was the best overall classifier; however, MD also showed significant results for the Chandra–Bhaga basin glaciers. MXL was previously shown to be a reliable algorithm in areas of spectral confusion [59], and to deliver accurate results [181]. Moreover, the entire processing and classification was performed on VNIR data, thus reiterating its utility in mapping facies in the absence of SWIR or thermal data [10,17]. Improvements over previous attempts included an attempt at mapping facies of two geographically distinct groups of glaciers (15 glaciers in total). With three atmospheric corrections, two pansharpening algorithms, and 12 classification algorithms, for a total of 15 glaciers, the current study evaluated an exhaustive 1620 thematic surface facies accuracy measurements.

Implementing a robust image-processing routine would aid in standardized preparation of satellite data and highlight the effects and anomalies that may result and promote another area of research. Accurate derivation of facies may also help calibrate distributed mass balance modelling [18]. Keeping this at the center, the study provided the following recommendations for further attempts. (1) the FLAASH algorithm would retrieve the best spectral reflectance, while being slightly sensitive to residual atmospheric effects in the blue band [142]. (2) HCS may enhance target detection of facies only if coupled with FLAASH for WV-2 data; however, Snehamani et al. [45] suggested considering the usage of pansharpening based on the value of time and cost. Here, GS and HCS subsets were the bulkiest and took the maximum processing time. As no significant improvement in overall accuracy was observed in this case, the study refrained from recommending it for future use. (3) Between the CC and AC methods, AC was the most computationally demanding. CC processing was faster and more accurate. ACE and MTTCIMF are recommended from the AC for future testing for improved mapping using larger training samples [157], and if possible, more spectral bands. Among the CC methods, MXL and MD are recommended for further use. (4) Different information extraction approaches such as machine learning [182], SVM [54], object-based mapping [99], band ratios [11], and multidataset/auxiliary layers [36] can be tasked in the future to comparatively map complex facies against the results achieved here. Analysis of denoised satellite imagery for future mapping of glacier facies can be compared with the baseline results of this study. Finer-resolution albedo products derived from in situ measurements or high-spatial-resolution satellite data can be assessed against facies' reflectance spectra. In addition to denoising, advanced classification methods such as object-based mapping have helped reduce the effect of salt-and-pepper noise, and thus open another pathway for a potential comparative study [183].

This study was the first of a three-part series that will present a complete account of image-processing routines, parameters, and their associated impacts on the thematic classification of glacier surface facies. The current study focused on image-processing routines and pixel-based classification techniques. The forthcoming studies will focus on more complex information extraction methods, the combined effects of processing parameters on the different classification techniques, and a band-by-band analysis of all these attempts at mapping facies with upcoming methods.

5. Conclusions

This study evaluated three atmospheric correction methods and two pansharpening methods for their impacts on glacier facies classification of five conventional and seven advanced classifiers. This was carried out using WV-2 data for glaciers in two separate cryosphere regions: Chandra-Bhaga basin, Himalayas; and Ny-Ålesund, Svalbard. The atmospheric correction methods included DOS, QUAC, and FLAASH. The pansharpening methods included GS and HCS. The conventional methods consisted of MHD, MXL, MD, SAM, and WTA. The advanced methods consisted of ACE, CEM, MF, MTMF, MTTCIMF, OSP, and TCIMF. The focus of the work was on testing the effects of variations in processing schemes on the resultant classification of surface facies using VHR WV-2 imagery, and not on the mapping of facies to the highest accuracy possible. This permitted the use of image-derived spectra and visual interpretation to assign validation points. FLAASH-derived spectral signatures were used as a reference for comparison against the literature, with good agreement. The lack of field data was not a hindrance, as the accuracy assessment focused on analyzing the deviation in performance between each processing scheme before cumulatively assigning the classifiers a reliability order/ranking. The FLAASH subsets delivered higher overall accuracies, followed by QUAC and DOS. The MXL classifier was the least variant across the three atmospheric corrections, delivering OA values of 0.78, 0.73, and 0.79 for DOS, FLAASH, and QUAC corrections, respectively. WTA classification of the DOS subsets resulted in the highest OA of 0.81, whereas the lowest OA (0.01) was delivered by MTTCIMF classification of the FLAASH subsets. Pansharpening did not improve performance, but rather caused a decrement in the derived reflectance, as well as in classifier performance. Based upon the average error rate of the classified GS and HCS subsets, the following order of reliability was derived: GS_DOS > HCS_FLAASH > HCS_QUAC > HCS_DOS > GS_FLAASH > GS_QUAC. Cumulatively, The HCS pansharpening delivered better results than the GS pansharpening. For the Chandra–Bhaga basin glaciers, crevasses and shadowed snow were best mapped by WTA (F1 scores of 0.89 and 0.95); glacier ice,

IMD, debris, and snow were best classified by MD (F1 scores of 0.96, 0.55, 0.62, and 0.98, respectively). For the Ny-Ålesund glaciers, dry snow, wet snow, melting snow, shadowed snow, glacier ice, melting glacier ice, dirty ice, and streams and crevasses were best mapped by MXL, with F1 scores of 1.00, 0.69, 0.68, 1.00, 0.74, 0.57, 0.73, and 0.81, respectively. Saturated snow was classified equally well by WTA and MXL (F1 score of 0.53). The final order of classifier performance, independent of atmospheric corrections and pansharpening, was: MXL > WTA > MHD > ACE > MD > CEM = MF > SAM > MTMF = TCIMF > OSP > MTTCIMF. The best CC method was the MXL, whereas the best AC method was the ACE. An assessment based on computational time suggested that FLAASH correction followed by MXL classification was the most efficient mechanism for supervised classification of surface facies. The experiment carried out here was an exhaustive assessment to decipher which method of image processing was the most efficient and accurate for surface facies classification. Future recommendations have been provided to test the robustness of the current results and potentially apply it across a larger scale. As an important indicator of a changing planet, accurate derivation of surficial glacier properties will play a key role in the broader analysis of environmental change. This study presented an important first phase in the development of an efficient mapping and monitoring system.

Supplementary Materials: The following are available online at https://www.mdpi.com/article/10.3390/rs14061414/s1, Figure S1: Example of polygonal training ROIs for classification displayed upon Samudra Tapu and Midtre Lovenbreen, respectively. The ROIs were assigned after visual and spectral analysis of the observable facies. Both the images are portrayed with a band combination of Red: NIR1 (B8), Green: Red (B5), and Blue: Green (B3). Figure S2: Data gaps in CGLS PROBA-V Surface Albedo v1.5.1. for the date of 24 October 2014. The inset is a highlight the global data product, leading to a zoomed inset of the Indian Himalayas showing the location of the Samudra Tapu (Chandra–Bhaga basin). Inset b displays the boundary of the Samudra Tapu glacier and isolated pixels with albedo data. Table S1: Nomenclature of processing schemes used in the current study. TP: samples are those that were in the positive class and were correctly classified, TN: samples that were correctly classified as negative, FP: samples that were not truly of the positive class but were incorrectly mapped as positive, FN: samples that were mapped as negative when they actually were positive [134]. Sheet S1: Excel sheet showing all the average derived values for each measure of accuracy. Each measure is presented with values for all the atmospheric corrections, pansharpening methods, classification algorithms, and averages across all processing schemes.

Author Contributions: Conceptualization and experiment setup, S.D.J. and S.F.W.; methodology and manuscript design, S.D.J. and S.F.W.; software and processing, S.D.J. and S.F.W.; validation, S.D.J. and S.F.W.; result analysis and discussion, S.D.J. and S.F.W.; resources, A.J.L. and K.B.; writing—original draft preparation, S.F.W. and S.D.J.; writing—review and editing, S.D.J., A.J.L. and K.B.; visualization, S.F.W.; supervision, S.D.J., A.J.L. and K.B.; project administration, S.D.J., A.J.L. and K.B. All authors have read and agreed to the published version of the manuscript.

Funding: The article processing charge (APC) for this publication was covered by the SIOS-Knowledge Center (Research Council of Norway project no. 291644) through the Early Career Researcher presentation award for S.F.W. during the SIOS Online conference on EO and RS, June 2020.

Data Availability Statement: Freely available data used in the current study- (1) ASTER GDEM v2. Downloaded from: Gdex.cr.usgs.gov/gdex/ (accessed on 2 February 2017). The data is now moved to GDEM v3: asterweb.jpl.nasa.gov/gdem.asp (reviewed on 12 March 2022) ASTER GDEM is a product of Japan's Ministry of Economy, Trade, and Industry (METI) and NASA. (2) Arctic DEM. Available online: Pgc.umn.edu/data/arcticdem/ (accessed on 21 January 2019). (3) VIIRS VNP43IA3 Albedo Product. Available online: https://lpdaac.usgs.gov/products/vnp43ia3v001/ (accessed on 25 February 2022). (4) CGLS PROBA-V Surface Albedo Data. Available online: https://land.copernicus.eu/global/products/sa (accessed on 25 February 2022).

Acknowledgments: S.F.W. would like to thank SIOS for awarding an Early Career Researcher award for his presentation in the SIOS online conference 2020 and participation in the SIOS terrestrial remote sensing training course in 2020. The authors would like to thank DigitalGlobe for providing high-resolution satellite data. Chandra–Bhaga basin imagery © 2014 Maxar; Ny-Ålesund imagery © 2016 Maxar. The authors thank the reviewers for their constructive criticism, which helped improve the manuscript. The authors also thank HOD, Civil Engineering, and Director, MIT, MAHE for their support.

Conflicts of Interest: The authors declare no conflict of interest.

References

1. Cisek, D.; Mahajan, M.; Brown, M.; Genaway, D. Remote sensing data integration for mapping glacial extents. In Proceedings of the 2017 New York Scientific Data Summit (NYSDS), IEEE Conference, New York, NY, USA, 6–9 August 2017; pp. 1–4.
2. Kargel, J.S.; Abrams, M.J.; Bishop, M.P.; Bush, A.; Hamilton, G.; Jiskoot, H.; Kääb, A.; Kieffer, H.H.; Lee, E.M.; Paul, F.; et al. Multispectral imaging contributions to global land ice measurements from space. *Remote Sens. Environ.* **2005**, *99*, 187–219. [CrossRef]
3. Naegeli, K.; Damm, A.; Huss, M.; Wulf, H.; Schaepman, M.; Hoelzle, M. Cross-Comparison of Albedo Products for Glacier Surfaces Derived from Airborne and Satellite (Sentinel-2 and Landsat 8) Optical Data. *Remote Sens.* **2017**, *9*, 110. [CrossRef]
4. Pope, E.L.; Willis, I.C.; Pope, A.; Miles, E.S.; Arnold, N.S.; Rees, W.G. Contrasting snow and ice albedos derived from MODIS, Landsat ETM+ and airborne data from Langjökull, Iceland. *Remote Sens. Environ.* **2016**, *175*, 183–195. [CrossRef]
5. Rabatel, A.; Dedieu, J.-P.; Vincent, C. Using remote-sensing data to determine equilibrium-line altitude and mass-balance time series: Validation on three French glaciers, 1994–2002. *J. Glaciol.* **2005**, *51*, 539–546. [CrossRef]
6. Parrot, J.F.; Lyberis, N.; Lefauconnier, B.; Manby, G. SPOT multispectral data and digital terrain model for the analysis of ice-snow fields on arctic glaciers. *Int. J. Remote Sens.* **1993**, *14*, 425–440. [CrossRef]
7. Kraaijenbrink, P.D.A.; Shea, J.M.; Litt, M.; Steiner, J.F.; Treichler, D.; Koch, I.; Immerzeel, W. Mapping Surface Temperatures on a Debris-Covered Glacier with an Unmanned Aerial Vehicle. *Front. Earth Sci.* **2018**, *6*, 64. [CrossRef]
8. Karimi, N.; Farokhnia, A.; Karimi, L.; Eftekhari, M.; Ghalkhani, H. Combining optical and thermal remote sensing data for mapping debris-covered glaciers (Alamkouh Glaciers, Iran). *Cold Reg. Sci. Technol.* **2012**, *71*, 73–83. [CrossRef]
9. Bhardwaj, A.; Joshi, P.; Snehmani; Sam, L.; Singh, M.; Singh, S.; Kumar, R. Applicability of Landsat 8 data for characterizing glacier facies and supraglacial debris. *Int. J. Appl. Earth Obs. Geoinf.* **2015**, *38*, 51–64. [CrossRef]
10. Pope, A.; Rees, G. Using in situ spectra to explore Landsat classification of glacier surfaces. *Int. J. Appl. Earth Obs. Geoinf.* **2014**, *27*, 42–52. [CrossRef]
11. Paul, F.; Winsvold, S.H.; Kääb, A.; Nagler, T.; Schwaizer, G. Glacier Remote Sensing Using Sentinel-2. Part II: Mapping Glacier Extents and Surface Facies, and Comparison to Landsat 8. *Remote Sens.* **2016**, *8*, 575. [CrossRef]
12. Williams, R.S.; Hall, D.K.; Benson, C.S. Analysis of glacier facies using satellite techniques. *J. Glaciol.* **1991**, *37*, 120–128. [CrossRef]
13. Barzycka, B.; Grabiec, M.; Błaszczyk, M.; Ignatiuk, D.; Laska, M.; Hagen, J.O.; Jania, J. Changes of glacier facies on Hornsund glaciers (Svalbard) during the decade 2007–2017. *Remote Sens. Environ.* **2020**, *251*, 112060. [CrossRef]
14. Brown, I.A.; Kirkbride, M.P.; Vaughan, R.A. Find the firn line! The suitability of ERS-1 and ERS-2 SAR data for the analysis of glacier facies on Icelandic icecaps. *Int. J. Remote Sens.* **1999**, *20*, 3217–3230. [CrossRef]
15. Yousuf, B.; Shukla, A.; Arora, M.K.; Jasrotia, A.S. Glacier facies characterization using optical satellite data: Impacts of radiometric resolution, seasonality, and surface morphology. *Prog. Phys. Geogr. Earth Environ.* **2019**, *43*, 473–495. [CrossRef]
16. Pope, A.; Rees, W.G. Impact of spatial, spectral, and radiometric properties of multispectral imagers on glacier surface classification. *Remote Sens. Environ.* **2014**, *141*, 1–13. [CrossRef]
17. Jawak, S.D.; Wankhede, S.F.; Luis, A.J. Explorative Study on Mapping Surface Facies of Selected Glaciers from Chandra Basin, Himalaya Using World, View-2 Data. *Remote Sens.* **2019**, *11*, 1207. [CrossRef]
18. Braun, M.; Schuler, T.V.; Hock, R.; Brown, I.; Jackson, M. Comparison of remote sensing derived glacier facies maps with distributed mass balance modelling at Engabreen, northern Norway. *IAHS Publ. Ser. Proc. Rep.* **2007**, *318*, 126–134.
19. Benson, C. *Stratigraphic Studies in the Snow and Firn of the Greenland Ice Sheet, No. RR70*; Cold Regions Research and Engineering Lab: Hanover, NH, USA, 1962. Available online: http://acwc.sdp.sirsi.net/client/en_US/search/asset/1001392;jsessionid=351D596A6CE87F45BAEB04E7B9ECE897.enterprise-15000 (accessed on 17 January 2018).
20. Braun, M.; Rau, F.; Saurer, H.; Gobmann, H. Development of radar glacier zones on the King George Island ice cap, Antarctica, during austral summer 1996/97 as observed in ERS-2 SAR data. *Ann. Glaciol.* **2000**, *31*, 357–363. [CrossRef]
21. Brown, I.A. Radar Facies on the West Greenland Ice Sheet: Comparison with AVHRR Albedo Data. In Proceedings of the 22nd Symposium of the European Association of Remote Sensing Laboratories, Prague, Czech, 4–6 June 2002. Available online: http://www.earsel.org/symposia/2002-symposium-Prague/pdf/050.pdf (accessed on 18 August 2020).
22. Barzycka, B.; Błaszczyk, M.; Grabiec, M.; Jania, J. Glacier facies of Vestfonna (Svalbard) based on SAR images and GPR measurements. *Remote Sens. Environ.* **2019**, *221*, 373–385. [CrossRef]
23. Anderson, L.S.; Armstrong, W.H.; Anderson, R.S.; Buri, P. Debris cover and the thinning of Kennicott Glacier, Alaska: In situ measurements, automated ice cliff delineation and distributed melt estimates. *Cryosphere* **2021**, *15*, 265–282. [CrossRef]
24. Alifu, H.; Johnson, B.A.; Tateishi, R. Delineation of Debris-Covered Glaciers Based on a Combination of Geomorphometric Parameters and a TIR/NIR/SWIR Band Ratio. *IEEE J. Sel. Top. Appl. Earth Obs. Remote Sens.* **2016**, *9*, 781–792. [CrossRef]
25. Bhardwaj, A.; Joshi, P.K.; Snehmani; Singh, M.; Sam, L.; Gupta, R. Mapping debris-covered glaciers and identifying factors affecting the accuracy. *Cold Reg. Sci. Technol.* **2014**, *106–107*, 161–174. [CrossRef]

26. Foster, L.A.; Brock, B.W.; Cutler, M.E.J.; Diotri, F. A physically based method for estimating supraglacial debris thickness from thermal band remote-sensing data. *J. Glaciol.* **2012**, *58*, 677–691. [CrossRef]
27. Zhang, Y.; Hirabayashi, Y.; Fujita, K.; Liu, S.; Liu, Q. Heterogeneity in supraglacial debris thickness and its role in glacier mass changes of the Mount Gongga. *Sci. China Earth Sci.* **2015**, *59*, 170–184. [CrossRef]
28. Pandey, A.; Rai, A.; Gupta, S.K.; Shukla, D.P.; Dimri, A. Integrated approach for effective debris mapping in glacierized regions of Chandra River Basin, Western Himalayas, India. *Sci. Total Environ.* **2021**, *779*, 146492. [CrossRef] [PubMed]
29. Winsvold, S.H.; Kaab, A.; Nuth, C. Regional Glacier Mapping Using Optical Satellite Data Time Series. *IEEE J. Sel. Top. Appl. Earth Obs. Remote Sens.* **2016**, *9*, 3698–3711. [CrossRef]
30. Dozier, J. Snow Reflectance from LANDSAT-4 Thematic Mapper. *IEEE Trans. Geosci. Remote Sens.* **1984**, *GE-22*, 323–328. [CrossRef]
31. Hall, D.; Ormsby, J.; Bindschadler, R.; Siddalingaiah, H. Characterization of Snow and Ice Reflectance Zones on Glaciers Using Landsat Thematic Mapper Data. *Ann. Glaciol.* **1987**, *9*, 104–108. [CrossRef]
32. De Angelis, H.; Rau, F.; Skvarca, P. Snow zonation on Hielo Patagónico Sur, Southern Patagonia, derived from Landsat 5 TM data. *Glob. Planet. Chang.* **2007**, *59*, 149–158. [CrossRef]
33. Jawak, S.D.; Wankhede, S.F.; Luis, A.J.; Pandit, P.H.; Kumar, S. Implementing an object-based multi-index protocol for mapping surface glacier facies from Chandra-Bhaga basin, Himalaya. *Czech Polar Rep.* **2019**, *9*, 125–140. [CrossRef]
34. Ali, I.; Shukla, A.; Romshoo, S. Assessing linkages between spatial facies changes and dimensional variations of glaciers in the upper Indus Basin, western Himalaya. *Geomorphology* **2017**, *284*, 115–129. [CrossRef]
35. Shukla, A.; Ali, I. A hierarchical knowledge-based classification for glacier terrain mapping: A case study from Kolahoi Glacier, Kashmir Himalaya. *Ann. Glaciol.* **2016**, *57*, 1–10. [CrossRef]
36. Yousuf, B.; Shukla, A.; Arora, M.K.; Bindal, A.; Jasrotia, A.S. On Drivers of Subpixel Classification Accuracy—An Example from Glacier Facies. *IEEE J. Sel. Top. Appl. Earth Obs. Remote Sens.* **2020**, *13*, 601–608. [CrossRef]
37. Rahimzadeganasl, A.; Alganci, U.; Goksel, C. An Approach for the Pan Sharpening of Very High Resolution Satellite Images Using a CIELab Color Based Component Substitution Algorithm. *Appl. Sci.* **2019**, *9*, 5234. [CrossRef]
38. Xu, Y.; Smith, S.E.; Grunwald, S.; Abd-Elrahman, A.; Wani, S.P. Incorporation of satellite remote sensing pan-sharpened imagery into digital soil prediction and mapping models to characterize soil property variability in small agricultural fields. *ISPRS J. Photogramm. Remote Sens.* **2016**, *123*, 1–19. [CrossRef]
39. Meng, X.; Shen, H.; Li, H.; Zhang, L.; Fu, R. Review of the pansharpening methods for remote sensing images based on the idea of meta-analysis: Practical discussion and challenges. *Inf. Fusion* **2019**, *46*, 102–113. [CrossRef]
40. Jawak, S.; Luis, A.J. A spectral index ratio-based Antarctic land-cover mapping using hyperspatial 8-band World, View-2 imagery. *Polar Sci.* **2013**, *7*, 18–38. [CrossRef]
41. Jawak, S.D.; Luis, A.J.; Fretwell, P.T.; Convey, P.; Durairajan, U.A. Semiautomated Detection and Mapping of Vegetation Distribution in the Antarctic Environment Using Spatial-Spectral Characteristics of World, View-2 Imagery. *Remote Sens.* **2019**, *11*, 1909. [CrossRef]
42. Padwick, C.; Deskevich, M.; Pacifici, F.; Smallwood, S. World, View-2 Pan-Sharpening. In Proceedings of the ASPRS 2010 Annual Conference, San Diego, CA, USA, 26 April 2010; pp. 1–14. Available online: http://www.asprs.org/wp-content/uploads/2013/08/Padwick.pdf (accessed on 12 May 2021).
43. Wyczałek, I.; Wyzcałek, E. Studies on parsharpening and object-based classification of World, View-2 multispectral image. *Arch. Photogramm. Cartogr. Remote Sens.* **2013**, 109–117.
44. Pushparaj, J.; Hegde, A.V. Evaluation of pan-sharpening methods for spatial and spectral quality. *Appl. Geomat.* **2016**, *9*, 1–12. [CrossRef]
45. Snehmani, A.G.; Ganju, A.; Kumar, S.; Srivastava, P.K.; Hari Ram, R.P. A comparative analysis of pansharpening techniques on Quick, Bird and World, View-3 images. *Geocarto Int.* **2016**, *32*, 1268–1284. [CrossRef]
46. Nikolakopoulos, K. Quality assessment of ten fusion techniques applied on Worldview-2. *Eur. J. Remote Sens.* **2015**, *48*, 141–167. [CrossRef]
47. Rayegani, B.; Barati, S.; Goshtasb, H.; Sarkheil, H.; Ramezani, J. An effective approach to selecting the appropriate pan-sharpening method in digital change detection of natural ecosystems. *Ecol. Inform.* **2019**, *53*, 100984. [CrossRef]
48. Wu, B.; Fu, Q.; Sun, L.; Wang, X. Enhanced hyperspherical color space fusion technique preserving spectral and spatial content. *J. Appl. Remote Sens.* **2015**, *9*, 097291. [CrossRef]
49. Du, Y.; Teillet, P.M.; Cihlar, J. Radiometric Normalization of Multi-temporal High Resolution Satellite Images with Quality Control for Land Cover Change Detection. *Remote Sens. Environ.* **2002**, *82*, 123–134. [CrossRef]
50. Chavez, P.S., Jr. An improved dark-object subtraction technique for atmospheric scattering correction of multispectral data. *Remote Sens. Environ.* **1988**, *24*, 459–479. [CrossRef]
51. Bernstein, L.S.; Jin, X.; Gregor, B.; Adler-Golden, S.M. Quick atmospheric correction code: Algorithm description and recent upgrades. *Opt. Eng.* **2012**, *51*, 111719. [CrossRef]
52. Cooley, T.; Anderson, G.; Felde, G.; Hoke, M.; Ratkowski, A.; Chetwynd, J.; Gardner, J.; Adler-Golden, S.; Matthew, M.; Berk, A.; et al. FLAASH, a MODTRAN4-based atmospheric correction algorithm, its application and validation. In Proceedings of the IEEE International Geoscience and Remote Sensing Symposium 2003, Toronto, ON, Canada, 24–28 June 2002; IEEE: Piscataway, NJ, USA; Volume 3, pp. 1414–1418. [CrossRef]

53. Adler-Golden, S.M.; Matthew, M.W.; Bernstein, L.S.; Levine, R.Y.; Berk, A.; Richtsmeier, S.C.; Acharya, P.K.; Anderson, G.P.; Felde, J.W.; Gardner, J.A.; et al. Atmospheric correction for shortwave spectral imagery based on MODTRAN4. *SPIE Proc.* **1999**, *3753*, 61–69. [CrossRef]
54. Marcello, J.; Eugenio, F.; Perdomo, U.; Medina, A. Assessment of Atmospheric Algorithms to Retrieve Vegetation in Natural Protected Areas Using Multispectral High Resolution Imagery. *Sensors* **2016**, *16*, 1624. [CrossRef] [PubMed]
55. Shi, L.; Mao, Z.; Chen, P.; Han, S.; Gong, F.; Zhu, Q. Comparison and evaluation of atmospheric correction algorithms of QUAC, DOS, and FLAASH for HICO hyperspectral imagery. *SPIE Proc.* **2016**, *9999*, 999917. [CrossRef]
56. Dewi, E.K.; Trisakti, B. Comparing Atmospheric Correction Methods for Landsat Oli Data. *Int. J. Remote Sens. Earth Sci.* **2016**, *13*, 105–120. [CrossRef]
57. Chakouri, M.; Lhissou, R.; El Harti, A.; Maimouni, S.; Adiri, Z. Assessment of the image-based atmospheric correction of multispectral satellite images for geological mapping in arid and semi-arid regions. *Remote Sens. Appl. Soc. Environ.* **2020**, *20*, 100420. [CrossRef]
58. Guo, Z.; Geng, L.; Shen, B.; Wu, Y.; Chen, A.; Wang, N. Spatiotemporal Variability in the Glacier Snowline Altitude across High Mountain Asia and Potential Driving Factors. *Remote Sens.* **2021**, *13*, 425. [CrossRef]
59. Albert, T.H. Evaluation of Remote Sensing Techniques for Ice-Area Classification Applied to the Tropical Quelccaya Ice Cap, Peru. *Polar Geogr.* **2002**, *26*, 210–226. [CrossRef]
60. Luis, A.J.; Singh, S. High-resolution multispectral mapping facies on glacier surface in the Arctic using World, View-3 data. *Czech Polar Rep.* **2020**, *10*, 23–36. [CrossRef]
61. Lee, K.H.; Yum, J.M. A Review on Atmospheric Correction Technique Using Satellite Remote Sensing. *Korean J. Remote Sens.* **2019**, *35*, 1011–1030. [CrossRef]
62. Gore, A.; Mani, S.; HariRam, R.P.; Shekhar, C.; Ganju, A. Glacier surface characteristics derivation and monitoring using Hyperspectral datasets: A case study of Gepang Gath glacier, Western Himalaya. *Geocarto Int.* **2017**, *34*, 23–42. [CrossRef]
63. Thakur, P.K.; Garg, V.; Nikam, B.; Singh, S.; Chouksey, A.; Dhote, P.R.; Aggarwal, S.P.; Chauhan, P.; Kumar, A.S. Jasmine Snow Cover and Glacier Dynamics Study Using C-And L-Band Sar Datasets in Parts of North West Himalaya. *ISPRS-Int. Arch. Photogramm. Remote Sens. Spat. Inf. Sci.* **2018**, *XLII-5*, 375–382. [CrossRef]
64. Hallikainen, M.; Pulliainen, J.; Praks, J.; Arslan, A. Progress and challenges in radar remote sensing of snow. In Proceedings of the Third International Symposium on Retrieval of Bio-and Geophysical Parameters from SAR Data for Land Applications, Sheffield, UK, 11–14 September 2001; ESA SP-475. ESA Publications Division: Noordwijk, The Netherlands; pp. 185–192, ISBN 92-9092-741-0.
65. Thakur, P.K.; Aggarwal, S.P.; Garg, P.K.; Garg, R.D.; Mani, S.; Pandit, A.; Kumar, S. Snow physical parameter estimation using space-based SAR. *Geocarto Int.* **2012**, *27*, 263–288. [CrossRef]
66. Dozier, J.; Painter, T.H. Multispectral and hyperspectral remote sensing of alpine snow properties. *Annu. Rev. Earth Planet. Sci.* **2004**, *32*, 465–494. [CrossRef]
67. Racoviteanu, A.; Williams, M.W. Decision Tree and Texture Analysis for Mapping Debris-Covered Glaciers in the Kangchenjunga Area, Eastern Himalaya. *Remote Sens.* **2012**, *4*, 3078–3109. [CrossRef]
68. Negi, H.S.; Thakur, N.K.; Ganju, A. Snehmani Monitoring of Gangotri glacier using remote sensing and ground observations. *J. Earth Syst. Sci.* **2012**, *121*, 855–866. [CrossRef]
69. Schuler, T.V.; Kohler, J.; Elagina, N.; Hagen, J.O.M.; Hodson, A.J.; Jania, J.A.; Kääb, A.M.; Luks, B.; Małecki, J.; Moholdt, G.; et al. Reconciling Svalbard Glacier Mass Balance. *Front. Earth Sci.* **2020**, *8*, 8. [CrossRef]
70. Svendsen, H.; Beszczynska-Møller, A.; Hagen, J.O.; Lefauconnier, B.; Tverberg, V.; Gerland, S.; Ørbæk, J.B.; Bischof, K.; Papucci, C.; Zajączkowski, M.; et al. The physical environment of Kongsfjorden–Krossfjorden, an Arctic fjord system in Svalbard. *Polar Res.* **2002**, *21*, 133–166. [CrossRef]
71. Isaksen, K.; Nordli, Ø.; Førland, E.J.; Lupikasza, E.; Eastwood, S.; Niedźwiedź, T. Recent warming on Spitsbergen—Influence of atmospheric circulation and sea ice cover. *J. Geophys. Res. Atmos.* **2016**, *121*, 121. [CrossRef]
72. Nuth, C.; Kohler, J.; König, M.; von Deschwanden, A.; Hagen, J.O.; Kääb, A.; Moholdt, G.; Pettersson, R. Decadal changes from a multi-temporal glacier inventory of Svalbard. *Cryosphere* **2013**, *7*, 1603–1621. [CrossRef]
73. Van Pelt, W.J.J.; Pohjola, V.; Reijmer, C. The Changing Impact of Snow Conditions and Refreezing on the Mass Balance of an Idealized Svalbard Glacier. *Front. Earth Sci.* **2016**, *4*, 4. [CrossRef]
74. Van Pelt, W.; Kohler, J. Modelling the long-term mass balance and firn evolution of glaciers around Kongsfjorden, Svalbard. *J. Glaciol.* **2015**, *61*, 731–744. [CrossRef]
75. Hambrey, M.J.; Murray, T.; Glasser, N.; Hubbard, A.; Hubbard, B.; Stuart, G.; Hansen, S.; Kohler, J. Structure and changing dynamics of a polythermal valley glacier on a centennial timescale: Midre Lovénbreen, Svalbard. *J. Geophys. Res. Earth Surf.* **2005**, *110*. [CrossRef]
76. Evans, D.J.; Strzelecki, M.; Milledge, D.; Orton, C. Hørbyebreen polythermal glacial landsystem, Svalbard. *J. Maps* **2012**, *8*, 146–156. [CrossRef]
77. Hamberg, A. En resa till norra Ishafet sommaren 1892. *J. Geol.* **1895**, *4*, 25–61.
78. Liestøl, O. The glaciers in the Kongsfjorden area, Spitsbergen. *Nor. Geogr. Tidsskr.-Nor. J. Geogr.* **1988**, *42*, 231–238. [CrossRef]
79. Pandey, P.; Venkataraman, G. Changes in the glaciers of Chandra–Bhaga basin, Himachal Himalaya, India, between 1980 and 2010 measured using remote sensing. *Int. J. Remote Sens.* **2013**, *34*, 5584–5597. [CrossRef]

80. Kaushik, S.; Joshi, P.K.; Singh, T. Development of glacier mapping in Indian Himalaya: A review of approaches. *Int. J. Remote Sens.* **2019**, *40*, 6607–6634. [CrossRef]
81. Mir, R.A.; Jain, S.K.; Saraf, A.K.; Goswami, A. Glacier changes using satellite data and effect of climate in Tirungkhad basin located in western Himalaya. *Geocarto Int.* **2013**, *29*, 293–313. [CrossRef]
82. Pandey, P.; Kulkarni, A.V.; Venkataraman, G. Remote sensing study of snowline altitude at the end of melting season, Chandra-Bhaga basin, Himachal Pradesh, 1980–2007. *Geocarto Int.* **2013**, *28*, 311–322. [CrossRef]
83. Sahu, R.; Gupta, R.D. Glacier mapping and change analysis in Chandra basin, Western Himalaya, India during 1971–2016. *Int. J. Remote Sens.* **2020**, *41*, 6914–6945. [CrossRef]
84. Jawak, S.D.; Wankhede, S.F.; Luis, A.J. Exploration of Glacier Surface Facies, Mapping Techniques Using Very High Resolution Worldview-2 Satellite Data. *Proceedings* **2018**, *2*, 339. [CrossRef]
85. Shukla, A.; Arora, M.; Gupta, R. Synergistic approach for mapping debris-covered glaciers using optical–thermal remote sensing data with inputs from geomorphometric parameters. *Remote Sens. Environ.* **2010**, *114*, 1378–1387. [CrossRef]
86. Shukla, A.; Gupta, R.; Arora, M. Estimation of debris cover and its temporal variation using optical satellite sensor data: A case study in Chenab basin, Himalaya. *J. Glaciol.* **2009**, *55*, 444–452. [CrossRef]
87. Raup, B.; Racoviteanu, A.; Khalsa, S.J.; Helm, C.; Armstrong, R.; Arnaud, Y. The GLIMS geospatial glacier database: A new tool for studying glacier change. *Glob. Planet. Chang.* **2007**, *56*, 101–110. [CrossRef]
88. Digital Globe Product Details. Available online: https://www.geosoluciones.cl/documentos/worldview/Digital,Globe-Core-Imagery-Products-Guide.pdf (accessed on 20 February 2020).
89. ASTER GDEM v2. Available online: Gdex.cr.usgs.gov/gdex/ (accessed on 2 February 2017).
90. Arctic DEM. Available online: Pgc.umn.edu/data/arcticdem/ (accessed on 21 January 2019).
91. Porter, C.; Morin, P.; Howat, I.; Noh, M.-J.; Bates, B.; Peterman, K.; Keesey, S.; Schlenk, M.; Gardiner, J.; Tomko, K.; et al. "ArcticDEM", Harvard Dataverse, V1. 2018. Available online: https://www.pgc.umn.edu/data/arcticdem/ (accessed on 13 March 2022).
92. Radiative Transfer Code. Available online: https://www.harrisgeospatial.com/docs/backgroundflaash.html (accessed on 20 November 2021).
93. Atmospheric Correction User Guide. Available online: https://www.l3harrisgeospatial.com/portals/0/pdfs/envi/Flaash_Module.pdf (accessed on 20 November 2021).
94. Kaufman, Y.; Wald, A.; Remer, L.; Gao, B.-C.; Li, R.-R.; Flynn, L. The MODIS 2.1-μm channel-correlation with visible reflectance for use in remote sensing of aerosol. *IEEE Trans. Geosci. Remote Sens.* **1997**, *35*, 1286–1298. [CrossRef]
95. Abreu, L.W.; Anderson, G.P. The MODTRAN 2/3 report and LOWTRAN 7 model. *Contract* **1996**, *19628*, 132.
96. Teillet, P.; Fedosejevs, G. On the Dark Target Approach to Atmospheric Correction of Remotely Sensed Data. *Can. J. Remote Sens.* **1995**, *21*, 374–387. [CrossRef]
97. Zhang, Z.; He, G.; Zhang, X.; Long, T.; Wang, G.; Wang, M. A coupled atmospheric and topographic correction algorithm for remotely sensed satellite imagery over mountainous terrain. *GIScience Remote Sens.* **2017**, *55*, 400–416. [CrossRef]
98. Rumora, L.; Miler, M.; Medak, D. Impact of Various Atmospheric Corrections on Sentinel-2 Land Cover Classification Accuracy Using Machine Learning Classifiers. *ISPRS Int. J. Geo-Inform.* **2020**, *9*, 277. [CrossRef]
99. Laben, C.A.; Brower, B.V. Process for Enhancing the Spatial Resolution of Multispectral Imagery Using Pan-Sharpening. U.S. Patent 6,011,875, 4 January 2000.
100. Rastner, P.; Bolch, T.; Notarnicola, C.; Paul, F. A Comparison of Pixel-and Object-Based Glacier Classification with Optical Satellite Images. *IEEE J. Sel. Top. Appl. Earth Obs. Remote Sens.* **2014**, *7*, 853–862. [CrossRef]
101. Description of CC and AC Algorithms. Available online: https://www.l3harrisgeospatial.com/Learn/Whitepapers/Whitepaper-Detail/ArtMID/17811/Article,ID/17299/Workflow-Tools-in-ENVI (accessed on 24 November 2021).
102. Mahmon, N.A.; Ya'Acob, N.; Yusof, A.L. Differences of image classification techniques for land use and land cover classification. In Proceedings of the 2015 IEEE 11th International Colloquium on Signal Processing & Its Applications (CSPA), Kuala Lumpur, Malaysia, 6–8 March 2015; Institute of Electrical and Electronics Engineers (IEEE): Piscataway, NJ, USA, 2015; pp. 90–94.
103. Hamill, P.; Giordano, M.; Ward, C.; Giles, D.; Holben, B. An AERONET-based aerosol classification using the Mahalanobis distance. *Atmos. Environ.* **2016**, *140*, 213–233. [CrossRef]
104. Doma, M.I.; Gomaa, M.S.; Amer, R.A. Sensitivity of Pixel-Based Classifiers to Training Sample Size in Case of High Resolution Satellite Imagery. *ERJ. Eng. Res. J.* **2014**, *37*, 365–370. [CrossRef]
105. Gao, Y.; Mas, J.F. A comparison of the performance of pixel-based and object-based classifications over images with various spatial resolutions. *Online J. Earth Sci.* **2008**, *2*, 27–35.
106. Gevana, D.; Camacho, L.; Carandang, A.; Camacho, S.; Im, S. Land use characterization and change detection of a small mangrove area in Banacon Island, Bohol, Philippines using a maximum likelihood classification method. *For. Sci. Technol.* **2015**, *11*, 197–205. [CrossRef]
107. Chandra Bhensle, A.; Raja, R. An efficient face recognition using PCA and Euclidean distance classification. *Int. J. Comput. Sci. Mob. Comput.* **2014**, *3*, 407–413.
108. Ahmed, A.; Muaz, M.; Ali, M.; Yasir, M.; Minallah, N.; Ullah, S.; Khan, S. Comparing pixel-based classifiers for detecting tobacco crops in north-west Pakistan. In Proceedings of the 2015 7th International Conference on Recent Advances in Space Technologies (RAST), Istanbul, Turkey, 16–19 June 2015; IEEE Conference: Piscataway, NJ, USA, 2015; pp. 211–216.

109. Cho, M.A.; Debba, P.; Mathieu, R.; Naidoo, L.; Van Aardt, J.; Asner, G. Improving Discrimination of Savanna Tree Species Through a Multiple-Endmember Spectral Angle Mapper Approach: Canopy-Level Analysis. *IEEE Trans. Geosci. Remote Sens.* **2010**, *48*, 4133–4142. [CrossRef]
110. Petropoulos, G.P.; Vadrevu, K.P.; Xanthopoulos, G.; Karantounias, G.; Scholze, M. A Comparison of Spectral Angle Mapper and Artificial Neural Network Classifiers Combined with Landsat TM Imagery Analysis for Obtaining Burnt Area Mapping. *Sensors* **2010**, *10*, 1967–1985. [CrossRef] [PubMed]
111. Chen, K.; Wang, L.; Chi, H. Methods of Combining Multiple Classifiers with Different Features and Their Applications to Text-Independent Speaker Identification. *Int. J. Pattern Recognit. Artif. Intell.* **1997**, *11*, 417–445. [CrossRef]
112. Mancini, A.; Frontoni, E.; Zingaretti, P. A Winner Takes All Mechanism for Automatic Object Extraction from Multi-Source Data. In Proceedings of the 2009 17th International Conference on Geoinformatics IEEE, Fairfax, VA, USA, 12–14 August 2009; pp. 1–6.
113. Ni, L.; Xu, H.; Zhou, X. Mineral Identification and Mapping by Synthesis of Hyperspectral VNIR/SWIR and Multispectral TIR Remotely Sensed Data with Different Classifiers. *IEEE J. Sel. Top. Appl. Earth Obs. Remote Sens.* **2020**, *13*, 3155–3163. [CrossRef]
114. Sukcharoenpong, A.; Yilmaz, A.; Li, R. An Integrated Active Contour Approach to Shoreline Mapping Using HSI and DEM. *IEEE Trans. Geosci. Remote Sens.* **2015**, *54*, 1586–1597. [CrossRef]
115. Zou, S.; Gader, P.; Zare, A. Hyperspectral tree crown classification using the multiple instance adaptive cosine estimator. *Peer J.* **2019**, *7*, e6405. [CrossRef] [PubMed]
116. Soul, M.E.; Broadwater, J.B. Featureless classification for active sonar systems. In Proceedings of the OCEANS'10 IEEE SYDNEY, Sydney, NSW, Australia, 24–28 May 2010; IEEE: Piscataway, NJ, USA, 2010; pp. 1–5.
117. Ren, H.; Du, Q.; Chang, C.-I.; Jensen, J. Comparison between constrained energy minimization based approaches for hyperspectral imagery. In Proceedings of the IEEE Workshop on Advances in Techniques for Analysis of Remotely Sensed Data, Greenbelt, MD, USA, 27–28 October 2003; IEEE: Piscataway, NJ, USA, 2003; pp. 244–248.
118. Du, Q.; Ren, H.; Chang, C.-I. A comparative study for orthogonal subspace projection and constrained energy minimization. *IEEE Trans. Geosci. Remote Sens.* **2003**, *41*, 1525–1529. [CrossRef]
119. Pour, A.B.; Park, T.-Y.S.; Park, Y.; Hong, J.K.; Pradhan, B. Application of Constrained Energy Minimization (CEM) Algorithm to ASTER Data for Alteration Mineral Mapping. In Proceedings of the IGARSS 2019–2019 IEEE International Geoscience and Remote Sensing Symposium, IEEE, Yokohama, Japan, 28 July–2 August 2019; pp. 6760–6763.
120. Gürsoy, Ö.; Atun, R. Investigating Surface Water Pollution by Integrated Remotely Sensed and Field Spectral Measurement Data: A Case Study. *Pol. J. Environ. Stud.* **2019**, *28*, 2139–2144. [CrossRef]
121. Harris, J.R.; Rogge, D.; Hitchcock, R.; Ijewliw, O.; Wright, D. Mapping lithology in Canada's Arctic: Application of hyperspectral data using the minimum noise fraction transformation and matched filtering. *Can. J. Earth Sci.* **2005**, *42*, 2173–2193. [CrossRef]
122. Funk, C.; Theiler, J.; Roberts, D.; Borel, C. Clustering to improve matched filter detection of weak gas plumes in hyperspectral thermal imagery. *IEEE Trans. Geosci. Remote Sens.* **2001**, *39*, 1410–1420. [CrossRef]
123. Mehr, S.G.; Ahadnejad, V.; Abbaspour, R.A.; Hamzeh, M. Using the mixture-tuned matched filtering method for lithological mapping with Landsat TM5 images. *Int. J. Remote Sens.* **2013**, *34*, 8803–8816. [CrossRef]
124. Williams, A.P.; Hunt, E.R. Estimation of leafy spurge cover from hyperspectral imagery using mixture tuned matched filtering. *Remote Sens. Environ.* **2002**, *82*, 446–456. [CrossRef]
125. Zadeh, M.; Tangestani, M.; Roldan, F.; Yusta, I. Mineral Exploration and Alteration Zone Mapping Using Mixture Tuned Matched Filtering Approach on ASTER Data at the Central Part of Dehaj-Sarduiyeh Copper Belt, SE Iran. *IEEE J. Sel. Top. Appl. Earth Obs. Remote Sens.* **2014**, *7*, 284–289. [CrossRef]
126. Singha, A.; Bhowmik, M.K.; Bhattacherjee, D. Akin-based Orthogonal Space (AOS): A subspace learning method for face recognition. *Multimedia Tools Appl.* **2020**, *79*, 35069–35091. [CrossRef]
127. Ren, H.; Chang, C.-I. A target-constrained interference-minimized filter for subpixel target detection in hyperspectral imagery. In Proceedings of the IGARSS 2000, IEEE 2000 International Geoscience and Remote Sensing Symposium, Taking the Pulse of the Planet: The Role of Remote Sensing in Managing the Environment, Honolulu, HI, USA, 24–28 July 2000; IEEE: Piscataway, NJ, USA, 2002; Volume 4, pp. 1545–1547.
128. Du, Q.; Ren, H. On the performance of CEM and TCIMF. In Proceedings of the Defense and Security, Algorithms and Technologies for Multispectral, Hyperspectral, and Ultraspectral Imagery XI, Orlando, FL, USA, 28 March 2005; SPIE: Bellingham, WA, USA, 2005; Volume 5806, pp. 861–869.
129. Millan, V.E.G.; Pakzad, K.; Faude, U.; Teuwsen, S.; Muterthies, A. Target Detection of Mine-RELATED flooded Areas Using AISA-Eagle Data. In Proceedings of the 2014 6th Workshop on Hyperspectral Image and Signal Processing: Evolution in Remote Sensing (WHISPERS), Lausanne, Switzerland, 24–27 June 2014; IEEE: Piscataway, NJ, USA, 2014; pp. 1–4.
130. Seyedain, S.A.; Valadan Zoej, M.J.; Maghsoudi, Y.; Janalipour, M. Improving the Classification Accuracy Using Combination of Target Detection Algorithms in Hyperspectral Images. *J. Geomatics Sci. Technol.* **2015**, *4*, 161–174. Available online: http://jgst.issge.ir/article-1-170-en.html (accessed on 18 September 2021).
131. Kumar, C.; Shetty, A.; Raval, S.; Champatiray, P.K.; Sharma, R. Sub-pixel mineral mapping using EO-1 hyperion hyperspectral data. *Int. Arch. Photogramm. Remote Sens. Spat. Inf. Sci.* **2014**, *40*, 455. [CrossRef]
132. Sidike, P.; Khan, J.; Alam, M.; Bhuiyan, S. Spectral Unmixing of Hyperspectral Data for Oil Spill Detection. In *Optics and Photonics for Information Processing*; SPIE Optical Engineering + Applications: San Diego, CA, USA, 2012; Volume 8498, p. 84981B. [CrossRef]

133. Jawak, S.D.; Devliyal, P.; Luis, A.J. A Comprehensive Review on Pixel Oriented and object Oriented Methods for Information Extraction from Remotely Sensed Satellite Images with a Special Emphasis on Cryospheric Applications. *Adv. Remote Sens.* **2015**, *4*, 177–195. [CrossRef]
134. Keshri, A.K.; Shukla, A.; Gupta, R.P. ASTER ratio indices for supraglacial terrain mapping. *Int. J. Remote Sens.* **2009**, *30*, 519–524. [CrossRef]
135. Richards, J.A. *Remote Sensing Digital Image Analysis*; Springer: Berlin/Heidelberg, Germany, 2013.
136. Maxwell, A.E.; Warner, T.A. Thematic Classification Accuracy Assessment with Inherently Uncertain Boundaries: An Argument for Center-Weighted Accuracy Assessment Metrics. *Remote Sens.* **2020**, *12*, 1905. [CrossRef]
137. Casacchia, R.; Lauta, F.; Salvatori, R.; Cagnati, A.; Valt, M.; Ørbæk, J.B. Radiometric investigation of different snow covers in Svalbard. *Polar Res.* **2001**, *20*, 13–22. [CrossRef]
138. Warren, S.G. Optical properties of snow. *Rev. Geophys.* **1982**, *20*, 67–89. [CrossRef]
139. Hinkler, J.; Ørbæk, J.B.; Hansen, B.U. Detection of spatial, temporal, and spectral surface changes in the Ny-Ålesund area 79°, N, Svalbard, using a low cost multispectral camera in combination with spectroradiometer measurements. *Phys. Chem. Earth Parts A/B/C* **2003**, *28*, 1229–1239. [CrossRef]
140. Gao, J.; Liu, Y. Applications of remote sensing, GIS and GPS in glaciology: A review. *Prog. Phys. Geogr. Earth Environ.* **2001**, *25*, 520–540. [CrossRef]
141. Zeng, Q.; Cao, M.; Feng, X.; Liang, F.; Chen, X.; Sheng, W. Study on spectral reflection characteristics of snow, ice and water of northwest China. *Sci. Sin. Ser. B* **1984**, *46*, 647–656.
142. Bernardo, N.; Watanabe, F.; Rodrigues, T.; Alcântara, E. Atmospheric correction issues for retrieving total suspended matter concentrations in inland waters using OLI/Landsat-8 image. *Adv. Space Res.* **2017**, *59*, 2335–2348. [CrossRef]
143. Roy, D.P.; Wulder, M.A.; Loveland, T.R.; Woodcock, C.E.; Allen, R.G.; Anderson, M.C.; Helder, D.; Irons, J.R.; Johnson, D.M.; Kennedy, R.; et al. Landsat-8: Science and product vision for terrestrial global change research. *Remote Sens. Environ.* **2014**, *145*, 154–172. [CrossRef]
144. Binding, C.E.; Jerome, J.H.; Bukata, R.P.; Booty, W.G. Suspended particulate matter in Lake Erie derived from MODIS aquatic colour imagery. *Int. J. Remote Sens.* **2010**, *31*, 5239–5255. [CrossRef]
145. Prieto-Amparan, J.A.; Villarreal-Guerrero, F.; Martinez-Salvador, M.; Manjarrez-Domínguez, C.; Santellano-Estrada, E.; Pinedo-Alvarez, A. Atmospheric and Radiometric Correction Algorithms for the Multitemporal Assessment of Grasslands Productivity. *Remote Sens.* **2018**, *10*, 219. [CrossRef]
146. Lu, D.; Mausel, P.; Brondizio, E.; Moran, E. Assessment of atmospheric correction methods for Landsat TM data applicable to Amazon basin LBA research. *Int. J. Remote Sens.* **2002**, *23*, 2651–2671. [CrossRef]
147. Gong, S.; Huang, J.; Li, Y.; Wang, H. Comparison of atmospheric correction algorithms for TM image in inland waters. *Int. J. Remote Sens.* **2008**, *29*, 2199–2210. [CrossRef]
148. Manakos, I.; Manevski, K.; Kalaitzidis, C.; Edler, D. Comparison between Atmospheric Correction Modules on the Basis of Worldview-2 Imagery and In Situ spectroradiometric Measurements. In Proceedings of the 7th EARSeL SIG Imaging Spectroscopy Workshop, Edinburgh, UK, 11–13 April 2011.
149. Saini, V.; Tiwari, R.; Gupta, R. Comparison of FLAASH and QUAC Atmospheric Correction Methods for Resourcesat-2 LISS-IV Data. In Proceedings of the SPIE, Earth Observing Missions and Sensors: Development, Implementation, and Characterization IV, New Delhi, India, 2 May 2016.
150. Jin, X.; Paswaters, S.; Cline, H. A comparative study of target detection algorithms for hyperspectral imagery. *SPIE Proc.* **2009**, *7*, 73341W. [CrossRef]
151. Jawak, S.D.; Luis, A. Very high-resolution satellite data for improved land cover extraction of Larsemann Hills, Eastern Antarctica. *J. Appl. Remote Sens.* **2013**, *7*, 73460. [CrossRef]
152. Schadt, E.E.; Linderman, M.D.; Sorenson, J.; Lee, L.; Nolan, G.P. Computational solutions to large-scale data management and analysis. *Nat. Rev. Genet.* **2010**, *11*, 647–657. [CrossRef] [PubMed]
153. Racoviteanu, A.; Paul, F.; Raup, B.; Khalsa, S.J.; Armstrong, R. Challenges and recommendations in mapping of glacier parameters from space: Results of the 2008 Global Land Ice Measurements from Space (GLIMS) workshop, Boulder, Colorado, USA. *Ann. Glaciol.* **2009**, *50*, 53–69. [CrossRef]
154. Robson, B.A.; Bolch, T.; Mac Donell, S.; Hölbling, D.; Rastner, P.; Schaffer, N. Automated detection of rock glaciers using deep learning and object-based image analysis. *Remote Sens. Environ.* **2020**, *250*, 112033. [CrossRef]
155. Lu, Y.; Zhang, Z.; Shangguan, D.; Yang, J. Novel Machine Learning Method Integrating Ensemble Learning and Deep Learning for Mapping Debris-Covered Glaciers. *Remote Sens.* **2021**, *13*, 2595. [CrossRef]
156. De Beer, C.M.; Sharp, M.J. Topographic influences on recent changes of very small glaciers in the Monashee Mountains, British Columbia, Canada. *J. Glaciol.* **2009**, *55*, 691–700. [CrossRef]
157. Petrou, Z.I.; Kosmidou, V.; Manakos, I.; Stathaki, T.; Adamo, M.; Tarantino, C.; Tomaselli, V.; Blonda, P.; Petrou, M. A rule-based classification methodology to handle uncertainty in habitat mapping employing evidential reasoning and fuzzy logic. *Pattern Recognit. Lett.* **2014**, *48*, 24–33. [CrossRef]
158. Benz, U.C.; Hofmann, P.; Willhauck, G.; Lingenfelder, I.; Heynen, M. Multi-resolution, object-oriented fuzzy analysis of remote sensing data for GIS-ready information. *ISPRS J. Photogramm. Remote Sens.* **2004**, *58*, 239–258. [CrossRef]

159. Ansari, R.A.; Buddhiraju, K.M. Noise Filtering in High-Resolution Satellite Images Using Composite Multiresolution Transforms. *PFG–J. Photogramm. Remote Sens. Geoinf. Sci.* **2018**, *86*, 249–261. [CrossRef]
160. Liang, H.; Li, N.; Zhao, S. Salt and Pepper Noise Removal Method Based on a Detail-Aware Filter. *Symmetry* **2021**, *13*, 515. [CrossRef]
161. Green, A.; Berman, M.; Switzer, P.; Craig, M. A transformation for ordering multispectral data in terms of image quality with implications for noise removal. *IEEE Trans. Geosci. Remote Sens.* **1988**, *26*, 65–74. [CrossRef]
162. Shimamura, Y.; Izumi, T.; Matsuyama, H. Evaluation of a useful method to identify snow-covered areas under vegetation–Comparisons among a newly proposed snow index, normalized difference snow index, and visible reflectance. *Int. J. Remote Sens.* **2006**, *27*, 4867–4884. [CrossRef]
163. ENVI User's Guide. 2009. Available online: https://www.tetracam.com/PDFs/Rec_Cite9.pdf (accessed on 1 March 2022).
164. Ramezan, C.A.; Warner, T.A.; Maxwell, A.E.; Price, B.S. Effects of Training Set Size on Supervised Machine-Learning Land-Cover Classification of Large-Area High-Resolution Remotely Sensed Data. *Remote Sens.* **2021**, *13*, 368. [CrossRef]
165. Schaaf, C.; Wang, Z. *MCD43A4 MODIS/Terra+Aqua BRDF/Albedo Nadir BRDF Adjusted Ref Daily L3 Global-500m V006*; NASA EOSDIS Land Processes DAAC: Sioux Falls, SD, USA, 2018.
166. Schaaf, C.; Wang, Z.; Zhang, X.; Strahler, A. *VIIRS/NPP BRDF/Albedo Albedo Daily L3 Global 500m SIN Grid V001*; NASA EOSDIS Land Processes DAAC: Sioux Falls, SD, USA, 2018. [CrossRef]
167. Schaaf, C.; Wang, Z.; Zhang, X.; Strahler, A. *VIIRS/NPP BRDF/Albedo Albedo Daily L3 Global 1km SIN Grid V001*; NASA EOSDIS Land Processes DAAC: Sioux Falls, SD, USA, 2018. [CrossRef]
168. Dierckx, W.; Sterckx, S.; Benhadj, I.; Livens, S.; Duhoux, G.; Van Achteren, T.; Francois, M.; Mellab, K.; Saint, G. PROBA-V mission for global vegetation monitoring: Standard products and image quality. *Int. J. Remote Sens.* **2014**, *35*, 2589–2614. [CrossRef]
169. Download of VIIRS VNP43IA3 Datasets. Available online: https://lpdaac.usgs.gov/products/vnp43ia3v001/ (accessed on 25 February 2022).
170. Download of CGLS PROBA-V Surface Albedo Data. Available online: https://land.copernicus.eu/global/products/sa (accessed on 25 February 2022).
171. Knap, W.H.; Reijmer, C.H.; Oerlemans, J. Narrowband to broadband conversion of Landsat TM glacier albedos. *Int. J. Remote. Sens.* **1999**, *20*, 2091–2110. [CrossRef]
172. Liang, S. Narrowband to broadband conversions of land surface albedo I: Algorithms. *Remote Sens. Environ.* **2000**, *76*, 213–238. [CrossRef]
173. Zhou, Y.; Wang, D.; Liang, S.; Yu, Y.; He, T. Assessment of the Suomi NPP VIIRS Land Surface Albedo Data Using Station Measurements and High-Resolution Albedo Maps. *Remote Sens.* **2016**, *8*, 137. [CrossRef]
174. Yang, Y.; Sun, L.; Zhu, J.; Wei, J.; Su, Q.; Sun, W.; Liu, F.; Shu, M. A simplified Suomi NPP VIIRS dust detection algorithm. *J. Atmos. Sol.-Terr. Phys.* **2017**, *164*, 314–323. [CrossRef]
175. Moroni, B.; Arnalds, O.; Dagsson-Waldhauserová, P.; Crocchianti, S.; Vivani, R.; Cappelletti, D. Mineralogical and Chemical Records of Icelandic Dust Sources Upon Ny-Ålesund (Svalbard Islands). *Front. Earth Sci.* **2018**, *6*, 6080–6097. [CrossRef]
176. Di Mauro, B.; Fava, F.; Ferrero, L.; Garzonio, R.; Baccolo, G.; Delmonte, B.; Colombo, R. Mineral dust impact on snow radiative properties in the European Alps combining ground, UAV, and satellite observations. *J. Geophys. Res. Atmos.* **2015**, *120*, 6080–6097. [CrossRef]
177. Baddock, M.C.; Mockford, T.; Bullard, J.E.; Thorsteinsson, T. Pathways of high-latitude dust in the North Atlantic. *Earth Planet. Sci. Lett.* **2017**, *459*, 170–182. [CrossRef]
178. Dagsson-Waldhauserova, P.; Magnusdottir, A.Ö.; Olafsson, H.; Arnalds, O. The Spatial Variation of Dust Particulate Matter Concentrations during Two Icelandic Dust Storms in 2015. *Atmosphere* **2016**, *7*, 77. [CrossRef]
179. Casey, K.A.; Kääb, A.; Benn, D.I. Geochemical characterization of supraglacial debris via in situ and optical remote sensing methods: A case study in Khumbu Himalaya, Nepal. *Cryosphere* **2012**, *6*, 85–100. [CrossRef]
180. Norwegian Climate Service Center: Observations and Weather Statistics. Available online: https://seklima.met.no/observations/ (accessed on 25 May 2021).
181. Close, O.; Benjamin, B.; Petit, S.; Fripiat, X.; Hallot, E. Use of Sentinel-2 and LUCAS Database for the Inventory of Land Use, Land Use Change, and Forestry in Wallonia, Belgium. *Land* **2018**, *7*, 154. [CrossRef]
182. Tsai, Y.-L.S.; Dietz, A.; Oppelt, N.; Kuenzer, C. Wet and Dry Snow Detection Using Sentinel-1 SAR Data for Mountainous Areas with a Machine Learning Technique. *Remote Sens.* **2019**, *11*, 895. [CrossRef]
183. Charoenjit, K.; Zuddas, P.; Allemand, P.; Pattanakiat, S.; Pachana, K. Estimation of biomass and carbon stock in Para rubber plantations using object-based classification from Thaichote satellite data in Eastern Thailand. *J. Appl. Remote Sens.* **2015**, *9*, 96072. [CrossRef]

remote sensing

MDPI

Article

Effect of Image-Processing Routines on Geographic Object-Based Image Analysis for Mapping Glacier Surface Facies from Svalbard and the Himalayas

Shridhar D. Jawak [1], Sagar F. Wankhede [2,*], Alvarinho J. Luis [3] and Keshava Balakrishna [2]

[1] Svalbard Integrated Arctic Earth Observing System (SIOS), SIOS Knowledge Centre, Svalbard Science Centre, P.O. Box 156, N-9171 Longyearbyen, Svalbard, Norway
[2] Department of Civil Engineering, Manipal Institute of Technology, Manipal Academy of Higher Education, Manipal 576104, India
[3] Earth System Sciences Organization, National Centre for Polar and Ocean Research (NCPOR), Ministry of Earth Sciences, Government of India, Headland Sada, Vasco-da-Gama, Goa 403804, India
* Correspondence: sagar.wankhede@learner.manipal.edu

Abstract: Advancements in remote sensing have led to the development of Geographic Object-Based Image Analysis (GEOBIA). This method of information extraction focuses on segregating correlated pixels into groups for easier classification. This is of excellent use in analyzing very-high-resolution (VHR) data. The application of GEOBIA for glacier surface mapping, however, necessitates multiple scales of segmentation and input of supportive ancillary data. The mapping of glacier surface facies presents a unique problem to GEOBIA on account of its separable but closely matching spectral characteristics and often disheveled surface. Debris cover can induce challenges and requires additions of slope, temperature, and short-wave infrared data as supplements to enable efficient mapping. Moreover, as the influence of atmospheric corrections and image sharpening can derive variations in the apparent surface reflectance, a robust analysis of the effects of these processing routines in a GEOBIA environment is lacking. The current study aims to investigate the impact of three atmospheric corrections, Dark Object Subtraction (DOS), Quick Atmospheric Correction (QUAC), and Fast Line-of-Sight Atmospheric Analysis of Hypercubes (FLAASH), and two pansharpening methods, viz., Gram–Schmidt (GS) and Hyperspherical Color Sharpening (HCS), on the classification of surface facies using GEOBIA. This analysis is performed on VHR WorldView-2 imagery of selected glaciers in Ny-Ålesund, Svalbard, and Chandra–Bhaga basin, Himalaya. The image subsets are segmented using multiresolution segmentation with constant parameters. Three rule sets are defined: rule set 1 utilizes only spectral information, rule set 2 contains only spatial and contextual features, and rule set 3 combines both spatial and spectral attributes. Rule set 3 performs the best across all processing schemes with the highest overall accuracy, followed by rule set 1 and lastly rule set 2. This trend is observed for every image subset. Among the atmospheric corrections, DOS displays consistent performance and is the most reliable, followed by QUAC and FLAASH. Pansharpening improved overall accuracy and GS performed better than HCS. The study reports robust segmentation parameters that may be transferable to other VHR-based glacier surface facies mapping applications. The rule sets are adjusted across the processing schemes to adjust to the change in spectral characteristics introduced by the varying routines. The results indicate that GEOBIA for glacier surface facies mapping may be less prone to the differences in spectral signatures introduced by different atmospheric corrections but may respond well to increasing spatial resolution. The study highlighted the role of spatial attributes for mapping fine features, and in combination with appropriate spectral features may enhance thematic classification.

Keywords: geographic object-based image analysis; atmospheric correction; pansharpening; WorldView-2; Ny-Ålesund; Chandra–Bhaga basin; glacier surface facies

Citation: Jawak, S.D.; Wankhede, S.F.; Luis, A.J.; Balakrishna, K. Effect of Image-Processing Routines on Geographic Object-Based Image Analysis for Mapping Glacier Surface Facies from Svalbard and the Himalayas. *Remote Sens.* **2022**, *14*, 4403. https://doi.org/10.3390/rs14174403

Academic Editor: Ulrich Kamp

Received: 8 July 2022
Accepted: 31 August 2022
Published: 4 September 2022

1. Introduction

Glacial systems are dynamic. This dynamicity revolves around the intricate dependence of the glacier's health on its immediate environment and long-term climate. Glaciers serve not only as sources of fresh water but also as drivers for socio-economic and industrial growth [1,2]. Continuous monitoring of these systems is not only important for their indications of climate variations [3], but also for their cascading impact on human life. Remote sensing (RS) is one of the most effective mechanisms for near-continuous monitoring of the ever-changing cryosphere. Satellite multispectral sensors have an advantage, as the derivable reflectance from these sensors enable the identification of various supraglacial features [4]. Glacier surface facies are easily distinguished based upon variations in surface reflectance [5]. These facies are formed by the natural aging and movement of snow and its melting, refreezing, and intermixing with dust and debris. The spatial distribution of facies can be incorporated into distributed mass balance models [6] and used for validating existing three-dimensional models. Development of a recent open-source cryospheric model, the Snow Multidata Mapping and Modeling (S3M) 5.1, may allow for assimilation of glacier surface facies [7].

Multispectral remote sensors capture scenes which necessitate processing routines to derive usable information from raw data. These processing routines exert an influential impact on the consequent information extraction methods [8]. Moreover, the method of information extraction plays an important role in developing a sustainable mapping mechanism. Traditional pixel-based image analysis (PBIA) focuses on assigning a thematic class to each pixel, while object-based image analysis (OBIA) groups homogenous pixels into objects for classification [9] and is reported to be superior for remote sensing classification [10]. Hay and Castilla [11] define the primary aim of Geographic Object-Based Image Analysis (GEOBIA) as a discipline to develop theory, methods, and tools to replicate human interpretation of RS imagery through automated/semi-automated mechanisms. GEOBIA differs from OBIA in that it is solely focused on RS of the Earth and its surface.

GEOBIA

The foundation of OBIA is image segmentation [12]. Segments are groups/objects of similar pixels determined by criteria of homogeneity. Segmentation decreases image detail, reduces spectral complexity, and enhances understanding of image content [13]. These criteria result in additional features such as mean values per band, minimum and maximum values, relationships to neighbor objects, spatial topology, and geometric descriptions [14]. Thus, segmentation operations and their resultant features are suggested to be comparable to building a database with information of the objects. The subsequent classification of the objects can be reviewed as a database query [15]. Studies testing the capacity of OBIA in various applications have resulted in intriguing research problems related to selection of segmentation parameters, classification methods, and validation of end results. For example, Kim et al. [16] investigated multi-scale and single-scale segmentation and incorporation of a texture-derived grey-level co-occurrence matrix (GLCM) at different quantization levels. Their results suggest that multi-scale segmentation offers the highest accuracy when combined with GLCM for classification of features in a salt marsh environment. Verhagen and Dragut [17] utilized OBIA to perform predictive archaeological mapping using a digital elevation model (DEM). They suggest that geomorphological analysis using additional contextual features may enable better mapping and highlight the necessity of expert knowledge for interpretation of image objects. This implies that expert knowledge may drive the success or failure of segmentation and sequential classification.

Among GEOBIA for glaciological applications, Höfle et al. [18] identified ice, firn, snow, and crevasses using airborne laser scanning point cloud and intensity data through segmentation and subsequently supervised classification. Rastner et al. [9] comparatively assessed pixel-based image analysis (PBIA) and OBIA for mapping clean snow, ice, and debris-covered ice. Robson et al. [19] classified clean ice and debris-covered ice, by combining Synthetic Aperture Radar (SAR) and optical and topographic data in an OBIA environment. Sharda and Shrivastava [20] mapped snow, ice, and a glacial lake on the Siachen Glacier using band ratios and thresholds on indices. Jawak et al. [21] comparatively assessed PBIA and OBIA for mapping glacier surface facies in the Chandra–Bhaga basin, Himalaya, using a multi-index approach. GEOBIA was also tested for change detection of seasonal snow cover [22]. More recently, Mitkari et al. [23] mapped debris cover using a multi-scale segmentation approach utilizing ancillary datasets for mapping glacial lakes, exposed ice, debris cones, rills, crevasses, snow, ice-mixed debris, and supraglacial debris.

Much of the research involving GEOBIA for glacier mapping involves multi-scale segmentation and incorporation of ancillary data for delineating features. However, little to no research is conducted on the impact of basic image-processing routines on GEOBIA classification. The influence of atmospheric corrections and image sharpening can induce variations in the final thematic outputs of surface facies maps [8]. This study is a subsequent part of Jawak et al. [8]. Therefore, the previous work is referred to as "paper 1" henceforth. Hence, the current study aims to fill this research gap by analyzing the effects of variable processing schemes on the resultant thematic GEOBIA classification of glacier surface facies in two distinct study areas: Ny-Ålesund, Svalbard, and Chandra–Bhaga basin, Himalaya.

2. Study Areas and Data Used

2.1. Study Sites

Svalbard lies between 75° to 82°N and 10° to 35°E. The rate of warming experienced by this region is almost twice that of the global average [24]. Ny-Ålesund houses a research hub for international researchers and some of the most well-studied glaciers. The glaciers selected for this study include Vestre Brøggerbreen (VB), Austre Lovénbreen (AL), Austre Brøggerbreen (AB), Midtre Lovénbreen (ML), Edithbreen (EB), Botnfjellbreen (BB), Pedersbreen (PB), and Uvérsbreen (UB). The second site is the Chandra–Bhaga basin, which falls in the Lahaul and Spiti district of Himachal Pradesh, India. It lies between 32°05′N to 32°45′N and 76°50′E to 77°50′E. This basin hosts India's Himalayan research base, Himansh. The glaciers selected are Samudra Tapu (ST), CB1, CB2, CB3, CB4, CB5, and CB6. Figure 1 highlights the geospatial location of the study sites, whereas Supplementary Table S1 highlights the area of each glacier and their Global Land Ice Measurements from Space (GLIMS) reference ID [25].

2.2. Satellite and Elevation Data

The primary data of this study were LV2A-processed images from Digital Globe (now Maxar Technologies), Inc., Westminster, CO, USA [26]. The Himalayan image was acquired on 16 October 2014 (WorldView-2 © 2014 Maxar Technologies, Pasadena, CA, USA). It has a multispectral (MSL) resolution of 2 m and a panchromatic (PAN) resolution of 0.5 m. The Svalbard image was acquired on 10 August 2016 (WorldView-3 © 2016 Maxar Technologies). This product has an at-nadir spatial resolution of 1.24 m, whereas the SWIR bands and PAN band had resolutions of 3.7 m and 0.31 m, respectively. The datasets had a radiometric resolution of 16 bits per pixel. The spectral resolution of WorldView-2 (WV-2) consisted of the bands PAN (0.45–0.80 μm), Coastal (0.40–0.45 μm), Blue (0.45–0.51 μm), Green (0.51–0.58 μm), Yellow (0.585–0.625 μm), Red (0.63–0.69 μm), Red Edge (0.705–0.745 μm), near-infrared 1 (NIR 1) (0.770–0.895 μm), and near-infrared 2 (NIR 2) (0.86–1.04 μm). The projection and datum of the Svalbard image are WGS 1984 UTM Zone 33N, and the Himalayan image are WGS 1984 UTM Zone 43N.

Elevation data consisted of 30 m Advanced Spaceborne Thermal Emission and Reflection Radiometer (ASTER) and Global Digital Elevation Model (GDEM) v2 [27] for the Chandra–Bhaga basin, and 5 m Arctic DEM [28,29] for Ny-Ålesund.

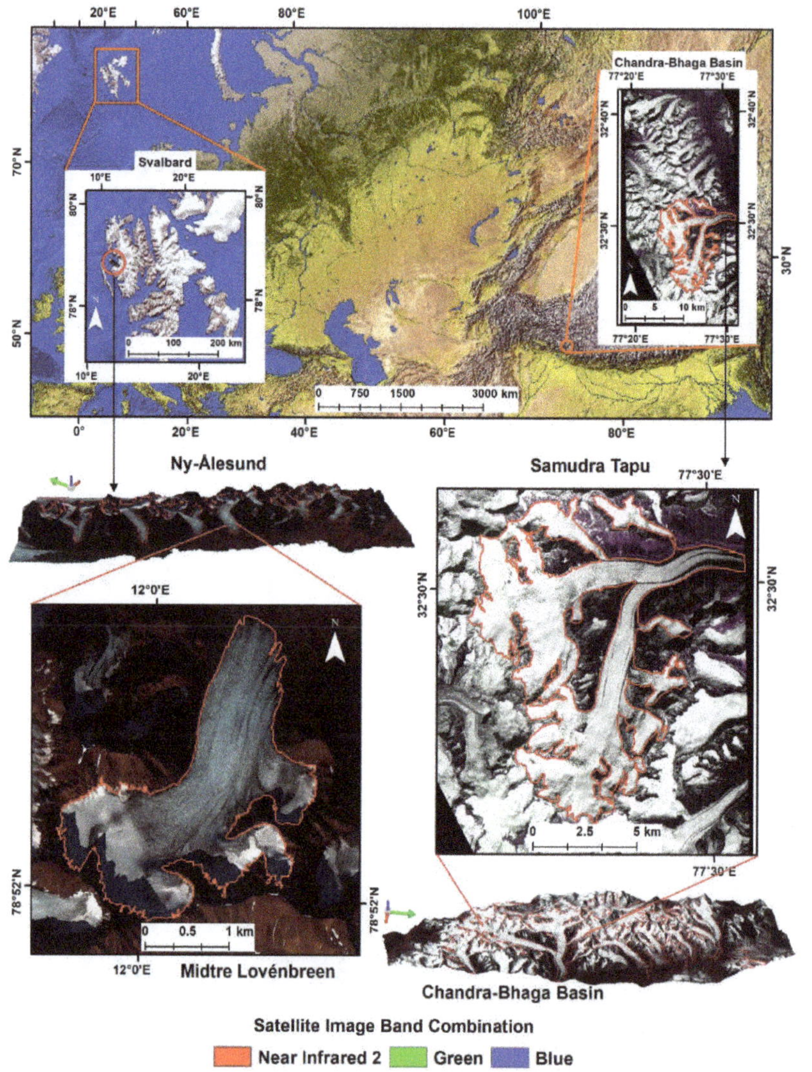

Figure 1. Geospatial location of the study areas. The 3D elevated surfaces are pansharpened images draped on digital elevation models. The insets of ML (WorldView-3 © 2016 Maxar Technologies) and ST (WorldView-2 © 2014 Maxar Technologies) are highlighted using the digitized boundary. The top image showing the global position of Svalbard and Chandra–Bhaga basin was prepared using Natural Earth (free vector and raster map data @ naturalearthdata.com).

3. Research Methodology

3.1. Experimental Setup

This study aims to map surface facies using a multi-rule set GEOBIA approach on selected glaciers in Ny-Ålesund, Svalbard, and Chandra–Bhaga basin, Indian Himalaya, using visible to near-infrared (VNIR) very-high-resolution (VHR) WV-2 data. Three atmospheric corrections, Dark Object Subtraction (DOS), Fast Line-of-Sight Atmospheric Analysis of Hypercubes (FLAASH), and Quick Atmospheric Correction (QUAC) are used to obtain reflectance, followed by pansharpening via Gram–Schmidt (GS) and Hyperspherical

Color Sharpening (HCS). Glacial extents are defined by digitizing over the pansharpened raised images as 3D surfaces using an ASTER GDEM v2 and Arctic DEM for the two areas. The results are then examined using error matrices and by comparison with published literature. The multiple rule sets and data processing are elaborated in the subsequent sections. In summary, we apply three rule sets in a GEOBIA domain to determine the variation induced by different processing routines on the classification of glacier surface facies. The nomenclature used in the study is described in Table 1 and the methodology is depicted in Figure 2.

3.2. Processing Routines

3.2.1. Deriving Reflectance: Radiometric and Atmospheric Corrections

Deriving reflectance from multispectral imagery entails a two-step routine: (a) conversion of digital numbers/brightness values to at-sensor spectral radiance; (b) application of an atmospheric correction model to calculate apparent spectral reflectance. The Environment for Visualizing Images (ENVI) 5.3 includes a calibration tool which was used to perform the first step. The radiometrically calibrated images were then subjected to three different atmospheric corrections. (1) The FLAASH correction is a dual-step module that simulates the atmosphere at the moment of image capture. This is performed by utilizing aerosol description and water column amount along with sensor and image data in an atmosphere model [30] and aerosol model [31]. According to Abreu and Anderson [32], the atmospheric model was set to tropical for the Himalayan image and subarctic summer for the Ny-Ålesund image. Similarly, the aerosol model was set as tropospheric for the Himalayan image and maritime for the Ny-Ålesund image [32]. Other input parameters such as pixel size, aerosol height, CO2 mixing ratio, water column multiplier, zenith angle, sensor altitude, and scene center location were computed automatically upon selection of image and sensor. (2) The DOS correction is built upon the principle that atmospheric scattering upwells path radiance in the dark pixels of an image [33]. Removal of this contribution to the path radiance can be performed using the value of a single dark pixel [34]. Following paper 1 and Rumora et al. [35], Top of Atmosphere (TOA) reflectance values of user-defined dark pixels were used as an input to the DOS correction. Supplementary Table S2 consists of the spectral-band-wise at-sensor reflectance values of selected dark pixels for the DOS correction. (3) Similar to DOS, the QUAC model is an in-scene approach, utilizing no model computations, rather using central wavelengths and the first step of sensor calibration [36]. The process involves direct input of the image into the QUAC module without any user intervention.

Table 1. Nomenclature of processing schemes and classifications used in the current study.

Nomenclature/Abbreviation	Description/Definition
DOS	DOS-corrected
FLAASH	FLAASH-corrected
QUAC	QUAC-corrected
GS_DOS	DOS followed by GS sharpening
GS_FLAASH	FLAASH followed by GS sharpening
GS_QUAC	QUAC followed by GS sharpening
HCS_DOS	DOS followed by HCS sharpening
HCS_FLAASH	FLAASH followed by HCS sharpening
HCS_QUAC	QUAC followed by HCS sharpening
DOS_Rule Set 1/2/3	DOS followed by classification by any of the three rule sets
FLAASH_ Rule Set 1/2/3	FLAASH followed by classification by any of the three rule sets
QUAC_ Rule Set 1/2/3	QUAC followed by classification by any of the three rule sets

Table 1. *Cont.*

Nomenclature/Abbreviation	Description/Definition
GS_DOS_ Rule Set 1/2/3	DOS followed by GS followed by classification by any of the three rule sets
GS_FLAASH_ Rule Set 1/2/3	FLAASH followed by GS followed by classification by any of the three rule sets
GS_QUAC_ Rule Set 1/2/3	QUAC followed by GS followed by classification by any of the three rule sets
HCS_DOS_ Rule Set 1/2/3	DOS followed by HCS followed by classification by any of the three rule sets
HCS_FLAASH_ Rule Set 1/2/3	FLAASH followed by HCS followed by classification by any of the three rule sets
HCS_QUAC_ Rule Set 1/2/3	QUAC followed by HCS followed by classification by any of the three rule sets

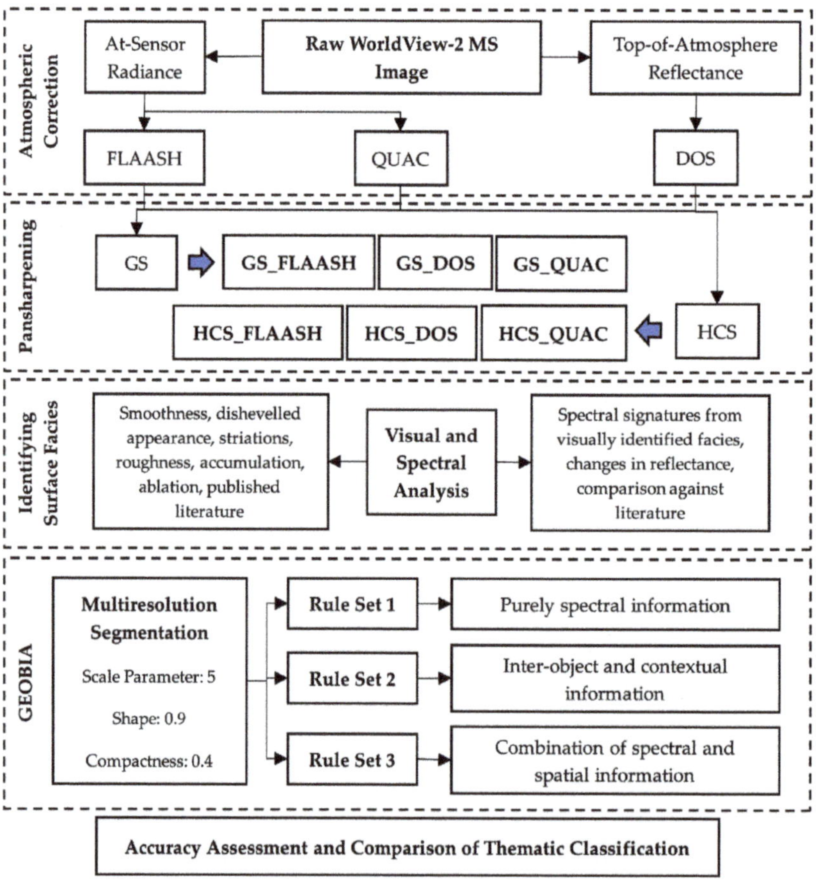

Figure 2. Methodology of the proposed study. MS: multispectral image; FLAASH: Fast Line-of-Sight Atmospheric Analysis of Hypercubes; QUAC: Quick Atmospheric Correction; DOS: Dark Object Subtraction; GS: Gram–Schmidt; HCS: Hyperspherical Color Sharpening; GEOBIA: Geographic Object-Based Image Analysis. The blue arrows from GS and HCS highlight the individual image subsets post pansharpening.

3.2.2. Pansharpening and Glacier Extent Delineation

HCS replaces the intensity component of MS data in the Hyperspherical color space with the intensity-matched form of the PAN band [37]. This was performed in ERDAS IMAGINE. GS predicts the panchromatic data based on the spectral response function of the sensor [38]. The suitability of generating a 3D surface to determine a glacier's boundaries and ice divides is demonstrated in previous studies [21,39]. Hence, the current study draped GS-pansharpened imagery over the Arctic DEM for Ny-Ålesund, and over the ASTER GDEM v2 for the Chandra–Bhaga basin to observe and digitize glacial extents.

3.3. *Mapping Facies in a GEOBIA Domain*

Surface facies were identified on the selected glaciers using visual and spectral characteristics. The characteristics of facies in the Chandra–Bhaga basin and in Ny-Ålesund are described in detail by Jawak et al. [21] and paper 1. The logic following the identification of facies is the reduction in reflectance due to mixing of dust, debris, and moisture, along with comparison against literature performed previously [21] (paper 1). Facies identified in the Ny-Ålesund image consist of dry snow, wet snow, melting snow, saturated snow, shadowed snow, glacier ice, melting glacier ice, dirty ice, and streams and crevasses. Facies observed in the Chandra–Bhaga basin image consist of snow, shadowed snow, glacier ice, ice mixed debris, debris, and crevasses.

3.3.1. Multiresolution Segmentation

To derive maximum spatial and spectral information from VNIR VHR imagery, the GEOBIA approach must first focus on identifying the optimal segmentation algorithm and subsequently its associated parameters. In this study, we utilize the multiresolution segmentation [40] algorithm loaded into eCognition Developer. Multiresolution segmentation is an iterative process that begins by aggregating highly correlated adjacent pixels into objects. This cycle repeats until the conditions set by the scale parameter, shape/color, and compactness are satisfied [41]. The scale parameter determines the size of the resultant objects, whereas shape regulates the influence of spectral characteristics [41]. Kim et al. [16] define optimal segmentation as that which generates objects of similar size to that of the target ground features.

In the current study, we utilized a trial-and-error approach to the scale, shape, and compactness parameters to determine which setting delivers the most meaningful objects. For image subsets of both study sites, the best scale parameter was found to be 5, shape was set at 0.9, and compactness at 0.4. Along with common parameters, the same layer weights were assigned during the segmentation process. NIR 1 was assigned the highest weight at 3, followed by Blue, Green, Red Edge and Yellow at 2, followed by Coastal, NIR 2, and Red at 1 each. NIR 1 was assigned the maximum weight as it is the most impacted by moisture, enabling greater spectral differentiation. The other layers were weighed one at a time in a repeated iterative process to determine the best weight to deliver meaningful objects. The segmentation parameters were consistently used to segment all image-processing schemes from both study sites. Figures 3 and 4 highlight pre and post segmentation as well as the post classification objects from the atmospherically corrected and pansharpened subsets, respectively.

3.3.2. Object Features and Rule Sets

Post segmentation, image objects provide much more than just spectral information. In this study, several object features were identified which were incorporated into rule sets for enabling classification. The features identified for classification are mean, quantile, standard deviation, min. pixel value, max. pixel value, edge contrast of neighbor pixels, number of overlapping thematic objects, relative border to, and customized ratios/arithmetic features. Table 2 describes each feature and the features per band used in the current study. The rule sets incorporating these features are developed based on a logic of testing the spectral and contextual features to determine how much information and which features are more suitable for mapping surface facies across different processing schemes.

1. Rule Set 1: Only object spectral information This rule set utilizes only spectral information from the mean values per band and customized ratios developed from the mean values to classify objects, the reasoning being to test the level of accuracy achievable when classifying objects using only spectral properties.
2. Rule Set 2: Inter-object and contextual information This rule set utilizes features that are not direct object spectral properties or ratioed spectral properties. This rule set will test classification of segmented objects without direct information on the spectral properties of the object.
3. Rule Set 3: Combination of spectral and contextual information This rule set will attempt to combine the features of both rule sets to achieve the best possible classification map. Supplementary Table S3 presents the exact rule set post segmentation for all the processing schemes. Table 3 highlights the best performing rule sets for the associated processing scheme for both study areas based on individual overall accuracy and F1 score.

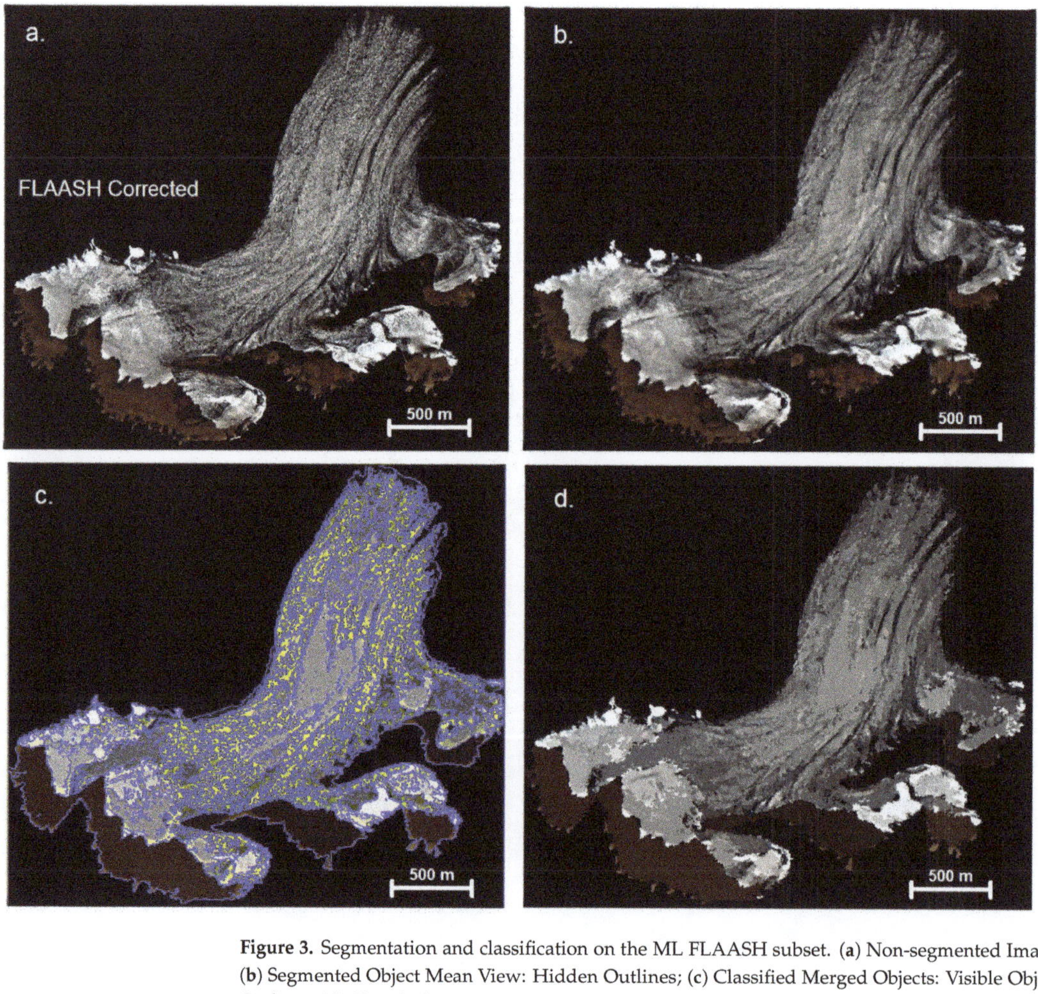

Figure 3. Segmentation and classification on the ML FLAASH subset. (**a**) Non-segmented Image; (**b**) Segmented Object Mean View: Hidden Outlines; (**c**) Classified Merged Objects: Visible Object Outlines; (**d**) Classified Merged Objects: Hidden Outlines. RGB: Coastal, Blue, Green.

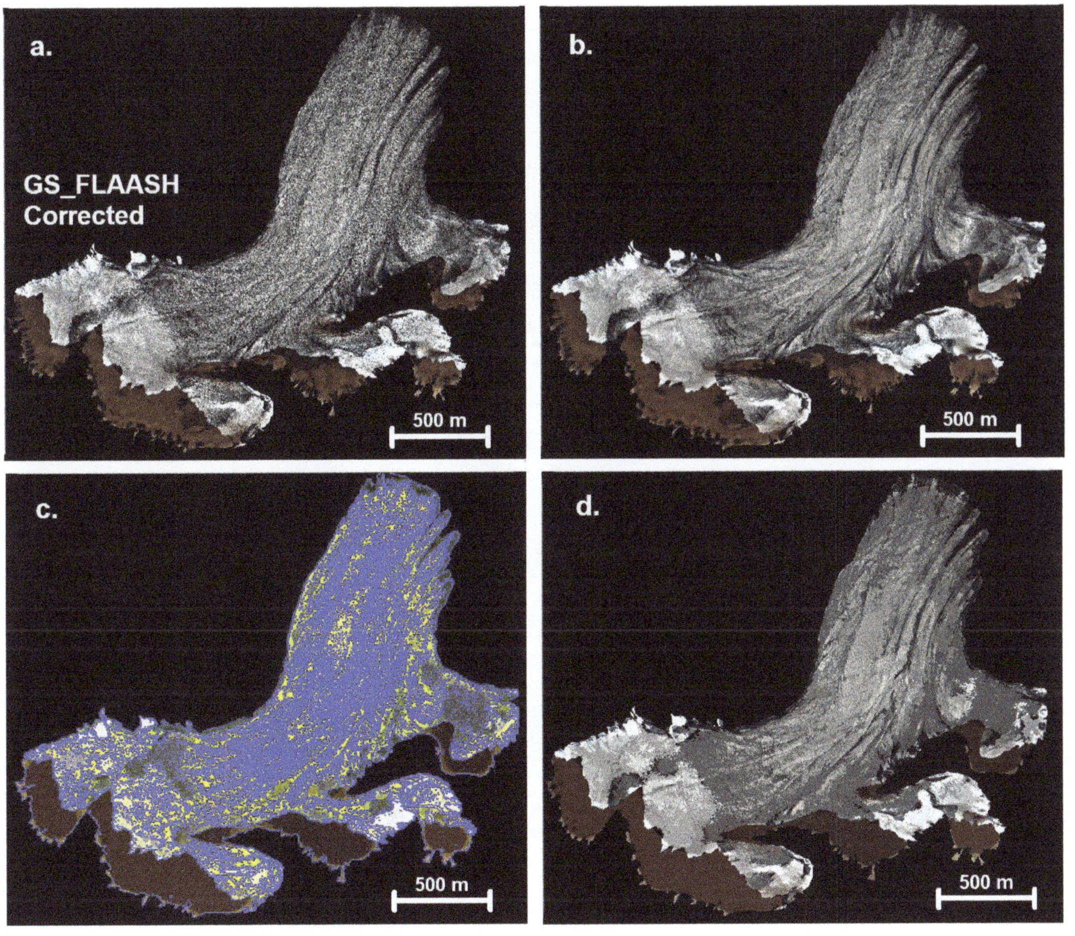

Figure 4. Segmentation and classification on the ML GS_FLAASH subset. (**a**) Non-segmented Image; (**b**) Segmented Object Mean View: Hidden Outlines; (**c**) Classified Merged Objects: Visible Object Outlines; (**d**) Classified Merged Objects: Hidden Outlines. RGB: Coastal, Blue, Green.

Table 2. Set of object, class, and customized features used to classify facies in this study. Description of the features in addition to the specific features used in the current study. Quantile [0.5] refers to the 50th percentile or middle quartile of the object data per spectral band.

Type of Feature	Feature Name	Description	Features Tested in This Study
Object Features: Layer Values	Mean Value per Layer/Band	The mean layer intensity value of an image object [42]	Coastal, Blue, Green, Yellow, Red, Red Edge, NIR 1, NIR 2, Brightness, Max. Difference
Object Features: Layer Values	Quantile	The feature value, where a specified percentage of image objects from the selected image object have a smaller feature value [42]	Quantile [0.5] (Coastal), Quantile [0.5] (Coastal), Quantile [0.5] (Blue), Quantile [0.5] (Green), Quantile [0.5] (Yellow), Quantile [0.5] (Red), Quantile [0.5] (Red Edge), Quantile [0.5] (NIR 1), Quantile [0.5] (NIR 2)

Table 2. *Cont.*

Type of Feature	Feature Name	Description	Features Tested in This Study
Object Features: Layer Values: Pixel-Based	Standard Deviation	The standard deviation of the feature value from all objects of the selected image object domain [42]	Coastal, Blue, Green, Yellow, Red, Red Edge, NIR 1, NIR 2
Object Features: Layer Values: Pixel-Based	Minimum Pixel Value	The value of the pixel with the minimum layer intensity value in the image object [42]	Coastal, Blue, Green, Yellow, Red, Red Edge, NIR 1, NIR 2
Object Features: Layer Values: Pixel-Based	Maximum Pixel Value	The value of the pixel with the maximum layer intensity value in the image object [42]	Coastal, Blue, Green, Yellow, Red, Red Edge, NIR 1, NIR 2
Object Features: Layer Values: Pixel-Based	Edge Contrast of Neighbor Pixels	Refers to the edge contrast of an image object to the surrounding volume of a given size	Coastal(3), Blue(3), Green(3), Yellow(3), Red(3), Red Edge(3), NIR 1(3), NIR 2(3)
Object Features: Thematic Attributes	Number of Overlapping Thematic Objects	The number of thematic objects that an image overlaps with if the scene contains a thematic layer [42]	Manual Digitized Layer of Shadowed Snow
Class-Related Features: Relations to Neighbor Objects	Relative Border To	Is the ratio of the common border length of an image object with a neighboring image object assigned to a defined class to the total border length [42]	Classified Objects
Object Features: Customized Features	Arithmetic Feature	Composed of existing features, variables, and constants, which are combined via arithmetic operations [42]	*Customized Ratios (using Mean Value)* R_RE = (Red/Red Edge) CB_CB = (Coastal − Blue)/(Coastal + Blue) G_C = (Red)/(Coastal) RC_RG = (Red/Coastal) × (Red/Green) Max_Min_RE = (Max. Pixel Value Red Edge − Min. pixel value Red Edge) Y_C = (Yellow/Coastal) C_G = (Coastal/Green) R_C = (Red/Coastal) C_N1 = (Coastal/NIR 1) G_RE = (Green/Red Edge) R_B = (Red/Blue) R_G = (Red/Green) N2_Y = (NIR 2/Yellow) N1_R = (NIR 1/Red) N1_N2 = (NIR 1/NIR 2) CN2_CN2 = (Coastal − NIR 2)/(Coastal + NIR 2) N1N2_N1N2 = (NIR 1 − NIR 2)/(NIR 1 + NIR 2)

Table 3. Best performing rule sets for respective processing scheme for each study site based on individual overall accuracy and F1 score. Detailed rule sets for all processing schemes are available in Supplementary Table S3.

Study Site	Rule Set and Processing Scheme		
	Rule Set 1: FLAASH	**Rule Set 2: DOS**	**Rule Set 3: QUAC**
Ny-Ålesund, Svalbard	*Shadowed Snow* R_RE $\geq$ 1.15 Rel. Border to Shadowed Snow > 0 *Dirty Ice* CB_CB < −0.17 Rel. border to Dirty Ice > 0.2 G_C $\geq$ 1.6 *Dry Snow* Mean NIR 2 > 0.6 Mean NIR 1 >0.4 Rel. Border to DS $\geq$ 0.4 *Wet Snow* Mean Blue $\geq$ 0.57 *Melting Snow* Mean Red Edge $\geq$ 0.5 *Saturated Snow* RC_RG $\geq$ 1.2 Rel. Border to SaS > 0.5 Mean Red $\leq$ 0.3 Rel. Border to SaS > 0.4 *Melting Glacier Ice* Mean Red $\leq$ 0.37 *Glacier Ice* Mean Coastal $\leq$ 0.5	*Shadowed Snow* Standard Deviation NIR 2 $\leq$ 0.006 Min. Pixel Value Blue $\leq$ 0.25 Rel. Border to Shadowed Snow > 0.1 *Dry Snow* Quantile [0.5] Red Edge $\geq$ 0.8 Quantile [0.5] Yellow $\geq$ 0.7 Quantile [0.5] NIR 1 $\geq$ 0.6 *Wet Snow* Min. Pixel Value Blue $\geq$ 0.46 *Dirty Ice* Quantile [0.5] Coastal $\leq$ 0.15 Max. Pixel Value Blue $\leq$ 0.25 Max. Pixel Value Green $\leq$ 0.23 *Melting Snow* Min. Pixel Value Yellow $\geq$ 0.32 *Streams and Crevasses* Standard Deviation NIR 2 $\geq$ 0.06 *Saturated Snow* Min. Pixel Value NIR 1 $\leq$ 0.12 Quantile [0.5] NIR 2 $\leq$ 0.15 *Glacier Ice* Min. Pixel Value Coastal $\geq$ 0.3 Min. Pixel Value Red Edge $\geq$ 0.28 *Melting Glacier Ice* Min. Pixel Value Red Edge < 0.28	*Shadowed Snow* G_RE $\geq$ 1.5 Rel. Border to Shadowed Snow $\geq$ 0.1 *Dirty Ice* R_C $\leq$ 1.6 *Dry Snow* Mean NIR 2 $\geq$ 0.4 *Wet Snow* Mean NIR 1 $\geq$ 0.36 *Melting Snow* Mean Red $\geq$ 0.3 Standard Deviation NIR 1 $\leq$ 0.09 *Streams and Crevasses* Standard Deviation Yellow $\geq$ 0.15 *Saturated Snow* Mean Green $\leq$ 0.17 Min. Pixel Value Coastal $\leq$ 0.05 *Melting Glacier Ice* Mean Red $\leq$ 0.3 *Glacier Ice* Min. Pixel Value Green > 0.14
	Rule Set 1: HCS_FLAASH	**Rule Set 2: HCS_QUAC**	**Rule Set 3: HCS_FLAASH**
Chandra–Bhaga basin, Himalaya	*Shadowed Snow* No. of Overlapping Objects: Shadowed Snow = 1 *Debris* R_B $\geq$ 7 *Ice Mixed Debris* R_B $\geq$ 5 N2_Y $\geq$ 1 *Snow* Mean Coastal $\geq$ 0.7 R_RE $\geq$ 0.8 *Crevasses* R_RE <= 0.8 Mean NIR 2 < 0.6 *Glacier Ice* Mean NIR 2 $\geq$ 0.6	*Shadowed Snow* No. of Overlapping Objects: Shadowed Snow = 1 *Snow* Quantile [0.5] (Red) $\geq$ 0.45 Quantile [0.5] (NIR 1) $\geq$ 0.45 Min. Pixel Value NIR 2 $\geq$ 0.45 *Crevasses* Standard Deviation Yellow $\geq$ 0.12 *Ice Mixed Debris* 0.15 < Max. Pixel Value Green > 0.05 *Debris* Max. Pixel Value NIR 1 < 0.15 *Glacier Ice* Quantile [0.5](Green) $\leq$ 0.35	*Shadowed Snow* No. of Overlapping Objects: Shadowed Snow = 1 *Debris* R_B $\geq$ 7 *Ice Mixed Debris* N2_Y $\geq$ 1 Quantile [0.5](Red Edge) $\leq$ 0.4 *Crevasses* Standard Deviation NIR 2 $\geq$ 0.2 *Snow* Quantile [0.5] (Coastal) $\geq$ 0.7 Quantile [0.5] (Blue) $\geq$ 0.7 R_RE $\geq$ 0.8 *Glacier Ice* Mean NIR 2 $\geq$ 0.6 Quantile [0.5] (Coastal) < 0.7

Figure 5 displays the classification results of the rule sets and processing schemes displayed in Table 3.

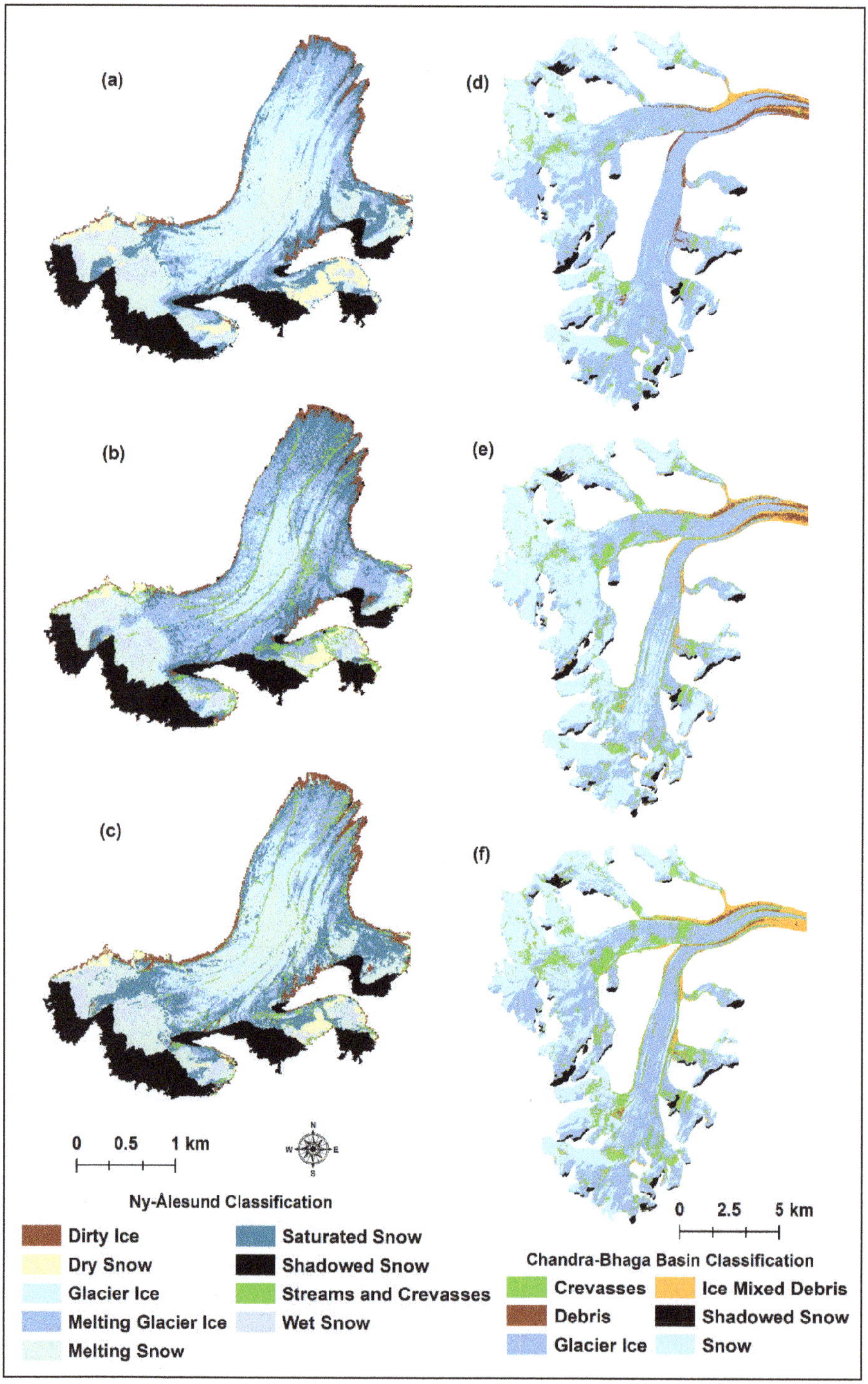

Figure 5. Classification results of the best rule sets from both study sites. Ny-Ålesund classification: (**a**) FLAASH_Rule Set 1; (**b**) DOS_Rule Set 2; (**c**) QUAC_Rule Set 3. Chandra-Bhaga Basin classification: (**d**) HCS_FLAASH_Rule Set 1; (**e**) HCS_QUAC_Rule Set 2; (**f**) HCS_FLAASH_Rule Set 3.

3.4. Accuracy Assessment

Logistical constraints in the field campaign to Svalbard, and harsh field conditions in the month of image acquisition in the Himalaya, prohibited ground truth collection. Avoiding bias in the manual assignment of reference data, a total of 1160 points were selected in an equalized random-sampling approach [42,43] to determine the accuracy of the thematic results. Error matrices were generated to calculate measures such as precision, recall, F1 score, overall accuracy (OA), error rate, and specificity [44]. While error rate is calculated as '100-OA' in terms of percentage, Supplementary Table S4 highlights the mathematical equations of the measures.

4. Results and Discussion

4.1. Quantitative Analysis of Surface Facies

4.1.1. Area of Surface Facies for Each Processing Scheme

The area of each facies is reported as an average across the rule sets for each individual processing scheme. Table 4 highlights the area per facies for glaciers ST and ML.

Table 4. Classified area of facies for each processing scheme for the glaciers ST and ML averaged across all rule sets are summarized here.

Study Sites	Facies	Atmospheric Correction			GS Pansharpening			HCS Pansharpening		
		DOS	QUAC	FLAASH	DOS	QUAC	FLAASH	DOS	QUAC	FLAASH
Samudra Tapu	Snow	30.83	26.01	27.55	28.13	27.23	27.63	27.66	27.51	27.60
	Shadowed Snow	2.23	2.24	2.18	2.22	2.21	2.20	2.21	2.21	2.21
	Ice-Mixed Debris	1.93	2.09	3.26	2.43	2.59	2.76	2.59	2.65	2.67
	Glacier Ice	34.72	40.18	35.72	36.87	37.59	36.73	37.06	37.13	36.97
	Debris	1.19	1.18	1.18	1.18	1.18	1.18	1.18	1.18	1.18
	Crevasses	5.11	4.31	6.11	5.17	5.20	5.49	5.29	5.33	5.37
Midtre Lovénbreen	Dirty Ice	0.28	0.30	0.30	0.28	0.36	0.31	0.28	0.32	0.29
	Dry Snow	0.20	0.29	0.26	0.25	0.19	0.22	0.22	0.20	0.22
	Glacier Ice	0.68	0.66	0.75	0.71	0.53	0.68	0.62	0.57	0.84
	Melting Glacier Ice	0.87	0.66	0.68	0.78	1.00	0.62	0.73	0.88	0.76
	Melting Snow	0.45	0.57	0.80	0.63	0.45	0.59	0.70	0.53	0.50
	Saturated Snow	0.90	0.91	0.72	0.80	0.88	1.02	0.97	0.89	0.92
	Shadowed Snow	0.78	0.74	0.77	0.73	0.78	0.76	0.67	0.75	0.72
	Streams and Crevasses	0.24	0.25	0.16	0.21	0.18	0.22	0.20	0.20	0.18
	Wet Snow	0.36	0.38	0.30	0.37	0.37	0.33	0.35	0.41	0.33

Glaciers in Ny-Ålesund are smaller in size when compared to glaciers in the Chandra–Bhaga basin. Owing to this, the variances in area of facies are greater for ST classification than the ML classification. Figure 6 presents the deviations from the average classified area for all facies across glaciers ML and ST using stacked bar graphs. For the ST glacier, cumulative variances across all processing schemes highlight that the atmospherically corrected, non-sharpened imageries (average of DOS, FLAASH, and QUAC) deliver the highest variance of 2.78 km^2. Among the individual subsets, FLAASH provides the least variance across all facies (0.76 km^2). The maximum variation was delivered by the QUAC subsets. The GS sharpened subsets provide moderate variance with an average of 2.54 km^2. The HCS subsets deliver the least variance in the classified area (13.68 km^2) among the facies for the ST glacier. Within the facies themselves, glacier ice observes the most cumulative variations in classified area, whereas shadowed snow varies the least.

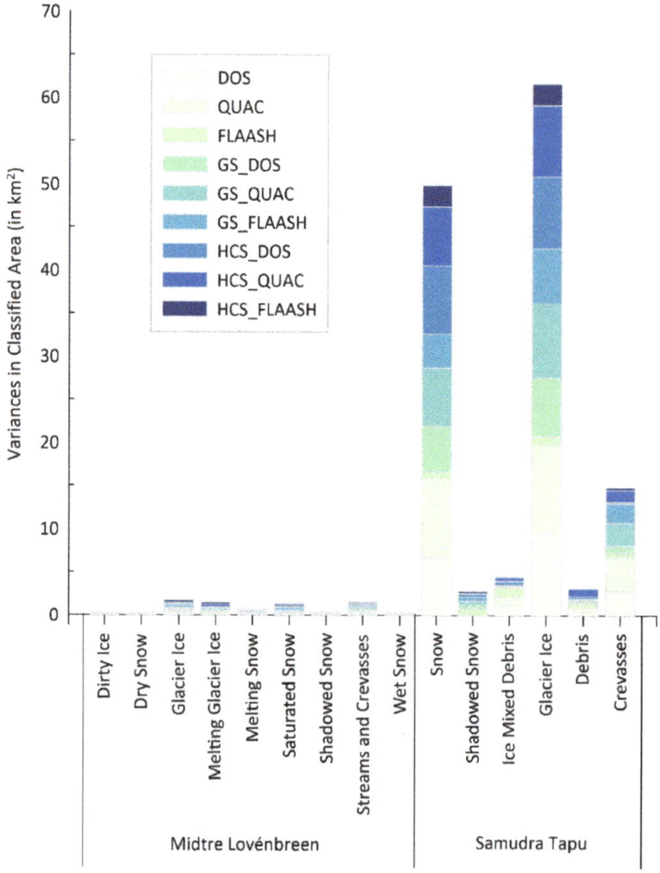

Figure 6. Variations in classified area of each facies for the representative glaciers ML and ST. Stacked columns highlight variances in the classified area between processing schemes for the same facies.

Unlike the ST classifications, the ML facies show an inverse trend of variability according to pansharpening. The pansharpened subsets report slightly higher cumulative average variances for the GS and HCS subsets (0.10 and 0.11 km^2) than the non-pansharpened subsets (0.09 km^2). Within the individual subsets the QUAC delivers least variability. Among the facies, dry snow retains maximum consistency across all processing schemes with a variance in classified area of 0.02 km^2. Glacier ice shows maximum variation at 0.19 km^2.

4.1.2. Accuracy Yielded by Each Rule Set

Complete accuracy assessment for the classification is detailed in Supplementary Sheet S1. Here, we aim to assess the classification results of the rule sets using the F1 score [44] as an average of all processing schemes/subsets. The aim is to utilize the robustness of the F1 score to determine reliability of the classified surface facies and rule sets. This is followed by the OA to delineate the order of reliable classification across both study areas and as an average of both study areas. Figure 7 highlights the F1 score of the classified facies averaged across all processing schemes for each rule set along with the respective deviation from the mean.

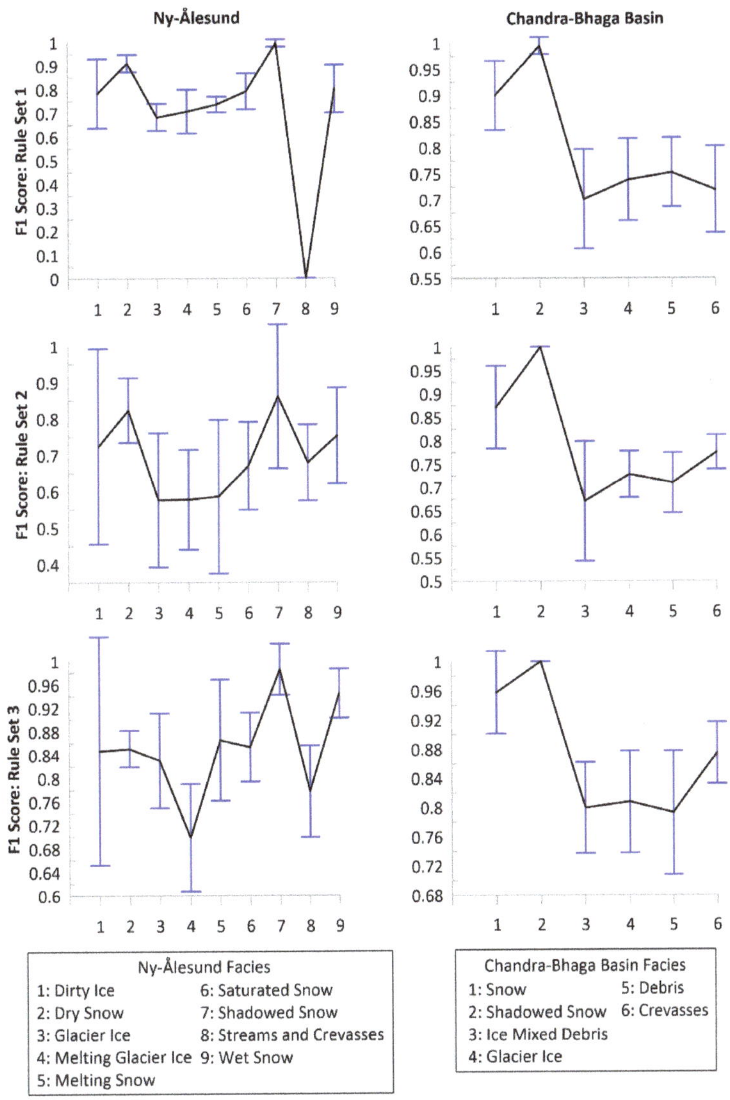

Figure 7. F1 score of each facies for the three rule sets as the mean of all processing schemes. The error bars in blue highlight deviations from the mean F1 score indicating reliability of the rule sets for the respective facies.

Overall observations suggest that rule set 3 delivered the best performance for both study sites. In the case of Ny-Ålesund classifications, it is observed that rule set 1 provides the least variance in mapped facies. However, the inability of spectral information to isolate fine features such as streams and crevasses is a major limiting factor. Rule set 2, which primarily utilizes spatial and contextual information, is limited by the large variations in performance across all processing schemes. An inference is possible here that different preprocessing routines may influence the spatial and contextual characteristics of objects to a greater extent than absolute spectral information. Unfortunately, this finding does not extend to facies in the Chandra–Bhaga basin. Here, rule set 2 delivered the median variance between rule set 1 and 3. The variance for the Chandra–Bhaga rule set 2 classification is less

(0.06) than the Ny-Ålesund rule set 2 classifications (0.16). The better performance of the Chandra–Bhaga classifications suggests that fewer classes with better spectral separability can result in better thematic classifications across all processing schemes by combining spatial–spectral attributes. Hence, the implication from these results highlights the decrease in impact of processing schemes on VHR imagery for facies with larger spectral separability and a lower number of classes and by utilizing spatial–spectral attributes. Table 5 summarizes the OA and the reliability orders for mapped facies and the study sites. The reliability orders highlight the best combination of features (Table 3 and Supplementary Table S3) that enabled mapping of the facies with the highest calculated accuracy.

Table 5. Overall accuracy (OA) of the classification of surface facies across both study sites using the three rule sets. The OA is calculated as an average across all processing schemes. Reliability orders based on the OA and variances represent the best rule set for mapping the individual facies.

Rule Sets		**Overall Accuracy (in %)**	
		Ny-Ålesund	*Chandra–Bhaga Basin*
Rule Set 1		75.03	80.00
Rule Set 2		69.51	79.08
Rule Set 3		85.06	87.09
Facies		**Reliability Order**	
Ny-Ålesund	Dirty Ice	rule set 3 > rule set 1 > rule set 2	
	Dry Snow	rule set 1 > rule set 3 > rule set 2	
	Glacier Ice	rule set 3 > rule set 1 > rule set 2	
	Melting Glacier Ice	rule set 1 > rule set 3 > rule set 2	
	Melting Snow	rule set 3 > rule set 1 > rule set 2	
	Saturated Snow	rule set 3 > rule set 1 > rule set 2	
	Shadowed Snow	rule set 1 = rule set 3 > rule set 2	
	Streams and Crevasses	rule set 3 > rule set 2 > rule set 1	
	Wet Snow	rule set 3 > rule set 1 > rule set 2	
Chandra–Bhaga Basin	Snow	rule set 3 > rule set 1 > rule set 2.	
	Shadowed Snow	rule set 2 = rule set 3 > rule set 1	
	Ice-Mixed Debris	rule set 3 > rule set 1 > rule set 2	
	Glacier Ice	rule set 3 > rule set 1 > rule set 2	
	Debris	rule set 3 > rule set 1 > rule set 2	
	Crevasses	rule set 3 > rule set 1 = rule set 2	

4.1.3. Variable Effect of Atmospheric Corrections

Figure 8 highlights the error rate yielded by each rule set for the atmospherically corrected image subsets. Rule set 3 delivers the most consistent performance, i.e., the lowest error rate except for the Himalaya FLAASH_Rule Set 1 classification. The common highest error rate was observed for both the Ny-Ålesund QUAC_Rule Set 1 and Ny-Ålesund FLAASH_Rule Set 2 classifications. When based upon variances in the performance of processing schemes, the reliability order of atmospheric corrections is DOS > FLAASH > QUAC.

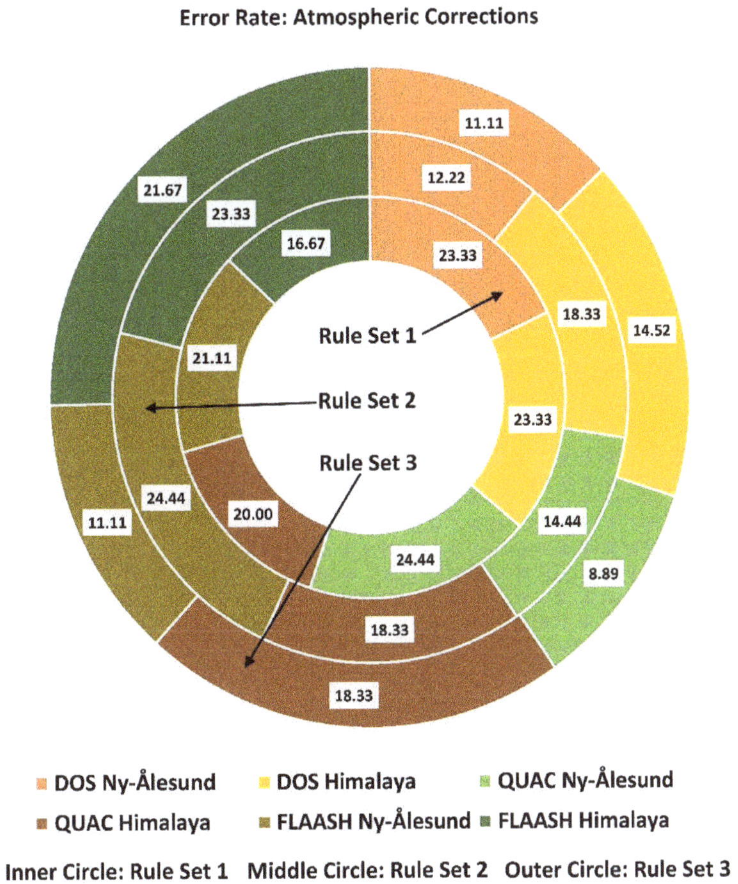

Figure 8. Error rates (in %) of the rule sets for the atmospheric corrections presented as a donut diagram. Inner circle: rule set 1, middle circle: rule set 2, and outer circle: rule set 3.

Rule set 3 delivers the most consistent performance, i.e., the lowest error rate except for the Himalaya FLAASH_Rule Set 1 classification. The common highest error rate was observed for both the Ny-Ålesund QUAC_Rule Set 1 and Ny-Ålesund FLAASH_Rule Set 2 classifications. When based upon variances in performance of processing schemes, the reliability order of atmospheric corrections is DOS > FLAASH > QUAC. The reliability of individual rule sets based upon the error rate is Ny-Ålesund QUAC_Rule Set 3 > Ny-Ålesund DOS_Rule Set 3 = Ny-Ålesund FLAASH_Rule Set 3 > Ny-Ålesund DOS_Rule Set 2 > Ny-Ålesund QUAC_Rule Set 2 > Himalaya DOS_Rule Set 3 > Himalaya FLAASH_Rule Set 1 > Himalaya DOS_Rule Set 2 = Himalaya QUAC_Rule Set 2 = Himalaya QUAC_Rule Set 3 > Himalaya QUAC_Rule Set 1 > Ny-Ålesund FLAASH_Rule Set 1 > Himalaya FLAASH_Rule Set 3 > Ny-Ålesund DOS_Rule Set 1 = Himalaya DOS_Rule Set 1 = Himalaya FLAASH_Rule Set 2 > Ny-Ålesund QUAC_Rule Set 1 = Ny-Ålesund FLAASH_Rule Set 2.

These findings imply that when combining spatial–spectral properties of objects, the DOS correction is sufficient for deriving reliable and accurate thematic classifications. FLAASH is most useful when the mapping procedure relies purely on spectral information. QUAC and DOS deliver similar performances for rule set 2, and this suggests that GEOBIA relying heavily on spatial properties between objects would benefit from simpler corrections. In a comparative assessment using GEOBIA for mapping benthic habitats, Siregar et al. [45]

suggested that the difference between FLAASH corrected and noncorrected classification is not significant. However, the profile used to highlight post correction spectral signatures highlights that the range of reflectance was not resampled from 0 to 1. Moreover, as the classification was performed using support vector machines (SVM) post segmentation, a comparison between rule-set-based classification is difficult. While the high performance of the DOS correction in the current study does not agree with the findings of Phiri et al. [46], it is likely because the current study utilizes rule sets rather than a classification algorithm (such as random forest) to classify segmented objects. Moreover, the superior performance of DOS is due to the combination of spatial and spectral properties, whereas FLAASH performs better for only spectral properties. This is important because where PBIA relies solely on spectral properties, GEOBIA can utilize spatial–spectral properties to overcome computational loads of complex atmospheric corrections.

4.1.4. Impact of Pansharpening

Table 6 depicts the average performance of pansharpened imagery across all three rule sets utilizing the OA as the assessing measure. Figure 9 highlights the OA for pansharpened, and non-sharpened images averaged across all the rule sets.

Table 6. Comparative performance of pansharpened and non-pansharpened subsets according to each rule set using the average overall accuracy across both study areas. Values of the best performing rule sets in each processing subset are emboldened and italicized.

Rule Sets	DOS	QUAC	FLAASH	GS			HCS		
				DOS	QUAC	FLAASH	DOS	QUAC	FLAASH
Rule Set 1	76.67	77.78	81.11	75.84	78.34	76.11	73.33	78.62	79.85
Rule Set 2	84.73	83.62	76.12	75.84	76.95	72.22	69.45	69.45	60.28
Rule Set 3	***87.19***	***86.39***	***83.61***	***85.00***	***85.00***	***89.45***	***87.50***	***88.62***	***81.95***

Considering rule set 1, neither GS nor HCS enhanced the OA for DOS and FLAASH subsets. HCS and GS added minor accuracy improvements to the QUAC subsets. Rule set 2 displayed the greatest loss in performance after pansharpening. The HCS_FLAASH reduced in performance by 15.84%. This is a significant decrease in OA. However, rule set 3 classifications showed enhancements in the OA of GS_FLAASH, HCS_DOS, and HCS_QUAC. Averaging across rule sets, the performance trend for the pansharpened subsets is GS_QUAC > GS_FLAASH > GS_DOS = HCS_QUAC > HCS_DOS > HCS_FLAASH. Overall comparison between pansharpening algorithms suggest that GS > HCS. Of all pansharpened image rule sets, rule set 3 delivered maximum performance and rule set 2 was the most inaccurate. This is surprising, as finer resolution can improve the contrast between image pixels, allowing for potentially better segregation of homogenous objects [47]. Rule set 2, which relies on the spatial distinction between objects, decreased in accuracy after pansharpening. Gavankar and Ghosh [48] utilized HCS sharpened Ikonos imagery to identify buildings in an urban setting. Buildings usually have consistent geometric shapes; this consistency can be leveraged using shape parameters such as circulometry and rectangulomtery [48]. A glacier's surface does not contain consistent geometric patterns, and this may indicate that segmentation of fine-resolution satellite data does not necessarily translate to improved classification. In a post-earthquake GEOBIA-based mapping application to identify damaged buildings and temporary relief sites, Omarzdeh et al. [49] found that damaged buildings which have lost their consistent structure are difficult to map. This inconsistency due to shape can also be detrimental for glacier facies mapping, as developing rules to accurately depict textural variations in observed facies is a challenge. Rule set 2 is testament to the challenges of spatial property-based mapping. The importance of combined spatial–spectral characteristics is evident, as segmentation can compensate the pansharpened image with high frequency information inserted during pansharpening [50].

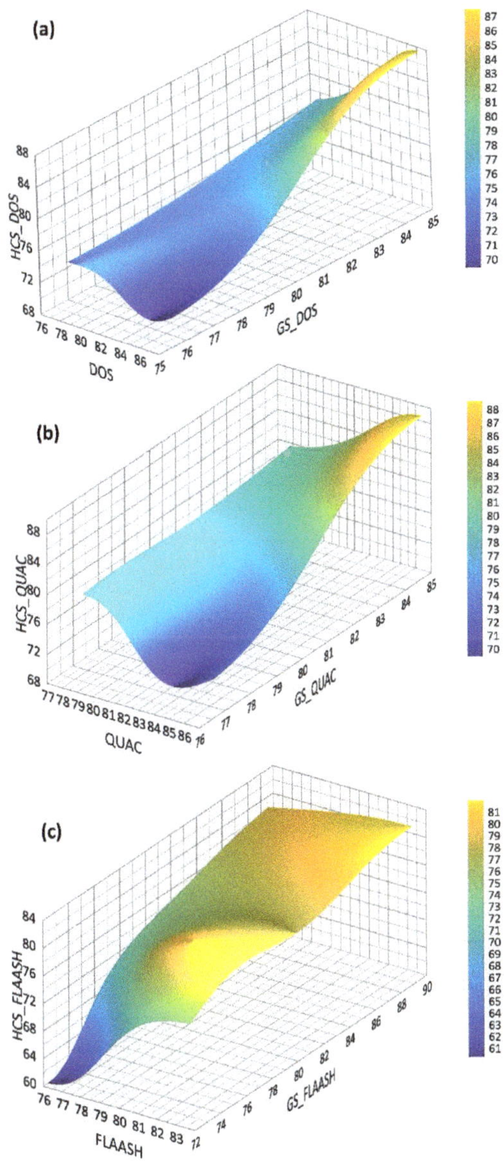

Figure 9. Overall-accuracy-based assessment of the performance of pansharpened vs. non-sharpened classifications. The 3D surface summarizes the OA from all three rule sets for (**a**) DOS vs. GS_DOS vs. HCS_DOS, (**b**) QUAC vs. GS_ QUAC vs. HCS_ QUAC, and (**c**) FLAASH vs. GS_FLAASH vs. HCS_FLAASH.

4.2. Discussion

In the quest of mapping earth features through GEOBIA, the focus is usually on multi-level/scale segmentation to highlight finer features. Glacier surfaces have a variety of undulating and disheveled features such as crevasses and debris which may necessitate such segmentation. However, in the current study, we focused on attempting to adapt the object features post segmentation to discern the visible surface facies. Utilizing only spectral information, streams and crevasses were not defined in any rule set 1 for Ny-Ålesund. In the case of Himalayan facies, crevasses were highlighted using all rule sets. This may be because the crevasses in Himalayan facies are larger and more clearly discernible than in Ny-Ålesund. Mitkari et al. [23] found some rills that were misclassified with periglacial debris in their multi-source GEOBIA approach on the Bara Shigri glacier. This is attributed to the coarser scale of segmentation. In the current study, however, the finer streams and crevasses were not classified using only spectral information. The contextual and spatial attributes of rule set 2 and rule set 3 enabled the identification of streams and crevasses. The standard deviation feature was most useful for discerning the minor streams and crevasses. This may point towards certain spatial attributes that can overcome the spectral constraints. Kim et al. [16] found that addition of the GLCM to multi-scale and single-scale segmentation improved classification performance. Moreover, the same constraint was found in the current study when mapping the same features using pansharpened images. Dabiri et al. [51] utilized Landsat images to perform landslide mapping using thresholds on indices and layer values in OBIA. However, due to acquisition of different images under different illumination conditions, each image was assigned thresholds independently. In the current study, differing processing routines cause a change in the derived reflectance [paper 1]. This therefore implies that the thresholds and object features utilized must be assigned accordingly. Each rule set for each processing scheme was assigned independently. It is observed that the maximum variation in rule sets is required in processing schemes for the Ny-Ålesund image, whereas most of the Himalayan images were classified with similar features, albeit with changes in the thresholds owing to changes in reflectance. This suggests that fewer features and larger targets can be extracted with more stable rule sets using changes in thresholds according to image and scene conditions. However, smaller features necessitate more rigorous processing. Rastner et al. [9] found that pixel-based mapping outperformed OBIA when detecting objects with single pixel size such as nunataks, narrow ridges, and couloirs. However, their analysis was performed on Landsat ETM+, ASTER, and Landsat TM imagery. In the case of the current study, the streams and crevasses in Ny-Ålesund were mapped using spatial attributes. A key finding is that improvement in spatial resolution did not necessitate a change in segmentation parameters. Furthermore, the same segmentation was applicable on all processing schemes, and this suggests that irrespective of the processing routines the combination of 'optimal' segmentation when combined with adjusted spatial and spectral attributes is sufficient for mapping glacier surface facies using VHR data.

In paper 1, we analyzed the variations induced by different processing routines in the thematic classification of facies using conventional and advanced pixel-based classification methods. In this study, we analyzed the impact of the same processing routines on GEOBIA mapping of surface facies. In paper 1, the FLAASH algorithm performed the best, whereas here, the DOS-based rule sets delivered superior performance. Pansharpening significantly reduced reliability of PBIA, whereas it selectively improved overall performance in OBIA. This highlights that segmentation and spatial–spectral attributes have a key advantage over hard pixel classification using VHR VNIR satellite data.

Significances and Challenges

GEOBIA classifications can be categorized into (a) operator driven (rule sets) and (b) operator assisted (classification algorithms). Johnson et al. [52] labels these groups as knowledge driven and automated classification. In the current study, we focus on rule sets because although semi-automated approaches are reported to be less subjective [52], rigorous analysis of individual object characteristics to identify meaningful segmentation and useful spatial–spectral features must be performed to gauge the quality of classification. Rule sets can be described as an attempt to replicate logical patterns of image interpretation from the human brain into classification operations [47]. The impact of processing routines on the modification of rule sets presents an opportunity to identify necessary changes in feature thresholds. The data of variation in thresholds, applicability of indices, and spectral information for each processing routine can also help drive the direction of semi-automated classification. Deep learning approaches such as SVM [45] and random forest (RF) [46] can be applied after OBIA to turn the classification into a semi-automatic process. However, the selection of the preprocessing method, the collection of training data, and the resultant goal of mapping facies using any GEOBIA method require a foundational test of the impact of these methods. Rule sets offer the best capacity for object-by-object analysis for many GEOBIA mapping applications and can be compared with results of semi-automatic methods. In this study, each rule set was defined by carefully testing all the parameters associated with the respective facies under consideration and the object features that can enable clear delineation of the target. For Ny-Ålesund, rule set 1 was not capable of differentiating streams and crevasses. This class was misclassified between dirty ice and saturated snow across all processing schemes. Enhancing spatial resolution did not improve mapping of fine features, although the overall classification did improve in cumulative accuracy. The variations in rule sets across processing schemes are found to be greater when the number of surface facies are greater, and the classes have a greater probability of spectral overlap. In such scenarios, scene-based adjustments are necessary for more accurate mapping [53]. However, as demonstrated in this study, spatial attributes may help overcome spectral complexities.

Shadowed areas in the Himalayan images were classified using an overlap of manually digitized vector data. This was constant throughout the experiment. However, the same issue did not arise in the classification of shadowed snow in Ny-Ålesund. Unfortunately, no index or mean threshold parameter was useful in delineating shadowed snow. This is a persistent problem in mountain glacier classification [54]. Figure 10 displays the spectral signatures of the shadowed areas on different points on the FLAASH subset of the glaciers. Figure 11 presents the variations in spectral reflectance of each of the processing schemes. In Figure 10, spectral pattern 2 shows a combination of saturated snow and dry/wet snow. Spectral patterns 3 and 4 are combinations of wet snow, melting snow, and saturated snow. Figure 10b highlights the variations in shadowed snow on the ST glacier. Figure 10b spectral pattern 2 does not even follow the same trend in reflectance. Patterns 1, 3, 4, and 5 vary in intensity, and these variations are significant when derived on other processing schemes (Figure 11). Using the atmospheric corrections in the current study, shadowed areas could not be divided into their constituent facies. Ryan et al. [55] and Leidman et al. [56] noted that shadowed areas can cause overestimation of darker classes. To avoid misclassification between streams and crevasses, dirty ice (the facies with the lowest reflectance), and the snow facies within the shadow, all shadowed areas were labelled as shadowed 'snow'. Shadows induced by larger features present challenges in the classification of smaller features lying within the shadow area. In such cases, the presence of more spectral bands and operator innovation is necessary for creative application of object features [49]. The aim of the current study is to focus on the impact of basic image-processing routines on GEOBIA mapping of glacier facies. The time needed for manual assignment of rules is immense. In this case, corrections for shadows were beyond scope. Moreover, differentiating between glacier features under shadowed areas is more problematic for large-scale mapping applications. Breaking down the component facies of shadowed areas on many glaciers would be an engaging challenge for future experiments. Ancillary layers have been used to supplement GEOBIA [19,23]. This study does not utilize multi-source datasets and highlights the capabilities of a combination of object features to map surface facies.

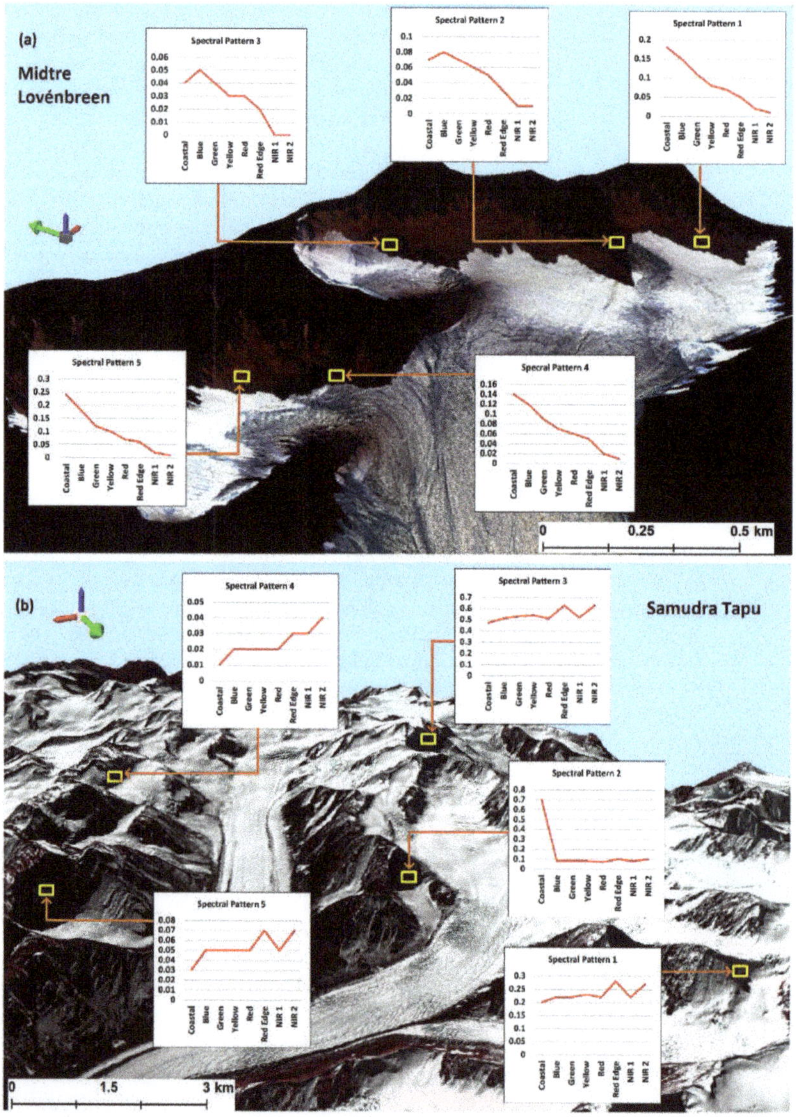

Figure 10. Complex spectral patterns of the shadowed snow at different points on the glacier. Base images are the FLAASH corrected images draped on the ArcticDEM for Ny-Ålesund and ASTER GDEM v2 for Chandra–Bhaga basin. Inset (**a**) 3D surface of glacier ML, (**b**) 3D surface of glacier ST.

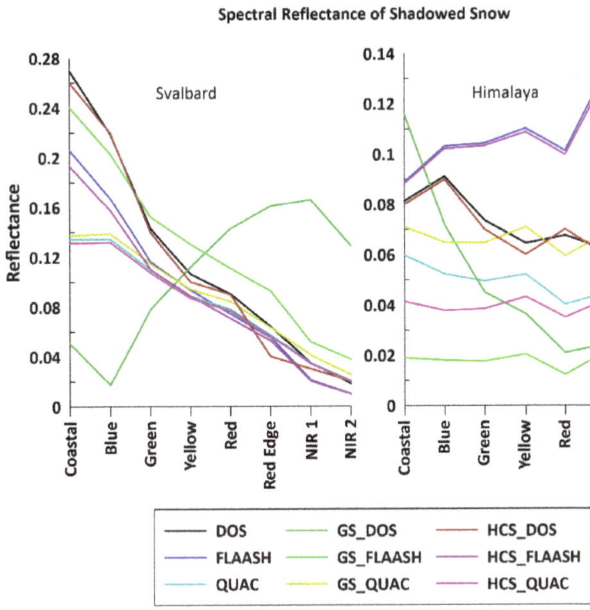

Figure 11. Spectral reflectance of shadowed snow derived across all processing schemes. The left graph shows the reflectance of shadowed snow in Svalbard, whereas the graph on the right shows the reflectance of shadowed snow in the Himalayas.

The Himalayan image shows a larger distribution of snow for its acquisition period. Unfortunately, free precipitation data are not available to corroborate this snow cover with a sudden precipitation event. Currently, free precipitation data for this region are only available from 1901 to 2002 [57]. Estimation of snow cover areas (SCA) and the trend of SCA variability for this region may help support any increase in snow cover during the period of acquisition. Prior to this, it is important to distinguish between SCA and surface facies mapping. SCA mapping is more often performed at a basin level or larger regional scale encompassing a coarser resolution for analyzing spatial, hypsographic, and temporal variability [58]. For example, the Moderate Resolution Imaging Spectroradiometer (MODIS) global snow cover product MOD10CM outlines SCA at a spatial resolution of 5 km [59]. Glacier surface facies are all the zones of snow and ice which can be distinguished based on visual and spectral properties. These facies are clearly visible at the end of the summer season, when almost all the seasonal snow has disappeared. While SCA operations can consider an entire glacier to be 'snow', facies mapping operations may divide snow into dry, wet, melting, saturated, etc. At the basin level and larger, SCA is reported to have substantially reduced [60]. However, at the sub-basin level, snow cover is reported to have greater dynamicity [61]. Rathore et al. [61] analyzed SCA variability and accumulation and ablation patterns using daily snow cover data from 2004 to 2014 for six sub-basins of the Chenab basin, which include the Chandra and Bhaga basins of the current study. The imagery used in the present study covers part of the Chandra and part of the Bhaga sub-basins. Hence, the name used here is the Chandra–Bhaga (sub)basin of the current study. Rathore et al. [61] observed an increasing trend in maximum snow cover during the accumulation period. Sub-basins such as Ravi and Warwan showed greater variability in the snow cover due to simultaneous accumulation and ablation, whereas the Chandra–Bhaga area did not display such variability. The larger distribution in snow cover in Chandra–Bhaga was attributed to the higher elevation of the sub-basin in comparison to the others. The authors even observed full coverage of snow even at temperatures of 13 °C. This is suggested to be because of melt causing a reduction in thickness, but not area. Moreover, precipitation data collected using an all-weather station (AWS) in the Bhaga basin suggests that rainfall (wet precipitation) is decreasing, whereas snowfall (dry precipitation) is not. SCA for the month of October also highlights a greater distribution of snow for the year 2014 than 2010–2013 [61]. Sahu and Gupta [62] report similar findings for seasonal analysis of SCA in the Chandra basin from 2000–2017. The authors observed that SCA begins increasing from September due to lower temperatures. This effectively corroborates the large distribution of snow observed and classified in the Chandra–Bhaga basin in the present study.

Unlike the Chandra–Bhaga basin, glaciers in Svalbard are experiencing accelerated thinning [63,64]. The ML glacier in Ny-Ålesund contains the longest record of observed data [65]. The Equilibrium Line Altitude (ELA) determines the transition from accumulation facies to ablation facies. Average elevation of the maximum snowline in summer can be used as a representative ELA [65]. The various snow facies in the ML glacier identified here comprise dry snow, wet snow, melting snow, saturated snow, and shadowed snow. The facies are based on visual assessments and spectral characteristics with comparisons against the literature [paper 1]. The image was acquired at the end of the ablation season in August 2016, and this implies that most of the seasonal snow has disappeared, leaving the full range of facies. As precipitation largely determines the ELA, it would be lower in Ny-Ålesund, as the region is closer to the coast [65,66]. The ELA estimated by Pelt et al. [65] is available until 2012. Recently, Garg et al. [67] mapped radar facies at the same site. They utilized the boundary of percolation/wet snow and clean ice as the ELA. Their analysis was performed on seasonal data for the years 2016 to 2020. The year 2016 was marked by an increase in winter season temperature and spring temperature. This invariably would cause an increase in the melt of accumulated snow, thus increasing the amount of wet snow adjacent to dry/perennial snow. This dry snow facies was not discernible in the observation made by Garg et al. [67]. However, in the current study, dry snow was characterizable using VHR multispectral characteristics and GEOBIA. The ELA is reported to be shifting upward due to increasing temperature and reduced precipitation [67]. Precipitation data for Ny-Ålesund for the month and year of image acquisition [68] show no sudden precipitation event prior to scene capture. This suggests that the facies maps and the rule sets are not impacted by any sudden increase in snow.

Increasing temperature in glaciated areas also raises the necessity for identifying and mapping permafrost. According to the National Snow and Ice Data Center (NSIDC), permafrost is a layer of soil/rock, at some depth below the surface, where the temperature has been continuously below 0 °C for a few years or where summer heat does not penetrate to the basal layer of frozen ground [69]. Dabski [70] describes the presence of permafrost in the periglacial and foreland area of a polythermal glacier. The author further highlights the difficulty in discerning between glacial and periglacial domains in high mountain and polar areas [71]. The primary mechanisms for identifying and measuring permafrost are through physical assessments in boreholes, geophysical sounding, temperature logging, and modelling [72]. As it is a subsurface phenomenon, permafrost can itself be difficult to observe using RS data [60]. The observable surface phenomenon is the active layer (lying above permafrost), which freezes, thaws, and refreezes seasonally. This layer can be measured either directly or by interpolation of 0 °C isotherm derived from borehole measurements [73,74]. Using data of mean annual ground temperature, depth of zero annual amplitude, and active layer thickness from the Global Terrestrial Network for Permafrost (GTN-P) database [75,76], Karjalainen et al. [77] predicted the risks of permafrost thaw. Much of the data within the GTN-P database comes from the Thermal State of Permafrost (TSP) Snapshot Borehole Inventory [78,79]. Westermann et al. [80] described two methods for characterizing permafrost using remote observations: (a) mapping landforms indicative of the presence of permafrost, and (b) characterizing and isolating physical variables that correspond to thermal subsurface phenomena. This requires penetrative radar to account for the freeze–thaw state and modelling of land surface temperature (LST) [81]. In the current experiment, we only rely on VNIR VHR multispectral data and thus can neither model LST nor identify subsurface conditions. Incorporating complex permafrost modelling into GEOBIA would be a challenge beyond the scope of the current study. Currently, global permafrost zonation maps are available at a spatial resolution of 1 km [81]. In this study, digitized glacial boundaries were used to extract the glaciers from the study sites, and a 3D visualization of ice divides was used to ensure maximum efficiency. Due to this, it was not necessary to isolate the glacial and periglacial margins. However, the process of permafrost extraction using a combination of RS datasets can be a promising, albeit challenging, direction.

We acknowledge the lack of field data for validation. However, the main aim of the experiment was to test the impact that various processing routines have on the final classification of glacier surface facies. The comparison between the SCA derived in the current study against previous findings along with the equalized random sampling approach and rigorous analysis of performances aid the robustness of the current study. As travel restrictions due to the COVID-19 pandemic did not permit a field campaign during the experiment, visual and spectral characteristics of facies were compared against the literature for their validity [21] (paper 1). A major limitation is the time needed to manually adjust rule sets for each processing scheme. Nevertheless, this works to our advantage in analyzing the implementation of GEOBIA for specific processing schemes. To the best of our knowledge, no study has yet tested the impacts of various processing schemes and rule sets on the thematic classification of glacier surface facies using GEOBIA. An important part of GEOBIA is the transferability of methods. Through the current study, we infer that even for the same sensor over the same glacier, the influence of processing routines dictates the thresholds of spectral features. However, the segmentation parameters tested here may be transferable, as they have consistently yielded optimal objects.

5. Conclusions

This study aimed to map glacier surface facies using a multi-rule set GEOBIA approach utilizing varying image-processing schemes to test the robustness of segmentation and reliability of rule sets. Three atmospheric corrections and two pansharpening algorithms were tested on VHR VNIR WV-2 for Ny-Ålesund and Chandra–Bhaga basin, Himalaya. The atmospheric correction methods included DOS, QUAC, and FLAASH. The pansharpening methods included GS and HCS. The selected segmentation parameters delivered consistent results across all processing schemes. A series of object features were identified to create three rule sets. Rule set 1 focused on only spectral information and rule set 2 relied on contextual and inter-object features, whereas rule set 3 combined both spatial and spectral information. Among the atmospheric corrections, DOS delivered the highest overall accuracy, followed by QUAC, and lastly FLAASH. Pansharpening improved overall performance with GS delivering greater accuracy than HCS. For the Ny-Ålesund glaciers, dirty ice, glacier ice, saturated snow, melting snow, wet snow, and streams and crevasses were best mapped by rule set 3, whereas dry snow and melting glacier ice were best mapped by rule set 1. For the Chandra–Bhaga basin glaciers, all facies were best mapped by rule set 3.

The current study highlights the requirement of selection of segmentation parameters. The incorporation of thresholds differed across rule sets, indicating a more processing-focused, in-scene, and image-specific approach for mapping applications of surface facies. Nevertheless, the segmentation parameters may be transferable to other VHR VNIR data. GEOBIA offers the capability of combining spatial–spectral attributes to glacier surface facies mapping, which can transcend the distortions in performance induced in PBIA by variable processing routines. Moreover, the classification is performed entirely devoid of ancillary data such as slope and temperature, enabling a much deeper understanding of the potential of VHR VNIR data for mapping glacier surface facies.

Supplementary Materials: The following supporting information can be downloaded at: https://www.mdpi.com/article/10.3390/rs14174403/s1, Table S1: The selected glaciers of the study, their areal extents, and GLIMS reference IDs. The extents were calculated from the delineated shapefiles using the geometry calculator in ArcGIS; Table S2: Spectral-band-wise at-sensor reflectance values of selected dark pixels for input into DOS correction module in ENVI 5.3. Table S3: Stepwise rule sets and threshold delineation for each processing scheme for both study sites. The ratios used here are highlighted in Table 2. Table S4: Measures of accuracy used in the current study. TP: samples are those that are in the positive class and are correctly classified, TN: samples that are correctly classified as negative, FP: samples that are not truly of the positive class but are incorrectly mapped as positive. FN: samples that are mapped as negative when they are positive. Sheet S1: Precision, recall, specificity, F1 score. Overall accuracy (OA), error rate (ER) for each processing scheme, rule sets, facies, and both study sites.

Author Contributions: Conceptualization and experiment setup, S.D.J. and S.F.W.; methodology and manuscript design, S.D.J. and S.F.W.; software and processing, S.D.J. and S.F.W.; validation, S.D.J. and S.F.W.; result analysis and discussion, S.D.J. and S.F.W.; resources, A.J.L. and K.B.; writing—original draft preparation, S.F.W. and S.D.J.; writing—review and editing, S.D.J., A.J.L. and K.B.; visualization, S.F.W.; supervision, S.D.J., A.J.L. and K.B.; project administration, S.D.J., A.J.L. and K.B. All authors have read and agreed to the published version of the manuscript.

Funding: This research received no external funding.

Data Availability Statement: Freely available data used in the current study—(1) ASTER GDEM v2. Downloaded from: Gdex.cr.usgs.gov/gdex/ (accessed on: 2 February 2017). The data are now moved to GDEM v3: (reviewed on: 12 March 2022) ASTER GDEM is a product of Japan's Ministry of Economy, Trade, and Industry (METI) and NASA. (2) Arctic DEM. Available online: Pgc.umn.edu/data/arcticdem/ (accessed on: 21 January 2019).

Acknowledgments: The authors would like to thank DigitalGlobe (now Maxar Technologies) for providing high-resolution satellite data. Chandra–Bhaga basin imagery WorldView-2 © 2014 Maxar Technologies; Ny-Ålesund imagery WorldView-3 © 2016 Maxar Technologies. The authors thank the reviewers for their constructive criticism, which helped improve the manuscript. The authors also thank HOD, Civil Engineering, and Director, MIT, MAHE for their support.

Conflicts of Interest: The authors declare no conflict of interest.

References

1. Huss, M. Present and future contribution of glacier storage change to runoff from macroscale drainage basins in Europe. *Water Resour. Res.* **2011**, *47*, W07511. [CrossRef]
2. Kulkarni, A.V.; Randhawa, S.S.; Rathore, B.P.; Bahuguna, I.M.; Sood, R.K. Snow and glacier melt runoff model to estimate hydropower potential. *J. Indian Soc. Remote Sens.* **2002**, *30*, 221–228. [CrossRef]
3. Gaddam, V.K.; Boddapati, R.; Kumar, T.; Kulkarni, A.V.; Bjornsson, H. Application of "OTSU"—An image segmentation method for differentiation of snow and ice regions of glaciers and assessment of mass budget in Chandra basin, Western Himalaya using Remote Sensing and GIS techniques. *Environ. Monit. Assess.* **2022**, *194*, 337. [CrossRef] [PubMed]
4. Bolch, T.; Menounos, B.; Wheate, R. Landsat-based inventory of glaciers in western Canada, 1985–2005. *Remote Sens. Environ.* **2010**, *114*, 127–137. [CrossRef]
5. Gore, A.; Mani, S.; HariRam, R.P.; Shekhar, C.; Ganju, A. Glacier surface characteristics derivation and monitoring using Hyperspectral datasets: A case study of Gepang Gath glacier, Western Himalaya. *Geocarto Int.* **2017**, *34*, 23–42. [CrossRef]
6. Braun, M.; Schuler, T.V.; Hock, R.; Brown, I.; Jackson, M. Comparison of remote sensing derived glacier facies maps with distributed mass balance modelling at Engabreen, northern Norway. *IAHS Publ. Ser. Proc. Rep.* **2007**, *318*, 126–134.
7. Avanzi, F.; Gabellani, S.; Delogu, F.; Silvestro, F.; Cremonese, E.; di Cella, U.M.; Ratto, S.; Stevenin, H. Snow Multidata Mapping and Modeling (S3M) 5.1: A distributed cryospheric model with dry and wet snow, data assimilation, glacier mass balance, and debris-driven melt. *Geosci. Model Dev.* **2022**, *15*, 4853–4879. [CrossRef]
8. Jawak, S.D.; Wankhede, S.F.; Luis, A.J.; Balakrishna, K. Impact of Image-Processing Routines on Mapping Glacier Surface Facies from Svalbard and the Himalayas Using Pixel-Based Methods. *Remote Sens.* **2022**, *14*, 1414. [CrossRef]
9. Rastner, P.; Bolch, T.; Notarnicola, C.; Paul, F. A Comparison of Pixel-and Object-Based Glacier Classification with Optical Satellite Images. *IEEE J. Sel. Top. Appl. Earth Obs. Remote Sens.* **2014**, *7*, 853–862. [CrossRef]
10. Guo, S.; Du, P.; Xia, J.; Tang, P.; Wang, X.; Meng, Y.; Wang, H. Spatiotemporal changes of glacier and seasonal snow fluctuations over the Namcha Barwa–Gyala Peri massif using object-based classification from Landsat time series. *ISPRS J. Photogramm. Remote Sens.* **2021**, *177*, 21–37. [CrossRef]

11. Hay, G.J.; Castilla, G. Geographic Object-Based Image Analysis (GEOBIA): A New Name for a New Discipline. In *Object-Based Image Analysis*; Springer: Berlin/Heidelberg, Germany, 2021; pp. 75–89.
12. Blaschke, T. Object based image analysis for remote sensing. *ISPRS J. Photogramm. Remote Sens.* **2010**, *65*, 2–16. [CrossRef]
13. Lang, S. Object-Based Image Analysis for Remote Sensing Applications: Modeling Reality–Dealing with Complexity. In *Object-Based Image Analysis*; Springer: Berlin/Heidelberg, Germany, 2008; pp. 3–27.
14. D'Oleire-Oltmanns, S.; Marzolff, I.; Tiede, D.; Blaschke, T. Detection of Gully-Affected Areas by Applying Object-Based Image Analysis (OBIA) in the Region of Taroudannt, Morocco. *Remote Sens.* **2014**, *6*, 8287–8309. [CrossRef]
15. Willhauck, G.; Schneider, T.; De Kok, R.; Ammer, U. Comparison of object-oriented classification techniques and standard image analysis for the use of change detection between SPOT multispectral satellite images and aerial photos. *Int. Arch. Photogramm. Remote Sens.* **2000**, *33 Pt B3*, 35–42.
16. Kim, M.; Warner, T.A.; Madden, M.; Atkinson, D.S. Multi-scale GEOBIA with very high spatial resolution digital aerial imagery: Scale, texture and image objects. *Int. J. Remote Sens.* **2011**, *32*, 2825–2850. [CrossRef]
17. Verhagen, P.; Drăguţ, L. Object-based landform delineation and classification from DEMs for archaeological predictive mapping. *J. Archaeol. Sci.* **2012**, *39*, 698–703. [CrossRef]
18. Höfle, B.; Geist, T.; Rutzinger, M.; Pfeifer, N. Glacier surface segmentation using airborne laser scanning point cloud and intensity data. *Int. Arch. Photogramm. Remote Sens. Spat. Inf. Sci.* **2007**, *36 Pt 3*, W52.
19. Robson, B.; Nuth, C.; Dahl, S.; Hölbling, D.; Strozzi, T.; Nielsen, P. Automated classification of debris-covered glaciers combining optical, SAR and topographic data in an object-based environment. *Remote Sens. Environ.* **2015**, *170*, 372–387. [CrossRef]
20. Sharda, S.; Srivastava, M. Classification of Siachen Glacier Using Object-Based Image Analysis. In Proceedings of the 2018 International Conference on Intelligent Circuits and Systems (ICICS), Phagwara, India, 19–20 April 2018.
21. Jawak, S.D.; Wankhede, S.F.; Luis, A.J. Explorative Study on Mapping Surface Facies of Selected Glaciers from Chandra Basin, Himalaya Using WorldView-2 Data. *Remote Sens.* **2019**, *11*, 1207. [CrossRef]
22. Farhan, S.B.; Kainat, M.; Shahzad, A.; Aziz, A.; Kazmi, S.J.H.; Shaikh, S.; Zhang, Y.; Gao, H.; Javed, M.N.; Zamir, U.B. Discrimination of Seasonal Snow Cover in Astore Basin, Western Himalaya using Fuzzy Membership Function of Object-Based Classification. *Int. J. Econ. Environ. Geol.* **2019**, *9*, 20–25.
23. Mitkari, K.V.; Arora, M.K.; Tiwari, R.K.; Sofat, S.; Gusain, H.S.; Tiwari, S.P. Large-Scale Debris Cover Glacier Mapping Using Multisource Object-Based Image Analysis Approach. *Remote Sens.* **2022**, *14*, 3202. [CrossRef]
24. Isaksen, K.; Nordli, Ø.; Førland, E.J.; Lupikasza, E.; Eastwood, S.; Niedźwiedź, T. Recent warming on Spitsbergen—Influence of atmospheric circulation and sea ice cover. *J. Geophys. Res. Atmos.* **2016**, *121*, 11913. [CrossRef]
25. Raup, B.; Racoviteanu, A.; Khalsa, S.J.; Helm, C.; Armstrong, R.; Arnaud, Y. The GLIMS geospatial glacier database: A new tool for studying glacier change. *Glob. Planet. Change* **2007**, *56*, 101–110. [CrossRef]
26. Digital Globe Product Details. Available online: https://www.geosoluciones.cl/documentos/worldview/DigitalGlobe-Core-Imagery-Products-Guide.pdf (accessed on 20 February 2020).
27. ASTER GDEM v2. Available online: https://asterweb.jpl.nasa.gov/gdem.asp (accessed on 7 July 2022).
28. Arctic DEM. Available online: Pgc.umn.edu/data/arcticdem/ (accessed on 21 January 2019).
29. Porter, C.; Morin, P.; Howat, I.; Noh, M.-J.; Bates, B.; Peterman, K.; Keesey, S.; Schlenk, M.; Gardiner, J.; Tomko, K.; et al. "ArcticDEM", Harvard Dataverse, V1. 2018. Available online: https://www.pgc.umn.edu/data/arcticdem/ (accessed on 13 March 2022).
30. Atmospheric Correction User Guide. Available online: https://www.l3harrisgeospatial.com/portals/0/pdfs/envi/Flaash_Module.pdf (accessed on 20 November 2021).
31. Kaufman, Y.; Wald, A.; Remer, L.; Gao, B.-C.; Li, R.-R.; Flynn, L. The MODIS 2.1-μm channel-correlation with visible reflectance for use in remote sensing of aerosol. *IEEE Trans. Geosci. Remote Sens.* **1997**, *35*, 1286–1298. [CrossRef]
32. Abreu, L.W.; Anderson, G.P. The MODTRAN 2/3 report and LOWTRAN 7 model. *Contract* **1996**, *19628*, 132.
33. Teillet, P.; Fedosejevs, G. On the Dark Target Approach to Atmospheric Correction of Remotely Sensed Data. *Can. J. Remote Sens.* **1995**, *21*, 374–387. [CrossRef]
34. Zhang, Z.; He, G.; Zhang, X.; Long, T.; Wang, G.; Wang, M. A coupled atmospheric and topographic correction algorithm for remotely sensed satellite imagery over mountainous terrain. *GIScience Remote Sens.* **2017**, *55*, 400–416. [CrossRef]
35. Rumora, L.; Miler, M.; Medak, D. Impact of Various Atmospheric Corrections on Sentinel-2 Land Cover Classification Accuracy Using Machine Learning Classifiers. *ISPRS Int. J. Geo-Inform.* **2020**, *9*, 277. [CrossRef]
36. Bernstein, L.S.; Jin, X.; Gregor, B.; Adler-Golden, S.M. Quick atmospheric correction code: Algorithm description and recent upgrades. *Opt. Eng.* **2012**, *51*, 111719. [CrossRef]
37. Pushparaj, J.; Hegde, A.V. Evaluation of pan-sharpening methods for spatial and spectral quality. *Appl. Geomat.* **2016**, *9*, 1–12. [CrossRef]
38. Laben, C.A.; Brower, B.V. Process for Enhancing the Spatial Resolution of Multispectral Imagery Using Pan-Sharpening. U.S. Patent 6,011,875, 4 January 2000.
39. Bhardwaj, A.; Joshi, P.; Snehmani; Sam, L.; Singh, M.; Singh, S.; Kumar, R. Applicability of Landsat 8 data for characterizing glacier facies and supraglacial debris. *Int. J. Appl. Earth Obs. Geoinf.* **2015**, *38*, 51–64. [CrossRef]

40. Baatz, M.; Schape, A. Multiresolution Segmentation: An Optimization Approach for High Quality Multi-Scale Image Segmentation. In *Angewandte Geographische Informations-Verarbeitung, XII*; Strobl, J., Blaschke, T., Griesbner, G., Eds.; Wichmann Verlag: Heidelberg/Karlsruhe, Germany, 2000; pp. 12–23.
41. Han, Y.; Javed, A.; Jung, S.; Liu, S. Object-Based Change Detection of Very High Resolution Images by Fusing Pixel-Based Change Detection Results Using Weighted Dempster–Shafer Theory. *Remote Sens.* **2020**, *12*, 983. [CrossRef]
42. Keshri, A.K.; Shukla, A.; Gupta, R.P. ASTER ratio indices for supraglacial terrain mapping. *Int. J. Remote Sens.* **2009**, *30*, 519–524. [CrossRef]
43. Richards, J.A. *Remote Sensing Digital Image Analysis*; Springer: Berlin/Heidelberg, Germany, 2013.
44. Maxwell, A.E.; Warner, T.A. Thematic Classification Accuracy Assessment with Inherently Uncertain Boundaries: An Argument for Center-Weighted Accuracy Assessment Metrics. *Remote Sens.* **2020**, *12*, 1905. [CrossRef]
45. Siregar, V.P.; Prabowo, N.W.; Agus, S.B.; Subarno, T. The effect of atmospheric correction on object based image classification using SPOT-7 imagery: A case study in the Harapan and Kelapa Islands. *IOP Conf. Ser. Earth Environ. Sci.* **2018**, *176*, 012028. [CrossRef]
46. Phiri, D.; Morgenroth, J.; Xu, C.; Hermosilla, T. Effects of pre-processing methods on Landsat OLI-8 land cover classification using OBIA and random forests classifier. *Int. J. Appl. Earth Obs. Geoinf.* **2018**, *73*, 170–178. [CrossRef]
47. Roganda, M.S.; Murti, S.H.; Widyatmanti, W. Mapping the distribution of natural ecosystems on peatlands through vegetation using the object-based image analysis (OBIA) method in Bangko district, Rokan Hilir regency, Riau. *IOP Conf. Ser. Earth Environ. Sci.* **2022**, *1047*, 012017. [CrossRef]
48. Gavankar, N.L.; Ghosh, S.K. Object based building footprint detection from high resolution multispectral satellite image using K-means clustering algorithm and shape parameters. *Geocarto Int.* **2019**, *34*, 626–643. [CrossRef]
49. Omarzadeh, D.; Karimzadeh, S.; Matsuoka, M.; Feizizadeh, B. Earthquake Aftermath from Very High-Resolution WorldView-2 Image and Semi-Automated Object-Based Image Analysis (Case Study: Kermanshah, Sarpol-e Zahab, Iran). *Remote Sens.* **2021**, *13*, 4272. [CrossRef]
50. Witharana, C.; Civco, D.L.; Meyer, T.H. Evaluation of pansharpening algorithms in support of earth observation based rapid-mapping workflows. *Appl. Geogr.* **2013**, *37*, 63–87. [CrossRef]
51. Dabiri, Z.; Hölbling, D.; Abad, L.; Prasicek, G.; Argentin, A.L.; Tsai, T.T. An Object-Based Approach for Monitoring the Evolution of Landslide-dammed Lakes and Detecting Triggering Landslides in Taiwan. *Int. Arch. Photogramm. Remote Sens. Spat. Inf. Sci.* **2019**, *42*, 103–108. [CrossRef]
52. Johnson, B.A.; Tateishi, R.; Hoan, N.T. A hybrid pansharpening approach and multiscale object-based image analysis for mapping diseased pine and oak trees. *Int. J. Remote Sens.* **2013**, *34*, 6969–6982. [CrossRef]
53. Shukla, A.; Ali, I. A hierarchical knowledge-based classification for glacier terrain mapping: A case study from Kolahoi Glacier, Kashmir Himalaya. *Ann. Glaciol.* **2016**, *57*, 1–10. [CrossRef]
54. Lu, Y.; Zhang, Z.; Shangguan, D.; Yang, J. Novel Machine Learning Method Integrating Ensemble Learning and Deep Learning for Mapping Debris-Covered Glaciers. *Remote Sens.* **2021**, *13*, 2595. [CrossRef]
55. Ryan, J.C.; Hubbard, A.; Stibal, M.; Irvine-Fynn, T.D.; Cook, J.; Smith, L.C.; Cameron, K.; Box, J. Dark zone of the Greenland Ice Sheet controlled by distributed biologically-active impurities. *Nat. Commun.* **2018**, *9*, 1065. [CrossRef]
56. Leidman, S.; Rennermalm, Å.K.; Lathrop, R.G.; Cooper, M. Terrain-Based Shadow Correction Method for Assessing Supraglacial Features on the Greenland Ice Sheet. *Front. Remote Sens.* **2021**, *2*, 690474. [CrossRef]
57. Free Precipitation Data for India. Available online: https://www.indiawaterportal.org/met_data (accessed on 14 August 2022).
58. Desinayak, N.; Prasad, A.K.; El-Askary, H.; Kafatos, M.; Asrar, G.R. Snow cover variability and trend over the Hindu Kush Himalayan region using MODIS and SRTM data. *Ann. Geophys.* **2022**, *40*, 67–82. [CrossRef]
59. Hall, D.K.; Riggs. *MODIS/Terra Snow Cover Monthly L3 Global 0.05Deg CMG, Version 61*; NASA National Snow and Ice Data Center Distributed Active Archive Center: Boulder, CO, USA, 2021. [CrossRef]
60. Poloczanska, E. The IPCC Special Report on Ocean and Cryosphere in a Changing Climate-a view from the mountain tops to the deepest depths. In Proceedings of the 2020 Ocean Sciences Meeting, San Diego, CA, USA, 16–21 February 2020. [CrossRef]
61. Rathore, B.P.; Singh, S.K.; Jani, P.; Bahuguna, I.M.; Brahmbhatt, R.; Rajawat, A.S.; Randhawa, S.S.; Vyas, A. Monitoring of snow cover variability in Chenab Basin using IRS AWiFS sensor. *J. Indian Soc. Remote Sens.* **2018**, *46*, 1497–1506. [CrossRef]
62. Sahu, R.; Gupta, R.D. Snow cover area analysis and its relation with climate variability in Chandra basin, Western Himalaya, during 2001–2017 using MODIS and ERA5 data. *Environ. Monit. Assess.* **2020**, *192*, 489. [CrossRef]
63. Kohler, J.; James, T.D.; Murray, T.; Nuth, C.; Brandt, O.; Barrand, N.E.; Aas, H.F.; Luckman, A. Acceleration in thinning rate on western Svalbard glaciers. *Geophys. Res. Lett.* **2007**, *34*, L18502. [CrossRef]
64. Wang, Z.; Lin, G.; Ai, S. How long will an Arctic mountain glacier survive? A case study of Austre Lovénbreen, Svalbard. *Polar Res.* **2019**, *38*, 3519. [CrossRef]
65. Van Pelt, W.J.; Kohler, J.; Liston, G.E.; Hagen, J.O.; Luks, B.; Reijmer, C.H.; Pohjola, V.A. Multidecadal climate and seasonal snow conditions in Svalbard. *J. Geophys. Res. Earth Surf.* **2016**, *121*, 2100–2117. [CrossRef]
66. Hagen, J.O.; Kohler, J.; Melvold, K.; Winther, J.G. Glaciers in Svalbard: Mass balance, runoff and freshwater flux. *Polar Res.* **2003**, *22*, 145–159. [CrossRef]
67. Garg, V.; Thakur, P.K.; Rajak, D.R.; Aggarwal, S.P.; Kumar, P. Spatio-temporal changes in radar zones and ELA estimation of glaciers in Ny-Ålesund using Sentinel-1 SAR. *Polar Sci.* **2022**, *31*, 100786. [CrossRef]

68. Free Precipitation Data for Ny-Ålesund. Available online: https://seklima.met.no/observations/ (accessed on 14 August 2022).
69. Concept of Permafrost from National Snow and Ice Data Center. Available online: https://nsidc.org/learn/parts-cryosphere/frozen-ground-permafrost (accessed on 23 August 2022).
70. Dąbski, M. Should Glaciers Be Considered Permafrost? *Geosciences* **2019**, *9*, 517. [CrossRef]
71. Berthling, I.; Schomacker, A.; Benediktsson, Í.Ö. The Glacial and Periglacial Research Frontier: Where from Here? In *Treatise on Geomorphology*; Shroder, J.F., Ed.; Elsevier: San Diego, CA, USA, 2013; pp. 479–499.
72. Haeberli, W.; Noetzli, J.; Arenson, L.; Delaloye, R.; Gärtner-Roer, I.; Gruber, S.; Isaksen, K.; Kneisel, C.; Krautblatter, M.; Phillips, M. Mountain permafrost: Development and challenges of a young research field. *J. Glaciol.* **2010**, *56*, 1043–1058. [CrossRef]
73. Christiansen, H.H.; Gilbert, G.L.; Demidov, N.; Guglielmin, M.; Isaksen, K.; Osuch, M.; Boike, J. Permafrost temperatures and active layer thickness in Svalbard during 2017/2018 (PermaSval). In *SESS Report 2019-The State of Environmental Science in Svalbard*; Svalbard Integrated Arctic Earth Observing System: Longyearbyen, Norway, 2020.
74. Burn, C.R. The active layer: Two contrasting definitions. *Permafr. Periglac. Processes* **1998**, *9*, 411–416. [CrossRef]
75. Biskaborn, B.K.; Lanckman, J.P.; Lantuit, H.; Elger, K.; Streletskiy, D.A.; Cable, W.L.; Romanovsky, V.E. The new database of the Global Terrestrial Network for Permafrost (GTN-P). *Earth Syst. Sci. Data* **2015**, *7*, 245–259. [CrossRef]
76. Global Terrestrial Network for Permafrost Database. Available online: http://gtnpdatabase.org/ (accessed on 23 August 2022).
77. Karjalainen, O.; Aalto, J.; Luoto, M.; Westermann, S.; Romanovsky, V.E.; Nelson, F.E.; Etzelmüller, B.; Hjort, J. Circumpolar permafrost maps and geohazard indices for near-future infrastructure risk assessments. *Sci. Data* **2019**, *6*, 190037. [CrossRef]
78. International Permafrost Association (IPA). *IPY 2007–2009 Thermal State of Permafrost (TSP) Snapshot Borehole Inventory, Version 1*; National Snow and Ice Data Center: Boulder, CO, USA, 2010.
79. Romanovsky, V.E.; Smith, S.L.; Christiansen, H.H. Permafrost thermal state in the polar Northern Hemisphere during the international polar year 2007–2009: A synthesis. *Permafr. Periglac. Processes* **2010**, *21*, 106–116. [CrossRef]
80. Westermann, S.; Duguay, C.R.; Grosse, G.; Kaab, A. Remote sensing of permafrost and frozen ground. In *Remote Sensing of the Cryosphere*; Wiley-Blackwell: Hoboken, NJ, USA, 2015; pp. 307–344. [CrossRef]
81. Obu, J.; Westermann, S.; Bartsch, A.; Berdnikov, N.; Christiansen, H.H.; Dashtseren, A.; Delaloye, R.; Elberling, B.; Etzelmüller, B.; Kholodov, A.; et al. Northern Hemisphere permafrost map based on TTOP modelling for 2000–2016 at 1 km^2 scale. *Earth-Sci. Rev.* **2019**, *193*, 299–316. [CrossRef]

remote sensing

MDPI

Article

Multispectral Characteristics of Glacier Surface Facies (Chandra-Bhaga Basin, Himalaya, and Ny-Ålesund, Svalbard) through Investigations of Pixel and Object-Based Mapping Using Variable Processing Routines

Shridhar D. Jawak [1], Sagar F. Wankhede [2,*], Alvarinho J. Luis [3] and Keshava Balakrishna [2]

[1] Svalbard Integrated Arctic Earth Observing System (SIOS), SIOS Knowledge Centre, Svalbard Science Centre, P.O. Box 156, N-9171 Longyearbyen, Norway

[2] Department of Civil Engineering, Manipal Institute of Technology, Manipal Academy of Higher Education, Manipal, Udupi 576104, Karnataka, India

[3] Earth System Sciences Organization, National Centre for Polar and Ocean Research (NCPOR), Ministry of Earth Sciences, Government of India, Headland Sada, Vasco-da-Gama 403804, Goa, India

* Correspondence: sagar.wankhede@learner.manipal.edu

Citation: Jawak, S.D.; Wankhede, S.F.; Luis, A.J.; Balakrishna, K. Multispectral Characteristics of Glacier Surface Facies (Chandra-Bhaga Basin, Himalaya, and Ny-Ålesund, Svalbard) through Investigations of Pixel and Object-Based Mapping Using Variable Processing Routines. *Remote Sens.* **2022**, *14*, 6311. https://doi.org/10.3390/rs14246311

Academic Editor: Peter Romanov

Received: 1 September 2022
Accepted: 8 December 2022
Published: 13 December 2022

Abstract: Fundamental image processing methods, such as atmospheric corrections and pansharpening, influence the signal of the pixel. This morphs the spectral signature of target features causing a change in both the final spectra and the way different mapping methods may assign thematic classes. In the current study, we aim to identify the variations induced by popular image processing methods in the spectral reflectance and final thematic maps of facies. To this end, we have tested three different atmospheric corrections: (a) Quick Atmospheric Correction (QUAC), (b) Dark Object Subtraction (DOS), and (c) Fast Line-of-Sight Atmospheric Analysis of Hypercubes (FLAASH), and two pansharpening methods: (a) Hyperspherical Color Sharpening (HCS) and (b) Gram–Schmidt (GS). WorldView-2 and WorldView-3 satellite images over Chandra-Bhaga Basin, Himalaya, and Ny-Ålesund, Svalbard are tested via spectral subsets in traditional (BGRN1), unconventional (CYRN2), visible to near-infrared (VNIR), and the complete available spectrum (VNIR_SWIR). Thematic mapping was comparatively performed using 12 pixel-based (PBIA) algorithms and 3 object-based (GEOBIA) rule sets. Thus, we test the impact of varying image processing routines, effectiveness of specific spectral bands, utility of PBIA, and versatility of GEOBIA for mapping facies. Our findings suggest that the image processing routines exert an extreme impact on the end spectral reflectance. DOS delivers the most reliable performance (overall accuracy = 0.64) averaged across all processing schemes. GEOBIA delivers much higher accuracy when the QUAC correction is employed and if the image is enhanced by GS pansharpening (overall accuracy = 0.79). SWIR bands have not enhanced the classification results and VNIR band combination yields superior performance (overall accuracy = 0.59). The maximum likelihood classifier (PBIA) delivers consistent and reliable performance (overall accuracy = 0.61) across all processing schemes and can be used after DOS correction without pansharpening, as it deteriorates spectral information. GEOBIA appears to be robust against modulations in atmospheric corrections but is enhanced by pansharpening. When utilizing GEOBIA, we find that a combination of spatial and spectral object features (rule set 3) delivers the best performance (overall accuracy = 0.86), rather than relying only on spectral (rule set 1) or spatial (rule set 2) object features. The multiresolution segmentation parameters used here may be transferable to other very high resolution (VHR) VNIR mapping of facies as it yielded consistent objects across all processing schemes.

Keywords: surface facies of glaciers; pixel-based image analysis; geographic object-based image analysis; atmospheric corrections; pansharpening; image processing routines

1. Introduction

Glaciological zones are areas of differentiated snow and compact ice that form because of the natural accumulation and ablation processes of a glacier. These zones, also known as facies, when viewed on the surface of the glacier, can be called glacier surface facies. Facies vary according to the season and local weather. A glacier reveals its complete set of facies at the very end of the melting season. Each facies varies in terms of its physical properties. The presence of moisture, stage of melt, mixing of impurities, dust, debris, and anthropogenic particulates all contribute to the kind of spectral response produced by each facies. Accurate spatial mapping of glacier facies can act as an input for calibrating distributed mass balance models [1]. Multispectral sensors which can detect the spectral response patterns of these facies provide an immense source of data at various levels of resolution. However, to utilize the potential of multispectral sensors for mapping glacier surface facies, a thorough evaluation of popular preprocessing routines is necessary to determine the appropriate methodology for the application. The current study aims to map glacier surface facies on selected glaciers of Chandra-Bhaga Basin, Himalaya, and Ny-Ålesund, Svalbard, using a variety of preprocessing routines and classification methods. The primary goal is to identify the variations in the final thematic mapping of glacier surface facies due to different processing methods. The broader purpose is to determine the best processing method for specific facies and spectral bands across processing routines and classification techniques using very high-resolution imagery.

Extraction of surface facies has been demonstrated using methods incorporating data from a single sensor [2–5] as well as multiple sensors and data products [6–11]. Many of the persistent problems of mapping mountain glaciers revolve around solving the spectral complexities of supraglacial debris [6,12–14]. Debris characterization utilizes the shortwave infrared (SWIR), visible to near infrared (VNIR), and thermal infrared (TIR) range of optical satellite data [15]. Elevation from Digital Elevation Models (DEMs) is also demonstrated to be useful for adjusting the topographic effects on observable debris properties [6]. However, the supraglacial surface including facies such as ice mixed debris, dirty ice, and debris have been mapped using only the VNIR spectrum [3,16]. Pope and Rees [16] mapped the glacier facies on the Midtre Lovénbreen glacier using Airborne Thematic Mapper (ATM) imagery, field spectra, and Landsat ETM+ imagery by developing linear combinations (LCs) of principal components (PCs) of the spectral signatures. While this study does utilize ATM data, much of the mapping is performed using the VNIR spectrum. In Pope and Rees [7], the authors assessed the effects of radiometric, spectral, spatial, and properties of various sensors using resampled ATM imagery. While they found that enhanced spatial resolution does not necessarily improve mapping, Paul et al. [17] suggested that improved resolution would deliver better final thematic products. However, Jawak et al. [18] tested the impact that enhanced spatial resolution can have on conventional and advanced pixel-based image analysis (PBIA) for mapping facies using very high resolution (VHR) Worldview-2 (WV-2) satellite data. Their results suggest that pansharpening does not improve end classification. Moreover, spectral signatures of facies derived from pansharpened methods showed a decrease in overall intensity of reflectance. VHR imagery based PBIA classification itself, however, is subject to the problems of misclassification, salt, and pepper effect, and even data redundancy [19].

Geographic object-based image analysis (GEOBIA) is demonstrated to overcome the problems associated with VHR imagery by segmenting the image into objects for improving feature separability based on the assigned scale factors [20]. Comparative analysis of PBIA and GEOBIA highlights the effectiveness of GEOBIA for retrieving highly accurate thematic classification [21]. GEOBIA also provides an efficient structure for adding ancillary information to extract maximum features from the primary data. As demonstrated by Mitkari et al. [22], debris cover is efficiently mapped through GEOBIA using a multi-scale segmentation approach using ancillary data for identifying glacial lakes, minor features such as debris cones, rills, and other features such as crevasses, exposed ice, snow, ice mixed debris, and supraglacial debris. GEOBIA permits arithmetic operations on its segmented objects; this enables application of band ratios [23] and customized spectral indices [3]. Every information extraction method relies on interpretable data derived from the primary imagery, which can be enhanced and/or supplemented by ancillary data. However, as demonstrated in Jawak et al. [3], the primary image when preprocessed using different atmospheric correction and pansharpening methods, yields significantly distinct results. While GEOBIA may overcome salt and pepper effects, the changes induced in the spectral response of each object due to varying processing routines are yet to be fully tested.

Accurate retrieval of spectral reflectance relies upon reliable removal of atmospheric effects [24]. With many atmospheric correction methods available today, selection of the most appropriate method trickles down to the specific application under focus [25]. Fast Line-of-Sight Atmospheric Analysis of Hypercubes (FLAASH) correction is a method which models the atmosphere along with correcting the scattering from adjacent pixels into the field-of-view (FOV) [26]. However, comparative studies between atmospheric corrections have found that, although FLAASH may be slightly more accurate than the quick atmospheric correction (QUAC), QUAC is more efficient and generalizable [27]. Dark Object Subtraction (DOS) is found to be more useful for features with low to medium reflectivity [28]. Glacier surfaces contain a range of features ranging from high to low reflectivity. The key question then here is: can DOS correction identify some facies better than others? Mandanici et al. [29] compared the QUAC, FLAASH, DOS, Empirical Line, and 6S methods and found that a single most accurate method could not be identified. Moreover, in their results, the methods performed variably across different classification methods. This indicates that the combination of an atmospheric correction and classification method is also crucial for accurate mapping of the target feature. Utilizing WV-2 imagery, Eugenio et al. [30] highlighted that model-based methods such as FLAASH and 6S are more useful in spectrally complex environments. However, glaciers have been mapped individually using QUAC [31], DOS [32], and FLAASH [33]. This indicates that there is a gap in the knowledge of how different atmospheric corrections and mapping/classification methods may impact the extraction of glacier facies. Moreover, pansharpening is not only performed to improve the spatial resolution of multispectral bands, but also to enhance spectral information by unmixing coarse multispectral data using the finer panchromatic (PAN) data [34]. This in turn emphasizes the need for appropriate pansharpening methods based on the targeted application. While the Gram–Schmidt (GS) has been shown to retain better spectral information [35], the Hyperspherical Color Sharpening (HCS) was developed for VHR WorldView-2 (WV-2) imagery. Furthermore, while the GS has been used to supplement boundary digitization for extracting glaciers [3], it has not been directly tested for mapping facies. Furthermore, the compounding effects of different atmospheric corrections and pansharpening methods have not been tested using different mapping methods for extracting facies. This poses an interesting area of research which is attempted by the current study.

The general recommendation for selection of atmospheric correction methods and pansharpening centers on the specific application of the study [25,36]. VNIR based PBIA using multiple processing routines is discussed in Jawak et al. [18] (henceforward referred to as Paper 1). In Paper 1, it was observed that FLAASH delivers consistent results across the processing schemes, whereas pansharpening by HCS and GS degraded the classification results. In Paper 2 [37], which focuses on VNIR based GEOBIA, it was observed that the DOS correction delivers superior results, and pansharpening improves the classification results, with the GS outperforming HCS. While these results can be tested for VHR VNIR based approaches, a more robust analysis would require testing the effects of SWIR bands and different spectral band combinations to clearly determine the best processing routines for mapping facies using VHR multispectral imagery. The current experiment builds upon Paper 1 and Paper 2 by conducting a comparative analysis of GEOBIA and PBIA using multiple image processing routines for identifying surface facies Chandra-Bhaga Basin, Indian Himalaya, and in Ny-Ålesund, Svalbard.

2. Study Area and Data Used

Study area 1 is the Norwegian archipelago of Svalbard which lies between 75° and 82°N. The rate of warming observed in this region is almost twice that of the global mean [38]. This group of islands contains Ny-Ålesund, an international research town which possesses some of the most well studied glaciers. The glaciers tested here include Austre Lovénbreen (AL), Vestre Brøggerbreen (VB), Austre Brøggerbreen (AB), Uvérsbreen (UB), Edithbreen (EB), Midtre Lovénbreen (ML), Pedersenbreen (PB), and Botnfjellbreen (BB). Study area 2 is the Chandra-Bhaga Basin, on the northern slopes of Pir-Panjal range of Himalaya, in the Lahaul-Spiti valley of Himachal Pradesh, India [39,40]. It lies between 32°05′ to 32°45′N. This basin contains India's Himalayan research base, Himansh. The glaciers selected are CB1, CB2, CB3, CB4, CB5, CB6, and Samudra Tapu (ST). Figure 1 highlights the geospatial location of the study sites, whereas Supplementary Table S1 highlights the area of each glacier and their Global Land Ice Measurements from Space (GLIMS) reference ID [41].

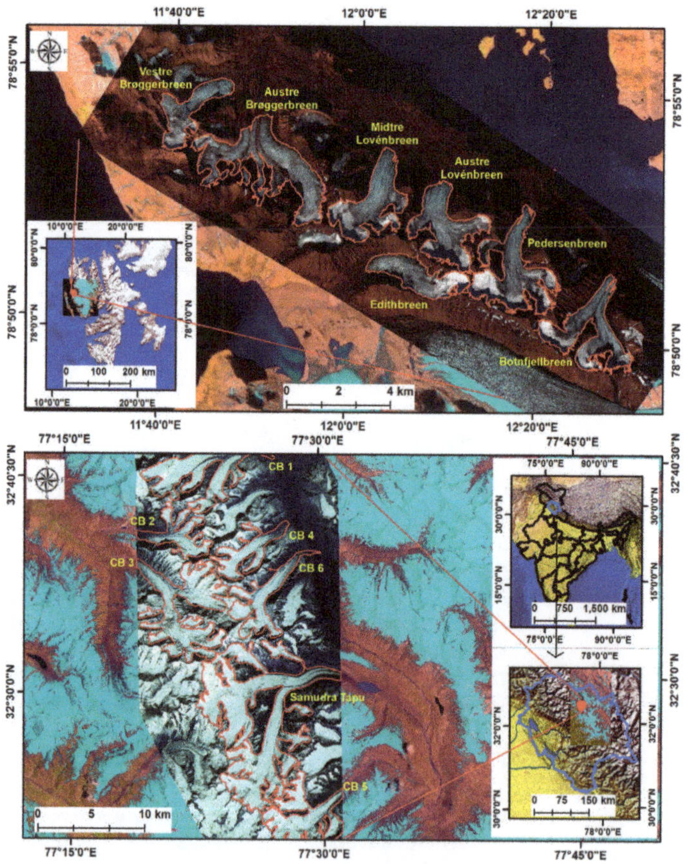

Figure 1. Geographic location of the selected glaciers and the local geomorphology. The VHR WorldView-2/3 imagery is imposed upon Sentinel 2A (10 m) imagery (Himalaya: Copernicus Sentinel data 2021, processed by ESA; Svalbard: Copernicus Sentinel data 2022, processed by ESA). WorldView-2/3 was obtained from Digital Globe, Inc., Westminster, CO, USA. Chandra–Bhaga Basin image: WorldView-2 © 2014 Maxar; Ny-Ålesund image: WorldView-3 © 2016 Maxar. Band display combination: Red—Band 2, Green—Band 3, Blue—Band 2. Inset maps outlining Himachal Pradesh, India, and Svalbard were prepared using Natural Earth (Free vector and raster map data @ naturalearthdata.com).

The primary data of this study are LV2A images from Digital Globe, Inc., Westminster, CO, USA [42]. Table 1 describes the information on the spectral bands and ground sampling distance (GSD) of the VHR images used in the present study.

Table 1. Spectral band names, wavelength ranges, ground sampling distances (GSD), and acquisition dates of the primary images of the current study.

Worldview-2 (WV-2): 16 October 2014			WorldView-3 (WV-3): 10 August 2016		
Name	**Wavelength (μm)**	**GSD**	**Name**	**Wavelength (μm)**	**GSD**
PAN	0.45–0.80	0.46 m	PAN	0.45–0.80	0.31 m
Coastal	0.40–0.45	1.84 m	Coastal	0.40–0.45	1.24 m
Blue	0.45–0.51		Blue	0.45–0.51	
Green	0.51–0.58		Green	0.51–0.58	
Yellow	0.58–0.62		Yellow	0.58–0.62	
Red	0.63–0.69		Red	0.63–0.69	
Red Edge	0.70–0.74		Red Edge	0.70–0.74	
NIR 1	0.77–0.89		NIR 1	0.77–0.89	
NIR 2	0.86–1.04		NIR 2	0.86–1.04	
			SWIR 1	1.19–1.22	3.7 m
			SWIR 2	1.55–1.59	
			SWIR 3	1.64–1.68	
			SWIR 4	1.71–1.75	
			SWIR 5	2.14–2.18	
			SWIR 6	2.18–2.22	
			SWIR 7	2.23–2.28	
			SWIR 8	2.29–2.36	

The projection and datum of the Svalbard image are WGS 1984 UTM Zone 33N, and the Himalayan image are WGS 1984 UTM Zone 43N. Elevation data consisted of 30 m Advanced Spaceborne Thermal Emission and Reflection Radiometer (ASTER) and Global Digital Elevation Model (GDEM) v2 [43] for the Chandra–Bhaga Basin, and 5 m Arctic DEM [44,45] for Ny-Ålesund.

3. Research Methodology

3.1. Summary of the Experiment

This study aimed to map glacier surface facies for selected glaciers in the Chandra–Bhaga Basin, Indian Himalayas, and Ny-Ålesund, Svalbard, using VNIR VHR WV-2/3 imagery. Three corrections, viz., DOS, FLAASH, and Quick Atmospheric Correction (QUAC), were used to calculate reflectance, followed by image fusion using Hyperspherical Color Sharpening (HCS) and Gram–Schmidt (GS). The processing schemes were then converted into subsets with selected spectral band combinations (Figure 2).

The first subset uses the Blue, Green, Red, and NIR 1 (BGRN1) spectral bands. The second subset uses the Coastal, Yellow, Red Edge, and NIR 2 (CYRN2) spectral bands. The third subset uses the complete VNIR range, whereas the fourth subset uses the entire set of spectral bands including the VNIR and SWIR (VNIR_SWIR). This resulted in 27 image subsets for a single glacier in the Chandra-Bhaga Basin, and 36 image subsets for a single glacier in Ny-Ålesund. The glacier boundaries were delineated over 3D surface images using Arctic DEM and ASTER GDEM v2, respectively. This was followed by visual and spectral analysis to identify the surface facies. This was used to determine regions of interest (ROI) for selecting training data as input into conventional and advanced PBIA classification workflows. GEOBIA classification was then performed by segmenting the subsets using common segmentation parameters. This was then followed by image classification using three rule sets (Figure 2). Thus, the current study utilizes three atmospheric corrections, two pansharpening methods, PBIA and GEOBIA classification algorithms to test the effects of atmospheric corrections, pansharpening, and information extraction methods on the classification of surface facies. Supplementary Table S2 summarizes the nomenclature used in the current study.

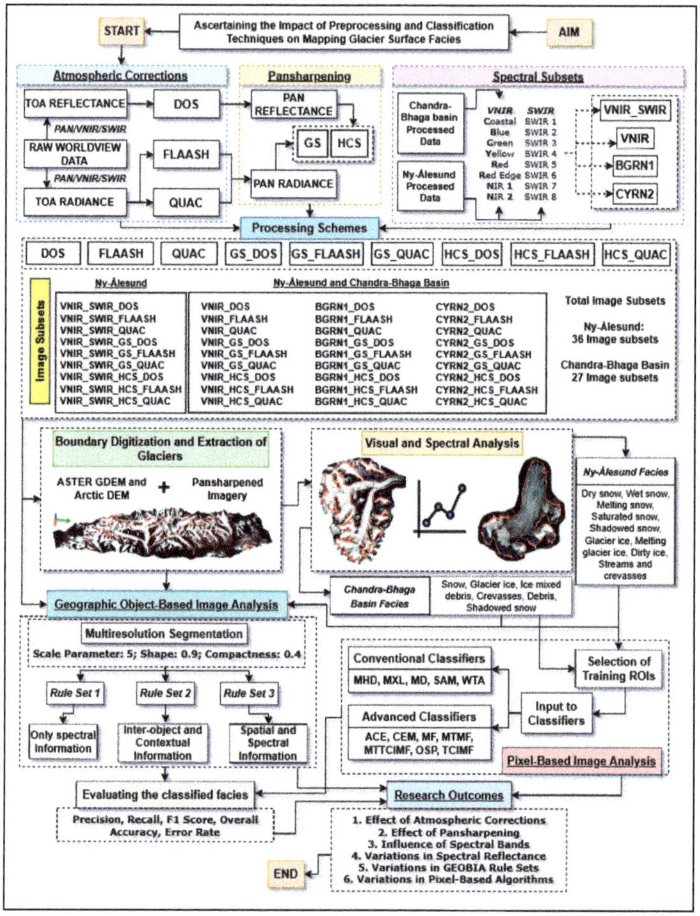

Figure 2. Highlights the broad design of the experiment. TOA: Top of Atmosphere; MSS: Multispectral; PAN: Panchromatic; DOS: Dark Object Subtraction; QUAC: Quick Atmospheric Correction; FLAASH: Fast-Line-of-Sight Atmospheric Analysis of Spectral Hypercubes; HCS: Hyperspherical Color Space; GS: Gram–Schmidt; SAM: Spectral Angle Mapper; MD: Minimum Distance; MXL: Maximum Likelihood; MHD: Mahalanobis Distance; WTA: Winner Takes All; MF: Matched Filtering; MTTCIMF: Mixture-Tuned Target-Constrained Interference-Minimized Filter; TCIMF: Target-Constrained Interference-Minimized Filter; ACE: Adaptive Coherence Estimator; CEM: Constrained Energy Minimization; OSP: Orthogonal Space Projection; MTMF: Mixture-Tuned Matched Filtering.

3.2. The Preprocessing Routines

3.2.1. Atmospheric Correction

Conversion of raw pixel brightness to apparent reflectance is a dual-step process, which comprises: (1) converting DNs to at-sensor radiance and (2) deriving apparent reflectance using atmospheric correction. The first step was performed through the radiometric calibration module in Environment for Visualizing Images (ENVI) 5.3. The three atmospheric correction models are described as follows. The FLAASH correction algorithm requires: (1) retrieval of atmospheric parameters which include aerosol description and water column amount; (2) using aerosol description and model atmosphere to calculate reflectance from radiance using the radiative transfer code [46]. Atmospheric model and aerosol description [47,48] were defined as prescribed by Abreu and Anderson [49]. Table 2 displays automatically computed and user-defined parameters which are necessary for executing FLAASH for the images of each study area.

Table 2. Input parameters for FLAASH atmospheric correction.

Parameter	Chandra–Bhaga Basin	Ny-Ålesund	Computation
Flight date	16 October 2014	10 August 2018	Imagery metadata
Scene center location	Lat: 32.5324 Long: 77.4175	Lat: 78.8816 Long: 12.0734	Automatic computation
GMT	5.6825	12.7456	User-defined
Sensor altitude (km)	770	770	Automatic computation
View zenith angle (degrees)	180.00	180.00	Automatic computation
Initial visibility (km)	40.00	40.00	User-defined
Atmospheric model	1 (Tropical)	4 (Subarctic Summer)	User-defined [49]
Aerosol model	6 (Tropospheric)	4 (Maritime)	User-defined [49]
Water column multiplier	1.00	1.00	Automatic computation
Pixel size (m)	2.00	0.90	Automatic computation
Aerosol scale height	1.50	1.50	Automatic computation
CO2 mixing ratio (ppm)	390.00	390.00	Automatic computation

The DOS correction functions upon the principle that atmospheric scattering upwells path radiance in the darkest pixels of an image [50]. Removal of this upwelling into the path radiance can be performed using the value of a single dark pixel [51]. Following Paper 1, Paper 2, and Rumora et al. [52], Top of Atmosphere (TOA) reflectance values of user defined dark pixels (Table 3) were used as input to the DOS correction. (3) Similar to DOS, the QUAC model is an in-scene approach, utilizing no model computations, instead using central wavelengths and the TOA radiance [53].

Table 3. TOA reflectance values of selected dark pixels for input into DOS module.

Wavelengths	Mean at-Sensor Reflectance of Selected Dark Pixels	
	Ny-Ålesund	Chandra–Bhaga Basin
Coastal	0.09	0.17
Blue	0.06	0.14
Green	0.04	0.11
Yellow	0.03	0.09
Red	0.03	0.08
Red Edge	0.02	0.08
NIR1	0.01	0.06
NIR2	0.01	0.06
SWIR 1	0.00	–
SWIR 2	0.00	–
SWIR 3	0.00	–
SWIR 4	0.00	–
SWIR 5	0.00	–
SWIR 6	0.00	–
SWIR 7	0.00	–
SWIR 8	0.00	–

3.2.2. Pansharpening and Glacier Boundary Digitization

HCS replaces the intensity constituent of MS data in the hyperspherical color space with the intensity form of the panchromatic band [54]. This was performed in ERDAS IMAGINE. GS predicts the PAN data from the spectral response function of the sensor [55]. The suitability of generating a 3D surface to determine a glacier's boundaries and ice divides is demonstrated in previous studies [3,56]. Hence, the current study draped pansharpened imagery over the ASTER GDEM v2 for the Chandra-Bhaga Basin, and Arctic DEM for Ny-Ålesund, to observe and digitize glacial boundaries. These boundaries were then used to extract the individual glaciers from the entire scene of the study areas.

3.2.3. Identifying Glacier Surface Facies

Surface facies were identified on the selected glaciers using visual and spectral characteristics. An in-depth analysis of the characteristics and extraction of VNIR spectral signatures is described in Jawak et al. [3], Paper 1 and Paper 2. The logic following the identification of facies is the reduction in reflectance due to mixing of dust, debris, and moisture, along with comparison against literature performed previously [3], Paper 1. Facies identified in Ny-Ålesund images consist of wet snow, melting snow, dry snow, saturated snow, shadowed snow, melting glacier ice, glacier ice, streams and crevasses, and dirty ice. Facies observed in the Chandra-Bhaga Basin image consist of snow, shadowed snow, ice mixed debris, glacier ice, crevasses, and debris.

In this paper, the spectral signatures are highlighted in Figures 3 and 4. For the Ny-Ålesund facies, inclusion of SWIR bands only added a marginal average of 0.20 in the reflectance of dry snow in the SWIR 1 band. This varied by 0.04 across all processing schemes. SWIR 4 added an average of 0.02, whereas SWIR 7 and SWIR 8 each delivered an average reflectance of 0.01 each for dry snow. Wet snow showed an average reflectance of 0.06, melting snow displayed 0.08, saturated snow delivered 0.03, glacier ice delivered 0.04, melting glacier ice showed 0.01, and streams and crevasses showed 0.01 in SWIR 1 across all processing schemes. SWIR bands 2 to 8 showed reflectance of 0.00 across all facies and processing schemes, whereas dirty ice and shadowed snow delivered no response in any SWIR band.

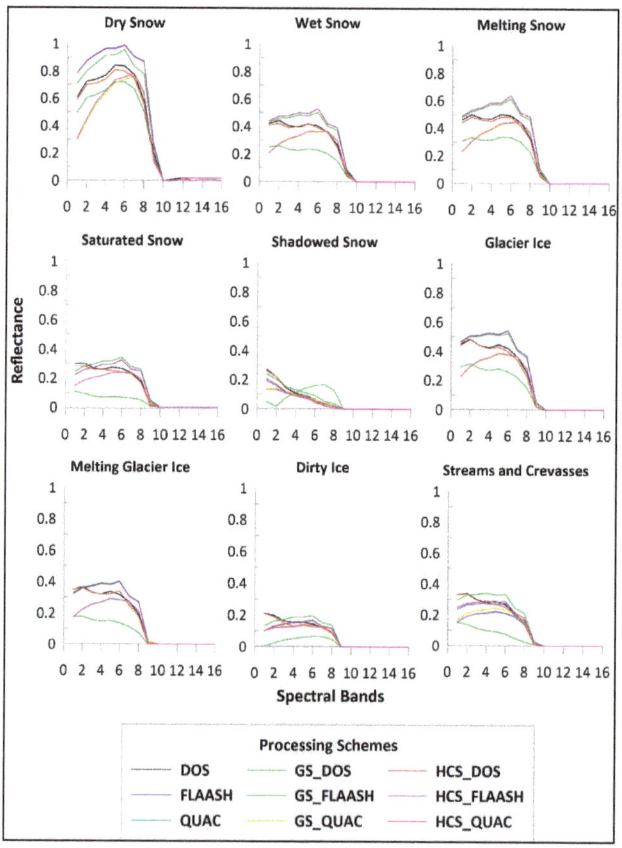

Figure 3. Difference in spectral response of facies in Svalbard for the processing schemes. 2: Band 2 = Blue; 4: Band 4 = Yellow; 6: Band 6 = Red Edge; 8: Band 8 = NIR 2; 10: Band 10 = SWIR 2; 12: Band 12 = SWIR 4; 14: Band 14 = SWIR 6; 16: Band 16 = SWIR 8.

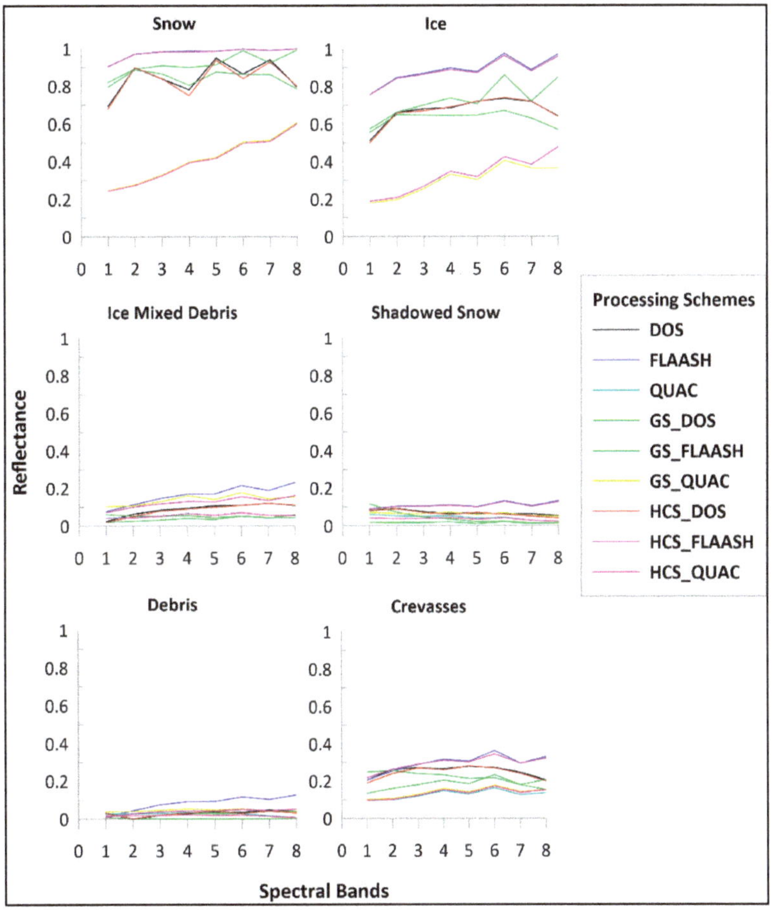

Figure 4. Difference in spectral response of facies in Chandra–Bhaga Basin for the processing schemes. 1: Band 1 = Coastal; 2: Band 2 = Blue; 3: Band 3 = Green; 4: Band 4 = Yellow; 5: Band 5 = Red; 6: Band 6 = Red Edge; 7: Band 7 = NIR 1; 8: Band 8 = NIR 2.

3.3. Mapping Facies Using Conventional and Advanced Pixel-Based Image Analysis (PBIA)

As highlighted in Paper 1, classifications such as that performed by Pope and Rees [16] provide an excellent basis for comparative analysis between the results of the current study and their findings. However, as they utilized the ISODATA classification algorithm, and an Airborne Thematic Mapper (ATM), their results need to be compared against other classifications of facies in a similar area, like the results of Luis and Singh [2]. Supervised classification algorithms of facies maps have been shown to yield high accuracies [3,57,58]. This study utilizes the Minimum Distance to Mean (MD), Maximum Likelihood (MXL), Spectral Angle Mapper (SAM), Mahalanobis Distance (MHD), and Winner Takes All (WTA), Constrained Energy Minimization (CEM), Mixture-Tuned Matched Filtering (MTMF), Mixture-Tuned Target-Constrained Interference-Minimized Filter (MTTCIMF), Matched Filtering (MF), Orthogonal Space Projection (OSP), Adaptive Coherence Estimator (ACE), and Target-Constrained Interference-Minimized Filter (TCIMF). All classifiers are available in ENVI under the Terrain Categorization (TERCAT) and Target Detection (TD) workflows. Paper 1 describes each classification algorithm in detail. The current paper improves upon previous work by testing the classification algorithms for their classification results using all image subsets and processing schemes.

PBIA necessitates: (a) Assigning training data; and (b) Input to classification algorithms. Training data were extracted through analysis of the visual and spectral characteristics of target facie using Polygonal regions of interest (ROIs) (Paper 1). The MTTCIMF, MF, and MTMF required a minimum noise fraction (MNF) transformation before classification. Algorithms that did not require this transformation were classified directly after insertion of the image and associated ROIs. Default parameters were left unchanged, and post-classification enhancement was avoided to negate unintentional bias. Hence, as performed by Jawak et al. [3], stretch (square root) and rule thresholds (0.4) were common throughout the PBIA process.

3.4. Mapping Facies Using a Geographic Object-Based Image Analysis (GEOBIA)

3.4.1. Multiresolution Segmentation

In this study, we utilize the multiresolution segmentation [59] algorithm in the eCognition Developer. Multiresolution segmentation is an iterative process that begins by aggregating highly correlated adjacent pixels into objects. This cycle repeats until the conditions set by the scale parameter, shape/color, and compactness are satisfied [60]. The scale parameter determines the size of the resultant objects, whereas shape regulates the influence of spectral characteristics [60]. The parameter values and layer weights used for segmentation are described in Table 4. NIR 1 was assigned the maximum weight as it is most impacted by moisture enabling greater spectral differentiation. The other layers were weighed one at a time in a repeated iterative process to determine the best weight to deliver meaningful objects. The same layer weights were used for all the image subsets regardless of the processing scheme. This invariably suggests that the scale parameter, shape, and compactness values set here have resulted in consistent objects across all processing schemes. This suggests that, for VHR imagery, layer weights may be less impactful for segmentation than the primary parameters. The parameters delivered consistent results across both study areas indicating the potential transferability to VHR images of other glaciated areas.

Table 4. Parameters used to implement multiresolution segmentation.

Layer Weights							
VNIR_SWIR	**Value**	**VNIR**	**Value**	**BGRN1**	**Value**	**CYRN2**	**Value**
Coastal	1	Coastal	1			Coastal	1
Blue	2	Blue	2	Blue	2		
Green	2	Green	2	Green	2		
Yellow	2	Yellow	2			Yellow	2
Red	1	Red	1	Red	1		
Red Edge	2	Red Edge	2			Red Edge	2
NIR 1	3	NIR 1	3	NIR 1	3		
NIR 2	1	NIR 2	1			NIR 2	1
SWIR 1	1						
SWIR 2	1						
SWIR 3	1						
SWIR 4	1						
SWIR 5	1						
SWIR 6	1						
SWIR 7	1						
SWIR 8	1						
Common Object Parameters					**Value**		
Scale					5		
Shape					0.9		
Compactness					0.4		

3.4.2. Object Features and Rule Sets

In this study, several object features were identified which were incorporated into rule sets for enabling classification. The features identified for classification are mean, quantile, standard deviation, min. pixel value, max. pixel value, edge contrast of neighbor pixels, number of overlapping thematic objects, relative border to, and customized ratios/arithmetic features [61]. The rule sets incorporating these features are developed based on a logic of testing the spectral and contextual features to determine how much information and which features are more suitable for mapping surface facies across different processing schemes. Three rule sets were defined, rule set 1 focused on incorporating only spectral information, rule set 2 focused on using only spatial inter-object information, and rule set 3 utilized both spatial and spectral attributes. Paper 2 describes the features and rule sets used in the current study using the VNIR approach. Essentially, the same features are used in the same rule sets for mapping the facies across all image subsets, with substitutions made in the required bands. Considering the combination of CYRN2, Coastal is used as a substitute for the blue band, yellow is used as a substitute for the green band, red edge is used as a substitute for the red band, and NIR 2 is used as a substitute for the NIR 1 band. The main contribution of SWIR bands was in the delineation of dirty ice. Only SWIR 1 was found to be useful; the other SWIR bands did not provide any discernible signal for target object identification. Table 5 contains the features used to classify objects into specific facies classes. The table is adapted from Paper 2 (focused on VNIR) with additions of the features used for BGRN1, CYRN2, and VNIR_SWIR spectral band combinations.

Table 5. Object features and the rule sets employing them for GEOBIA classification in the current study.

Rule Sets	Type of Feature	Feature Name	Features Tested in This Study
Rule Set 1 and 3	Object Features: Layer Values	Mean Value per Layer/Band	Coastal, Blue, Green, Yellow, Red, Red Edge, NIR 1, NIR 2, SWIR 1, Brightness, Max. difference
Rule Set 2 and 3	Object Features: Layer Values	Quantile (50th percentile)	Quantile (Coastal), Quantile (Coastal), Quantile (Blue), Quantile (Green), Quantile (Yellow), Quantile (Red), Quantile (Red Edge), Quantile (NIR 1), Quantile (NIR 2), Quantile (NIR 2), Quantile (SWIR 1)
Rule Set 2 and 3	Object Features: Layer Values: Pixel-Based	Standard Deviation	Coastal, Blue, Green, Yellow, Red, Red Edge, NIR 1, NIR 2, SWIR 1
Rule Set 2 and 3	Object Features: Layer Values: Pixel-Based	Minimum Pixel Value	Coastal, Blue, Green, Yellow, Red, Red Edge, NIR 1, NIR 2, SWIR 1
Rule Set 2 and 3	Object Features: Layer Values: Pixel-Based	Maximum Pixel Value	Coastal, Blue, Green, Yellow, Red, Red Edge, NIR 1, NIR 2, SWIR 1
Rule Set 2 and 3	Object Features: Layer Values: Pixel-Based	Edge Contrast of Neighbor Pixels	Coastal (3), Blue (3), Green (3), Yellow (3), Red (3), Red Edge (3), NIR 1 (3), NIR 2 (3), SWIR 1 (3)
All three rule sets	Object Features: Thematic Attributes	Number of Overlapping Thematic Objects	Manual Digitized Layer of Shadowed Snow
Rule Set 2 and 3	Class-Related Features: Relations to Neighbor Objects	Relative Border To	Classified Objects
Rule Set 1 and 3	Object Features: Customized Features	Arithmetic Feature	*Customized Ratios (using Mean Value)* R_RE = (Red/Red Edge) CB_CB = (Coastal − Blue)/(Coastal + Blue) G_C = (Green)/(Coastal) RC_RG = (Red/Coastal) * (Red/Green) Max_Min_RE = (Max. pixel value Red Edge − Min. pixel value Red Edge) Y_C = (Yellow/Coastal) C_G = (Coastal/Green) R_C = (Red/Coastal) C_N1 = (Coastal/NIR 1) G_RE = (Green/Red Edge) R_B = (Red/Blue) R_G = (Red/Green) N2_Y = (NIR 2/Yellow) N1_R = (NIR 1/Red) N1_N2 = (NIR 1/NIR 2) CN2_CN2 = (Coastal − NIR 2)/(Coastal + NIR 2) N1N2_N1N2 = (NIR 1 − NIR 2)/(NIR 1 + NIR 2) BN1_BS1 = (Blue − NIR 1)/(Blue + SWIR 1) GN1_GS1 = (Green − NIR 1)/(Green + SWIR 1) N1S1 = (NIR 1 − SWIR 1)/(NIR 1 + SWIR 1)

3.5. Accuracy Assessment

Logistical constraints in the field campaign to Svalbard, and harsh field conditions in the month of image acquisition in the Himalaya prohibited ground truth collection. Avoiding bias in the manual assignment of reference data, a total of 1160 points were selected in an equalized random-sampling approach [5,62] to determine the accuracy of the thematic results. Error matrices were generated to calculate measures such as precision, recall, F1 score, overall accuracy (OA), error rate and specificity [63]. While error rate is calculated as '1-OA', Supplementary Table S2 highlights the mathematical equations of the measures.

4. Results and Discussion

4.1. Pixel-Based Image Analysis

We first present the PBIA results of all band combinations across processing schemes using the F1 score as the harmonic mean of recall and precision. The performance of the classifiers is presented as an average across both study areas and processing schemes in Table 6.

Table 6. Performance of PBIA in the current study averaged across all processing schemes for each subset using the F1 score. The values of the best performing classifiers are highlighted in bold and italicized.

Classifier	Spectral Band Combinations				Average
	BGRN1	CYRN2	VNIR	VNIR_SWIR	
ACE	0.35	0.34	0.37	0.36	0.36
CEM	0.23	0.22	0.23	0.22	0.23
MF	0.13	0.22	0.25	0.23	0.21
MTMF	0.10	0.12	0.15	0.14	0.13
MTTCIMF	0.10	0.10	0.12	0.12	0.11
OSP	0.14	0.09	0.12	0.12	0.12
TCIMF	0.39	0.13	0.16	0.16	0.21
MHD	***0.48***	0.38	0.42	0.41	0.42
MXL	0.29	***0.47***	***0.50***	***0.49***	0.44
MD	0.22	0.27	0.31	0.30	0.28
SAM	0.44	0.21	0.25	0.25	0.29
WTA	0.44	0.43	0.46	0.45	***0.45***

For the BGRN1 combination, the best performing classifier was the MHD, whereas the worst performing classifiers were the MTMF and MTTCIMF. For the CYRN2 subsets, the MXL was the best performing classifier, whereas the OSP delivered the lowest F1 score. For the VNIR subsets, MXL delivered the highest F1 score of 0.50, whereas the lowest score of 0.12 was obtained by MTTCIMF and OSP. For the VNIR_SWIR subsets, MXL delivered the highest F1 score, while MTTCIMF and OSP delivered the lowest scores. On averaging the scores across all band combinations, the MXL achieved the highest accuracy, whereas the MTTCIMF achieved the lowest accuracy. The trend of PBIA classifier performance across all processing schemes and image subsets is WTA > MXL > MHD > ACE > SAM > MD > CEM > TCIMF = MF > MTMF > OSP > MTTCIMF. The trend of accuracy among image subsets based on average performance across all classifiers is VNIR = BGRN1 > VNIR_SWIR > CYRN2.

4.1.1. Effect of Atmospheric Corrections on PBIA

The effect of atmospheric correction algorithms on the PBIA classifications of facies using each band combination is presented here as an average of both study areas. The OA is used here as a comparative measure. Table 7 displays the OA for all PBIA classifiers across each image subset based on atmospheric corrections. For the DOS correction, ACE has delivered an average OA of 0.61 across all image subsets. CEM and MF have produced an OA of 0.53, whereas MTMF has delivered an OA of 0.23. MTTCIMF has delivered an OA of 0.02. OSP has yielded an OA of 0.28. TCIMF has performed at an OA of 0.22. MHD produced an OA of 0.70, whereas MXL delivered an average OA of 0.76. MD and SAM have produced OAs of 0.62 and 0.45. WTA yielded the maximum OA of 0.79.

Table 7. Performance of the classifiers according to each atmospheric correction across all image subsets using OA as the evaluative measure. The values of the best performing classifiers are highlighted in bold and italicized.

Classifier	Comparison between Atmospheric Corrections Using Overall Accuracy											
	BGRN1			CYRN2			VNIR			VNIR_SWIR		
	DOS	*FLAASH*	*QUAC*	*DOS*	*FLAASH*	*QUAC*	*DOS*	*FLAASH*	*QUAC*	*DOS*	*FLAASH*	*QUAC*
ACE	0.60	0.47	0.55	0.60	0.46	0.55	0.63	0.47	0.59	0.62	0.46	0.58
CEM	0.53	0.36	0.35	0.51	0.35	0.35	0.55	0.38	0.37	0.52	0.37	0.36
MF	0.53	0.36	0.31	0.52	0.36	0.30	0.55	0.38	0.31	0.53	0.37	0.31
MTMF	0.22	0.15	0.05	0.21	0.14	0.05	0.24	0.17	0.07	0.23	0.16	0.06
MTTCIMF	0.01	0.01	0.14	0.01	0.01	0.14	0.02	0.01	0.16	0.02	0.01	0.15
OSP	0.28	0.11	0.17	0.27	0.10	0.16	0.29	0.13	0.19	0.28	0.13	0.19
TCIMF	0.21	0.11	0.22	0.20	0.10	0.19	0.24	0.12	0.25	0.22	0.11	0.24
MHD	0.70	0.61	0.66	0.68	0.59	0.64	0.70	0.61	0.66	0.70	0.59	0.65
MXL	0.75	0.66	***0.76***	0.75	0.67	***0.77***	0.78	***0.73***	***0.79***	0.77	***0.73***	***0.79***
MD	0.62	0.61	0.46	0.61	0.60	0.45	0.64	0.64	0.48	0.62	0.63	0.48
SAM	0.44	0.33	0.37	0.43	0.32	0.36	0.46	0.35	0.38	0.45	0.34	0.37
WTA	***0.79***	***0.72***	0.71	***0.77***	***0.73***	0.71	***0.81***	***0.73***	0.73	***0.78***	***0.73***	0.72

For the FLAASH atmospheric correction, ACE delivered an average OA of 0.47 across all image subsets. CEM and MF delivered OAs of 0.37 each, whereas MTMF delivered an OA of 0.16. MTTCIMF yielded the lowest OA of 0.01, while OSP achieved an OA of 0.12. TCIMF yielded an average OA of 0.11, and MHD delivered an OA of 0.60. MXL achieved an OA of 0.70, whereas MD delivered 0.62. WTA delivered the highest OA of 0.73, whereas SAM achieved 0.34. Among the QUAC corrected image subsets, ACE delivered an average OA of 0.57, whereas CEM yielded an OA of 0.36. MF and MTMF achieved OAs of 0.31 and 0.06 each. MTTCIMF and OSP delivered OAs of 0.15 and 0.18 respectively. TCIMF yielded an OA of 0.23. MXL delivered the highest OA of 0.78, whereas WTA yielded 0.72. MHD and MD delivered OAs of 0.65 and 0.47, whereas SAM yielded an OA of 0.37, respectively. Among the DOS subsets, all PBIA classifiers varied by 0.01. However, among the FLAASH subsets, MXL delivered a maximum variance of 0.03, and MD yielded 0.02. A variance of 0.01 was achieved by CEM, MF, MTMF, OSP, TCIMF, MHD, and SAM. ACE, MTTCIMF and WTA have delivered no variance in the resultant OAs. Among the QUAC subsets, ACE and TCIMF varied in performance by 0.02, whereas MF yielded no variance among the classified subsets. A common variance of 0.01 was achieved by CEM, MTMF, MTTCIMF, OSP, MHD, MXL, MD, SAM, and WTA.

In summary, the trend of PBIA classifiers among the DOS subsets is WTA > MXL > MHD > MD > ACE > CEM = MF > SAM > OSP > MTMF > TCIMF > MTTCIMF. The trend of PBIA performance among the QUAC subsets is MXL > WTA > MHD > ACE > MD > SAM > CEM > MF > TCIMF > OSP > MTTCIMF > MTMF. The trend of classifier performance for the FLAASH subsets is WTA > MXL > MD > MHD > ACE > CEM = MF > SAM > MTMF > OSP > TCIMF > MTTCIMF. Among the atmospheric corrections, the trend of performance based on average OAs across all processing schemes is DOS > QUAC > FLAASH.

4.1.2. Effect of Pansharpening on PBIA

The effect of pansharpening algorithms on the PBIA classifications of facies using each band combination is presented here as an average of both study areas. The ER is used here as a comparative measure. Table 8 displays the ER for all PBIA classifiers across each image subset based on methods of pansharpening.

Table 8. Effect of pansharpening across each image subset based on average error rate. NP: Not pansharpened, GS: Gram–Schmidt, HCS: Hyperspherical Color Sharpening. All the values are calculated as averages across the atmospheric corrections.

Classifier	Comparison between Pansharpening Methods Using Error Rate											
	BGRN1			CYRN2			VNIR			VNIR_SWIR		
	NP	*GS*	*HCS*	*NP*	*GS*	*HCS*	*NP*	*GS*	*HCS*	*NP*	*GS*	*HCS*
ACE	0.46	0.74	0.66	0.46	0.74	0.67	0.44	0.72	0.63	0.45	0.72	0.65
CEM	0.59	0.81	0.76	0.60	0.81	0.78	0.57	0.80	0.74	0.59	0.81	0.74
MF	0.60	0.82	0.73	0.61	0.82	0.74	0.59	0.80	0.71	0.60	0.80	0.71
MTMF	0.86	0.85	0.81	0.87	0.86	0.81	0.85	0.82	0.78	0.85	0.83	0.79
MTTCIMF	0.95	0.91	0.73	0.95	0.92	0.74	0.94	0.89	0.71	0.94	0.90	0.72
OSP	0.82	0.87	0.84	0.83	0.89	0.85	0.80	0.86	0.82	0.80	0.86	0.83
TCIMF	0.82	0.87	0.77	0.84	0.88	0.78	0.80	0.85	0.75	0.81	0.86	0.76
MHD	0.35	0.66	0.56	0.36	0.67	0.58	0.35	0.66	0.56	0.36	0.67	0.58
MXL	0.28	0.62	0.60	0.27	0.62	0.61	0.24	0.59	0.57	0.24	0.60	0.58
MD	0.44	0.78	0.76	0.45	0.78	0.77	0.42	0.76	0.77	0.43	0.77	0.78
SAM	0.62	0.81	0.83	0.63	0.82	0.83	0.61	0.80	0.81	0.62	0.81	0.82
WTA	0.26	0.62	0.63	0.27	0.61	0.60	0.25	0.62	0.61	0.26	0.63	0.62

Among the BGRN1 subsets, for the ACE classification, GS increased in error by 0.28, whereas CEM and MF increased in error by 0.22. Both OSP and TCIMF decreased performance by 0.05. MXL and MD increased in error by 0.34, whereas MHD and SAM showed an increase of 0.31 and 0.19, respectively. WTA has increased in error by 0.36. However, MTMF and MTTCIMF decreased in error by 0.01 and 0.04. HCS resulted in an increase in error by a maximum of 0.37 for WTA, whereas MXL and MD decreased in performance by 0.34 each. MHD and SAM showed an increase in error by 0.21, respectively. ACE and OSP increased in error by 0.2, whereas CEM and MF decreased in performance by 0.17 and 0.13. MTTCIMF and TCIMF improved in performance by 0.22 and 0.05. Among the CYRN2 image subsets, for the ACE classification, GS and HCS caused an increase in error by 0.28 and 0.21, respectively. For the CEM classification, GS reduces performance by 0.21, and HCS by 0.18. For the MF classification, GS reduces performance by 0.21 and HCS by 0.13. For the MTMF classification, GS and HCS improve performance by 0.01 and 0.06. MTTCIMF classes show an improvement in accuracy by 0.03 and 0.21 through the GS and HCS sharpening. OSP classification reduced in error by 0.06 and 0.02 via the GS and HCS sharpening, respectively. For the TCIMF classification, HCS improved accuracy by 0.06; however, GS reduced by 0.04. MHD classification reduces in performance by 0.31 and 0.22 using GS and HCS sharpening, respectively. For the MXL classification, GS and HCS reduced performance by 0.35 and 0.34. For the MD classification, GS and HCS reduced performance by 0.33 and 0.32. For the SAM classification, GS and HCS reduced accuracy by 0.19 and 0.20, respectively. For the WTA classification, GS and HCS decreased performance by 0.34 and 0.33, respectively. Among the VNIR subsets, for the ACE classification, GS and HCS reduced performance by 0.28 and 0.19 each. For the CEM classification, GS and HCS reduced performance by 0.23 and 0.17, respectively. For the MF classes, GS and HCS reduced accuracy by 0.21 and 0.12, respectively. MTMF classes showed an increase in performance of 0.03 and 0.07 via the GS and HCS, respectively. MTTCIMF classes improved the resultant accuracy by 0.05 and 0.23 for GS and HCS sharpening, respectively. OSP classification showed a decrease in accuracy by 0.06 and 0.02. For the TCIMF classification, GS resulted in an increase in error by 0.05; however, it has resulted in an improvement in accuracy of 0.05 using the HCS sharpening. MHD, MXL, and MD showed a reduction in performance by 0.31, 0.35, and 0.34 using the GS sharpening. However, using the HCS sharpening, the same classifiers showed a decrease in performance by 0.21, 0.33, and 0.35, respectively. WTA and SAM classification reduced in accuracy by 0.37 and 0.19 using the GS sharpening. However, the HCS sharpening reduced the performance by 0.36 and 0.2.

Among the VNIR_SWIR image subsets, MTMF and MTTCIMF showed an increase in performance by 0.02 and 0.04 using GS, and 0.06 and 0.22 using HCS sharpening, respectively. ACE classification showed an increase in error by 0.27 and 0.2 using the GS and HCS, respectively. The CEM classification reduced in performance by 0.22 and 0.15 using GS and HCS, respectively. MF classification reduced in performance by 0.2 and 0.11 using the GS and HCS, whereas the OSP classification reduced in performance by 0.06 and 0.03 using the GS and HCS, respectively. TCIMF classification showed an increase in performance using the HCS (0.05) and reduction using GS (0.05). MHD classification resulted in an increase in error by 0.31 and 0.22 using GS and HCS each. For the MXL classification, GS and HCS reduced in accuracy by 0.36 and 0.34, respectively. For the MD classifications, GS and HCS reduced in accuracy by 0.34 and 0.35, respectively. Among the SAM classification, GS and HCS reduced in accuracy by 0.19 and 0.2, respectively. WTA classification showed a decrease in performance by 0.37 and 0.36, using GS and HCS, respectively. For the BGRN1 subsets, the trend of classifier performance based upon averaging GS and HCS classification error rates is MHD = MXL > WTA > ACE > MD > MF > CEM > SAM = TCIMF = MTTCIMF > MTMF > OSP. For the CYRN2 subsets, the trend of average classifier performance is WTA > MXL > MHD > ACE > MF = MD > CEM> SAM = TCIMF = MTTCIMF > MTMF > OSP. For the VNIR subsets, the trend of average classifier performance is MXL > MHD > WTA > ACE > MF > CEM = MD > MTMF > MTTCIMF > TCIMF > SAM > OSP. For the VNIR_SWIR subsets, the trend of average classifier performance is MXL > WTA = MHD > ACE > MF > CEM = MD > MTMF = MTTCIMF = TCIMF > SAM > OSP. Among the image subsets, the trend of performance averaged across all classifiers is VNIR_HCS > VNIR_SWIR_HCS = BGRN1_HCS > CYRN2_HCS > VNIR_GS> VNIR_SWIR_GS > BGRN1_GS > CYRN2_GS. The trend of pansharpening algorithm reliability is HCS > GS.

4.2. Geographic Object-Based Image Analysis

We first present the GEOBIA classification results of all band combinations across processing schemes using the F1 score as the harmonic mean of recall and precision. The performance of the classifiers is presented as an average across both study areas and processing schemes in Table 9. Rule Set 3 achieved the highest F1 score across all image subsets, followed by rule set 1 and lastly rule set 2. Thus, the performance of the spectral band combinations is ranked as follows: BGRN1 = VNIR > VNIR_SWIR = CYRN2.

Table 9. Performance of GEOBIA in the current study averaged across all processing schemes for each subset using the F1 score. The values of the best performing classifiers are highlighted in bold and italicized.

Rule Sets	Spectral Band Combinations				Average
	BGRN1	CYRN2	VNIR	VNIR_SWIR	
Rule Set 1	0.75	0.73	0.77	0.73	0.75
Rule Set 2	0.73	0.72	0.75	0.72	0.73
Rule Set 3	***0.86***	***0.85***	***0.86***	***0.85***	***0.85***

4.2.1. Effect of Atmospheric Corrections on GEOBIA

The effect of atmospheric correction algorithms on the GEOBIA classifications of facies using each band combination is presented here as an average of both study areas. The OA is used here as a comparative measure. Table 10 displays the OA for all GEOBIA classifiers across each image subset based on atmospheric corrections.

For all subsets, rule set 3 delivered the maximum OA. For the DOS and QUAC classifications, rule set 1 delivered the lowest OA, whereas, for the FLAASH classifications, rule set 2 yielded the minimum OA. The trend of rule set reliability for DOS and QUAC is rule set 3 > rule set 2 > rule set 1. The trend of rule set reliability for FLAASH is rule set 3 > rule set 1 > rule set 2. For the image subsets, the reliability of atmospheric corrections is QUAC > DOS > FLAASH.

Table 10. Performance of the GEOBIA rule sets according to each atmospheric correction across all image subsets using OA as the evaluative measure. The values of the best performing classifiers are highlighted in bold and italicized.

Classifier	Comparison between Atmospheric Corrections Using Overall Accuracy											
	BGRN1			CYRN2			VNIR			VNIR_SWIR		
	DOS	*FLAASH*	*QUAC*	*DOS*	*FLAASH*	*QUAC*	*DOS*	*FLAASH*	*QUAC*	*DOS*	*FLAASH*	*QUAC*
Rule Set 1	0.73	0.81	0.77	0.72	0.79	0.76	0.77	0.81	0.78	0.72	0.79	0.76
Rule Set 2	0.83	0.75	0.84	0.82	0.76	0.83	0.85	0.77	0.84	0.82	0.76	0.83
Rule Set 3	***0.87***	***0.84***	***0.87***	***0.86***	***0.84***	***0.87***	***0.87***	***0.84***	***0.87***	***0.86***	***0.84***	***0.87***

4.2.2. Effect of Pansharpening on GEOBIA

The effect of pansharpening algorithms on the GEOBIA classifications of facies using each band combination is presented here as an average of both study areas. The ER is used here as a comparative measure. Table 11 displays the ER for all GEOBIA classifiers across each image subset based on methods of pansharpening.

Table 11. Effect of Pansharpening across each image subset based on average error rate. NP: Not pansharpened, GS: Gram–Schmidt, HCS: Hyperspherical Color Sharpening. All the values are calculated as averages across the atmospheric corrections.

Classifier	Comparison between Pansharpening Methods Using Error Rate											
	BGRN1			CYRN2			VNIR			VNIR_SWIR		
	NP	*GS*	*HCS*	*NP*	*GS*	*HCS*	*NP*	*GS*	*HCS*	*NP*	*GS*	*HCS*
Rule Set 1	0.24	0.24	0.24	0.25	0.24	0.26	0.25	0.24	0.23	0.25	0.24	0.26
Rule Set 2	0.20	0.26	0.33	0.20	0.27	0.35	0.20	0.25	0.32	0.20	0.27	0.35
Rule Set 3	0.15	0.14	0.16	0.15	0.14	0.15	0.15	0.14	0.14	0.15	0.14	0.15

Among the BGRN1 subsets, GS and HCS deliver cumulative ERs of 0.24 for rule set 1. For rule set 2, GS and HCS deliver ERs of 0.26 and 0.33, respectively. For rule set 3, GS and HCS yield ERs of 0.14 and 0.16, respectively. Among the CYRN2 subsets, GS and HCS deliver ERs of 0.24 and 0.26 for rule set 1, 0.27 and 0.35 for rule set 2, and 0.14 and 0.15 for rule set 3. Among the VNIR subsets, GS and HCS deliver ERs of 0.24 and 0.23 for rule set 1, 0.25 and 0.32 for rule set 2, and a common 0.14 for rule set 3. Among the VNIR_SWIR subsets, GS delivered ERs of 0.24, 0.27, and 0.14 for rule sets 1, 2, and 3. However, HCS delivered ERs of 0.26, 0.35, and 0.15 each for rule sets 1, 2, and 3, respectively.

Hence, averaging across rule sets, the reliability of pansharpening performance for GEOBIA in the current study is GS > HCS.

4.3. Discussion

4.3.1. Manifestation of Facies

Although the total number and type of facies are discernible at the end of summer, it is the mapping method which ultimately characterizes pixels/objects to a thematic class. In Figure 5, we observe the facies of ML for the VNIR_DOS processing scheme classified by the MXL algorithm. Figure 6 highlights the mapped facies using the VNIR_DOS scheme characterized by MXL for the Samudra Tapu (ST) glacier. The highest elevation of the glacier is shadowed due to illumination conditions at the time of screen capture. In lower latitudes, higher solar elevation angle may reduce the provenance of shadow due to illumination, but the steep terrain of mountain glaciers causes shadows to be a persistent problem [64]. While some facies such as wet snow, melting snow, and saturated snow can be visually identified within the shadowed area due to varying intensity of the shadow, the same classes provide complex spectral properties within the overall shadowed area. This adds to the difficulty in isolating specific classes within shadows. Hence, the entire shadowed area is termed as shadowed snow as many of the different snow classes are encompassed within it (Paper 2). Distribution of facies follows the natural accumulation and ablation cycle with dry snow at the highest elevation, followed by wet snow, and melting snow. Similarly, the ablation region is characterized by glacier ice, melting glacier ice, dirty ice, and streams and crevasses. Saturated snow is an interesting class in the current study which is the transition facies between accumulation and ablation. This is classified here as a 'snow' class.

Garg et al. [65] utilized end of ablation SAR data to map dry snow, wet snow/middle percolation, percolation refreeze, superimposed ice, and clean ice zones for the same glaciers in Ny-Ålesund. The authors utilized the boundary between wet snow and clean ice to delineate the ELA. In the current study, the region characterized by Garg et al. [65] as wet snow is further differentiated into wet snow and melting snow based on reduced reflectance [10]. Moreover, dry snow was not identified using SAR on the same glaciers [65]. VHR optical sensors may have an advantage here as dry snow in the present study is found limited to the highest elevation and could be characterized based on its maximum reflectance indicative of no infiltration of water Paper 1, [66]. Dry snow is bright and therefore highly reflective [7]. Visual identification of dry and wet snow can be made based on the reduction in reflectance and slightly darker appearance (Paper 1). Spectral variations highlight the drop in intensity of reflectance of wet snow when compared to dry snow in all processing schemes (Figure 3). This indicates a broad separability of dry snow with all processing schemes. In the current study, dry and wet snow are part of several facies on the glaciers in Ny-Ålesund. However, depending on the method of mapping and broad goal of the study, the context and definition of wet snow appear to be different. This also stems from the purpose of snow-covered area (SCA) mapping, which is typically performed for large regions and considers the entire glacier as 'snow' and is not necessarily focused on identifying facies (Paper 2).

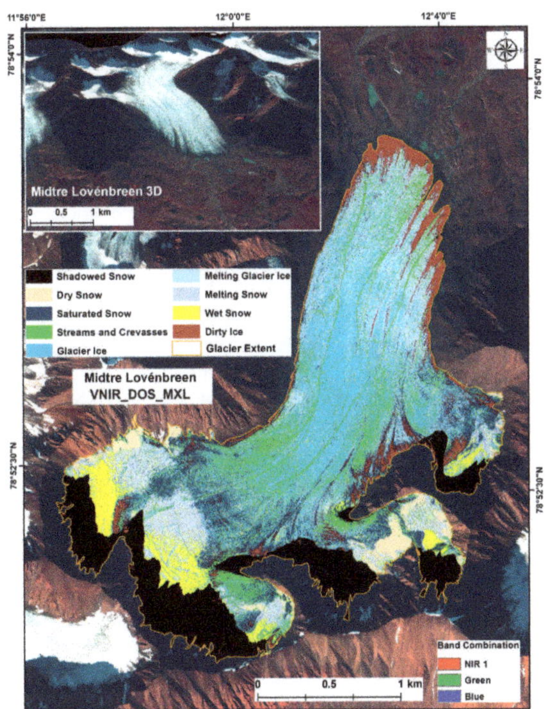

Figure 5. MXL classification of VNIR_DOS processing scheme of the ML glacier. Background image: WorldView-3 © 2016 Maxar.

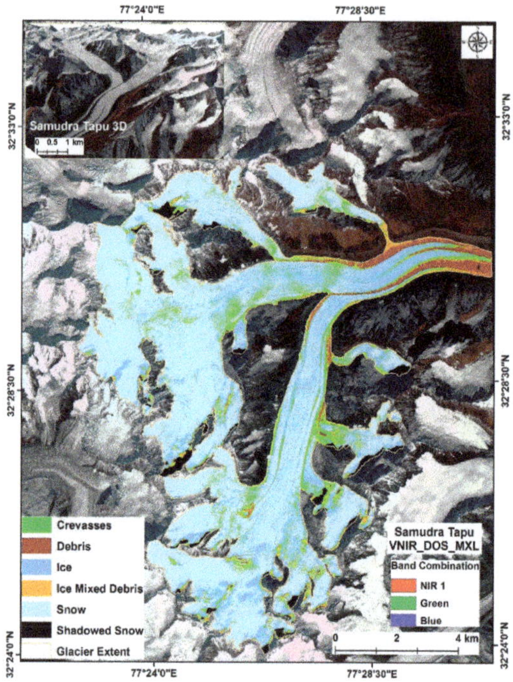

Figure 6. MXL classification of VNIR_DOS processing scheme of the ST glacier. Background image: WorldView-2 © 2014 Maxar.

For example, Snapir et al. [67] combined MODIS SCA maps with Sentinel-1 composites to distinguish between dry and wet snow for two large reservoir catchments in the Himalayas with a total area of 55,000 km^2. In such a case, the purpose of wet snow detection is to supplement monitoring of snowmelt and improve watershed management. Moreover, the authors found that SAR underestimates wet snow when the pixel contains a mix of dry and wet snow near the snow line. VHR optical sensors can improve discrimination of these classes due to fine resolution, if accompanied by the appropriate mapping method. Karbou et al. [68] compared Sentinel-1 wet snow extent and Sentinel-2 snow products for a temporal analysis to assist determination of melt-out. Over glacier areas, Sentinel-1 displayed an underestimation of wet snow extent. Although this underestimation is partially due to geometric distortions and differences in sensor properties, it may indicate toward greater applicability of optical sensors for wet snow extent over glacier areas. Nagajothi et al. [69] mapped dry snow, wet snow, and moraine covered ice using Sentinel 2A images for the Miyar glacier. The distinction between dry and wet snow was performed using an NIR threshold. Interestingly, the authors highlighted that wet snow and glacial ice have similar reflectance properties and can be distinguished on optical images through the shape and occurrence. In the current study, we also observe similar reflectance patterns between ice and wet snow (Figure 3). However, both PBIA and GEOBIA were able to differentiate between the two classes without misclassification. While in GEOBIA the customized ratios and manual thresholding is the cause for successful mapping, PBIA seems effective in differentiating these classes too. However, it is important to note that this is true only for the MXL and WTA classification using VNIR_DOS without any pansharpening. GEOBIA is clearly more accurate (Section 4.2.); however, PBIA can be effective if the image processing and mapping methods are selected carefully. Recently, Yousuf et al. [70] mapped wet snow along with other available facies using AWiFS and Sentinel-2A. The authors utilized ancillary data such as elevation, band ratios, and variances derived from principal components to perform sub-pixel classification using a pair wise classification strategy in a support vector machine (SVM). Sub-pixel classification in this case was accomplished by blending the pixels of higher resolution Sentinel-2A imagery with coarser resolution AWiFS data. The spatial resolution of the current imagery is 0.31 m at its finest (pansharpened WV-3). Hard pixel supervised classification in the current study was capable of mapping all the facies without ancillary data. However, GEOBIA rule sets in the current study are defined based on various object features described in Table 5. These objects were formed based on aggregation of homogenous pixels using multiresolution segmentation. Currently, these object features used cannot be considered as ancillary data as they are products solely created from the object properties themselves. Hence, the current study using both PBIA and GEOBIA was able to map glacier facies across different spectral band combinations, with VNIR achieving the highest accuracy.

Melting snow is a surface facies not usually mapped in the broad characterization of glacier facies. Often, melting snow is either monitored as wet snow in the mapping of dry and wet snow to simplify facies characterization for purposes such as snow cover mapping [71] and seasonal melt coupled with precipitation runoff dynamics [72] or it is considered to be glacial melt with the objective of overall glacial discharge modelling via SCA and watershed modelling [73]. From this, it is possible to surmise that snowmelt (run-off) is distinct from the melting snow surface facies in that it is the sum release of water from the melting of snow/ice and its transport via supraglacial streams and englacial and/or subglacial channels. Liang et al. [74] mapped melting surfaces on the Antarctic Ice sheet while monitoring freeze-thaw dynamics using time series analysis via Google Earth Engine (GEE) and Sentinel-1 SAR. The authors used this to observe spatiotemporal changes in surface melt activity. The facies mapped were dry snow, melting surfaces, and refrozen surfaces. Drops in the intensity of backscatter echo during summer were used to detect presence of liquid in the snow. In the broad context of freeze-thaw dynamics, penetration of water in the snowpack can be characterized generally as a melting surface/wet snow. However, in the specific context of glacier facies, wet snow and melting snow may be different surface facies based on the reduction in reflectance, elevation, and visible properties. Facies are individualistic to glaciers and some facies may exist on some but not on others. In this study, we are focused on identifying the maximum range of facies and testing the variability in the final thematic by modulating the image processing and mapping method. Hence, while it is possible to merge wet snow and melting snow, the utility of conventional PBIA with VHR VNIR data to track subtle variations in facies is key for robust mapping operations. GEOBIA is most accurate, but PBIA is more efficient. Mendes Jr. et al. [75] mapped glacier radar facies in Antarctica to retrieve snowmelt by monitoring the wet snow zone. The authors also mapped dry snow, frozen percolation, and bare ice zones. The authors highlighted that radar zones/facies are based on the surface properties of glacial bodies, whereas classical/glaciological facies depend on the properties of the entire glacier due to the previous accumulation [76]. Pope and Rees [16] highlighted the same with optical facies suggesting that optical mapping of facies can be surface classes rather than glaciological facies. Nonetheless, at the end of ablation season, when most of the seasonal now has melted, the maximum variations of facies are discernible. Unless a sudden precipitation event occurs prior to the date of image capture, the occurrence of facies should not differ. We found no recorded precipitation event before the date of image acquisition in the current study (Paper 1). However, without in situ monitoring of the subsurface phenomena, a direct and accurate characterization of glacier facies is not advisable. Due to this, in our previous work [3], we utilized the name surface facies to refer to the end of ablation facies mapped by optical imagery. In the current study, melting snow differed from wet snow in that it exhibited lower reflectance and is showed higher reflectance than saturated snow (Paper 1 and Figure 3). This slight variation is only effectively characterized by the reflectance obtained from VNIR_DOS and VNIR_FLAASH image subsets. This suggests that pansharpening does not enhance separability of closely matching but distinct surface facies. Moreover, the higher accuracy of MXL and WTA suggest that only in the combination of VNIR_DOS/FLAASH and either MXL or WTA can PBIA accurately map these facies. GEOBIA rule set 3 overcomes the effects of image processing and in fact improves after pansharpening. However, rule sets are analyst driven. The development of each rule set permits a detailed evaluation of objects, object features, and the individual exertions of each feature for each facies. This will lead to a highly accurate thematic map, but it is achieved at the expense of efficiency. However, the saturated snow class presents an interesting case because, if the ELA was characterized previously using the border between wet snow and clean ice [65], in the present case, it would be the border between melting snow and saturated snow. Pope and Rees [16] found two snow classes on the ML glacier using unsupervised classification of Landsat ETM+ imagery. The extent of snow class 6 identified by Pope and Rees [16] is similar to the saturated snow extent identified here. Moreover, the facies was named saturated snow as its reflectance resembled that of the same class identified by Hinkler et al. [77]. All the facies are characterized by observing similar reflectance properties with previous literature (Paper 1).

The ice facies in the current study correspond to glacial ice occurring below the ELA for the respective glaciers. Glacial ice is generally 'clean' with ice mixed with debris or dirty ice being characterized separately. While in SCA mapping ice and snow facies may often be considered into the overall glacier body, in facies mapping, they are usually categorized independently. Bhardwaj et al. [6] aggregated ice and snow into one class considering it as the accumulation region while mapping supraglacial debris and ice mixed debris on two small glaciers in the Chandra-Bhaga basin. The mapping was performed using band ratios of Landsat TM and ETM+ images, utilizing an ASTER DEM for deriving slope and curvature. Due to the coarser resolution of Landsat ETM+ images, slope and curvature derived from the DEM was necessary to delineate glacier boundaries. In the current study, VHR images visualized as 3D surfaces with base heights derived from the ASTER GDEM v2 were efficient in identifying all the glacier boundaries in Ny-Ålesund, as well as some of the ice divides in the Chandra-Bhaga basin [3]. Moreover, the greater range of accumulation and ablation surface facies are discernible using VHR VNIR_SWIR data. Bhardwaj et al. [56] mapped snow and clean ice independently in addition to slush, crevasses, dirty ice, and supraglacial debris using Landsat 8 OLI. Interestingly, the authors utilized TOA reflectance and brightness temperature. Atmospheric correction algorithms convert TOA reflectance to apparent surface reflectance, thereby providing a comparison against established spectral signatures of identified facies. This allowed for a comparison with the spectral characteristics of ice in this study against previous extractions of spectral reflectance for the same facies (3, Paper 1). Moreover, when compared against different image processing methods, the VNIR_DOS stands out as the most reliable. The spectral reflectance pattern of ice in the current study resembles that from Prieur et al. [78] extracted using Landsat 8 OLI. Keshri et al. [5] distinguished snow and ice after separating debris and ice mixed debris utilizing a stepwise implementation of Normalized Difference Snow Index (NDSI), Normalized Difference Glacier Index (NGDI), and Normalized Difference Snow Ice Index (NDSII). This approach is similar in implementation to GEOBIA rule sets. GEOBIA rule sets are formed after a careful evaluation of object features and customized features followed by a stepwise assignment of thresholds, and categorization into thematic classes. In Paper 2, we provided all the rule sets used for mapping. For example, in Ny-Ålesund, in all the rule sets, ice is classified at the end after all the other facies are extracted. Only rule set 2, for the VNIR_DOS subset, necessitated that ice be classified before melting glacier ice using the min. pixel value of the red edge band. In the case of rule set 1 for VNIR_FLAASH, ice was mapped using the mean reflectance of the coastal band. In rule set 3, using the VNIR_QUAC, ice was mapped using the min. pixel value of the green band. For the Chandra-Bhaga basin, using rule set 1 for the HCS_FLAASH, ice was mapped using the mean reflectance of the NIR band. In rule set 2 using the HCS_QUAC subset, ice was characterized using the quantile of the green band. In rule set 3, using HCS_FLAASH, ice was categorized using the mean reflectance of NIR2 and the quantile of the coastal band. All facies are characterizable using each of the spectral band combinations and each image processing scheme. VNIR is most accurate, however, GEOBIA permits mapping due to the intervention of an analyst molding the rule sets according to the image processing scheme. Nonetheless, even PBIA via MXL is capable of mapping ice. Azoni et al. [79] mapped ice using the MXL algorithm on the Forni glacier using VHR UAV orthophotos (0.15 m spatial resolution) with a high precision of 0.91. The order of facies characterization is important as outlined by the procedure adopted by Keshri et al. [5]. This process, though time consuming and not as straight forward as supervised PBIA, is effective in exploiting the limitations introduced by limited spectral bands. Alifu et al. [80] utilized a TIR/NIR/SWIR band ratio to map supraglacial debris by separating the rest of the glacier as clean ice using a density slice approach. The authors also utilized a similar geomorphometric approach as Bhardwaj et al. [6] to separate the glacier from the non-glacier area. In the current study, however, ice was mappable without the use of thermal bands in the satellite imagery, by simple PBIA and by mean reflectance and object features in GEOBIA. In a machine learning-based classification of glacier surface classes, MXL, SVM, and RF classifiers were tested using a combination of Landsat TM and OLI with a range of normalized difference indices to support the characterization [81]. All three methods delivered greater than 99% accuracy, with the SVM only slightly outperforming the other methods. Although the facies themselves were limited to three, the performance of the MXL is attributed to its robust capacity [82].

The dirty ice and alluvial/depositional fan conjunction is important for accurately mapping glacier extents. Dirty ice in the current study was characterized previously as off-glacier [16]. The current imagery highlights water bodies outside the glacial terminus in the alluvial fan (Figure 5). Manual digitization of the glacier terminus using the 3D elevation surface (upper left inset, Figure 5) aided in avoiding possible confusion between the alluvial/depositional fan and dirty ice. Dirty ice is also important because this facies is suggested to have a higher melt rate than other facies [83]. Dirty ice is defined as partially debris-covered ice [83,84]. The Ostrem curve suggests that debris cover lesser than the 'critical thickness' of debris increases melt [85,86]. Moreover, areas with thin and partial debris cover are discontinuous [83] leading to difficulty in effective mapping. Mapping of patchy and thin debris cover is necessary to track spatial variations in ablation. VHR optical sensors may enable efficient mapping of dirty ice by improved visibility due to enhanced spatial resolution (Paper 1) and tracking of reflectance variability due to sensitivity in the NIR region [16]. Accurate mapping of dirty ice would improve spatially distributed ablation modeling. In the current study, digitization of the glacier supported by 3D surface analysis enabled accurate extraction of the glacier body. Dirty ice is also observed in the upper regions of the glacier (Figure 5) intermixed with saturated snow and streams and crevasses. This may be attributed to misclassification of streams and crevasses into dirty ice due to some spectral overlap in the visible range (Figure 3). While temporal analysis movement of debris may permit a detailed explanation of the occurrence of dirty ice at higher elevations, it is beyond the scope of the current study. The intermixing of saturated snow, streams, and crevasses and dirty ice may be explained by the emergence of debris from fine transverse crevasses which are then dislodged by melt and supraglacial streams and transported into the dirty ice region [83]. The occurrence of dirty ice at the edges of the glacier may be due to the entrainment of debris and dirt along its margins [87,88]. Haq et al. [87] mapped dirty ice using Hyperion hyperspectral data and Random Forest classification in the Indian Himalayas. In the spectral reflectance drawn from Hyperion imagery, dirty ice shows maximum spectral separability in the visible range. Between 750–800 nm, dirty ice shows an overlap with ice mixed debris, after which the SWIR range contains minimal variations between all facies [87]. Thus, in the absence of hyperspectral imagery, VHR VNIR imagery may hold a key advantage for mapping patchy and discontinuous debris cover. In the current study, the VNIR subset yielded better performance among all atmospheric corrections and pansharpening methods. Florath et al. [89] (2022) mapped a range of facies including dirty ice in a comparative assessment of unsupervised and supervised classification methods. Their results highlight the efficacy of supervised approaches such as Random Forest (RF) in mapping facies. However, their results were based on mapping carried out on Sentinel-2A images. In the current study, we evaluated several spectral band combinations and found that the best mapping of not only dirty ice, but all facies was without the SWIR range. Hence, the VNIR spectrum still holds maximum potential for executing PBIA and GEOBIA spatial-spectral approaches for mapping patchy debris. However, both Florath et al. [89] and the current study highlight the utility of supervised approaches for mapping glacier facies when ancillary data are not available. Separability of spectral signatures is key for identification and mapping of facies. Dirty ice can be confusing because of the confounding spectral signatures with streams and crevasses. Advanced classification methods are poor at generalizing spectral signatures into constituent classes (Paper 1), in such cases, simpler algorithms such as MXL and MHD may classify better. While GEOBIA rule set 3 is much more accurate than PBIA, the time needed to develop rule sets can be detrimental when faster albeit slightly less accurate methods can be utilized.

Ice mixed debris (IMD) is a mixture of debris of varying sizes with ice [87,90]. Spectrally, it is of slightly higher reflectance than debris [3] and lower reflectance than ice [91] and is most prevalent in the ablation zone. Accurate mapping of IMD like dirty ice and debris is important because the conversion of IMD to debris can be a direct assessment of the impact of climatic variations on the health of the glacier [92]. A reliance on thermal bands to isolate ice mixed debris and debris from the overall glacier body is well reported in literature [5,70,92–94]. In the current study, we lack thermal bands in the sensor. Moreover, as the present focus is on mapping facies without the ancillary inputs from other datasets, we have relied purely on the visual and spectral properties of the facies arising from the imagery and previous literature to guide the classification strategy. The spectral signatures of IMD arising from DOS and FLAASH corrections are most useful for comparative purposes and have a better contrast against debris. This is most important for PBIA as only non-pansharpened subsets led to accurate extraction by MXL. However, GEOBIA overcomes the distortions in spectral signatures induced by all image processing schemes. In the absence of thermal bands, elevation-based information derived from DEMs, band ratioing, and/or spectral indices are necessary for mapping IMD [95]. In a comparative assessment of a hierarchical knowledge-based classification (HKBC) utilizing ancillary information from DEMs, ratios, and indices against MXL, the HKBC classified IMD better with a user's accuracy of 81.82% [91]. Parallels against the current study can be drawn by considering the superior performance of GEOBIA against PBIA. However, we have only utilized image data to generate objects and their features. Rule sets themselves are knowledge-driven on account of stepwise development. Even sub-pixel mapping of IMD utilizing SVM has relied on ancillary information from both DEMs and thermal bands when using medium-to-high resolution imagery [70]. Our results point toward the efficacy of VHR imagery in congruence with appropriate mapping methods for effective characterization of debris. Extraction of IMD or debris via hyperspectral images require significant preprocessing for removal of noise, bad bands, segregation of pure spectra, and atmospheric correction. Although such extensive processing yields accurate results, it is time consuming [87]. In such cases, the current study may prove useful as the lack of spectral bands may be overcome by utilizing VHR data and spatial-spectral GEOBIA rule sets. In a recent GEOBIA based approach for mapping debris cover in addition to other facies, Mitkari et al. [22] utilized thermal infrared (TIR) band (Landsat TM) and a NIR2/Yellow index (Table 5) from WV-2 to map IMD. In the current study, we have also utilized the same index for mapping IMD for the CYRN2, VNIR, and VNIR_SWIR rule set 1 and 3 subsets. However, for the BGRN1, the R_B index (Table 5) was sufficient but not ideal. For rule set 2, IMD was characterized using the max. pixel value of the objects in the green band. The average threshold range for this was 0.05–0.15 (Paper 2). However, this value changed based on atmospheric correction and pansharpening.

Supraglacial debris cover occurs by the entrainment and deposition of bedrock material in the ablation zone due to the interplay of summer ablation and the glacier's movement. Based on the size, deposition, and intermixing with ice, the surface facies may transition from dirty ice, to IMD, to debris cover with ice now beneath a continuous debris cover. Other sources of debris input consist of rockfalls and mass movements from the surrounding valley, as well as deposition via strong winds. Debris-covered glaciers are defined as a glacier, wherein a portion of the ablation region is overlain by a continuous cover of debris across its width [96,97]. The varying influence of the spatiotemporal dimensions of debris on the ablation of the glacier has been the core of several investigations [83,92,96,98–101]. In the current study, we observed the debris class only on the Chandra-Bhaga Basin glaciers as the Himalayas are known to possess a cover of debris in the ablation zone [97,102]. Glaciers in Ny-Ålesund are known to be relatively clean [7]; therefore, the facies at the end of the ablation zone was characterized as dirty ice. Pratibha and Kulkarni [99] mapped temporal changes in supraglacial debris by first classifying snow and snow-free areas of the glaciers using MXL, and subsequently applying band ratios derived from Landsat OLI, ETM, and TM imagery to characterize debris. Although the study did not utilize thermal bands, it relied upon images of three different Landsat sensors to successfully map debris. Nonetheless, it does highlight the utility of simple processing methods to extract debris cover. Mölg et al. [64] created an inventory of the debris cover in High Mountain Asia using a threshold on band ratios on Landsat TM and ETM+ imagery. However, Azzoni et al. [79] suggest that debris mapping is more reliable via VHR imagery because of the capacity to distinguish between debris cover of various distributions, and smaller features such as cones which may be misclassified in coarser resolution data. Kaushik et al. [103] utilized a deep neural network (DNN) to map debris with a combination of optical, SAR, and elevation data. The authors trained the DNN framework over different test sites in the Himalayas and then applied it to a different region. This highlights the potential transferability of the authors' method. However, this DNN structure can misclassify IMD or dirty ice into debris class as the method is not trained for a range of facies, but rather specifically for variations in supraglacial debris of different glaciers. Robson et al. [9] combined optical, SAR coherence, and elevation data in a GEOBIA domain to distinguish clean ice, glacial lakes, and debris. The authors utilized a three-step hierarchical segmentation approach to separate the classes. In the current study, the same multiresolution segmentation parameters were applicable for mapping debris in all the different processing schemes (Paper 2). In a GEOBIA based mapping of supraglacial debris in Antarctica using WV-2 imagery, debris was mapped with an overall accuracy of 93% [104]. However, the authors noted that the geomorphology of Antarctica induces spectral similarity between many of the features. In the current study, the spectral contrast between debris, snow, ice, and crevasses is quite large. The separability with IMD is appreciable, with shadowed snow being the most confusing class. This may highlight why PBIA using MXL here provided better results than that observed in Antarctica (Paper 1, [104]). We extracted debris using the max. pixel value of NIR1/2 for rule sets 2 and 3, the R_B ratio for BGRN1 and the mean reflectance of NIR2 for VNIR and VNIR_SWIR in rule sets 1 and 3. Mitkari et al. [22] utilized brightness temperature and slope via GEOBIA to separate supraglacial and periglacial debris. Often, boundaries of debris-covered glaciers are difficult to ascertain without the use of ancillary data due to confounding properties with the surrounding geology. Manual delineation of glacier boundaries is therefore a practical approach to ascertain minimal error of misclassification [3,56,99].

Crevasses are elongated deep cracks or fractures in the ice [22] resulting from the type of flow of the glacier [105]. Owing to this shape, crevasses may be easily detected in GEOBIA via the appropriate segmentation parameter. The spectral signature of a crevasse is usually of lower intensity than ice but maintains the same trend [3]. This can lead to misclassification of crevasses. Although crevasses are not considered as classical/glaciological facies, they are important surface facies due to their hazard potential. As crevasses represent the dynamicity of the glacier, identifying their extent and shape is useful for understanding mass balance processes [106]. In the Chandra-Bhaga basin, we observe that the crevasses are clearly discernible and mostly benefit from GEOBIA GS_QUAC_VNIR_Rule sets 2 and 3. This is because rule set 2 utilizes mostly spatial and contextual information, wherein the standard deviation object feature of the Yellow (CYRN2/VNIR/VNIR_SWIR subsets) and NIR1/2 (BGRN1/VNIR VNIR/VNIR_SWIR subsets) were applied successfully. The same features were utilized in rule set 3 as it combines both spatial and spectral information. In rule set 1, where only direct spectral information was used, the R_RE, N2_Y, and mean reflectance of the red band (Table 5) ratios were used to delineate crevasses. In Ny-Ålesund, streams and crevasses were categorized into one class, as the width of these facies is much smaller than that of the Chandra-Bhaga Basin crevasses. Moreover, supraglacial streams transport water via crevasse channels and categorizing them into one facies helped in ensuring spectral separability of the other facies. Moreover, the disheveled appearance of crevasses in Chandra-Bhaga Basin aided in their visual identification alongside enhancing the contextual capabilities of GEOBIA. A U-Net-based deep-learning model for mapping crevasses and supraglacial streams highlighted the necessity of mapping fine stripe crevasses which are only visible at fine spatial resolutions [107]. The VHR GEOBIA method employed here can be used to supplement Ground Penetrating Radar (GPR) based investigations of crevasses [108,109] to highlight risky zones.

4.3.2. Variations in the Best Performance

The striking factor of the segmentation parameters used here is the wide applicability across all the image processing methods for both study areas. This contradicts the assertions by Hao et al. [110], who suggest that because objects generated in VHR imagery are molded by a larger number of pixels [111] and different image targets will require different scale parameters. This supports the necessity of multiscale segmentation for different image targets. However, the authors [110] also observed that the image objects' size/area is affected by more factors than just the scale parameter. These factors include spectral, spatial, and geometric properties. Multiresolution segmentation permits the analyst to exert greater control over the resultant objects by modulating the segmentation criteria [112]. This causes the determination of 'optimal' segmentation parameters to be subjective, time-consuming, and challenging [113]. Built up areas with buildings offers crisp geometric shapes clearly discernible in VHR imagery which can be useful for extracting these features. However, natural land covers may often have image targets with close spectral and spatial properties. Multiscale image segmentation has been employed to map glacier cirques [114], volcanic landforms [115], Antarctic supraglacial debris [104], glacier facies [9,22], and glacial landforms [116]. However, single scale segmentation has also proven useful in delineating objects for mapping landslide-dammed lakes [117], snow cover [118], supraglacial ponds [119], and icebergs [120]. Ice marginal lakes were extracted using two different segmentation parameters for Sentinel-1 and Sentinel-2 imagery [121]. Although it is not an optical vs. optical image comparison, the same lakes required two different scales with two different sensors. Spatiotemporal analysis of forest cover found that pixels in Corona and Landsat TM/OLI imagery can be segmented using similar scale parameters [122]. This highlights that both single and multiscale parameters can be effectively utilized for segmenting objects of varying sizes. According to our knowledge, no study has evaluated the impact of image processing routines on the resultant characterization of glacier surface facies using GEOBIA or PBIA. This is significant because the selection of the basic process to derive reflectance introduces changes in the spectral signatures of image targets. Moreover, pansharpening increases the number of pixels aggregated into objects. Therefore, the necessity to evaluate modulations needed for subsets of each processing scheme in segmentation and rule set development is non-trivial. In the current study, we find that no change in scale parameter is needed when segmenting any of the various subsets. This suggests that the parameters applied here may be transferred to other VHR imagery for segmenting glacier surfaces. However, every subset required individual manual modulations to test the applicable ratios/features, the thresholds for each feature, and the order of extracting facies. Multiscale segmentation utilizes different object sizes to match different image targets. It follows the logic that object features may have overlapping properties for various facies. This is extended to rule sets as well because, even if a single segmentation level was sufficient to delineate objects for all image targets, object features must be evaluated and extracted individually. As each processing scheme modifies the spectral properties of the pixels in those subsets (Figures 3 and 4), rule sets utilizing features based on spectral properties will require reconfiguration. This extends to pansharpened subsets as well because GS and HCS also exert differential impacts on the overall spectral and spatial quality of the image. Each rule set was created methodically after segmenting based on step-by-step observations for individual subsets. Thus, GEOBIA rule set 3 delivers high accuracy and robust effectivity but is exhaustive in applicability.

The technical calculations described in Section 4.2. highlight the high accuracies obtained by GEOBIA rule set 3. The modifications in this rule set for each processing scheme also resulted in changes in the final thematic maps. Figure 7 highlights the differences in mapped facies on the Samudra Tapu glacier from rule set 3 for the VNIR_GS_DOS, VNIR_GS_FLAASH, and VNIR_GS_QUAC subsets, respectively. GS_FLAASH overestimates snow in the ablation region but simultaneously shows a patchy distribution in the well-known accumulation area [123], whereas GS_DOS only slightly underestimates snow in the accumulation zone. Both GS_DOS and GS_FLAASH overestimate ice in the accumulation zone due to the misclassification of some areas into ice. GS_ QUAC yields a better overall coverage of snow and ice. The classification of debris from the GS_QUAC subset closely follows the known occurrence of the same facies [94]. However, GS_DOS underestimates and GS_FLAASH overestimates debris when compared with GS_QUAC. Crevasses are well mapped by all three processing schemes, highlighting the efficacy of GEOBIA in extracting these minor features. IMD shows a large variability between all three subsets. GS_QUAC appears to have characterized IMD more efficiently against the overestimation by GS_FLAASH and underestimation by GS_QUAC. In recent years, the overall glacier area in the Chandra-Bhaga Basin is found to be larger than other basins in the Himalayas [124]. This is attributed to the higher elevation of this basin and the reduction in wet precipitation with dry precipitation being consistent [124]. The month and year of image acquisition in the Chandra-Basin is reported to have a larger distribution of SCA [124,125]. Thus, distribution of snow and ice in the current imagery (Figure 7) observed from the GS_QUAC subset is consistent with recent findings. These variations are important because selection of the most appropriate image processing routine for any imagery would ideally yield the most realistic and accurate thematic classification. This is also evident from the fact that in PBIA pansharpening reduced the performance, whereas, in GEOBIA, it enhanced the result. To the best of our knowledge, no other study has compared the impact of variations in image processing in GEOBIA for mapping facies; this makes a direct comparison with the current results difficult.

4.4. Comparative Analysis

In this study, we have attempted to compile the results of Paper 1 and Paper 2 with additions of the variations of spectral band combinations into the processing routine based GEOBIA and PBIA analysis of glacier surface facies.

It is observed that the literature for selection of atmospheric correction and pansharpening algorithms described in Paper 1 often delivers a recommendation that is application specific [25]. The addition of SWIR bands to the classification in both PBIA and OBIA did not significantly enhance the results. For pixel-based classification, it was found that the bands most affected by atmospheric effects (Coastal and Blue) were the main bands used to decipher atmospheric information [126,127]. Moreover, NIR 1 and NIR 2 are strongly influenced by water absorption and dispersion of suspended particles [128]. The FLAASH algorithm which utilizes these bands to predict the atmospheric effects would then be most affected by detrimental effects of pansharpening on the spectral information (Paper 1). In PBIA based on VNIR data (Paper 1), it was found that the FLAASH atmospheric correction performed better. However, when tested across various spectral band combinations in PBIA, the DOS correction yields superior rigor. This implies that the DOS may be more suitable for large data volume analysis using PBIA. Furthermore, as the DOS algorithm is simplistic in its configuration Paper 1 [129], it may promote faster analysis. In the GEOBIA domain, however, the QUAC algorithm delivered superior results. Karimi et al. [31] mapped debris covered glaciers using the QUAC algorithm as the image correction method with efficient results. In the current case, this study is the first of its kind to test image correction methods for the application of glacier surface facies. The FLAASH correction generally delivers the most realistic spectral reflectance [130,131] and is consistent using VNIR imagery. However, across image processing mechanisms, we find it delivers inferior results when compared to DOS and QUAC.

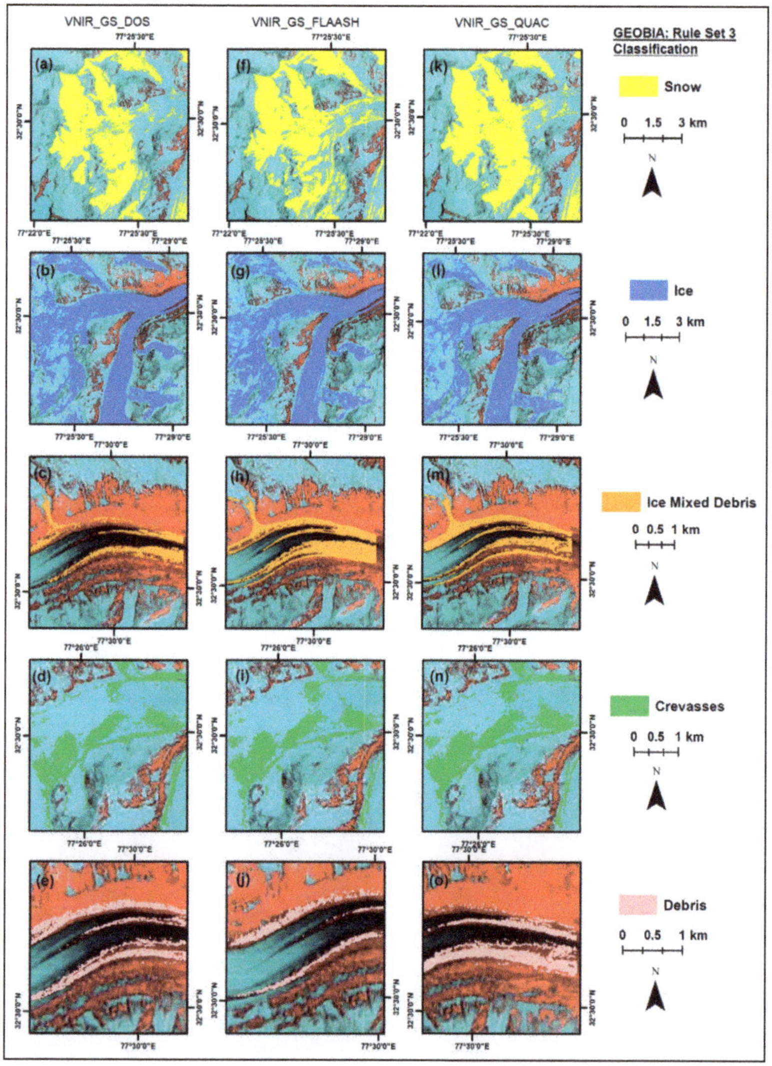

Figure 7. VNIR_GEOBIA_Rule Set 3 classifications of surface facies of the Samudra Tapu glacier. Background image: WorldView-2 © 2014 Maxar. Subfigures (**a**–**e**) correspond to the snow, ice, ice mixed debris, crevasses, and debris classified from the VNIR_GS_DOS processing scheme. Subfigures (**f**–**j**) correspond to the snow, ice, ice mixed debris, crevasses, and debris classified from the VNIR_GS_FLAASH processing scheme. Subfigures (**k**–**o**) correspond to the snow, ice, ice mixed debris, crevasses, and debris classified from the VNIR_GS_QUAC processing scheme.

The nature of classification algorithm and workflows sometimes necessitate specification of Red, Blue, Green, and NIR bands. For example, the TERCAT workflow in ENVI requires this information before proceeding to classification. It would make intuitive sense that substitution of these bands by manipulating spectral band subsets will result in variations in classification using the same image. Here we find that the VNIR and BGRN1 subsets deliver superior results when compared to CYRN2 and VNIR_SWIR. This is true for both PBIA and GEOBIA processes. While GEOBIA delivers superior performance across all atmospheric corrections and pansharpening methods, it appears that conventional spectral bands retain maximum capacity for targeted information extraction. SWIR bands can aid in the mapping of glacier debris and even help define ratios for mapping dust and supraglacial mineral composition [132]. This was found to be true for the GEOBIA rule sets. SWIR bands were most useful in the delineation of dirty ice. However, object features based on VNIR imagery were also capable for mapping dirty ice. Moreover, the image subsets with the best GEOBIA based mapping in the current study are BGRN1 and VNIR; this may again point towards the utility of conventional spectral bands for mapping facies. Moreover, as this mapping can be accomplished without SWIR, the focus on VNIR data for mapping glacier facies and supraglacial debris can be further explored. GEOBIA across all image subsets did not require a change in the segmentation parameters of scale parameter, shape, and size. In comparison to Paper 2, much of the rule set logic is maintained. We were able to map fine features that have proven difficult in the past [133].

As observed in Paper 1, PBIA greatly suffers from the detrimental effects of pansharpening. However, we also find that MTTCIMF and TCIMF observe a modest increase in overall accuracy consistent across both pansharpening methods and image subsets. Even though this increase does not improve the overall reliability, it still points towards the possibility of improving MTTCIMF based mapping at finer resolution (Paper 1). GEOBIA, however, improves with pansharpening, with the GS delivering better results than the HCS (Paper 2). When tested across large datasets, the suitability of GS may hold true [55].

Comparative Analysis of WV-2/3 Spectral Performance

WV-3 imagery has been termed as super-spectral [134] and hyper-spatial [135]. The spectral characteristics of both WV-2 and WV-3 have been tested across many applications. Some of the main applications are agriculture (crop yield, crop productivity, crop type), mineral mapping, cryosphere (glacial lakes, glacier facies, snow cover extent), urban material characterization, flood monitoring, disaster analysis, biomass estimation, etc. Here, we discuss previous literature and then compare our own results to infer cross-application robustness (if any) of the spectral bands.

In an analysis of WV-2 for mapping specific minerals, FLAASH corrected SWIR bands were found to be most useful for alteration and hydrocarbon detection using SVM [136]. Ye at al. [136] found that the most characterizable absorption features were found in the SWIR region. Sun et al. [137] distinguished between hydroxyl, Al-OH, Mg-OH, Fe-OH, iron stained, carbonates, kaolinites, calcites, siderite, and jarosite alterations using a combination of PCA, mineral indices and SAM classification on WV-3 imagery. Their indices utilized SWIR 1, SWIR 3, SWIR 5, SWIR 6, SWIR 7, SWIR 8, green, and red edge bands. SWIR bands enhanced the results of petroleum (Hydrocarbons) characterization by Asadzadeh et al. [138]. Their results highlight the absence of hydrocarbon features in the VNIR bands and consequently the extreme reliance on SWIR bands. WV-3 SWIR bands are reported to possess better spatial consistency of mineralogical units [139]. Kruse et al. [140] compared pre-launch simulated WV-3 spectral bands with AVIRIS and post-launch WV-3 image data to map minerals using the MTMF classifier. Their results suggest that WV-3 performs better for calcite and muscovite than buddingtonite, kalinite, and silica. Most confusion was found in alluvial fan areas, which is a mixing zone for these minerals. Although glacier surface facies can be spectrally distinct, we find that spectral overlap does exist between facies, especially in the visible region (Figures 3 and 4). However, most of the overlap is observed for facies in the Chandra-Bhaga Basin due to the large coverage of snow. This suggests that timing of image capture for end of ablation facies mapping is extremely crucial for observing all of a glacier's available facies.

Hunt Jr. et al. [141] found spectral indices created using SWIR bands 2, 3, and 4 were superior in estimating leaf water content through WV-3 imagery, whereas SWIR 1 was not useful. In the current study, however, only SWIR 1 was useful. Eckert [142] found that WV-2 red edge and NIR bands have the highest correlation with field data for aboveground biomass estimation. Without field data presently, we cannot definitively suggest similar results. Nevertheless, looking at the accuracies, it is safe to assume that the NIR bands are most useful for mapping facies. The red edge band can be swapped by the red band without many changes to the final output. Of course, at least with GEOBIA, it depends on the rule sets defined by the operator. Sibanda et al. [143] found that the red edge spectrum improved grassland mapping accuracy by 14%. Pu et al. [144] also described the potential of the red edge band for detecting reduction in chlorophyll content. The efficacy of red edge is not replicable in the present study. While using a class-based sensor independent spectral band NDVI, Upadhyay et al. [145] found the Yellow, Red, Red Edge, NIR 1, and NIR 2 bands to be most applicable for mapping crops. Immitzer et al. [146] analyzed the conventional (BGRN1), additional bands (CYRN2), and full 8-band (VNIR) range of the WV-2 sensor for comparative classification of tree species using RF and GEOBIA. Like the manual digitization used to remove shadowed areas in the current study, Immitzer et al. [146] focused only on the sunlit portions of tree crowns. Akin to the present results, VNIR bands yielded maximum accuracy through GEOBIA, and the utility of the non-conventional bands was species specific. By and large, the conventional BGRN1 subsets delivered higher accuracies than the CYRN2 bands. Marshall et al. [147] found that additional bands do not improve spectral separability while testing the influence of additional bands (CYRN2) of WV-2 using MTMF for mapping invasive grass species. In an analysis of mangrove species and non-mangrove area classification, Heenkenda et al. [148] utilized spectral band combinations of pansharpened and non-sharpened imagery. They highlight the utility of pansharpened VNIR imagery for effective classification using GEOBIA. The authors suggest that rule sets can be easily transferred to other areas as they are made of image variables rather than set numerical values. Unfortunately, this is not consistent with glacier facies mapping. This may be because facies themselves vary across different glaciated areas and must first be robustly analyzed to determine their occurrence, followed by spectral characterization and then development of rule sets. However, we do believe that the segmentation parameters developed in Paper 2 and implemented here are robust as they have been applicable without any deviation. The rule sets themselves need to be assessed carefully as they are a logical progression of operator intuition and skill. Here, the rule sets applicable in Ny-Ålesund were not applicable in the Chandra-Bhaga Basin; moreover, the same rule sets were not even transferable across different processing schemes. This highlights the importance of accurate selection of image processing routines.

In a comparative analysis between DOS and FLAASH corrected images for bathymetry mapping, DOS delivered superior performance with the most useful bands being coastal, green, yellow, and NIR 2 [149]. Malinowski et al. [150] analyzed patterns of localized flooding using 12 different decision tree (DT) methods comprising of different PBIA and GEOBIA approaches on WV-2 imagery. DOS delivered more accurate reflectance for water than ATCOR. The authors highlight that NIR 2 is the most water sensitive band in the WV-2 arsenal. This could have potential for mapping saturated snow facies, but in the current study, we observe that all facies can be mapped by BGRN1 with better accuracies than CYRN2. Collin et al. [149] compared Hyperion and WV-3 data using spectral combinations of BGR (3 wavelengths), BGRNIR (4 wavelengths), CBGYR (5 wavelengths), BGRNIRs (8 wavelengths), and full 16 band spectrums in an Artificial Neural Network (ANN) structure for mapping coastal systems. The authors suggest that most of the over-reflecting discrepancies between the sensors are a result of the coarser resolution of Hyperion when compared to WV-3. Between the spectral subsets, however, the 16-band full spectrum subset delivered maximum accuracy for WV-3. In the current study, however, the VNIR

(8 wavelength) subset delivered highest accuracy across all processing schemes and classification methods. One reason may be the little to no target information contained in the SWIR bands from SWIR 2 to SWIR 8. Only SWIR 1 could be used for delineating dirty ice, melting snow, and melting glacier ice. However, this was performed to utilize the SWIR band in the combination and not because it could not be alternated by a VNIR alternative. When matching field spectra to WV-2 bands to map asphalt roads, blue, green, red, and NIR 2 reported maximum usability [151]. Variations in roofing materials were distinguishable from the yellow band onwards using pansharpened WV-2 imagery [152].

Karimi et al. [31] mapped clean ice, periglacial debris cover, and supraglacial debris cover using 4 band WV-2 (CBGR) and Landsat-TM imagery. The authors performed an inter-sensor pansharpening by enhancing Landsat-TM using PAN WV-2 data, and QUAC was used to derive reflectance. The 4 band WV-2 imagery is reported to be effective in classifying ice due to its high reflectivity. The authors also differentiated the temperatures of shadowed areas, periglacial, and supraglacial debris using Landsat-TM. The goal of the current study is to only utilize the attributes of the optical imagery to determine the variations in output according to different processing schemes and classification methods. This allows for an independent robust analysis of the image itself in varying scenarios. Moreover, manual digitization of the glacier boundary in the current study did not necessitate distinction between peri- and supraglacial debris. Tiwari et al. [153] utilized WV-2 imagery solely for data interpretation purpose to identify features in the ablation zone of the Bara Shigri glacier to facilitate supraglacial debris mapping using ASTER thermal, GDEM, and TerraSAR-X imagery. Buhler et al. [154] identified wet snow and wind transported snow using WV-2 imagery. However, their analysis was performed on non-atmospherically corrected data. The NIR 2 band which was suggested to enable greater differentiation between snow surfaces did not individually enhance classification in the current study but aided in the differentiation between dry and wet snow (whenever available). Jawak et al. [155] extracted snow cover using a comparative analysis of 14 spectral index ratios on principle component sharpened and GS sharpened WV-2 imagery. Their spectral indices using the additional bands yellow, red edge, and NIR 2 were most significant in extracting snow cover. However, in the current study, only NIR 2 was moderately effective. Yellow and red edge did not add any significant improvement and could be easily replaced by green or red band. Gray et al. [156] highlighted an interesting application of WV-2/3 imagery for mapping green and red snow algae. The imagery was converted to reflectance using the 6S radiative transfer model. Algal distributions documented on field were used to train SAM classifier to distinguish between green, red, dirty, and clean snow. The authors highlighted that processing routines can induce biases in the final algal distributions. This is true even in the case of the current study. Modulations in spectral reflectance due to changing atmospheric corrections, resolution improvement by pansharpening, selection of classification algorithm (PBIA), and even operator skill (GEOBIA) play a large role in the final thematic classifications.

4.5. Final Inferences and Recommendations

The statistical evaluation tabulated and described in Sections 4.1 and 4.2. is summarized and illustrated to outline our findings in Supplementary Figures S1–S4. Among atmospheric corrections (Supplementary Figure S1), we find that DOS (OA = 0.64) is the most reliable correction across all image subsets. When analyzed between PBIA and GEOBIA, DOS is preferred for PBIA, whereas QUAC is most suitable for GEOBIA. Among the PBIA classification methods, MXL and WTA work best with DOS. In GEOBIA, rule sets 2 and 3 work best with QUAC, whereas rule set 1 performs well with FLAASH. Between the spectral band combinations classified by PBIA, VNIR_DOS is the most accurate, whereas in GEOBIA, BGRN1_QUAC, VNIR_DOS, and VNIR_QUAC yielded maximum OA of 0.83. Among pansharpening methods (Supplementary Figure S2), HCS (0.64) delivers an overall better performance than GS. However, this is due to the aggregated extremely poor performance of GS in PBIA. In GEOBIA, both GS and HCS improve accuracy, with GS delivering the maximum OA of 0.79. Among the GEOBIA rule sets, rule set 3 shows the maximum improvement in accuracy with GS scoring an OA of 0.86. Among the PBIA classifiers, MHD yields the highest OA with HCS. Between the spectral band combinations, the VNIR_HCS combination is the best, but with a very low OA of 0.29. The VNIR_GS and BGRN1_GS combination via GEOBIA classification yields the maximum OA of (0.79). A comparison between GEOBIA rule sets and PBIA classifiers (Supplementary Figure S3) outlines that GEOBIA is the most accurate with rule set 3 delivering the highest OA of 0.86. Within the PBIA algorithms, MXL delivers the maximum OA of 0.61. Overall findings of the spectral band combinations (Supplementary Figure S4) suggest that VNIR is the best performing subset with an aggregated OA of 0.59. The same trend is maintained among the PBIA and GEOBIA classifications with the VNIR subsets achieving OA of 0.37 and 0.81, respectively.

In Supplementary Figure S5, we provide processing methodologies which may be followed when using specific spectral band combinations with the tested image processing and classification methods. The most accurate method is the VNIR_VHR subset atmospherically corrected by QUAC, pansharpened by GS, and classified in GEOBIA by a combination of spatial and spectral object features using multiresolution segmentation and rule set 3 (Paper 2). The most efficient method is found to be the VNIR_VHR subset atmospherically corrected by DOS and classified by MXL.

4.6. Significances and Limitations

In the current study, we find that the DOS atmospheric correction performs best for VNIR imagery using PBIA. When utilizing GEOBIA, the QUAC correction may yield satisfactory results, but the DOS correction outperforms all in the overall analysis. Pansharpening does not improve PBIA but causes a significant improvement in GEOBIA classifications. Rule set definition must include both spatial and spectral attributes. SWIR data were the most useful for mapping dirty ice, but ultimately did not improve the classification process. Scene-based adjustments will always be profound when trying to extract maximum information from satellite data (Paper 2 [60]). Moreover, shadowed snow has continued to remain a problem (Paper 2 [9]) even with the addition of SWIR bands. Mixed pixel classifications in PBIA are a persistent phenomenon in an image with highly diverse spectral [152]. Overlap of spectral signatures can also cause misclassifications [140]. In glacier facies mapping, the main driver is the accurate characterization of spectral features into consequent classes. This places an onus on the operator for effective identification and comparison with literature (in absence of field data) and with field spectra (whenever available). PBIA can also result in fractured nature of classification due to spectral complexity on VHR imagery [157]. This can be reduced by GEOBIA [150]. The key for effective GEOBIA is segmentation. The multiresolution algorithm parameters selected here were applicable across all image subsets and processing schemes. This was constant even when the thresholds and features used for mapping facies were not consistent. The 'optimal' segmentation parameters selected here may be transferable and need to be tested further. The current study acknowledges the lack of field data for verification. However, the equalized random sampling approach utilized here [5] ensures a robust analysis. The classification of facies and their manifestation in the region was compared with previous studies to qualitatively assess the mapping methods. Furthermore, the focus of the study is the compounding effect of preprocessing routines on the resultant classification of glacier facies using GEOBIA and PBIA, not the most accurate map of glacier facies. The major limitation is the time needed to manually design rule sets. Moreover, the cognitive skill and performance of an analyst may play a role in the accuracy of rule set-based classification. While an existing rule set may be tested exactly for its transferability, development of a rule set for the same facies may be determined according to the differing knowledge, skills, and psychological variability of the analyst [31]. To the best of our knowledge, this is the first study to test the effects of GEOBIA and PBIA using various image processing techniques for mapping glacier surface facies using VHR satellite data. While the current results are tested on VHR VNIR_SWIR data, the results of the study must be tested on medium resolution satellite data for asserting the scalability of each rule set and classifier. We have not utilized an ancillary data as in-scene methods utilizing the properties of the available image to the maximum are necessary for understanding how facies can be mapped in different scenarios as well as reducing the reliance on ancillary data for accurate results. Moreover, it is well known that scene-based adjustments are necessary for implementing the same mapping method on different satellite images over different areas. Our results highlight that, in addition to sensor- and glacier-based adjustments, image processing-based adjustments are also necessary due to the impact of different methods of atmospheric correction and pansharpening on the properties of the pixels/objects of the same image.

Future Directions

The SVM classifier, although not tested here, is shown to have excellent capacity for detailed information extraction [136]. In this study, all indices for GEOBIA were created through a trial-and-error approach, while this is effective for an exhaustive study testing imagery and processing routines without field data. Shorter applications benefiting from field spectra can utilize an optimum index factor (OIF), which filters spectral bands based on standard deviation and correlation coefficients to determine optimal wavelengths [158,159]. Samsudin et al. [160] used the OIF to select spectral bands for creating indices for mapping degraded roof materials through WV-3 imagery. Alternatively, Shahi et al. [151] used a stepwise discriminant analysis (DA) to select bands for creating spectral indices for identifying asphalt roads. Resolution enhancement in the current study performed established and tested methods of pansharpening. Vivone and Chanussot [161] outline the potential of hypersharpening, which fuses a multispectral or hyperspectral image with a higher resolution multispectral/hyperspectral image. This could be potentially explored for multilevel resolution enhancements to characterize supraglacial features. To produce facies maps similar in detail to USGS mineral maps, we need a combination of hyperspectral shape matching [162], field spectra, laboratory spectra [163], and an expert mechanism of multispectral and hyperspectral wavelength feature characterization [140]. Papers 1, 2, and the current work account for the image processing and classification approaches which may be most suitable for facies mapping. Next, we intend to expand the methodology and examine classifications of SVM, random forest (RF), and neural network classifiers (NNC) in continuation of the current work to deliver a comprehensive machine learning based assessment of glacier facies mapping for the study sites—beyond which the current work will involve field investigations to produce detailed maps and rigorous comparison against all previous attempts.

5. Conclusions

This study focused on analyzing the impact of pre-processing routines and information extraction methods for mapping glacier surface facies across two different study areas. Three atmospheric corrections and two pansharpening algorithms were tested on VHR VNIR WV-2/3 for Ny-Ålesund, and Chandra-Bhaga Basin, Himalaya. The atmospheric correction methods included DOS, QUAC, and FLAASH. The pansharpening methods included GS and HCS. Following pansharpening, all processing schemes were divided into four spectral band subsets. Subset 1 is BGRN1, subset 2 is CYRN2, subset 3 is the VNIR range, whereas subset 4 is the entire VNIR_SWIR range. These subsets were then subjected to PBIA and GEOBIA classification methods. The PBIA process involved the utilization of conventional classifiers such as MHD, MXL, MD, SAM, and WTA; and advanced classifiers such as ACE, CEM, MF, MTMF, MTTCIMF, OSP, and TCIMF. The GEOBIA process involved multiresolution segmentation followed by development of three rule sets. Rule set 1 was developed using only object spectral information, rule set 2 was developed using only object contextual and spatial features, and rule set 3 was developed using a combination of both spatial and spectral features. The parameters used for segmentation consist of a scale parameter set at 5, shape set at 0.9, and compactness at 0.4. The features identified for classification are mean, quantile, standard deviation, min. pixel value, max. pixel value, edge contrast of neighbor pixels, number of overlapping thematic objects, relative border to, and customized ratios/arithmetic features. Among the PBIA classifications, the MXL classification on VNIR subsets achieved the highest F1 score of 0.50. MTTCIMF is the worst classifier yielding the lowest accuracy. However, MTTCIMF increased in performance when pan sharpened using the HCS method, whereas all other classifiers show a decrease in accuracy when using pan sharpened data. For the PBIA classifications, the best performing image subset is the VNIR. The trend of reliable performance among the atmospheric corrections for PBIA is DOS (OA = 0.48) > QUAC (OA = 0.40) > FLAASH (OA = 0.38). Pansharpening appears to be detrimental to VHR PBIA of surface facies and is not recommended. In the case of the GEOBIA classifications, the reliability trend among atmospheric corrections is QUAC (OA = 0.82) > DOS (OA = 0.81) > FLAASH (OA = 80). Among the pansharpening methods, GS (OA = 0.79) is more suitable than HCS (OA = 0.76) for the GEOBIA classifications. Rule set 3 delivers the best classification result, whereas the subsets with the highest accuracy (common OA = 0.79) are the VNIR and BGRN1 subset. This highlights the effectivity of the conventional spectral bands for characterizing glacier facies through both PBIA and GEOBIA. While GEOBIA delivers an overall greater performance than PBIA, the latter is more efficient. GEOBIA is limited by the human intervention for developing the rule sets. However, the current study highlights that GEOBIA may be better suited to map glacier surface facies using VHR VNIR data through a combination of spatial and spectral attributes. The segmentation parameters were consistent across all processing schemes and subsets and may be transferable to other VHR VNIR based facies mapping applications. The mapping procedures outlined in the current study may be applied on medium resolution satellite data. However, sub-pixel classification may be necessary to enhance the characterization of facies in coarser resolutions [70]. The direct alternative would be to utilize fine spatial resolution imagery [82] such as that used in the current study combined with the proposed methodology to map facies.

Supplementary Materials: The following supporting information can be downloaded at: https://www.mdpi.com/article/10.3390/rs14246311/s1, Table S1: The selected glaciers of the study, their areal extents, and GLIMS reference IDs. The extents were calculated from the delineated shapefiles using the geometry calculator in ArcGIS, Table S2: Abbreviations/nomenclature of processing schemes used in the current study. AC: Advanced Classifiers; CC: Conventional Classifiers. Table S3: Measures of accuracy used in the current study. TP: Samples are those that are in the positive class and are correctly classified, TN: Samples that are correctly classified as negative, FP: Samples that are not truly of the positive class but are incorrectly mapped as positive, FN: Samples that are mapped as negative when they are positive. Figure S1: Highlights the impact of atmospheric corrections progressively at each level of image processing, Figure S2: Highlights the impact of pansharpening progressively at each level of image processing, Figure S3: Displays the variations in overall accuracy of GEOBIA rule sets and PBIA algorithms. Figure S4: Recommendations of the most practical processing strategies if using any of the image processing routines and mapping methods.

Author Contributions: Conceptualization and experiment setup, S.D.J. and S.F.W.; methodology and manuscript design, S.D.J. and S.F.W.; software and processing, S.D.J. and S.F.W.; validation, S.D.J. and S.F.W.; result analysis and discussion, S.D.J. and S.F.W.; resources, A.J.L. and K.B.; writing—original draft preparation, S.F.W. and S.D.J.; writing—review and editing, S.D.J., A.J.L. and K.B.; visualization, S.F.W.; supervision, S.D.J., A.J.L. and K.B.; project administration, S.D.J., A.J.L. and K.B. All authors have read and agreed to the published version of the manuscript.

Funding: This research received no external funding.

Data Availability Statement: Freely available data used in the current study-(1) ASTER GDEM v2. Downloaded from: Gdex.cr.usgs.gov/gdex/ (accessed on 2 February 2017). The data are now moved to GDEM v3: (reviewed on 12 March 2022) ASTER GDEM is a product of Japan's Ministry of Economy, Trade, and Industry (METI) and NASA. (2) Arctic DEM. Available online: Pgc.umn.edu/data/arcticdem/ (accessed on 21 January 2019). Sentinel 2A (10 m)-Himalaya: Copernicus Sentinel data 2021, processed by ESA; Svalbard: Copernicus Sentinel data 2022, processed by ESA). Natural Earth: Free vector and raster map data @ naturalearthdata.com.

Acknowledgments: The authors would like to thank DigitalGlobe/MAXAR for providing high-resolution satellite data. Chandra–Bhaga Basin WorldView-2 © 2014 Maxar; Ny-Ålesund WorldView-3 © 2016 Maxar. The authors also thank HOD, Civil Engineering, and Director, MIT, MAHE for their support. S.D.J and A.J.L would like to thank M. Ravichandran, Secretary, MoES and Thamban Meloth, Director NCPOR for their encouragement and support. The authors thank the anonymous reviewers for their constructive criticism which helped develop and improve the manuscript.

Conflicts of Interest: The authors declare no conflict of interest.

References

1. Braun, M.; Schuler, T.V.; Hock, R.; Brown, I.; Jackson, M. Comparison of remote sensing derived glacier facies maps with distributed mass balance modelling at Engabreen, Northern Norway. *IAHS Publ. Ser. Proc. Rep.* **2007**, *318*, 126–134.
2. Luis, A.J.; Singh, S. High-resolution multispectral mapping facies on glacier surface in the Arctic using World, View-3 data. *Czech Polar Rep.* **2020**, *10*, 23–36. [CrossRef]
3. Jawak, S.D.; Wankhede, S.F.; Luis, A.J. Explorative Study on Mapping Surface Facies of Selected Glaciers from Chandra Basin, Himalaya Using World, View-2 Data. *Remote Sens.* **2019**, *11*, 1207. [CrossRef]
4. Jawak, S.D.; Wankhede, S.F.; Luis, A.J.; Pandit, P.H.; Kumar, S. Implementing an object-based multi-index protocol for mapping surface glacier facies from Chandra-Bhaga basin, Himalaya. *Czech Polar Rep.* **2019**, *9*, 125–140. [CrossRef]
5. Keshri, A.K.; Shukla, A.; Gupta, R.P. ASTER ratio indices for supraglacial terrain mapping. *Int. J. Remote Sens.* **2009**, *30*, 519–524. [CrossRef]
6. Bhardwaj, A.; Joshi, P.; Snehmani; Singh, M.; Sam, L.; Gupta, R. Mapping debris-covered glaciers and identifying factors affecting the accuracy. *Cold Reg. Sci. Technol.* **2014**, *106–107*, 161–174. [CrossRef]
7. Pope, A.; Rees, W.G. Impact of spatial, spectral, and radiometric properties of multispectral imagers on glacier surface classification. *Remote Sens. Environ.* **2014**, *141*, 1–13. [CrossRef]
8. Kundu, S.; Chakraborty, M. Delineation of glacial zones of Gangotri and other glaciers of Central Himalaya using RISAT-1 C-band dual-pol SAR. *Int. J. Remote Sens.* **2015**, *36*, 1529–1550. [CrossRef]
9. Robson, B.; Nuth, C.; Dahl, S.; Hölbling, D.; Strozzi, T.; Nielsen, P. Automated classification of debris-covered glaciers combining optical, SAR and topographic data in an object-based environment. *Remote Sens. Environ.* **2015**, *170*, 372–387. [CrossRef]
10. Gore, A.; Mani, S.; Shekhar, C.; Ganju, A. Glacier surface characteristics derivation and monitoring using Hyperspectral datasets: A case study of Gepang Gath glacier, Western Himalaya. *Geocarto Int.* **2017**, *34*, 23–42. [CrossRef]
11. Yousuf, B.; Shukla, A.; Arora, M.K.; Jasrotia, A.S. Glacier facies characterization using optical satellite data: Impacts of radiometric resolution, seasonality, and surface morphology. *Prog. Phys. Geogr. Earth Environ.* **2019**, *43*, 473–495. [CrossRef]
12. Foster, L.A.; Brock, B.W.; Cutler, M.E.J.; Diotri, F. A physically based method for estimating supraglacial debris thickness from thermal band remote-sensing data. *J. Glaciol.* **2012**, *58*, 677–691. [CrossRef]
13. Zhang, Y.; Hirabayashi, Y.; Fujita, K.; Liu, S.; Liu, Q. Heterogeneity in supraglacial debris thickness and its role in glacier mass changes of the Mount Gongga. *Sci. China Earth Sci.* **2015**, *59*, 170–184. [CrossRef]
14. Pandey, A.; Rai, A.; Gupta, S.K.; Shukla, D.P.; Dimri, A. Integrated approach for effective debris mapping in glacierized regions of Chandra River Basin, Western Himalayas, India. *Sci. Total Environ.* **2021**, *779*, 146492. [CrossRef]
15. Winsvold, S.H.; Kaab, A.; Nuth, C. Regional Glacier Mapping Using Optical Satellite Data Time Series. *IEEE J. Sel. Top. Appl. Earth Obs. Remote Sens.* **2016**, *9*, 3698–3711. [CrossRef]
16. Pope, A.; Rees, G. Using in situ spectra to explore Landsat classification of glacier surfaces. *Int. J. Appl. Earth Obs. Geoinf.* **2014**, *27*, 42–52. [CrossRef]
17. Paul, F.; Winsvold, S.H.; Kääb, A.; Nagler, T.; Schwaizer, G. Glacier Remote Sensing Using Sentinel-2. Part II: Mapping Glacier Extents and Surface Facies, and Comparison to Landsat 8. *Remote Sens.* **2016**, *8*, 575. [CrossRef]
18. Jawak, S.D.; Wankhede, S.F.; Luis, A.J.; Balakrishna, K. Impact of Image-Processing Routines on Mapping Glacier Surface Facies from Svalbard and the Himalayas Using Pixel-Based Methods. *Remote Sens.* **2022**, *14*, 1414. [CrossRef]
19. Jensen, J.R.; Lulla, K. Introductory digital image processing: A remote sensing perspective. *Geocarto Int.* **1987**, *2*, 65. [CrossRef]
20. Arbiol, R.; Zhang, Y.; i Comellas, V.P. Advanced classification techniques: A review. *Rev. Catalana Geogr.* **2007**, *12*, 31.
21. Weih, R.C.; Riggan, N.D. Object-based classification vs. pixel-based classification: Comparative importance of multi-resolution imagery. *Int. Arch. Photogram. Remote Sens. Spat. Inf. Sci.* **2010**, *38*, C7.
22. Mitkari, K.V.; Arora, M.K.; Tiwari, R.K.; Sofat, S.; Gusain, H.S.; Tiwari, S.P. Large-Scale Debris Cover Glacier Mapping Using Multisource Object-Based Image Analysis Approach. *Remote Sens.* **2022**, *14*, 3202. [CrossRef]
23. Sharda, S.; Srivastava, M. Classification of Siachen Glacier Using Object-Based Image Analysis. In Proceedings of the 2018 International Conference on Intelligent Circuits and Systems (ICICS), Phagwara, India, 19–20 April 2018.
24. Gao, B.C.; Davis, C.; Goetz, A. A review of atmospheric correction techniques for hyperspectral remote sensing of land surfaces and ocean color. In Proceedings of the 2006 IEEE International Symposium on Geoscience and Remote Sensing, Denver, CO, USA, 31 July–4 August 2006; pp. 1979–1981. [CrossRef]
25. Lee, K.H.; Yum, J.M. A Review on Atmospheric Correction Technique Using Satellite Remote Sensing. *Korean J. Remote Sens.* **2019**, *35*, 1011–1030. [CrossRef]

26. Gao, B.C.; Montes, M.J.; Davis, C.O.; Goetz, A.F. Atmospheric correction algorithms for hyperspectral remote sensing data of land and ocean. *Remote Sens. Environ.* **2009**, *113*, S17–S24. [CrossRef]
27. Guo, Y.; Zeng, F. Asstmospheric correction comparison of SPOT-5 image based on model FLAASH and model QUAC. *Int. Arch. Photogramm. Remote Sens. Spat. Inf. Sci.* **2012**, *39*, 21–23.
28. Nazeer, M.; Nichol, J.E.; Yung, Y.K. Evaluation of atmospheric correction models and Landsat surface reflectance product in an urban coastal environment. *Int. J. Remote Sens.* **2014**, *35*, 6271–6291. [CrossRef]
29. Mandanici, E.; Franci, F.; Bitelli, G.; Agapiou, A.; Alexakis, D.; Hadjimitsis, D.G. Comparison between empirical and physically based models of atmospheric correction. In Proceedings of the Third International Conference on Remote Sensing and Geoinformation of the Environment (RSCy2015), Paphos, Cyprus, 16–19 March 2015; Volume 9535, pp. 110–119. [CrossRef]
30. Eugenio, F.; Marcello, J.; Martin, J.; Rodríguez-Esparragón, D. Benthic Habitat Mapping Using Multispectral High-Resolution Imagery: Evaluation of Shallow Water Atmospheric Correction Techniques. *Sensors* **2017**, *17*, 2639. [CrossRef]
31. Karimi, N.; Farokhnia, A.; Karimi, L.; Eftekhari, M.; Ghalkhani, H. Combining optical and thermal remote sensing data for mapping debris-covered glaciers (Alamkouh Glaciers, Iran). *Cold Reg. Sci. Technol.* **2012**, *71*, 73–83. [CrossRef]
32. Albert, T.H. Evaluation of Remote Sensing Techniques for Ice-Area Classification Applied to the Tropical Quelccaya Ice Cap, Peru. *Polar Geogr.* **2002**, *26*, 210–226. [CrossRef]
33. Guo, Z.; Geng, L.; Shen, B.; Wu, Y.; Chen, A.; Wang, N. Spatiotemporal Variability in the Glacier Snowline Altitude across High Mountain Asia and Potential Driving Factors. *Remote Sens.* **2021**, *13*, 425. [CrossRef]
34. Garzelli, A.; Nencini, F.; Alparone, L.; Aiazzi, B.; Baronti, S. Pan-sharpening of multispectral images: A critical review and comparison. In Proceedings of the IGARSS 2004. 2004 IEEE International Geoscience and Remote Sensing Symposium, Anchorage, AK, USA, 20–24 September 2004. [CrossRef]
35. Xu, Q.; Zhang, Y.; Li, B. Recent advances in pansharpening and key problems in applications. *Int. J. Image Data Fusion* **2014**, *5*, 175–195. [CrossRef]
36. Snehmani, A.G.; Ganju, A.; Kumar, S.; Srivastava, P.K.; Hari Ram, R.P. A comparative analysis of pansharpening techniques on Quick, Bird and World, View-3 images. *Geocarto Int.* **2016**, *32*, 1268–1284. [CrossRef]
37. Jawak, S.D.; Wankhede, S.F.; Luis, A.J.; Balakrishna, K. Effect of Image-Processing Routines on Geographic Object-Based Image Analysis for Mapping Glacier Surface Facies from Svalbard and the Himalayas. *Remote Sens.* **2022**, *14*, 4403. [CrossRef]
38. Isaksen, K.; Nordli, Ø.; Førland, E.J.; Lupikasza, E.; Eastwood, S.; Niedźwiedź, T. Recent warming on Spitsbergen—Influence of atmospheric circulation and sea ice cover. *J. Geophys. Res. Atmos.* **2016**, *121*, 121. [CrossRef]
39. Pandey, P.; Ali, S.N.; Ramanathan, A.L.; Venkataraman, G. Regional representation of glaciers in Chandra Basin region, western Himalaya, India. *Geosci. Front.* **2017**, *8*, 841–850. [CrossRef]
40. Pandey, P.; Venkataraman, G. Changes in the glaciers of Chandra–Bhaga basin, Himachal Himalaya, India, between 1980 and 2010 measured using remote sensing. *Int. J. Remote Sens.* **2013**, *34*, 5584–5597. [CrossRef]
41. Raup, B.; Racoviteanu, A.; Khalsa, S.J.; Helm, C.; Armstrong, R.; Arnaud, Y. The GLIMS geospatial glacier database: A new tool for studying glacier change. *Glob. Planet. Chang.* **2007**, *56*, 101–110. [CrossRef]
42. Digital Globe Product Details. Available online: https://www.geosoluciones.cl/documentos/worldview/Digital,Globe-Core-Imagery-Products-Guide.pdf (accessed on 20 February 2020).
43. ASTER GDEM v2. Available online: Gdex.cr.usgs.gov/gdex/ (accessed on 2 February 2017).
44. Arctic DEM. Available online: www.pgc.umn.edu/data/arcticdem/ (accessed on 21 January 2019).
45. Porter, C.; Morin, P.; Howat, I.; Noh, M.-J.; Bates, B.; Peterman, K.; Keesey, S.; Schlenk, M.; Gardiner, J.; Tomko, K.; et al. "ArcticDEM", Harvard Dataverse, V1. 2018. Available online: https://www.pgc.umn.edu/data/arcticdem/ (accessed on 13 March 2022).
46. Radiative Transfer Code. Available online: https://www.harrisgeospatial.com/docs/backgroundflaash.html (accessed on 17 February 2017).
47. Atmospheric Correction User Guide. Available online: https://www.l3harrisgeospatial.com/portals/0/pdfs/envi/Flaash_Module.pdf (accessed on 17 February 2017).
48. Kaufman, Y.; Wald, A.; Remer, L.; Gao, B.-C.; Li, R.-R.; Flynn, L. The MODIS 2.1-μm channel-correlation with visible reflectance for use in remote sensing of aerosol. *IEEE Trans. Geosci. Remote Sens.* **1997**, *35*, 1286–1298. [CrossRef]
49. Abreu, L.W.; Anderson, G.P. The MODTRAN 2/3 report and LOWTRAN 7 model. *Contract* **1996**, *19628*, 132.
50. Teillet, P.; Fedosejevs, G. On the Dark Target Approach to Atmospheric Correction of Remotely Sensed Data. *Can. J. Remote Sens.* **1995**, *21*, 374–387. [CrossRef]
51. Zhang, Z.; He, G.; Zhang, X.; Long, T.; Wang, G.; Wang, M. A coupled atmospheric and topographic correction algorithm for remotely sensed satellite imagery over mountainous terrain. *GISci. Remote Sens.* **2017**, *55*, 400–416. [CrossRef]
52. Rumora, L.; Miler, M.; Medak, D. Impact of Various Atmospheric Corrections on Sentinel-2 Land Cover Classification Accuracy Using Machine Learning Classifiers. *ISPRS Int. J. Geo-Inform.* **2020**, *9*, 277. [CrossRef]
53. Bernstein, L.S.; Jin, X.; Gregor, B.; Adler-Golden, S.M. Quick atmospheric correction code: Algorithm description and recent upgrades. *Opt. Eng.* **2012**, *51*, 111719. [CrossRef]
54. Pushparaj, J.; Hegde, A.V. Evaluation of pan-sharpening methods for spatial and spectral quality. *Appl. Geomat.* **2016**, *9*, 1–12. [CrossRef]
55. Laben, C.A.; Brower, B.V. Process for Enhancing the Spatial Resolution of Multispectral Imagery Using Pan-Sharpening. U.S. Patent 6,011,875, 4 January 2000.

56. Bhardwaj, A.; Joshi, P.; Snehmani; Sam, L.; Singh, M.; Singh, S.; Kumar, R. Applicability of Landsat 8 data for characterizing glacier facies and supraglacial debris. *Int. J. Appl. Earth Obs. Geoinf.* **2015**, *38*, 51–64. [CrossRef]
57. De Angelis, H.; Rau, F.; Skvarca, P. Snow zonation on Hielo Patagónico Sur, Southern Patagonia, derived from Landsat 5 TM data. *Glob. Planet. Chang.* **2007**, *59*, 149–158. [CrossRef]
58. Shukla, A.; Ali, I. A hierarchical knowledge-based classification for glacier terrain mapping: A case study from Kolahoi Glacier, Kashmir Himalaya. *Ann. Glaciol.* **2016**, *57*, 1–10. [CrossRef]
59. Baatz, M.; Schape, A. Multiresolution Segmentation: An Optimization Approach for High Quality Multi-Scale Image Segmentation. In *Angewandte Geographische Informations-Verarbeitung, XII*; Strobl, J., Blaschke, T., Griesbner, G., Eds.; Wichmann Verlag: Heidelberg/Karlsruhe, Germany, 2000; pp. 12–23.
60. Han, Y.; Javed, A.; Jung, S.; Liu, S. Object-Based Change Detection of Very High Resolution Images by Fusing Pixel-Based Change Detection Results Using Weighted Dempster–Shafer Theory. *Remote Sens.* **2020**, *12*, 983. [CrossRef]
61. Trimble GmbH. *eCognition Developer 9.0 User Guide*; Trimble Germany GmbH: Munich, Germany, 2014.
62. Richards, J.A. *Remote Sensing Digital Image Analysis*; Springer: Berlin/Heidelberg, Germany, 2013.
63. Maxwell, A.E.; Warner, T.A. Thematic Classification Accuracy Assessment with Inherently Uncertain Boundaries: An Argument for Center-Weighted Accuracy Assessment Metrics. *Remote Sens.* **2020**, *12*, 1905. [CrossRef]
64. Mölg, N.; Bolch, T.; Rastner, P.; Strozzi, T.; Paul, F. A consistent glacier inventory for Karakoram and Pamir derived from Landsat data: Distribution of debris cover and mapping challenges. *Earth Syst. Sci. Data* **2018**, *10*, 1807–1827. [CrossRef]
65. Garg, V.; Thakur, P.K.; Rajak, D.R.; Aggarwal, S.P.; Kumar, P. Spatio-temporal changes in radar zones and ELA estimation of glaciers in Ny-Ålesund using Sentinel-1 SAR. *Polar Sci.* **2022**, *31*, 100786. [CrossRef]
66. Casacchia, R.; Lauta, F.; Salvatori, R.; Cagnati, A.; Valt, M.; Ørbæk, J.B. Radiometric investigation of different snow covers in Svalbard. *Polar Res.* **2001**, *20*, 13–22. [CrossRef]
67. Snapir, B.; Momblanch, A.; Jain, S.K.; Waine, T.W.; Holman, I.P. A method for monthly mapping of wet and dry snow using Sentinel-1 and MODIS: Application to a Himalayan river basin. *Int. J. Appl. Earth Obs. Geoinf.* **2019**, *74*, 222–230. [CrossRef]
68. Karbou, F.; Veyssière, G.; Coleou, C.; Dufour, A.; Gouttevin, I.; Durand, P.; Gascoin, S.; Grizonnet, M. Monitoring Wet Snow Over an Alpine Region Using Sentinel-1 Observations. *Remote Sens.* **2021**, *13*, 381. [CrossRef]
69. Nagajothi, V.; Geetha, P.M.; Sharma, P.; Krishnaveni, D. Classification of Dry/Wet Snow Using Sentinel-2 High Spatial Resolution Optical Data. In *Intelligent Data Engineering and Analytics*; Springer: Singapore, 2021; pp. 1–9. [CrossRef]
70. Yousuf, B.; Shukla, A.; Arora, M.K. Temporal Variability of the Satopanth Glacier Facies at Sub-pixel Scale, Garhwal Himalaya, India. In *Mountain Landscapes in Transition*; Springer: Cham, Switzerland, 2022; pp. 207–218. [CrossRef]
71. Ji, X.; Chen, Y.; Tong, L.; Jia, M.; Tan, L.; Fan, S. Area retrieval of melting snow in alpine areas. In Proceedings of the 2014 IEEE Geoscience and Remote Sensing Symposium, Quebec City, QC, Canada, 13–18 July 2014; pp. 3991–3993. [CrossRef]
72. Vickers, H.; Malnes, E.; Eckerstorfer, M. A Synthetic Aperture Radar Based Method for Long Term Monitoring of Seasonal Snowmelt and Wintertime Rain-On-Snow Events in Svalbard. *Front. Earth Sci.* **2022**, *10*, 2296–6463. [CrossRef]
73. Aggarwal, S.P.; Thakur, P.K.; Nikam, B.R.; Garg, V. Integrated approach for snowmelt run-off estimation using temperature index model, remote sensing and GIS. *Curr. Sci.* **2014**, *106*, 397–407.
74. Liang, D.; Guo, H.; Zhang, L.; Cheng, Y.; Zhu, Q.; Liu, X. Time-series snowmelt detection over the Antarctic using Sentinel-1 SAR images on Google Earth Engine. *Remote Sens. Environ.* **2021**, *256*, 112318. [CrossRef]
75. Mendes, C.W., Jr.; Arigony, N.J.; Hillebrand, F.L.; De Freitas, M.W.D.; Costi, J.; Simões, J.C. Snowmelt retrieval algorithm for the Antarctic Peninsula using SAR imageries. *An. Acad. Bras. Cienc.* **2022**, *94*, e20210217. [CrossRef]
76. Paterson, W.S.B. *The Physics of Glaciers*; Elsevier: Amsterdam, The Netherlands, 1994.
77. Hinkler, J.; Ørbæk, J.B.; Hansen, B.U. Detection of spatial, temporal, and spectral surface changes in the Ny-Ålesund area 79° N, Svalbard, using a low cost multispectral camera in combination with spectroradiometer measurements. *Phys. Chem. Earth Parts A/B/C* **2003**, *28*, 1229–1239. [CrossRef]
78. Prieur, C.; Rabatel, A.; Thomas, J.-B.; Farup, I.; Chanussot, J. Machine Learning Approaches to Automatically Detect Glacier Snow Lines on Multi-Spectral Satellite Images. *Remote Sens.* **2022**, *14*, 3868. [CrossRef]
79. Azzoni, R.S.; Fugazza, D.; Zerboni, A.; Senese, A.; D'Agata, C.; Maragno, D.; Carzaniga, A.; Cernuschi, M.; Diolaiuti, G.A. Evaluating high-resolution remote sensing data for reconstructing the recent evolution of supra glacial debris: A study in the Central Alps (Stelvio Park, Italy). *Prog. Phys. Geogr. Earth Environ.* **2018**, *42*, 3–23. [CrossRef]
80. Alifu, H.; Johnson, B.A.; Tateishi, R. Delineation of Debris-Covered Glaciers Based on a Combination of Geomorphometric Parameters and a TIR/NIR/SWIR Band Ratio. *IEEE J. Sel. Top. Appl. Earth Obs. Remote Sens.* **2016**, *9*, 781–792. [CrossRef]
81. Ambinakudige, S.; Intsiful, A. Estimation of area and volume change in the glaciers of the Columbia Icefield, Canada using machine learning algorithms and Landsat images. *Remote Sens. Appl. Soc. Environ.* **2022**, *26*, 100732. [CrossRef]
82. Lu, D.; Weng, Q. A survey of image classification methods and techniques for improving classification performance. *Int. J. Remote Sens.* **2007**, *28*, 823–870. [CrossRef]
83. Fyffe, C.L.; Woodget, A.S.; Kirkbride, M.P.; Deline, P.; Westoby, M.J.; Brock, B.W. Processes at the margins of supraglacial debris cover: Quantifying dirty ice ablation and debris redistribution. *Earth Surf. Process. Landf.* **2020**, *45*, 2272–2290. [CrossRef]
84. Chandler, D.M.; Alcock, J.D.; Wadham, J.L.; Mackie, S.L.; Telling, J. Seasonal changes of ice surface characteristics and productivity in the ablation zone of the Greenland Ice Sheet. *Cryosphere* **2015**, *9*, 487–504. [CrossRef]

85. Østrem, G. Ice melting under a thin layer of moraine, and the existence of ice cores in moraine ridges. *Geogr. Ann.* **1959**, *41*, 228–230. [CrossRef]
86. Østrem, G. Problems of dating ice-cored moraines. *Geogr. Ann. Ser. A Phys. Geogr.* **1965**, *47*, 1–38. [CrossRef]
87. Haq, M.A.; Alshehri, M.; Rahaman, G.; Ghosh, A.; Baral, P.; Shekhar, C. Snow and glacial feature identification using Hyperion dataset and machine learning algorithms. *Arab. J. Geosci.* **2021**, *14*, 1525. [CrossRef]
88. Croot, D.G.; Sharp, R.P. Living ice. Understanding glaciers and glaciation. *Geogr. J.* **2006**, *155*, 410. [CrossRef]
89. Florath, J.; Keller, S.; Abarca-del-Rio, R.; Hinz, S.; Staub, G.; Weinmann, M. Glacier Monitoring Based on Multi-Spectral and Multi-Temporal Satellite Data: A Case Study for Classification with Respect to Different Snow and Ice Types. *Remote Sens.* **2022**, *14*, 845. [CrossRef]
90. Bennett, M. *Glaciers and Glaciation*; Benn, D.I., Evans, D.J.A., Eds.; Boreas: London, UK, 2011. [CrossRef]
91. Pandey, A.; Rai, A.; Gupta, S.K.; Shukla, D.P. Hierarchical Knowledge Based Classification (Hkbc) On Sentinel-2a Data for Glacier Mapping of Bhaga River Basin, Northwest Himalaya. *Red* **2017**, *10*, 665.
92. Ali, I.; Shukla, A.; Romshoo, S. Assessing linkages between spatial facies changes and dimensional variations of glaciers in the upper Indus Basin, western Himalaya. *Geomorphology* **2017**, *284*, 115–129. [CrossRef]
93. Shukla, A.; Gupta, R.; Arora, M. Estimation of debris cover and its temporal variation using optical satellite sensor data: A case study in Chenab basin, Himalaya. *J. Glaciol.* **2009**, *55*, 444–452. [CrossRef]
94. Shukla, A.; Arora, M.; Gupta, R. Synergistic approach for mapping debris-covered glaciers using optical–thermal remote sensing data with inputs from geomorphometric parameters. *Remote Sens. Environ.* **2010**, *114*, 1378–1387. [CrossRef]
95. Ghosh, S.; Pandey, A.C.; Nathawat, M.S. Mapping of debris-covered glaciers in parts of the Greater Himalaya Range, Ladakh, western Himalaya, using remote sensing and GIS. *J. Appl. Remote Sens.* **2014**, *8*, 083579. [CrossRef]
96. Fleischer, F.; Otto, J.C.; Junker, R.R.; Hölbling, D. Evolution of debris cover on glaciers of the Eastern Alps, Austria, between 1996 and 2015. *Earth Surf. Process. Landf.* **2021**, *46*, 1673–1691. [CrossRef]
97. Kirkbride, M.P. Debris-Covered Glaciers. In *Encyclopedia of Snow, Ice and Glaciers*; Singh, V.P., Singh, P., Haritashya, U.K., Eds.; Springer: Dordrecht, The Netherlands, 2011. [CrossRef]
98. Shrestha, R.; Kayastha, R.B.; Kayastha, R. Effect of debris on seasonal ice melt (2016−2018) on Ponkar Glacier, Manang, Nepal. *Sci. Cold Arid. Reg.* **2020**, *12*, 261–271. [CrossRef]
99. Pratibha, S.; Kulkarni, A.V. Decadal change in supraglacial debris cover in Baspa basin, Western Himalaya. *Curr. Sci.* **2018**, *114*, 792–799. [CrossRef]
100. Gibson, M.J.; Glasser, N.F.; Quincey, D.J.; Mayer, C.; Rowan, A.V.; Irvine-Fynn, T.D. Temporal variations in supraglacial debris distribution on Baltoro Glacier, Karakoram between 2001 and 2012. *Geomorphology* **2017**, *295*, 572–585. [CrossRef]
101. Nicholson, L.I.; McCarthy, M.; Pritchard, H.D.; Willis, I. Supraglacial debris thickness variability: Impact on ablation and relation to terrain properties. *Cryosphere* **2018**, *12*, 3719–3734. [CrossRef]
102. Racoviteanu, A.E.; Nicholson, L.; Glasser, N.F. Surface composition of debris-covered glaciers across the Himalaya using linear spectral unmixing of Landsat 8 OLI imagery. *Cryosphere* **2021**, *15*, 4557–4588. [CrossRef]
103. Kaushik, S.; Singh, T.; Bhardwaj, A.; Joshi, P.K.; Dietz, A.J. Automated Delineation of Supraglacial Debris Cover Using Deep Learning and Multisource Remote Sensing Data. *Remote Sens.* **2022**, *14*, 1352. [CrossRef]
104. Jawak, S.D.; Jadhav, A.; Luis, A.J. Object-oriented feature extraction approach for mapping supraglacial debris in Schirmacher Oasis using very high-resolution satellite data. In *Land Surface and Cryosphere Remote Sensing III*; SPIE: Bellingham, WA, USA, 2016; Volume 9877, pp. 337–345. [CrossRef]
105. Bennett, M.M.; Glasser, N.F. (Eds.) *Glacial Geology: Ice Sheets and Landforms*; John Wiley & Sons: Hoboken, NJ, USA, 2011.
106. Colgan, W.; Rajaram, H.; Abdalati, W.; McCutchan, C.; Mottram, R.; Moussavi, M.S.; Grigsby, S. Glacier crevasses: Observations, models, and mass balance implications. *Rev. Geophys.* **2016**, *54*, 119–161. [CrossRef]
107. Chen, F. Comparing Methods for Segmenting Supra-Glacial Lakes and Surface Features in the Mount Everest Region of the Himalayas Using Chinese GaoFen-3 SAR Images. *Remote Sens.* **2021**, *13*, 2429. [CrossRef]
108. Singh, K.K.; Negi, H.S.; Ganju, A.; Kulkarni, A.V.; Kumar, A.; Mishra, V.D.; Kumar, S. Crevasses detection in Himalayan glaciers using ground-penetrating radar. *Curr. Sci.* **2013**, *105*, 1288–1295.
109. Taurisano, A.; Tronstad, S.; Brandt, O.; Kohler, J. On the use of ground penetrating radar for detecting and reducing crevasse-hazard in Dronning Maud Land, Antarctica. *Cold Reg. Sci. Technol.* **2006**, *45*, 166–177. [CrossRef]
110. Hao, S.; Cui, Y.; Wang, J. Segmentation Scale Effect Analysis in the Object-Oriented Method of High-Spatial-Resolution Image Classification. *Sensors* **2021**, *21*, 7935. [CrossRef]
111. Hossain, M.D.; Chen, D.M. Segmentation for Object-based Image analysis (OBIA): A review of algorithm and challenges from remote sensing perspective. *ISPRS J. Photogram. Remote Sens.* **2019**, *150*, 115–134. [CrossRef]
112. Arifjanov, A.M.; Akmalov, S.B.; Apakhodjaeva, T.U.; Tojikhodjaeva, D.S. Comparison Of Pixel To Pixel And Object Based Image Analysis with using Worldview-2 Satellite Images of Yangiobod Village of Syrdarya Province. Интеркарто. Интергис **2020**, *26*, 313–321. [CrossRef]
113. Kucharczyk, M.; Hay, G.J.; Ghaffarian, S.; Hugenholtz, C.H. Geographic Object-Based Image Analysis: A Primer and Future Directions. *Remote Sens.* **2020**, *12*, 2012. [CrossRef]
114. Arundel, S.T. Pairing semantics and object-based image analysis for national terrain mapping—A first-case scenario of cirques. In Proceedings of the GEOBIA 2016: Solutions and synergies, Enschede, The Netherlands, 14–16 September 2016; University of Twente Faculty of Geo-Information and Earth Observation (ITC): Enschede, The Netherlands, 2016. [CrossRef]

115. Feizizadeh, B.; Garajeh, M.K.; Blaschke, T.; Lakes, T. An object based image analysis applied for volcanic and glacial landforms mapping in Sahand Mountain, Iran. *Catena* **2021**, *198*, 105073. [CrossRef]
116. Robb, C.; Willis, I.; Arnold, N.; Guðmundsson, S. A semi-automated method for mapping glacial geomorphology tested at Breiðamerkurjökull, Iceland. *Remote Sens. Environ.* **2015**, *163*, 80–90. [CrossRef]
117. Dabiri, Z.; Hölbling, D.; Abad, L.; Prasicek, G.; Argentin, A.L.; Tsai, T.T. An Object-Based Approach for Monitoring the Evolution of Landslide-dammed Lakes and Detecting Triggering Landslides in Taiwan. *Int. Arch. Photogramm. Remote Sens. Spat. Inf. Sci.* **2019**, *42*, 103–108. [CrossRef]
118. Farhan, S.B.; Kainat, M.; Shahzad, A.; Aziz, A.; Kazmi, S.J.H.; Shaikh, S.; Zhang, Y.; Gao, H.; Javed, M.N.; Zamir, U.B. Discrimination of Seasonal Snow Cover in Astore Basin, Western Himalaya using Fuzzy Membership Function of Object-Based Classification. *Int. J. Econ. Environ. Geol.* **2019**, *9*, 20–25. [CrossRef]
119. Kraaijenbrink, P.D.A.; Shea, J.M.; Pellicciotti, F.; De Jong, S.M.; Immerzeel, W.W. Object-based analysis of unmanned aerial vehicle imagery to map and characterise surface features on a debris-covered glacier. *Remote Sens. Environ.* **2016**, *186*, 581–595. [CrossRef]
120. Podgórski, J.; Pętlicki, M. Detailed Lacustrine Calving Iceberg Inventory from Very High Resolution Optical Imagery and Object-Based Image Analysis. *Remote Sens.* **2020**, *12*, 1807. [CrossRef]
121. Dabiri, Z.; Hölbling, D.; Abad, L.; Guðmundsson, S. Comparing the Applicability of Sentinel-1 and Sentinel-2 for Mapping the Evolution of Ice-marginal Lakes in Southeast Iceland. *GI_Forum* **2021**, *9*, 46–52. [CrossRef]
122. Rendenieks, Z.; Nita, M.D.; Nikodemus, O.; Radeloff, V.C. Half a century of forest cover change along the Latvian-Russian border captured by object-based image analysis of Corona and Landsat TM/OLI data. *Remote Sens. Environ.* **2020**, *249*, 112010. [CrossRef]
123. Pandey, P.; Kulkarni, A.V.; Venkataraman, G. Remote sensing study of snowline altitude at the end of melting season, Chandra-Bhaga basin, Himachal Pradesh, 1980–2007. *Geocarto Int.* **2013**, *28*, 311–322. [CrossRef]
124. Rathore, B.P.; Singh, S.K.; Jani, P.; Bahuguna, I.M.; Brahmbhatt, R.; Rajawat, A.S.; Randhawa, S.S.; Vyas, A. Monitoring of snow cover variability in Chenab Basin using IRS AWiFS sensor. *J. Indian Soc. Remote Sens.* **2018**, *46*, 1497–1506. [CrossRef]
125. Sahu, R.; Gupta, R.D. Snow cover area analysis and its relation with climate variability in Chandra basin, Western Himalaya, during 2001–2017 using MODIS and ERA5 data. *Environ. Monit. Assess.* **2020**, *192*, 489. [CrossRef] [PubMed]
126. Bernardo, N.; Watanabe, F.; Rodrigues, T.; Alcântara, E. Atmospheric correction issues for retrieving total suspended matter concentrations in inland waters using OLI/Landsat-8 image. *Adv. Space Res.* **2017**, *59*, 2335–2348. [CrossRef]
127. Roy, D.P.; Wulder, M.A.; Loveland, T.R.; Woodcock, C.E.; Allen, R.G.; Anderson, M.C.; Helder, D.; Irons, J.R.; Johnson, D.M.; Kennedy, R.; et al. Landsat-8: Science and product vision for terrestrial global change research. *Remote Sens. Environ.* **2014**, *145*, 154–172. [CrossRef]
128. Binding, C.E.; Jerome, J.H.; Bukata, R.P.; Booty, W.G. Suspended particulate matter in Lake Erie derived from MODIS aquatic colour imagery. *Int. J. Remote Sens.* **2010**, *31*, 5239–5255. [CrossRef]
129. Chakouri, M.; Lhissou, R.; El Harti, A.; Maimouni, S.; Adiri, Z. Assessment of the image-based atmospheric correction of multispectral satellite images for geological mapping in arid and semi-arid regions. *Remote Sens. Appl. Soc. Environ.* **2020**, *20*, 100420. [CrossRef]
130. Saini, V.; Tiwari, R.; Gupta, R. Comparison of FLAASH and QUAC Atmospheric Correction Methods for Resourcesat-2 LISS-IV Data. In Proceedings of the SPIE, Earth Observing Missions and Sensors: Development, Implementation, and Characterization IV, New Delhi, India, 2 May 2016.
131. Marcello, J.; Eugenio, F.; Perdomo, U.; Medina, A. Assessment of Atmospheric Algorithms to Retrieve Vegetation in Natural Protected Areas Using Multispectral High Resolution Imagery. *Sensors* **2016**, *16*, 1624. [CrossRef]
132. Casey, K.A.; Kääb, A.; Benn, D.I. Geochemical characterization of supraglacial debris via in situ and optical remote sensing methods: A case study in Khumbu Himalaya, Nepal. *Cryosphere* **2012**, *6*, 85–100. [CrossRef]
133. Rastner, P.; Bolch, T.; Notarnicola, C.; Paul, F. A Comparison of Pixel-and Object-Based Glacier Classification with Optical Satellite Images. *IEEE J. Sel. Top. Appl. Earth Obs. Remote Sens.* **2014**, *7*, 853–862. [CrossRef]
134. Longbotham, N.; Pacifici, F.; Malitz, S.; Baugh, W.; Camps-Valls, G. Measuring the spatial and spectral performance of WorldView-3. In *Hyperspectral Imaging and Sounding of the Environment*; Optica Publishing Group: Washington, DC, USA, 2015; p. HW3B-2. [CrossRef]
135. Collin, A.; Andel, M.; James, D.; Claudet, J. The superspectral/hyperspatial worldview-3 as the link between spaceborne hyperspectral and airborne hyperspatial sensors: The case study of the complex tropical coast. *ISPRS-Int. Arch. Photogramm. Remote Sens. Spat. Inf. Sci.* **2019**, *42*, 1849–1854. [CrossRef]
136. Ye, B.; Tian, S.; Ge, J.; Sun, Y. Assessment of WorldView-3 Data for Lithological Mapping. *Remote Sens.* **2017**, *9*, 1132. [CrossRef]
137. Sun, Y.; Tian, S.; Di, B. Extracting mineral alteration information using WorldView-3 data. *Geosci. Front.* **2017**, *8*, 1051–1062. [CrossRef]
138. Asadzadeh, S.; de Souza Filho, C.R. Investigating the capability of WorldView-3 superspectral data for direct hydrocarbon detection. *Remote Sens. Environ.* **2016**, *173*, 162–173. [CrossRef]
139. Mars, J.C. Mineral and lithologic mapping capability of WorldView 3 data at Mountain Pass, California, using true-and false-color composite images, band ratios, and logical operator algorithms. *Econ. Geol.* **2018**, *113*, 1587–1601. [CrossRef]
140. Kruse, F.A.; Baugh, W.M.; Perry, S.L. Validation of DigitalGlobe WorldView-3 Earth imaging satellite shortwave infrared bands for mineral mapping. *J. Appl. Remote Sens.* **2015**, *9*, 096044. [CrossRef]

141. Hunt Jr, E.R.; Daughtry, C.S.; Li, L. Feasibility of estimating leaf water content using spectral indices from WorldView-3's near-infrared and shortwave infrared bands. *Int. J. Remote Sens.* **2016**, *37*, 388–402. [CrossRef]
142. Eckert, S. Improved Forest biomass and carbon estimations using texture measures from WorldView-2 satellite data. *Remote Sens.* **2012**, *4*, 810–829. [CrossRef]
143. Sibanda, M.; Mutanga, O.; Rouget, M. Testing the capabilities of the new WorldView-3 space-borne sensor's red-edge spectral band in discriminating and mapping complex grassland management treatments. *Int. J. Remote Sens.* **2017**, *38*, 1–22. [CrossRef]
144. Pu, R.; Gong, P.; Biging, G.S.; Larrieu, M.R. Extraction of red edge optical parameters from Hyperion data for estimation of forest leaf area index. *IEEE Trans. Geosci. Remote Sens.* **2003**, *41*, 916–921. [CrossRef]
145. Upadhyay, P.; Ghosh, S.K.; Kumar, A.; Roy, P.S.; Gilbert, I. Effect on specific crop mapping using WorldView-2 multispectral add-on bands: Soft classification approach. *J. Appl. Remote Sens.* **2012**, *6*, 063524. [CrossRef]
146. Immitzer, M.; Atzberger, C.; Koukal, T. Tree Species Classification with Random Forest Using Very High Spatial Resolution 8-Band WorldView-2 Satellite Data. *Remote Sens.* **2012**, *4*, 2661–2693. [CrossRef]
147. Marshall, V.; Lewis, M.; Ostendorf, B. Do additional bands (coastal, NIR-2, red-edge and yellow) in WorldView-2 multispectral imagery improve discrimination of an Invasive Tussock, Buffel Grass (Cenchrus Ciliaris). *Int. Arch. Photogramm. Remote Sens. Spat. Inf. Sci.* **2012**, *39*, B8. [CrossRef]
148. Heenkenda, M.K.; Joyce, K.E.; Maier, S.W.; Bartolo, R. Mangrove Species Identification: Comparing WorldView-2 with Aerial Photographs. *Remote Sens.* **2014**, *6*, 6064–6088. [CrossRef]
149. Collin, A.; Hench, J.L. Towards Deeper Measurements of Tropical Reefscape Structure Using the WorldView-2 Spaceborne Sensor. *Remote Sens.* **2012**, *4*, 1425–1447. [CrossRef]
150. Malinowski, R.; Groom, G.; Schwanghart, W.; Heckrath, G. Detection and Delineation of Localized Flooding from WorldView-2 Multispectral Data. *Remote Sens.* **2015**, *7*, 14853–14875. [CrossRef]
151. Shahi, K.; Shafri, H.Z.; Taherzadeh, E.; Mansor, S.; Muniandy, R. A novel spectral index to automatically extract road networks from WorldView-2 satellite imagery. *Egypt. J. Remote Sens. Space Sci.* **2015**, *18*, 27–33. [CrossRef]
152. Abriha, D.; Kovács, Z.; Ninsawat, S.; Bertalan, L.; Balázs, B.; Szabó, S. Identification of roofing materials with Discriminant Function Analysis and Random Forest classifiers on pan-sharpened WorldView-2 imagery–a comparison. *Hung. Geogr. Bull.* **2018**, *67*, 375–392. [CrossRef]
153. Tiwari, R.K.; Gupta, R.P.; Gens, R.; Prakash, A. Use of optical, thermal and microwave imagery for debris characterization in Bara-Shigri glacier, Himalayas, India. In Proceedings of the 2012 IEEE International Geoscience and Remote Sensing Symposium, Munich, Germany, 22–27 July 2012; pp. 4422–4425. [CrossRef]
154. Bühler, Y.; Meier, L.; Meister, R. Continuous, high resolution snow surface type mapping in high alpine terrain using WorldView-2 data. *Digit. Globe* **2011**, 6. Available online: https://www.researchgate.net/profile/Roland-Meister-3/publication/267859153_Continuous_high_resolution_snow_surface_type_mapping_in_high_alpine_terrain_using_WorldView-2_data/links/547370c10cf2d67fc0373851/Continuous-high-resolution-snow-surface-type-mapping-in-high-alpine-terrain-using-WorldView-2-data.pdf (accessed on 16 August 2022).
155. Jawak, S.D.; Khopkar, P.S.; Jadhav, S.P.; Luis, A.J. Customization of Normalized Difference Snow Index for Extraction of Snow Cover from Cryospheric Surface Using WorldView-2 Data. In Proceedings of the AGSE International Conference, Ahmedabad, India, 16–19 December 2013; pp. 16–19. Available online: https://www.researchgate.net/profile/Shridhar-Jawak/publication/270890440_Customization_of_Normalized_Difference_of_Snow_Index_NDSI_for_extraction_of_snow_andor_ice_cover_from_cryospheric_surface_using_WorldView-2_data/links/55279bb80cf229e6d6362dd3/Customization-of-Normalized-Difference-of-Snow-Index-NDSI-for-extraction-of-snow-and-or-ice-cover-from-cryospheric-surface-using-WorldView-2-data.pdf (accessed on 16 August 2022).
156. Gray, A.; Krolikowski, M.; Fretwell, P.; Convey, P.; Peck, L.S.; Mendelova, M.; Smith, A.G.; Davey, M.P. Remote sensing phenology of Antarctic green and red snow algae using WorldView satellites. *Front. Plant Sci.* **2021**, *12*, 877. [CrossRef]
157. Arroyo, L.A.; Johansen, K.; Armston, J.; Phinn, S. Integration of LiDAR and QuickBird imagery for mapping riparian biophysical parameters and land cover types in Australian tropical savannas. *For. Ecol. Manag.* **2010**, *259*, 598–606. [CrossRef]
158. Gardin, S.; Van Laere, S.M.J.; Van Coillie, F.M.B.; Anseel, F.; Duyck, W.; De Wulf, R.R.; Verbeke, L.P.C. Remote sensing meets psychology: A concept for operator performance assessment. *Remote Sens. Lett.* **2011**, *2*, 251–257. [CrossRef]
159. Shafri, H.Z.M.; Anuar, M.I.; Saripan, M.I. Modified vegetation indices for Ganoderma disease detection in oil palm from field spectroradiometer data. *J. Appl. Remote Sens.* **2009**, *3*, 033556. [CrossRef]
160. Samsudin, S.H.; Shafri, H.Z.; Hamedianfar, A. Development of spectral indices for roofing material condition status detection using field spectroscopy and WorldView-3 data. *J. Appl. Remote Sens.* **2016**, *10*, 025021. [CrossRef]
161. Vivone, G.; Chanussot, J. Fusion of short-wave infrared and visible near-infrared WorldView-3 data. *Inf. Fusion* **2020**, *61*, 71–83. [CrossRef]
162. Clark, R.N.; Swayze, G.A.; Livo, K.E.; Kokaly, R.F.; Sutley, S.J.; Dalton, J.B.; McDougal, R.R.; Gent, C.A. Imaging spectroscopy: Earth and planetary remote sensing with the USGS Tetracorder and expert systems. *J. Geophys. Res. Planets* **2003**, *108*, E12. [CrossRef]
163. Swayze, G.A.; Clark, R.N.; Goetz, A.F.; Livo, K.E.; Breit, G.N.; Kruse, F.A.; Sutley, S.J.; Snee, L.W.; Lowers, H.A.; Post, J.L.; et al. Mapping advanced argillic alteration at Cuprite, Nevada, using imaging spectroscopy. *Econ. Geol.* **2014**, *109*, 1179–1221. [CrossRef]

remote sensing

Article

Combined Use of Aerial Photogrammetry and Terrestrial Laser Scanning for Detecting Geomorphological Changes in Hornsund, Svalbard

Małgorzata Błaszczyk [1,*], Michał Laska [1], Agnar Sivertsen [2] and Shridhar D. Jawak [3]

1 Institute of Earth Sciences, University of Silesia in Katowice, Bedzinska 60, 41-200 Sosnowiec, Poland; michal.laska@us.edu.pl
2 NORCE Norwegian Research Center AS, Sykehusvn 21, 9019 Tromsø, Norway; agsi@norceresearch.no
3 Svalbard Integrated Arctic Earth Observing System (SIOS), SIOS Knowledge Centre, Svalbard Science Centre, P.O. Box 156, N-9171 Longyearbyen, Norway; shridhar.jawak@sios-svalbard.org
* Correspondence: malgorzata.blaszczyk@us.edu.pl

Abstract: The Arctic is a region undergoing continuous and significant changes in land relief due to different glaciological, geomorphological and hydrogeological processes. To study those phenomena, digital elevation models (DEMs) and highly accurate maps with high spatial resolution are of prime importance. In this work, we assess the accuracy of high-resolution photogrammetric DEMs and orthomosaics derived from aerial images captured in 2020 over Hornsund, Svalbard. Further, we demonstrate the accuracy of DEMs generated using point clouds acquired in 2021 with a Riegl VZ®-6000 terrestrial laser scanner (TLS). Aerial and terrestrial data were georeferenced and registered based on very reliable ground control points measured in the field. Both DEMs, however, had some data gaps due to insufficient overlaps in aerial images and limited sensing range of the TLS. Therefore, we compared and integrated the two techniques to create a continuous and gapless DEM for the scientific community in Svalbard. This approach also made it possible to identify geomorphological activity over a one-year period, such as the melting of ice cores at the periglacial zone, changes along the shoreline or snow thickness in gullies. The study highlights the potential for combining other techniques to represent the active processes in this region.

Keywords: structure-from-motion; terrestrial laser scanning; digital elevation model; Svalbard; SIOS

Citation: Błaszczyk, M.; Laska, M.; Sivertsen, A.; Jawak, S.D. Combined Use of Aerial Photogrammetry and Terrestrial Laser Scanning for Detecting Geomorphological Changes in Hornsund, Svalbard. *Remote Sens.* **2022**, *14*, 601. https://doi.org/10.3390/rs14030601

Academic Editor: Michael Lim

Received: 23 December 2021
Accepted: 25 January 2022
Published: 26 January 2022

1. Introduction

The Hornsund area in southern Spitsbergen (Svalbard) is the focus of wide glaciological [1–4], hydrological [5–9], snow [9], permafrost [10,11], geomorphometric [12–14] and biological [15] studies performed due to intensive alterations in the surveyed terrain. Digital mapping of such an environment and its related phenomena has become a fundamental requirement to keep track of all such alterations [16]. The most up-to-date, accurate and precise digital elevation models (DEM) for the study area are generated based on high-resolution satellites and aerial photos [17]. From these, the DEM with the highest spatial resolution (2 m) and accuracy (standard deviation = 0.6 m) is derived from aerial photographs taken in 2011. However, to monitor climate-induced change in the territory, there is a need for more frequent and accurate data on terrain elevation [17].

The aim of the paper is to examine and combine two DEMs produced from aerial image sets and terrestrial laser scanning (TLS). All data created within the project are further available to the scientific community through the Svalbard Integrated Arctic Earth Observing System (SIOS). Aerial images for the studied area were provided by SIOS through a dedicated call of proposals (https://sios-svalbard.org/AirborneRS (accessed on 23 December 2021)). One of SIOS's missions is to reduce the environmental footprint of scientific observations in Svalbard [18]. To achieve this mission, SIOS supports and coordinates the usage of unmanned aerial vehicles (UAVs) and a crewed Dornier aircraft to acquire imagery and hyperspectral data for the Svalbard research community to support scientific projects (https://sios-svalbard.org/AirborneRemoteSensing (accessed on 23 December 2021)). The project started in 2020 and so far, no publications have presented the accuracy of the data collected within the framework of the project. In this study, we examine and present the accuracy of products such as orthomosaics and DEMs obtained during the mission. The aerial data over two Hornsund zones, the Fuglebergsletta and Werenskioldbreen areas, were collected in June 2020 (Figure 1). The mission took place at the beginning of the ablation season, when some snow cover was present in the numerous gullies. Additionally, the flight level limited the sidelap of images, causing data gaps in the DEM. To address these limitations, a long-range TLS campaign was performed in August 2021. Long-range terrestrial laser scanning is an emerging method for the monitoring of complex and rough terrain such as mountain slopes, outcrops and deformations [16,19–23]. Here, we used the ultra-long laser scanner Riegl VZ®-6000, which has been successfully used in cryosphere studies such as glacier mass balance measurements in China [24,25]; the mass balance of very small glaciers in the Swiss Alps [26]; snow distribution at a glacier located in the Ötztal Alps, Austria [27]; glacier snowline determination [28]; or relationships between different climate forcing and flows of Helheim Glacier, Greenland [29]. However, at this point there are not many studies on the accuracy of ultra-long laser scans [23,26,30], especially over 2 km [27]. Further, to our knowledge there has not yet been a study published concerning the quality control of TLS-collected data with the Riegl VZ®-6000 instrument in the polar region. To fill this gap, we analyzed the accuracy of the relative and absolute registration of four scans collected over the complex Fuglebergsletta area, the most studied area near the Polish Polar Station.

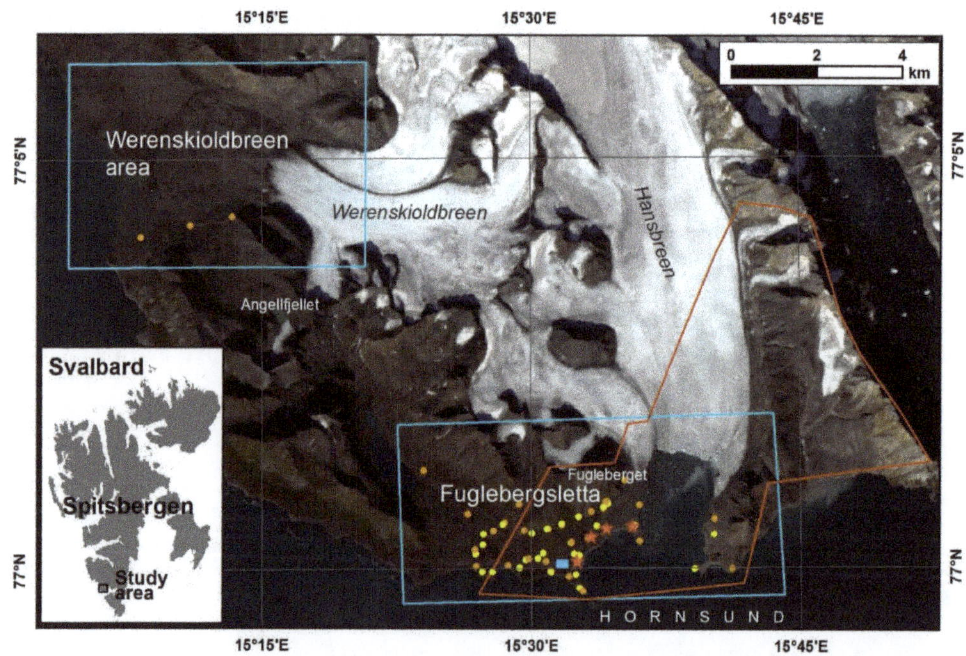

Figure 1. Location of the study area. The blue outlines represent the extent of aerial imagery acquired during the crewed aircraft campaign in 2020; the red polygon presents the range of data taken by terrestrial scanner Riegl VZ-6000 in 2021; the red asterisks illustrate the positions of the laser scanner; the orange dots represent ground control points and the yellow dots represent checkpoints measured by GPS in 2021; the blue rectangle presents the location of the Polish Polar Station, Hornsund.

Terrestrial laser scanning (TLS) was next applied to complement the aerial photogrammetry. The integration of 3D modeling techniques is advantageous for obtaining the most complete and useful object coverage for many application areas [19]. Integration of aircraft-based, helicopter-based and UAV photogrammetry with TLS is popular, especially at large or inaccessible sites [21,23,30–32]. In this paper, digital aerial photogrammetry and TLS were combined to produce a continuous DEM with the highest possible resolution and accuracy. Thanks to that, this DEM, based on aerial photos and scanning (TLS), can be successfully applied to many environmental applications, such as hydrology modeling, glacier change detection, quantifying depositional and erosional processes in dynamic and complex fluvial systems, the evolution of the landscape, geomorphology and the monitoring of landslide displacement [13,33].

2. Study Area

The study area is located in the southern part of Spitsbergen, Svalbard (Figure 1). The area is characterized by diverse surface coverage such as relatively flat topography, mountains, tidewater glaciers and land-based glaciers. Narrow coastal plains with raised marine terraces surround the fjord shores [34]. Further inland, mountain massifs range in elevation up to 763 m above sea level (a.s.l.). The steep rock walls are cut by a system of deep chutes and gullies [13]. Two regions are analyzed in this paper. The eastern part is Fuglebergsletta, containing the Fugleberget massif and Hansbreen tidewater glacier; the retreat of the Hansbreen caused the exposure of wide lateral moraine ridges with buried glacial ice [35]. The western studied part is the Werenskioldbreen area; the land-based Werenskioldbreen glacier's foreground is flat with active glaciofluvial landforms and moraine. The vicinity of the Polish Polar Station, located on the shore, makes these two areas the focus of numerous environmental studies.

3. Materials and Methods

3.1. Aerial Imagery

3.1.1. Data Preprocessing

Imagery acquisition over the research area was carried out during a SIOS crewed aircraft campaign on 22 June 2020. The aircraft, a Dornier DO228, is fitted with an RGB camera (Phasone IXU-150, Schneider LS 55 mm f/2.8) and a hyperspectral imager (VNIR-1800, Norsk elektrooptikk; [18]). The RGB camera and the hyperspectral imager can acquire images with a ground resolution of 0.1 m and 0.3 m from a flight altitude of 1000 m, respectively. During the Hornsund campaign, 622 high-resolution RGB photos were acquired. Out of these, 326 were related to the Fuglebergsletta area (covering Polish Polar Station and Hansbreen surroundings) and 296 to the foreground and frontal part of Werenskioldbreen (Figure 1). The images were all acquired at a flight altitude of 1000 m a.s.l., or approximately 935 m above ground level, resulting in a ground resolution of approximately 0.085 m per pixel. Due to the variable lighting conditions during the mission, preliminary exposure compensation was performed (Figure 2). The entire dataset was converted from the RAW format (48 bit IIQ) to JPEG (24-bit).

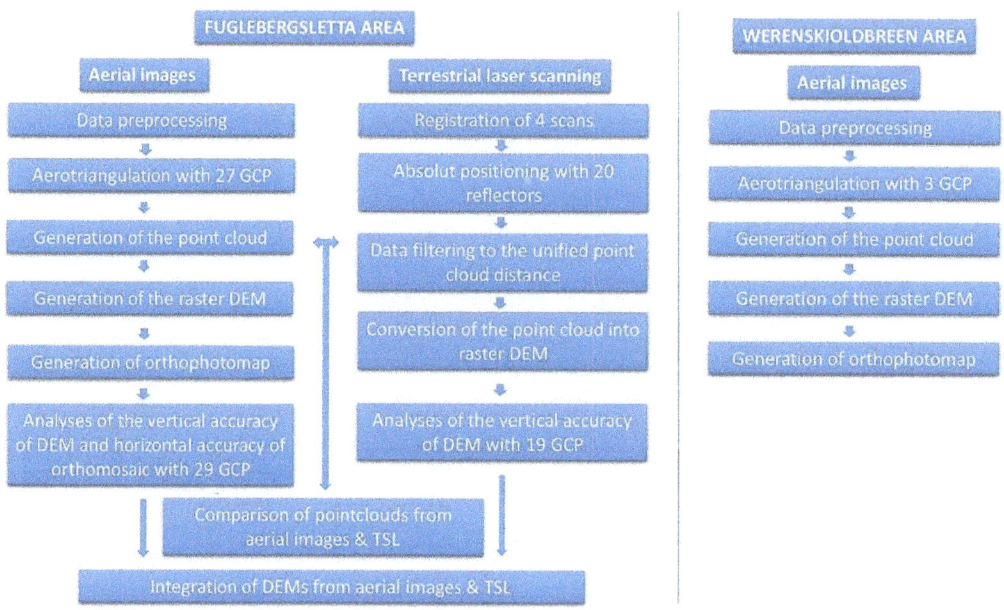

Figure 2. Processing workflow for airborne data and terrestrial laser scanning (TLS).

3.1.2. Ground Control Points and Checkpoints

There was no opportunity to place dedicated ground control points (GCPs) and checkpoints before the flight campaign. Therefore, natural points were measured during the fieldwork in 2021, between 13–18 August (Figure 1). The vast majority concerned centroids of the characteristic large boulders which could have been correctly defined in the previously taken aerial photos. Fifty-nine points were measured in WGS 84 UTM 33X projection by Leica GX1230GG receivers. Measurements were performed at GNSS post-processing mode (30 min sessions, Phase Fixed solution) and RTK-GNSS mode, with corrections from the reference station at the Polish Polar Station (PPS). The GCPs at Fuglebergsletta were within the 4 km range from the PPS. The GCPs at the Werenskioldbreen area were 11–12.5 km from the PPS. The average horizontal and vertical accuracy related to the post-processing measurements was 0.0003 m and 0.0008 m, respectively. The average horizontal and vertical accuracy related to the RTK mode was 0.01 m and 0.024 m, respectively.

3.1.3. Data Processing and Quality

The attitude and position of the camera were recorded for each image, using an onboard navigation system (Applanix POS AV 410). The flight data, together with the GNSS correction measurements from the Polish Polar Station, were postprocessed using Applanix PosPac MMS 8 software. The total accuracy of the postprocessed camera position and orientation were within 0.025 m and 0.01 degrees, respectively. The aerial images were processed in Agisoft Metashape 1.7.5 software (https://www.agisoft.com/ (accessed on 23 December 2021)), using the postprocessed camera position and orientation for each image. The structure-from-motion method was used to obtain high-resolution DEMs and orthomosaics (Figure 2). Structure from motion (SfM) is a photogrammetric technique for estimating three-dimensional models from two-dimensional images, collected in proper overlap and coupled with local motion signals [36]. It is widely used in modern photogrammetry, especially low-level, where UAVs are applied [37].

At the following stages of data processing, the images in full resolution (8280 × 6208 pixels) were applied. Next, 30 GCPs were used in the aerotriangulation process (Figure 1). In the

Fuglebergsletta area, the selected 27 GCPs were characterized by an appropriate distribution, presented in Figure 1. The GCP density was 1.17 GCP/km^2. In the Werenskioldbreen area, only three GCPs were measured, mainly located in the southern part of the study. The GCP density was only 0.11 GCP/km^2. The aggressive depth filtering was used to generate dense point clouds, which produced the best results among the available levels. The remaining artefacts were removed manually. Due to the negligible number of vertical objects and the use of very high-quality images, estimated by the software at above 0.80, the total number of removed points was insignificant, below 0.1%.

The remaining 29 points measured by GPS, which did not participate in the aerotriangulation process, were used as checkpoints (Figure 1) to validate the vertical accuracy of DEM and horizontal accuracy of the orthomosaic generated for Fuglebergsletta. The point cloud data were exported into the grid format (*.tif) and compared against the GCPs in ArcGIS software.

3.2. Terrestrial Laser Scanning (TLS)

3.2.1. Long-Range Terrestrial Laser Scanning

The TLS survey was carried out on 15 August 2021, using the Riegl VZ®-6000 long-range terrestrial laser scanner (Figure 1). The scanner has an effective maximum range of around 6 km, operates at a wavelength of 1064 nm and uses class 3B laser beams, so can be used in snow and glacier studies thanks to high rates of reflection (80%) from snow- and ice-covered terrain [23,24,29]. The result of a single laser scan is a large quantity of 3D data, usually of the order of several million point measurements, each with an x, y, z and intensity value. This dataset is termed a point cloud and is the raw 'product' of scanning [38]. The laser footprint size estimated from the beam divergence (0.12 mrad for Riegl VZ®-6000) gives spot sizes of 15 mm at 60 m, 120 mm at 1000 m, 240 mm at 2000 m and up to 720 mm at distances of 6000 m [39].

Although the range of 6 km theoretically covers most of the area of interest, in order to increase the density of scans in the far field and reduce shadowing problems [23], a total of four scans were acquired from various positions to cover the area as uniformly as possible. The scanning frequency was set to 30 kHz and column and line resolutions to 0.002°. Riegl instrument uses laser light, so is relatively slow compared to the phase difference scanners [40]. Therefore, the number of the positions and scan resolutions were chosen to take advantage of good atmospheric conditions, which is not common in harsh Arctic conditions. Associated phenomena affecting long-distance estimations are air temperature, pressure and humidity. Therefore, we used meteorological data from a nearby meteorological station at the Polish Polar Station during measurements.

3.2.2. Point Cloud Registration

Relative and absolute positioning of scans [38] (Figure 2) was performed using RiSCAN Pro 2.12.1 (http://www.riegl.com/products/software-packages/riscan-pro/ (accessed on 23 December 2021)). The maximum measured distance by TLS within the study area was about 7.8 km. However, to minimize the effect of the laser footprint size for long distances, during the registration process we used only parts of point clouds within the range of 3 km from the laser scanner position. The data were also filtered to remove noises, and the unstable area of the glacier was omitted during the registration of the scan.

Fine cloud-to-cloud registration was performed by a Multi-Station Adjustment (MSA) tool in RiSCAN Pro [32]. The software identifies common plane patches from different scan positions, links them together and minimizes the errors between all these linked plane patches by using an iterative matching algorithm [41]. The standard procedure in MSA uses a plane patch filter [32]; however, our study used triangulated polydata to increase the number of planes in the registration process. As a result, all point clouds from the four stations were merged into a single point cloud.

An absolute registration was performed in the last step to fit the merged point clouds into the reflectors scanned during the field campaign. Five Riegl flat reflectors (50 mm diameters) were distributed spatially as evenly as possible around each of the four scanner positions and measured using RTK-GNSS in the UTM 33X zone. Leica GX1230GG receivers operating in the RTK mode used corrections from the reference station at the Polish Polar Station. The average horizontal and vertical accuracy of RTK GNSS measurements of all 20 reflectors was 0.01 m and 0.02 m, respectively.

3.2.3. Validation of TLS

Absolute positioning allows laser scans to be integrated with the other registered data [38]. However, before such integration, validation of the data should be performed. The densities of TLS point clouds strongly depend on the distance to the scanning system and the chosen scan frequency [23]. Therefore, to uniform the data, point clouds were first filtered with the Octree algorithm in RiSCAN Pro 2.12.1 and then exported to the LAS format. Next, the data were imported to ArcGIS, converted into the raster format and compared against the 19 GCPs and checkpoints measured in 2021 (see Chapter 3.1.2) which were captured during scanning.

3.3. Integration of Aerial and TLS Based Data

To assess the difference between aerial-based and TLS-based DEMs, we used the freely available Multiscale Model-to-Model Cloud Comparison (M3C2) plugin for open-source CloudCompare software [42]. M3C2 is an algorithm used for multitemporal point cloud distance calculations. It estimates local positions in two input point clouds by using the surface normal vectors to determine the median point within a cylinder of defined radius [42,43]. For the point-cloud-based strategy, the M3C2 algorithm is chosen as the best-established method, especially in earth sciences, when dealing with irregular surfaces [44]. Here, we used the algorithm to compare the data and find areas where differences between both point clouds were significantly larger than the estimated accuracy of the DEMs generated from those point clouds. Buildings and noises were removed from the point clouds. Normal calculations were performed at a fixed scale (D = 2 m) using the core point file.

Both DEMs contain data gaps; therefore, we combined both DEMs to create a continuous gapless product with the best possible accuracy and resolution. Missing data in aerial-based DEM stem from too low a sidelap of images. TLS data gaps arise from two primary sources: a line-of-sight obstacle resulting in occlusion (holes in the data as some foreground interfered with the scanner's line of sight) and a dropout resulting from a specular reflective or absorbent surface, preventing the energy from a given laser pulse from returning to the TLS instrument [38,45]. Furthermore, TLS data has inhomogeneous point density throughout the area [16,22,30,42,46]. Thus, we used the aerial-based DEM with relatively homogeneous point distribution as the reference dataset. Next, the data gaps and areas covered by snow were replaced by the data from TLS. In a few small zones without any data, TLS data were interpolated using the Inverse Distance Weighting (IDW) algorithm [16].

4. Results

4.1. Digital Elevation Model and Orthomosaics Based on Aerial Imageries

4.1.1. Fuglebergsletta

As a result of aligning the images, 325 from 326 were correctly aligned. Camera locations and image overlaps are presented in Figure 3A. Details on data processing are shown in Table 1. The final resolution of the DEM equals 0.169 m and the point density was 35.2 points/m^2. The root mean square error for all 27 GCP locations calculated by the software was 0.0018 m. Based on the improved dense clouds, digital elevation models were generated. The DEM was exported to raster format with 0.169 m resolution, the best possible resolution with this software, to increase the possibility of mapping even the smallest geomorphological features [16]. The vertical quality of the DEM was assessed based on 29 GCPs that were not used in the aerotriangulation process (Figure 4A). The standard deviation equaled 0.14 m, with a mean value of −0.22 m and a median value of −0.19 m. The maximal vertical error of the DEM equaled 0.54 m. There is a small systematic error in the DEM. In general, the elevation of the DEM generated in Agisoft is slightly higher than the measured elevation of the checkpoints.

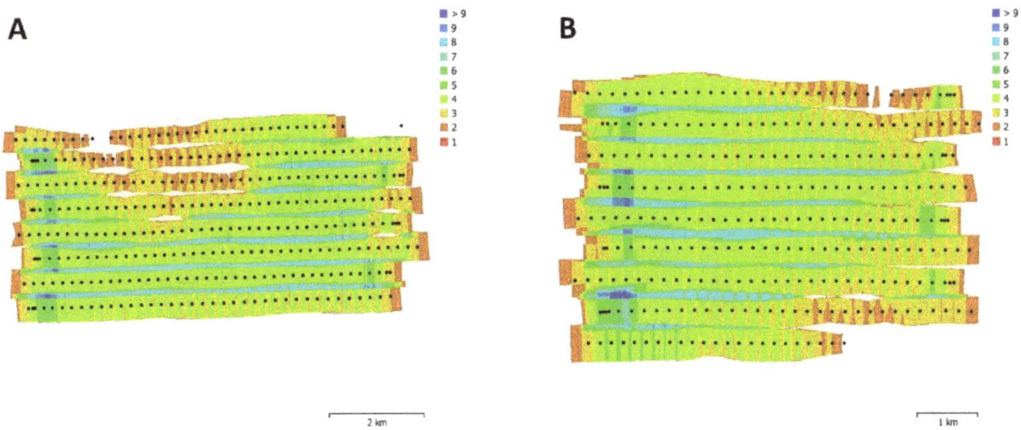

Figure 3. Camera locations (black dots) and image numbers and overlaps over Fuglebergsletta (**A**) and Werenskioldbreen area (**B**).

Table 1. Processing settings and final resolution of the products.

	Image Alignment		Dense Cloud		Depth Maps Filtering	3D Model		DEM		Orthomosaic	
	Accuracy	Tie Points	Quality	Points		Quality	Faces	Size	Resolution	Size	Resolution
Fuglebergsletta	High	356,693	High	1,074,237,705	Aggressive	High	213,379,209	73,519 × 38,769	0.169 m	106,313 × 50,381	0.0843 m
Werenskioldbreen	High	323,830	High	959,690,194	Aggressive	High	191,045,203	61,565 × 45,592	0.174 m	81,027 × 53,511	0.087 m

After the final colour calibration of the entire dataset, an orthomosaic was produced (Figure 4B). The resolution of the orthomosaic was two times higher than the DEMs, which were the basis for its creation, and equalled 0.0843 m. The horizontal accuracy of the orthomosaic was estimated based on the same GCP points as those used for the DEM accuracy assessment. The standard deviation of the horizontal accuracy equalled 0.10 m, with a mean value of −0.12 m and a median value of −0.10 m.

In the low-image-overlap areas, the insufficient color blending over merged images may be noticeable. Even using color and white balance calibration in postprocessing, this effect was difficult to eliminate. For flat areas located below c. 140 m a.s.l., the applied flight parameters were sufficient to assemble the models. However, the mountain slopes and peaks on both study sites were characterized by either insufficient or a complete lack of coverage (Figure 3A). This resulted in broad blank areas (Figure 4) in the DEM and orthomosaic. Parts of the model composed only of a pair of images occasionally presented coarse graining or a deep generalization of elevation data (Figure 5A). Similar effects occur even with multiple overlaps on watercourses, surface water (Figure 5A) and the sea. For this reason, the DEM shows negative point values, reaching –15 m. Artefacts were also generated by moving objects, such as animals (Figure 5B).

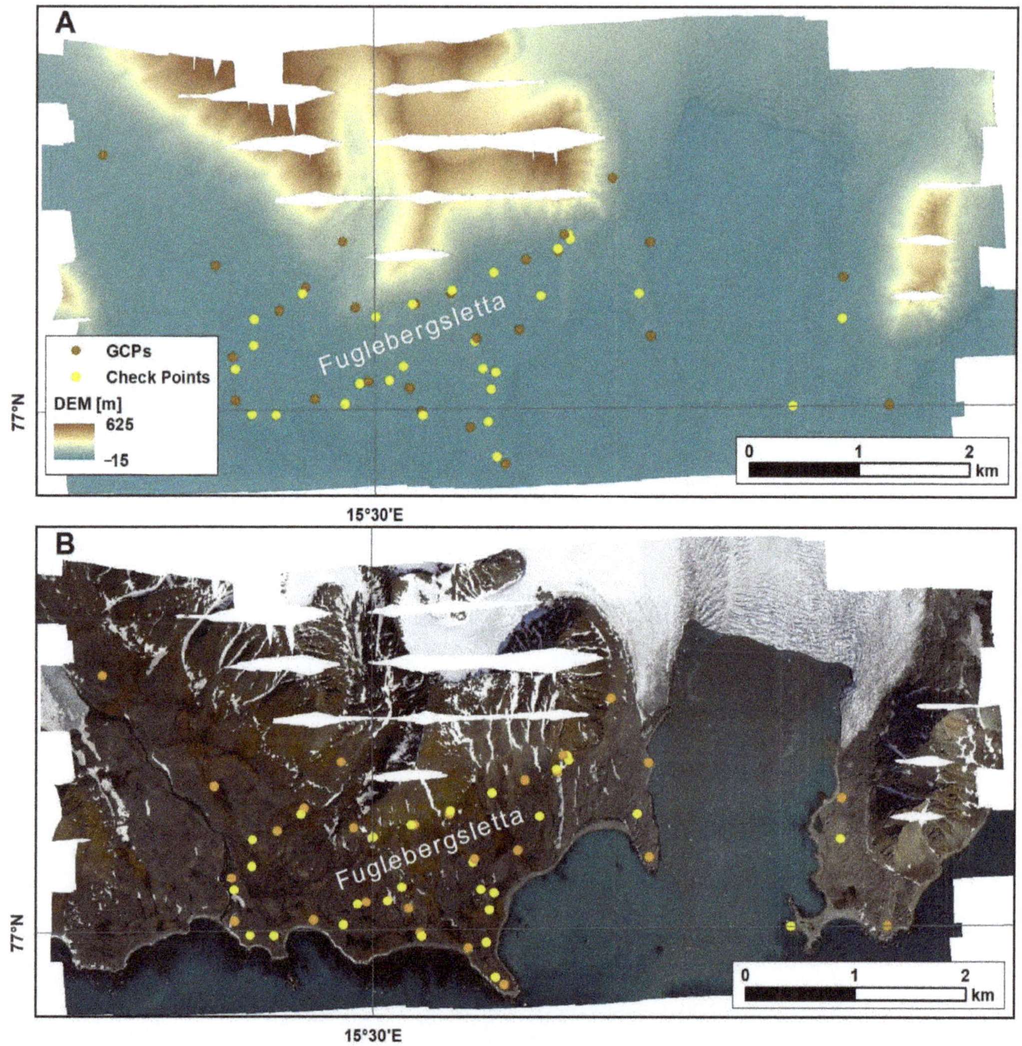

Figure 4. Digital Elevation Model (**A**) and orthomosaic (**B**) for the Fuglebergsletta area. Orange points represent GCPs used during processing in Agisoft, while yellow points served as checkpoints to estimate the quality of the products.

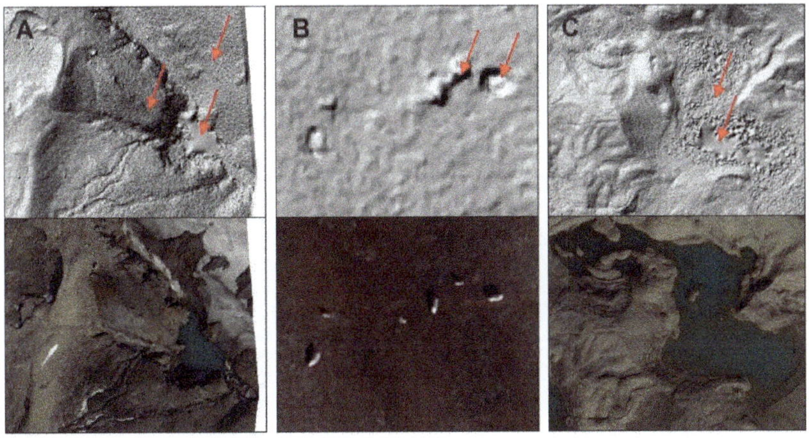

Figure 5. Artefacts described in the text: coarse graining or a deep generalization of elevation data (**A**), artefacts generated by reindeers (**B**) and noises over water bodies (**A**,**C**).

4.1.2. Werenskioldbreen Area

As a result of aligning the images, 285 out of 296 photos for Werenskioldbreen were aligned correctly. Both the automatic and manual alignment functions failed to merge ten images representing the southern part of the glacier. It was caused by low overlap over the Angellfjellet ridge (594 m a.s.l.) and partial cloud cover over this area, making it difficult to detect tie points correctly. Camera locations and image overlaps for the second area are presented in Figure 3B. Details on data processing are described in Table 1. The final resolution of the DEM equals 0.174 m and the point density is 33 points/m^2. The root mean square error for three points used in aerotriangulation was 0.0011 m. The resolution of the DEM in raster format and orthomosaics generated for the zone was 0.174 and 0.087 m, respectively (Figure 6). The resolution difference between orthomosaics for Werenskioldbreen and Fuglebergsletta was caused by the variation between mean ground level and flight altitude. Similar to the DEM for Fuglebergsletta, noises over water bodies (e.g., Figure 5C) are also present for the Werenskioldbreen area.

No checkpoints were applied to assess the final absolute vertical accuracy of the DEM and horizontal accuracy of the orthomosaic. However, we compared the DEM and orthomosaic generated using three points in aerotriangulation and without using any GCP. Results show significant improvement of the former products. The DEM accuracy measured over the three GCPs increased from 1.77 m to 0.22 m. The horizontal accuracy of the orthomosaic increased from 1.29 m to around 0.1 m (the orthomosaic was shifted towards the southeast).

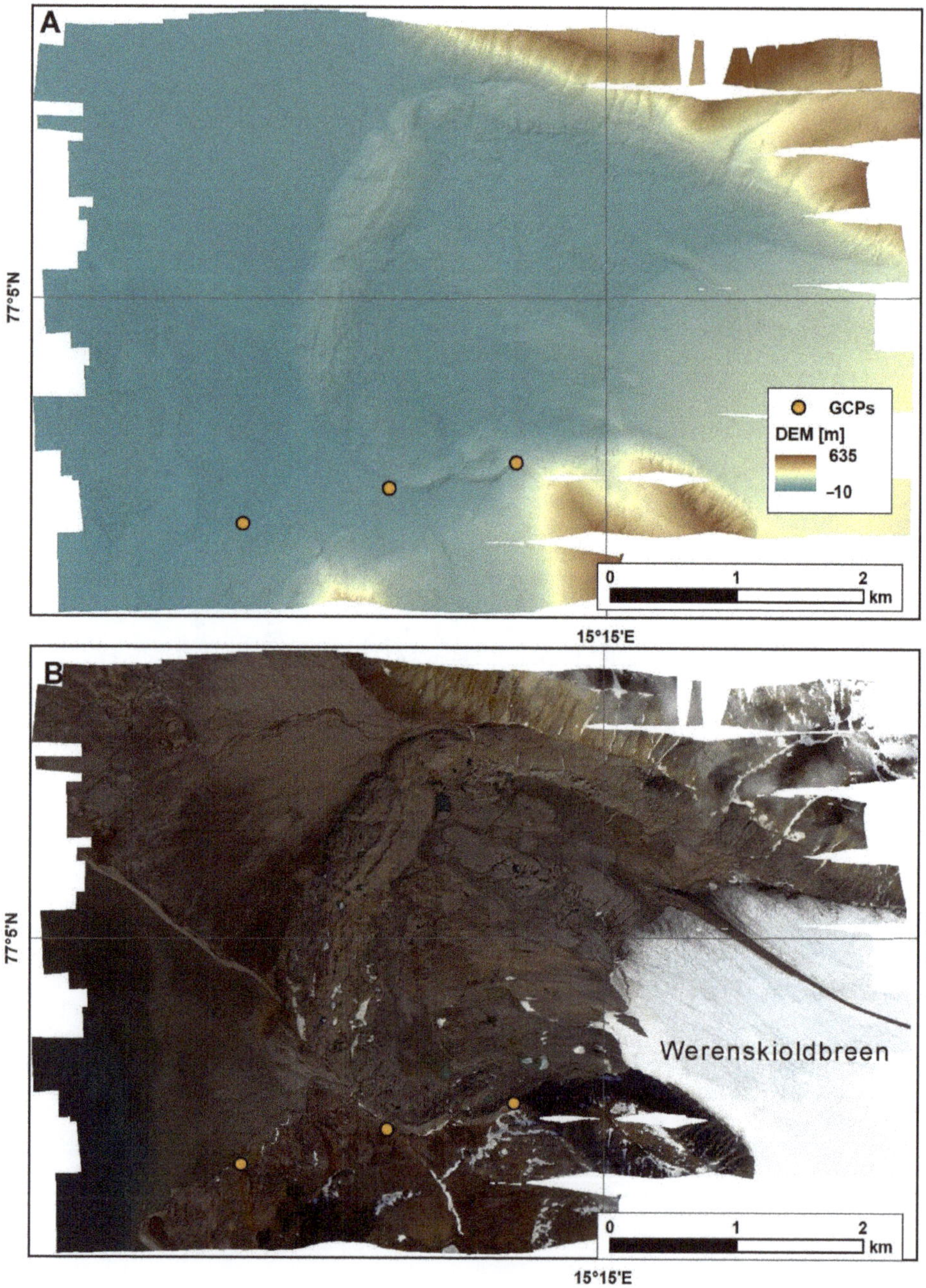

Figure 6. Digital Elevation Model (**A**) and orthomosaic (**B**) for the Werenskioldbreen area. Orange points represent GCPs used in processing in Agisoft.

4.2. TLS

In this study, we only used point clouds within 3 km from the laser scanner position (Figure 7; see Chapter 3.2.2). The four scans were registered using nearly 540,000 plane patches detected throughout the area (Figure 7). The final point cloud had over 117 million points, with a density of around 500–600 thousand points per m^2 in the near-field region of the scans and just a few points per m^2 at the distance of 3 km from the scanner. The RMS error obtained during the relative registration of all four point clouds was 0.09 m.

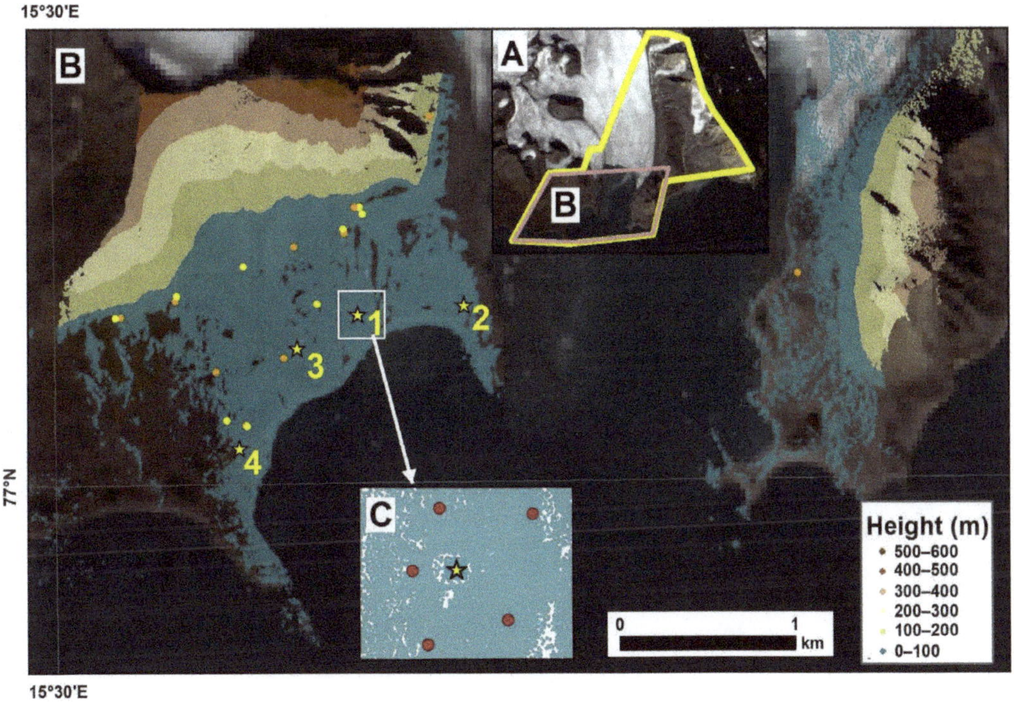

Figure 7. Terrestrial laser scanning data over the study area. The yellow polygon in (**A**) indicates the area covered by all scans, while the brown polygon presents the data range used in the registration procedure. Yellow asterisks in (**B**) illustrate the positions of the laser scanner. Orange and yellow dots represent GCPs and checkpoints that served to estimate the quality of the raster DEM generated from TLS data. Red dots in (**C**) present the example positioning of the RIEGL reflectors regarding the scanner position.

The final point cloud was then georeferenced utilizing 20 RIEGL reflectors (Figure 7C), with an RMS error of 0.13 m. The lowest accuracy in the northerly direction (0.2 to 0.3 m) was noticed for the two lateral scans, numbers 2 and 4. The lowest accuracy in the easterly direction was for scan number 1 (0.21 to 0.25 m). Height accuracy was high for all scans (of order 0.01–0.05 m).

The georeferenced cloud point was next filtered using an Octree algorithm in RiSCAN Pro 2.12.1 with a cell length of 0.16 m and converted into the raster format in ArcGIS (Esri, California) with 0.16 m resolution. As the areas of DEM interpolated in the shadow are characterized by lower accuracy, NoData values were assigned to the raster where holes in the data were present. Therefore, the DEM contains data gaps and keeps information only at spots measured directly by the laser scanner. The vertical quality of the TLS-based DEM was assessed based on the 19 independent points (Figure 7B). The standard deviation equalled 0.31 m and median value −0.19 m. The maximal vertical error of the DEM was 0.93 m.

4.3. Integration of Aerial DEM with TLS

The resulting M3C2-based point cloud distances are presented in Figure 8. Generally, the differences between both point clouds over the surveyed area were in the range of the vertical accuracy of the product (SD equalled 0.14 m and 0.31 m for aerial and TLS data, respectively). The map of surface differences shows local changes ranging from −4 m to 2 m (Figure 8A). The largest recorded differences, up to −4 m, are noted over the gullies and depressions filled by snow in 2020. Therefore, before integrating both DEMs, all areas covered by snow were removed from the aerial-based point cloud (Figure 8B,C). We also documented the erosion of lateral moraine of Hansbreen (around −0.7 m), which consists of an ice core covered with debris, as well as erosion along the shoreline. There are elevation changes of about 1 m concentrated over the mountain slope in the western part of the analyzed area. Numerous large boulders probably affect the quality and accuracy of the TLS-based DEM in that zone. As the scanner was located on marine terraces, around 300 m below, the laser beam was directed significantly upwards, and reflected off the wall of these boulders, rather than the top. Regardless of these local differences, we combined both DEMs into one continuous product. The final dataset (Figure 8D) is composed of a mosaic of two DEMs in raster format. All data gaps and snow areas in the aerial-based DEM were filled with the TLS-based DEM, with a blending option over a two-meter seamline. Moreover, we interpolated DEM values for a few remaining holes, where no data from both campaigns were present and interpolation of the DEM was made under the building and infrastructure of PPS (Figure 8D).

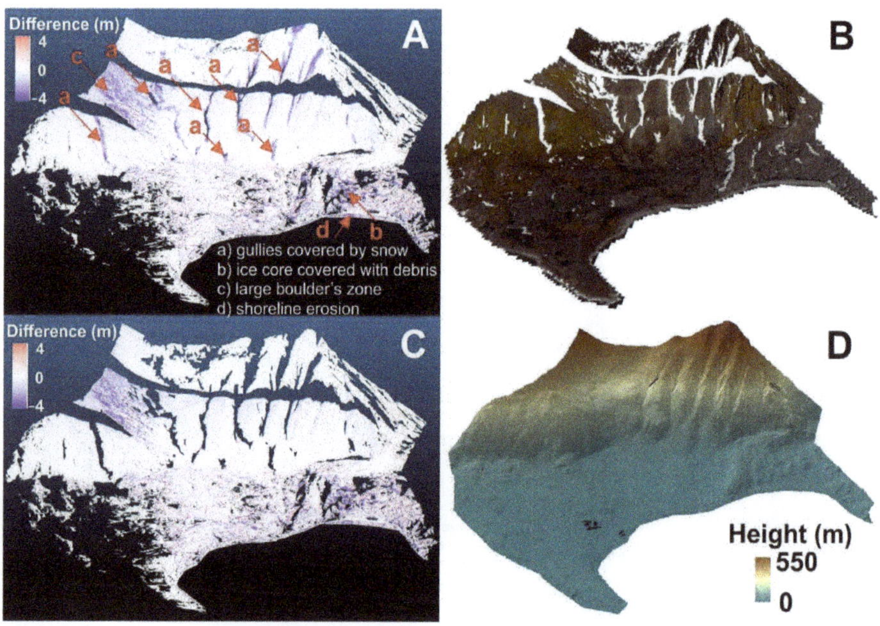

Figure 8. Comparison of the multitemporal point clouds: aerial-based point cloud from images taken in 2020 and TLS-based point cloud collected in 2021. (**A**) M3C2-calculated distance between aerial-based and TLS dataset; (**B**) orthomosaic presenting data gaps and snow cover over the land in 2020; (**C**) vertical difference of the point clouds with snow cover area eliminated from further data integration; (**D**) the final DEM integrated from both DEMs. Black polygons present the areas where DEM was interpolated.

5. Discussion

The two approaches to acquiring the terrain data discussed in this paper are helpful in studying the landform topography and different environmental processes in the Hornsund area. However, both technologies have their limitations. The best vertical accuracy was noted for the aerial-based DEM. However, the limited sidelap of the mission caused data gaps in DEM over mountain slopes. Such settings should be considered when planning subsequent missions in the future, especially when the area of interest is land relief of the mountain regions.

Further, as GCPs and checkpoints were not established prior to the photograph surveys, we chose large, outsize boulders to serve as calibration and validation points [47,48]. Using such existing 'stable' features is a limitation of studies that use photogrammetric methods to produce DEMs [48]. For the first studied zone, Fuglebergsletta, we used enough GCPs to generate the DEM and orthomosaic and assess their accuracy. For the Werenskioldbreen area, only three points would have been used to create the DEM and orthomosaic. Thus, more significant distortions in the northern part of the study must be expected due to the lack of GCPs in that part of the model. Notwithstanding this, we generated the best products with the best resolution and accuracy for the Hornsund area. Further, it is possible to supplement them with better-representing GCPs and recalculate the dataset in the future. We would also like to underline here that regardless of the vertical accuracy of DEMs, the very high resolution of the DEMs gives the possibility to map even the smallest geomorphological features, compute geomorphometric indices and carefully estimate geomorphometric parameters [16].

The accuracy of the TLS-based DEM is slightly lower than the aerial-based product. One of the properties that strongly influenced the long-range TLS was an inhomogeneous point density throughout the area of interest and positional uncertainty due to the laser beam width [22,30,42,46]. Laser footprint size at long 3 km distances equals around 0.36 m. Furthermore, on inclined surfaces the laser footprint becomes deformed to an ellipse [22]. The uncertainty caused by these large footprints are not yet quantified, but is expected to be in the order of decimeters and strongly depends on the distance from the scanner to the surface [27]. Laser scanner data are susceptible to data gaps in locations not in the direct line-of-sight of the scanner, resulting in 'range shadows' which inherently added uncertainty to the derived product [22,38,42,45,49]. Further, the configuration of the scan position and Riegl flat reflectors [30] was limited by terrain conditions, which would increase the error in relative alignment and absolute registration of scans. Considering all these limitations, the final vertical and horizontal accuracy of the TLS-based DEM of order 0.30 m is not unexpected to be in the lower-decimeter range [27] and is the best possible for the complex natural terrain. The DEM at interpolated areas has a quality lower than estimated in the study. In the future, viewshed analysis before TLS fieldwork could remove the shadowing effect [50,51] and improve the accuracy of the data.

The snow cover present in the aerial imagery taken in 2020 and data gaps in both point clouds led us to integrate both datasets [32,52]. Usually, within point-cloud-based strategies, the data acquired in at least two epochs are used to determine the geometric changes between them [30,43,44,52]. Here, apart from the accuracy assessment, we compared and combined the data acquired within one year. The distances between both point clouds were distributed unevenly over the whole area. The differences lower than the vertical accuracy of DEM (SD = 0.31 m) indicated the stable area. The final DEM for Fuglebergsletta was composed of the aerial-based DEM, with data gaps and the snow-covered regions filled by TLS data. One needs to keep in mind that the expected accuracy of such a product is lower than singular point clouds acquired over two years, as the combined DEM comes from the datasets collected in 2020 and 2021. Merging two DEMs from two different years is challenging, as the study area is undergoing continuous changes in land relief. However, temporal changes in most of the area are below the accuracy of both DEMs, and therefore the DEM can be successfully applied to hydrological, glaciological, biological and geomorphometric studies, when observed changes are larger than the accuracy of the

DEM. In other words, the merged DEM that we have generated represents the topography of the region's independent annual surface elevation changes, since the incorporated data were collected during the following two years. In the process of merging two datasets, the annual elevation changes may not reflect in the resulting DEM because of suppression of high-density elevation data recorded from TLS and photogrammetric methods [53,54].

To conclude, the final point clouds, DEMs and orthomosaic are the best terrain products with the highest spatial resolution for the Hornsund area and can be used for further environmental studies. They allowed us to identify local geomorphic processes, such as melting of ice cores at the periglacial area or shoreline erosion. However, for studies aiming to identify changes within a specified time range or slow geomorphometric processes on slopes, we recommend using data from the same types of sources (aerial/TLS) or stemming from a more extended period, such as DEMs generated from aerial imagery taken in 2011 [17]. In addition, we would like to underline here that all the DEMs should be first co-registered [55,56] and the final error must be defined to differentiate between real changes and data noise [22]. This is especially important for the DEM of the Werenskioldbreen area, as only three GCPs located along an almost-straight line were used in the data processing. This suggests that despite the high relative accuracy of the DEM, the absolute accuracy could be lower and decrease with the distance from measured GCPs; therefore, co-registration is needed when comparing with the other data.

6. Conclusions

The Hornsund area is the site in the focus of several studies due to the vicinity of the Polish Polar Station. It is a complex terrain under constant changes, although it lacks repeated datasets with spatial resolution high enough to map even the smallest geomorphological features. In the present study we derived aerial-based DEMs, TLS-based DEMs and orthomosaics, and assessed their accuracy and usefulness for further environmental studies.

Aerial data provided by SIOS were used to generate the DEMs and orthomosaics for the Fuglebergsletta and Werenskioldbreen areas. The products lack gaps over the mountain slopes due to the low sidelap during the flight. Nevertheless, the derived DEM has sufficient quality to study different geomorphometric features. Long-range TLS, despite limitations such as data gaps, complex terrain which limits scan positions and large laser footprint sizes at long distances, can also serve for future analysis of the different geomorphological processes. In particular, the area close to the scanner position with very dense point clouds can be used for future studies of 3D displacements and minor geomorphological features.

This work demonstrates that combining other techniques could be an option, especially in remote polar areas, when data acquisition depends on many factors. In order to remove the aforementioned holes in the datasets from aerial and TLS surveys, we integrated both DEMs over the area close to the Polish Polar Station. Comparison of the point clouds revealed a few small zones under local geomorphic processes, such as the melting of ice cores at the periglacial zone and changes along the shoreline. The differences in both products over the rest of the area were generally lower than the accuracy of both DEMs. This allowed us to combine the data and create the final continuous DEM without gaps for the Fuglebergsletta area. After the proper co-registration, all point clouds and DEMs can be successfully applied in further studies of landform evolution and hydrological, geomorphometric and other processes in the region.

Author Contributions: Conceptualization, M.B.; methodology, M.B., M.L., A.S. and S.D.J.; flight planning, mission execution and aerial data acquisition A.S., S.D.J., M.L. and M.B.; data curation, M.B. and M.L.; writing—original draft preparation, M.B. and M.L.; writing—review and editing, S.D.J. and A.S. All authors have read and agreed to the published version of the manuscript.

Funding: The fieldwork has been financed by the Centre for Polar Studies of the University of Silesia, Poland. The SIOS KC would like to acknowledge the support from the Research Council of Norway (RCN) through projects No. 291644 & No. 269927 and the Norwegian Space Agency (NoSA) through contract No. NIT 12.20.5. The Hyperspectral sensor onboard the Dornier aircraft is funded by the SIOS-InfraNor project (No. 269927).

Informed Consent Statement: Not applicable.

Data Availability Statement: Publicly available datasets were analyzed in this study. This data can be found here: DEM for Fuglebergsletta area from Dornier images: https://ppdb.us.edu.pl/geonetwork/srv/eng/catalog.search#/metadata/69987fe3-f0ed-4fdc-b597-41dd5ae7d21a (accessed on 23 December 2021), Orthomosaic for Fuglebergsletta area from Dornier images: https://ppdb.us.edu.pl/geonetwork/srv/eng/catalog.search#/metadata/9b980572-2bcb-401f-a2e6-2a7a16428dd7 (accessed on 23 December 2021), DEM for Werenskioldbreen area from Dornier images: https://ppdb.us.edu.pl/geonetwork/srv/eng/catalog.search#/metadata/87a556ed-6b8f-4d06-9c20-1a5b6b23d280 (accessed on 23 December 2021), Orthomosaic for Werenskioldbreen area from Dornier images: https://ppdb.us.edu.pl/geonetwork/srv/eng/catalog.search#/metadata/c99085cb-7e4c-43dd-866b-f2f909b44937 (accessed on 23 December 2021), Terrestrial laser scanner (TLS) data for Fuglebergsletta area: https://ppdb.us.edu.pl/geonetwork/srv/eng/catalog.search#/metadata/8bb03c41-35de-46b5-87c1-bd6c82fa2914 (accessed on 23 December 2021), Combined DEM for Fuglebergsletta area from Dornier images and terrestrial laser scanner (TLS): https://ppdb.us.edu.pl/geonetwork/srv/eng/catalog.search#/metadata/eec5dd69-20c0-46c6-aa16-8bb9cece3e85 (accessed on 23 December 2021).

Acknowledgments: The authors would like to acknowledge the Institute of Geophysics PAS and project POLARPOL for enabling the use of terrestrial laser scanner Riegl VZ®-6000. The studies were carried out as part of the scientific activity of the Centre for Polar Studies (University of Silesia in Katowice) with the use of research and logistic equipment (GNSS receivers and supporting equipment) of the Polar Laboratory of the University of Silesia in Katowice. The authors would like to thank Jacek Jania and participants of the polar expedition of the University of Silesia in Katowice 2021 for assistance with the fieldwork. Thanks are directed to the staff of the Institute of Geophysics PAS and the Polish Polar Station, Hornsund for assistance during the field campaign. All the authors acknowledge the SIOS Knowledge Centre (SIOS-KC) for funding and conducting flight campaigns over the study area.

Conflicts of Interest: The authors declare no conflict of interest.

References

1. Grabiec, M.; Puczko, D.; Budzik, T.; Gajek, G. Snow distribution patterns on Svalbard glaciers derived from radio-echo soundings. *Pol. Polar Res.* **2011**, *32*, 393–421. [CrossRef]
2. Grabiec, M.; Jania, J.A.; Puczko, D.; Kolondra, L.; Budzik, T. Surface and bed morphology of Hansbreen, a tidewater glacier in Spitsbergen. *Pol. Polar Res.* **2012**, *33*, 111–138. [CrossRef]
3. Ziaja, W.; Ostafin, K. Landscape–seascape dynamics in the isthmus between Sørkapp Land and the rest of Spitsbergen: Will a new big Arctic island form? *AMBIO* **2015**, *44*, 332–342. [CrossRef]
4. Błaszczyk, M.; Jania, J.A.; Ciepły, M.; Grabiec, M.; Ignatiuk, D.; Kolondra, L.; Kruss, A.; Luks, B.; Moskalik, M.; Pastusiak, T.; et al. Factors Controlling Terminus Position of Hansbreen, a Tidewater Glacier in Svalbard. *J. Geophys. Res. Earth Surf.* **2021**, *126*, e2020JF005763. [CrossRef]
5. Majchrowska, E.; Ignatiuk, D.; Jania, J.; Marszałek, H.; Wąsik, M. Seasonal and interannual variability in runoff from the Werenskioldbreen catchment, Spitsbergen. *Pol. Polar Res.* **2015**, *36*, 197–224. [CrossRef]
6. Stachnik, Ł.; Majchrowska, E.; Yde, J.C.; Nawrot, A.P.; Cichała-Kamrowska, K.; Ignatiuk, D.; Piechota, A. Chemical denudation and the role of sulfide oxidation at Werenskioldbreen, Svalbard. *J. Hydrol.* **2016**, *538*, 177–193. [CrossRef]
7. Wawrzyniak, T.; Osuch, M.; Nawrot, A.; Napiorkowski, J.J. Run-off modelling in an Arctic unglaciated catchment (Fuglebekken, Spitsbergen). *Ann. Glaciol.* **2017**, *58*, 36–46. [CrossRef]
8. Osuch, M.; Wawrzyniak, T.; Nawrot, A. Diagnosis of the hydrology of a small Arctic permafrost catchment using HBV conceptual rainfall-runoff model. *Hydrol. Res.* **2019**, *50*, 459–478. [CrossRef]
9. Kępski, D.; Luks, B.; Migała, K.; Wawrzyniak, T.; Westermann, S.; Wojtuń, B. Terrestrial Remote Sensing of Snowmelt in a Diverse High-Arctic Tundra Environment Using Time-Lapse Imagery. *Remote Sens.* **2017**, *9*, 733. [CrossRef]
10. Dolnicki, P.; Grabiec, M.; Puczko, D.; Gawor, Ł.; Budzik, T.; Klementowski, J. Variability of temperature and thickness of permafrost active layer at coastal sites of Svalbard. *Pol. Polar Res.* **2013**, *34*, 353–374. [CrossRef]
11. Glazer, M.; Dobiński, W.; Marciniak, A.; Majdański, M.; Błaszczyk, M. Spatial distribution and controls of permafrost development in non-glacial Arctic catchment over the Holocene, Fuglebekken, SW Spitsbergen. *Geomorphology* **2020**, *358*, 107128. [CrossRef]
12. Zagórski, P.; Rodzik, J.; Moskalik, M.; Strzelecki, M.C.; Lim, M.; Błaszczyk, M.; Promińska, A.; Kruszewski, G.; Styszyńska, A.; Malczewski, A. Multidecadal (1960–2011) shoreline changes in Isbjørnhamna (Hornsund, Svalbard). *Pol. Polar Res.* **2015**, *36*, 369–390. [CrossRef]

13. Senderak, K.; Kondracka, M.; Gądek, B. Talus slope evolution under the influence of glaciers with the example of slopes near the Hans Glacier, SW Spitsbergen, Norway. *Geomorphology* **2017**, *285*, 225–234. [CrossRef]
14. Strzelecki, M.C.; Szczuciński, W.; Dominiczak, A.; Zagórski, P.; Dudek, J.; Knight, J. New fjords, new coasts, new landscapes: The geomorphology of paraglacial coasts formed after recent glacier retreat in Brepollen (Hornsund, southern Svalbard). *Earth Surf. Process. Landf.* **2020**, *45*, 1325–1334. [CrossRef]
15. Sułowicz, S.; Bondarczuk, K.; Ignatiuk, D.; Jania, J.A.; Piotrowska-Seget, Z. Microbial communities from subglacial water of naled ice bodies in the forefield of Werenskioldbreen, Svalbard. *Sci. Total Environ.* **2020**, *723*, 138025. [CrossRef] [PubMed]
16. Marotta, F.; Teruggi, S.; Achille, C.; Vassena, G.P.; Fassi, F. Integrated Laser Scanner Techniques to Produce High-Resolution DTM of Vegetated Territory. *Remote Sens.* **2021**, *13*, 2504. [CrossRef]
17. Błaszczyk, M.; Ignatiuk, D.; Grabiec, M.; Kolondra, L.; Laska, M.; Decaux, L.; Jania, J.; Berthier, E.; Luks, B.; Barzycka, B.; et al. Quality Assessment and Glaciological Applications of Digital Elevation Models Derived from Space-Borne and Aerial Images over Two Tidewater Glaciers of Southern Spitsbergen. *Remote Sens.* **2019**, *11*, 1121. [CrossRef]
18. Jawak, S.D.; Andersen, B.N.; Pohjola, V.A.; Godøy, Ø.; Hübner, C.; Jennings, I.; Ignatiuk, D.; Holmén, K.; Sivertsen, A.; Hann, R.; et al. SIOS's Earth Observation (EO), Remote Sensing (RS), and Operational Activities in Response to COVID-19. *Remote Sens.* **2021**, *13*, 712. [CrossRef]
19. Buckley, S.J.; Schwarz, E.; Terlaky, V.; Howell, J.A.; Arnott, R.W.C. Terrestrial laser scanning combined with photogrammetry for digital outcrop modelling. *Int. Arch. Photogramm. Remote Sens. Spat. Inf. Sci.* **2009**, *38*, 75–80.
20. Prokop, A.; Delaney, C. A high resolution approach to defining spatial snow height distribution in avalanche release zones for dynamic avalanche modeling. In Proceedings of the ISSW 2010, Lake Tahoe, CA, USA, 17–22 October 2010; pp. 839–845.
21. Sima, A.A.; Buckley, S.J.; Schneider, D.; Howell, J.A. An improved workflow for image- and laser-based virtual geological outcrop modelling. *Int. Arch. Photogramm. Remote Sens. Spat. Inf. Sci.* **2010**, *3*, 115–119.
22. Fey, C.; Wichmann, V. Long-range terrestrial laser scanning for geomorphological change detection in alpine terrain—Handling uncertainties. *Earth Surf. Process. Landf.* **2017**, *42*, 789–802. [CrossRef]
23. Müller, D.; Walter, T.R.; Schöpa, A.; Witt, T.; Steinke, B.; Gudmundsson, M.T.; Dürig, T. High-Resolution Digital Elevation Modeling from TLS and UAV Campaign Reveals Structural Complexity at the 2014/2015 Holuhraun Eruption Site, Iceland. *Front. Earth Sci.* **2017**, *5*, 59. [CrossRef]
24. Xu, C.; Li, Z.; Wang, F.; Li, H.; Wang, W.; Wang, L.I.N. Using an ultra-long-range terrestrial laser scanner to monitor the net mass balance of Urumqi Glacier No. 1, eastern Tien Shan, China, at the monthly scale. *J. Glaciol.* **2017**, *63*, 792–802. [CrossRef]
25. Wang, F.; Xu, C.; Li, Z.-Q.; Anjum, M.N.; Wang, L. Applicability of an ultra-long-range terrestrial laser scanner to monitor the mass balance of Muz Taw Glacier, Sawir Mountains, China. *Sci. Cold Arid Reg.* **2018**, *10*, 47–54.
26. Fischer, M.; Huss, M.; Kummert, M.; Hoelzle, M. Application and validation of long-range terrestrial laser scanning to monitor the mass balance of very small glaciers in the Swiss Alps. *Cryosphere* **2016**, *10*, 1279–1295. [CrossRef]
27. Voordendag, A.B.; Goger, B.; Klug, C.; Prinz, R.; Rutzinger, M.; Kaser, G. Automated and permanent long-range terrestrial laser scanning in a high mountain environment: Setup and first results. *ISPRS Ann. Photogramm. Remote Sens. Spat. Inf. Sci.* **2021**, *2*, 153–160. [CrossRef]
28. Prantl, H.; Nicholson, L.; Sailer, R.; Hanzer, F.; Juen, I.F.; Rastner, P. Glacier Snowline Determination from Terrestrial Laser Scanning Intensity Data. *Geosciences* **2017**, *7*, 60. [CrossRef]
29. LeWinter, A.; Finnegan, D.C.; Hamilton, G.S.; Stearns, L.A.; Gadomski, P.J. Continuous Monitoring of Greenland Outlet Glaciers Using an Autonomous Terrestrial LiDAR Scanning System: Design, Development and Testing at Helheim Glacier. In Proceedings of the AGU Fall Meeting, San Fracnisco, CA, USA, 15–16 December 2014.
30. Pfeiffer, J.; Zieher, T.; Bremer, M.; Wichmann, V.; Rutzinger, M. Derivation of Three-Dimensional Displacement Vectors from Multi-Temporal Long-Range Terrestrial Laser Scanning at the Reissenschuh Landslide (Tyrol, Austria). *Remote Sens.* **2018**, *10*, 1688. [CrossRef]
31. Dietrich, A.; Krautblatter, M. Deciphering controls for debris-flow erosion derived from a LiDAR-recorded extreme event and a calibrated numerical model (Roßbichelbach, Germany). *Earth Surf. Process. Landf.* **2019**, *44*, 1346–1361. [CrossRef]
32. Pagano, M.; Palma, B.; Ruocco, A.; Parise, M. Discontinuity Characterization of Rock Masses through Terrestrial Laser Scanner and Unmanned Aerial Vehicle Techniques Aimed at Slope Stability Assessment. *Appl. Sci.* **2020**, *10*, 2960. [CrossRef]
33. Milenković, M.; Pfeifer, N.; Glira, P. Applying Terrestrial Laser Scanning for Soil Surface Roughness Assessment. *Remote Sens.* **2015**, *7*, 2007. [CrossRef]
34. Błaszczyk, M.; Jania, J.A.; Kolondra, L. Fluctuations of tidewater glaciers in Hornsund Fjord (Southern Svalbard) since the beginning of the 20th century. *Pol. Polar Res.* **2013**, *34*, 327–352. [CrossRef]
35. Jania, J. *Dynamiczne Procesy Glacjalne na Południowym Spitsbergenie (w Świetle Badań Fotointerpretacyjnych i Fotogrametrycznych). (Dynamic Glacial Processes in South Spitsbergen [in Light of Photo Interpretation and Photogrammetric Research]);* Wydawnictwo Uniwersytetu Śląskiego: Katowice, Poland, 1988.
36. Westoby, M.J.; Brasington, J.; Glasser, N.F.; Hambrey, M.J.; Reynolds, J.M. 'Structure-from-Motion' photogrammetry: A low-cost, effective tool for geoscience applications. *Geomorphology* **2012**, *179*, 300–314. [CrossRef]
37. Hann, R.; Altstädter, B.; Betlem, P.; Deja, K.; Dragańska-Deja, K.; Ewertowski, M.; Hartvich, F.; Jonassen, M.; Lampert, A.; Laska, M.; et al. Scientific Applications of Unmanned Vehicles in Svalbard. In *Moreno-Ibáñez et al (eds) SESS Report 2020;* Svalbard Integrated Arctic Earth Observing System: Longyearbyen, Svalbard and Jan Mayen, 2021.

38. Buckley, S.J.; Howell, J.A.; Enge, H.D.; Kurz, T.H. Terrestrial laser scanning in geology: Data acquisition, processing and accuracy considerations. *J.Geol. Soc.* **2008**, *165*, 625. [CrossRef]
39. Riegl. Available online: http://www.riegl.com/nc/products/terrestrial-scanning/produktdetail/product/scanner/33/ (accessed on 16 December 2021).
40. Kersten, T.P.; Mechelke, K.; Lindstaedt, M.; Sternberg, H. Methods for Geometric Accuracy Investigations of Terrestrial Laser Scanning Systems. *Photogramm. Fernerkund. Geoinf.* **2009**, *4*, 301–315. [CrossRef] [PubMed]
41. Buchroithner, M.F.; Gaisecker, T. Ice surface changes in Eisriesenwelt (Salzburg, Austria) based on LIDAR measurements between 2017 and 2020. *Die Höhle* **2020**. Available online: http://www.riegl.com/media-events/projects/terrestrial-scanning/project/manfred-f-buchroithner-thomas-gaisecker-ice-surface-changes-in-eisriesenwelt-salzburg-austria/ (accessed on 23 December 2021).
42. Lague, D.; Brodu, N.; Leroux, J. Accurate 3D comparison of complex topography with terrestrial laser scanner: Application to the Rangitikei canyon (N-Z). *ISPRS J. Photogramm. Remote Sens.* **2013**, *82*, 10–26. [CrossRef]
43. Anders, K.; Marx, S.; Boike, J.; Herfort, B.; Wilcox, E.J.; Langer, M.; Marsh, P.; Höfle, B. Multitemporal terrestrial laser scanning point clouds for thaw subsidence observation at Arctic permafrost monitoring sites. *Earth Surf. Process. Landf.* **2020**, *45*, 1589–1600. [CrossRef]
44. Holst, C.; Janßen, J.; Schmitz, B.; Blome, M.; Dercks, M.; Schoch-Baumann, A.; Blöthe, J.; Schrott, L.; Kuhlmann, H.; Medic, T. Increasing Spatio-Temporal Resolution for Monitoring Alpine Solifluction Using Terrestrial Laser Scanners and 3D Vector Fields. *Remote Sens.* **2021**, *13*, 1192. [CrossRef]
45. O'Banion, M.S.; Olsen, M.J.; Hollenbeck, J.P.; Wright, W.C. Data Gap Classification for Terrestrial Laser Scanning-Derived Digital Elevation Models. *ISPRS Int. J. Geo-Inform.* **2020**, *9*, 749. [CrossRef]
46. Lichti, D.D.; Gordon, S.J. Error propagation in directly georeferenced terrestrial laser scanner point clouds for cultural heritage recording. In Proceedings of the FIG Working Week 2004, Athens, Greece, 22–27 May 2004.
47. Staines, K.E.H.; Carrivick, J.L.; Tweed, F.S.; Evans, A.J.; Russell, A.J.; Jóhannesson, T.; Roberts, M. A multi-dimensional analysis of pro-glacial landscape change at Sólheimajökull, southern Iceland. *Earth Surf. Process. Landf.* **2015**, *40*, 809–822. [CrossRef]
48. Tonkin, T.N.; Midgley, N.G.; Cook, S.J.; Graham, D.J. Ice-cored moraine degradation mapped and quantified using an unmanned aerial vehicle: A case study from a polythermal glacier in Svalbard. *Geomorphology* **2016**, *258*, 1–10. [CrossRef]
49. Nesbit, P.R.; Hugenholtz, C.H. Enhancing UAV–SfM 3D Model Accuracy in High-Relief Landscapes by Incorporating Oblique Images. *Remote Sens.* **2019**, *11*, 239. [CrossRef]
50. Starek, M.J.; Chu, T.; Mitasova, H.; Harmon, R.S. Viewshed simulation and optimization for digital terrain modelling with terrestrial laser scanning. *Int. J. Remote Sens.* **2020**, *41*, 6409–6426. [CrossRef]
51. Heritage, G.L.; Milan, D.J.; Large, A.R.G.; Fuller, I.C. Influence of survey strategy and interpolation model on DEM quality. *Geomorphology* **2009**, *112*, 334–344. [CrossRef]
52. Belmonte, A.; Sankey, T.; Biederman, J.; Bradford, J.; Goetz, S.; Kolb, T. UAV-Based Estimate of Snow Cover Dynamics: Optimizing Semi-Arid Forest Structure for Snow Persistence. *Remote Sens.* **2021**, *13*, 1036. [CrossRef]
53. Jawak, S.D.; Luis, A.J. Synergistic use of multitemporal RAMP, ICESat and GPS to construct an accurate DEM of the Larsemann Hills region, Antarctica. *Adv. Space Res.* **2012**, *50*, 457–470. [CrossRef]
54. Jawak, S.D.; Luis, A.J. Synergetic merging of Cartosat-1 and RAMP to generate improved digital elevation model of Schirmacher Oasis, east Antarctica. *Int. Arch. Photogramm. Remote Sens. Spat. Inf. Sci.* **2014**, *40*, 517. [CrossRef]
55. Berthier, E.; Arnaud, Y.; Kumar, R.; Ahmad, S.; Wagnon, P.; Chevallier, P. Remote sensing estimates of glacier mass balances in the Himachal Pradesh (Western Himalaya, India). *Remote Sens. Environ.* **2007**, *108*, 327–338. [CrossRef]
56. Nuth, C.; Schuler, T.V.; Kohler, J.; Altena, B.; Hagen, J.O. Estimating the long-term calving flux of Kronebreen, Svalbard, from geodetic elevation changes and mass-balance modeling. *J. Glaciol.* **2012**, *58*, 119–133. [CrossRef]

remote sensing

MDPI

Article

Airborne Validation of ICESat-2 ATLAS Data over Crevassed Surfaces and Other Complex Glacial Environments: Results from Experiments of Laser Altimeter and Kinematic GPS Data Collection from a Helicopter over a Surging Arctic Glacier (Negribreen, Svalbard)

Ute C. Herzfeld [1,*], Matthew Lawson [1], Thomas Trantow [1] and Thomas Nylen [2]

1 Geomathematics, Remote Sensing and Cryospheric Sciences Laboratory, Department of Electrical, Computer and Energy Engineering University of Colorado Boulder, Boulder, CO 80309-0574, USA; Matthew.Lawson@colorado.edu (M.L.); Thomas.Trantow@colorado.edu (T.T.)
2 DTU Space, Danish Technical University (DTU), Elektrovej, Building 328, Kgs, 2800 Lyngby, Denmark; thony@space.dtu.dk
* Correspondence: ute.herzfeld@colorado.edu

Citation: Herzfeld, U.C.; Lawson, M.; Trantow, T.; Nylen, T. Airborne Validation of ICESat-2 ATLAS Data over Crevassed Surfaces and Other Complex Glacial Environments: Results from Experiments of Laser Altimeter and Kinematic GPS Data Collection from a Helicopter over a Surging Arctic Glacier (Negribreen, Svalbard). *Remote Sens.* **2022**, *14*, 1185. https://doi.org/10.3390/rs14051185

Academic Editors: Shridhar Jawak, Andreas Kääb, Veijo Pohjola, Hiroyuki Enomoto, Geir Moholdt, Kjell Arild Høgda, Małgorzata Błaszczyk, Bo N. Andersen, Ann Mari Fjæraa, Bartłomiej Luks, Roberto Salzano and Frode Dinessen

Received: 4 October 2021
Accepted: 30 January 2022
Published: 27 February 2022

Abstract: The topic of this paper is the airborne evaluation of ICESat-2 Advanced Topographic Laser Altimeter System (ATLAS) measurement capabilities and surface-height-determination over crevassed glacial terrain, with a focus on the geodetical accuracy of geophysical data collected from a helicopter. To obtain surface heights over crevassed and otherwise complex ice surface, ICESat-2 data are analyzed using the density-dimension algorithm for ice surfaces (DDA-ice), which yields surface heights at the nominal 0.7 m along-track spacing of ATLAS data. As the result of an ongoing surge, Negribreen, Svalbard, provided an ideal situation for the validation objectives in 2018 and 2019, because many different crevasse types and morphologically complex ice surfaces existed in close proximity. Airborne geophysical data, including laser altimeter data (profilometer data at 905 nm frequency), differential Global Positioning System (GPS), Inertial Measurement Unit (IMU) data, on-board-time-lapse imagery and photographs, were collected during two campaigns in summers of 2018 and 2019. Airborne experiment setup, geodetical correction and data processing steps are described here. To date, there is relatively little knowledge of the geodetical accuracy that can be obtained from kinematic data collection from a helicopter. Our study finds that (1) Kinematic GPS data collection with correction in post-processing yields higher accuracies than Real-Time-Kinematic (RTK) data collection. (2) Processing of only the rover data using the Natural Resources Canada Spatial Reference System Precise Point Positioning (CSRS-PPP) software is sufficiently accurate for the sub-satellite validation purpose. (3) Distances between ICESat-2 ground tracks and airborne ground tracks were generally better than 25 m, while distance between predicted and actual ICESat-2 ground track was on the order of 9 m, which allows direct comparison of ice-surface heights and spatial statistical characteristics of crevasses from the satellite and airborne measurements. (4) The Lasertech Universal Laser System (ULS), operated at up to 300 m above ground level, yields full return frequency (400 Hz) and 0.06–0.08 m on-ice along-track spacing of height measurements. (5) Cross-over differences of airborne laser altimeter data are -0.172 ± 2.564 m along straight paths, which implies a precision of approximately 2.6 m for ICESat-2 validation experiments in crevassed terrain. (6) In summary, the comparatively light-weight experiment setup of a suite of small survey equipment mounted on a Eurocopter (Helicopter AS-350) and kinematic GPS data analyzed in post-processing using CSRS-PPP leads to high accuracy repeats of the ICESat-2 tracks. The technical results (1)–(6) indicate that direct comparison of ice-surface heights and crevasse depths from the ICESat-2 and airborne laser altimeter data is warranted. Numerical evaluation of height comparisons utilizes spatial surface roughness measures. The final result of the validation is that ICESat-2 ATLAS data, analyzed with the DDA-ice, facilitate surface-height determination over crevassed terrain, in good agreement with airborne data, including spatial characteristics, such as surface roughness, crevasse spacing and depth, which are key informants on the deformation and dynamics of a glacier during surge.

Keywords: ICESat-2; laser altimetry; kinematic GPS experiments; glaciology; surge glaciers; svalbard; density dimension algorithm for ice surfaces; airborne validation of satellite data

1. Introduction

With the Advanced Topographic Laser Altimeter System (ATLAS), NASA's ICESat-2, launched on 15 September 2018, carries the first space-borne multi-beam micro-pulse photon-counting laser altimeter system [1,2]. The ATLAS system records returns from every single photon in the 532 nm range of the sensor, which facilitates determination of along-track surface heights at the 0.7 m resolution of the sensor (under clear-sky atmospheric conditions), using the Density Dimension Algorithm for ice surfaces (DDA-ice) [3]. This capability of high-density measurements constitutes a better than 200-fold improvement in resolution over the 173-m spacing of the Geoscience Laser Altimeter System (GLAS) of the ICESat Mission, which operated from 2003 to 2009 [4,5].

Such increased density of measurements requires a field validation with high geodetical accuracy to establish that the surface heights determined from ICESat-2 accurately represent the actual surface heights and morphologies of an ice surface, especially for the complex morphology of heavily crevassed ice surfaces. Capturing crevassed surfaces accurately is important, because crevassing occurs as a sign of fast-moving and accelerating glaciers. Glacial acceleration is one of the largest sources of uncertainty in sea-level-rise assessment, according to the Fifth Assessment Report (AR5) of the Intergovernmental Panel of Climate Change (IPCC) [6]. The different spatial characteristics of crevasse fields, including crevasse spacing, depth and surface roughness, yield information on the deformation characteristics and ice dynamics during glacial acceleration [7–11].

To assess the surface-height determination capabilities of ICESat-2 ATLAS, we carried out an airborne geophysical evaluation campaign over Negribreen, Svalbard, during surge in summer 2019. The objective of this paper is a geodetical and glaciological evaluation using laser altimeter, image and geodetical data collected during this campaign. Negribreen accelerated to more than 200 times its quiescent-phase ice speed in July 2017, with a maximal speed of 22 m/day at the height of the surge acceleration [3]. The glacier's rapid mass transfer to the Arctic Ocean during the surge provides a prime example of the relationship between dramatic mass loss through calving and the surge process [12]. A surge in a Svalbard glacier typically lasts 7–10 years [13]. Maximal velocities in summer 2019 were still around 7 = 6 m/day, based on our analysis of Sentinel-1 Synthetic Aperture Radar (SAR) data.

Negribreen during surge is an ideal validation region, because many types of crevasses and other complex surface types exist in close proximity (Figure 1). This heavily crevassed environment provides a set of challenges for surface-height measurement and surface-height determination, whose solution will indicate that surface heights can be determined accurately in other crevassed and smooth glaciated regions as well. Because crevasses are relatively small features, a high geodetical accuracy and a close repeat of satellite track by airborne tracks is important, both will be analyzed in this paper. As there is not much literature about kinematic GPS data collection from a helicopter for the purpose of satellite data evaluation, the mathematical geodetical results of this paper are expected to be of use for future airborne experiments.

Figure 1. Negribreen during surge in 2017–2019. The multitude of different surface types indicates that a glacier during surge provides an ideal test situation for evaluation of height determination over crevassed and other complex ice surfaces. Photographs by Ute Herzfeld. (**a**) Negribreen surge July 2017, overview. (**b**) Wavy crevasses and folded moraines in central Negribreen, July 2017. (**c**) Several different crevasse provinces in central Negribreen, July 2017. (**d**) Calving front, ice fingers advancing into the ocean (front right), with smooth surface of non-surging Ordonannsbreen in the back left, July 2017. (**e**) Crevasse province near calving front July 2018. (**f**) Blocky crevasses near RGT 594 region, July 2018. (**g**) Ordonannsbreen-Negribreen boundary near calving front, August 2019. (**h**) Folded moraines and surge crevasses, August 2019.

The precision of ICESat-2 ATL06 (version 003) [14,15] ice surface height data has been determined as better than 7.2 cm, with a bias of less than 3.3 cm for smooth and flat ice/snow surfaces at 88° Southern latitude in Antarctica, which is the circle of convergence of the ICESat-2 ground tracks [16]. Ref [17] conducted a field evaluation along the Chinese Antarctic Expedition Route (CHINARE, a route driven with a Pisten Bully) near Amery Ice Shelf, using GNSS, in December 2019 to February 2020, finding that ICESat-2 ATL06 (version 003 data) are precise to 1.5 cm with 9.1 cm precision and surface heights of ATL03 photon events [18,19] are accurate to 4.3 cm with 8.5 cm precision. Since this is a route that was driven with a Pisten Bully, it is all smooth, uncrevassed ice. However, the ICESat-2 ATL06 data product is a 40 m along-track ice-surface height product with 20 m postings, which generally does not report crevassed and otherwise morphologically complex glaciated terrain, and the photon classification on the ATL03 product often fails over such terrain [3]. The study in this paper complements such determinations of measurement accuracy over smooth and flat ice surfaces by investigations of surface-height determination over crevassed and otherwise complex regions, derived using the DDA-ice. The surface-height-determination problem has been addressed in [3], here, we focus on the geodetical component of ICESat-2 validation campaign data over crevassed terrain.

In this paper, we describe the ICESat-2 ATLAS data, experimental and instrumental setup of the airborne geophysical field campaigns, the process for geodetical correction of the airborne data using GPS/GNSS data and IMU data, and the resultant accuracies. Results will include (1) the capability to repeat satellite tracks during a helicopter-based airborne validation campaign, (2) the type of geodetical corrections and associated accuracy, (3) spatial characteristics of airborne data (on-ice spacing of resultant altimeter data), (4) determination of cross-over heights from airborne laser altimeter data after correction as an indicator of vertical precision of the field data, (5) the resultant accuracy of surface-height determination and variability, derived from ICESat-2 ATLAS data using the DDA-ice, and (6) the resultant ability to derive spatial characterizations of the morphological properties of the ice surface, exemplified by a surface-roughness parameter.

2. ICESat-2 ATLAS Data

2.1. ATLAS Instrumentation, Data Collection Geometry, Basic Corrections and ATL03 Data Product

ICESat-2 ATLAS data are collected along-track, for 3 pairs of 2 beams, a strong beam and a weak beam per pair, where the weak beam has a quarter of the energy of the strong beam [1,2]. Across-track separation of the beams on the Earth's surface is 3.3 km between the centers of adjacent pairs and 90 m for the beams within each pair; ICESat-2 beam pattern and track geometry are illustrated in Figure 2. The sensor operates in the 532 nm wavelength (green light) with a pulse-repetition rate (PRF) of 10 kHz. The PRF results in a nominal 0.7 m spacing of the laser pulses on the Earth's surface, under clear-sky atmospheric conditions. ATLAS has a footprint diameter of less than 17.4 m at 85% encircled energy. The Field of View (FOV) of the receiver telescope is 83.8 μrad, equivalent to 45 m FOV on the surface of the Earth. Observatory performance has been assessed in [20], where the actual footprint is characterized as closer to 10 m in diameter. As the FOV moves along the satellite ground track and returns from every single photon (in the 532 nm wavelength domain of the sensor) are recorded, surface structures at much higher resolution than footprint diameter can be resolved (see Section 2.2 and [3]). The instrumentation and derivation of the recorded photon point cloud are described in [2], and any technical component of the instrument and data set necessary for understanding the work in this paper is found in [3].

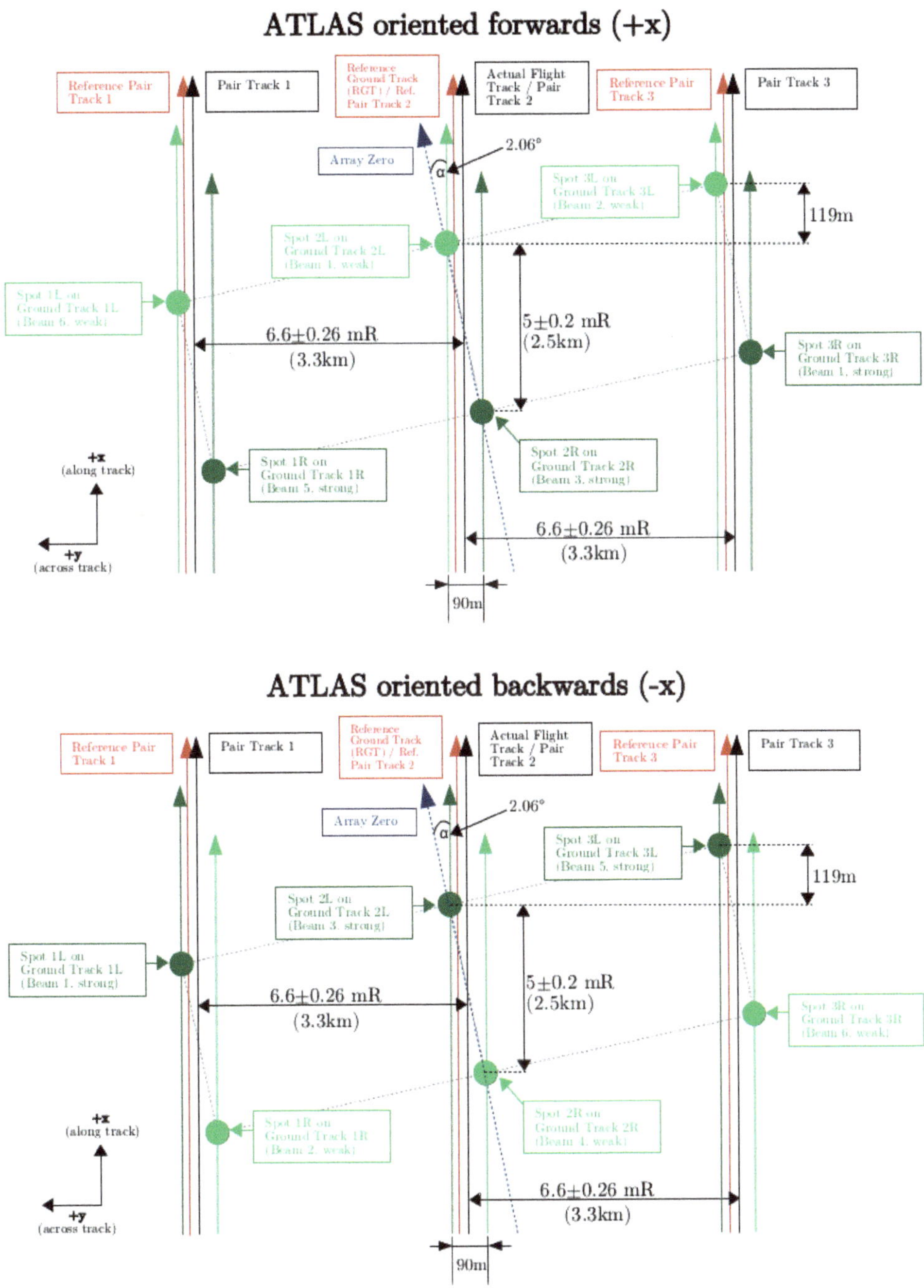

Figure 2. ICESat-2 beam pattern and track geometry. For observatory orientations of ATLAS flying forward ("weak beams leading") and ATLAS flying backward ("strong beams leading"). Beam numbering is a feature of the transmitter array, while spot and ground track numbering is a feature of the ATLAS data products, see Table 1. (Figure from [3], Figure 3).

ATLAS records returns from every photon in the 532 nm wavelength domain of the sensor. This results in a so-called photon point cloud. The photon point cloud includes photon events from ambient light (background) and photon events that results from the active laser signal of the instrument; these types of photons are initially not distinguishable. Classification of photons into signal and background requires a sophisticated mathematical algorithm, since the classification problem is mathematically ill-posed. For the solution of the classification problem and the determination of surface heights, we apply the density-dimension algorithm for ice surfaces, the DDA-ice [3,21], briefly introduced in Section 2.2. Data in this paper stem from the ICESat-2 ATL03 data product (version 003) [19], described fully in the Algorithm Theoretical Base Document (ATBD) for Geolocated Photon Points [18]. There is a classification of photons provided on ATL03, however, the work in our paper utilizes the classification based on the DDA-ice (the ATL03 classification often fails over crevassed surfaces). Note that technically, the data sets (ICESat-2 ATLAS data products) are referred to by version number (here: version 3), while the algorithms used to create the data products are referred to by release number (here: release 3). Because the geodetical corrections have not changed significantly in a way that would affect the ATL03 data geolocation and especially the representation of crevasse characteristics, the field validation presented in this paper is expected to be valid for subsequent releases as well (to release 5, the last release at publication date of this paper) [22].

Of relevance for the airborne validation campaign of the ICESat-2 ATLAS data is the concept of Reference Ground Tracks (RGTs). ATLAS ground tracks are referenced to the RGTs, which are calculated ahead of time, prior to the mission. There are a total of 1392 RGTs. Actual ATLAS ground tracks are derived after data collection and following detailed geolocation.

The ICESat-2 beam pattern and track geometry depend on the orientation of the observatory, which can be flown in two modes, forwards (+x) and backwards (−x) (Figure 2). All data in this paper were collected during an "ATLAS oriented backwards" phase in July and August 2019. The orientation of the observatory changes which ground track stems from a weak beam and which one from a strong beam (see, Table 1). Therefore, for each track used in this paper "strong beam" or "weak beam" are specified (tracks are labeled gt1r, gt1l, gt2r, gt2l, gt3r, gt3l where the number stands for the beam and the letter for "left" or "right". Beam 2 is the center beam and the RGT splits the distance between gt2r and gt2l by design. Reference pair tracks split the distance between (gt1r, gt1l) and (gt3r, gt3l). Both weak beam and strong beam data were evaluated during our validation campaign. For shortness, we use the terminology "predicted ATLAS tracks" for ground tracks calculated ahead of time from the RGTs and "actual ATLAS tracks" for ground tracks derived from ATLAS observations, corrected as summarized in the next paragraph.

After data collection, the geolocation process results in the association of latitude, longitude and height for each telemetered photon event in the point cloud. It uses Precision Pointing Determination (PPD) of the laser and Precision Orbit Determination (POD). ATLAS carries a Laser Reference System (LRS), two GPS antennas and Spacecraft Star Trackers. The process is described in short in [3], in more detail in [2] and in full in the geodetical part of the ATL03 ATBD for Receive Photon Geolocation ATL03g [22], see also [23]. The location of the "bounce point" of the laser altimeter on the ice surface is given in the International Terrestrial Reference Frame (ITRF, Petit and Luzum [24]) ITRF 2014 (ITRF14) as latitude, longitude and height w.r.t. the World Geodetical System WGS84 in realization G1150 with ellipsoid constants (a_e = 6,378,137 m, $\frac{1}{f}$ = 298.257223563) [2]. We will utilize the same referencing for the airborne laser altimeter data of our campaign. After geolocation, several geophysical corrections are applied to the photon points [2,3].

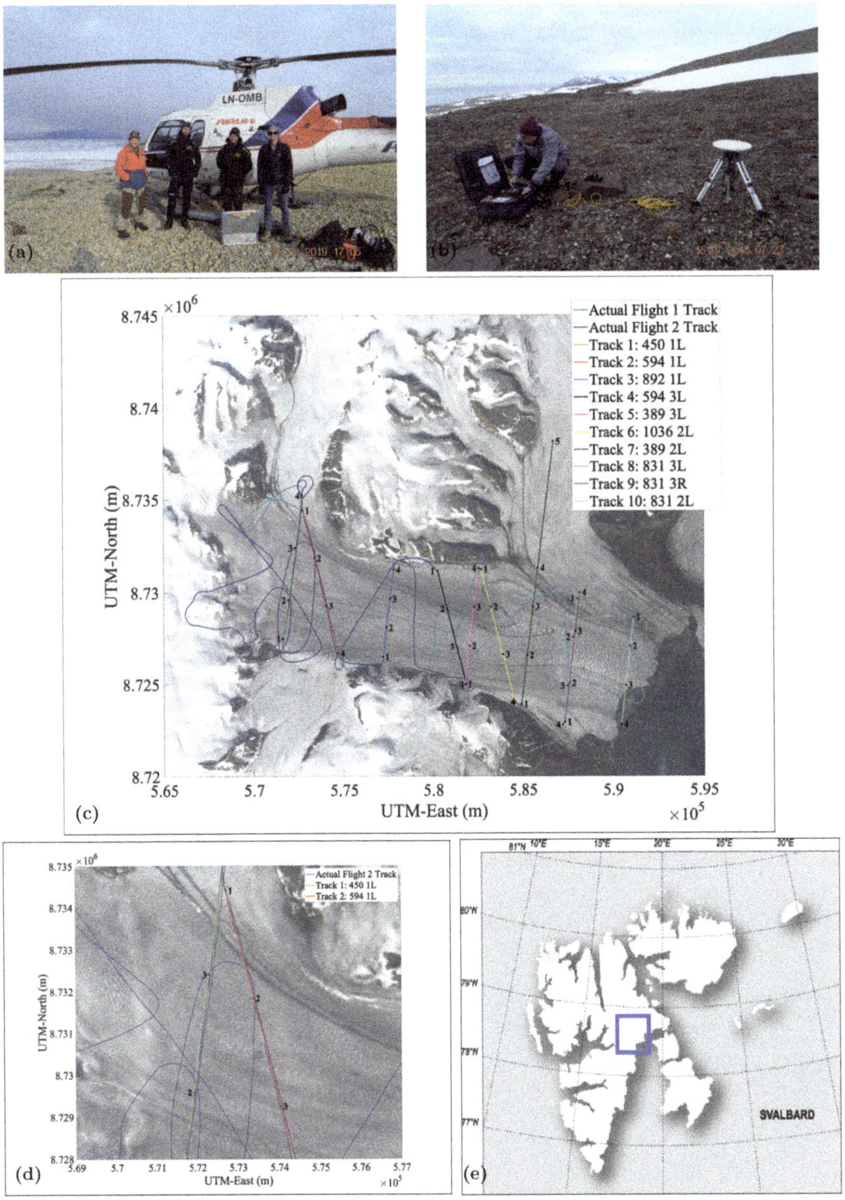

Figure 3. Negribreen ICESat-2 Airborne Validation Campaign 2019: Experiment setup, track maps, location map and satellite image. (**a**) Field team 2019 (f.l.t.r. Ute Herzfeld, Harald Sandal (pilot) and Gustav Svanström (technician), Matthew Lawson, in front of the Airlift/ NPI helicopter; photograph by Thomas Trantow), on the side of Negribreen, flight 1, 12 August 2019, (**b**) GPS base station on a ridge above Negribreen, 2018 (Connor Myers), photograph by Ute Herzfeld. (**c**) Planned and actual flight tracks and predicted ICESat-2 ATLAS ground tracks. Flight 1, 12 August 2019; Flight 2: 13 August 2019. Background Landsat-8 data (5 August 2019) [LC08_L1TP_215003_20190805_20190820_01_T1]. (**d**) Zoom into (**c**) for area of RGT 594 and RGT 450 analyzed in this paper. (**e**) Location map. (Figure panels (**c**–**e**) after [3]).

Table 1. Identifiers for ATLAS orientations. (After [3], Table 1).

ATLAS Flying Forwards							
PCE	**Strength**	**ATLAS Spot**	**Pair**	**GT**	**ATL03 Beam**	**ATL04/09 Atmosphere**	**Orientation**
1	Strong	1	Right	gt3r	5	profile_3	1
1	Weak	2	Left	gt3l	6	profile_3	1
2	Strong	3	Right	gt2r	3	profile_2	1
2	Weak	4	Left	gt2l	4	profile_2	1
3	Strong	5	Right	gt1r	1	profile_1	1
3	Weak	6	Left	gt1l	2	profile_1	1
ATLAS Flying Backwards							
PCE	**Strength**	**ATLAS Spot**	**Pair**	**GT**	**ATL03 Beam**	**ATL04/09 Atmosphere**	**Orientation**
1	Strong	1	Right	gt1l	5	profile_1	0
1	Weak	2	Left	gt1r	6	profile_1	0
2	Strong	3	Right	gt2l	3	profile_2	0
2	Weak	4	Left	gt2r	4	profile_2	0
3	Strong	5	Right	gt3l	1	profile_3	0
3	Weak	6	Left	gt3r	2	profile_3	0

2.2. The Density-Dimension Algorithm for Ice-Surface-Height Determination (DDA-Ice) in Crevassed and Other Complex Terrain—Description in a Nutshell

The analysis of the ATLAS data utilizes the Density-Dimension Algorithm for ice surfaces (DDA-ice) [3,21]. The DDA-ice is an algorithm specifically developed for surface-height determination from micro-pulse photon-counting laser altimeter data, especially those of the ICESat-2 ATLAS instrument and similar sensors. The DDA-ice is introduced in [21] for data from the Slope Imaging Multi-polarization Photon-counting Lidar (SIMPL) [25,26], an airborne micro-pulse photon-counting laser altimeter system that was employed as an ICESat-2 predecessor instrument in test flights, and has been first applied to ICESat-2 ATLAS data in [3]. The data from the ATLAS instrument have the form of a point cloud of geolocated photons [2]. The DDA-ice utilizes a Gaussian radial basis function (*rbf*) for data aggregation. The *rbf* is evaluated for every single photon in the point cloud; this operation results in a density field. In the next step, an auto-adaptive threshold function is applied to separate signal and background photons; this function automatically adapts to the different background properties of daytime, twilight and nighttime, apparent surface reflectivity (ASR) and some instrument characteristics, such as the afterpulse of the lidar signal. The algorithm is termed density-dimension algorithm, because density is used as an additional dimension in the signal-background classification step. The resultant surface heights have the same resolution as the sensor, which is nominally 0.7 m along-track under clear-sky atmospheric conditions. The DDA-ice uses its own cloud flag algorithm (it does not rely on the cloud flag in the atmospheric data product ATL04 [27,28]). The last step of the DDA-ice is the ground follower. The ground follower is a piece-wise linear interpolator, whose segmentation automatically adapts to surface roughness, resulting in 2.5 m spacing for crevassed (or generally, rough) surfaces and 5 m for uncrevassed (or generally, smooth) surfaces. The DDA-ice is controlled by a set of so-called algorithm-specific parameters, introduced to facilitate quick adjustment of the algorithm should the performance characteristics of the ATLAS lasers change. The values of the algorithm-specific parameters employed in the analysis in this study are given in Table 2. Input data for the DDA-ice are geolocated photons, as given in the publicly available ICESat-2 ATLAS data product ATL03 "Global Geolocated Photons", version 3 [19], described in Neumann et al. [2,29]. The

DDA-ice does not utilize the signal classification given in ATL03, but signal-background classification is an intrinsic step of the DDA-ice (for a comparison, see [3]).

Table 2. DDA-ice parameters. Strong beam, weak beam—ATLAS beams. Parameters are the same as used in [3], parameters used for strong beam are the same as parameters optimized for analysis of SIMPL_green (532nm parallel polarized) channels. Variables Q and S are now algorithm-specific parameters for DDA-ice. Resolution of the ground follower is 2.5 m over rough/ crevassed surfaces and 5 m over smooth surfaces. See also [21].

Symbol	Meaning	Strong Beam ATLAS (Actual)	Weak Beam ATLAS (Actual)
s	standard deviation (m)	3	4
u	cutoff (number of standard deviations)	1	1
a	anisotropy	1	1
q	threshold quantile	0.5	0.6
k	threshold bias offset	1	0.2
l	slab thickness (m)	200	200
R	resolution of ground follower (m)	5	5
r	factor to reduce the R parameter	2	1
-	resolution of ground follower for rough surfaces (m)	2.5	5
Q	crevasse depth quantile	0.5	0.5
S	standard deviation threshold of thresholded signal to trigger small step size in ground follower (m)	1.75	1.75

3. Experiment Setup for the Negribreen ICESat-2 Airborne Geophysical Validation Campaign

3.1. Geographic Region and Ice-Surface Types during Surge

The Negribreen Glacier System (NGS) is located on Spitsbergen in Svalbard, it forms an outlet of the Filchnerfønna ice cap (see, Figure 3). The calving front of Negribreen was located at approximately (78.573° N, 19.083° E) in 2019. An overview of main areas of the Negribreen Glacier System, including Negribreen (in center), Ordonnansbreen (the glacier that joins Negribreen close to the calving front) and Akademikerbreen (the glacier that joins upper Negribreen from the North) is seen in the background Landsat-8 image in Figure 3c.

Ice surface types during the surge include many different crevasse classes (Figure 1). Figure 1a shows smooth ice of Ordonannsbreen contrasting with the highly fractured ice of Negribreen, which calves into the bay, part of the Arctic Ocean. Wavy crevasses and folded moraines, the indicators that a glacier is a surge-type glacier, are seen in Figure 1b. Figure 1c illustrates several types of crevasse provinces, delineated from neighboring provinces by relatively clear boundaries. Heavily crevassed ice that advances rapidly into the Arctic Ocean indicates the rapid mass loss during the surge in July 2017 (Figure 1d) and July 2018 (Figure 1e). Figure 1f shows a typical surge crevasse type, characterized by blocky crevasses with subordinate, thin multidirectional crevasses in the region of RGT594. The surface of slow-moving Ordonannsbreen (left) is only crevassed near the boundary to surging Negribreen and otherwise smooth (Figure 1g). Lastly, Figure 1h presents a picture of multi-generational crevasses resultant from several deformation events in an area where the folded moraines exist, taken in August 2019. While maximal ice speeds during the height of acceleration in July 2017 reached 22 m per day, the region of crevassed ice expanded during the evolution of the surge and the maximal speeds decreased. Ice-surface velocity magnitudes derived from analysis of Sentinel-1 SAR data in August of 2019 were about 6 m (5.96 m) near the front of the glacier and 1–2 m in the region of the surveys analyzed in this paper (Figure 4a). High-resolution satellite imagery from Planet SkySat, collected on 2019, August 18, illustrates the type of complex crevassing that exists near the ice front, in the fastest-moving region of the glacier, over 3 years after the start of the surge in spring 2016 (Figure 4b).

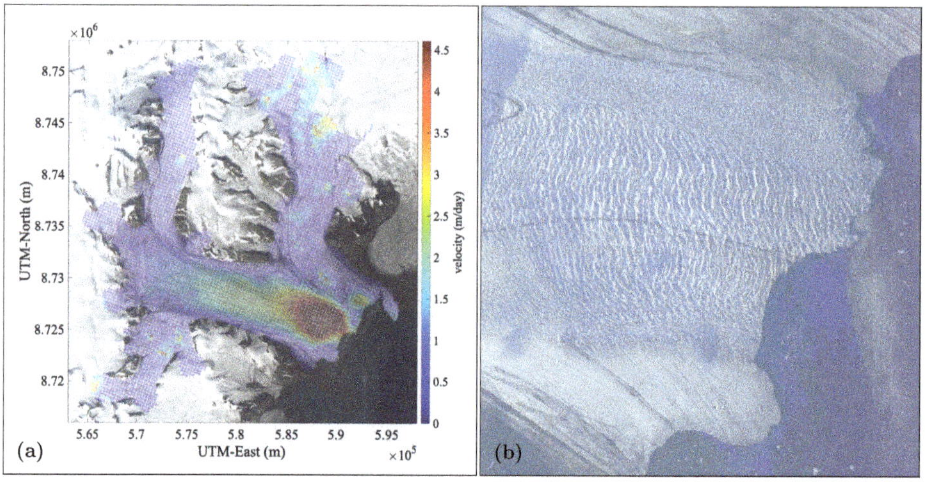

Figure 4. Negribreen velocity magnitude and crevassing at time of validation campaign in August 2019. (**a**) Velocity magnitude, calculated from ESA Copernicus Mission Sentinel-1 SAR data for 9 August 2019 to 21 August 2019. Highest velocity magnitude was 5.96 m/day near the calving front. (**b**) Crevasse types in the fastest-moving region near the front, Planet SkySat image, collected 18 August 2019. [20190818_150858_ssc9_u0002_panchromatic_dn.tif]. (Figure panel (**b**) after [3]).

3.2. Experiment Setup

We conducted airborne geophysical observation campaigns over Negribreen during surge in the summers of 2017, 2018 and 2019. The campaigns over Negribreen had two objectives, (1) study of the state of the surge in the Negribreen Glacier System and (2) validation of ICESat-2 data. The 2018 campaign was carried out in preparation of ICESat-2 validation, prior to launch of ICESat-2 (which was on 15 September 2018), while the 2019 campaign obtained synoptical airborne and satellite observations. Airborne data were collected from a helicopter (Eurocopter AS 350), operated by Airlift under the auspices of the Norwegian Polar Institute, chartered for our campaign by the University of Colorado Geomathematics Group (Figure 3a).

During the 2019 field validation campaign, airborne geophysical data were collected along predicted ICESat-2 ATLAS ground tracks (see, Figure 3c) on 12 August 2019 (Flight 1) and 13 August 2019 (Flight 2). The locations of ground tracks from our flight campaign are referred to as flight tracks, or flight ground tracks (with coordinates calculated as described in this paper). The analysis in this paper utilizes predicted ATLAS ground tracks, actual ATLAS ground tracks and flight ground tracks. A zoom into Figure 3c, seen in Figure 3d, shows the tracks along RGT 450 and RGT 594, which will be analyzed in this paper, their relative location and crevasse types surveyed. Planet SkySat satellite image data, specifically collected to support this project, provide additional information suitable for evaluation of ICESat-2 measurements. Figure 3f shows crevasse provinces near the calving front in a Planet SkySat image taken on 18 August 2019. Extensional wavy crevasses, indicative of the area of highest velocities at time of image capture, dominate the northern central part of Negribreen, north of the trace of a medial moraine in upstream Negribreen. The fast-moving region is flanked by shear crevasses in its north and south, then (in an outward direction from the center) more closely spaced crevasses in the north, followed by smooth, slow-moving ice provinces in the north and south.

3.3. Instrumentation

During the 2018 and 2019 airborne geophysical campaigns, we collected laser altimeter data and image data, which were geolocated using coincidently collected Global Positioning System (GPS) and Inertial Measurement Unit (IMU) data. The same instruments were used in both years, except for the GPS Rover, and included the following:

(1) *Laser altimetry:* Universal Laser System (ULS), LaserTech Inc. (Denver, Colorado, U.S.A.), mounted on helicopter (400 Hz data);
(2) *Global Positioning System (GPS) Receivers (Base and Rover):*
    (2.1) Kinematic GPS
        (2.1.1) GPS Base Station placed on the side of Negribreen (Trimble NetR9)
        (2.1.2) GPS Rover, mounted on a skid of helicopter (Septentrio APS-3G in 2018, Trimble R10 in 2019) (10 Hz data)
    (2.2) "Button GPS", a simple GPS system placed on the dashboard of the helicopter under the window (1 Hz data)
(3) *Attitude correction:* Inertial Measurement Unit (IMU), LORD 3DM-GX5-15 Vertical Reference Unit (mounted inside the helicopter, taped to the floor) (1 Hz data)
(4) *Nadir-pointed time-lapse imagery:* GoPro Hero5 Camera, mounted outside on the helicopter (2 Hz data)
(5) *Photographs:* Several handheld cameras (NIKON D5100 and other Nikon cameras)

(1) Laser Altimetry. The LaserTech ULS is a class-1, eye-safe laser that operates at 905 nm (Near-Infrared) with a configurable pulse-repetition frequency (PRF), set to 4000 Hz with 10-point averaging to yield an effective PRF of 400 Hz during the Negribreen campaigns. Beam divergence is 3 mrad, which results in a 30 cm footprint at 100 m range (see, Table 3). The ULS was integrated by members of the University of Colorado Geomathematics, Remote Sensing and Cryospheric Sciences Laboratory (the first three authors) for altimetry use from small manned and unmanned aircraft, utilizing a so-called engineering kit provided by LaserTech. An earlier integration is described in [30], for use aboard the NASA Sierra UAV during the Characterization of Arctic Sea Ice Experiment (CASIE) in 2009. Findings from geostatistical analysis of the CASIE data are reported and related to sea-ice modeling in [31].

(2) GPS data. Two separate GPS units were used for redundancy. The first, called "Button GPS", is a low cost, plug-and-play type used for making quick, low-accuracy plots and serves as a backup for the second system. Button GPS was placed on the dashboard of the helicopter, where it had good sky view through the large front window. The second system is a high-cost high-accuracy two-part GPS system, which consists of a base station reference receiver and a helicopter mounted kinematic GPS unit (rover). The base station collects observations including atmospheric delay parameters and orbit information that allow for post-processing that increases the position accuracy of the kinematic GPS unit mounted on the helicopter. Sky view of the base station was unobstructed, because the base station was positioned on an elevated moraine edge or near a cliff edge on the southern side of Negribreen, several hundred meters above the glacier (Figure 3a,b). There were no large mountains obstructing the sky view of the base station. The rover was mounted on one of the skids of the helicopter, and thus the body of the helicopter and the helicopter rotors possibly interfered with sky view of the rover. The instrumentation for the differential GPS data collection was provided by UNAVCO in both years.

(3) IMU data. IMU data record the attitude of the helicopter during the flights. These data are collected to allow correction for attitude and vibration. The IMU is also owned by our Geomathematics Laboratory and integrated into the survey system. Our post-processing includes coincidently collected IMU data to remove helicopter vibration with a low-pass filter, and to correct for roll, pitch, and yaw with a transformation matrix, taking into account the spatial offset between the mounting positions of the laser altimeter system and the GPS rover on the helicopter.

Table 3. Technical specifications of the Universal Laser System (ULS) and settings during the Negribreen airborne geophysical observation and validation campaigns in 2017, 2018 and 2019. FDA—United States of America, Federal Food and Drug Administration, CFR—Code of Federal Regulations Title 21, see https://www.accessdata.fda.gov/scripts/cdrh/cfdocs/cfcfr/cfrsearch.cfm?FR=1040.10 (accessed on 3 October 2021).

Manufacturer	Laser Technology, Inc.
Model	Universal Laser System (ULS)
Wavelength	905 nm (IR)
Beam Divergence	3 mrad (~30 cm diameter footprint at 100 m range)
Exit Aperture	42 mm
Pulse Repetition Frequency (PRF)	10–4500 Hz, configurable, 4000 Hz for Negribreen campaign
Averaging	10 points for Negribreen campaign (400 Hz effective PRF)
Maximum Range Over Ice Targets	500 m
Eye Safety	Class 1, 7mm eye safe (FDA,CFR21)
Accuracy	±2 cm
Resolution	1 mm
Dimensions	5.3 in L × 4.75 in W × 2.5 in H
Weight	1.75 lbs (0.8 kg)

(4) GoPro time-lapse-imagery was recorded at two images per second in linear imaging mode. The linear mode was selected, because it yields the least distortion of the imagery. Time-lapse imagery is better suited for comparison with discrete altimeter data than video data, results in much smaller data volumes and reduces battery energy usage.

(5) Photographs were taken using handheld cameras to provide overviews of the glacier system during surge, especially from high altitudes above ground level (a.g.l.), to collect documentation of crevasse provinces, calving events and other features of special interest to the study of the surge process. Several thousand images were taken during each flight.

4. Comparison of Different Approaches to GPS-Data Collection and Processing: Real-Time Kinematic (RTK) Compared to Kinematic GPS Data Collection with Differential Correction Using Base and Rover Data

4.1. Scientific Motivation and Experiments

In preparation of the ICESat-2 launch (15 September 2018), in summer 2018 we carried out experiments aimed at optimizing GPS data collection and resultant repeat accuracy of satellite ground tracks during airborne validation flights of ICESat-2 measurement capabilities. These include experiments with Real-Time Kinematic (RTK) data collection, which are motivated as described in the following paragraphs.

There is relatively little knowledge of the geodetical accuracy that can be obtained from kinematic data collection from a fast-moving platform, especially from a helicopter that is susceptible to wind and turbulence. In an approach that utilizes a base and rover with post-processed solutions, the challenge is to acquire sufficient satellite views to maintain a fixed solution in the GPS post-processing for at least 10 min, while the helicopter is flying. The faster the vehicle/rover moves, and the less linearly due to turbulence, the harder it is to find solutions. The mounting location of the GPS receiver and, for a helicopter, interference with rotor blade occultation, create additional unknowns [32].

The problem is compounded by the limited mounting options for a GPS rover on a helicopter. In our Svalbard experiments, the rover had to be mounted on one of the skids for safety reasons, where it had limited sky view due to the body of the helicopter and the rotors. Therefore, we included experiments of RTK data collection in our 2018 experiments, as an alternative to the post-processed base-rover GPS data analysis.

Combining the 2018 pre-launch experiments with 2019 experiments along near-time ICESat-2 tracks, we have results from three types of GPS experiments:

(1) Kinematic GPS data collection with differential correction using base and rover data (summer 2018)
(2) Real-time kinematic (RTK) GPS data collection using base and rover data (summer 2018)
(3) Kinematic GPS data collection with differential correction of rover data using Natural Resources Canada Spatial Reference System Precise Point Positioning (CSRS-PPP, [33,34]), (summer 2019)

Type 1 GPS experiments involve setting up a GPS base station at the side of the glacier on an elevated moraine or near a cliff edge for the duration of the airborne survey, typically two hours over the Negribreen region, and collecting kinematic data with a rover GPS mounted to the helicopter. During these experiments, the pilot follows a predetermined flight path. Experiments of best flight path segmentation were carried out. After the field experiment, base station data, recorded at 10Hz, are used to differentially correct rover data, also collected at 10 Hz. Because of regulations of the governor of Svalbard, the GPS base cannot be left in the glacier region. While a 24 h recording time for the base would be preferred, the length of time of about 2 h has been found to be sufficient for the accuracy of the project objectives.

Type 2 experiments require a base station and a rover as well. But in contrast to type-1 experiments, during RTK data collection, the correction of the rover position is carried out during the flight, using information sent over by the base station, and displayed on a blue-tooth-enabled device. This enables a correction of the flight path during flight.

In 2019, the GPS base station malfunctioned or was set incorrectly, recording only 15-second data, thus the differential correction applied to the 2019 GPS rover data uses Spatial Reference System Precise Point Positioning (CSRS-PPP), a service provided by Canadian Geodetic Survey, Natural Resources Canada, for post processing [33,34]. This situation yields an unintentional third type of experiment, whose results may be of interest for flights where a base station cannot be placed.

4.2. Results

(1) Kinematic GPS data collection with differential correction of rover data with base station data in post-processing yields the highest accuracy of positioning, of our three types of experiments. Experiments on flight-path segmentation indicate that quarter-crossings of the glacier work best, aiming to fly a straight line across a 2–3 km distance. Full glacier crossings are similarly good with respect to resultant accuracy of track repeats and require fewer waypoints to be entered manually by the pilot into his on-board system. Notably, the entire survey setup requires good collaboration and communication between the pilot, the experiment lead (the PI) and the survey team who continuously monitors instrument functionality and height (altimeter range). As a result, full glacier crossings of about 10–20 km flight distance depending on location were used in 2019.

(2) Experiments with RTK indicated that the added accuracy from the real-time-kinematic correction is more than offset by the inaccuracies introduced by the continued reactions of the pilot to adjust the flightpath and thus the aircraft attitude to the updated position information (from the RTK system). Flying straight towards points, predetermined by the expedition lead/ scientist with 2–4 points for a glacier crossing of 10–20 km track length and shared with the pilot prior to the flight, yields data from a more "quiet" helicopter flight and more consistent, linearly interpolatable data. Thus, the repeat accuracy of a pre-determined satellite track is not limited by the accuracy of the GPS solution, but instead by the maneuvering characteristics of a helicopter. The pilot cannot fly a straight path when the quick changes in the RTK-GPS directions call for unpractically frequent changes of the flight direction of the helicopter. A helicopter turn necessitates leaning the helicopter to the side, which then introduces additional distortions to the flight path and results in off-nadir laser pointing.

(3) Results from CSRS-PPP [33,34] are sufficiently accurate for validation of ICESat-2 surface height determination and morphological characteristics, as will be detailed in the following sections, while most likely significantly less accurate than positions differentially corrected with a base station. The CSRS post-processing uses precise satellite orbits, clock and bias corrections from a global network of receivers. Because of the much shorter base line resultant from a placing a base station in the field area, differential corrections using a locally placed base station are known to yield highest accuracy [35].

(4) Despite the problem of interference of the rotor blades and the helicopter body with view of GPS satellites, the amount of GPS rover data collected was found to be sufficient for accurate geolocation of the ULS data, despite the mounting location of the GPS rover on the skids.

(5) Outlook: Based on the 2018 and 2019 experiments (and previous experiments in similar types of field surveys), highest accuracy will be achieved by using differential GPS with a base station on the side of the glacier (as shorter baselines yield higher position accuracy) and a GPS rover on the aircraft. It may be easiest, especially for students new to GPS data collection, to utilize a rover and a base GPS instrument from the same brand, while not necessary to process the data (which can be converted to RINEX format independently of the receiver brand). Trimble has a proprietary format, but RINEX conversion remains a possibility.

The summer 2018 data are not used in the remainder of this paper, because they were collected prior to the launch of ICESat-2 on 15 September 2018.

5. Surface-Height Determination from Synchronization of Laser and GPS and IMU Corrections

5.1. Processing of 2019 GPS Data

Processing of 2019 GPS data was carried out using the following steps:

(1) GPS data were downloaded from the rover and the base station. In the 2019, flight 2 GPS recordings, the base station recorded only 15s-data. The rover recorded at 10Hz. In the differential data analysis, using the rover/base data, we were not able to obtain a fixed solution in the Trimble Business Center (TBC), only a float solution. We attribute this to the likely possibility that the rover was not tracking a sufficient number of satellite vehicles (SVs) to get a fixed solution. Therefore, the following steps were undertaken to obtain a solution.

(2) GPS data from the rover (Trimble R10) *(rover_data_set_v1.T02)* were converted from Trimble format (.T02) to RINEX (Receiver INdependent EXchange format) *(rover_data_set_v2.RINEX)*. This is performed using Trimble's Convert to RINEX utility (accessed on 3 October 2021).

(3) The RINEX *(rover_data_set_v2.RINEX)* file is submitted to the Spatial Reference System Precise Point Positioning (CSRS-PPP), provided by Natural Resources of Canada. CSRS-PPP is an online application for global navigation satellite systems (GNSS) data post-processing. It uses precise satellite orbit, clock and bias corrections derived from a global network of receivers to determine accurate user positions anywhere on the globe, regardless of proximity to reference stations (from: https://www.nrcan.gc.ca/maps-tools-and-publications/tools/geodetic-reference-systems/data/10923 (accessed on 3 October 2021)). Upon submission of RINEX files, enhanced positioning precisions in the North American Datum of 1983 of the Canadian Spatial Reference System (NAD83(CSRS)) or the International Terrestrial Reference System (ITRS) in realization ITRF14 are returned.

In essence, the CSRS software was able to provide a fixed solution for the rover data. To best match with the ICESat-2 data and their coordinate reference, we selected ITRF14 as the reference frame for the CSRS-PPP solutions. The coordinates used in this analysis are given as *(rover_data_set_PPP_v3)*, in csv format (human readable, including lat, lon, elevation, time, reference frame) with reference to ITRF14. The Canadian Geodetic Survey uses the GRS80 ellipsoid (a_e = 6,378,137 m, $\frac{1}{f}$ = 298.257222101) for the realization of ITRF14 in CSRS-PPP (Brian Donahue, Canadian Geodetic Survey, friendly pers. comm. 19 January

2022). For practical purposes, the ITRF14 corresponds to WGS84 (conversion from ITRF14 to WGS84 results in identical coordinates, for a more precise description, see Section 2). In summary, practically the same coordinate reference is utilized for the airborne data and the ICESat-2 ATLAS data.

5.2. Comparison of Button GPS and Corrected Rover GPS Data

In addition to the rover GPS, GPS data were also collected using a simple instrument, the Button GPS, located on board the helicopter. The Button GPS is a plug-and-play type inexpensive instrument, that is placed under the windshield of the helicopter and has proven to be very reliable during several years of airborne campaigns. The Button GPS records location at 1 Hz, while the rover GPS records location at 10 Hz. This Button GPS has lower accuracy, especially of height data, and lower frequency than the rover GPS, but is directly connected to the ULS laser system and records on the same computer. The rover GPS recorded only time-from-start-of-rover-time, it records on a GPS-internal computer and hence its times are independent of the ULS. Thus, we are using the time stamps from the Button GPS to align the corrected rover-GPS data with the ULS measurements.

This process is illustrated in Figure 5, which shows heights and times from the two GPS receivers during flight 2. Button GPS recordings appear to be "ahead of" the rover GPS with respect to distance, an effect that results from starting the Button GPS first. Since the Button GPS and ULS laser were being controlled from the same computer, we have exact time synchronization between them. In contrast, the rover GPS was controlled through its internal computer (and a handheld controller), which results in timing independent from that of the ULS laser. The first correction step involved time-shifting the rover GPS to align with the Button GPS/ULS laser (Figure 5).

5.3. Derivation of Surface Height Using ULS and Differential GPS or Button GPS

Ice-surface height is calculated by subtracting the laser-retrieved height from the GPS-retrieved height. Two apparent surface height values result from pairing ULS data with rover GPS data and Button GPS data (Figure 3b). In addition to the time lags between the rover GPS and the ULS laser, the change in height of the aircraft and the surface roughness of the glacier contribute to the apparent differences in height. For example, when the GPS starts to move down (aircraft lowers), the laser-retrieved height would normally be lower as well. However, in our case, the laser is ahead of the GPS, so it has already started recording lower heights. The different start times contribute a small part of the large apparent height difference values between the rover GPS-ULS and the Button GPS-ULS data, the main source of the apparent height difference is the poor height determination of the Button GPS. ULS data with ranges greater than 500 m are excluded from the analysis, as the ULS does not provide enough returns at ranges larger than 500 m (see Section 7).

5.4. Synchronization of Laser and Differential GPS and Button GPS

Crevasse locations are used to further improve the determination of a time shift, performed visually to match. Time accuracy is 0.1 s. Results in Figure 5c show that crevasse morphology lines up in the recordings of the Button GPS and the differential GPS, thus using crevasse morphology to aid in determination of a time shift is a feasible approach. There are still height differences on the order of tens of meters, which are due to the inherent properties of the low-resolution Button GPS. Because absolute heights from Button GPS data are less accurate than heights from corrected rover GPS data, heights from corrected rover GPS data are used in the following analysis.

5.5. IMU Correction

An Inertial Measurement Unit (IMU) records the attitude of the survey platform, here the helicopter, during flight, at (in our case) a frequency of 1 Hz. As the laser operates from the helicopter, it actually measures the range from the helicopter to the Earth's surface, rather than the height above the surface, and along a vector to the ground that is not normal to the ground surface. The IMU correction is a change of reference system, from that centered at the moving helicopter to an Earth-centric system. Prior to the actual correction, a low pass filter (moving average filter) is applied to remove helicopter vibration from the IMU data. The attitude reported by the IMU sensor is utilized in the form of Euler angles for correcting lidar laser range returns, and an orientation matrix for correcting the location of the range measurement in (x,y) coordinates. Formally, the IMU correction utilizes a quaternion multiplication (introduced e.g., in [36]). The IMU correction is applied to the GPS-corrected ULS data (with GPS corrections as described in Sections 5.1–5.4). It should be noted that the IMU corrections do not only yield corrected locations on the ground, but also corrections of the range values (this is a matrix multiplication applied to all components of the ULS data).

Figure 6a,b shows the magnitudes of the components (roll, pitch and yaw) of the IMU corrections. Figure 6c,d illustrates the magnitude of range corrections. At the scale of absolute range, this is not resolved (in Figure 6c,d). The correction magnitude (Figure 6c,d) is less than 5 m on average, with some spikes up to 10 m. These usually occur when the aircraft is turning or during heavy winds. We experienced heavy winds near the end of flight 2, which explains the largest difference.

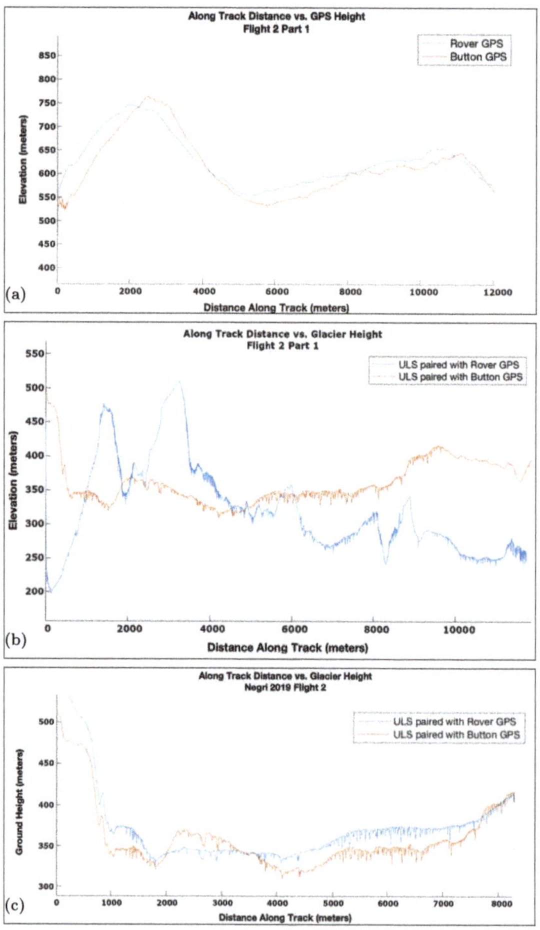

Figure 5. Illustration of the process of surface-height determination from synchronization of ULS and GPS data. Negribreen validation campaign. Flight 2, 13 August 2019. Corrected GPS data are used

(see Equation (1)). (**a**) Comparison of GPS heights and times from two GPS receivers. Button GPS (red), recorded at 1 Hz, started first, together with ULS and recorded on the same computer as the ULS (exact time synchronization of ULS and Button GPS). Rover GPS (blue), recorded at 10 Hz, started second, time recorded on the GPS-internal computer. The first correction step is time-shifting the rover GPS to align with the Button GPS/ULS laser. (**b**) ULS apparent surface heights, derived when pairing ULS data with Rover GPS data (blue) and Button GPS data (red). (**c**) ULS surface heights, after time-shift correction between ULS laser and the rover GPS. Now both elevation measurements and crevasse morphologies resemble each other.

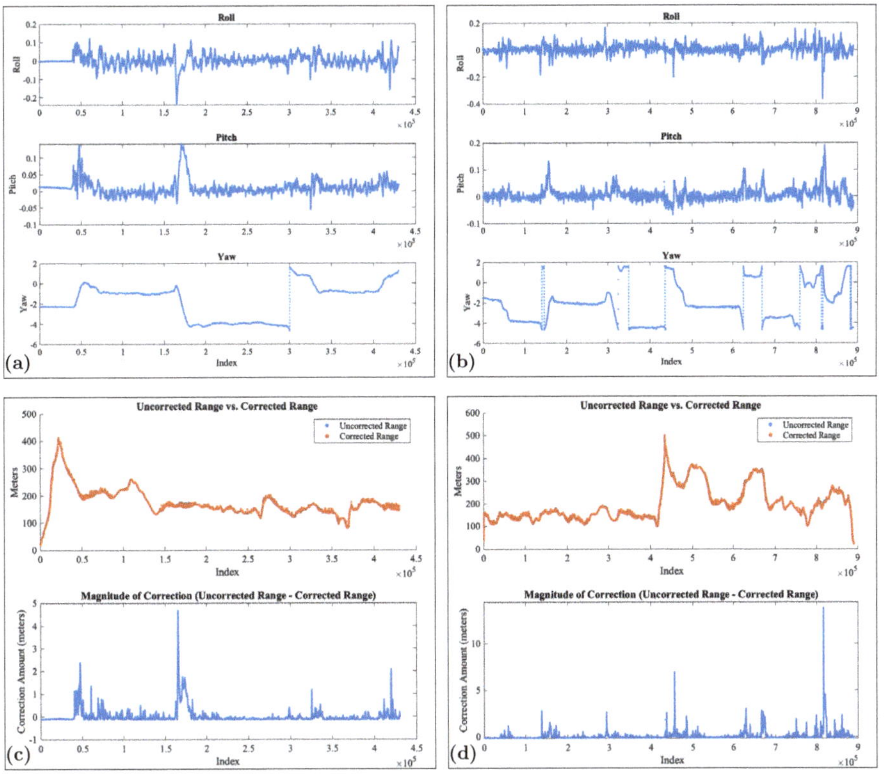

Figure 6. IMU values and IMU-corrected ULS ranges. Negribreen validation campaign. Flight 2, 13 August 2019. (**a,b**) IMU values (roll, pitch and yaw) for flight2, part1 (**a**) and flight2, part2 (**b**). (**c,d**) Range correction (total corrected and uncorrected range and magnitude of correction) for flight2, part1 (**c**) and flight2, part2 (**d**).

6. Analysis of Airborne Geophysical Data and Results: Ice-Surface Heights, Technical Information and Crossover Analysis

In this section, we analyze the airborne geophysical data (without comparison to ICESat-2 data) to derive ULS range and ULS-based ice surface heights along flight tracks (Section 6.1). Section 6.2 is devoted to technical information that may be useful to design future airborne validation campaigns from small aircraft, including helicopter velocity, optimal ULS range and aircraft altitude above ground, resultant data frequency and on-ice spacing. These results generalize to other types of manned aircraft and to some extent, unmanned aircraft. Precision of the ULS height data in crevassed terrain is assessed using a crossover analysis (Section 6.3).

6.1. Results Part 1: ULS Range and Ice-Surface Heights Based on GPS and IMU-Corrected Data

Following the GPS and IMU corrections as described in Section 5, we now have the correctly geolocated coordinates for the ULS data. After application of the IMU and GPS geodetical corrections to the ULS data, we obtain ULS data with geographical coordinates

$$x_{geogcorr} = (lat_{corr}, lon_{corr}, z_{corr}), \tag{1}$$

where z_{corr} is the corrected surface height. As described above, the GPS coordinates are given in IRTF14 (as output from the CSRS-PPP software). The bounce point of the ATLAS signal is also given in the IRTF14 reference frame as latitude, longitude and height w.r.t. the WGS-84 (G1150) ellipsoid for each point in the photon cloud reported on the ATL03 data product [2,3,18,22] (Section 2.1). Therefore, corrected ULS data from the validation campaign are directly comparable to ATL03 data.

Lastly, ULS data $x_{geogcorr}$ given with reference to the geographic coordinate system are converted to ULS data with reference to the Universal Transverse Mercator (UTM) coordinate system

$$x_{utmcorr} = (north_{corr}, east_{corr}, up_{corr}) \tag{2}$$

with reference to the WGS84 ellipsoid and UTM zone 33 (central meridian 15° E). We use $up_{corr} = z_{corr}$. We are now able to determine the range of the ULS to the ice surface at the laser measurement rate (Figure 7a), which combined with the aircraft position gives the surface height of the glacier (Figure 7b). The corrected surface heights can be used in several glaciological applications, including monitoring of surface elevation change during the surge, indicative of mass transfer during the surge, and derivation of the spatial characteristics of the crevassed ice surface (see, Section 7).

6.2. Results Part 2: Technical Information: Helicopter Velocity, ULS Range, Data Frequency and On-Ice-Spacing

The range of the ULS is an important experiment parameter to control, as the ULS functions for altitudes up to about 500 m and works best for shorter ranges up to about 300 m. Helicopter safety requires that the instrument be flown at sufficient distance above the surface, dependent on (a) visibility and surface reflectance, and (b) topographic relief, morphology and crevassity of the terrain. Following [11], we use the term "crevassity" as a measure of the void fraction in glacier ice as a material, akin to porosity. Crevassity is utilized here as a general term to describe the magnitude and complexity of the overall deformation at a given location, which for a surge glacier manifests itself in the crevasse characteristics.

All ICESat-2 validation flight lines were flown at this optimal height range, with tracks in the upper central Negribreen (inlcuding those for RGT 450 and RGT 594) collected with a range of 100-200 m and tracks traversing towards the upper inflows flown at 200–350 m range (Figure 7a). The ULS is originally designed for measurements at much shorter ranges and there is no practical lower limit of the measurement range (we test the instrument in the laboratory, for distances down to 1 m). The practical lower limit of the range is constrained by flight operations safety, as the helicopter pilot needs to maintain a safe height above the highly reflective, sloping and crevassed ice surface, at a flight speed that is efficient for data collection.

Flight velocity magnitudes, evaluated after GPS and IMU corrections, are given in Figure 7c and after along-track smoothing in Figure 7d. Speeds for the validation tracks (RGT 450 and RGT 594) are generally between 25 m/s (90 km/h) and 35 m/s (126 km/h), slower over steep ice falls. Averaged helicopter velocities were 17 m/s (averaged for the whole survey flight) and 30–35 m/s while flying over the ICESat-2 segments. Mean velocities were 34.426 m/s over the RGT 450 segment and 32.0451 m/s over the RGT 594 segment. The ULS functions well at higher velocities as well, as supported by the fact that frequency of recorded data is generally close to the 400 Hz nominal frequency of ULS setting used during our experiments (Figure 7e, for ULS settings, see Table 3). Resultant on-ice along-track spacing was 0.06–0.08 m for the validation tracks (Figure 7f).

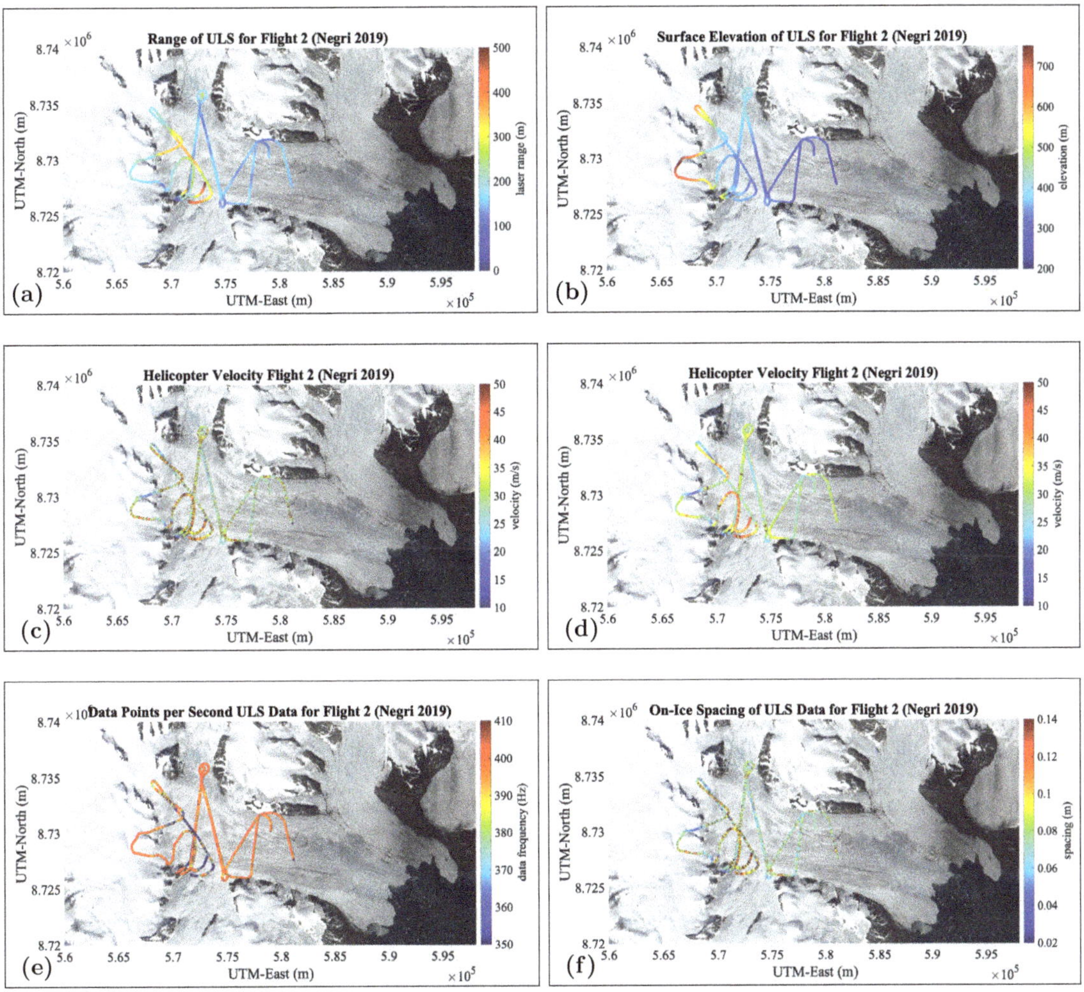

Figure 7. Results from ULS data collection. Negribreen validation campaign. Flight 2, 13 August 2019. The scale does not allow for identification of crevassed regions. (**a**) ULS range aircraft to ice surface. (**b**) ULS surface height (elevation). (**c**) Aircraft velocity magnitude. (**d**) Aircraft velocity magnitude, smoothed. (**e**) Data points per second. (**f**) On-ice spacing of ULS data. Background: Landsat-8 image from 5 August 2019 [LC08_L1TP_215003_20190805_20190820_01_T1].

Flight altitudes above ground are further constrained by the other instruments: Higher altitudes allow collection of better imagery, i.e., imagery that shows larger crevassed regions. Flight altitudes of 10 km or more are flown for collection of overview imagery. Altitude in transit from one observation target to another can be higher than 300 m and still yield altimeter data useful for science. For example, the blue stretch in Figure 7e is flown at 400 m above ground, resulting in lower data frequency and somewhat larger on-ice spacing of 0.08–0.14 m. ULS data with ranges greater than 500 m are excluded from the analysis, as the ULS does not work properly at this range, and the situation of large ranges only occurs when we flew high to collect overview imagery and the ULS was not turned off. In conclusion, the ULS can be operated with full recording frequency for altitudes up to 300 m above ground level (a.g.l.). The advantage of flying at lower altitudes a.g.l. is that the on-ground spacing is smaller, as the spacing is a direct function of altitude above ground. Between 300 m and 500 m, the recording frequency can drop somewhat, but still yield useful science data. Lower altitudes of 100 m can be used for flights with unmanned aerial systems over sea ice, where risk is lower and the surface is flat, as during the Characterization of Arctic Sea Ice Experiment (CASIE) [30,31].

6.3. Results Part 3: Crossover Analysis of ULS Heights

The crossover analysis is performed on the corrected height estimates for the 2019 flight data in the upper glacier (see Table 4). Crossover height differences are calculated after averaging all points within a 10 m radius of the intersection location for each crossing flight segment, resultant in 14 crossover locations with a mean elevation difference of 0.111 ± 3.056 m. Similar values are achieved when the averaging radius is 1 m and 0.1 m. Largest differences (red and dark blue points in Figure 8) occur at locations where the aircraft is turning significantly, where laser ranges are large and IMU corrections are most relied upon, and at locations of large crevasses where height variations on the order of tens of meters occur. However, most of the crossover differences are near zero (green points) when the aircraft is flying straight over non-crevassed areas or areas with beginning crevassing. If we remove the 3 crossover locations that occur on significant turns, we obtain a mean elevation difference of -0.172 ± 2.564 m for the remaining 11 points. Since underflights of ICESat-2 tracks follow straight flight paths, the estimated the precision of our measurements is approximately 2.6 m in crevassed terrain. Accuracy estimates would require ground elevation truths independent of and coincident with our flight data, however collection of ground measurements on a glacier during surge is not feasible (dangerous). Crevasse depths ranging to 45 m were observed with ICESat-2 and determined with the DDA-ice [3].

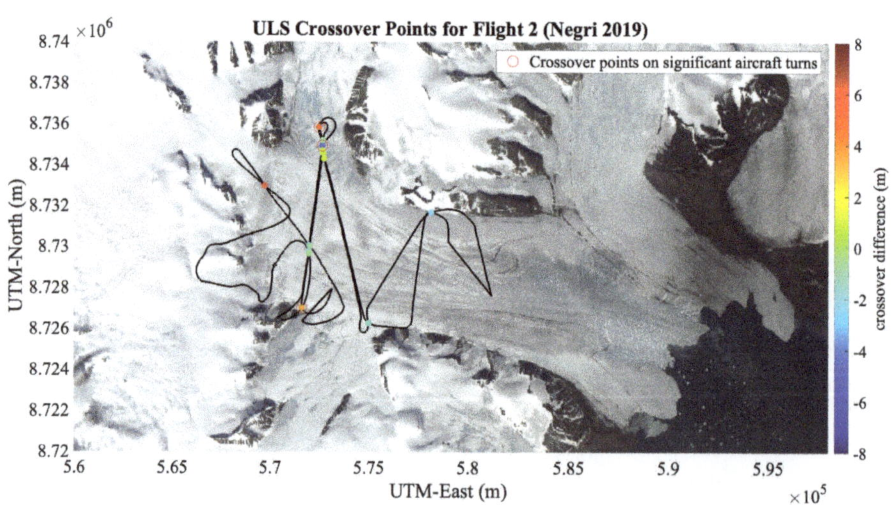

Figure 8. Crossover elevations from ULS data collection. Negribreen validation campaign. Flight 2, 13 August 2019. There are 14 crossover locations. Removing the three crossover points on large turns, for the remaining 11 points the mean difference is -0.172 ± 2.564 m. Background: Landsat-8 image from 5 August 2019 [LC08_L1TP_215003_20190805_20190820_01_T1].

Table 4. Crossover analysis for Flight 2, 13 August 2019, Negribreen ICESat-2 Validation Campaign 2019. Estimates are given when using all crossover locations (seen in Figure 8) and for locations only along relatively straight flight segments, i.e., where the helicopter was not performing a significant turning maneuver. Crossover estimates are averaged within a radius specified in the last column of the table.

	Mean Elevation Difference (m)	Standard Deviation of Elevation Difference (m)	Number of Points	Crossover Radius (m)
All Crossover Locations	0.111	3.056	14	10
Crossover Locations Excluding Turns	−0.172	2.564	11	10

7. Analysis of Geolocated ULS and ICESat-2 ATLAS Data

7.1. Overview

Following the geodetical and technical results in the last sections, we now have established the technical basis needed to (1) quantitatively investigate the capability of a light-weight experiment using a small helicopter to repeat satellite tracks accurately and (2) compare surface heights and crevasse morphologies from ULS and ATLAS measurements. Results relating to the first topic will be summarized in Section 8 and results regarding the second topic will be reported in Section 9. Here, the analysis is carried out as a case study for each of five sets of tracks, to facilitate understanding the relationships between flight-repeat capability of our experiment setup dependent on a number of factors including wind, resultant track separation, crevasse representation in ULS data and in ATLAS data for weak and strong beams, crevasse morphology, crevasse characteristics and surface roughness.

ICESat-2 tracks for RGT 450 and RGT 594 were repeated, including both strong and weak beams, resultant tracks for ULS and ATLAS are shown in Figure 9a,b and listed in Table 5. A useful tool in this context is plotting the results in 1-km along-track segments, as shown in Figure 9. Figure 9a,b provides the 1-km segments for flight 2, parts 1 and 2, respectively, carried out on 13 August 2019 over Negribreen.

Data from flight 1 are not used in the analysis in this paper, because the GPS rover did not record data. Data from the Button GPS are still useful for analysis of surge evolution, reported elsewhere.

Flight 2 has two parts, each part includes an overpass of RGT 594 (gt1 l and gt1r) and RGT 450 (gt1l and gt1r) tracks. The track of the RGT 594 weak beam (gt1r) was overflown twice (both passes in part 2 of flight 2) in order to obtain information on the quantitative measures listed above from a repeat experiment. However, because the weather worsened significantly during flight 2, the first repeat occurred during approximately normal weather conditions and the second one during unusually strong winds of over 30 knots (60 km/h). The increasing winds led the pilot to end the data collection for this day soon after. The three tracks for RGT 594 were flown in the order (1) "RGT 594, gt1r, part 1", (2) "RGT 594, gt1l", (3) "RGT 594, gt1r, part 2", with other flight lines in between, hence wind speed increases in this order. Labels in the figures are as follows: "RGT 594, gt1r, part 1" refers to overpass 1 (less wind). "RGT594, gt1r, part 2" refers to overpass 2 (stronger wind). During strong winds, the pilot has to hold the helicopter at an angle relative to the direction of flight and tilted, which makes following the prescribed tracks a lot harder and increases the IMU values. The IMU still functions for this situation (see Figure 6). Resultant track separations for RGT 594, gt1r, part 2 are higher than for the other tracks, but crevasse comparisons are still useable for evaluation of ICESat-2 tracks, as the analysis will show.

The tracks of the RGT 594 gt1r and gt1l are the location of analysis of 2018 prelaunch Negribreen airborne campaign data and first post-launch ICESat-2 data from early winter 2018 reported elsewhere, which indicated that ICESat-2 measures summer's crevasses through winter's snow cover. Because the identification of crevasses under snow was based on non-synoptic airborne and satellite data with several months separation, an additional objective for the 2019 validation campaign was the collection of synoptic satellite and airborne data for the same crevasse field surveyed in summer 2018 with the ULS and in winter 2018 with ICESat-2 ATLAS.

In the following subsections, we analyze each of the tracks (1) RGT 450, gt1l (strong beam) (Figure 10), (2) RGT 450, gt1r (weak beam) (Figure 11), (3) RGT 594, gt1l (strong beam) (Figure 12), (4) RGT 594, gt1r (weak beam), overpass 1 (Figure 13), and (5) RGT 594, gt1r (weak beam), overpass 2 (Figure 14).

For each example, we show the location of the ground track sections, indicated by red and green tracks, superimposed on a Landsat image from 5 August 2019 (collected only 8 days before flight 2), in panel (a) of Figures 10–14. Zooming into the area, with the flight tracks plotted over the aforementioned SkySat image from 18 August 2019, gives and impression of the crevasse provinces under the flight tracks, and visually illustrates the across-track distances between the predicted and actual ATLAS tracks and the actual ULS tracks (panel (b) in Figures 10–14). The distances are then given in a graph that illustrates the variability of the distance between the ULS track and the actual ATLAS track, with the distance between the predicted and actual ATLAS tacks added in for comparison (panel (c) of Figures 10–14). Note that at time of flight, only the predicted ATLAS tracks were available, thus we aimed to underfly the predicted tracks for each beam. Track variability is exaggerated by the over-heightened plot style. Variability of the ATLAS-predicted to ATLAS-actual distance is shown in better resolution in panel (d) of Figures 10–14. In summary, capability for track repeat ranges from an average of 11.32 m to 24.87 m, with an outlier of 60m during strong storm conditions. Differences between ICESat-2 predicted and actual range from 6.51 m to 15.65 m for the average difference of a glacier crossing.

Figure panels (e) and (f) in Figures 10–14 aid in the comparative analysis of the surface heights of crevassed regions. Segment lengths in (a), (c), (d) and (e) are the same within each figure and of approximately 2–3 km length, while segment lengths in (b) and (f) show subsets for better representation of the crevasse morphologies. To further aid the comparison of panels (e) and (f), we have added a blue bar in panels (e) that shows the location and coverage of the DDA-ice results seen in figure panels (f) of the same figure. Surface heights of ICESat-2 ATLAS data are derived by application of the DDA-ice, which was briefly described in Section 2.2.

Table 5. Average distances of ATLAS predicted vs. ATLAS actual and ATLAS actual vs. ULS actual. Values calculated for entire glacier crossings. Note that Figures 9–13 highlight 2–3 km sections for each crossing.

RGT_Beam	ATLAS Predicted vs. ATLAS Actual (m)	ATLAS Actual vs. ULS Actual (m)	Figure
450_gt1l (strong)	9.28	20.40	Figure 10
450_gt1r (weak)	6.51	11.32	Figure 11
594_gt1l (strong)	15.65	60.38	Figure 12
594_gt1r (weak) (part 1)	9.16	24.87	Figure 13
594_gt1r (weak) (part 2)	9.16	24.35	Figure 14

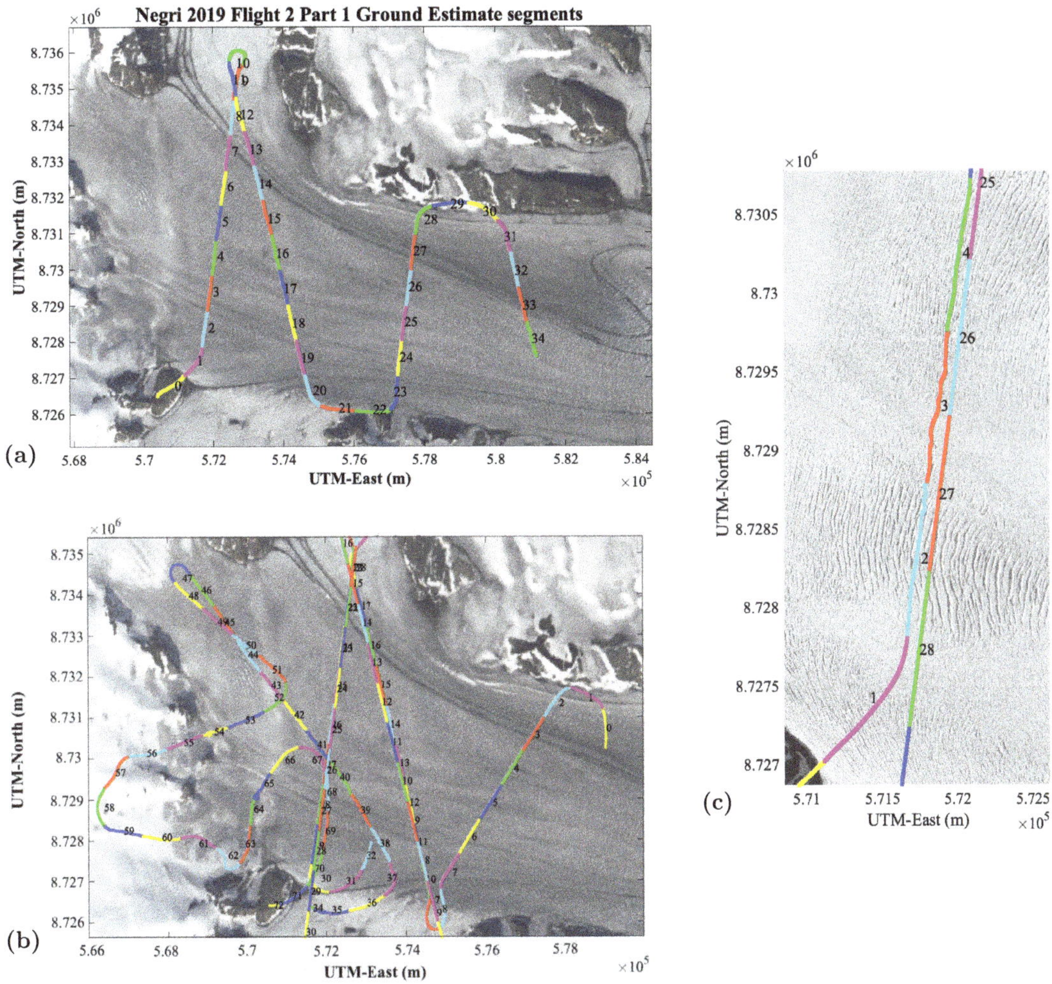

Figure 9. Flight path segments, resultant from ULS data collection, and co-location with ATLAS ground tracks over crevasse fields. Negribreen validation campaign. Flight 2, 13 August 2019. (**a**,**b**) Flight paths, colored in 1-km segments to match the along-track distance vs. elevation plots in Figures 9–13. Background: Landsat-8 image from 5 August 2019 [LC08_L1TP_215003_20190805_20190820_01_T1]. (**a**) Flight 2, part 1. (**b**) Flight 2, part 2. (**c**) Example of co-located airborne ground track (ULS), left, and ATLAS ground track (RG 450 1l), right. Even with an 18 meter average separation, the airborne ground track (ULS) and the ATLAS ground track cross the same crevasse provinces. Background: Planet SkySat image from 18 August 2019 [20190818_150858_ssc9_u0002_panchromatic_dn.tif].

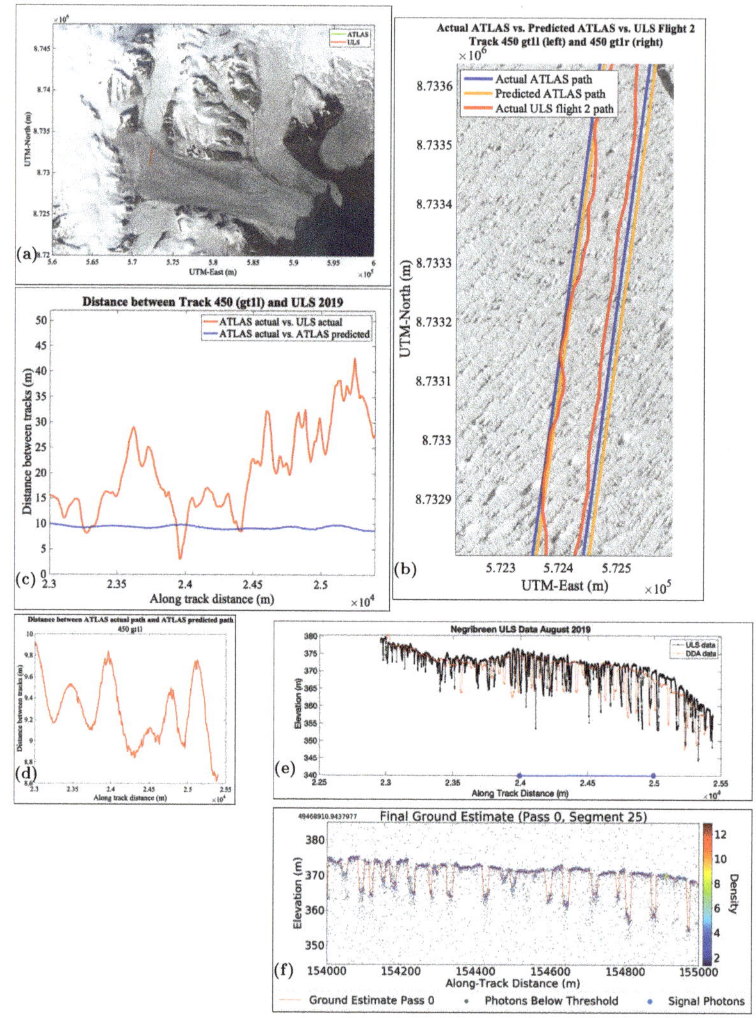

Figure 10. ICESat-2 ATLAS vs. ULS data. Track repeat accuracy and results. RGT 450, gt1l (strong). Negribreen validation campaign. Flight 2, 13 August 2019. (**a**) Location of ATLAS segments over Negribreen and closest ULS data (note the tracks are so close that the green line disappears under the red line). For segment numbering, see Figure 9 (ATLAS DDA segments). Background: Landsat-8 image from 5 August 2019 [LC08_L1TP_215003_20190805_20190820_01_T1]. Segment lengths in (**a**,**c**–**e**) are the same, segments in (**b**,**f**) are subsets. (**b**) Track separation between our ULS flight path, the ATLAS predicted flight path and the ATLAS actual flight path for granule: ATL03_20190727132129_04500405_003_01.h5. Background: Planet SkySat image from 18 August 2019 [20190818_150858_ssc9_u0002_panchromatic_dn.tif]. (**c**) Separations between ATLAS Track 450 (gt1l) actual and ULS flight path, and between ATLAS Track 450 (gt1l) actual and ATLAS Track 450 (gt1l) predicted. Calculated for actual ATLAS and ULS flight paths from data shown in (**b**). (**d**) Track separation between predicted and actual ATLAS 450 (gt1l) flight paths. (**e**) ULS surface heights and ICESat-2 ATLAS DDA-ice surface heights superimposed, after GPS and IMU correction of ULS paths (ground locations). (**f**) ICESat-2 ATLAS surface heights over crevassed region derived using DDA-ice, RGT 450, gt1l, segment 25. DDA-ice parameters as in Table 2. ATL03 data set given in caption for (**c**). Rel. 03 data.

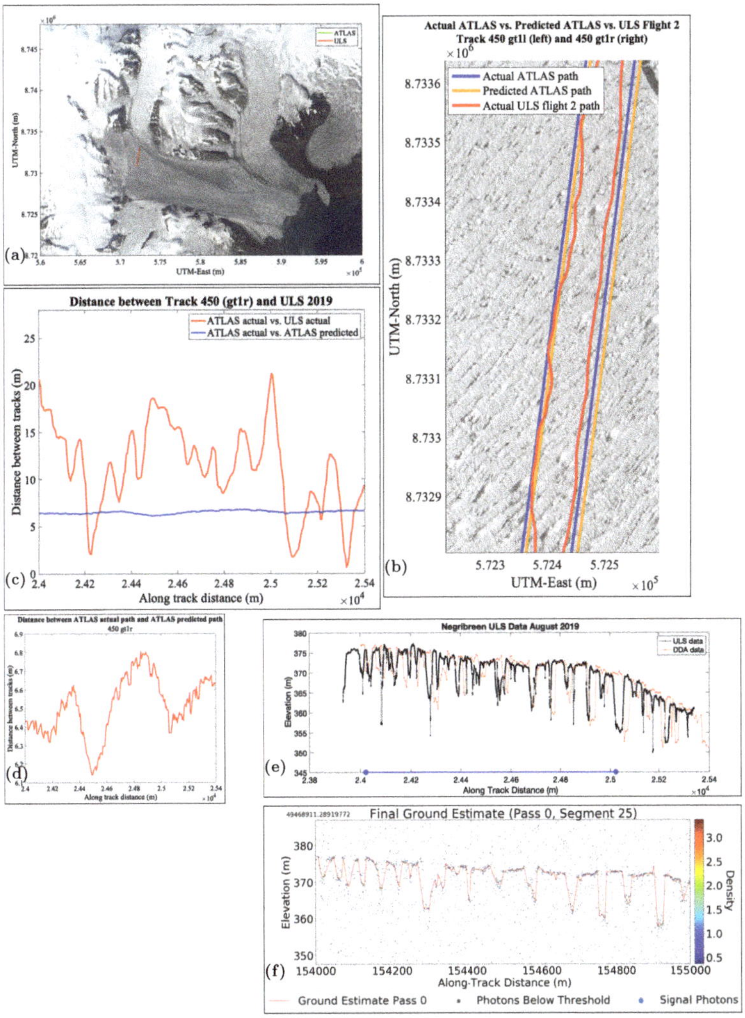

Figure 11. ICESat-2 ATLAS vs. ULS data. Track repeat accuracy and results. RGT 450, gt1r (weak). Negribreen validation campaign. Flight 2, 13 August 2019. (**a**) Location of DDA segments over Negribreen and closest ULS data (note the tracks are so close that the green line disappears under the red line). For segment numbering, see Figure 9 (ATLAS DDA segments). Background: Landsat-8 image from 5 August 2019 [LC08_L1TP_215003_20190805_20190820_01_T1]. Segment lengths in (**a**,**c**–**e**) are the same, segments in (**b**,**f**) are subsets. (**b**) Track separation between our ULS flight path, the ATLAS predicted flight path and the ATLAS actual flight path for granule: ATL03_20190727132129_04500405_003_01.h5. Background: Planet SkySat image from 18 August 2019 [20190818_150858_ssc9_u0002_panchromatic_dn.tif]. (**c**) Separations between ATLAS Track 450 (gt1r) actual and ULS flight path, and between ATLAS Track 450 (gt1r) actual and ATLAS Track 450 (gt1r) predicted. Calculated for actual ATLAS and ULS flight paths from data shown in (**b**). (**d**) Track separation between predicted and actual ATLAS 450 (gt1r) flight paths. (**e**) ULS surface heights and ICESat-2 ATLAS DDA-ice surface heights superimposed, after GPS and IMU correction of ULS paths (ground locations). Blue line: Section covered in f. (**f**) ICESat-2 ATLAS surface heights over crevassed region derived using DDA-ice, RGT 450, gt1l, segment 25. DDA-ice parameters as in Table 2. ATL03 data set given in caption for (**c**). Rel. 03 data.

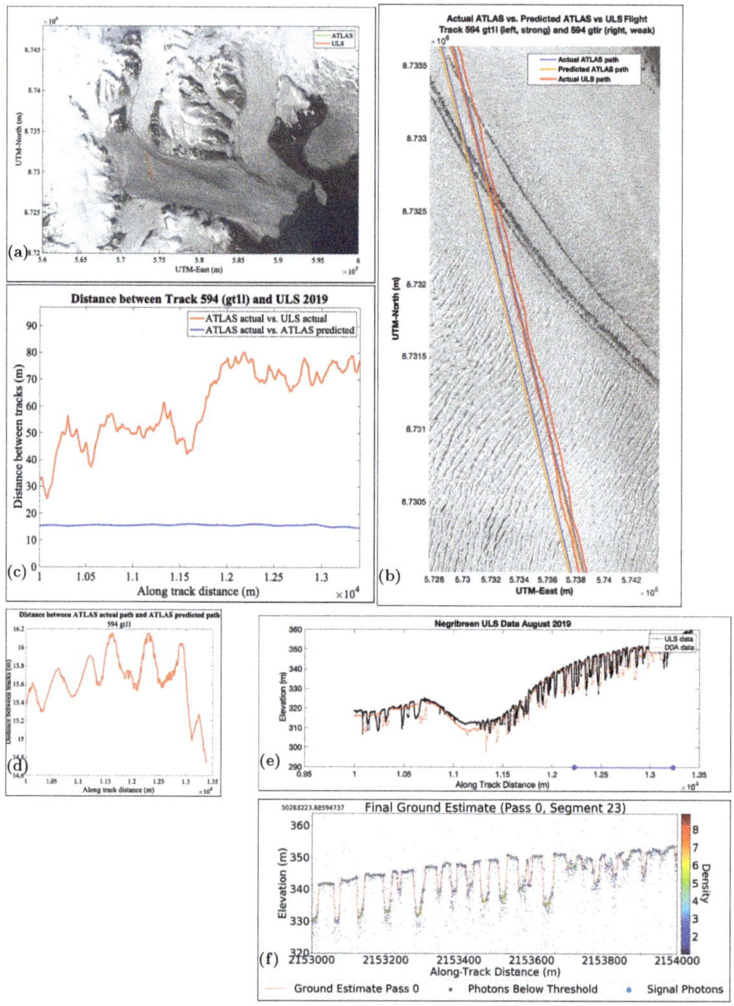

Figure 12. ICESat-2 ATLAS vs. ULS data. Track repeat accuracy and results. RGT 594, gt1l (strong). Negribreen validation campaign. Flight 2, 13 August 2019. (**a**) Location of DDA segments over Negribreen and closest ULS data. For segment numbering, see Figure 9 (ATLAS DDA segments). Background: Landsat-8 image from 5 August 2019 [LC08_L1TP_215003_20190805_20190820_01_T1]. Segment lengths in (**a**,**c**–**e**) are the same, segments in (**b**,**f**) are subsets. (**b**) Track separation between our ULS flight path, the ATLAS predicted flight path, and the ATLAS actual flight path for granule: [ATL03_20190805232841_05940403_002_01.h5]. Background: Planet SkySat image from 18 August 2019 [20190818_150858_ssc9_u0002_panchromatic_dn.tif]. (**c**) Separations between ATLAS Track 594 (gt1l) actual and ULS flight path, and between ATLAS Track 594 (gt1l) actual and ATLAS Track 594 (gt1l) predicted. Calculated for actual ATLAS and ULS flight paths from data shown in (**b**). (**d**) Track separation between predicted and actual ATLAS 594 (gt1l) flight paths; same segment length as in (**a**,**c**). (**e**) ULS surface heights and ICESat-2 ATLAS DDA-ice surface heights superimposed, after GPS and IMU correction of ULS paths (ground locations). Blue line: Section covered in f. (**f**) ICESat-2 ATLAS surface heights over crevassed region derived using DDA-ice, RGT 594, gt1l, segment 23. DDA-ice parameters as in Table 2. ATL03 data set given in caption for (**c**). Rel. 03 data (see, [3]).

7.2. Concepts of Crevasse Provinces and Surface Roughness

Important concepts in the analysis of ice dynamics, especially for glacial accelerations which lead to heavy crevassing as is the case during a surge, are crevasse provinces and ice-surface roughness. The analysis of crevasse provinces allows study of the deformation characteristics of a glacier during surge and provides information on several aspects of ice dynamics, which can be modeled [7,9,11,37]. The mathematically easiest way to distinguish crevassity is through calculation of ice-surface roughness [8,11].

The concepts are illustrated in Figure 9c. A crevasse province is defined as an area that is homogeneous with respect to crevasse type, or ice deformation, and maximal with this homogeneity property [8]. For example, a province of slightly curved, wide surge crevasses extends for almost exactly segment 2 of the flight track and half of segments 27 and 28 of the satellite track. A different province of more narrowly spaced, less open and partly snow-filled crevasses extends northwards starting at segment 26 of the satellite ground track.

Figure 9c gives an example of co-geolocated ground tracks, with the airborne tracks on the left and the satellite ground track on the right (note that segment numbering is computed automatically and independently for each data set). The tracks in this section have an average spacing of about 18 m. The underlying Planet SkySat image, which was collected on 18 August 2019, only 5 days after flight 2 shows that the same types of crevasses are crossed by the satellite and the airborne track. Notably, with a separation of 18 m, the same crevasse provinces are crossed by the ULS tracks and the ICESat-2 tracks. This example indicates that a separation of 18 m is sufficiently small to capture the same characteristics and, specifically, function for a validation of ICESat-2 surface heights and crevasse characteristics.

7.3. Analysis of Data for RGT 450, gt1l (Strong Beam)

Predicted ICESat-2 tracks, actual ICESat-2 tracks and actual ULS flight tracks, projected to the ground, are shown in Figure 10b, plotted over a Planet SkySat image from a near-time SkySat acquisition, for beam pair 1 of RGT 450. For the strong beam (gt1l, left in Figure 10b), the ULS ground tracks are often coincide with the predicted ATLAS tracks, at the resolution of this figure. The distance between the ULS tracks and the actual ATLAS tracks is given in Figure 10c, the average distance is 18 m for the section and 20.40 m for the glacier crossing (Table 5); this distance characterizes the height comparisons. Notably, the distance between the actual and the predicted ICESat-2 track is 9.28 m (Figure 10d), which indicates that with an the 18-20m distance the track repeat capability of the airborne campaign is very good for a helicopter-based operation. Typically, repeat accuracy of satellite tracks (ATLAS actual versus ATLAS predicted) can be expected to be much better than that of a flight campaign (ULS actual versus ATLAS predicted). For the weak beam (gt1r, right set of tracks in Figure 10b, see Figure 11), the distance between actual ICESat-2 and ULS tracks is on the same order as the distance between the predicted and actual ICESat-2 tracks (and sometimes better, for example in the southern part of the plot, near 8,732,900 UTM-North).

The Planet SkySat image indicates that ULS tracks and predicted and actual ICESat-2 tracks cross the same crevasse types and in most cases the same individual crevasses. This observation is confirmed by the comparison of heights from the ULS data and ATLAS data, processed with the DDA-ice, seen in Figure 10e,f. It should be noted that the ULS operates at 905 nm (NIR), while ATLAS operates at 532 nm (green), and thus the comparison figure reflects the different penetration properties of the two altimeters, the small location differences of the ground tracks and the separation in time (the glacier moves at several meters per day in this region, because of the surge). Along-track distances of the ULS data were shifted a few meters to better align crevasses in the ULS and ATLAS data, to account for glacier movement during the separation times. Flight 2 was carried out on 13 August 2019, ATLAS data along RGT 450 were collected 27 July 2019, resulting in a 17-day time separation. Penetration differences of red and green lidars are analyzed in [21], using ICESat-2 simulator data.

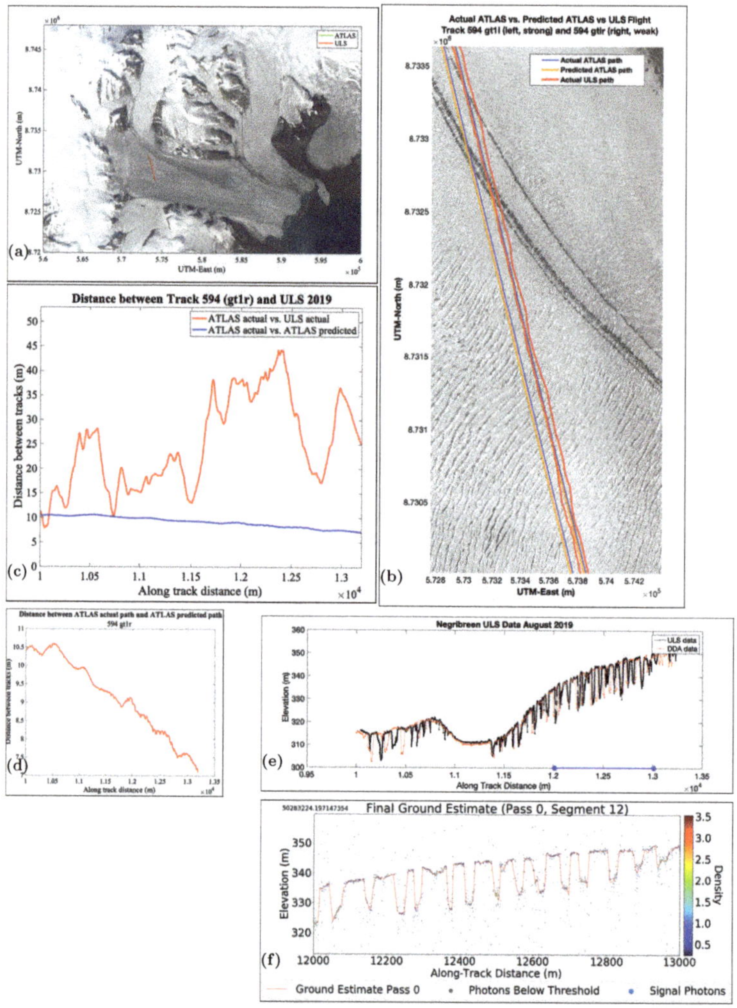

Figure 13. ICESat-2 ATLAS vs. ULS data. Track repeat accuracy and results. RGT 594, gt1r (weak), overpass 1. Negribreen validation campaign. Flight 2, 13 August 2019. (**a**) Location of DDA segments over Negribreen and closest ULS data (note the tracks are so close that the green line disappears under the red line). For segment numbering, see Figure 9 (ATLAS DDA segments). Background: Landsat-8 image from 5 August 2019 [LC08_L1TP_215003_20190805_20190820_01_T1]. Segment lengths in (**a**,**c**–**e**) are the same, segments in (**b**,**f**) are subsets. (**b**) Track separation between our ULS flight path, the ATLAS predicted flight path, and the ATLAS actual flight path for granule: [ATL03_20190805232841_05940403_002_01.h5]. Background: Planet SkySat image from 18 August 2019 [20190818_150858_ssc9_u0002_panchromatic_dn.tif]. (**c**) Separations between ATLAS Track 594 (gt1r) actual and ULS flight path, and between ATLAS Track 594 (gt1r) actual and ATLAS Track 594 (gt1r) predicted. Calculated for actual ATLAS and ULS flight paths from data shown in (**b**). (**d**) Track separation between predicted and actual ATLAS 594 (gt1r) flight paths. (**e**) ULS surface heights and ICESat-2 ATLAS DDA-ice surface heights superimposed, after GPS and IMU correction of ULS paths (ground locations). Blue line: Section covered in f. (**f**) ICESat-2 ATLAS surface heights over crevassed region derived using DDA-ice, RGT 594, gt1r, segment 12. DDA-ice parameters as in Table 2. ATL03 data set given in caption for (**c**). Rel. 03 data.

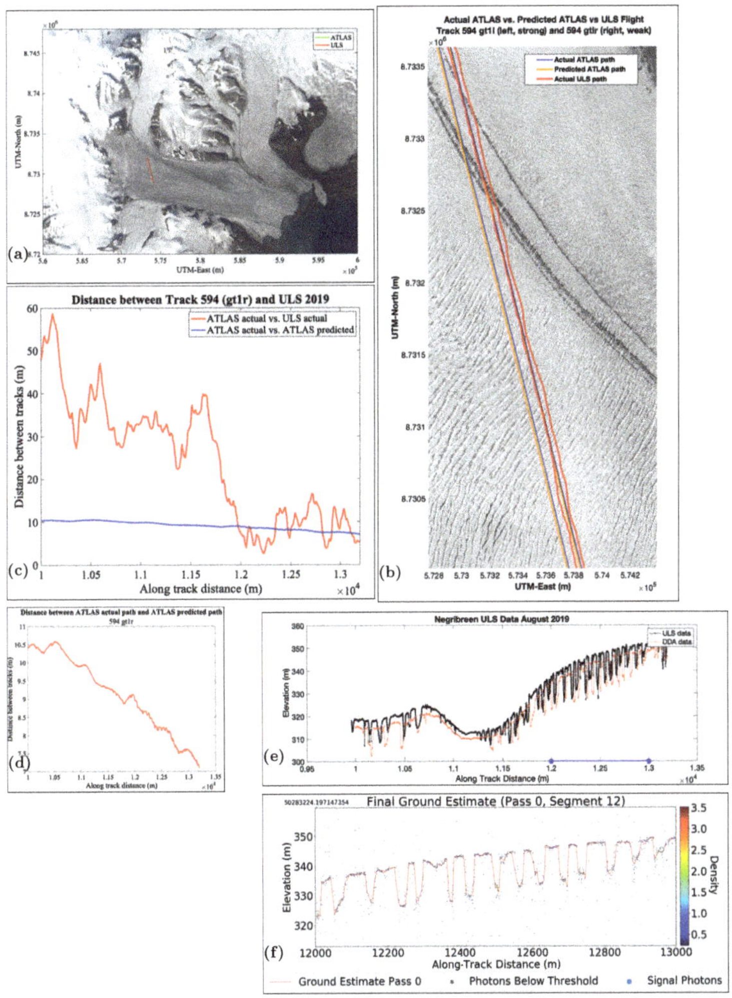

Figure 14. ICESat-2 ATLAS vs. ULS data. Track repeat accuracy and results. RGT 594, gt1r (weak), overpass 2 (high winds of 60 km/h). Negribreen validation campaign. Flight 2, 13 August 2019. (**a**) Location of DDA segments over Negribreen and closest ULS data (note the tracks are so close that the green line disappears under the red line). For segment numbering, see Figure 9 (ATLAS DDA segments). Background: Landsat-8 image from 5 August 2019 [LC08_L1TP_215003_20190805_20190820_01_T1]. Segment lengths in (**a**,**c**–**e**) are the same, segments in (**b**,**f**) are subsets. (**b**) Track separation between our ULS flight path, the ATLAS predicted flight path, and the ATLAS actual flight path for granule: [ATL03_20190805232841_05940403_002_01.h5]. Background: Planet SkySat image from 18 August 2019 [20190818_150858_ssc9_u0002_panchromatic_dn.tif]. (**c**) Separations between ATLAS Track 594 (gt1r) actual and ULS flight path, and between ATLAS Track 594 (gt1r) actual and ATLAS Track 594 (gt1r) predicted. Calculated for actual ATLAS and ULS flight paths from data shown in (**b**). (**d**) Track separation between predicted and actual ATLAS 594 (gt1r) flight paths. (**e**) ULS surface heights and ICESat-2 ATLAS DDA-ice surface heights superimposed, after GPS and IMU correction of ULS paths (ground locations). Blue line: Section covered in f. (**f**) ICESat-2 ATLAS surface heights over crevassed region derived using DDA-ice, RGT 594, gt1r, segment 12. DDA-ice parameters as in Table 2. ATL03 data set given in caption for (**c**). Rel. 03 data.

In summary, agreement of morphologies and heights is very good, and thus we conclude that the experiment setup of our airborne geophysical validation campaign serves the purpose of a validation campaign. In consequence, the resultant ICESat-2 surface heights, as calculated with the DDA-ice from the ATL03 photon data, represent the heights of the crevassed surface and the interior of the crevasses. These surface heights and crevasse morphologies are thus suited for geophysical and glaciological analyses [3]. Details in the DDA-ice results (Figure 10f) show the typical morphology of surge crevasses with relatively sharp edges to the ice surface and largely straight crevasse walls. The latter is due to the general absence of points that could define a complex crevasse walls, as are detected in the case of Jakobshav Isbræ (Ilulissat Ice Stream) crevasses [3].

7.4. Analysis of Data for RGT 450, gt1r (Weak Beam)

Figure 11 shows the same type of analyses as Figure 10, now for the weak beam of RGT 450, gt1r. For the weak beam (gt1r, right set of tracks in Figure 11b), the distance between actual ICESat-2 and ULS tracks is on the same order as the distance between the predicted and actual ICESat-2 tracks (and sometimes better, for example in the southern part of the plot, near 8,732,900 UTM-North). Average separation of the ATLAS (actual) and ULS (actual) tracks for this glacier crossing is only 11.32 m. In this case, the distance to the predicted tracks is larger than to the actual tracks. As in the previous example (Figure 10), along-track distances were shifted to account for the 17-day time separation between ATLAS and ULS data along the RGT 450 track.

All conclusions regarding evaluation capabilities hold as stated for the example of the RGT 450 strong beam. In conclusion, we are able to validate the ICESat-2 ATLAS surface-height determination capability for the weak beam as well, with exemplary results over crevassed terrain shown in Figure 11f.

7.5. Analysis of Data for RGT 594, gt1l (Strong Beam)

Figure 12 presents the results of RGT 594 with its characteristics surge crevasses, with clear-cut edges, almost regular spacing and depths of 15 m (for details, see [3]). Track separation ranges from 20 m to 80 m with strong fluctuations. As the SkySat image (Figure 12b) and the comparison plot ULS-ICESat-2 indicates, crevasse provinces still line up correctly, which means that crevasses are identified where they exist.

Absolute surface heights show a difference in some paces, because of the track separation in time and location. Flight 2 was carried out on 13 August 2019, ATLAS data along RGT 594 were collected 5 August 2019, resulting in an 8-day time separation. For the example from gt1l (RGT 594), a distance-shift of ULS data was not applied, as it was not needed.

As in the example of the RGT 450 data, the ULS data for this track (RGT 594, gt1l) can be used for validation of crevasse detection and characterization of morphological types and their spatial characteristics. The crevasse profile, analyzed with the DDA-ice (Figure 12f) shows the clear-cut edges of recently opened crevasses that have not been eroded yet (where erosion and weathering would result in rounded edges, see [38]), especially in the section 2,153,000–2,153,600 m along-track distance. In contrast, the remainder of the profile segment 23 shows eroded crevasses (2,153,600–2,154,000). This example also indicates that relative dating of crevasse opening may be possible using ATLAS data and the DDA-ice.

7.6. Analysis of Data for RGT 594, gt1r (Weak Beam), Overpass 1

In this example, we find a close match of individual crevasses, in addition to the match of spatial crevasse characteristics. Notably this best matching of crevasses between ULS and ICESat-2 data exists for a track with separation of 10 to 45 m, and the locations of lowest separation do not line up with locations of best crevasse matches between the two instruments, airborne and satellite altimeters (Figure 13c). DDA-ice surface heights over crevassed terrain are retrieved correctly for the weak beam in this case as well. The results of the DDA-ice analysis (Figure 12f) show the typical square-cut blocky profile of a field of recently opened crevasses. Remarkably this type of morphological detail is reflected in the ATLAS data and their analysis. A distance shift was applied to the ULS data, to align crevasse morphologies with those in the ATLAS RGT 594 gt1r (weak beam) data.

7.7. Analysis of Data for RGT 594, gt1r (Weak Beam), Overpass 2

The plan for overpass 2 was to underfly the strong beam again, but we ended up closer to the weak beam, for which analyses are presented (Figure 14). During this part of the experiment, the helicopter drifted in high winds from the track location of the strong beam to that of the weak beam, ending up within 10 m from the weak beam. The ULS-ICESat-2 comparison has a larger apparent error than for overpass 1 of the same track. This is most likely attributable to an increased error in the IUM correction, due to larger and more variable IMU values during high winds, which in turn was caused by more helicopter attitude changes than during the earlier parts of flight 2.

As seen in Figure 14e, the crevasse characteristics are similar, despite the existing offset in absolute surface height values between ULS and ATLAS data. This example demonstrates particularly well that a crevasse characterization based on surface-roughness analysis is more robust than absolute surface-height determination.

8. Results Part 4: Track-Repeat Accuracies (ULS versus ICESat-2 ATLAS-Predicted and ATLAS-Actual)

The "actual" and "predicted" track repeat capabilities of our experiment setup are summarized in Table 5, which gives the average distances between (a) the ATLAS-predicted and the ATLAS-actual tracks and (b) the ATLAS-actual and ULS-actual tracks.

By design of the experiments, we aimed to repeat the predicted ATLAS tracks, which are based on the RGTs and the type of ATLAS track available during the flight experiments. The track-repeat capability for predicted ATLAS ground tracks is typically 18–20 m; this distance characterizes the combined capability of flight planning, helicopter flights in changing weather conditions and GPS and IMU corrections to underfly a (planned/ predicted) satellite track. Notably, the distance between predicted and actual ATLAS ground tracks has an average of 6.51 m to 15.65 m, with approximately 9 m for three out of five case studies. Therefore the track repeat distance between airborne and predicted satellite tracks (repeat capability) is only twice as large as the ATLAS-predicted to ATLAS-actual tracking capability, which is an impressive value for helicopter-based experiments.

The distance that characterizes the height comparisons between the ULS data and the ATLAS surface heights is the distance between actual-ATLAS and ULS (actual) ground tracks. Average distances between ATLAS-actual and ULS ground tracks is generally better than 25 m, with a minimum of 11.32 m and an outlier of 60 m track separation. In a kinematic experiment, it is not possible to achieve the co-locational accuracy of a static experiment, and observation of surface heights and crevasses on a surging glacier is not feasible (it is dangerous). The distances that we have achieved are sufficiently small to allow the surface-height validation.

It is sometimes, but not always, possible to match the location and height of individual crevasses. The next section will focus on the surface-height comparison and the role of summative quantitative characteristics, as well as on crevasse morphology.

9. Results Part 5: Comparison of Surface Heights and Crevasse Characteristics from ULS and ICESat-2 ATLAS Data. Surface Roughness

9.1. Surface Heights and Crevasse Morphologies

The numerical evaluation of of ICESat-2 surface heights, compared to heights from airborne observations, is carried out by an analysis of spatial surface structures, here crevasse characteristics (analysis of spatial surface roughness), rather than by point-wise height differencing. The primary reason for this approach to evaluation is that the evolution of the surge, and of glacial acceleration in general, is captured in crevasse characteristics. Comparison of absolute surface heights is less meaningful, for the following three reasons: (1) Due to the surge, the glacier moves at 1–2 m per day in the survey area in upper

Negribreen, according to a velocity analysis based on Sentinel-1 SAR data (Figure 4a) of the European Space Agency's (ESA's) Copernicus Mission. (2) There is a spatial track separation of zero to typically 20-25 m between the actual ATLAS ground tracks and the ULS ground tracks (see, Table 5) over a sloping glacier. (3) Time separation between flight 2 (13 August 2019) and ICESat-2 ATLAS data collection is 8 days for RGT 594 (5 August 2019) and 17 days for RGT450 (27 July 2019). Depending on track orientation, slope and velocity vector, these factors can cancel each other out or increase actual surface height differences.

Planet SkySat imagery was collected to support the ICESat-2 validation objective and the study of the surge. Comparison with a Planet SkySat image from 18 August 2019 (only 5 days after flight 2 on 13 August 2019) supports the finding that the same crevasse types are observed in the ULS data and ICESat-2 ATLAS data for all glacier crossings of our validation experiments. This matches the result from the analysis of track distances (Result Part 4, Section 8). Comparisons of ULS surface-height profiles and ATLAS surface-height profiles show the same characteristics with respect to presence/absence of crevasse fields, crevasse morphologies and crevasse spacing (for the latter, see Section 9.2). The crevasse profiles, analyzed with the DDA-ice, clearly show the clear-cut edges of recently opened surge crevasses in some fresh crevasse fields, contrasting to older crevasses that have been eroded by weathering, melting and snow deposition.

Differences in absolute height of the ice-surface between crevasses are attributed to the separation distance of the tracks over a sloping glacier and to the surge-related fast movement of the glacier during to the time separation between the airborne and satellite observations. Along-track distances were shifted a few meters to better align crevasses in the ULS and ATLAS data, in some experiments, to account for glacier movement between the times of the airborne experiments (13 August 2019) and ICESat-2 data collection (RGT 450: 27 July 2019, 17-day separation; RGT 594: 5 August 2019, 8-day separation). The ULS operates at 905 nm, which yields different penetration characteristics especially in slush and water.

The detail of the ICESat-2 DDA-ice surface heights in crevassed and smooth terrain and the comparison to airborne data indicates that surface heights, measured by ICESat-2, are accurately represented in the DDA-ice analysis.

9.2. *Crevasse Spacing, Crevasse Depth and Surface Roughness*

Crevasse spacing and depths for ATLAS segment 23 (RGT 594 gt1l) (Figure 11f) and matching ULS data are characterized by the values in the following table (Table 6), cited after [3]. These values match well, considering that the green light (532 nm) of the ATLAS sensor and the NIR light (905 nm) of the ULS have different penetration depths into snow, firn and slush and thus interact differently with the cryospheric materials at the bottom of the crevasses.

Table 6. Crevasse spacing and depths for Negribreen evaluation profile "segment 23 (RGT 594) 5 August 2019". Comparison of results from ULS airborne laser altimeter data and DDA-ice applied to ICESat-2 ATLAS data ATL03_20190805232841_05940403_002_01.h5. (From [3], Table 3).

	Mean Crevasse Spacing (m)	Maximum Crevasse Depth (m)	Mean Crevasse Depth (m) (>5 m)
DDA-ice	52.12	16.01	10.95
ULS	58.82	13.96	10.18

Surface roughness is analyzed for both the ULS data and the DDA-ice ATLAS surface heights, employing the $pond_{res}$ parameter of geostatistical characterization [31,39,40].

The $pond_{res}$ parameter is defined as the maximal value of the first-order residual vario function. First-order vario functions are used to mathematically summarize the spatial surface roughness in a given area. Vario functions are defined in a discrete mathematics framework, formally akin to the variogram of geostatistics (which requires the assumption that the data set may be considered a realization of a spatial random function that outlines the intrinsic hypothesis [41,42]). Other than the variogram, vario functions exist always, which makes them particularly useful for applications in computational geosciences. The first-order vario function is defined as follows [31,39]:

$$v_1(h) = \frac{1}{2n} \sum_{i=1}^{n} [z(x_i) - z(x_i + h)]^2 \tag{3}$$

for pairs of points $(x_i, z(x_i)), (x_i + h, z(x_i + h)) \in \mathcal{D}$, where $\mathcal{D}$ is a region in $\mathcal{R}^2$ (case of survey profiles) or $\mathcal{R}^3$ (case of survey areas) and n is the number of pairs separated by h; the distance value h is also termed "lag". The function $v_1(h)$ is called the *first-order vario function*. This function exists always and has a finite value. In situations where a regional trend underlies the data, the residual vario function is often more useful to analyze roughness than the vario function. Using

$$m(h) = \frac{1}{n} \sum_{i=1}^{n} [z(x_i) - z(x_i + h)], \tag{4}$$

the residual vario function is defined as:

$$res_1(h) = v_1(h) - \frac{1}{2} m(h)^2. \tag{5}$$

The results of the roughness analysis using the $pond_{res}$ parameter are shown in Figure 15, overlain on the Planet SkySat image. Surface roughness characteristics of both ATLAS and ULS surface profiles line up with the location of crevasse fields.

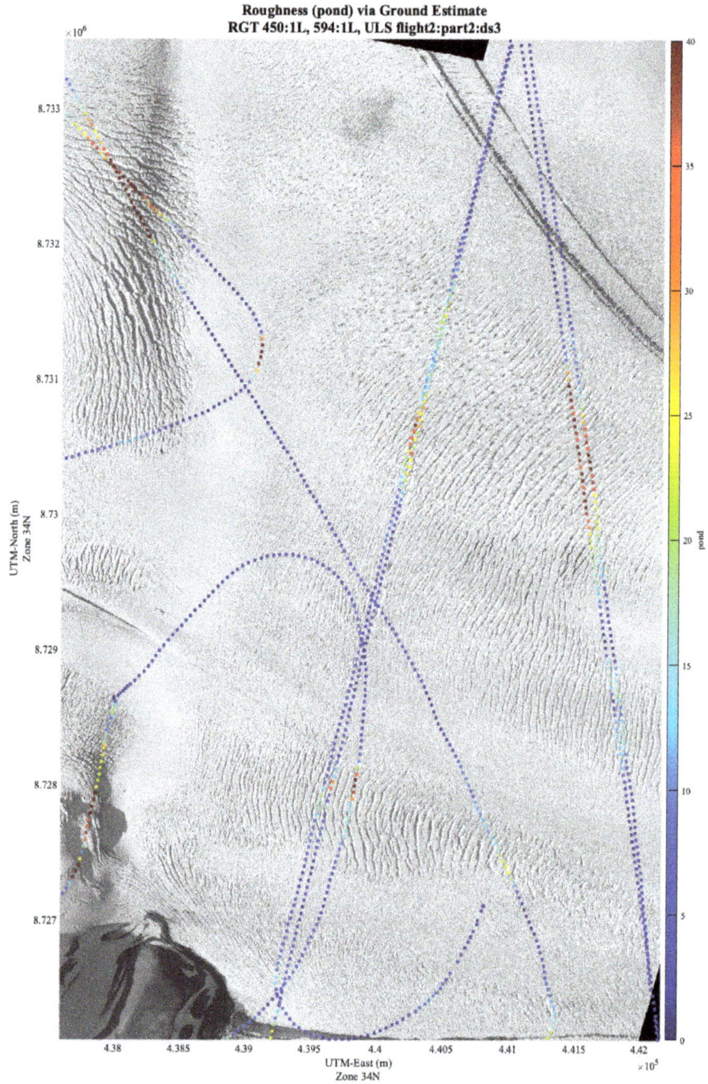

Figure 15. Ice-surface roughness in upper Negribreen during surge in 2019, from airborne altimeter data (ULS) data and ICESat-2 ATLAS Data, compared to Planet SkySat image data. ICESat-2 ATLAS Granules: [ATL03_20190805232841_05940403_002_01.h5] [RGT 594 gtl1 5 August 2019] and [ATL03_20190727132129_04500405_002_01.h5] [RGT 450 gtl1]. SkySat Image [20190818_150858_ssc9_u0002_panchromatic_dn.tif]. SkySat ssc9 data have 0.72 m pixel size. Roughness measure $pond_{res}$ calculated for ICESat-2 ATLAS data, analyzed with the DDA-ice, and for ULS data. ICESat-2 RGT594 and matching ULS tracks are the right almost straight set; ICESat-2 RGT450 and matching ULS tracks are the left almost straight sets. For RGT 594, the airborne track (ULS) is to the east (right) of the satellite track. For RGT 450, the airborne track and the satellite track are almost in the same location. Generally, surface roughness from parallel tracks and crossover locations matches between ICESat-2 and ULS. The $pond_{res}$ measure increases with intensity of crevassing evident in the SkySat image.

10. Summary and Conclusions

The topic of this paper is the airborne evaluation of ICESat-2 Advanced Topographic Laser Altimeter System (ATLAS) measurement capabilities and surface-height-determination over crevassed glacial terrain, with a focus on the geodetical accuracy of geophysical data collected from a helicopter. Negribreen, Svalbard, during surge in 2019 provided an ideal situation for the validation of ICESat-2 ATLAS measurements and surface height-determination over crevassed and otherwise complex ice surfaces, using the the density-dimension algorithm for ice surfaces (DDA-ice). As a result of the ongoing surge, different types of morphologically complex ice surface types, including many different types of wet and dry crevasses, were found in close proximity. Airborne geophysical data were collected during two campaigns in summers of 2018 and 2019. Airborne experiment setup, geodetical correction and data processing steps are described in this study, focussing on the 2019 data in the correction steps.

Experiment setup. The experiment setup consisted of a suite of small geophysical instruments mounted on a helicopter (an AS-350 "Eurocopter"), a LaserTech Universal Laser System (ULS) (a laser profilometer operating at 905 nm frequency), a Global Positioning System (GPS) receiver for collection of kinematic "rover" data, an Inertial Measurement Unit (IMU) for collection of attitude data of the platform and a GoPro Hero 5 for recording of on-board-time-lapse imagery. A GPS base station was placed on the side of the glacier for each flight experiment (and collected at the end of the flight). Photographs from handheld cameras complemented the surveys. Flight tracks of the airborne experiments were planned to repeat predicted ATLAS ground tracks, which are calculated based on the ICESat-2 Reference Ground Tracks (RGTs) and the only type of track data available prior to ICESat-2 overpasses. Following ICESat-2 data collection, actual ATLAS ground tracks become available after a few weeks following detailed geodetical correction. Planet SkySat data (0.72 m pixel size) were collected to support the validation campaign, the imagery closest in time to the 2019 flights stems from 8 August 2019 and has a 5-day separation to the 2019 flight data analyzed here.

ICESat-2 ATLAS Algorithm: DDA-ice. To obtain surface heights over crevassed and otherwise complex ice surfaces, ICESat-2 data are analyzed using the density-dimension algorithm for ice surfaces (DDA-ice). The DDA-ice is an algorithm that was specifically designed for the analysis of micro-pulse photon-counting laser altimeter data as collected by ATLAS. The DDA-ice utilizes the radial basis function for data aggregation into a density field, which adds a dimension, and an auto-adaptive threshold function for separation of signal from background. The DDA-ice yields surface heights at the nominal 0.7 m along-track spacing of ATLAS data (under clear-sky atmospheric conditions) and 2.5 m piece-wise linear interpolated surface heights and crevasse morphologies. In that this paper substantiates the geodetical results and provides detailed and accurate information of the validation campaign, the results here further validate the surface-height determination over crevassed and otherwise morphologically complex terrain using the DDA-ice, adding to the results reported in [3].

Geodetical Results. The study contributes to the relatively small amount of knowledge of the geodetical accuracy that can be obtained from kinematic data collection from a helicopter and thus may be of interest for future airborne satellite validation campaigns. We analyzed and compared the geodetical accuracy of (a) kinematic GPS data, collected by a rover on a helicopter and a base station on the side of the glacier, corrected in post-processing, (b) real-time kinematic (RTK) GPS data, where the updated position is calculated during flight and communicated to the pilot for adjustment of the flight track, and (c) kinematic GPS data, collected by a rover and post-processed without using base-station data. Findings of our study regarding these different GPS-data collection and correction modes include the following: (1) Kinematic GPS data collection with correction in post-processing yields higher accuracies than RTK data collection, because the frequent adjustments of the flight tracks lead to inaccuracies that more than offset the across-track distances resultant from straight lines flown across the glacier. (2) Processing of only the rover data using the Natural Resources Canada Spatial Reference System Precise Point Positioning (CSRS-PPP [34]) software is sufficiently accurate for the sub-satellite validation purpose of our study. (3) Distances between ICESat-2 ground tracks and airborne ground tracks were generally better than 25 m, while distances between predicted and actual ICESat-2 ground track were on the order of 9 m, which allows direct comparison of ice-surface heights and spatial statistical characteristics of crevasses from the satellite and airborne measurements.

Capability for track repeat ranges from an average of 11.32 m to 24.87 m for the distance between ICESat-2 ATLAS-actual and ULS (actual) ground tracks, with an outlier of 60 m. Differences between ICESat-2 ATLAS-predicted ground tracks (based on the RGTs) and ATLAS-actual ground tracks (determined after ICESat-2 data collection and geodetical correction) range from 6.51 m to 15.65 m for the average difference of a glacier crossing, with an approximately 9 m average difference as the most common value in the experiments in our study.

Technical Measurement Results. Following geodetical correction, the 2019 airborne campaign data over Negribreen are analyzed with respect to measurement accuracies and precisions. (4) The Lasertech Universal Laser System (ULS), operated at up to 300 m above ground level, yields full return frequency (400 Hz) and 0.06–0.08 m on-ice along-track spacing of height measurements. (5) Cross-over differences of airborne laser altimeter data are -0.172 ± 2.564 m along straight paths over generally crevassed terrain, which implies a precision of approximately 2.6 m for ICESat-2 validation experiments. (6) In summary, the comparatively light-weight experiment setup of a suite of small survey equipment mounted on a Eurocopter (Helicopter AS-350) and kinematic GPS data analyzed in post-processing using CSRS-PPP leads to high-accuracy repeats of ICESat-2 tracks.

Comparison of Surface Heights and Crevasse Characteristics from ULS and ICESat-2 ATLAS Data. Surface Roughness. As a result of the geodetical analysis and the technical measurement analysis, the data collected during the Negribreen airborne geophysical validation campaigns can be utilized for validation of ICESat-2 ATLAS data, analyzed with the DDA-ice. A detailed analysis of airborne and satellite data for and from RGT 450 and RGT 594, which cross different crevasse provinces in upper Negribreen, is presented. The analysis includes data from the weak beam and the strong beam of ATLAS (which differ in strength of the transmitted laser signal by a factor of 4) and a repeat experiment for one beam.

It is often, but not always, possible to match individual crevasses between the airborne height data and the ICESat-2 data. Differences are caused by the fast movement of the glacier combined with the distance between the airborne and satellite ground tracks. Instead of obtaining an always perfect match between individual crevasses, the morphological characteristics of a crevasse field are ultimately the objective of observation and analysis. Therefore, the comparison between airborne ULS data and ICESat-2 ATLAS data includes surface heights, crevasse spacing and depth, crevasse morphology (shape of the crevasse profile) and surface roughness. All parameters are found to be in good agreement. In conclusion, ICESat-2 surface heights, determined with the DDA-ice, are suitable for geophysical studies of fast-moving glaciers and glacial acceleration, especially surging.

Geophysical conclusions. The glaciological motivation for collection of data over a surge glacier is the investigation of the surge itself, including the evolution of the ice dynamics, glacial hydrology, surface height changes and mass transfer within the glacier system, as well as calving leading to mass transfer from the glacier system to the ocean as a contributing factor to sea-level change. In all of these contexts, the morphological characteristics of crevasse fields and ice fronts are more relevant than measurements of individual crevasse shapes. The validation of surface height measurements in this study supports the finding that the morphological characteristics, including crevasse depth, spacing and surface roughness of crevassed and uncrevassed areas, are in close agreement between the airborne and the satellite data (when analyzed with the DDA-ice). The latter finding was already reported in [3], this paper provides the geodetical substantiation of the geophysical results in the referenced paper and a larger base of analyzed examples from the field experiments. This paper is also a validation of the capabilities of the DDA-ice to identify crevasses and complex morphological shapes in ICESat-2 ATLAS data.

The final result of the validation is that ICESat-2 ATLAS data, analyzed with the DDA-ice, facilitate surface-height determination over crevassed terrain, in good agreement with airborne data, including spatial characteristics, such as surface roughness, crevasse spacing and depth, which are key informants on the deformation and dynamics of a glacier during surge.

Author Contributions: U.C.H. designed the study, carried out part of the analysis and wrote the paper. M.L. and T.T. carried out most of the data analysis and contributed to writing the paper. T.N. carried out an essential part of the GPS data analysis and contributed to writing the paper. All authors have read and agreed to the published version of the manuscript.

Funding: Research and data collection were supported by the U.S. National Aeronautics and Space Administration (NASA) Earth Sciences Division under awards 80NSSC20K0975, 80NSSC18K1439 and NNX17AG75G and by the U.S. National Science Foundation (NSF) under awards OPP-1745705 and OPP-1942356 (Principal Investigator for all awards: Ute Herzfeld). The data collection was also partly supported through a 2018 Access Pilot Project (2017_0010) of the Svalbard Integrated Observing System (SIOS). All this support is gratefully acknowledged.

Institutional Review Board Statement: Not applicable.

Informed Consent Statement: Not applicable.

Data Availability Statement: (1) ICESat-2 data products are available under https://earthdata.nasa.gov/ (accessed 3 October 2021) provided by the National Snow and Ice Data Center (NSIDC). The ICESat-2 ATLAS data product ATL03, and any other ICESat-2 ATLAS data product cited in this paper, can be identified by their digital object identifiers (dois), as given in the references. (2) Data collected as part of the Negribreen Airborne Geophysical Campaigns, collected by the first author and her team, in 2018 are available through the NSF Arctic Data Center and can be accessed at https://arcticdata.io/data/10.18739/A2QF8JK7T/ (accessed on 3 October 2021). The data sets are referenced as follows: Ute Herzfeld and Thomas Trantow, Airborne Laser Altimeter, Global Positioning System (GPS), Inertial Measurement Unit (IMU) and Imagery Campaign of the Surging Negribreen Glacier, Svalbard, in July 2017 and July 2018. Arctic Data Center, doi:/data/10.18739/A2QF8JK7T. The project landing page is: https://arcticdata.io/catalog/view/doi%3A10.18739%2FA2QF8JK7T (accessed on 3 October 2021). (3) Planet SkySat data are in the process of becoming available through the NASA Commercial SmallSat Data Acquisition Project (CSDAP) https://earthdata.nasa.gov/esds/csdap (accessed on 3 October 2021). Results from the Pilot Project of the first author are included in the Report, accessible at https://cdn.earthdata.nasa.gov/conduit/upload/14180/CSDAPEvaluationReport_Apr20.pdf (accessed on 3 October 2021) (4) Landsat-8 data are freely available through the U.S. Geological Survey, see https://www.usgs.gov/core-science-systems/nli/landsat/landsat-data-access?qt-science_support_page_related_con=0#qt-science_support_page_related_con (accessed on 3 October 2021) Landsat data products held in the USGS archives can be searched and downloaded at no charge from a variety of data portals. This page provides information about accessing Landsat Collection 2 and Landsat Collection 1 Level-1 and Level-2 data products (as quoted from the cited webpage).

Acknowledgments: Thanks are due to colleagues and staff at SIOS, NPI, UNIS and Airlift and captain and crew of R/V Lance, especially to our pilots and technicians Anders Bjørghum, Gunnar Nordahl, Tor Andre Vaksdal, Harald Sandal and Gustav Svanström (Airlift), to Jack Kohler, Havard Hansen, Harvey Goodwin, Geir Ove Aspnes, Jørn Dybdahl, Norwegian Polar Institute, and Chris Borstad, University Centre in Svalbard (UNIS), now University of Montana, to Inger Jennings and Heikki Lihavainen (SIOS), to Kristin Woxholth, Longyearbyen, for help with logistical support, to the ICESat-2 Project, especially Thomas Neumann, Kaitlin Harbeck, David Hancock and Anthony Martino for collaboration and support regarding ICESat-2, to Annie Zaino, Joe Pettit, Spencer Niebuhr (UNAVCO Boulder) for help with GPS data collection and correction, to Brian Donahue, Canadian Geodetic Survey, for information on CSRS-PPP to Compton Tucker and Manil Maskey (NASA) and Paris Good, Planet, for coordination of SkySat data acquisition under the NASA Commercial SmallSat Data Acquisition Project (CSDAP), to Laurie Padman and Susan Howard (Earth&Space Research, Corvallis, Oregon) for comments on the study, and to Connor Myers and Sam Bennetts,

Geomathematics Group, University of Colorado Boulder, for assistance with data collection over Negribreen in 2018 and 2017, respectively, and last not least to Shridhar Jawak, UNIS, Longyearbyen, Svalbard for inviting us to contribute to the Virtual Remote-Sensing Conference. The helicopter was provided by the Norwegian Polar Institute and operated by Airlift. Collection of airborne data was conducted with permission of the National Security Authority of Norway, the Civil Aviation Authority of Norway and the Governor of Svalbard, registered as Research in Svalbard Project RIS-10827 "NEGRIBREEN SURGE" (2017–2019). All this support is gratefully acknowledged.

Conflicts of Interest: The authors declare no conflict of interest.

References

1. Markus, T.; Neumann, T.; Martino, A.; Abdalati, W.; Brunt, K.; Csatho, B.; Farrell, S.; Fricker, H.; Gardner, A.; Harding, D.; et al. The Ice, Cloud, and land Elevation Satellite-2 (ICESat-2): Science requirements, concept, and implementation. *Remote. Sens. Environ.* **2017**, *190*, 260–273. [CrossRef]
2. Neumann, T.A.; Martino, A.J.; Markus, T.; Bae, S.; Bock, M.R.; Brenner, A.C.; Brunt, K.M.; Cavanaugh, J.; Fernandes, S.T.; Hancock, D.W.; et al. The Ice, Cloud, and Land Elevation Satellite–2 mission: A global geolocated photon product derived from the Advanced Topographic Laser Altimeter System. *Remote. Sens. Environ.* **2019**, *233*, 111325. [CrossRef] [PubMed]
3. Herzfeld, U.; Trantow, T.; Lawson, M.; Hans, J.; Medley, G. Surface heights and crevasse types of surging and fast-moving glaciers from ICESat-2 laser altimeter data—Application of the density-dimension algorithm (DDA-ice) and validation using airborne altimeter and planet skysat data. *Sci. Remote. Sens.* **2021**, *3*, 100013. [CrossRef]
4. Schutz, B.; Zwally, H.; Shuman, C.; Hancock, D.; DiMarzio, J. Overview of the ICESat Mission. *Geophys. Res. Lett.* **2005**, *32*. [CrossRef]
5. Zwally, H.; Schutz, B.; Abdalati, W.; Abshire, J.; Bentley, C.; Brenner, A.; Bufton, J.; Dezio, J.; Hancock, D.; Harding, D.; et al. ICESat's laser measurements of polar ice, atmosphere, ocean, and land. *J. Geodyn.* **2002**, *34*, 405–445. [CrossRef]
6. Stocker, T.F.; Qin, H.; Plattner, G.-K.; Tignor, M.; Allen, S.K.; Boschung, J.; Nauels, A.; Xia, Y.; Bex, V.; Midgley, P.M.E. *Climate Change 2013: The Physical Science Basis. Contribution of Working Group I to the Fifth Assessment Report of the Intergovernmental Panel on Climate Change*; Cambridge University Press: Cambridge, UK, 2013.
7. Herzfeld, U.C.; Mayer, H. Surge of Bering Glacier and Bagley Ice Field, Alaska: An update to August 1995 and an interpretation of brittle-deformation patterns. *J. Glaciol.* **1997**, *43*, 427–434. [CrossRef]
8. Herzfeld, U.C.; McDonald, B.; Wallin, B.F.; Krabill, W.; Manizade, S.; Sonntag, J.; Mayer, H.; Yearsley, W.A.; Chen, P.A.; Weltman, A. Elevation changes and dynamic provinces of Jakobshavn Isbræ, Greenland, derived using generalized spatial surface roughness from ICESat GLAS and ATM data. *J. Glaciol.* **2014**, *60*, 834–848. [CrossRef]
9. Mayer, H.; Herzfeld, U. Structural glaciology of the fast-moving Jakobshavn Isbræ, Greenland, compared to the surging Bering Glacier, Alaska, USA. *Ann. Glaciol.* **2000**, *30*, 243–249. [CrossRef]
10. Mayer, H.; Herzfeld, U. A structural segmentation, kinematic analysis and dynamic interpretation of Jakobshavns Isbræ, West Greenland. *Z. Gletsch. Glazialgeol.* **2001**, *37*, 107–124.
11. Trantow, T.; Herzfeld, U.C. Crevasses as indicators of surge dynamics in the Bering Bagley Glacier System, Alaska: Numerical experiments and comparison to image data analysis. *J. Geophys. Res. Earth Surf.* **2018**, *123*, 1615–1637. [CrossRef]
12. Herzfeld, U.; Trantow, T.; Bennetts, S. Surge-forced structural disintegration, enhanced calving and resultant rapid mass loss of a large Arctic fjord glacier (Negribreen, Svalbard). *GRL* **2018**, *submitted*.
13. Murray, T.; Strozzi, T.; Luckman, A.; Jiskoot, H.; Christakos, P. Is there a single surge mechanism? Contrasts in dynamics between glacier surges in Svalbard and other regions. *J. Geophys. Res.* **2003**, *108*, 2237. [CrossRef]
14. Smith, B.; Fricker, H.; Gardner, A.; Siegfried, M.; Adusumilli, S.; Csatho, B.; Holschuh, N.; Paolo, F. *ICESat-2 L3a Land Ice Height, Version 3*; NASA ICESat-2 Project; NASA: Boulder, CO, USA, 2020. [CrossRef]
15. Smith, B.; Hancock, D.; Harbeck, K.; Roberts, L.; Neumann, T.; Brunt, K.; Fricker, H.; Gardner, A.; Siegfried, M.; Adusumilli, S.; et al. *ICESat-2 Algorithm Theoretical Basis Document for Land Ice Along-Track Height Product, February 20, 2020*; NASA ICESat-2 Project; NASA: Greenbelt, MD, USA, 2020; 107p. [CrossRef]
16. Brunt, K.M.; Smith, B.E.; Sutterly, T.C.; Kurtz, N.T.; Neumann, T.A. Comparisons of Satellite and Airborne Altimetry With Ground-Based Data From the Interior of the Antarctic Ice Sheet. *Geophys. Res. Lett.* **2020**, *48*, e2020GL090572. [CrossRef]
17. Cui, H.; Li, R.; Li, H.; Hao, T.; Qiao, G.; He, Y.; Hai, G.; Xie, H.; Cheng, Y.; Li, B. Field validation of ICESat-2 data along CHINARE Route in East Antarctica. In *The International Archives of the Photogrammetry, Remote Sensing and Spatial Information Sciences*; XLIII-B3-2021 (XXIV ISPRS Congress); Copernicus: Göttingen, Germany, 2021; pp. 443–448. [CrossRef]
18. Neumann, T.; Brenner, A.; Hancock, D.; Robbins, J.; Saba, J.; Harbeck, K.; Gibbons, A.; Lee, J.; Luthcke, S.; Rebold, T. *ICESat-2 Algorithm Theoretical Basis Document for Global Geolocated Photons, 15 October 2019*; NASA ICESat-2 Project; NASA: Greenbelt, MD, USA, 2019; 202p. [CrossRef]
19. Neumann, T.; Brenner, A.; Hancock, D.; Robbins, J.; Saba, J.; Harbeck, K.; Gibbons, A.; Lee, J.; Luthcke, S.; Rebold, T. *ATLAS/ICESat-2 L2A Global Geolocated Photon Data, Version 3*; NASA ICESat-2 Project; NASA: Greenbelt, MD, USA, 2020. [CrossRef]
20. Magruder, L.; Neumann, T.; Kurtz, N. ICESat-2 early mission synopsis and observatory performance. *Earth Space Sci.* **2021**, *8*, e2020EA001555. [CrossRef]

21. Herzfeld, U.; Trantow, T.; Harding, D.; Dabney, P. Surface-height determination of crevassed glaciers—Mathematical principles of an Auto-Adaptive Density-Dimension Algorithm and validation using ICESat-2 Simulator (SIMPL) data. *IEEE Trans. Geosci. Remote. Sens.* **2017**, *55*, 1874–1896. [CrossRef]
22. Luthcke, S.B.; Pennington, T.; Rebold, T.; Thomas, T. *Algorithm Theoretical Basis Document (ATBD) for ATL03g ICESat-2 Receive Photon Geolocation, 28 October 2019*; NASA ICESat-2 Project; NASA: Greenbelt, MD, USA, 2019; 63p.
23. Luthcke, S.; Zelensky, N.; Rowlands, D.; Lemoine, F.; Williams, T. The 1-centimeter orbit: Jason-1 precision orbit determination using gps, slr, doris, and altimeter data special issue: Jason-1 calibration/validation. *Mar. Geod.* **2003**, *26*, 399–421. [CrossRef]
24. Petit, G.; Luzum, B. *IERS Conventions (2010) IERS*; Technical Note No. 36; International Earth Rotation and Reference Systems Service: Frankfurt, Germany, 2010.
25. Dabney, P.W.; Harding, D.J.; Abshire, J.; Huss, T.; Jodor, G.; Machan, R.; Marzouk, J.; Rush, K.; Seas, A.; Shuman, C.; et al. The Slope Imaging Multi-polarization Photon-counting Lidar: Development and performance results. In *Geoscience and Remote Sensing Symposium (IGARSS), 2010 IEEE International*; IEEE: Piscataway, NJ, USA, 2010; pp. 653–656. Available online: http://ieeexplore.ieee.org/xpls/abs_all.jsp?arnumber=5650862 (accessed on 3 October 2021).
26. Harding, D.; Dabney, P.; Valett, S.; Yu, A.; Vasilyev, A.; Kelly, A. Airborne polarimetric, two-color laser altimeter measurements of lake ice cover: A pathfinder for nasa's icesat-2 spaceflight mission. In *Geoscience and Remote Sensing Symposium (IGARSS), 2011 IEEE international*; IEEE: Piscataway, NJ, USA, 2011; pp. 3598–3601.
27. Palm, S.; Yang, Y.; Herzfeld, U. *ICESat-2 Algorithm Theoretical Basis Document for the Atmosphere, Part I: Level 2 and 3 Data Products, February 14, 2020*; NASA ICESat-2 Project; NASA: Greenbelt, MD, USA, 2020; 104p. [CrossRef]
28. Palm, S.; Yang, Y.; Herzfeld, U.; Hancock, D.; Barbieri, K.; Wimert, J. *ATLAS/ICESat-2 L2A Normalized Relative Backscatter Profiles, Version 3*; NASA ICESat-2 Project; NASA: Greenbelt, MD, USA, 2020. [CrossRef]
29. Neumann, T.; Brenner, A.; Hancock, D.; Robbins, J.; Saba, J.; Harbeck, K.; Gibbons, A.; Lee, J.; Luthcke, S.; Rebold, T. *ICESat-2 Algorithm Theoretical Basis Document for Global Geolocated Photons, April 2020*; NASA ICESat-2 Project; NASA: Greenbelt, MD, USA, 2020; 207p. [CrossRef]
30. Crocker, R.I.; Maslanik, J.A.; Palo, S.E.; Adler, J.J.; Herzfeld, U.C.; Emery, W.J. A sensor package for ice surface characterization using small unmanned aircraft systems. *IEEE Trans. Geosc. Remote Sens.* **2011**, *49*, 1033–1047. [CrossRef]
31. Herzfeld, U.; Hunke, E.; McDonald, B.; Wallin, B. Sea Ice Deformation in Fram Strait—Comparison of CICE simulations with analysis and classification of airborne remote-sensing data. *Cold Reg. Sci. Technol.* **2015**, *117*, 19–33. [CrossRef]
32. Brodin, G.; Cooper, J.; Stevens, J. Measuring the effect of helicopter rotors on GPS reception. *Aeronaut. J. New Ser.* **2007**, *111*, 561–570. [CrossRef]
33. Banville, S.; Hassen, E.; Lamothe, P.; Farinaccio, J.; Donahue, B.; Mireault, Y.; Goudarzi, M.A.; Collins, P.; Ghoddousi-Fard, R.; Kamali, O. Enabling ambiguity resolution in CSRS-PPP. *Navigation* **2021**, *68*, 433–451. [CrossRef]
34. Tétreault, P.; Kouba, J.; Héroux, P.; Legree, P. CSRS-PPP: an internet service for GPS user access to the Canadian Spatial Reference Frame. *Geomatica* **2005**, *59*, 17–28.
35. Langley, R.B. Rtk gps. *Gps World* **1998**, *9*, 70–76.
36. Parent, R. *Computer Animation, Algorithms and Techniques*; Elsevier: Amsterdam, The Netherlands, 2012. [CrossRef]
37. Herzfeld, U.C.; Clarke, G.K.C.; Mayer, H.; Greve, R. Derivation of deformation characteristics in fast-moving glaciers. *Comput. Geosci.* **2004**, *30*, 291–302. [CrossRef]
38. Herzfeld, U.C.; McDonald, B.; Stachura, M.; Hale, R.G.; Chen, P.; Trantow, T. Bering Glacier surge 2011: Analysis of laser altimeter data. *Ann. Glaciol.* **2013**, *54*, 158–170. [CrossRef]
39. Herzfeld, U.C. Vario functions of higher order–definition and application to characterization of snow surface roughness. *Comput. Geosci.* **2002**, *28*, 641–660. [CrossRef]
40. Herzfeld, U.C. Master of the obscure—Automated geostatistical classification in presence of complex geophysical processes. *Math. Geosci.* **2008**, *40*, 587–618. [CrossRef]
41. Matheron, G. Principles of geostatistics. *Econ. Geol.* **1963**, *58*, 1246. [CrossRef]
42. Matheron, G. The intrinsic random functions and their applications. *Adv. Appl. Probab.* **1973**, *5*, 439–468. [CrossRef]

remote sensing

MDPI

Article

Fifty Years of Tidewater Glacier Surface Elevation and Retreat Dynamics along the South-East Coast of Spitsbergen (Svalbard Archipelago)

Jan Kavan [1,2,*], Guy D. Tallentire [3], Mihail Demidionov [1], Justyna Dudek [4] and Mateusz C. Strzelecki [2]

1 Polar-Geo Lab, Faculty of Science, Masaryk University, 61137 Brno, Czech Republic; demidionovmihail@gmail.com
2 Institute of Geography and Regional Development, University of Wroclaw, 50-137 Wroclaw, Poland; mateusz.strzelecki@uwr.edu.pl
3 Geography and Environment, Loughborough University, Loughborough LE11 3TU, UK; g.d.tallentire@lboro.ac.uk
4 Institute of Geography and Spatial Organization, Polish Academy of Sciences, 00-818 Warsaw, Poland; j.dudek@geopan.torun.pl
* Correspondence: jan.kavan.cb@gmail.com

Abstract: Tidewater glaciers on the east coast of Svalbard were examined for surface elevation changes and retreat rate. An archival digital elevation model (DEM) from 1970 (generated from aerial images by the Norwegian Polar Institute) in combination with recent ArcticDEM were used to compare the surface elevation changes of eleven glaciers. This approach was complemented by a retreat rate estimation based on the analysis of Landsat and Sentinel-2 images. In total, four of the 11 tidewater glaciers became land-based due to the retreat of their termini. The remaining tidewater glaciers retreated at an average annual retreat rate of 48 m year^{-1}, and with range between 10–150 m year^{-1}. All the glaciers studied experienced thinning in their frontal zones with maximum surface elevation loss exceeding 100 m in the ablation areas of three glaciers. In contrast to the massive retreat and thinning of the frontal zones, a minor increase in ice thickness was recorded in some accumulation areas of the glaciers, exceeding 10 m on three glaciers. The change in glacier geometry suggests an important shift in glacier dynamics over the last 50 years, which very likely reflects the overall trend of increasing air temperatures. Such changes in glacier geometry are common at surging glaciers in their quiescent phase. Surging was detected on two glaciers studied, and was documented by the glacier front readvance and massive surface thinning in high elevated areas.

Keywords: tidewater glaciers; surface elevation changes; glacier geometry; Svalbard

Citation: Kavan, J.; Tallentire, G.D.; Demidionov, M.; Dudek, J.; Strzelecki, M.C. Fifty Years of Tidewater Glacier Surface Elevation and Retreat Dynamics along the South-East Coast of Spitsbergen (Svalbard Archipelago). *Remote Sens.* **2022**, *14*, 354. https://doi.org/10.3390/rs14020354

Academic Editors: Veijo Pohjola, Shridhar D. Jawak, Andreas Kääb, Hiroyuki Enomoto, Geir Moholdt, Kjell Arild Høgda, Małgorzata Błaszczyk, Bo N. Andersen, Ann Mari Fjæraa, Bartłomiej Luks, Roberto Salzano and Frode Dinessen

Received: 15 November 2021
Accepted: 10 January 2022
Published: 13 January 2022

1. Introduction

Glaciers cover almost 60% of Svalbard [1], constituting the predominant part of the local ecosystem. The interior of the archipelago is covered by vast ice caps often flowing directly into the ocean. This is also the case of the eastern coast of Svalbard, where a large part of the coastline is formed by fronts of marine terminating glaciers. Only small areas are formed by steep cliffs where glaciers are not present. Glaciers are essential stores of water in the Arctic, releasing the stored water during a relatively short melting period, usually between three to four months on the south-east coast of Svalbard [2]. Tidewater glaciers present on Svalbard's east coast are often considered as biological hotspots providing the adjacent marine environment with essential nutrients and other mineral material. Many animal species rely heavily on this element of the ecosystem, typically marine mammals or birds in Svalbard [3]. The inflow of nutrients also supports primary production near glacier calving fronts [4,5].

Blaszczyk et al. [6] identified 163 tidewater glaciers present on the Svalbard archipelago with a total length of 860 km calving ice-cliffs, while, Nuth et al. [1] identified 740 km of tidewater glacier terminus width, only a few years later. Blaszczyk et al. [6] also stressed the importance of calving, which contributes as much as 21% to the overall mass loss from Svalbard glaciers. Hagen et al. [7] estimated the calving of tidewater glaciers to contribute slightly less (approximately 15%). Martín-Moreno et al. [8] reported a significant loss of glacier area, namely in the south and south-east of Svalbard, where a reduction of approximately 16% was estimated since the end of the Little Ice Age.

There is widespread agreement that Svalbard's glaciers have been losing mass since the Little Ice Age, with over 5000 km^2 of ice lost since the glaciers and ice caps of the archipelago were at their most recent maxima [8]. In this period, the greatest retreat rates were observed for tidewater glaciers which occurred in response to oceanic warming [9]. The last 50 years have seen an increase in the rate of mass loss of Svalbard's glaciers, showing an even greater negative mass balance since 2000, a result of Svalbard being one of the fastest warming regions on Earth [10]. Much of this understanding comes from in-situ mass balance measurements carried out over four decades on some of the region's glaciers, and is supported by remote sensing observations which also indicate increased glacier frontal retreat and mass loss [1,11]. Despite the recent advancement in knowledge about the response of Svalbard's glaciers to climate change, derived from these datasets, very little is known about the evolution of glaciers in Svalbard in the early 20th century [12]. Most of the mass balance observations and measurements on Svalbard have been carried out on small land-based glaciers in the west and central regions of the archipelago [7]. The areas most difficult to access, including the eastern and the southern coasts of Spitsbergen, are among the least studied parts of the Svalbard archipelago [13–16]. Direct mass balance measurements cover less than 0.5% of the glaciated area of Svalbard. The reference glaciers are on the west of the archipelago, and the climatic differences between the regions of study and a lack of similar in situ measurements on the eastern coast make it difficult to predict glacier evolution in this area during the same period [17].

Precise historic geographic data are rather sparse in the high Arctic and most of the information on glacier dynamics is usually limited to the last 20 or 30 yearswhen remote sensing data has become more readily available [18–22], or are limited to a single glacier where past direct observational data are available [23–25]. Members of the glaciological community are trying to bridge the gap by using different modelling approaches [12,26,27].

The aim of this paper is to describe the retreat rate and surface elevation changes of the eleven tidewater glaciers, as well as the overall development of glaciers on the east coast of Svalbard. To achieve this, we use historic satellite and aerial images, and a 1970 digital elevation model (DEM), which represents a unique opportunity to study long-term glacier dynamics.

2. Material and Methods

2.1. Study Site

The study site is the south-eastern region of Spitsbergen, the largest island of the Svalbard archipelago, and covers 77.5 km of coastline, with 11 glaciers terminating into the Barents Sea in 1970 (Figure 1). The total glacier area as defined in 2019 is 283 km^2. Note that the 1970 DEM does not cover the complete area of all the glaciers studied. The basic characteristics of all glaciers studied are summarized in Table 1.

The archipelago of Svalbard intersects three large water bodies: the Greenland Sea to the west, the Barents Sea to the east, and the Arctic Ocean to the north. The temperature of the surrounding water masses influences the climate of the archipelago, which is milder than that of other areas at the same latitude and, at the same time, more sensitive to changes relating to atmospheric fronts [28]. The eastern part of Svalbard experiences a cold climate characterized by relatively low temperatures as a result of the cooler waters of the East Spitsbergen Current flowing along its coastline, which often brings sea ice from the interior of the Arctic. During the Little Ice Age and in the 19th century, the eastern shores of Spitsbergen were often blocked by pack ice even during summer [14].

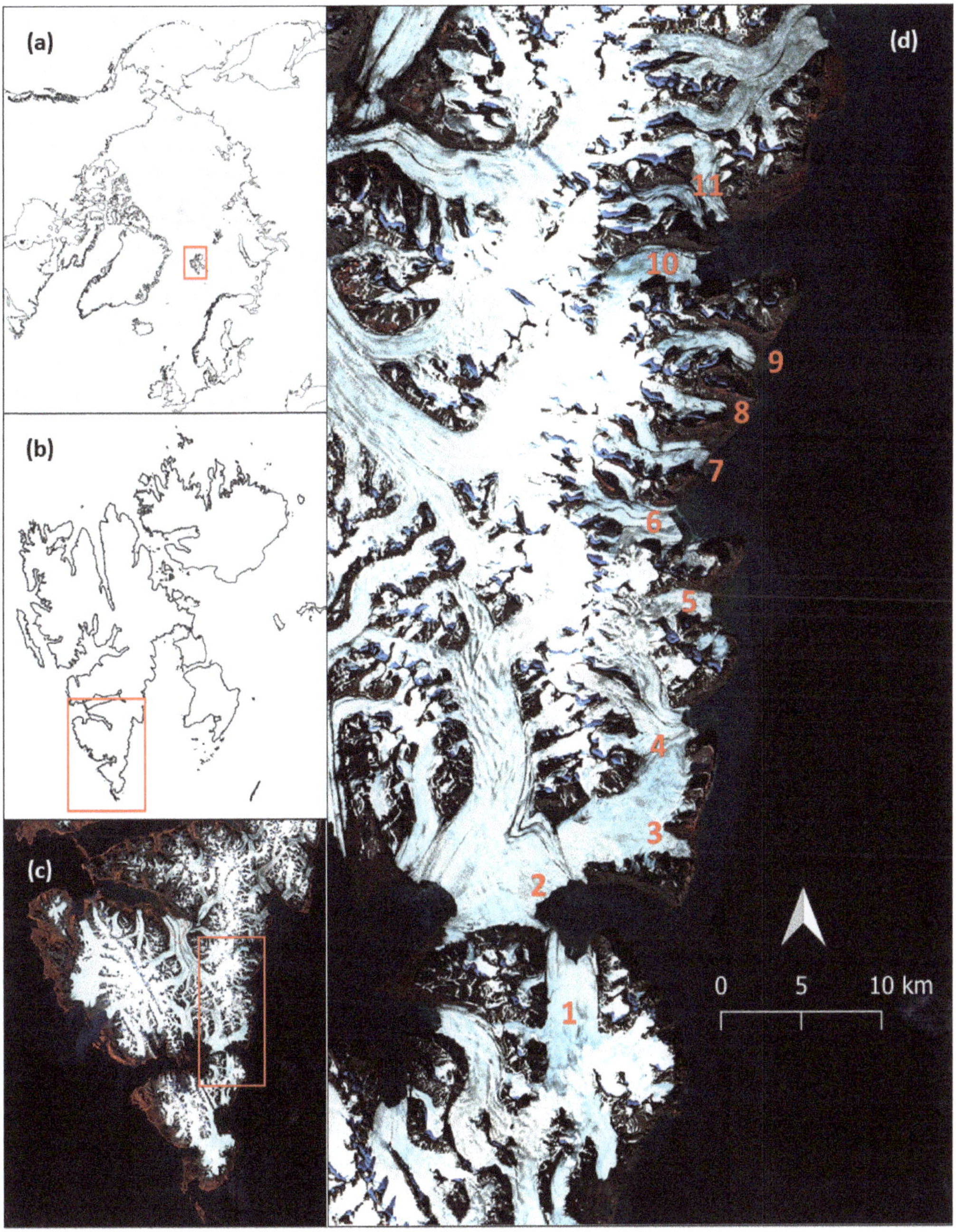

Figure 1. Study site; (**a**) the Arctic with location of Svalbard; (**b**) Svalbard with close up of the southern Spitsbergen island; (**c**) Sentinel-2 satellite image of southern Spitsbergen in false color; (**d**) the glaciers used in this study: 1—Sykorabreen, 2—Hambergbreen, 3—Staupbreen, 4—Markhambreen, 5—Crollbreen, 6—Davisbreen, 7—Skimmebreen, 8—Bellingbreen, 9—Anna Margrethebreen, 10—Emil'janovbreen, 11—Spaelbreen.

Table 1. Glacier characteristics (area and length derived from 2017 Sentinel images; elevation data based on 1970 DEM).

Glacier	Id (Figure 1)	Area (km^2)	Length (km)	Mean Elevation (m)	Maximum Elevation (m)	Present Terminus
Sykorabreen	1	54.6	16.7	248	550	Marine
Hambergbreen	2	16.6	13.2	189	360	Marine
Staupbreen	3	20.2	7.7	257	610	Land
Markhambreen	4	44	9.2	278	640	Marine
Crollbreen	5	20.4	8.8	279	630	Marine
Davisbreen	6	31.2	12.1	252	570	Marine
Skimmebreen	7	14.6	8.1	261	580	Land
Bellingbreen	8	6.7	6.6	266	570	Land
Anna Margrethbreen	9	14	8.4	208	550	Land
Emil'janovbreen	10	33.7	16.4	352	710	Marine
Spaelbreen	11	26.9	9.9	291	670	Marine

This stands in contrast with the western coast, where the West Spitsbergen Current brings warmer waters (due to its connection with the Gulf Stream) flowing from the south, restricting the duration of sea ice cover to a shorter temporal period [29,30]. These climatic differences are also well reflected in the presence of vegetation cover on the western coast compared to the eastern coast which has minimal vegetation (see Figure 1c) [31]. Such climatic characteristics result in the presence of vast ice caps and numerous glaciers which are marine terminating, making tidewater glaciers the prevalent glacier type on the east coast of Svalbard. Nordli et al. [32,33] reported the long-term air temperature series showing a distinct increasing trend since the 1970s, despite the timeseries being a composite for the whole of Svalbard. The eastern coast of Svalbard is not permanently monitored [28,29], but it is suggested that the air temperature has similar temporal patterns as the composite series shown by e.g., Sinnhuber [34] for the north-east region of Svalbard. The morphology of the terrain is heterogeneous with steep cliffs in the coastal areas reaching up to 500 m a.s.l., which are carved by large glaciers flowing down from the ice caps and plateaus towards the ocean. The highest peaks in the accumulation zone reach up to 800 m a.s.l.

The first modern comprehensive inventory of Svalbard glaciers was compiled by Hagen et al. [28], which constitutes the basis to describe glaciers included in this study. Blaszczyk et al. [6] specifically described Svalbard's tidewater glaciers with emphasis on flow velocity, calving fluxes, and mass loss. Many tidewater glaciers in Svalbard have experienced surge events during the last 150 years [13,35–40]. Such activity is described for most of the glaciers in this study [6,13], even though there is only one reported surge event on Markhambreen from around 2010. Noormets et al. [41] reported two major surge events prior to the study period at Hambergbreen. The eastern coast of Svalbard is inaccessible and thus lacks any long-term direct mass balance monitoring, with most of our knowledge based on remote sensing data often with restricted spatial and temporal resolution [10].

2.2. Detection of Surface Elevation Changes

Two digital elevation models (DEMs) were used to demonstrate the evolution of the glaciers' geometry. The first DEM was produced by Norwegian Polar Institute (NPI) and is derived from aerial images taken in the summer of 1970. The spatial resolution of the model is 20 m. The second DEM, produced by the Polar Geospatial Data Center at University of Minnesota (ArcticDEM), has a 2 m resolution [42]. The final Mosaic product (a compilation of several individual tiles) was used as the study site covered a relatively large area and no individual overflight (or series of overflights from one single year) was detected as covering the study site completely. This has obviously limited the precise quantification of the surface elevation changes, as no precise date could be attributed to the ArcticDEM. Through comparison with Sentinel-2 derived front positions of the glaciers, the ArcticDEM is most likely averaged to the state in the year 2017.

The ArcticDEM is projected to the National Snow and Ice Data Center (NSIDC) Sea Ice Polar Stereographic North and referenced to the ellipsoidal WGS84 horizontal datum

(EPSG:3413). The 1970 NPI DEM is projected using the European Terrestrial Reference System 1989 using the GRS 1980 ellipsoid (local reference system UTM-zone 33). As a result, the ArcticDEM shows a systematic shift in altitude of 28.5 m compared to the local reference system. The discrepancy between the two projection systems was solved by correction of the ArcticDEM. The shift of 28.5 m was estimated with use of 20 ground control points spread evenly throughout the study area. Comparison of the elevation values was made on flat surfaces in the study area in order to avoid errors resulting from the different resolutions of the DEMs. The standard deviation of the elevation shift was 1.03 m. The differences are likely to be more obvious in areas with high slope or heterogeneous terrain, which has to be considered when interpreting the results. Therefore, the quantification of the surface elevation changes serves mostly for identifying the spatial patterns of the changes and not for precise calculations.

The surface elevation changes were obtained as a simple overlap of the two DEMs, excluding the non-glacierized areas. The resulting surface elevation changes were used to define the zero-surface elevation change altitude as a parameter describing mass loss and gain. Glacier boundaries were defined using Sentinel-2 images from 2017 (frontal and lateral zones of the glacier) and ArcticDEM derived boundaries in the accumulation zone (based on flow accumulation function). The longitudinal profiles for each of the glaciers were derived in the centrelines and consisted of the profile derived from the 1970 DEM and ArcticDEM.

2.3. Glacier Front Positions

A series of Sentinel-2 and Landsat 5, 7, and 8 images were used to detect the positions of glacier fronts in 2000, 2010, and 2019. Positions of glacier fronts in 1990 were derived directly from the digital map database of the Norwegian Polar Institute. These were calculated based on a 1990 aerial mapping survey. Overall, two glaciers not covered by the photogrammetric overflight in 1990—Sykorabreen and Hambergbreen—and were instead mapped using Landsat imagery acquired in the summer of the same year. Glacier front positions in 1970 were derived directly from the 1970 DEM as described above. The satellite images were downloaded via the Sentinel Hub EO browser. The images were already georeferenced, and the TIFF files were directly processed in the QGIS environment. The positions of the glacier fronts were delimited manually for each of the years studied. Images obtained at the end of the summer seasons were used for the detection of front positions to keep consistency in the detection process and to avoid irregularities caused by seasonal variations in glacier front positions (winter advance/summer retreat).

The average retreat rate was calculated as a ratio of the area delimited between the glacier front positions studied and the length of the glacier centerline between the two front positions.

For more information on data sources refer to Supplementary Table S1.

3. Results

3.1. Glacier Retreat

All of the glaciers studied have experienced retreat since 1970 (Table 2). The average retreat rate was 48 m year^{-1}, with the maximum retreat rate found at Hambergbreen (up to 150 m year^{-1}). In total, four of the 11 glaciers studied have changed from marine terminating to land-based glaciers, which has also affected (i.e., slowed) their retreat rate (Figure 2). Staupbreen, Bellingbreen, and Anna Margrethbreen became land-based between 1970 and 1990, whereas Skimmebreen made this transition between 1990 and 2000. The high retreat rate of the Hambergbreen and Sykorabreen glaciers resulted in the separation of the original glacier into two almost independent glaciers and consequently also increased the width of the calving front of the two glaciers. Breakup of the original vast tidewater glacier into separate glacier termini is a relatively common feature and can also be observed in the case of Emil'janovbreen and Spaelbreen, where the two glaciers became separated around 2005. Similarly, Davisbreen and Skimmebreen were terminating directly into the sea in one glacier terminus until around 1990.

Table 2. Surface elevation changes and retreat rates of the glaciers studied; average surface elevation change uncertainty is ±1.1 m, the annual change uncertainty corresponds to ±0.023 m/year; ice front retreat rate uncertainty is ±0.8 m.

Glacier	Id (Figure 1)	Average Surface Elevation Change (m)	Annual Change (m/year)	Ice Front Retreat Rate (m/year)
Sykorabreen *	1	−12.7	−0.27	149.0
Hambergbreen *	2	−39.2	−0.83	149.0
Staupbreen	3	−29.9	−0.64	16.3
Markhambreen	4	−30.5	−0.65	24.5
Crollbreen	5	−10.7	−0.23	32.7
Davisbreen	6	−11	−0.23	89.8
Skimmebreen	7	−19.5	−0.41	21.4
Bellingbreen	8	−13.8	−0.29	20.4
Anna Margrethbreen	9	−23.8	−0.51	10.2
Emil'janovbreen *	10	−45.2	−0.96	71.4
Spaelbreen *	11	−24.1	−0.51	71.4

* one value for both glaciers as they share the common outlet.

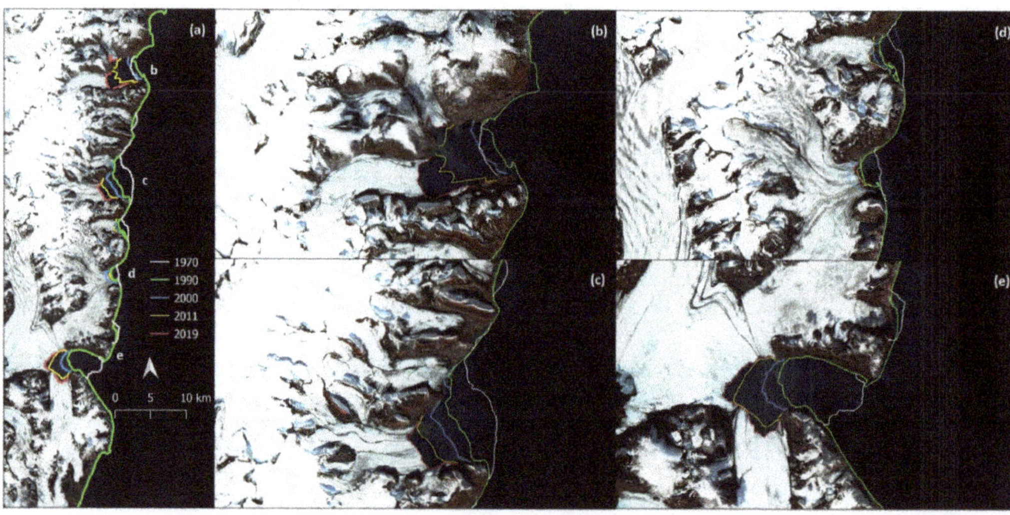

Figure 2. Glacier front positions in 1970, 1990, 2000, 2010, and 2019 of the tidewater glaciers that remained marine terminating in 2019; the whole study site (**a**) with close up views of four major areas where the retreat was quantified. (**b**) Emil'janovbreen and Spaelbreen, (**c**) Davisbreen and Skimmebreen, (**d**) Markhambreen and Crollbreen, (**e**) Hambergbreen and Sykorabreen.

The irregularities found in the retreat rate can be attributed to natural variability or occasional surge events. A minor readvancement was recorded in the case of a common outlet of Emil'janovbreen and Spaelbreen between 1970 and 1990; although we do not have any written record of a surge event, the high surface elevation loss would suggest that this took place. In the case of Markhambreen, a surge event was recorded around 2010, and is quite clear from the glacier front positions illustrated in detail in Figure 3.

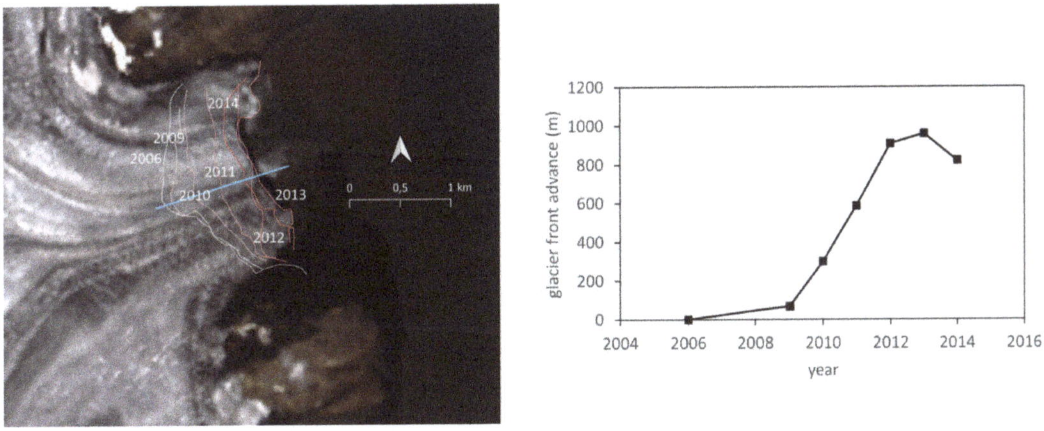

Figure 3. Glacier front positions during the surge event on Markhambreen with the advance rate in graphics, Landsat image from 2013 as a background image.

3.2. *Surface Elevation Changes*

All of the glaciers studied experienced similar development in terms of changes in their surface elevation between 1970 and the present. The common feature that can be observed is the surface lowering in their ablation areas, which is apparent up to an altitude of approximately 200–300 m a.s.l. The higher reaches of the glaciers experienced no clear change or gain in their mass. However, this trend is disturbed by occasional surge events as reported, for example, from Markhambreen since 2008. The readvancing of the glacier front is visible in Figure 3, and the disturbed post surge geometry is also well documented in Figure 4. Similarly, Emil'janovbreen experienced a surge event between 1970 and 1990, as documented by the readvancing glacier front (Figure 2) and massive surface lowering reaching up to 120 m and propagating towards the glacier terminus (Figure 4). Markhambreen experienced severe loss in mass in the whole northern branch of the glacier i.e., surface lowering in the accumulation zone. The two minor branches of the glacier system did not undergo surging and saw slight mass gain during this time. The surface elevation changes are summarized in Table 2.

3.3. *Zero Surface Elevation Change Altitude*

The altitudes which distinguish mass loss and gain vary among the glaciers studied from 240 to 530 m a.s.l. (Table 3). The estimated zero-surface elevation change altitude (ZSECA) could be considered as the long-term average for the whole study period. There are clear differences among the glaciers, but the general trend is dominant. The lower parts of the glaciers have lost substantial mass, whereas the upper parts remained relatively stable or even gained some mass-up to 20 m in some cases (Figure 5). The two exceptions were the glaciers that experienced surges during the study period, i.e., Markhambreen and Emil'janovbreen, where the whole glacier surface experienced thinning. Some of the glaciers were not fully covered by the 1970 DEM. It was therefore not possible to estimate the ZSECA for Hambergbreen. We can only suggest that all the glacier is currently below the ZSECA as all the surface covered by 1970 DEM is thinning, and the small part not covered by the 1970 DEM lies below 200 m a.s.l.

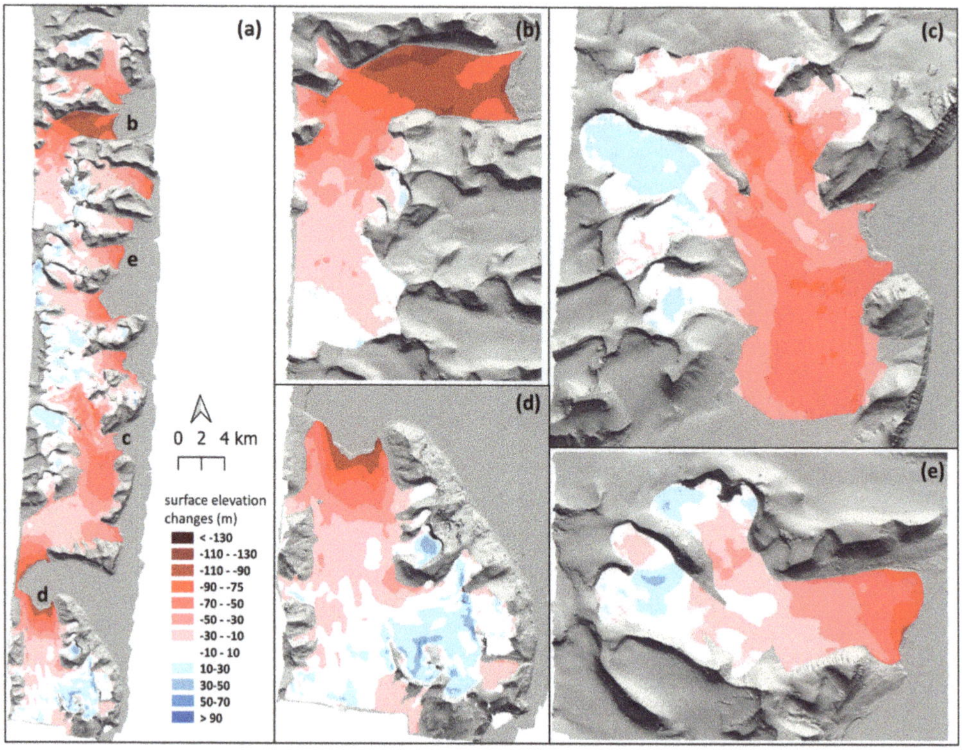

Figure 4. Surface elevation changes between 1970 and recent ArcticDEM expressed as elevation change (**a**); the detailed views represent the glaciers frequently mentioned in the text i.e., the two glaciers that underwent surge (**b**) Emil'janovbreen, (**c**) Markhambreen; (**d**) Sykorabreen as a part of the Hambergbreen-Sykorabreen glacier system with the highest retreat rate and (**e**) Skimmebreen as an example of a glacier that turned into land-based system; background is the ArcticDEM, glacier outlines delimited from 2017 Sentinel-2 images.

Table 3. Zero surface elevation change altitude (ZSECA) and glacier above ZSECA ratio; the uncertainty of area above ZSECA values arisen from the elevation uncertainty are reported in brackets.

Glacier	Id (Figure 1)	ZSECA (m a.s.l.)	Glacier above ZSECA (%)	Comment
Sykorabreen	1	230	46.8 (42.2–49.7)	
Hambergbreen	2	No	0	The whole glacier is losing mass
Staupbreen	3	480	1.2 (0.8–1.3)	
Markhambreen	4	350	8.5 (6.6–15.3)	Southern part only, northern losing mass
Crollbreen	5	240	45.4 (39.7–50.6)	
Davisbreen	6	250	45.7 (42.1–47.6)	
Skimmebreen	7	300–400	18.6 (15.9–21.2)	Different altitude on different glacier branches
Bellingbreen	8	370	17.7 (16.2–21.9)	
Anna Margrethbreen	9	250	23.9 (21.8–25.9)	
Emil'janovbreen	10	530	9 (7.3–11.2)	Southern part only
Spaelbreen	11	No	0.2 (0–0.6)	Small discontinuous areas of mass gain

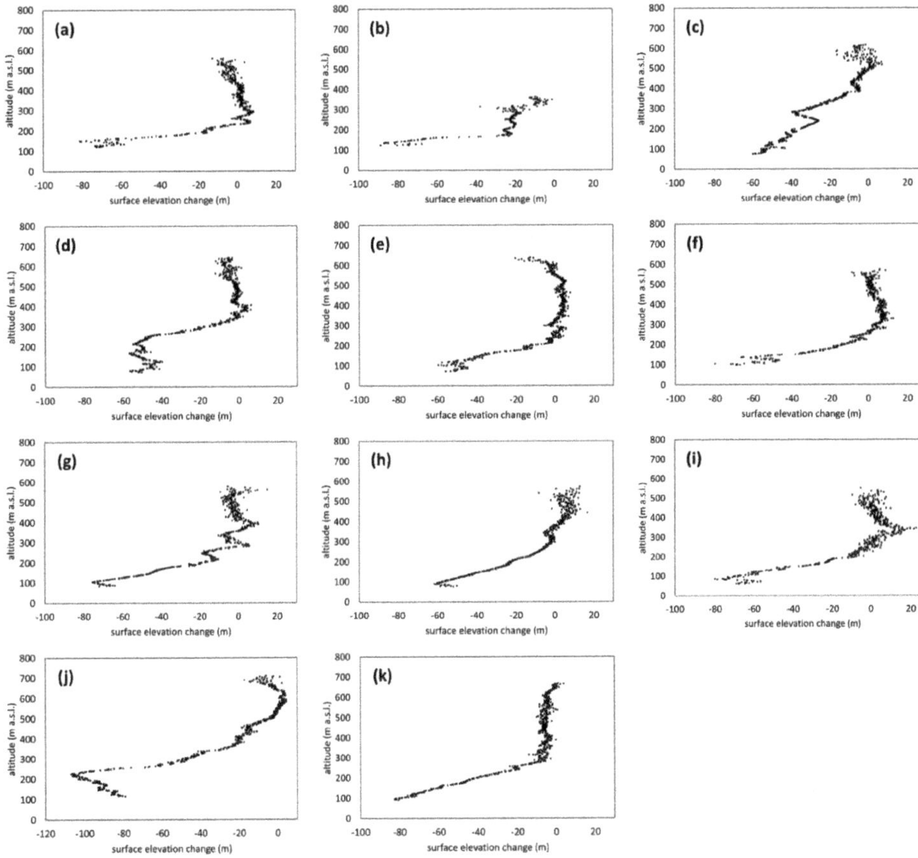

Figure 5. Surface elevation changes in relation to glacier altitude (1970 DEM); (**a**) Sykorabreen, (**b**) Hambergbreen, (**c**) Staupbreen, (**d**) Markhambreen, (**e**) Crollbreen, (**f**) Davisbreen, (**g**) Skimebreen, (**h**) Bellingbreen, (**i**) Anna Margrethebreen, (**j**) Emil'janovbreen, (**k**) Spaelbreen.

3.4. Glacier Geometry

The changes in glacier thickness combined with massive retreat of the glacier fronts resulted in significant change in glacier geometry. A number of the glaciers retreated up to 7 km (e.g., Hambergbreen) and significantly reduced in thickness at their calving front. On the contrary, most of the glaciers experienced mass gain in the highest parts of their accumulation areas, which has led to an increase in slope. In general, the present glaciers have steeper surfaces, are shorter in length, are thinner in their ablation zones, and have thicker accumulation areas. The exceptions are Markhambreen and Emil'janovbreen. The northern branch of Markhambreen experienced a surge around 2010, and the entire northern part of the glacier has lost large portion of its mass, even in the high elevation areas. The same is true for Emil'janovbreen, which has lost a large amount of mass across its whole area and therefore subsequently experienced dramatic thinning. The comparison of centerline profiles between the 1970 DEM and ArcticDEM is illustrated in Figure 6 with examples from several of the studied glaciers. It is clearly visible that the lower parts of the glaciers are thinning, whereas the upper parts are gaining mass or do not change considerably. The exceptions of Emil'janovbreen and Markhambreen are also well illustrated. The central branch of Markhambreen experienced mass gain in its upper reaches, whereas the northern branch has lost much of its mass due to surging.

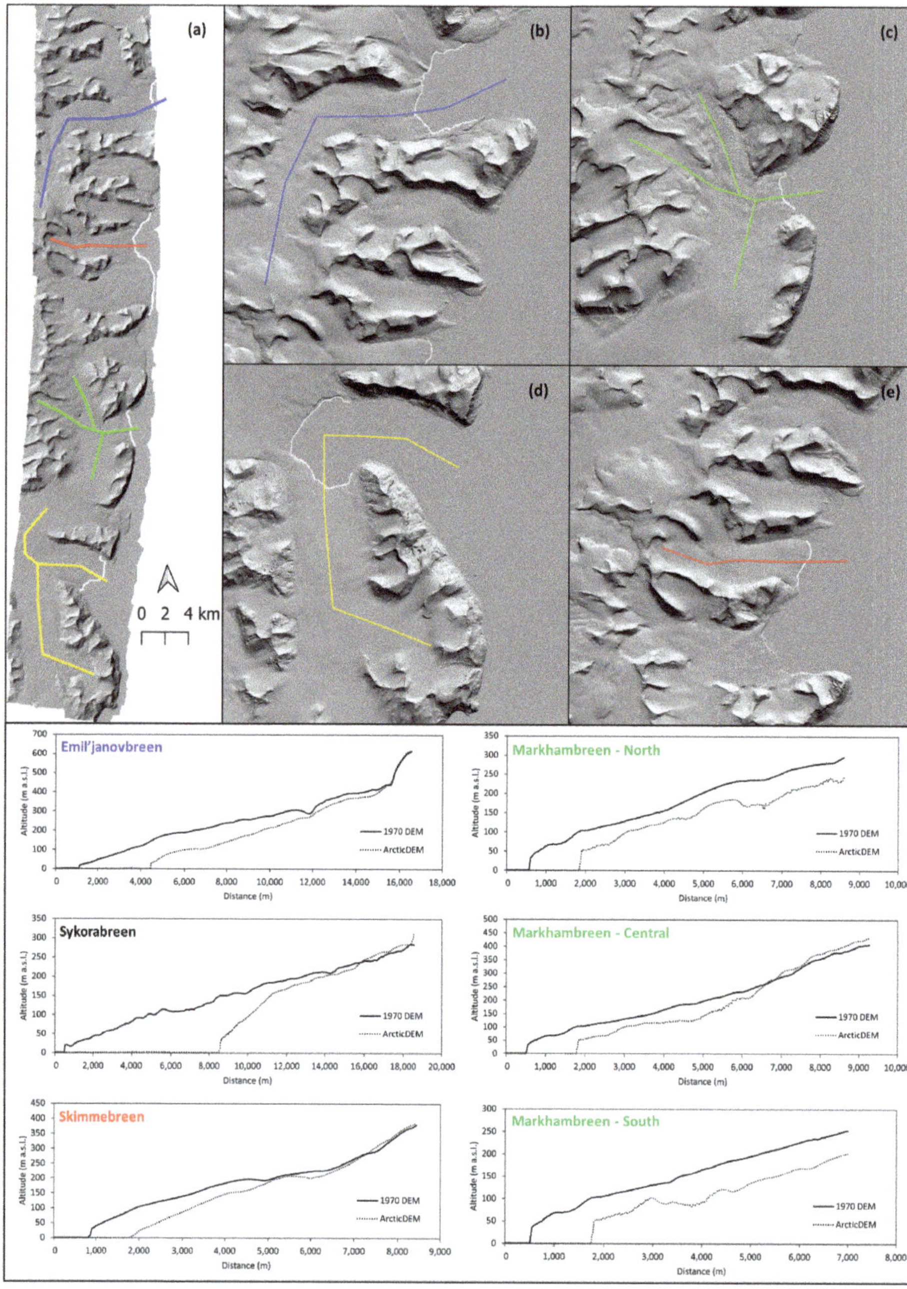

Figure 6. Longitudinal profiles (colored lines) of glaciers studied from DEM in 1970 and ArcticDEM; the overview of the whole study area (**a**); Emil'janovbreen (**b**); Markhambreen (**c**); Sykorabreen (**d**); Skimmebreen (**e**); selected glaciers shown in this figure correspond to the glaciers from Figure 4; The overview figure with 1970 DEM as background, the detailed views of profile lines with ArcticDEM as a background; The captions of the longitudinal profiles are in the same color as on the map to ease the orientation.

4. Discussion

The massive retreat rate of tidewater glaciers goes hand in hand with findings of glacier behavior from other parts of Svalbard ([10] for recent glacier behavior or Martín-Moreno et al. [8] for post Little Ice Age development), and specifically for other tidewater glaciers in the archipelago [6]. There are very few studies which have presented results of glacier recession that this could be related to. Historical analysis of glacier changes in this area showed that frontal retreat of Hambergbreen (ca. 16 km) was accompanied by its surface thinning of 60–100 m during the period of 1901–2000 [43]. The most rapid retreat rate (150 m year^{-1}) identified in the case of Hambergbreen is similar to that of Hornbreen, flowing towards the other side of the glaciated peninsula [44,45]. In fact, despite a recent slowdown of retreat in both Hornbreen and Hambergbreen, it is expected that the waterway between the east and west coast will open around 2055–2065 [43,46]. The retreat rate is lower in the case of glaciers that have already retreated and transformed into a land-based systems. In such cases, the lowering of the glacier surface is the dominant process as the frontal area is relatively thick. This behavior is similar to that of land-based glaciers right after their maximum extent at the end of the Little Ice Age [47].

All of the glaciers have lost large parts of their mass despite a slight surface elevation gain in the accumulation zones. This corresponds well with the increasing air temperature trend [32,33] that began in the 1970s across the whole of the Svalbard archipelago. The mass gain in the glaciers' upper reaches could be attributed to increased precipitation. Førland et al. [48] identified significant increase in precipitation at all weather stations used for assessing the precipitation trends between 1975–2011. This trend was even more pronounced in more recent years. The increase in precipitation does not compensate the increase in melt rates, and the annual surplus from the accumulation area will take some time to get the advected ice down to the ablation area. This is the case of most of the glaciers studied. This stems well with observations of Blaszczyk et al. [19] and Schuler et al. [10], who reported high velocities only on Hambergbreen and the lower part of Sykorabreen—the glacier with largest surface elevation losses. The mass loss across all parts of Hambergbreen, despite its snow line of 398 m a.s.l. as identified by Laska et al. [49], might also be explained by high flow velocities moving the ice mass downglacier from the accumulation area. The redistribution of snow due to the combination of terrain morphology and prevailing wind may also be an important factor [21,50]. This may be the cause for mass loss in the uppermost parts of Staupbreen and Crollbreen, for example (Figure 4). The comparison of glacier front positions in 1936 (as seen from the oblique aerial images) and 1970 (the DEM used in this study) suggested that the retreat rate was considerably lower before 1970, which may be explained by relatively stable air temperatures, including a short period of cooling during the 1960s [33], and the intensification of warm water inflow in the last decades which has resulted in increased ocean temperatures and a decline in sea ice extent [51]. The geometry of the glaciers probably led firstly to surface lowering, followed by retreat of the frontal area later, as reported for the first half of the 20th century in central Svalbard [46]. A similar change in glacier geometry was also reported by Marlin et al. [52] in the case of Austre Lovénbreen on the west coast of Svalbard. Moreover, the generally low altitude of the glaciers studied could have contributed to their massive ablation and could be a result of high sensitivity to air temperature variations [53]. Blaszczyk et al. [19] reported surface lowering (−1.6 m year^{-1} between 2012–2017) at Hornbreen, the neighbouring glacier to Hambergbreen, in an area of the glacier where similar surface lowering was detected by this study. Strozzi et al. [54] detected the same spatial pattern of surface elevation changes in the case of Stonebreen tidewater glacier on the Edgeøya island (eastern Svalbard) when comparing a similar time period (1971–2011). Pälli et al. [43] also reported thinning in almost all of the studied glaciers since 1900. Surprisingly, Sykorabreen was gaining mass in the upper parts in the early 20th century.

Using the two DEMs which were acquired with different techniques (i.e., aerial photographs and satellite images) can bring some constraints to the precision of the data especially when comparing the two DEMS. However, regarding the long time span between the two DEMs and large differences in the elevation changes, only the general spatial patterns of change can be drawn. Similarly, the methods used in Pälli et al. [42] for estimating the changes on local glaciers have their limitations, resulting from the accuracy issues with historic mapping.

The described geometry changes in the glaciers (retreat and steepening of the accumulation zone) are classic signs usually detected before surge events [55]. Unlike land-based glaciers, where the surge usually begins in the accumulation zone, the surge of a tidewater glacier in Svalbard often starts at the terminus and propagates upwards [56]. This is very likely the case of the Markhambreen surge described here. The glacier has advanced approximately 900 m between 2009 and 2012, resulting in a mean annual advance of about 300 m. Such retreat and advance rates are in good correspondence with previously reported rates for example at Blomstrandbreen, in the northwest of Svalbard [57] when accounting for mass loss due to calving, which is estimated at around 20% [58]. This may have led to the flow velocity being in the correct window of reported values for most Svalbard glacier surges—around 1 m day^{-1} [59]. Interestingly, the surge of Markhambreen coincides with the surge of Nathorstbreen (flowing to the north-east from the accumulation area neighboring the glaciers studied) despite the two glaciers not being directly linked. Sund et al. [58] reported that Nathorstbreen started surging after October 2008, and resulted in up to 50 m surface lowering in some areas. The mass loss in the accumulation area of Markhambreen was even more extreme, with certain areas of the glacier surface reducing by more than 70 m. Nuth et al. [60] used the example of the Nathorstbreen surge to demonstrate the impact of warming in the Arctic on the surging activity of local glaciers. Surprisingly, the thinning of the glacier tongue and subsequent basal freezing may enhance storage of subglacial meltwater in the upper parts of the glacier and thus lead to more surging. This may be the case of the glaciers studied, as all of them display the described parameters. Surging events were recorded also prior to the study period for example, at Hambergbreen [13], where recently Noormets et al. [41] detected two surges around 1900 and 1957. It is likely that the glacier front in 1970 (i.e., the basis for our comparison) represents the maximum extent after the surge starting in 1957. This may also explain the fastest recorded retreat rate (150 m year^{-1}) of the glaciers studied.

Tidewater glaciers do not only represent a dominant landscape feature on the eastern coast of Svalbard, but also constitute one of the most important ecological hotspots where the marine ecosystem is fed by nutrients and other compounds from the terrestrial environment [4,5,61]. The subglacial meltwater discharge also modifies the general circulation of water in front of the glacier terminus, usually leading to water upwelling and increased mixing of the fjord water [62,63]. This enhances the biological importance of such sites especially for marine mammals, or for foraging birds [3,64]. In this light, the shift of four glaciers from marine terminating to land-based can be seen as an important loss of biodiversity, as the fjord circulation at the glacier fronts will likely have changed, probably causing loss of certain species directly connected to the activity that occurs in this area (e.g., bearded seals). Similar development is also expected across other parts of Svalbard [65].

The observed glacier retreat also led to another significant change in the environment—opening of new bays and formation of pristine coastal zones. Such new coasts usually evolve from young unconsolidated glacial sediments left along the shores of fjords by retreating glaciers [45]. For instance, the opening of the Hambergbukta exposed over 10 km of fresh paraglacial coasts, where lateral moraines, crevasse-squeeze ridges, and eskers evolved into new narrow beaches, cuspate barriers, spits, and lagoons similar to those observed in the Brepollen on the opposite side of the Hornbreen-Hambergbreen glacier saddle [45]. Formation of the new embayment between the retreating Emil'janovbreen and Spaelbreen resulted in development of over 12 km of new shorelines filled with a number of well-developed accumulative coastal landforms, including one of the longest spit systems in this part of the archipelago, which formed along the northern banks of the new bay.

Overall, the coastal systems developing in front of retreating tidewater glaciers on the southeastern coast of Spitsbergen are one of the most geomorphologically dynamic paraglacial coastal environments in the entire Svalbard archipelago.

5. Conclusions

The tidewater glaciers in eastern Svalbard have retreated at a considerable rate (mean retreat rate 48 m year^{-1}) between 1970 and 2019, with a maximum retreat rate recorded of 150 m year^{-1} at Hambergbreen. The retreat was so extreme that four of the 11 glaciers studied had completely lost their connection with the sea and became land-based glacier systems retreating from their respective fjords. The slowdown of the retreat at Crollbreen and Spaelbreen and appearance of new land in their lateral marginal areas suggests that these two glaciers may be the next to become land-based systems in the study area. It is highly probable that such behavior will be observed in many of the remaining tidewater glaciers in Svalbard in the near future. The shift from marine terminating to land-based glaciers may also have serious implications for biodiversity, as a number of species rely directly on processes at the glacier terminus i.e., calving, whilst many others benefit from fjord water circulation induced by meltwater discharge at a glacier's grounding line.

There was a common spatial pattern in the geometry of all of the studied glaciers with thinning in the ablation areas of the glaciers, and the gaining of mass in the upper accumulation areas. This together with the massive retreat has completely changed the geometry resulting in shorter and steeper glaciers. Such dynamics are often observed in glaciers in their pre-surge period. This may be the case of the glaciers studied, as it was also demonstrated at Markhambreen, which experienced an important surge around 2010. The surge event was also imprinted in the general surface lowering of all the northern part of the glacier and episodic readvance of its calving front. A similar behavior was observed at Emil'janovbreen, where surging occurred between 1970 and 1990. However, it is unlikely that all the glaciers will surge as the reduced glacier flow velocities may explain such geometry changes as well. The accumulation rate in the upper parts of the glaciers is far below the overall ablation rate. The glacier geometry changes and retreat rate reflected changes in climate well—higher air temperatures leading to surface lowering in the frontal zones and higher precipitation (together with reduced flow velocities) causing the mass gain in upper reaches.

The paper has shown the importance and great value of historic spatial data such as the 1970 DEM that has enabled us to assess the changes of glacier dynamics over almost half a century. The massive retreat of the glaciers has also exposed large number of new coasts that are currently being reshaped by coastal processes, and which require further work to understand their impact on southeast Spitsbergen and more broadly across the archipelago of Svalbard.

Supplementary Materials: The following supporting information can be downloaded at: https://www.mdpi.com/article/10.3390/rs14020354/s1, Table S1: Source data used in the analyses.

Author Contributions: Conceptualization, J.K.; methodology, J.K. and M.D.; software, J.K. and M.D.; validation, J.K.; formal analysis, J.K.; investigation, J.K.; resources, J.K. and M.C.S.; data curation, J.K.; writing—original draft preparation, J.K., G.D.T., J.D. and M.C.S.; writing—review and editing, J.K. and G.D.T.; visualization, J.K. and M.D.; supervision, M.C.S.; project administration, J.K.; funding acquisition, J.K. and M.C.S. All authors have read and agreed to the published version of the manuscript.

Funding: This research was funded by the Masaryk University project ARCTOS MU (MUNI/G/1540/2019) and MUNI/A/1570/2020. The research was also funded through the Norwegian Financial Mechanism 2014–2021: SVELTA—Svalbard Delta Systems Under Warming Climate (UMO-2020/37/K/ST10/02852) based at the University of Wroclaw.

Data Availability Statement: ArcticDEM data were provided by the Polar Geospatial Center under NSF-OPP awards 1043681, 1559691, and 1542736. NPI 1970 DEM was provided by the Norwegian Polar Institute.

Acknowledgments: This work was supported by the Masaryk University project ARCTOS MU (MUNI/G/1540/2019) and MUNI/A/1570/2020. The research leading to these results has received funding from the Norwegian Financial Mechanism 2014–2021: SVELTA—Svalbard Delta Systems Under Warming Climate (UMO-2020/37/K/ST10/02852) based at the University of Wroclaw. ArcticDEM were provided by the Polar Geospatial Center under NSF-OPP awards 1043681, 1559691, and 1542736. We would like to thank to Veijo Pohjola and four anonymous reviewers for their valuable comments and suggestions.

Conflicts of Interest: The authors declare no conflict of interest.

References

1. Nuth, C.; Kohler, J.; König, M.; von Deschwanden, A.; Hagen, J.O.; Kääb, A.; Moholdt, G.; Pettersson, R. Decadal changes from a multi-temporal glacier inventory of Svalbard. *Cryosphere* **2013**, *7*, 1603–1621. [CrossRef]
2. Rotschky, G.; Schuler, T.V.; Haarpaintner, J.; Kohler, J.; Isaksson, E. Spatio-temporal variability of snowmelt across Svalbard during the period 2000–2008 derived from QuikSCAT/SeaWinds scatterometry. *Polar Res.* **2011**, *30*, 5963. [CrossRef]
3. Lydersen, C.; Assmy, P.; Falk-Petersen, S.; Kohler, J.; Kovacs, K.M.; Reigstad, M.; Steen, H.; Strøm, H.; Sundfjord, A.; Varpe, Ø.; et al. The importance of tidewater glaciers for marine mammals and seabirds in Svalbard, Norway. *J. Mar. Syst.* **2014**, *129*, 452–471. [CrossRef]
4. Hopwood, M.J.; Carroll, D.; Browning, T.J.; Meire, L.; Mortensen, J.; Krisch, S.; Achterberg, E.P. Non-linear response of summertime marine productivity to increased meltwater discharge around Greenland. *Nat. Commun.* **2018**, *9*, 3256. [CrossRef] [PubMed]
5. Vonnahme, T.R.; Persson, E.; Dietrich, U.; Hejdukova, E.; Dybwad, C.; Elster, J.; Chierici, M.; Gradinger, R. Early spring subglacial discharge plumes fuel under-ice primary production at a Svalbard tidewater glacier. *Cryosphere* **2021**, *15*, 2083–2107. [CrossRef]
6. Blaszczyk, M.; Jania, J.A.; Hagen, J.O. Tidewater glaciers of Svalbard: Recent changes and estimates of calv-ing fluxes. *Pol Polar Res.* **2009**, *30*, 85–142.
7. Hagen, J.O.; Kohler, J.; Melvold, K.; Winther, J.-G. Glaciers in Svalbard: Mass balance, runoff and freshwater flux. *Polar Res.* **2003**, *22*, 145–159. [CrossRef]
8. Martín-Moreno, R.; Álvarez, F.A.; Hagen, J.O. 'Little Ice Age' glacier extent and subsequent retreat in Svalbard archipelago. *Holocene* **2017**, *27*, 1379–1390. [CrossRef]
9. Luckman, A.; Benn, D.; Cottier, F.; Bevan, S.; Nilsen, F.; Inall, M. Calving rates at tidewater glaciers vary strongly with ocean temperature. *Nat. Commun.* **2015**, *6*, 8566. [CrossRef] [PubMed]
10. Schuler, T.V.; Kohler, J.; Elagina, N.; Hagen, J.O.M.; Hodson, A.J.; Jania, J.A.; Kääb, A.M.; Luks, B.; Małecki, J.; Moholdt, G.; et al. Reconciling Svalbard Glacier Mass Balance. *Front. Earth Sci.* **2020**, *8*, 156. [CrossRef]
11. Nuth, C.; Moholdt, G.; Kohler, J.; Hagen, J.O.; Kääb, A. Svalbard glacier elevation changes and contribution to sea level rise. *J. Geophys. Res.* **2010**, *115*, F01008. [CrossRef]
12. Möller, M.; Kohler, J. Differing Climatic Mass Balance Evolution Across Svalbard Glacier Regions Over 1900–2010. *Front. Earth Sci.* **2018**, *6*, 128. [CrossRef]
13. Lefauconnier, B.; Hagen, J.O. Surging and calving glaciers in eastern Svalbard. *Nor. Polarinst. Medd.* **1991**, *116*, 130.
14. Ziaja, W.; Maciejowski, W.; Ostafin, K. *Northeastern Sørkappland Landscape Dynamics (Spitsbergen, Svalbard)*; Jagiellonian University Press: Krakow, Poland, 2007; ISBN 978-83-233-2317-4.
15. Ziaja, W.; Ostafin, K. Landscape–Seascape dynamics in the isthmus between Sørkapp Land and the rest of Spitsbergen: Will a new big Arctic island form? *Ambio* **2015**, *44*, 332–342. [CrossRef] [PubMed]
16. Cygankiewicz-Truś, A.; Ziaja, W. From Glaciated Landscape to Unglaciated Seascape: Transformation of the Hambergbreen–Hambergbukta Area, SE Spitsbergen, 1900–2017. *Ann. Am. Assoc. Geogr.* **2021**, *111*, 1949–1966. [CrossRef]
17. Hanssen-Bauer, I.; Førland, E.J.; Hisdal, H.; Mayer, S.; Sandø, A.B.; Sorteberg, A. *Climate in Svalbard 2100—A Knowledge Base for Climate Adaptation*; Norwegian Centre for Climate Services (NCCS) for Norwegian Environment Agency: Trondheim, Norway, 2019.
18. Moholdt, G.; Nuth, C.; Hagen, J.O.; Kohler, J. Recent elevation changes of Svalbard glaciers derived from ICESat laser altimetry. *Remote Sens. Environ.* **2010**, *114*, 2756–2767. [CrossRef]
19. Błaszczyk, M.; Ignatiuk, D.; Grabiec, M.; Kolondra, L.; Laska, M.; Decaux, L.; Jania, J.; Berthier, E.; Luks, B.; Barzycka, B.; et al. Quality Assessment and Glaciological Applications of Digital Elevation Models Derived from Space-Borne and Aerial Images over Two Tidewater Glaciers of Southern Spitsbergen. *Remote Sens.* **2019**, *11*, 1121. [CrossRef]
20. Rastner, P.; Prinz, R.; Notarnicola, C.; Nicholson, L.; Sailer, R.; Schwaizer, G.; Paul, F. On the Automated Mapping of Snow Cover on Glaciers and Calculation of Snow Line Altitudes from Multi-Temporal Landsat Data. *Remote Sens.* **2019**, *11*, 1410. [CrossRef]
21. Kavan, J.; Haagmans, V. Seasonal dynamics of snow ablation on selected glaciers in central Spitsbergen derived from Sentinel-2 satellite images. *J. Glaciol.* **2021**, *67*, 961–966. [CrossRef]
22. Schellenberger, T.; Van Wychen, W.; Copland, L.; Kääb, A.; Gray, L. An Inter-Comparison of Techniques for Determining Velocities of Maritime Arctic Glaciers, Svalbard, Using Radarsat-2 Wide Fine Mode Data. *Remote Sens.* **2016**, *8*, 785. [CrossRef]

23. Małecki, J. Elevation and volume changes of seven Dickson Land glaciers, Svalbard, 1960–1990–2009. *Polar Res.* **2013**, *32*, 18400. [CrossRef]
24. Ziaja, W.; Dudek, J.; Ostafin, K. Landscape transformation under the Gåsbreen glacier recession since 1899, southwestern Spitsbergen. *Pol. Polar Res.* **2016**, *37*, 155–172. [CrossRef]
25. Wójcik, K.A.; Sobota, I. Spatial and temporal changes in ablation, distribution and evolution of glacial zones on Irenebreen, a small glacier of the High Arctic, Svalbard. *Polar Sci.* **2020**, *23*, 100503. [CrossRef]
26. Østby, T.I.; Schuler, T.V.; Hagen, J.O.; Hock, R.; Kohler, J.; Reijmer, C.H. Diagnosing the decline in climatic mass balance of glaciers in Svalbard over 1957–2014. *Cryosphere* **2017**, *11*, 191–215. [CrossRef]
27. van Pelt, W.J.J.; Schuler, T.V.; Pohjola, V.A.; Pettersson, R. Accelerating future mass loss of Svalbard glaciers from a multi-model ensemble. *J. Glaciol.* **2021**, *67*, 485–499. [CrossRef]
28. Hagen, J.O.; Liestøl, O.; Roland, E.; Jørgensen, T. Glacier Atlas of Svalbard and Jan Mayen. *Nor. Pola-Rinstitutt Medd.* **1993**, *129*, 144.
29. Maciejowski, W.; Michniewski, A. Variations in weather on the East and West coasts of South Spitsber-gen, Svalbard. *Pol. Polar Res.* **2007**, *28*, 123–136.
30. Sulikowska, A.; Wypych, A.; Mitka, K.; Maciejowski, W.; Ostafin, K.; Ziaja, W. Summer weather condi-tions in 2005 and 2016 on the western and eastern coasts of south Spitsbergen. *Pol. Polar Res.* **2018**, *39*, 127–144. [CrossRef]
31. Szymański, W.; Maciejowski, W.; Ostafin, K.; Ziaja, W.; Sobucki, M. Impact of parent material, vegetation cover, and site wetness on variability of soil properties in proglacial areas of small glaciers along the northeastern coast of Sørkappland (SE Spitsbergen). *Catena* **2019**, *183*, 104209. [CrossRef]
32. Nordli, Ø.; Przybylak, R.; Ogilvie, A.E.; Isaksen, K. Long-term temperature trends and variability on Spitsbergen: The extended Svalbard Airport temperature series, 1898–2012. *Polar Res.* **2014**, *33*, 21349. [CrossRef]
33. Nordli, Ø.; Wyszyński, P.; Gjelten, H.M.; Isaksen, K.; Łupikasza, E.; Niedźwiedź, T.; Przybylak, R. Revisiting the extended Svalbard Airport monthly temperature series, and the compiled corresponding daily series 1898–2018. *Polar Res.* **2020**, *39*, 3614. [CrossRef]
34. Sinnhuber, B.-M. Historic temperature observations on Nordaustlandet, north-east Svalbard. *Polar Res.* **2021**, *40*, 7564. [CrossRef]
35. Sund, M.; Eiken, T.; Hagen, J.O.; Kääb, A. Svalbard surge dynamics derived from geometric changes. *Ann. Glaciol.* **2009**, *50*, 50–60. [CrossRef]
36. Sund, M.; Błaszczyk, M.; Eiken, T.; Jania, J.A. The Implications of Surge and Tidewater Glacier Dynamics Related to Climate Change Response of Svalbard Glaciers. Ph.D. Thesis, Universitet i Oslo, Oslo, Norway, 2011.
37. Flink, A.E.; Noormets, R.; Kirchner, N.; Benn, D.I.; Luckman, A.; Lovell, H. The evolution of a submarine landform record following recent and multiple surges of Tunabreen glacier, Svalbard. *Quat. Sci. Rev.* **2015**, *108*, 37–50. [CrossRef]
38. Sevestre, H.; Benn, D.I. Climatic and geometric controls on the global distribution of surge-type glaciers: Implications for a unifying model of surging. *J. Glaciol.* **2015**, *61*, 646–662. [CrossRef]
39. Lovell, H.; Fleming, E.J.; Benn, D.I.; Hubbard, B.; Lukas, S.; Rea, B.R.; Noormets, R.; Flink, A.E. Debris entrainment and landform genesis during tidewater glacier surges. *J. Geophys. Res. Earth Surf.* **2015**, *120*, 1574–1595. [CrossRef]
40. Farnsworth, W.R.; Ingólfsson, Ó.; Retelle, M.; Schomacker, A. Over 400 previously undocumented Svalbard surge-type glaciers identified. *Geomorphology* **2016**, *264*, 52–60. [CrossRef]
41. Noormets, R.; Flink, A.; Kirchner, N. Glacial dynamics and deglaciation history of Hambergbukta reconstructed from submarine landforms and sediment cores, SE Spitsbergen, Svalbard. *Boreas* **2021**, *50*, 29–50. [CrossRef]
42. Porter, C.; Morin, P.; Howat, I.; Noh, M.-J.; Bates, B.; Peterman, K.; Keesey, S.; Schlenk, M.; Gardiner, J.; Tomko, K.; et al. ArcticDEM. *Earth Environ. Sci.* **2018**. Available online: https://www.pgc.umn.edu/data/arcticdem/ (accessed on 11 May 2021). [CrossRef]
43. Pälli, A.; Moore, J.C.; Jania, J.; Glowacki, P. Glacier changes in southern Spitsbergen, Svalbard, 1901–2000. *Ann. Glaciol.* **2003**, *37*, 219–225. [CrossRef]
44. Błaszczyk, M.; Jania, J.A.; Kolondra, L. Fluctuations of tidewater glaciers in Hornsund Fjord (Southern Svalbard) since the beginning of the 20th century. *Pol. Polar Res.* **2013**, *34*, 327–352. [CrossRef]
45. Strzelecki, M.C.; Szczuciński, W.; Dominiczak, A.; Zagórski, P.; Dudek, J.; Knight, J. New fjords, new coasts, new landscapes: The geomorphology of paraglacial coasts formed after recent glacier retreat in Brepollen (Hornsund, southern Svalbard). *Earth Surf. Process Landf.* **2020**, *45*, 1325–1334. [CrossRef]
46. Grabiec, M.; Ignatiuk, D.; Jania, J.A.; Moskalik, M.; Głowacki, P.; Błaszczyk, M.; Budzik, T.; Walczowski, W. Coast formation in an Arctic area due to glacier surge and retreat: The Hornbreen-Hambergbreen case from Spistbergen. *Earth Surf. Process Landf.* **2017**, *43*, 387–400. [CrossRef]
47. Kavan, J. Early twentieth century evolution of Ferdinand glacier, Svalbard, based on historic photographs and structure-from-motion technique. *Geogr. Ann.* **2020**, *102*, 57–67. [CrossRef]
48. Førland, E.J.; Benestad, R.E.; Hanssen-Bauer, I.; Haugen, J.E.; Skaugen, T.E. Temperature and Precipitation Development at Svalbard 1900–2100. *Adv. Meteorol.* **2011**, *2011*, 893790. [CrossRef]
49. Laska, M.; Barzycka, B.; Luks, B. Melting Characteristics of Snow Cover on Tidewater Glaciers in Hornsund Fjord, Svalbard. *Water* **2017**, *9*, 804. [CrossRef]

50. Małecki, J. Snow accumulation on a small high-arctic glacier svenbreen: Variability and topographic controls. *Geogr. Ann.* **2015**, *97*, 809–817. [CrossRef]
51. Asbjørnsen, H.; Årthun, M.; Skagseth, Ø.; Eldevik, T. Mechanisms Underlying Recent Arctic Atlantification. *Geophys. Res. Lett.* **2020**, *47*, e2020GL088036. [CrossRef]
52. Marlin, C.; Tolle, F.; Griselin, M.; Bernard, E.; Saintenoy, A.; Quenet, M.; Friedt, J.-M. Change in geometry of a high Arctic glacier from 1948 to 2013 (Austre Lovénbreen, Svalbard). *Geogr. Ann.* **2017**, *99*, 115–138. [CrossRef]
53. Noël, B.; Jakobs, C.L.; Van Pelt, W.J.J.; Lhermitte, S.; Wouters, B.; Kohler, J.; Hagen, J.O.; Luks, B.; Reijmer, C.H.; Van De Berg, W.J.; et al. Low elevation of Svalbard glaciers drives high mass loss variability. *Nat. Commun.* **2020**, *11*, 4597. [CrossRef]
54. Strozzi, T.; Kääb, A.; Schellenberger, T. Frontal destabilization of Stonebreen, Edgeøya, Svalbard. *Cryosphere* **2017**, *11*, 553–566. [CrossRef]
55. Murray, T.; Stuart, G.W.; Miller, P.J.; Woodward, J.; Smith, A.M.; Porter, P.R.; Jiskoot, H. Glacier surge propagation by thermal evolution at the bed. *J. Geophys. Res.* **2000**, *105*, 13491–13507. [CrossRef]
56. Sevestre, H.; Benn, D.I.; Luckman, A.; Nuth, C.; Kohler, J.; Lindbäck, K.; Pettersson, R. Tidewater Glacier Surges Initiated at the Terminus. *J. Geophys. Res. Earth* **2018**, *123*, 1035–1051. [CrossRef]
57. Burton, D.J.; Dowdeswell, J.; Hogan, K.A.; Noormets, R. Marginal Fluctuations of a Svalbard Surge-Type Tidewater Glacier, Blomstrandbreen, Since the Little Ice Age: A Record of Three Surges. *Arctic Antarct. Alp. Res.* **2016**, *48*, 411–426. [CrossRef]
58. Sund, M.; Lauknes, T.R.; Eiken, T. Surge dynamics in the Nathorstbreen glacier system, Svalbard. *Cryosphere* **2014**, *8*, 623–638. [CrossRef]
59. Murray, T.; Strozzi, T.; Luckman, A.; Jiskoot, H.; Christakos, P. Is there a single surge mechanism? Contrasts in dynamics between glacier surges in Svalbard and other regions. *J. Geophys. Res.* **2003**, *108*, 2237. [CrossRef]
60. Nuth, C.; Gilbert, A.; Köhler, A.; McNabb, R.; Schellenberger, T.; Sevestre, H.; Weidle, C.; Girod, L.; Luckman, A.; Kääb, A. Dynamic vulnerability revealed in the collapse of an Arctic tidewater glacier. *Sci. Rep.* **2019**, *9*, 5541. [CrossRef] [PubMed]
61. Meire, L.; Mortensen, J.; Meire, P.; Juul-Pedersen, T.; Sejr, M.K.; Rysgaard, S.; Nygaard, R.; Huybrechts, P.; Meysman, F.J.R. Marine-terminating glaciers sustain high productivity in Greenland fjords. *Glob. Chang. Biol.* **2017**, *23*, 5344–5357. [CrossRef] [PubMed]
62. Mankoff, K.D.; Straneo, F.; Cenedese, C.; Das, S.B.; Richards, C.G.; Singh, S. Structure and dynamics of a subglacial discharge plume in a Greenlandic fjord. *J. Geophys. Res. Oceans* **2016**, *121*, 8670–8688. [CrossRef]
63. Jackson, R.H.; Shroyer, E.L.; Nash, J.; Sutherland, D.A.; Carroll, D.; Fried, M.J.; Catania, G.A.; Bartholomaus, T.C.; Stearns, L.A. Near-glacier surveying of a subglacial discharge plume: Implications for plume parameterizations. *Geophys. Res. Lett.* **2017**, *44*, 6886–6894. [CrossRef]
64. Bertrand, P.; Strøm, H.; Bêty, J.; Steen, H.; Kohler, J.; Vihtakari, M.; van Pelt, W.; Yoccoz, N.; Hop, H.; Harris, S.; et al. Feeding at the front line: Interannual variation in the use of glacier fronts by foraging black-legged kittiwakes. *Mar. Ecol. Prog. Ser.* **2021**, *677*, 197–208. [CrossRef]
65. Torsvik, T.; Albretsen, J.; Sundfjord, A.; Kohler, J.; Sandvik, A.D.; Skarðhamar, J.; Lindbäck, K.; Everett, A. Impact of tidewater glacier retreat on the fjord system: Modeling present and future circulation in Kongsfjorden, Svalbard. *Estuarine Coast. Shelf Sci.* **2019**, *220*, 152–165. [CrossRef]

remote sensing

MDPI

Article

Seasonal InSAR Displacements Documenting the Active Layer Freeze and Thaw Progression in Central-Western Spitsbergen, Svalbard

Line Rouyet [1,2,3,*], Lin Liu [4], Sarah Marie Strand [3,5], Hanne Hvidtfeldt Christiansen [3], Tom Rune Lauknes [1,2] and Yngvar Larsen [1]

1 NORCE Norwegian Research Centre AS, P.O. Box 6434, 9294 Tromsø, Norway; tlau@norceresearch.no (T.R.L.); ynla@norceresearch.no (Y.L.)
2 Department of Geosciences, The Arctic University of Norway (UiT), P.O. Box 6050, Langnes, 9037 Tromsø, Norway
3 Arctic Geology Department, The University Centre in Svalbard (UNIS), P.O. Box 156, 9171 Longyearbyen, Norway; sarahs1@unis.no (S.M.S.); hannec@unis.no (H.H.C.)
4 Earth System Science Programme, Faculty of Science, The Chinese University of Hong Kong, Central Ave, Hong Kong 999077, China; liulin@cuhk.edu.hk
5 Department of Geosciences, University of Oslo (UiO), P.O. Box 1047, Blindern, 0316 Oslo, Norway
* Correspondence: liro@norceresearch.no

Abstract: In permafrost areas, the active layer undergoes seasonal frost heave and thaw subsidence caused by ice formation and melting. The amplitude and timing of the ground displacement cycles depend on the climatic and ground conditions. Here we used Sentinel-1 Synthetic Aperture Radar Interferometry (InSAR) to document the seasonal displacement progression in three regions of Svalbard. We retrieved June–November 2017 time series and identified thaw subsidence maxima and their timing. InSAR measurements were compared with a composite index model based on ground surface temperature. Cyclic seasonal patterns are identified in all areas, but the timing of the displacement progression varies. The subsidence maxima occurred later on the warm western coast (Kapp Linné and Ny-Ålesund) compared to the colder interior (Adventdalen). The composite index model is generally able to explain the observed patterns. In Adventdalen, the model matches the InSAR time series at the location of the borehole. In Kapp Linné and Ny-Ålesund, larger deviations are found at the pixel-scale, but km or regional averaging improves the fit. The study highlights the potential for further development of regional InSAR products to represent the cyclic displacements in permafrost areas and infer the active layer thermal dynamics.

Keywords: permafrost; active layer; InSAR; time series; ground displacement; ground temperature; displacement progression; thaw progression; Arctic; Svalbard

Citation: Rouyet, L.; Liu, L.; Strand, S.M.; Christiansen, H.H.; Lauknes, T.R.; Larsen, Y. Seasonal InSAR Displacements Documenting the Active Layer Freeze and Thaw Progression in Central-Western Spitsbergen, Svalbard. *Remote Sens.* **2021**, *13*, 2977. https://doi.org/10.3390/rs13152977

Academic Editor: Francesco Casu

Received: 21 May 2021
Accepted: 22 July 2021
Published: 28 July 2021

1. Introduction

Permafrost, defined as ground that remains at or below 0 °C for at least two consecutive years, is an essential component of the terrestrial cryosphere that is sensitive to climate change [1]. Permafrost degradation contributes to global warming by releasing greenhouse gases previously trapped in the frozen ground [2] and has direct impacts on infrastructure [3] and ecosystems [4]. The upper part of the ground, the active layer (AL), is seasonally frozen and thawed, and determines a vast set of ecological and hydrological processes occurring in permafrost landscapes [5,6]. Permafrost thermal state and AL thickness (ALT) are the two components of the Permafrost Essential Climate Variable (ECV). These variables are typically measured by in-situ techniques [7], but the scarce network of monitoring sites makes remote and large polar regions difficult to comprehensively document. This leads to large uncertainties in the estimate of the current permafrost state and future projections [8]. However, surface changes documented by satellite remote sensing allow us to indirectly investigate permafrost dynamics. The exploitation of optical [9], radar [10] and thermal [11] imagery for this purpose has significantly increased these past decades [12].

Seasonal AL freezing and thawing induces cyclic heave and subsidence of the ground surface due to ice formation and melting [5,6]. The variability of the ground thermal regime, water content and physical AL properties lead to an uneven amplitude, distribution and temporal variability of these displacements [13,14]. Satellite Synthetic Aperture Radar Interferometry (InSAR) allows for documenting line-of-sight (LOS) ground surface displacements between radar images taken at different times [15]. InSAR-based displacement maps can cover extensive areas and document the spatial distribution of thaw subsidence [16]. InSAR has also been used to estimate the ALT [17] and map areas with high content of excess ground ice at the top of permafrost [18]. Multi-temporal InSAR techniques allow for the retrieval of displacement time series, valuable for studying the seasonal progression of the ground displacements related to freeze and thaw cycles [19–22], as well as the interannual changes of surface elevation [23–26].

In Svalbard, InSAR time series highlighted that the temporal patterns of the seasonal subsidence and heave match the AL thermal variations measured in boreholes in Adventdalen and Endalen [21]. Based on Global Positioning System Interferometric Reflectometry (GPS-IR) in an Alaskan site, Hu et al. [27] showed that a composite index model is able to characterize the cyclic patterns. The model estimates the subsidence and heave based on assumptions of homogeneous soil water content and using the Stefan equation, to calculate the variable freezing/thawing front depth based on air or ground temperature records [28,29]. Both in Svalbard and Alaska, the results suggest that displacement time series can indirectly document the AL thermal regime and thus complement and upscale traditional point-based field monitoring. The results indicate that the timing of the maximal subsidence can be used as a proxy for the transition between the thawing period and the freeze-back onset. However, other studies conversely concluded that the displacement progression described by the Stefan equation does not reproduce observations, due to the unconsidered impacts of hydrologic controls [30–32] and the potential time lag between temperature and displacements [19,33]. Further research is thus necessary to compare measured and modelled displacement time series, and to study the importance of the temperature control on AL displacement patterns in different environmental settings. In addition, with the development of national to multi-national InSAR mapping services based on freely available images from the Copernicus Sentinel-1 satellite mission [34,35], the ability to map ground movement at large scales is dramatically increasing.

However, the operational exploitation of InSAR technology for the monitoring of ground dynamics in extensive permafrost areas still needs to be demonstrated. The currently applied processing strategies are mostly designed for moving areas with relatively constant displacement directions and the InSAR parameter chosen for mapping purpose is usually the mean annual ground velocity, which has limited applicability in areas affected by cyclic patterns. In Svalbard and similar polar environments, dedicated products that take the complex seasonal ground dynamics into account are required.

Here we study to what degree seasonal displacements time series are controlled by temperature variations and discuss the value of novel InSAR products designed for mapping cyclic ground dynamics in Svalbard. Based on pre-existing techniques, the novelty of our study is to propose a simple procedure to summarize and compare complex displacement time series in extensive polar regions, and to couple InSAR and modelling for the interpretation of the active layer dynamics in permafrost landscapes. Specifically, our objectives are to (1) to develop Sentinel-1 InSAR products documenting the spatial variability and timing of the seasonal thaw subsidence maxima in three regions of Svalbard characterized by different geomorphological and climatic conditions; (2) compare the displacement time series with a composite index based on temperature and evaluate how a simple model can explain the progression of subsidence and heave patterns in the study areas; (3) discuss the potential and limitations of using the timing of the maximal subsidence as a proxy for the end of the thawing season and suggest ideas for the development of alternative InSAR products in polar areas characterized by cyclic patterns.

2. Study Areas

Svalbard is characterized by a polar-tundra climate [36] influenced by the West Spitsbergen oceanic current, which warms in particular the western parts of the archipelago [37]. The Spitsbergen island experiences a large climatic gradient with higher temperature and greater precipitation in the west compared to the more continental interior [38]. The periglacial land area has continuous permafrost, varying from approximately 100 m in thickness in valley bottoms and coastal areas to 500 m in the mountains [39]. The seasonal and inter-annual meteorological variations, as well as the diversity of local environmental conditions (water content, ground material properties, snow cover and vegetation) largely influence the ground thermal regime and consequently the dynamics of periglacial landforms [40–42]. Monitoring of ground temperature and ALT indicates that permafrost is warming and ALT is increasing [43–45]. Projections for the twenty-first century suggest similar future trends following climate change scenarios [38,46]. The study focuses on three different and well-studied areas of central and western Spitsbergen: Adventdalen, Kapp Linné and Ny-Ålesund (Figure 1). All three areas have permafrost observation sites as part of the Svalbard Integrated Arctic Earth Observing System (SIOS) [44]. The selected areas vary from 121 to 307 km^2 in size (Table 1).

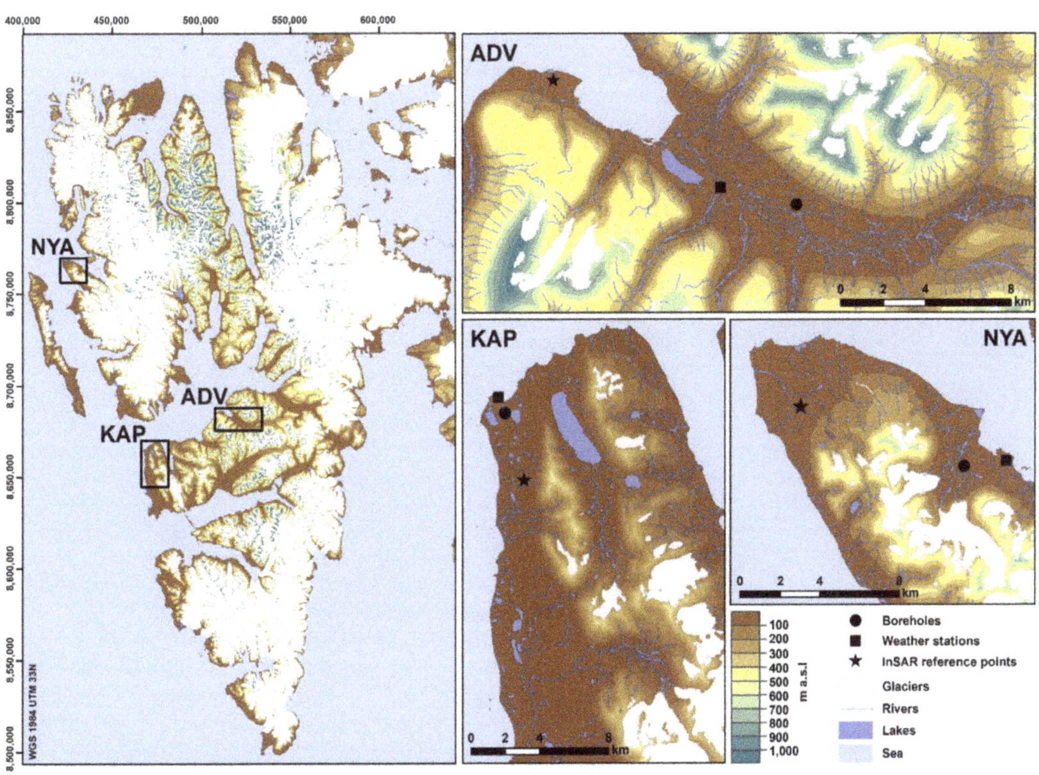

Figure 1. Location of the three study areas in central and western Spitsbergen, Svalbard: Adventdalen (ADV), Kapp Linné (KAP) and Ny-Ålesund (NYA). Digital elevation model and topographical information are from the Norwegian Polar Institute [47,48].

Table 1. Information about the study areas and the InSAR datasets. Locations of the InSAR reference points are shown in Figure 1. The interferogram networks are shown in Supplements S1–S3.

	Land Area [km^2]	Selected Number of Interferograms	Line-of-Sight (LOS) Incidence Angle (I_a)/Compass Direction (D_i)	Reference Points (UTM 33N)
Adventdalen (ADV)	307	90	I_a: 37.3° / D_i: 69.5°	8,685,931 511,282
Kapp Linné (KAP)	288	88	I_a: 34.0° / D_i: 67.8°	8,551,011 469,616
Ny-Ålesund (NYA)	121	84	I_a: 34.3° / D_i: 66.0°	8,765,916 423,918

The Adventdalen (ADV) area is dominated by a SE−NW oriented valley tributary to the large Isfjord system. Longyearbyen, Svalbard's main settlement and airport, is located in the western part of the study area (black star, Figure 1). ADV has a complex topography with mountain tops up to approximately 1050 m a.s.l. and glacially eroded valleys down to sea level, carved into flat-lying sedimentary rocks, consisting primarily of sandstones and shales [49]. Following regional deglaciation, fluvial and periglacial activities have further developed the ADV geomorphology. The valley floors are infilled by fluvial, alluvial and eolian (loess) deposits [50]. The permafrost thickness is about 100 m near the coast and increases up-valley [51]. The accumulation of eolian deposits on alluvial terraces led to the upward growth of syngenetic permafrost, underlain by epigenetic permafrost formed by the downward freezing of fluvial and marine deposits [52]. Ground ice distribution is variable [52,53], but higher contents are generally found in the top few metres of the syngenetic permafrost, especially in eolian deposits, and at the top of the underlying epigenetic permafrost [52]. Permafrost ECV observations have been carried out in this area since 2000 [45]. In 2016−2017, the mean annual temperature of the permafrost surface varied between approximatively −0.5 and −3.2 °C in two SIOS boreholes. ALT was between 0.9 and 1.8 m [44].

The Kapp Linné (KAP) area is located in the westernmost part of the Nordenskiöld peninsula and is greatly influenced by the North Atlantic maritime regime [54]. The region is characterized by the northwest-striking West Spitsbergen Fold Belt [55], and a low-lying Precambrian bedrock platform—the strandflat—mantled by raised beach deposits [56]. The Isfjord Radio weather station is situated in the northwestern part of this strandflat complex (black square, Figure 1). The Griegfjellet ridgeline, composed of pre-Cambrian phyllite [57], reaches up to approximately 780 m a.s.l. and separates the coastal strandflat from the Linné valley and its proglacial lake Linnévatnet [58]. The strandflat geomorphology is characterized by a complex assemblage of dry, coarse-grained, raised marine beach ridges and exposed weathered bedrock interspersed with thermokarst lakes, organic-rich shallow bogs with small palsas, ice-wedge polygons and sorted/unsorted circles [59,60]. Permafrost ECV observations have been carried out in this area since 2008 [43]. In 2016−2017, the mean annual temperature of the permafrost surface varied between approximatively −1.5 and −1.8 °C in two SIOS boreholes. ALT was between 1.8 and 3.0 m [44].

The Ny-Ålesund (NYA) area is located along Kongfjorden in northwestern Spitsbergen. It includes a former coal mining village, converted into a research station (black square, Figure 1). The region is characterized by a strandflat area in the outer part of the Brøggerhalvøya peninsula and steep topography with the highest peaks at 790 m a.s.l in the southeastern part. In Brøggerhalvøya, late Paleozoic to early Triassic sedimentary sequences are overlain by Paleocene coal-rich sediments [61]. The bedrock is covered by Quaternary terrestrial and coastal sediments, consisting of till, colluvial, fluvial and raised beach deposits [62]. Fine-scale variabilities of surface temperature [63], snow cover [64] and active layer thickness [65] have been documented in the intensively studied Bayelva area (black circle, Figure 1). Here a permafrost research site collects long-term environmental data series since 1998 [66]. In 2016–2017, the mean annual temperature of the permafrost surface was approximatively −2.7 °C in the two SIOS boreholes in the area. ALT was between 1.5 and 2.0 m [44].

3. Data and Methods

3.1. Sentinel-1 SBAS Time Series

InSAR results are based on SAR images from the Sentinel-1 satellites of the European Commission Copernicus Programme. The selected scenes have been acquired with the Interferometric Wide swath mode in an ascending geometry (track 14). The sensor looks obliquely downward (LOS incidence angle I_a, Table 1), towards ENE (LOS compass direction D_i, Table 1). The same period (22 June to 25 November 2017) was selected for each study area to provide comparable time series. The chosen start and end dates are expected to reduce the risk of significant decorrelation due to extensive snow cover. It should, however, be noted that ground thawing is expected to start before 22 June (typically late May–early June in ADV) [21,45], which leads to an underestimation of the total seasonal subsidence if ground ice melts in the upper part of the active layer. This has no major implication considering that the scope of the study is to document the timing of the transition from thaw subsidence to frost heave, occurring later within the season, and discuss the relative temporal variability of the displacement patterns.

InSAR results were processed using the NORCE GSAR software [67]. Single Look Complex (SLC) images were co-registered and multi-looked using a range/azimuth factor of 8 × 2. Interferometric image pairs (interferograms) were generated with a maximal temporal baseline of 48 days. After removal of strongly decorrelated interferograms (mean coherence <0.5) due to fast movement, snow and moist surface, the effective temporal baseline is 6 to 12 days at the beginning and end of the time series. Longer temporal baselines have been included in the middle of the time series due to more stable ground conditions. The final selection includes 84 to 90 interferograms depending on the study area (Table 1 and Supplements S1–S3). The ADV time series is continuous. Acquisition 27.08.2017 in KAP is affected by major ionospheric effect, while acquisition 08.10.2017 in NYA is affected by snow. They have thus been discarded, introducing a gap in the time series. The constrained spatial baselines (maximum value: 188 m) lead to small topographical phase components that have been estimated and removed using a 20 m Digital Elevation Model (DEM) [47]. The noise-level was reduced in all interferograms by applying a spatially adaptive coherence-dependent Goldstein filter [68,69]. The contribution from the stratified atmosphere was mitigated by a data driven approach where we fit a linear relationship between residual phase and topography [70] using the available DEM [47]. Based on the redundant sets of interferograms, we further corrected the stratified delay per scene using a network-based approach [71]. The remaining turbulent component was mitigated by averaging all pairs centred on common acquisitions and using the redundancy to iteratively estimate the atmospheric contribution of each scene [72]. Pixels affected by low signal stability due to snow (e.g., glaciers, perennial patches) or water (e.g., lakes, rivers) in most of the pairs were removed by applying a coherence-based filter (0.5 in 50% of the selected interferograms).

The unwrapping was performed using the SNAPHU software [73]. InSAR is a spatially relative technique, which means that the results must be calibrated to a reference location. We tested different reference points and chose references in areas assumed to be stable. For all study areas, reference points are on low-inclined surface (≤2°) with high mean coherence (≥0.8), located either on infrastructure or on visible rock outcrops based on aerial imagery [74] (Table 1). We estimated ground displacement time series using the Small Baseline Subset (SBAS) method [75]. The inversion is performed using an L1-norm-based cost function, which is more robust than L2-norm with respect to unwrapping errors [76]. Atmospheric filtering used a spatial filter of 500 m and a temporal filter of 18 days.

The initial InSAR measurements correspond to one-dimensional sensor-to-ground distance changes along the LOS, expressed in mm. The displacement times series are temporally relative to the first scene of the dataset (22 June 2017) and spatially relative to the reference points (Table 1). The results were geocoded using the 20 m DEM [47] and have a 40 m spatial resolution. The time series have a six-day temporal resolution.

3.2. InSAR Post-Processing and Identification of the Thaw Subsidence Maxima

We propose a relatively simple workflow to generate high level products based on seasonal SBAS time series (inputs from Section 3.1). We applied conservative filters to remove unreliable and irrelevant information when focusing on the timing of the thaw subsidence maxima. The procedure follows five main steps to filter the SBAS results, convert the LOS values to vertical displacements, identify the maximal subsidence and extract the acquisition date of the maxima. These steps are illustrated in Figure 2 (observed displacements) and summarized hereafter:

A. Pre-filtering of the initial SBAS results:

- Criterion 1 "Ambiguity": InSAR signal becomes aliased when the displacement gradient between adjacent pixels is higher than a quarter of the wavelength during the selected time interval. When using Sentinel-1 (5.6 cm wavelength), if the displacement is over 14 mm between the acquisitions used to build interferograms, there is a higher probability of spatial phase unwrapping error [69]. We therefore filtered out the results likely to be affected by a phase unwrapping error by filtering out pixels including a displacement gradient over 14 mm between successive acquisitions of the time series. If the displacement difference is over the ambiguity threshold, for example between the first acquisition (June 22) and the second (28 June), the pixel is discarded.
- Criterion 2 "Slope angle": Creep processes on slopes are likely to mask out the transition from subsidence to heave due to a gradual and continuous downslope displacement component. Based on a 20 m DEM [47], we computed the slope angle using ArcGIS (©ESRI). We discarded all pixels with slope angle >1.5°, to focus on flat areas (Supplement S4). Solifluction can occur on low-inclined surfaces, and has been reported on 2° slopes [77]. The conservative threshold of 1.5° was used to account for the relatively low DEM resolution, likely to underestimate local slope variabilities.
- Criterion 3 "Coherence": Decorrelation sources due to snow, ground moisture and vegetation may affect the quality of the displacement estimates [78]. We applied a secondary coherence thresholding more conservative than at the processing stage (Section 3.1). Pixels with mean coherence <0.55 based on the selected interferograms (Table 1; Supplement S1–S3) were discarded.

B. Vertical conversion: InSAR measurements are intrinsically one-dimensional, along the oblique LOS (Table 1) and therefore provide ambiguous information in complex topography, especially if the movement orientation is spatially heterogeneous and/or temporally variable. As we focus here on flat areas, we can assume that all displacements occur vertically (subsidence and heave). Although some local areas (e.g., coastal areas affected by erosion) may slightly deviate from this general assumption, we expect that the dominant ground behaviour at the landscape scale is along the vertical. Therefore, we converted all results from LOS to vertical displacement using the following equation:

$$V_{disp} = \frac{LOS_{disp}}{\cos(I_a)} \quad (1)$$

where V_{disp} is the vertical displacement, LOS_{disp} is the LOS-projected displacement, and I_a is the incidence angle of the radar beam (Table 1). V_{disp} documents a subsidence (positive value) or a heave (negative value), relatively to the first acquisition date.

C. Subsidence maxima identification: For each time series, the maximal value was identified, and its corresponding Day of Year (DOY) was extracted. It should be noted that the DOY identification is based on the subsidence maximum only and does not take into account the entire pattern of the displacement progression, which may lead to erroneous value if the ground level flattens at the end of the thawing season.

D. Post-filtering of the selected time series:
   - Criterion 4 "Cyclicity": Pixels with DOY corresponding to the first or the last acquisition of the series (i.e., without any subsidence/heave pattern) were discarded, as they do not document a cyclic process. We assume that these pixels correspond to remaining low-inclined areas affected by downslope creep. In combination with Criterion 2 "Slope angle", Criterion 4 is an additional way to ensure that we focus the analysis on flat areas dominated by vertical patterns.
   - Criterion 5 "Maxima": All pixels with a maximal subsidence <10 mm were additionally discarded, as we assume that the transition between subsidence to heave in areas with low displacement amplitude is likely to be masked out by noise sources (e.g., atmospheric effects, bias due to snow or ground moisture). The temporal resolution of the DOY product is 6 days (12 days when there is one missing acquisition), corresponding the repeat-pass interval of the Sentinel-1 mission.

E. InSAR outputs: For analysing the spatial distribution of the maximal subsidence, we used the results after the four first steps of filtering (Criteria 1−4), while all five criteria are used to map the DOY. For further comparison with the temperature-based model (Sections 3.3 and 3.4), we focused on time series at three different scales (local, intermediate and regional) by:
   - Extracting the nearest pixels to the boreholes;
   - Averaging the series for the pixels within 1 km^2 around the boreholes;
   - Averaging the pixels with a DOY of the subsidence maxima within the interquartile range of all results, as we assume that they are representative of the ground behaviour at the regional scale.

   When averaging time series, the dispersion of the selected values can be represented by the standard deviation around the mean of the selected pixels for each acquisition date.

3.3. Air and Ground Surface Temperature

To interpret the InSAR measurements and evaluate the importance of the thermal control on the ground dynamics, we used daily averaged air temperature from three weather stations and ground surface temperature from three boreholes in all study areas (Table 2; Figure 1).

3.4. Composite Index Model of Seasonal Time Series

The SBAS time series were compared to a simple composite index model based on temperature (input from Section 3.3). The calculation of the composite index is explained in detail by Hu et al. [27]. The main steps of the procedure are schematically summarized in Figure 2 (modelled displacements) and can be described as follows:

A. For a given time (t) in the season, the thawing (freezing) depth (Z) of the ground can be modelled using the Stefan equation [28,29], as follows:

$$Z(t) = \sqrt{\frac{2knA(t)}{\rho\theta L}} \tag{2}$$

where k is the thermal conductivity of soil (units: $W \cdot m^{-1} \cdot K^{-1}$), n is the n-factor ratio accounting for the offset between the air temperature and the ground surface temperature, influenced by the surface conditions (e.g., snow, vegetation) [80], and $A(t)$ is the accumulated degree days of thawing or freezing (units: °C days). ρ is the soil bulk density (units: kg/m^3), θ is the volumetric water content (units: m^3/m^3), and L is the specific latent heat of fusion for water (units: J/kg).

The phase change of water causes volume change of the ground medium within the thawed (frozen) layer. We assume that the 1D ground medium has homogenous and constant thermal properties and water content. All the water pores within the active layer are assumed to be affected by phase change, which causes a ~9% volume decrease and increase leading to subsidence and heave. Based on these assumptions and following the steps detailed in Hu et al. [27], we can express the subsidence (*s*) and heave (*h*) as a simple function of the variables *A(t)* and the time-invariant coefficients *E*:

$$s(t) = E_T\sqrt{A_T(t)} \text{ and } h(t) = E_F\sqrt{A_F(t)} \quad (3)$$

where *A(t)* is the time-variant accumulated degree days of thawing (*T*) (units: °C days) or freezing (*F*) (units: °C days), based on the daily averaged air and ground surface temperatures from the sites introduced in Section 3.3. E_F and E_T are time-invariant coefficients based on ground/water properties of the frozen (*F*) and thawed (*T*) ground (the soil bulk density, volumetric water content, water/ice density, latent heat of fusion for water, thermal conductivity, n-factors) [27].

B. The two seasonal coefficients E_F and E_T can be related by a scaling factor α:

$$\alpha = \frac{E_F}{E_T} = \sqrt{\frac{k_F n_F}{k_T n_T}} \quad (4)$$

Where k_F and k_T are the thermal conductivities of the frozen (*F*) and thawed (*T*) ground and n_F and n_T are the *n*-factors n_F and n_T as described in Equation (2).

The scaling factor applied by Hu et al. [27] is 1.44 based on the ground properties of the Utqiaġvik site (Anchorage, AK, USA), and using air temperature series for the calculation of the composite index. In our study, the *k* and *n* coefficients remain undefined, due to variable and partly unknown ground properties in the study areas. We therefore focused on their relative difference (ratio between frozen and thawed ground) represented by the scaling factor α. When using ground surface temperature, the *n*-factors are 1, so α reduces to $\sqrt{\frac{k_F}{k_T}}$, i.e., the difference of thermal responses between thawed and frozen ground. We tested and compared the results using five α values (1, 1.2, 1.4, 1.6 and 1.8).

C. Based on Equations (3) and (4), the composite index I_c can be expressed as:

$$I_c(t) = \sqrt{A_T(t)} - \alpha\sqrt{A_F(t)} \quad (5)$$

We set the composite index to be zero until the start of the thawing season. The initiation of the calculation starts when the first daily averaged air and ground surface temperatures above 0 °C are recorded at the weather stations or in the boreholes.

D. Because we are only interested in characterizing the temporal pattern of ground displacements, we normalized the composite index with its maximum value and rescaled it by multiplying the index by the maximal value of the SBAS displacement time series:

$$d(t) = d_s\bar{I}_c(t) \quad (6)$$

where *d(t)* is the rescaled composite index, directly comparable with the observed displacement (Section 3.2), d_s is the maximum seasonal subsidence based on the SBAS time series (units: mm) and $\bar{I}_c$ is the composite index (Equation (5)) normalized with its maximum value.

The comparisons between the SBAS time series and the model (normalized and rescaled composite index) consisted of:

- Comparing the timing of the transition between the subsidence and the heave from the observations and the thawing and freezing periods from the models;
- Evaluating the goodness of the fit between the observations and the models by documenting the proportion of the variance of the seasonal SBAS displacements that is explained by the normalized index (R^2);
- Analysing the temporal variations of the entire observed displacement time series with respect to the rescaled composite index;
- Discussing the results' differences when using air or ground surface temperature, as well as single pixels closest to the boreholes, 1-km^2 or regional averaged displacement time series.

We finally interpreted the fit/deviation between the observation and the model by discussing the limitations of the SBAS products and the validity/invalidity of the assumptions behind the simplified composite index model. The findings were related to the results from Section 3.2 to discuss the potential of DOY maxima products to document the cyclic dynamics of the active layer in permafrost landscapes.

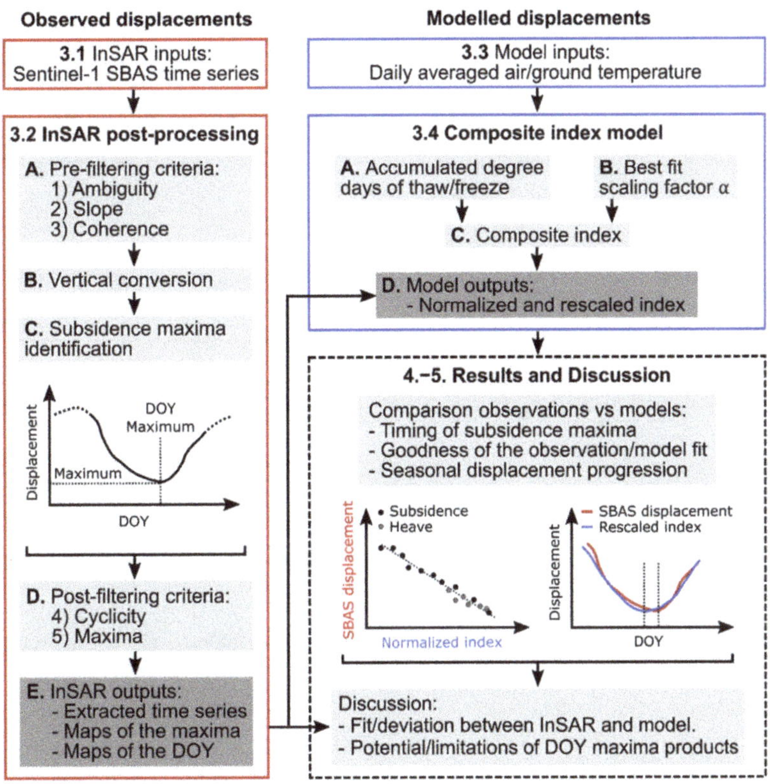

Figure 2. Schematic illustration of the study workflow. The boxes are labelled according to the numbers of Sections 3.1–3.4 and the processing steps listed in the text. Red boxes show the workflow of the InSAR procedure (observed displacements). Blue boxes show the workflow of the modelling procedure (modelled displacements). Dashed black box highlights the procedure to compare and discuss the observed and modelled displacements. Light grey areas show the main processing/analysis steps. Dark grey areas show the main outputs.

Table 2. Information about the weather stations, the boreholes and the documented air and ground temperature time series in Table 1.

	Station and Data Types	Coordinates (UTM 33N)	Altitude [m a.s.l]	Site Information and Reference
Adventdalen (ADV)	Weather station. Air temperature.	8,681,070 N 518,966 E	15	Adventdalen station 99870. Reference: NCCS, 2021 [79].
	Borehole. Ground temperature.	8,680,294 N 522,504 E	17	Ice-wedge polygons in eolian deposit. The area is affected by long-term subsidence, exposing the upper sensors closer to the surface. Data from logger at −23 cm is therefore used in this study, assuming to be representative of the ground surface conditions. The borehole is part of the UNIS monitoring network and temperature data has previously been compared with InSAR in Rouyet et al., 2019 [21].
Kapp Linné (KAP)	Weather station. Air temperature.	8,665,721 N 468,119 E	7	Isfjord Radio station 99790. Reference: NCCS, 2021 [79].
	Borehole. Ground temperature.	8,664,808 N 468512 E	21	Beach ridge on strandflat composed of coarse-grained beach sediment. Sensor at ground surface. GTN-P and NORPERM references: NO 36/KL-B-2. Reference: Christiansen et al., 2010; 2021 [43,44].
Ny-Ålesund (NYA)	Weather station. Air temperature.	434,216 N 8,763,255 E	8	Ny-Ålesund station 99910. Reference: NCCS, 2021 [79].
	Borehole. Ground temperature.	8,762,985 N 432,118 E	25	Diamict surface with fine-grained glaciofluvial sediments. Profile C. Sensor at −1 cm. GTN-P reference: NO GE 60. Reference: Boike et al., 2018 [66].

4. Results

4.1. Thaw Subsidence Maxima

The final SBAS results (Section 3.2) are presented on maps showing the distribution of the subsidence maxima and their corresponding DOY (Figures 3–5). The five criteria used for filtering significantly reduce the exploitable observations to 3–5% of the initially documented pixels (Supplement S5) but still provide a good coverage in the flat areas, mostly in the valley bottoms in ADV, and in the strandflat areas in KAP and NYA, with total pixel numbers of 14,547 in ADV, 21,198 in KAP and 6021 in NYA. The observed patterns are described by analysing the results at the landscape scale, as well as within selected km^2 areas around the three boreholes and over landforms experiencing a behaviour that deviates from the regional trend. Orthophoto imagery of the corresponding locations can be found in supporting material (Supplement S6).

In ADV, displacements occurring between June 22 and the day of the subsidence maxima are mostly over 20 mm (Figure 3A). The displacement distribution is geomorphologically-controlled [21]. Large displacements are detected on landforms assumed to have high water content and composed of fine-grained frost-susceptible sediments. Greatest subsidence up to approximately 100 mm are found in the outer part of alluvial fans and in the eolian terraces surrounding the ADV braided river. The timing of the subsidence maxima is quite homogenous in the valley bottom (Figure 3B), with values between 251 and 269 (mid–late September). The maxima are identified later on the valley sides, e.g., where the borehole is located (Figure 3B, area C; Figure 6C), suggesting different ground behaviour between the terraces (Supplement S6C) and the fluvial riverbed (Supplement S6D). Most ADV time series show a clear cyclic displacement pattern, with a distinct and quick transition between subsidence and heave that allows for a mostly unambiguous extraction of the DOY maxima (Figure 6C,D). The subsidence maxima are identified earlier on the SW blockfield plateau (Figure 3B, area E; Figure 6E), suggesting earlier freeze-back due to higher elevations (400–500 m a.s.l). The end of the time series in November is often affected by a stabilization or lowering effect (Figure 6C,E). This presumed artefact, further discussed in Section 5.1, especially affects some alluvial fans in the inner part of the valley (eastern side in Figure 3B) where outliers with late DOY maxima are identified.

In KAP, the variability of the subsidence maxima matches the complexity of the landform assemblage in the area (Figure 4A), especially in the northern part (Supplement S6, area F). Low subsidence values (typically lower than 20 mm) are detected in areas composed of exposed bedrock or coarse-grained beach ridges, but maxima up to 120 mm are found in organic-rich and fine-grained sediments in beach ridge depressions. A spatial gradation with earlier DOY along the eastern slope compared to the coastal part highlights the impact of the maritime influence within the area (Figure 4B). The timing of the subsidence maxima is mostly homogenous in the northern part, e.g., where the borehole is located (Figure 4B, area F). The subsidence maxima occur considerably later than in ADV (287 to 305, i.e., in mid–late October). The transition between subsidence and heave is clear on series from the northern part (Figure 4B, area F; Figure 6F) but becomes more ambiguous towards the South, where earlier DOY are detected (Figure 4B, areas G–H; Figure 6G,H). The flattening of the displacement curves in mid–late summer shows that the ground level stabilizes for a long period (Figure 6G,H). The different behaviour observed on the beach ridges in the southeastern part of the strandflat may indicate specific ground conditions at the mouth of the Orustdalen valley (Figure 3B, area H). This is further discussed in Section 5.1.

In NYA, the documented surfaces are mostly composed of coarse beach and diamict deposits. The relatively little frost-susceptibility of the material leads to subsidence values generally lower than for the two other study areas, with maxima mostly under 40 mm (Figure 5A). Maximal subsidence, up to 80 mm, are found primarily in the western part of the peninsula, in the alluvial deposits surrounding the Kvadehukelva riverbed. Beach ridge depressions on the strandflat show relatively homogenous DOY values with subsidence maxima occurring approximatively at the same time as in the KAP area (287 to 299, in mid-October) (Figure 5B, e.g., in area J).

The northwestern part of the strandflat has considerably earlier DOY (Figure 5B, area K) compared to the surroundings (Figure 5B, area J), suggesting a different ground behaviour in the exposed coastal part of the landscape. This area, in the outer part of alluvial fans, experiences rather stable surface position, with only minor subsidence through the documented summer period, followed by quick and large heaving when freeze-back occurs (Figure 5B, area K; Figure 6K). In the inner part of Kongsfjorden (NE in Figure 5), some altitudinal variations and geographical zonation are visible, with earlier DOY compared to the outer part of the peninsula. Around the borehole, a heave pattern is detected in late August to mid-September, prior to the main one starting in mid-October (Figure 6I). We further discuss the potential causes of this pattern in relation with the composite index model (Sections 4.2 and 5.1).

When comparing the results for the three areas, clear differences are identified (Figure 7). The median DOY of the subsidence maxima is earlier in ADV compared to KAP and NYA. Even considering the interquartile variability of the distribution, the values of ADV show no overlap with the two other areas (Figure 7A). KAP and NYA DOY maxima have overlapping distributions, but the detected subsidence values are lower in NYA (Figure 7A). Cyclic patterns are visible in all regionally averaged time series (Figure 8B), but the results highlight differences between the three study areas that are further discussed in Section 5.1. Time series with maxima below/over the first and third quartiles of the entire area (Figure 8A,C) exemplify the deviation from the most common pattern. Due to long thawing periods, the averaged time series in KAP and NYA show that the ground tends to remain stable during several months in mid–late summer (Figure 8), which likely affects the reliability of the automatic extraction of the subsidence maxima. This issue is further discussed in Section 5.2.

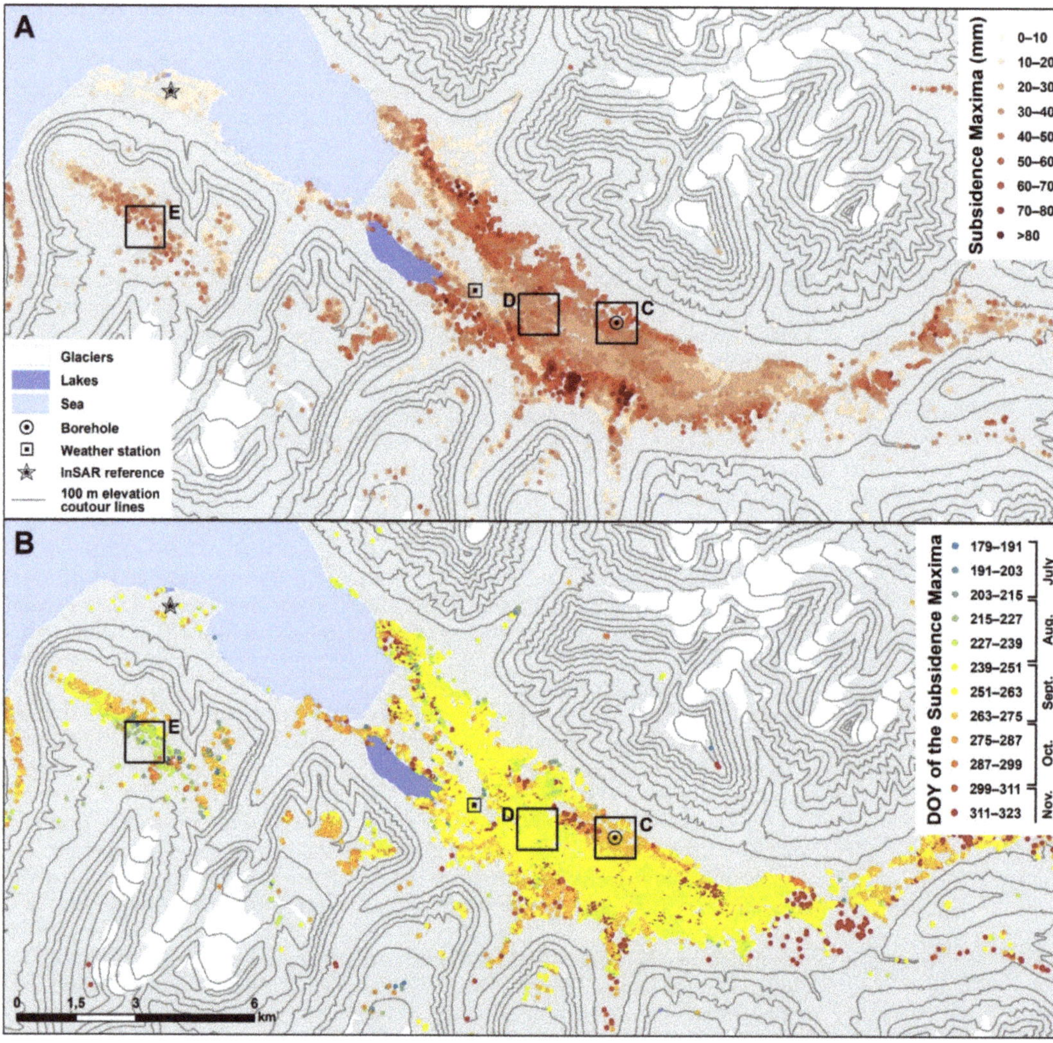

Figure 3. Spatial distribution of filtered SBAS results in Adventdalen (ADV) based on 22.06.2017–25.11.2017 time series. (**A**) Subsidence maxima for the pixels selected after the four first filtering criteria (Section 3.2). (**B**) Day of Year (DOY) of the subsidence maxima for the pixels selected after the five filtering criteria (Section 3.2). Black squares (C–E) show the km^2 extents used to average the time series (Figure 6C–E). Abbreviations refer to August (Aug.), September (Sept.), October (Oct.) and November (Nov.). Orthophoto imagery of the corresponding locations are shown in Supplement S6. One hundred metres of elevation contour lines, glacier, sea and lake layers are from the Norwegian Polar Institute [47,48].

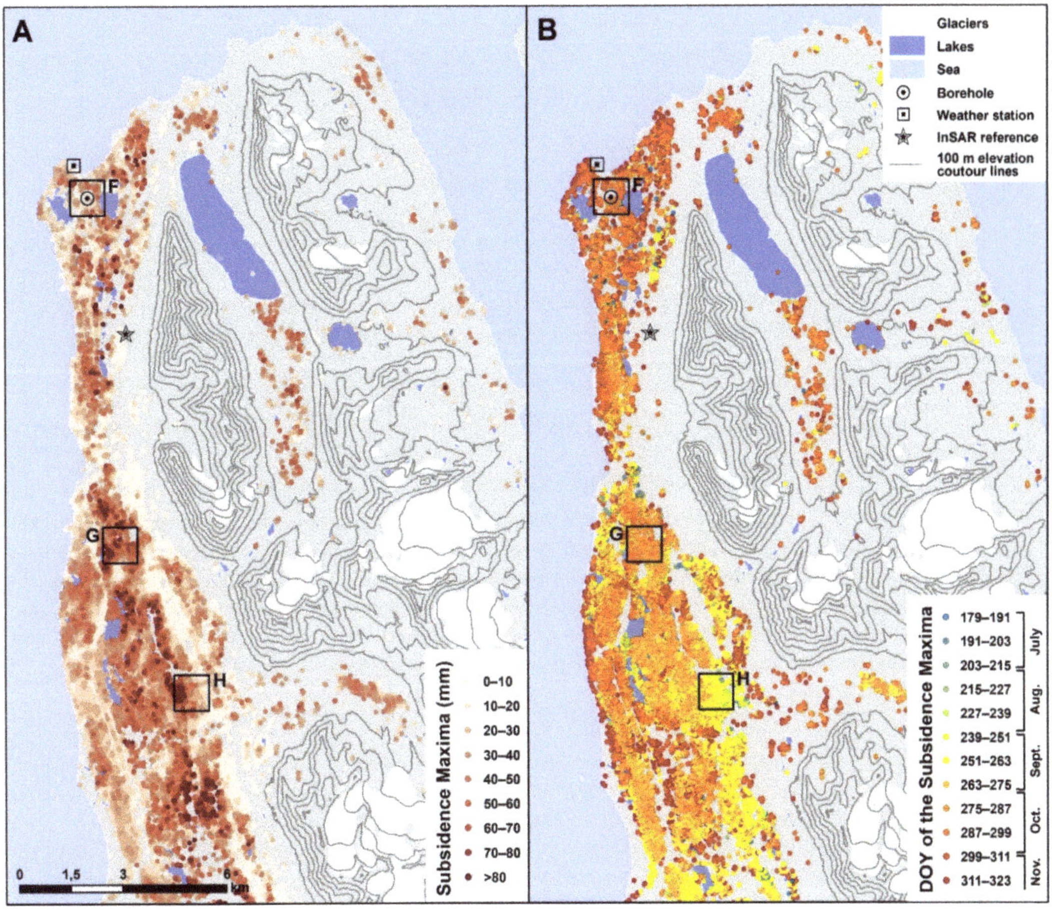

Figure 4. Spatial distribution of filtered SBAS results in Kapp Linné (KAP) based on 22.06.2017–25.11.2017 time series. (**A**) Subsidence maxima for the pixels selected after the four first filtering criteria (Section 3.2). (**B**) Day of Year (DOY) of the subsidence maxima for the pixels selected after the five filtering criteria (Section 3.2). Black squares (F–H) show the km^2 extents used to average the time series (Figure 6F–H). Abbreviations refer to August (Aug.), September (Sept.), October (Oct.) and November (Nov.). Orthophoto imagery of the corresponding locations are shown in Supplement S6. One hundred metres of elevation contour lines, glacier, sea and lake layers are from the Norwegian Polar Institute [47,48].

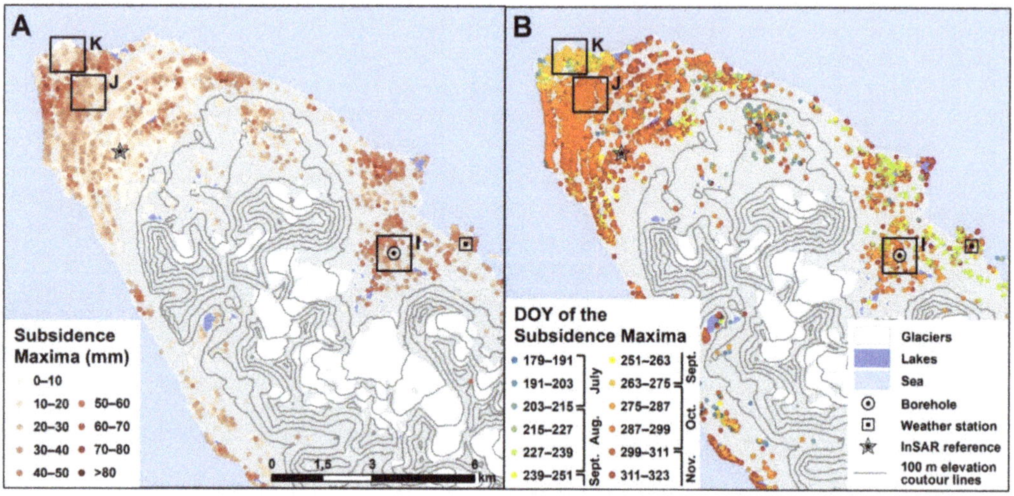

Figure 5. Spatial distribution of filtered SBAS results in Ny-Ålesund (NYA) based on 22.06.2017–25.11.2017 time series. (**A**) Subsidence maxima for the pixels selected after the four first filtering criteria (Section 3.2). (**B**) Day of Year (DOY) of the subsidence maxima for the pixels selected after the five filtering criteria (Section 3.2). Black squares (I–K) show the km^2 extents used to average the time series (Figure 6I–K). Abbreviations refer to August (Aug.), September (Sept.), October (Oct.) and November (Nov.). Orthophoto imagery of the corresponding locations are shown in Supplement S6. One hundred metres of elevation contour lines, glacier, sea and lake layers are from the Norwegian Polar Institute [47,48].

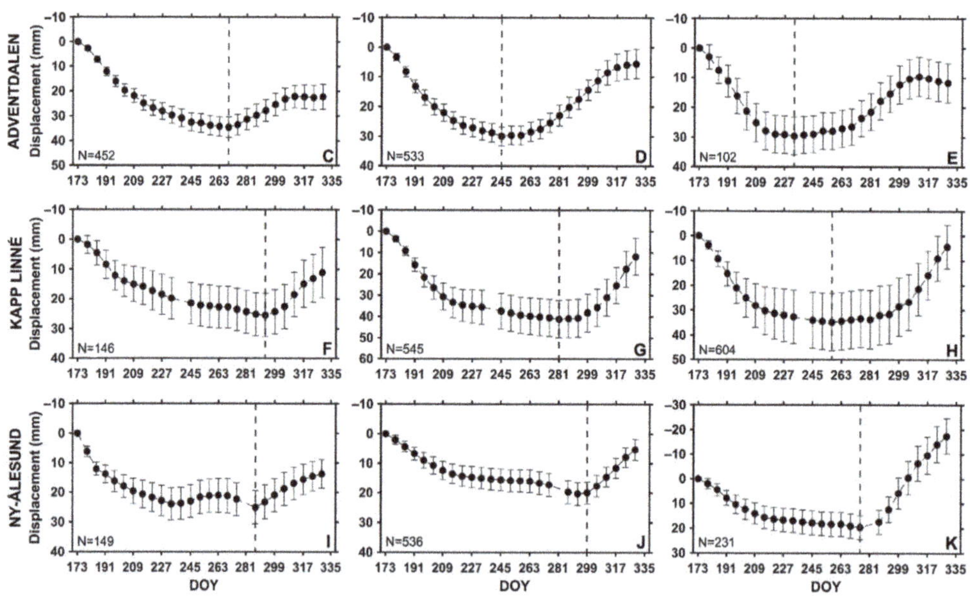

Figure 6. Examples of the spatial variability of the seasonal displacement progression and the detected DOY of the subsidence maxima for time series averaged over 1 km^2. Locations (C–K) are shown in Figures 3–5 (black squares). Orthophoto imagery of the areas are shown in Supplement S6. The vertical bars show the standard deviation of the displacement values around the mean for each acquisition date. The vertical dashed lines show the averaged detected DOY of the subsidence maxima. N values document the number of pixels that have been averaged for each area. Note that the y-axis scale varies for each graph.

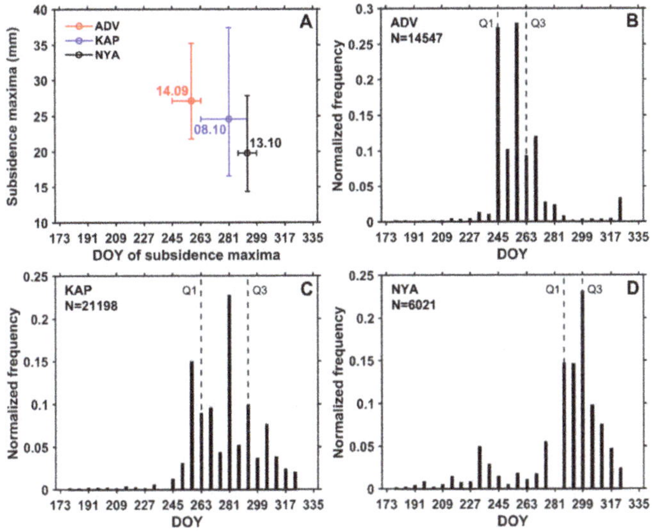

Figure 7. (**A**) Median and interquartile range of the subsidence maxima (>10 mm) and associated DOY. Circles and dates show the median values. The bars show the interquartile range. (**B–D**) Distribution of the DOY of the subsidence maxima in Adventdalen (ADV), Kapp Linné (KAP) and Ny-Ålesund (NYA). Vertical dashed lines Q1 and Q3 show the first and third quartiles of the distribution. N values are the numbers of total pixels after filtering.

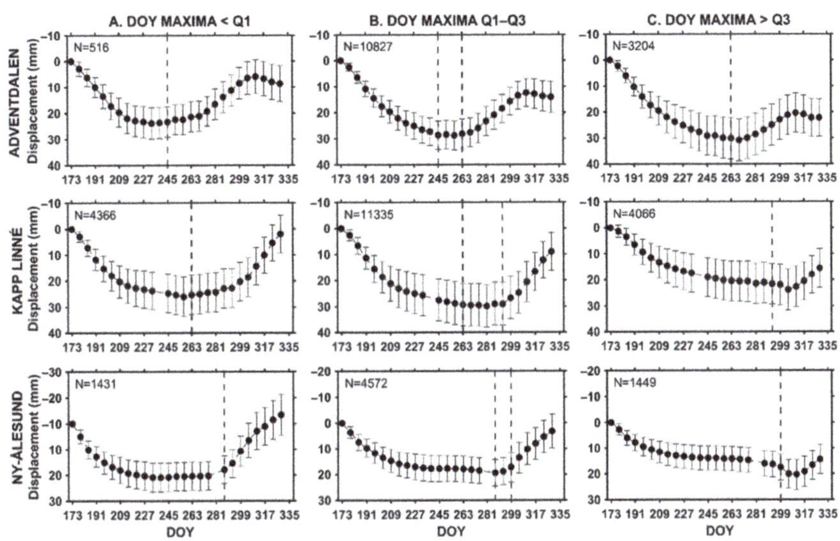

Figure 8. Averaged time series for each study area classified according to the DOY of the subsidence maxima. The vertical bars show the standard deviation of the displacement values around the mean for each acquisition date. Vertical dashed lines Q1 and Q3 show the first and third quartiles of the entire distribution. N values document the number of pixels that have been averaged in each category. (**A**) Average of the series with a detected DOY earlier than the first quartile (Q1) of the distribution. (**B**) Average of the series with a detected DOY within the interquartile range, assumed to be representative of ground dynamics at the regional scale. (**C**) Average of the series with a detected DOY later than the third quartile (Q3) of the distribution.

4.2. Composite Index Model of Seasonal Time Series

The daily averaged temperature series highlight differences between the air and ground surface temperatures measured at weather stations and in boreholes in each of the three study areas (Table 2). Short-term fluctuations are reduced in the time series from the boreholes (Figure 9A), mostly due to the insulating winter and spring effects of the snow and the thermal diffusivity of the upper soil layer. Especially at the beginning of the thawing season in ADV and NYA, a thermal offset and a time lag are observable between the two series. In NYA, the ground surface temperature during snow melting remained stable around 0 °C during three weeks after the air temperature became positive (Figure 9A). The start of the InSAR observation window on 22 June (measurement period, vertical black lines in Figure 9A) approximately fits the onset of ground thawing in the NYA borehole (20 June). In ADV and KAP boreholes, the ground thawing started on June 13 in ADV borehole and June 4 in KAP. This is 9 days (ADV) to 18 days (KAP) before the InSAR measurement period, which highlights the importance of interpreting the subsidence maxima values in a relative manner, due to the underestimation of the total seasonal subsidence.

The composite index was calculated based on air and ground temperature data from the three weather stations and boreholes (Section 3.3) (Supplements S7–S8). After testing five scaling factors α (Section 3.4), the results showing the best fit between the observed and modelled time series have been selected (Supplements S9–S10). At the regional scale, the pixels with a DOY of the subsidence maxima within the interquartile range of all series (Figure 8B) are compared with the composite index. The results show that for all study areas, the composite index models based on ground surface temperature from the boreholes provide a better fit with the observed subsidence and heave than when using air temperature as input (Figure 9B and Supplement S11). This can be explained by the lag between air and ground temperature, especially at the thaw onset. For further analysis, we therefore used ground surface temperature as input to the models at all scales and for all areas.

In ADV, at the borehole location, both the timing of the transition from subsidence to heave and the whole displacement progression is well represented by the composite index (Figure 10A). Similarly, the match is good for the intermediate scale (km^2 average) (Figure 10B). At the regional average, the general shape of the curve is well represented by the model, but the measurements are shifted compared to the modelled displacements, especially during heaving (Figure 10C). The observed subsidence maxima and the following heave occur earlier than the model. This shows that the borehole is located in an area that deviates from the interquartile regional averaged series, which is also visible when examining the spatial distribution of the DOY of the subsidence maxima (Figure 3B).

In KAP, the pixel closest to the borehole shows noisy patterns, especially at the beginning and the end of the documented period (Figure 11A). The progression of the displacements is not well represented by the model, but the timing of the transition matches the composite index. The km^2 average allows for removing some variability and improves the fit between measurements and the model (Figure 11B, left), although the whole displacement progression still appears to be controlled by other factors than temperature only (Figure 11B, right). The regional average reduces short-term variability, especially during late subsidence, which is well represented by the model. However, as for ADV, the timing of the transition is slightly shifted (Figure 11C). The natural variability in this wide and geomorphologically heterogeneous area does not allow for representing averaged regional displacements based on temperature from one single borehole site located on a dry beach ridge (Supplement S6F).

In NYA, the comparison between the composite index and the InSAR time series closest to the borehole shows a clear shift (earlier for the InSAR series) and a low R^2, showing that the processes occurring at this specific location can not only be explained by a simple temperature-based model (Figure 12A). With a km^2 average, the InSAR series highlight an ambiguous pattern, with a small heave occurring prior to the main one

(Figure 12B). It approximately occurs when the temperature drops and fluctuates around 0 °C for about a week (Figure 9A). The pattern is only visible in the highest part of the landscape (Figures 5I and 6I). Figure 12 generally highlights a poor correspondence between the measurements and the model for the acquisition dates close to the subsidence maxima. Nevertheless, when averaging series at a regional scale, the match between the observation and the model increases considerably (Figure 12C).

The comparison of the three areas leads to contrasting conclusions. In ADV, the time series have a clear cyclic pattern that is well represented by the model at local and km^2 scales. At larger scale, the fit between the observation and the model decreases. The area where the borehole is located appears to be not representative of the regional behaviour dominated by earlier DOY in the fluvial sediments of the brained river plain. In KAP, the km^2 average provides a better fit with the model than a single pixel. The timing of the subsidence-to-heave transition generally fits the temperature records, but the whole displacement progression is not fully explained by the model. At the regional scale, inversely, the displacement progression is well represented by the model, but the subsidence-to-heave transition is somewhat shifted. In NYA, the observation at local scale is not explained by the model. However, by averaging the series within a km^2 or at regional scale, the match improves considerably. The differences between the three areas are further discussed in Section 5.1 and used to identify the advantages and limitations of the proposed method and applied model in Section 5.2.

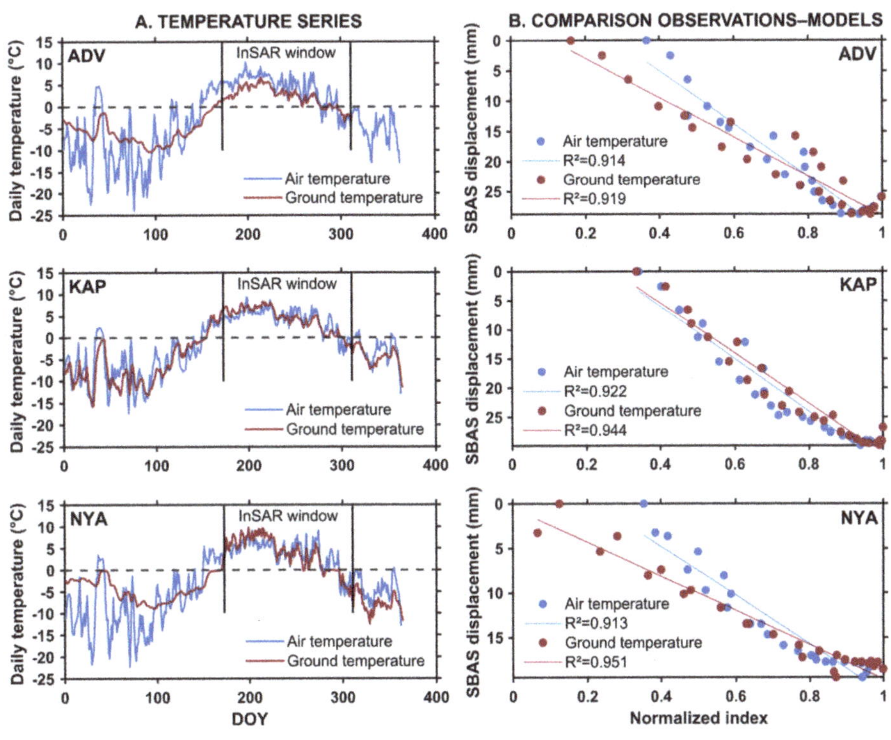

Figure 9. (**A**) Daily air and ground surface temperatures at the weather stations and in boreholes in Adventdalen (ADV), Kapp Linné (KAP) and Ny-Ålesund (NYA) (Table 2). (**B**) Comparisons between normalized composite index and SBAS displacement at the regional scale (DOY maxima Q1–Q3, Figure 8B). Comparisons between the SBAS time series and the rescaled normalized index are shown in Supplement S11. Information about temperature data is summarized in Table 2.

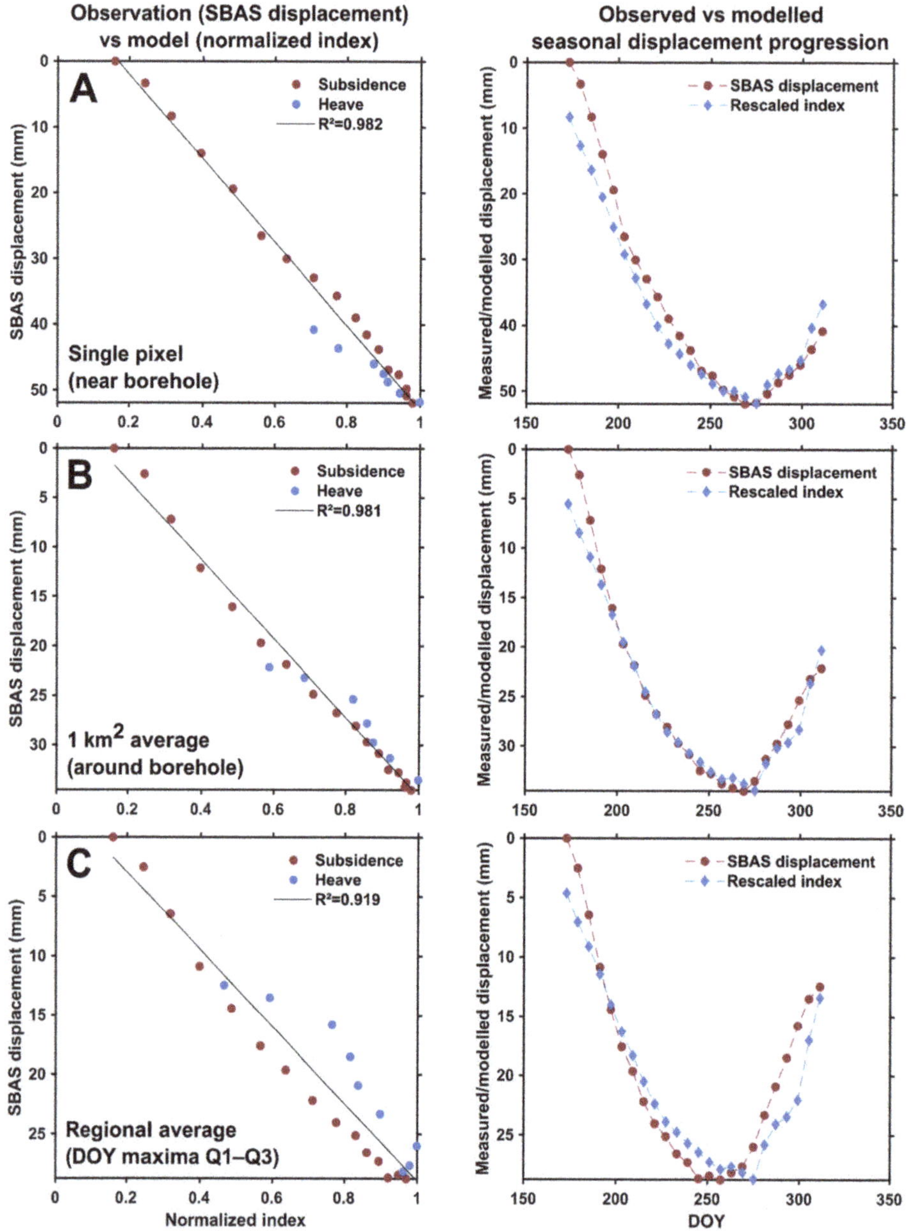

Figure 10. Comparison between the InSAR time series and the composite index at three scales in the Adventdalen (ADV) study area. **Left:** Goodness of the fit between SBAS displacement and normalized composite index. **Right:** Seasonal progression of the SBAS time series and the rescaled composite index. (**A**) The selected time series corresponds to the pixel where the borehole is located. (**B**) 1 km^2 average around the borehole (452 pixels). Black square C in Figure 3. (**C**) Regional average of the time series with a DOY within the interquartile of all the pixels (Figure 8B). Note that the y-axis scale varies for each graph. Information about temperature data is summarized in Table 2.

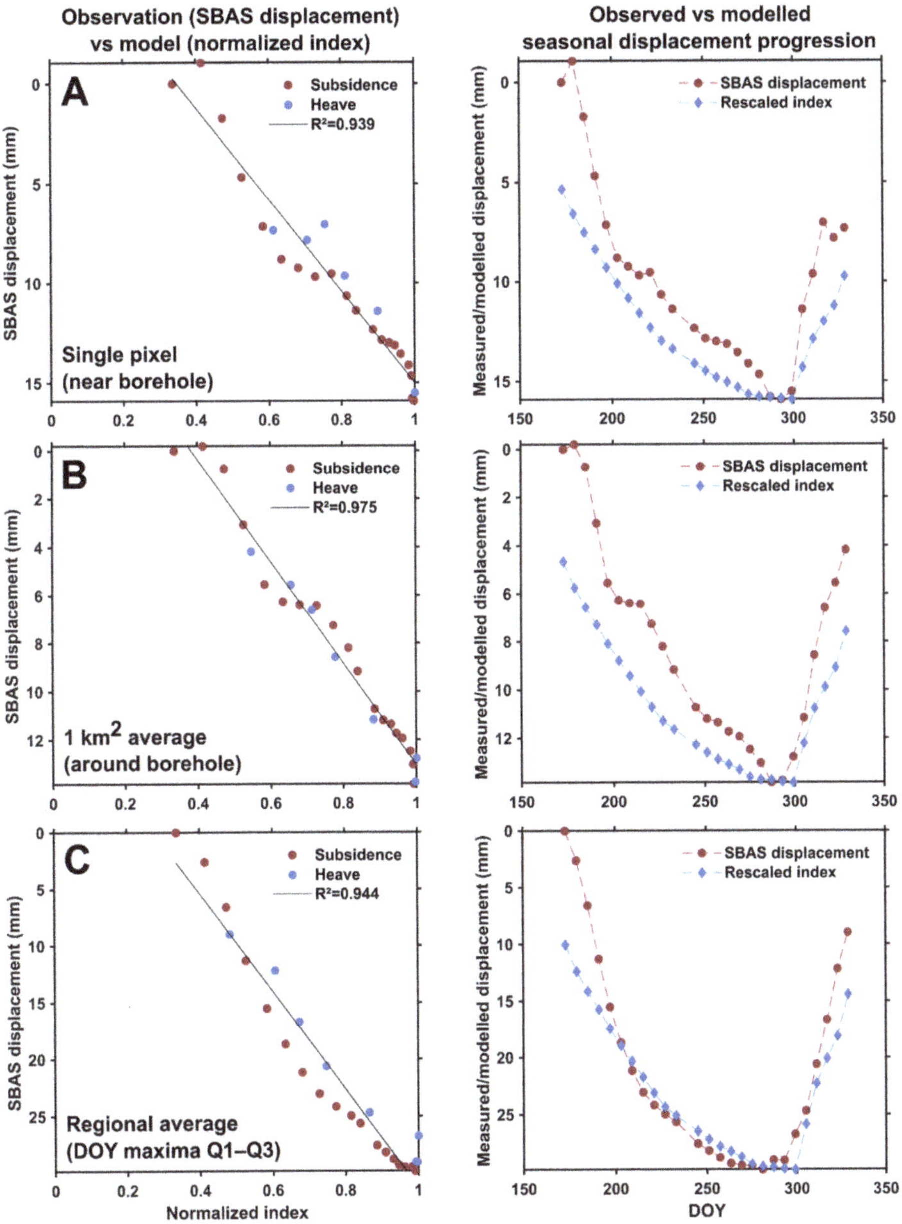

Figure 11. Comparison between the InSAR time series and the composite index at three scales in the Kapp Linné (KAP) study area. **Left:** Goodness of the fit between SBAS displacement and normalized composite index. **Right:** Seasonal progression of the SBAS time series and the rescaled composite index. (**A**) The pixel at the location of the borehole has been filtered out and we therefore used the nearest available time series (pixel centre is 62 m away from the borehole, black circle in Figure 4). (**B**) One km^2 average around the borehole (146 pixels). Black square F in Figure 4. (**C**) Regional average of the time series with a DOY within the interquartile of all the pixels (Figure 8B). Note that the *y*-axis scale varies for each graph. Information about temperature data is summarized in Table 2.

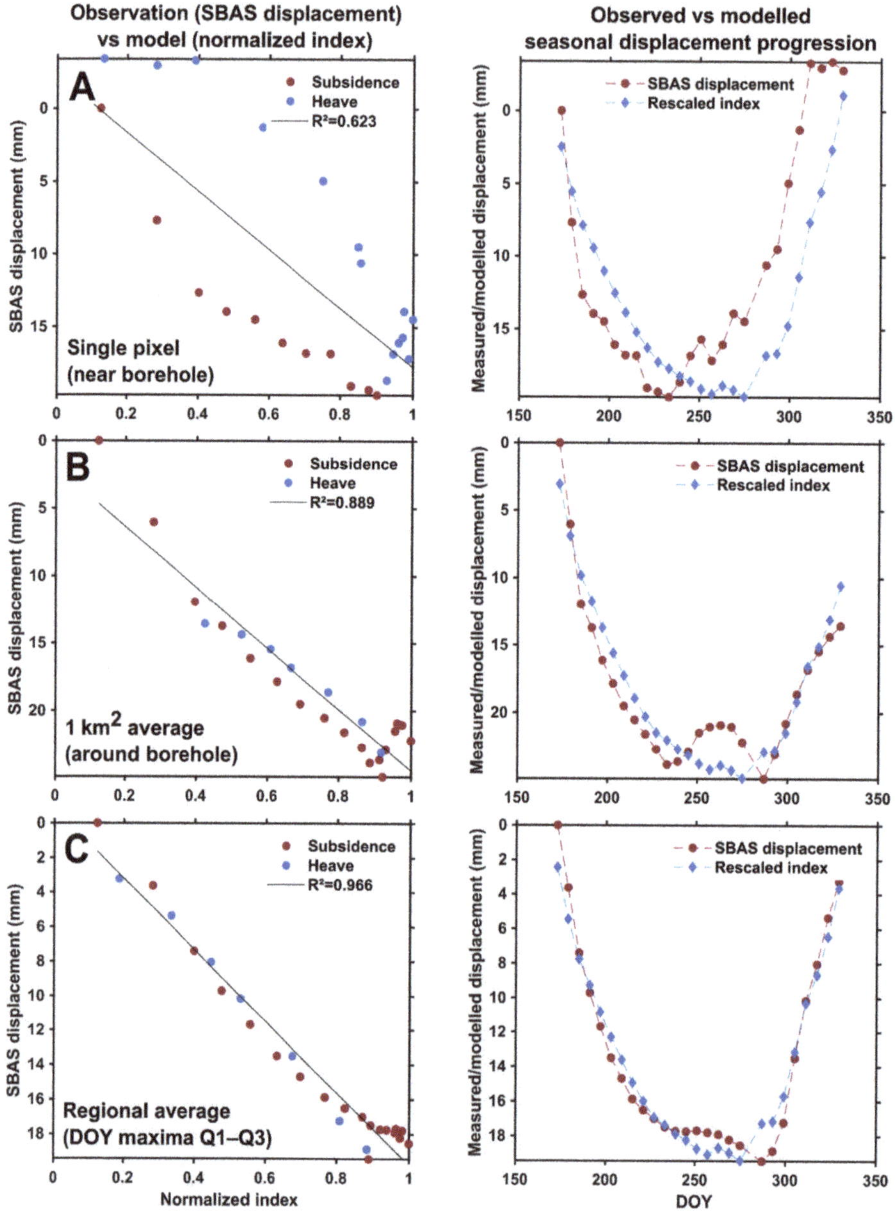

Figure 12. Comparison between the InSAR time series and the composite index at three scales in the Ny-Ålesund (NYA) study area. **Left:** Goodness of the fit between SBAS displacement and normalized composite index. **Right:** Seasonal progression of the SBAS time series and the rescaled composite index. (**A**) The pixel at the location of the borehole has been filtered out and we therefore used the nearest available time series (pixel centre is 43 m away from the borehole, black circle in Figure 5). (**B**) One km^2 average around the borehole (149 pixels). Black square I in Figure 5. (**C**) Regional average of the time series with a DOY within the interquartile of all the pixels (Figure 8B). Note that the *y*-axis scale varies for each graph. Information about temperature data is summarized in Table 2.

5. Discussion

5.1. Seasonal Displacement Patterns

The documented subsidence maxima in the three areas of central and western Spitsbergen (mostly within 15–35 mm, with maximum up to 120 mm) are in the commonly reported ranges of seasonal cyclic displacements in continuous permafrost areas [13,25,27] and generally agree with other observations made in Svalbard [81–83]. However, it is important to consider that the applied InSAR measurement period is underestimating the total subsidence, especially in areas where the thaw onset starts significantly earlier than the InSAR observation window (Figure 10A). Based on the thaw depth progression from in-situ probing in the University Centre in Svalbard Circumpolar Active Layer Monitoring network grid (UNISCALM) [42,45], early summer is indeed known to experience a quick thawing rate, which may induce an undocumented subsidence before the InSAR measurement period if ground ice melts in the upper part of the active layer. The underestimation of the subsidence is likely even greater in the warmer KAP area. The NYA area shows nevertheless clearly lower subsidence values than ADV and KAP (Figure 7A), which is most likely caused by a generally thinner and more coarse-grained sediment cover [62]. The large variability of the displacement values within and between the study areas is assumed to be controlled by the active layer thickness and the ground conditions (ice type/content, frost-susceptibility of the material). The relationship between landform types and InSAR displacement rates has been discussed in a former Svalbard study [21] and by several authors in other continuous permafrost environments of the Northern Hemisphere, such as Siberia, Alaska and Canada [16,20,84,85].

The InSAR time series after filtering highlight clear cyclic patterns, but the timing of the displacement progression varies between the three study areas. Distinct transitions from subsidence to heave are found in ADV, while more ambiguous patterns are observed in parts of the KAP and NYA areas. The extracted DOY of the subsidence maxima vary between the three regions. The results especially highlight a clear difference between the KAP and NYA areas located on the West coast and ADV in central Spitsbergen. This emphasizes the maritime influence of the warm sea in KAP and NYA, compared to ADV, one of the most continental parts of Svalbard [40,54].

In ADV, the mean annual air temperature was −2.8 °C in 2017 [79]. Ground surface temperature at borehole location showed relatively cold conditions with only 356 °C days of thaw in 2017 (Supplement S8). The area is characterized by early subsidence maxima (DOY median values: September 14) (Figure 7A). At the location of the borehole, the maximum occurs later (September 26), which approximately concurs with the ground surface temperature series (Figure 10A). The difference between the fine-grained terraces and the coarse sediments surrounding the riverbed in the valley bottom, visible on DOY maxima maps (Figure 3B) as well as on the entire time series (Figure 6C), highlight a contrasting response of the ground to temperature fluctuations. It suggests that the ground reacts quickly to the first recorded negative temperature in coarse fluvial sediments, while the process is delayed on the loess terraces. The different behaviour at the location of the borehole, compared to most of the valley bottom, is also documented by in-situ measurements. At the UNISCALM site, the thaw onset typically occurs between late May and early June [42,45], while thawing is delayed at the location of the selected borehole (first positive temperature at ground surface on 13 June 2017) (Figure 9, Supplement S8). InSAR time series with subsidence maxima above the first third quartile (Figure 8C) appear to be affected by a systematic bias at the end of the documented period. No clear source has been identified, but we hypothesize that the stabilization or lowering effect at the end of time series (November) may be caused by complex scattering processes due to snow or surface icing [78]. Time series with maxima below the first quartile of the entire area (Figure 8A) are mostly located in elevated parts of the landscape (blockfield plateaus) and suggest a natural earlier freezing onset due to colder, drier and/or well-drained conditions. Although less obvious, similar gradients seem to affect the two other areas. In KAP, later subsidence maxima are detected in the western part, most exposed to the maritime influence, compared to the more protected eastern part of the study area (Figure 4B). In NYA, early subsidence maxima are detected in the NE part (Kongsfjord interior), compared to the exposed strandflat in the outer part of the peninsula (Figure 5B).

In KAP, a large heterogeneity is highlighted both in terms of maximal subsidence values and timing of displacement progression (Figures 4 and 6F–H), which is assumed to illustrate the complexity of the landform assemblage in the wide area [59,60]. KAP has a warmer climate compared to ADV, with a mean annual air temperature of −1.5 °C in 2017 [79]. The study area experiences early ground thawing and late freezing leading to 674 °C days of thaw based on ground surface temperature (Supplement S8). The area is characterized by later subsidence maxima (DOY median value: October 8) (Figure 7A), compared to ADV (DOY median values: September 14). At the location of the KAP borehole, the detected DOY (20 October) approximatively fits the end of thawing season recorded by the temperature sensors (Figure 11A). Especially in the southern part (Figure 6G,H), the ground level stabilizes for a long period in mid–late summer, likely due to low ice content in the lower active layer. Noise sources are more likely to impact the displacement trends in periods with very few variations of ground level, which may induce uncertainties in the automatic identification of the subsidence maxima. KAP time series with DOY of the subsidence maxima above the first third quartile (Figure 8C) appear to be affected by late summer subsidence that may indicate the thaw front reached the ice-rich top of the permafrost. A similar pattern is also visible in the NYA study area.

NYA had a mean annual air temperature of −2.9 °C in 2017 [79]. The ground surface temperature from the borehole shows 570 °C days of thaw in 2017 (Supplement S8), which corresponds to an intermediate case compared to ADV and KAP. Observations in NYA generally highlight more ambiguous results, assumed to be caused by two main elements. First, snowfall occurred in early October 2017, which led to large decorrelation in all interferograms connected to the 8 October Sentinel-1 acquisition. This data gap reduced the temporal resolution to 12 days in a critical period for documenting the detection of the subsidence maxima. Second, the air and ground surface temperature series show that a long period is affected by oscillations close to the freezing point from September until the clear decrease of temperatures in late October (Figure 9A). These fluctuations are recorded both on time series from the weather station and the borehole, but the stabilization close to 0 °C is especially visible on ground temperature data (Figure 10A). A slight but consistent summer heave is detected at the end of the thawing season in the time series surrounding the NYA borehole (Figure 12B). A short period in mid-September shows a drop of both air and ground surface temperature close to or under 0 °C (Figure 9A, Supplement S8), but the AL ground thermal data [66] do not show any significant refreezing able to fully explain the displacement pattern. One possible explanation is that the km^2 average is dominated by an upward effect from surficial icing occurring around the riverbed, in the meltwater plain surrounding the hill where the borehole is located. The well-developed active layer may also have favoured water migration towards the frozen layer leading to ice formation at the permafrost table [86].

In general, the misfit between the InSAR observations and the model at the location of the borehole (Figure 12A) suggests that a simple model only based on surface temperature is not able to represent all mechanisms occurring in this area. Similarly, an early ground stabilization and heave pattern is detected in the western part of the Brøggerhalvøya peninsula (Figure 5B, area K, Figure 6K). This exposed coast was likely snow-free long before the start of the InSAR series (22 June), which possibly led to some subsidence prior to the documented period. The ambiguous pattern at this location may also suggest complex hydrological processes, variable water flux within the active layer and potential late summer ice segregation at the top of the permafrost [30–32]. Similar processes may also explain the early pattern detected in the southern part of KAP (Figures 6F and 8A). Located at the mouth of the Orustdalen valley occupied by large glacial systems in its upper part, this area may be subject to large water outflow variations.

5.2. InSAR Products as Proxy of the Active Layer Thermal Regime

The study has been designed with the general objective to develop InSAR products able to infer the active layer thermal regime in flat permafrost areas. This is based on the assumptions that (1) at the seasonal timescale, the subsidence and heave temporal patterns are mostly controlled by the AL thermal variations; (2) the subsidence maximum can be used as a proxy of the AL thaw maximum. The comparisons between the InSAR displacements and the composite index models show that the observations are generally well represented by the models, even if exclusively based on accumulated degree days of thawing and freezing, which confirms that the seasonal changes of surface level are mostly determined by the ground thermal conditions. Models based on ground surface temperature performed better than when using air temperature, which concurs with several studies documenting the thermal offset and time lag between the atmosphere and the ground, in particular due to snow cover [87–89]. At the local scale in ADV (Figure 10A) and for km or regional averages in KAP and NYA (Figures 11B,C and 12B,C), the temperature-based model well represents the observed displacement pattern (Figure 13, scenario 1), and the acquisition dates of the maximal InSAR displacement generally match the timing of transition between the thawing and freezing periods (Figure 13, scenario A). In these cases, the results show that the documented subsidence maxima can be appropriately used as a proxy of the end of the AL thawing period.

However, we also highlight different inferior scenarios when (1) the model fails to provide similar displacement series as InSAR due to oversimplistic assumptions, e.g., excluding the impact of variable water content, the water flux within the active layer and/or the potential late summer ice formation at the top of the permafrost (Figure 12, scenarios 2 and 3); (2) the extracted DOY of the subsidence maxima may not correspond to the end of the thawing period, due to ground level stabilization over long periods and short-term fluctuations in the time series (Figure 13, scenarios B and C). In certain cases, the model fails to explain the whole displacement progression but the timing of the transition between subsidence and heave fits with the temperature records (Figure 13, scenario 2). This is typically the case at the local scale in KAP (Figure 11A).

In other cases, the whole index is shifted compared to the measurements, such as in NYA (Figure 12A). Moreover, independently to the model validity, the extracted DOY may be erroneous. It typically occurs if the displacement time series has large variability due to summer heave or measurement noise (Figure 13B,C), and/or if the curve flattens due late summer ground stabilisation caused by low ice content in the lower AL (Figures 6F and 7A). In these cases, the DOY may be detected earlier than the actual end of the thawing season. Discontinuous series can also lead to shifted DOY identification if the natural transition occurs around the date of a discarded acquisition, such as in NYA (Figure 12C). When considering several thousand pixels in hundreds of km^2 areas, the major issue is to discriminate what causes the model to fail and/or the DOY extraction to misrepresent the targeted transition time.

The identified limitations highlight potential for further research to exploit InSAR time series and document the temporal dynamics of the active layer in permafrost areas. The developed products and applied model are based on simple assumptions and may require adjustments to be exploited systematically at a larger scale. We suggest that future developments can be achieved by considering the following elements:

- InSAR processing: The InSAR procedure is currently based on a site-dependent selection of interferograms that include several manual steps (Section 3.1). The variability of the snow cover is the main challenge that can lead to spatially and temporally discontinuous coherent interferometric signals (such as in NYA). Applying automated adaptive filtering, possibly based on a combination of SAR backscatter, InSAR coherence and external meteorological information would be valuable to upscale the procedure, for example to the entire glacier-free land of the Svalbard archipelago. Instead of exploiting similar acquisitions in areas with variable climatic conditions, adaptive InSAR observation windows would allow for the selection of locally relevant periods, starting from the first snow-free scene after thaw onset, and thus avoiding an underestimation of the total seasonal subsidence values (such as in ADV and KAP, Figure 9A).

- DOY extraction: To identify more robustly the timing of the transition between thawing and freezing seasons and solve the issue related to the late-summer flattening of the displacement curves visible in some time series (Figure 13, scenario C), more sophisticated procedures could be tested, for instance by fitting a polynomial function to the entire time series and/or by analysing the displacement gradients between acquisitions, in addition or instead of simply considering the maximal value of the InSAR time series. Scenarios where primary and secondary maxima are identified could also be valuable to further study the cases of summer heave patterns (Figure 13, scenario B).
- Time series averaging: While single time series may be affected by errors or unrepresentative local phenomena (Figures 11A and 12A), the results in KAP and NYA suggest that averaging reduces the noise level and dampens the effect of specific small-scale effects to focus on the main climate-controlled trends. Kilometric averaging may be favoured in future product development, to keep documenting spatial variability while providing more robust information about the general seasonal pattern. At this scale, InSAR processing could be performed with larger multi-looking factors to provide more robust phase statistics for each pixel. Kilometric averaged displacement time series can easily be compared and coupled with transient modelling of thermal conditions based on remotely sensed surface temperature at a similar resolution [11,90]. Comparing InSAR with modelled temperature would have the advantage of increasing the measurement density compared to weather stations and boreholes and could provide new insights on the causes of the spatial-temporal variability of the time series. Moreover, InSAR products could complement and potentially constrain the current permafrost models by indirectly documenting the variability of the ground ice content and the timing of active layer freeze/thaw at a fine resolution and a large scale.
- Time series modelling: To further evaluate the factors controlling the seasonal progression of the displacements in permafrost regions, InSAR observations could be compared with modelled time series based on a variable set of parameters. The composite index model applied in this study is based on the Stefan equation, with simplistic assumptions regarding the ground properties that can explain that the model fails to represent the measurements in several cases (Figure 13, scenarios 2 and 3). As discussed by Gruber [30], one main issue is related to the assumption of constant water content and absence of liquid water in the frozen layer. The model also assumes that the heave is caused by the volumetric change of pore water turning into pore ice (in-situ water freezing). It does not consider the ice segregation (formation of ice lenses), which is known to be an important factor causing frost heave [5,32,91]. Other formulations using the Leybenzon equation, the Kudryavtsev's or the Gold and Lachbruch's models [1,29] could be implemented and compared with InSAR time series. Frost heave models taken into account ice lens formation [92] could also be used to further interpret the InSAR-based displacement patterns and their controls.

By implementing the previously listed suggestions, it may become possible to further interpret the spatio-temporal variability of the ground dynamics, and compare the results with geological and hydrological variables based on field mapping and in-situ measurements, following previously applied methodologies in similar environments [84,85]. Remote sensing products documenting the ground dynamics can be used to monitor the active layer in Svalbard and other polar regions. ALT is one of the two components of the permafrost ECV, which is mainly obtained by field probing or borehole measurements [6,45]. ALT is often based on measurements of the maximal thaw depth. When the field measurements are discontinuous, e.g., through biweekly/monthly/seasonal probing, selecting the right time to measure the thaw depth can be a challenge. Providing systematic regional remotely sensed information about the active layer dynamics with a six-day resolution would help field scientists to target the best time of their in-situ ALT measurements. Based on one single season (June–November 2017), our study primarily aimed to test the feasibility of applying a simple procedure to generate InSAR maps and time series designed for identifying the cyclic ground dynamics in permafrost regions. The results suggest that the strategy could be valuably integrated into future operational monitoring systems, to document the long-term evolution of the active layer at large scale. If generated systematically, for each season and at a larger scale, such datasets can also enhance the comparison with other remotely sensed environmental variables, such as vegetation phenology and snow cover time series [93,94]. These products could also be combined with top-of-permafrost excess ground ice maps, estimated from late-summer InSAR subsidence, as developed by Zwieback & Meyer [18].

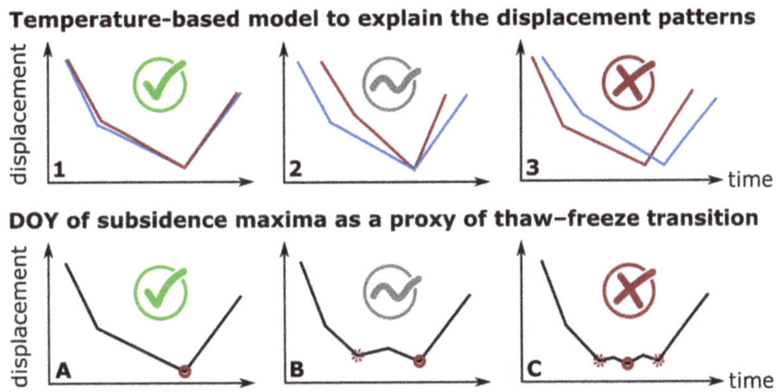

Figure 13. Examples of fit and deviation between the InSAR observations and the composite index model based on the Stefan equation (scenarios 1–3), and potential and limitations of DOY of subsidence maxima products as a proxy of the transition between the thawing and freezing season (scenarios (**A**–**C**)). Blue lines represent the rescaled index. Red lines are the measured InSAR time series. Red solid circles are the identified DOY of subsidence maxima. Dotted circles represent possible alternative solutions that lead to large uncertainties in the extraction of the targeted transition time.

6. Conclusions

The study first aimed to develop Sentinel-1 InSAR products documenting the spatial variability and timing of the seasonal thaw subsidence maxima in three regions of central and western Svalbard characterized by different geomorphological and climatic conditions. We show that the maximal subsidence and its timing vary between the three regions. The maxima occurred earlier in Adventdalen (mid-September) compared to Kapp Linné and Ny-Ålesund (early–mid October), both located along the warmer west coast. The identified maximal subsidence values vary within and between the regions and are assumed to depend on the active layer thickness, water/ice content and frost-susceptibility of the ground material. The results show clear cyclic patterns in all study areas and at the three considered scales (single pixels, km^2 and regional averages). The variable displacement progression is assumed to represent the natural variability of the climatic parameters and ground conditions. Some time series have long periods characterized by ground stabilization in late summer, which makes the identification of the subsidence maxima uncertain. Ambiguous displacement patterns are observed in Ny-Ålesund, including a secondary heave in a relatively cold period in late summer.

Then, we compared the displacement time series with a composite index based on air and ground surface temperature and evaluated how a simple model can explain the progression of subsidence and heave patterns in the study areas. Comparisons between the InSAR observations and the model show that the composite index based on ground surface temperature from the boreholes performs better than the air temperature index due to thermal offset and time lag between the atmosphere and the ground, especially at the thaw onset. In Adventdalen, the model explains well the displacement progression extracted at the location of the borehole. In Kapp Linné and in Ny-Ålesund, larger deviations are found at pixel-scale, likely due to site-specific factors and complex hydrological effects occurring in the active layer. However, km or regional averaging allows for improving the match between the models and measurements, which suggests that at this larger scale, the displacement patterns are primarily controlled by the thermal response of the active layer to atmospheric forcing.

Finally, we discussed the potential and limitations of using the timing of the maximal subsidence as a proxy for the end of the thawing season and suggested ideas for the development of alternative InSAR products in polar areas characterized by cyclic patterns. The findings show that dense and frequent InSAR measurements of thaw subsidence and frost heave have the potential to upscale the documentation of the active layer thermal regime over wide permafrost areas. Limitations related to the proposed InSAR products and the simple assumptions behind the composite index model are discussed and identify potentials for follow-up research to couple InSAR with modelling and design future operational remote sensing strategies in Svalbard.

Supplementary Materials: The following supplements are available online at https://www.mdpi.com/article/10.3390/rs13152977/s1: Supplements S1–S3: Baseline plots of the Sentinel-1 interferometric pairs. Supplement S4: Slope map and mask for the filtering criterion 2 "Slope angle". Figure S5: Percent of documented pixels after the five steps of filtering. Supplement S6: Orthophoto imagery of the 1-km^2 areas. Supplements S7 and S8: Air and ground temperature, calculated ADDT/ADDF and composite index. Supplements S9–S10: Coefficient of determination R^2 or proportion of the variance of the InSAR displacements that is explained by the composite index, based on air and ground temperature, after having tested five scaling factors α. Supplement S11: Measured and modelled displacements based on air temperature.

Author Contributions: Conceptualization: L.R. and L.L.; methodology and software: L.R., L.L., Y.L. and T.R.L.; formal analysis, investigation and visualization: L.R.; data curation: L.R., S.M.S., H.H.C. and T.R.L.; project funding and administration: L.R. and T.R.L.; writing—original draft preparation: L.R.; writing—review and editing: L.L., H.H.C., S.M.S., T.R.L. and Y.L. All authors have read and agreed to the published version of the manuscript.

Funding: The study is part of Rouyet's Ph.D. research project FrostInSAR (2017–2021) funded by the Space Research Programme of the Research Council of Norway (grant 263005). The development of the GSAR processing chain and previous research applying InSAR in the Norwegian periglacial environment have been supported by the Norwegian Space Centre, the European Space Agency and the Research Council of Norway. L. Liu was supported by the Hong Kong Research Grants Council (CUHK14305618).

Data Availability Statement: The filtered InSAR time series, the subsidence maxima and corresponding DOY are available in Zenodo: https://doi.org/10.5281/zenodo.4775398 (created on 20 May 2021). Additional data sources used in this study are listed in the references, included in the figures and tables, or in the supporting information associated with this publication.

Acknowledgments: We acknowledge the five judges from the editorial board of SIOS's special issue who awarded a related presentation at the SIOS's online conference (4–5 June 2020) at the origin of the research presented in the current manuscript. We acknowledge the valuable comments of two anonymous reviewers and the academic editor. We thank the helpful support of the managing editor Aurora Wang.

Conflicts of Interest: The authors declare no conflict of interest. The funders had no role in the design of the study; in the collection, analyses, or interpretation of data; in the writing of the manuscript, or in the decision to publish the results.

References

1. French, H.M. *The Periglacial Environment*, 3rd ed.; John Wiley & Sons: Hoboken, NJ, USA, 2007.
2. Schuur, E.A.; McGuire, A.D.; Schädel, C.; Grosse, G.; Harden, J.W.; Hayes, D.J.; Hugelius, G.; Koven, C.D.; Kuhry, P.; Lawrence, D.M. Climate change and the permafrost carbon feedback. *Nature* **2015**, *520*, 171–179. [CrossRef]
3. Hjort, J.; Karjalainen, O.; Aalto, J.; Westermann, S.; Romanovsky, V.E.; Nelson, F.E.; Etzelmüller, B.; Luoto, M. Degrading permafrost puts arctic infrastructure at risk by mid-century. *Nat. Commun.* **2018**, *9*, 1–9. [CrossRef]
4. Jorgenson, M.T.; Racine, C.H.; Walters, J.C.; Osterkamp, T.E. Permafrost degradation and ecological changes associated with a warming climate in Central Alaska. *Clim. Chang.* **2001**, *48*, 551–579. [CrossRef]
5. Rempel, A.W. Formation of ice lenses and frost heave. *J. Geophys. Res. Earth Surf.* **2007**, *112*, F02S21. [CrossRef]
6. Bonnaventure, P.P.; Lamoureux, S.F. The active layer: A conceptual review of monitoring, modelling techniques and changes in a warming climate. *Prog. Phys. Geogr.* **2013**, *37*, 352–376. [CrossRef]
7. Biskaborn, B.K.; Smith, S.L.; Noetzli, J.; Matthes, H.; Vieira, G.; Streletskiy, D.A.; Schoeneich, P.; Romanovsky, V.E.; Lewkowicz, A.G.; Abramov, A. Permafrost is warming at a global scale. *Nat. Commun.* **2019**, *10*, 1–11. [CrossRef] [PubMed]
8. Meredith, M.; Sommerkorn, M.; Cassotta, S.; Derksen, C.; Ekaykin, A.; Hollowed, A.; Kofinas, G.; Mackintosh, A.; Melbourne-Thomas, J.; Muelbert, M.M.C. Chapter 3. Polar regions. In *IPCC Special Report on the Ocean and Cryosphere in a Changing Climate;* The Intergovernmental Panel on Climate Change: Geneva, Switzerland, 2019.
9. Nitze, I.; Grosse, G.; Jones, B.M.; Romanovsky, V.E.; Boike, J. Remote sensing quantifies widespread abundance of permafrost region disturbances across the Arctic and Subarctic. *Nat. Commun.* **2018**, *9*, 1–11. [CrossRef] [PubMed]
10. Zhang, T.; Barry, R.G.; Armstrong, R.L. Application of satellite remote sensing techniques to frozen ground studies. *Polar Geogr.* **2004**, *28*, 163–196. [CrossRef]
11. Obu, J.; Westermann, S.; Bartsch, A.; Berdnikov, N.; Christiansen, H.H.; Dashtseren, A.; Delaloye, R.; Elberling, B.; Etzelmüller, B.; Kholodov, A. Northern hemisphere permafrost map based on TTOP modelling for 2000–2016 at 1 km^2 scale. *Earth Sci. Rev.* **2019**, *193*, 299–316. [CrossRef]
12. Philipp, M.; Dietz, A.; Buchelt, S.; Kuenzer, C. Trends in satellite earth observation for permafrost related analyses—A review. *Remote Sens.* **2021**, *13*, 1217. [CrossRef]
13. Romanovsky, V.E.; Marchenko, S.S.; Daanen, R.; Sergeev, D.O.; Walker, D.A. Soil climate and frost heave along the permafrost/ecological North American arctic transect. In Proceedings of the 9th International Conference on Permafrost, Fairbanks, Alaska, 29 June–3 July 2008; Volume 2, pp. 1519–1524.
14. Shiklomanov, N.I.; Streletskiy, D.A.; Little, J.D.; Nelson, F.E. Isotropic thaw subsidence in undisturbed permafrost landscapes. *Geophys. Res. Lett.* **2013**, *40*, 6356–6361. [CrossRef]
15. Bamler, R.; Hartl, P. Synthetic aperture radar interferometry. *Inverse Probl.* **1998**, *14*, R1. [CrossRef]
16. Short, N.; LeBlanc, A.-M.; Sladen, W.; Oldenborger, G.; Mathon-Dufour, V.; Brisco, B. RADARSAT-2 D-InSAR for ground displacement in permafrost terrain, validation from Iqaluit airport, Baffin Island, Canada. *Remote Sens. Environ.* **2014**, *141*, 40–51. [CrossRef]
17. Liu, L.; Zhang, T.; Wahr, J. InSAR measurements of surface deformation over permafrost on the north slope of Alaska. *J. Geophys. Res. Earth Surf.* **2010**, *115*, F03023. [CrossRef]
18. Zwieback, S.; Meyer, F.J. Vulnerable top-of-permafrost ground ice indicated by remotely sensed late-season subsidence. *Cryosphere* **2020**, *15*, 2041–2055. [CrossRef]
19. Daout, S.; Doin, M.-P.; Peltzer, G.; Socquet, A.; Lasserre, C. Large-Scale InSAR monitoring of permafrost freeze-thaw cycles on the Tibetan plateau. *Geophys. Res. Lett.* **2017**, *44*, 901–909. [CrossRef]
20. Bartsch, A.; Leibman, M.; Strozzi, T.; Khomutov, A.; Widhalm, B.; Babkina, E.; Mullanurov, D.; Ermokhina, K.; Kroisleitner, C.; Bergstedt, H. Seasonal progression of ground displacement identified with satellite radar interferometry and the impact of unusually warm conditions on permafrost at the Yamal Peninsula in 2016. *Remote Sens.* **2019**, *11*, 1865. [CrossRef]
21. Rouyet, L.; Lauknes, T.R.; Christiansen, H.H.; Strand, S.M.; Larsen, Y. Seasonal dynamics of a permafrost landscape, Adventdalen, Svalbard, investigated by InSAR. *Remote Sens. Environ.* **2019**, *231*, 111236. [CrossRef]
22. Reinosch, E.; Buckel, J.; Dong, J.; Gerke, M.; Baade, J.; Riedel, B. InSAR time series analysis of seasonal surface displacement dynamics on the Tibetan plateau. *Cryosphere* **2020**, *14*, 1633–1650. [CrossRef]
23. Liu, L.; Jafarov, E.E.; Schaefer, K.M.; Jones, B.M.; Zebker, H.A.; Williams, C.A.; Rogan, J.; Zhang, T. InSAR detects increase in surface subsidence caused by an arctic tundra fire. *Geophys. Res. Lett.* **2014**, *41*, 3906–3913. [CrossRef]
24. Strozzi, T.; Antonova, S.; Günther, F.; Mätzler, E.; Vieira, G.; Wegmüller, U.; Westermann, S.; Bartsch, A. Sentinel-1 SAR interferometry for surface deformation monitoring in low-land permafrost areas. *Remote Sens.* **2018**, *10*, 1360. [CrossRef]
25. Antonova, S.; Sudhaus, H.; Strozzi, T.; Zwieback, S.; Kääb, A.; Heim, B.; Langer, M.; Bornemann, N.; Boike, J. Thaw subsidence of a Yedoma landscape in Northern Siberia, measured in situ and estimated from TerraSAR-X interferometry. *Remote Sens.* **2018**, *10*, 494. [CrossRef]
26. Abe, T.; Iwahana, G.; Efremov, P.V.; Desyatkin, A.R.; Kawamura, T.; Fedorov, A.; Zhegusov, Y.; Yanagiya, K.; Tadono, T. Surface displacement revealed by L-Band InSAR analysis in the Mayya area, Central Yakutia, underlain by continuous permafrost. *Earth Planets Space* **2020**, *72*, 1–16. [CrossRef]
27. Hu, Y.; Liu, L.; Larson, K.M.; Schaefer, K.M.; Zhang, J.; Yao, Y. GPS interferometric reflectometry reveals cyclic elevation changes in thaw and freezing seasons in a permafrost area (Barrow, Alaska). *Geophys. Res. Lett.* **2018**, *45*, 5581–5589. [CrossRef]

28. Stefan, J. Über die theorie der eisbildung, insbesondere über die eisbildung im polarmeere. *Ann. Phys. Chem.* **1891**, *42*, 269–286. [CrossRef]
29. Riseborough, D.W.; Shiklomanov, N.; Etzelmüller, B.; Gruber, S.; Marchenko, S. Recent advances in permafrost modelling. *Permafr. Periglac. Process.* **2008**, *19*, 137–156. [CrossRef]
30. Gruber, S. Ground subsidence and heave over permafrost: Hourly time series reveal interannual, seasonal and shorter-term movement caused by freezing, thawing and water movement. *Cryosphere* **2020**, *14*, 1437–1447. [CrossRef]
31. Romanovsky, V.E.; Osterkamp, T.E. Thawing of the active layer on the coastal plain of the Alaskan arctic. *Permafr. Periglac. Process.* **1997**, *8*, 1–22. [CrossRef]
32. Thomas, H.R.; Cleall, P.; Li, Y.-C.; Harris, C.; Kern-Luetschg, M. Modelling of cryogenic processes in permafrost and seasonally frozen soils. *Géotechnique* **2009**, *59*, 173–184. [CrossRef]
33. Zhao, R.; Li, Z.-W.; Feng, G.-C.; Wang, Q.-J.; Hu, J. Monitoring surface deformation over permafrost with an improved SBAS-InSAR algorithm: With emphasis on climatic factors modeling. *Remote Sens. Environ.* **2016**, *184*, 276–287. [CrossRef]
34. Dehls, J.F.; Larsen, Y.; Marinkovic, P.; Lauknes, T.R.; Stødle, D.; Moldestad, D.A. INSAR.No: A national insar deformation mapping/monitoring service in Norway–from concept to operations. In Proceedings of the IGARSS 2019—IEEE International Geoscience and Remote Sensing Symposium, Yokohama, Japan, 28 July–2 August 2019; pp. 5461–5464.
35. Larsen, Y.; Marinkovic, P.; Dehls, J.F.; Bredal, M.; Bishop, C.; Jøkulsson, G.; Gjøvik, L.-P.; Frauenfelder, R.; Salazar, S.E.; Vöge, M.; et al. European Ground Motion Service: Service Implementation Plan and Product Specification Document. Copernicus Land Monitoring Service. 2020. Available online: https://land.copernicus.eu/user-corner/technical-library/egms-specification-and-implementation-plan (accessed on 1 March 2021).
36. Peel, M.C.; Finlayson, B.L.; McMahon, T.A. Updated world map of the Köppen-Geiger climate classification. *Hydrol. Earth Syst. Sci.* **2007**, *11*, 1633–1644. [CrossRef]
37. Harland, W.B. Svalbard. In *The Geology of Svalbard*; Geological Society: London, UK, 1997; Volume 17, pp. 3–15.
38. Hanssen-Bauer, I.; Førland, E.J.; Hisdal, H.; Mayer, S.; AB, S.; Sorteberg, A. *Climate in Svalbard 2100*; NCCS Report no. 1/2019; Norwegian Centre for Climate Services (NCCS): Trondheim, Norway; Norwegian Environment Agency (Miljødirektoratet): Trondheim, Norway, 2019.
39. Humlum, O.; Instanes, A.; Sollid, J.L. Permafrost in Svalbard: A review of research history, climatic background and engineering challenges. *Polar Res.* **2003**, *22*, 191–215. [CrossRef]
40. Christiansen, H.H.; Humlum, O.; Eckerstorfer, M. Central Svalbard 2000–2011 meteorological dynamics and periglacial landscape response. *Arct. Antarct. Alp. Res.* **2013**, *45*, 6–18. [CrossRef]
41. Harris, C.; Kern-Luetschg, M.; Christiansen, H.H.; Smith, F. The role of interannual climate variability in controlling solifluction processes, Endalen, Svalbard. *Permafr. Periglac. Process.* **2011**, *22*, 239–253. [CrossRef]
42. Schuh, C.; Frampton, A.; Christiansen, H.H. Soil moisture redistribution and its effect on inter-annual active layer temperature and thickness variations in a dry loess terrace in Adventdalen, Svalbard. *Cryosphere* **2017**, *11*, 635–651. [CrossRef]
43. Christiansen, H.H.; Etzelmüller, B.; Isaksen, K.; Juliussen, H.; Farbrot, H.; Humlum, O.; Johansson, M.; Ingeman-Nielsen, T.; Kristensen, L.; Hjort, J. The thermal state of permafrost in the Nordic area during the international polar year 2007–2009. *Permafr. Periglac. Process.* **2010**, *21*, 156–181. [CrossRef]
44. Christiansen, H.H.; Gilbert, G.L.; Neumann, U.; Demidov, N.; Guglielmin, M.; Isaksen, K.; Osuch, M.; Boike, J. Ground ice content, drilling methods and equipment and permafrost dynamics in Svalbard 2016–2019 (PermaSval). In *The State of Environmental Science in Svalbard (SESS Report 2020)*; Svalbard Integrated Arctic Earth Observing System: Svalbard, Norway, 2021.
45. Strand, S.M.; Christiansen, H.H.; Johansson, M.; Åkerman, J.; Humlum, O. Active layer thickening and controls on interannual variability in the Nordic arctic compared to the circum-arctic. *Permafr. Periglac. Process.* **2020**, *32*, 47–58. [CrossRef]
46. Etzelmüller, B.; Schuler, T.V.; Isaksen, K.; Christiansen, H.H.; Farbrot, H.; Benestad, R. Modeling the temperature evolution of Svalbard permafrost during the 20th and 21st Century. *Cryosphere* **2011**, *5*, 67–79. [CrossRef]
47. NPI Terrengmodell Svalbard (S0 Terrengmodell) [Dataset]. Norwegian Polar Institute. 2014. Available online: https://www.doi.org/10.21334/npolar.2014.dce53a47 (accessed on 1 March 2021).
48. NPI Kartdata Svalbard 1:100 000 (S100 Kartdata)/Map Data [Dataset]. Norwegian Polar Institute. 2014. Available online: https://www.doi.org/10.21334/npolar.2014.645336c7 (accessed on 1 March 2021).
49. Major, H.; Haremo, P.; Dallmann, W.K.; Andresen, A.; Salvigsen, O. *Geological Map of Svalbard 1: 100,000, Sheet C9G Adventdalen*; Norwegian Polar Institute: Tromsø, Norway, 2001.
50. Tolgensbakk, J.; Sørbel, L.; Høgard, K. *Geomorphological and Quaternary Geological Map of Svalbard 1: 100,000. Sheet C9Q Adventdalen. Temakart 31*; Norwegian Polar Institute: Tromsø, Norway, 2001.
51. Humlum, O. Holocene permafrost aggradation in Svalbard. *Geol. Soc. Lond. Spec. Publ.* **2005**, *242*, 119–129. [CrossRef]
52. Gilbert, G.L.; O'Neill, H.B.; Nemec, W.; Thiel, C.; Christiansen, H.H.; Buylaert, J.-P. Late quaternary sedimentation and permafrost development in a Svalbard Fjord-Valley, Norwegian high arctic. *Sedimentology* **2018**, *65*, 2531–2558. [CrossRef]
53. Cable, S.; Elberling, B.; Kroon, A. Holocene permafrost history and cryostratigraphy in the high-arctic Adventdalen Valley, Central Svalbard. *Boreas* **2018**, *47*, 423–442. [CrossRef]
54. Retelle, M.; Christiansen, H.H.; Hodson, A.; Nikulina, A.; Osuch, M.; Poleshuk, K.; Romashova, K.; Roof, S.; Rouyet, L.; Strand, S.M. Environmental monitoring in the Kapp Linne-Gronfjorden Region (KLEO). In *The State of Environmental Science in Svalbard (SESS Report 2019)*; Svalbard Integrated Arctic Earth Observing System: Longyearbyen, Norway, 2020.

55. Dallmann, W.K. *Geoscience Atlas of Svalbard*; Norwegian Polar Institute: Tromsø, Norway, 2015.
56. Landvik, J.Y.; Landvik, J.Y.; Salvigsen, O. The late Weichselian and Holocene shoreline displacement on the West-Central coast of Svalbard. *Polar Res.* **1987**, *5*, 29–44. [CrossRef]
57. Ohta, Y.; Hjelle, A. *Geological Map Svalbard 1: 100,000. Isfjorden: Spitsbergen. Sheet B9G*; Norwegian Polar Institute: Oslo, Norway, 1992.
58. Snyder, J.A.; Werner, A.; Miller, G.H. Holocene cirque glacier activity in Western Spitsbergen, Svalbard: Sediment records from proglacial Linnévatnet. *Holocene* **2000**, *10*, 555–563. [CrossRef]
59. Åkerman, H.J. Relations between slow slope processes and active-layer thickness 1972–2002, Kapp Linné, Svalbard. *Nor. Geogr. Tidsskr. Nor. J. Geogr.* **2005**, *59*, 116–128. [CrossRef]
60. Eckerstorfer, M.; Malnes, E.; Christiansen, H.H. Freeze/Thaw conditions at periglacial landforms in Kapp Linné, Svalbard, investigated using field observations, in situ, and radar satellite monitoring. *Geomorphology* **2017**, *293*, 433–447. [CrossRef]
61. Hjelle, A. *Geological Map of Svalbard 1: 100,000: Sheet A7G Kongsfjorden*; Norwegian Polar Institute: Tromsø, Norway, 1999.
62. Miccadei, E.; Piacentini, T.; Berti, C. Geomorphological features of the Kongsfjorden area: Ny-Ålesund, Blomstrandøya (NW Svalbard, Norway). *Rend. Lincei* **2016**, *27*, 217–228. [CrossRef]
63. Westermann, S.; Langer, M.; Boike, J. Spatial and temporal variations of summer surface temperatures of high-arctic tundra on Svalbard—Implications for MODIS LST based permafrost monitoring. *Remote Sens. Environ.* **2011**, *115*, 908–922. [CrossRef]
64. Boike, J.; Roth, K.; Ippisch, O. Seasonal snow cover on frozen ground: Energy balance calculations of a permafrost site near Ny-Ålesund, Spitsbergen. *J. Geophys. Res. Atmos.* **2003**, *108*, ALT-4. [CrossRef]
65. Westermann, S.; Wollschläger, U.; Boike, J. Monitoring of active layer dynamics at a permafrost site on Svalbard using multi-channel ground-penetrating radar. *Cryosphere* **2010**, *4*, 475–487. [CrossRef]
66. Boike, J.; Juszak, I.; Lange, S.; Chadburn, S.; Burke, E.; Overduin, P.P.; Roth, K.; Ippisch, O.; Bornemann, N.; Stern, L. A 20-year record (1998–2017) of permafrost, active layer and meteorological conditions at a high arctic permafrost research site (Bayelva, Spitsbergen). *Earth Syst. Sci. Data* **2018**, *10*, 355–390. [CrossRef]
67. Larsen, Y.; Engen, G.; Lauknes, T.R.; Malnes, E.; Høgda, K.A. A generic differential interferometric SAR processing system, with applications to land subsidence and snow-water equivalent retrieval. In Proceedings of the ESA Fringe Workshop, Frascati, Italy, 28 November–2 December 2005.
68. Baran, I.; Stewart, M.P.; Kampes, B.M.; Perski, Z.; Lilly, P. A modification to the Goldstein radar interferogram filter. *IEEE Trans. Geosci. Remote Sens.* **2003**, *41*, 2114–2118. [CrossRef]
69. Goldstein, R.M.; Werner, C.L. Radar interferogram filtering for geophysical applications. *Geophys. Res. Lett.* **1998**, *25*, 4035–4038. [CrossRef]
70. Cavalié, O.; Doin, M.-P.; Lasserre, C.; Briole, P. Ground motion measurement in the lake Mead area, Nevada, by differential synthetic aperture radar interferometry time series analysis: Probing the lithosphere rheological structure. *J. Geophys. Res. Solid Earth* **2007**, *112*, B03403. [CrossRef]
71. Lauknes, T.R. InSAR tropospheric stratification delays: Correction using a small baseline approach. *IEEE Geosci. Remote Sens. Lett.* **2011**, *8*, 1070–1074. [CrossRef]
72. Tymofyeyeva, E.; Fialko, Y. Mitigation of atmospheric phase delays in InSAR data, with application to the Eastern California shear zone. *J. Geophys. Res. Solid Earth* **2015**, *120*, 5952–5963. [CrossRef]
73. Chen, C.W.; Zebker, H.A. Phase unwrapping for large SAR interferograms: Statistical segmentation and generalized network models. *IEEE Trans. Geosci. Remote Sens.* **2002**, *40*, 1709–1719. [CrossRef]
74. NPI Data and Services. Svalbard Orthophoto—WMTS Basemap Service. Norwegian Polar Institute. 2021. Available online: https://geodata.npolar.no/ (accessed on 10 March 2021).
75. Berardino, P.; Fornaro, G.; Lanari, R.; Sansosti, E. A new algorithm for surface deformation monitoring based on small baseline differential SAR interferograms. *IEEE Trans. Geosci. Remote Sens.* **2002**, *40*, 2375–2383. [CrossRef]
76. Lauknes, T.R.; Zebker, H.A.; Larsen, Y. InSAR deformation time series using an L1-norm small-baseline approach. *IEEE Trans. Geosci. Remote Sens.* **2010**, *49*, 536–546. [CrossRef]
77. Matsuoka, N. Solifluction rates, processes and landforms: A global review. *Earth Sci. Rev.* **2001**, *55*, 107–134. [CrossRef]
78. Zwieback, S.; Liu, X.; Antonova, S.; Heim, B.; Bartsch, A.; Boike, J.; Hajnsek, I. A statistical test of phase closure to detect influences on DInSAR deformation estimates besides displacements and decorrelation noise: Two case studies in high-latitude regions. *IEEE Trans. Geosci. Remote Sens.* **2016**, *54*, 5588–5601. [CrossRef]
79. NCCS. 2017 Daily Air Temperature. Adventdalen Station 99870. Isfjord Radio Station 99790. Ny-Ålesund Station 99910. Norwegian Centre for Climate Services. 2021. Available online: https://seklima.met.no/observations/ (accessed on 30 March 2021).
80. Riseborough, D.W. Thawing and freezing indices in the active layer. In Proceedings of the 8th International Conference on Permafrost, Zurich, Switzerland, 21–25 July 2003; Volume 2, pp. 953–958.
81. Hallet, B.; Allard, M. Measurement of soil motion in sorted circles, Western Spitsbergen. In Proceedings of the 7th International Conference on Permafrost, Yellowknife, Canada, 23–27 June 1998; pp. 415–420.
82. Matsuoka, N.; Hirakawa, K. Solifluction resulting from one-sided and two-sided freezing: Field data from Svalbard. *Polar Geosci.* **2000**, *13*, 187–201.
83. Watanabe, T.; Matsuoka, N.; Christiansen, H.H. Mudboil and ice-wedge dynamics investigated by electrical resistivity tomography, ground temperatures and surface movements in Svalbard. *Geogr. Ann. Ser. A Phys. Geogr.* **2012**, *94*, 445–457. [CrossRef]

84. Chen, J.; Wu, Y.; O'Connor, M.; Cardenas, M.B.; Schaefer, K.; Michaelides, R.; Kling, G. Active layer freeze-thaw and water storage dynamics in permafrost environments inferred from InSAR. *Remote Sens. Environ.* **2020**, *248*, 112007. [CrossRef]
85. Rudy, A.C.A.; Lamoureux, S.F.; Treitz, P.; Short, N.; Brisco, B. Seasonal and multi-year surface displacements measured by DInSAR in a high arctic permafrost environment. *Int. J. Appl. Earth Obs. Geoinf.* **2018**, *64*, 51–61. [CrossRef]
86. Mackay, J.R. Downward water movement into frozen ground, Western Arctic Coast, Canada. *Can. J. Earth Sci.* **1983**, *20*, 120–134. [CrossRef]
87. Romanovsky, V.E.; Sazonova, T.S.; Balobaev, V.T.; Shender, N.I.; Sergueev, D.O. Past and recent changes in air and permafrost temperatures in Eastern Siberia. *Glob. Planet. Chang.* **2007**, *56*, 399–413. [CrossRef]
88. Throop, J.; Lewkowicz, A.G.; Smith, S.L. Climate and ground temperature relations at sites across the continuous and discontinuous permafrost zones, Northern Canada. *Can. J. Earth Sci.* **2012**, *49*, 865–876. [CrossRef]
89. Way, R.G.; Lewkowicz, A.G. Environmental controls on ground temperature and permafrost in Labrador, Northeast Canada. *Permafr. Periglac. Process.* **2018**, *29*, 73–85. [CrossRef]
90. Westermann, S.; Peter, M.; Langer, M.; Schwamborn, G.; Schirrmeister, L.; Etzelmüller, B.; Boike, J. Transient modeling of the ground thermal conditions using satellite data in the Lena River Delta, Siberia. *Cryosphere* **2017**, *11*, 1441–1463. [CrossRef]
91. Smith, M.W. Observations of soil freezing and frost heave at Inuvik, Northwest Territories, Canada. *Can. J. Earth Sci.* **1985**, *22*, 283–290. [CrossRef]
92. Yanagiya, K.; Furuya, M. Post-wildfire surface deformation near Batagay, Eastern Siberia, detected by L-Band and C-Band InSAR. *J. Geophys. Res. Earth Surf.* **2020**, *125*, e2019JF005473. [CrossRef]
93. Karlsen, S.R.; Elvebakk, A.; Høgda, K.A.; Grydeland, T. Spatial and temporal variability in the onset of the growing season on Svalbard, Arctic Norway—Measured by MODIS-NDVI satellite data. *Remote Sens.* **2014**, *6*, 8088–8106. [CrossRef]
94. Vickers, H.; Karlsen, S.R.; Malnes, E. A 20-year MODIS-based snow cover dataset for Svalbard and its link to phenological timing and sea ice variability. *Remote Sens.* **2020**, *12*, 1123. [CrossRef]

remote sensing

MDPI

Article

Disturbance Mapping in Arctic Tundra Improved by a Planning Workflow for Drone Studies: Advancing Tools for Future Ecosystem Monitoring

Isabell Eischeid [1,2,3,*], Eeva M. Soininen [1], Jakob J. Assmann [4], Rolf A. Ims [1], Jesper Madsen [3], Åshild Ø. Pedersen [2], Francesco Pirotti [5], Nigel G. Yoccoz [1] and Virve T. Ravolainen [2]

1 Department of Arctic and Marine Biology, UiT The Arctic University of Norway, 9037 Tromsø, Norway; eeva.soininen@uit.no (E.M.S.); rolf.ims@uit.no (R.A.I.); nigel.yoccoz@uit.no (N.G.Y.)
2 Fram Centre, Norwegian Polar Institute, 9296 Tromsø, Norway; ashild.pedersen@npolar.no (Å.Ø.P.); virve.ravolainen@npolar.no (V.T.R.)
3 Department of Ecoscience, Aarhus University, 8410 Rønde, Denmark; jm@ecos.au.dk
4 Department of Biology—Ecoinformatics and Biodiversity, Aarhus University, 8000 Aarhus C, Denmark; j.assmann@bios.au.dk
5 CIRGEO Interdepartmental Research Center of Geomatics, TESAF Department, University of Padova, Viale dell'Università 16, 35020 Legnaro, Italy; francesco.pirotti@unipd.it
* Correspondence: isabell.eischeid@uit.no

Abstract: The Arctic is under great pressure due to climate change. Drones are increasingly used as a tool in ecology and may be especially valuable in rapidly changing and remote landscapes, as can be found in the Arctic. For effective applications of drones, decisions of both ecological and technical character are needed. Here, we provide our method planning workflow for generating ground-cover maps with drones for ecological monitoring purposes. The workflow includes the selection of variables, layer resolutions, ground-cover classes and the development and validation of models. We implemented this workflow in a case study of the Arctic tundra to develop vegetation maps, including disturbed vegetation, at three study sites in Svalbard. For each site, we generated a high-resolution map of tundra vegetation using supervised random forest (RF) classifiers based on four spectral bands, the normalized difference vegetation index (NDVI) and three types of terrain variables—all derived from drone imagery. Our classifiers distinguished up to 15 different ground-cover classes, including two classes that identify vegetation state changes due to disturbance caused by herbivory (i.e., goose grubbing) and winter damage (i.e., 'rain-on-snow' and thaw-freeze). Areas classified as goose grubbing or winter damage had lower NDVI values than their undisturbed counterparts. The predictive ability of site-specific RF models was good (macro-F1 scores between 83% and 85%), but the area of the grubbing class was overestimated in parts of the moss tundra. A direct transfer of the models between study sites was not possible (macro-F1 scores under 50%). We show that drone image analysis can be an asset for studying future vegetation state changes on local scales in Arctic tundra ecosystems and encourage ecologists to use our tailored workflow to integrate drone mapping into long-term monitoring programs.

Keywords: classifier; disturbance; drone; ecological monitoring; GLCM; herbivore; random forest; Svalbard; winter climate effect; grubbing

Citation: Eischeid, I.; Soininen, E.M.; Assmann, J.J.; Ims, R.A.; Madsen, J.; Pedersen, Å.Ø.; Pirotti, F.; Yoccoz, N.G.; Ravolainen, V.T. Disturbance Mapping in Arctic Tundra Improved by a Planning Workflow for Drone Studies: Advancing Tools for Future Ecosystem Monitoring. *Remote Sens.* **2021**, *13*, 4466. https://doi.org/10.3390/rs13214466

Academic Editor: Shridhar D. Jawak

Received: 23 September 2021
Accepted: 2 November 2021
Published: 6 November 2021

1. Introduction

Rapid climate change is altering abiotic and biotic disturbance processes in the Arctic and could therefore lead to ecosystem state changes in tundra ecosystems [1–4]. Abiotic disturbances to the Arctic tundra may increase due to summer warming, leading to, e.g., permafrost collapse and erosion, whereas changes in the winter climate cause, for instance, rain-on-snow events and ground ice, permafrost thaw and changes of hydrological regimes [2,5–11]. In addition, climate change can also modify the intensity and frequency of biotic disturbances to tundra, e.g., via herbivore grazing [12,13]. The cumulative effect of changes in multiple drivers can thus alter the state of tundra ecosystems [14]. Ecosystem state changes can be detected by mapping vegetation or tundra surface types, quantifying, for example, increases in forest or shrub cover [15,16], or assessing variables such as vegetation type composition and percentage of bare ground-cover [17,18]. While these examples illustrate that some ecosystem state changes have happened in the Arctic, it remains to be documented by means of long-term monitoring, which ecosystem states continue to develop in an Arctic landscape that is facing rapid environmental change [14].

High Arctic tundra, such as in Svalbard, represents an ecosystem in high demand for monitoring [3,13], as new biotic and abiotic conditions are evolving fast with rapid climate change. In particular, the winter climate in Svalbard has, since the late 1990s, seen a regime shift towards more frequent rain-on-snow events due to mild temperatures and, consequently, more pronounced formation of basal ice [19] that encapsulates and damages plants [20]. This commonly occurs in low-land tundra with, e.g., dwarf shrub vegetation [21]. Additionally, increasing herbivore populations (e.g., goose and reindeer) can cause biotic disturbances [12,22]. In Svalbard, a mode of herbivory with particular importance to the tundra is pink-footed goose (*Anser brachyrhynchus*) grubbing, which disrupts the vegetation by removing plants and the moss layer [23–25]. The population of pink-footed geese has increased dramatically since the 1990s [26], and heavy grubbing can remove enough vegetation to cause patches of bare ground, potentially leading to soil erosion, destabilizing permafrost and changes in the soil-carbon cycle [14,27–29]. Warmer summers can destabilize the permafrost further but simultaneously also lead to higher primary production and above-ground biomass [30]. Effects of summer warming, changed winter climate and altered herbivory are expected to change the current state of this high Arctic ecosystem [14]. The disturbances described above commonly cover only small areas but occur frequently across the landscape, resulting in a heterogeneous tundra where the vegetation is interspersed with different types of disturbed patches. Detecting such disturbances at the landscape scale requires new, improved monitoring methods [14].

Drones have become more accessible in recent years and allow researchers to cover larger spatial extents at multiple temporal scales [31]. Drone images provide spatial grain sizes that allow studying ecological processes at local scales in highly heterogeneous landscapes. They can help transferring from local and detailed knowledge to broad-scale environments with more spatial and temporal complexity and can improve the interpretation of satellite imagery [32]. In the Arctic, drones have been used for vegetation mapping [33–35], measurements of cryosphere characteristics [36–45], observations of permafrost thaw [46,47] and to help bridge the gap between field- and satellite-derived data [32,35,48]. Long-term monitoring is stated as a goal in many recently published drone studies (e.g., [49,50]), but as the technology is quite new, few current studies have compared results from drone data between years or even within the same season [32,35].

Ecological monitoring is based on comparable, repeated measurements and requires robust methodology at all steps from data collection to statistical inference. Monitoring that is based on comparisons of drone images across time (seasons, years) or space (between different areas) faces methodological challenges for obtaining consistent and comparable data [51–53]. Recent studies have developed guidelines for drone data acquisition, geometric processing and radiometric calibrations [31,53,54]. This helps ecologists obtain high-quality images that would allow for temporal and spatial comparison. There has been less attention on how to systematically analyze these images to extract ecological variables relevant for monitoring an ecosystem state's changes resulting from environmental drivers. Arctic ecological monitoring programs (e.g., [55–58]), networks [59] and drone ecologists, in general, are tasked with finding the balance between their scientific interests and technical remote sensing abilities. The use of drones as a tool in ecology will therefore benefit from shared protocols that go beyond the first steps of image acquisition and processing. To better capture ecosystem changes, we need to include further steps in the overall workflow, such as variable selection, layer resolution, ground-cover class selection and model development, including ground calibration and validation [31,53].

The overall aim of our study is to test ways of using drone imagery to collect variables that will be important to describe vegetation state changes resulting from abiotic and biotic disturbance factors that currently frequently occur at our study sites in Svalbard. For this purpose, we develop a method planning workflow with a focus on the steps after image acquisition (Table 1) to serve as guidance for the use of drone images in tundra ecological monitoring.

Table 1. Method planning workflow for drone flight campaigns and image classifications in a tundra landscape. Studies for the steps where guidance has been published are referenced. Asterisks indicate the steps we included in our study.

Topic	Solution
(1) Flight planning	
What to consider when planning field work?	Choose appropriate image overlap and camera angles for desired final product. *
	Ensure that the type of the drone, the camera and flight speeds integrate well with one another to obtain high-quality images of suitable resolution and avoid blurring due to slow rolling shutter speeds [60]. *
	Follow appropriate radiometric calibration guidelines. See guidelines from Aasen et al. [31], Assmann et al. [54] and Tmušic et al. [53] with advice for choices of flight line, (image) overlap, camera type, drone type, weather and sun, radiometric calibration, geolocation, ground control points and ground truthing. * Ensure accurate geolocation of images and groundtruthing data. *
(2) Variable selection	
Which variables to derive from the drone images?	A priori knowledge of the landscape is important to select appropriate data layers and resolutions that represent ecologically explainable heterogeneity in the terrain. *
How to assess which of the available variables to include in a classifier?	Variable importance can be ranked in a preliminary classifier using a subset of the available data [61,62]. *
Which variables best discriminate between ground-cover classes?	Exploratory data analysis via visual inspection of data via plotting to detect patterns important for the classification. *
What neighborhood size to select?	Neighborhood size can be ranked in a preliminary classifier using a subset of the available data [61,62]. *
	Appropriate neighborhood sizes for secondary layers can be computationally derived using a minimum entropy approach [63].
	Computational limits may define the minimum resolution or maximum possible neighborhood sizes [64]. *

Table 1. *Cont.*

Topic	Solution
(3) Ground-cover class selection	
How to define the first choice of ground-cover classes for the classifier?	Data-driven—Cluster analysis to see the separability of data without human input (unsupervised classification).
	Research-driven—Considering data-driven results define classes that are present in the area of interest for monitoring or expected to change over time (supervised classifications). *
How to choose between classifier robustness and ground-cover class detail?	Pre-define the ecological context of the ground-cover classes to determine which ones are meaningful to merge due to ecological similarities. *
How to increase transparency on class selection and its effects on classification accuracy?	Define documentation of how the final classes, explain class merges and the research consequences of mixing classes. *
(4) Classifier development and validation	
How to choose groundtruthing points?	Choose areas that are representative for the ground-cover classes and large enough sample sizes [65]. *
	Aim for a training data cover of approximately 0.25% of the study site [66] (recommendation based on medium-coarse grain satellite data).
	Avoid spatial autocorrelation of groundtruthing points by stratified random sampling in a blocked design [67]
How to avoid overfitting the classifier?	Split the dataset into training and validation datasets (such as K-fold mechanism or using random resampling) [64]. *
How to assess classifier robustness?	Use an independent validation dataset for external validation [68]
	If an independent dataset is not available, repeat runs of the classifier though multiple K-fold runs or repeat sampling of training and validation dataset [68]. *
	Additional map validation by local experts can help discover issues that go undetected by classifier evaluation statistics. This can be conducted, for example, by visually comparing the classified map with the drone images, pictures or revisits to the site. *
(5) Transferability of classifier	
How to assess the potential for transferability?	Exploratory data analysis via visual inspection of data via plotting to detect trends/shifts in values across sites or time intervals. *
How to test the transferability of the classifier?	Repetition of data collection in new area, creation of independent classifier and test in other area. *
How to improve the transferability of the classifier?	Tree pruning, simplification of the classifier, transferability functions [69,70]

2. Material and Methods

2.1. Study System

Svalbard is a high Arctic archipelago (74°–81°N, 10°–35°E), where only 15% of the land is vegetated [71,72]. Our three study sites (Janssonhaugen (JAN), Sassendalen (SAS) and Todalen (TOD)) are located in Nordenskiöld Land in the center of the main island, Spitsbergen (Figure 1). The area is characterized by glacial valleys with continuous vegetation at low elevations and sparse to no vegetation at higher elevations and on mountain slopes [73]. The study sites have similar plant community compositions, and each spans a gradient from moist-mesic tundra mires to dry-mesic ridges [74]. More information on the plant communities can be found in Appendix A. Annual precipitation is around 190 mm [75], and soils range from hyperskeletic cryosols in the moist moss tundra, haplic cryosols in the dry ridges and turbic cryosols where thaw-freeze cycles perturbate the ground (based on descriptions of Migala et al. [76]).

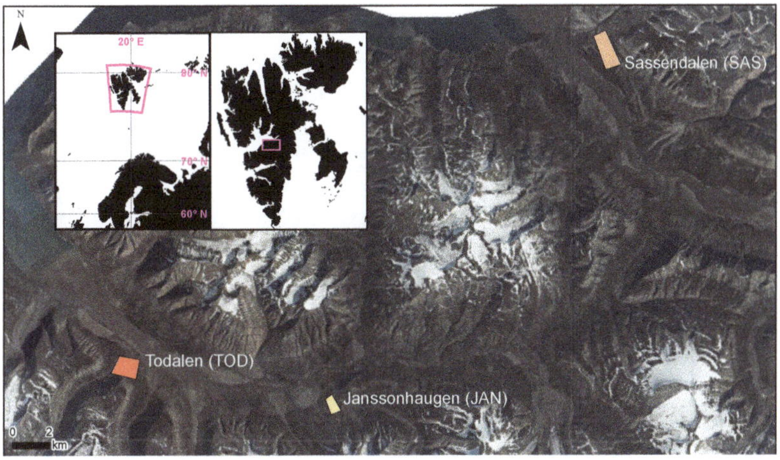

Figure 1. Location of drone flights of the case study in Central Spitsbergen, JAN: 78.16996°N, 16.300685°E, 0.3 km^2; SAS: 78.331412°N, 16.974537°E, 1 km^2; TOD: 78.188762°N, 15.82335°E, 1 km^2. Ground sampling distances (GSD) are listed for each site and camera (AeriaX RGB/Sequoia+ 4Band): JAN: 1.9 cm/9.8 cm; SAS: 2.2 cm/13.2 cm; TOD: 1.2 cm/10.3 cm. Aerial orthophoto provided by the Norwegian Polar Institute.

2.2. Study Preparation

We developed a workflow (see Table 1) to systematically plan the data collection and analysis steps of our study.

2.3. Data Collection

Our data collection choices are based on considerations presented in Table 1—(1) *Flight planning*. We captured aerial images during plant biomass peak season (20–28 July 2019) at the three study sites (Figure 1). We used a fixed wing drone (eBeeX by Sensefly) rigged with an AeriaX RGB camera and Sequoia+ four band camera (red, near infrared, red edge and green bands). As the Sequoia+ camera model is radiometrically calibrated automatically [77], we only took pictures of a spectral calibration target (Zenith Lite—SphereOptics) as backup. Image overlap was 70% or higher for the AeriaX RGB camera, while for the Sequoia+, side overlap was 60% and horizontal overlap 80%. We flew in perpendicular lines to the main slope of the terrain and kept a constant height over the ground (between 70 and 100 m above ground, depending on the camera and site) and speeds between 8–15 m/s.

Ground sampling distance (GSD) varied slightly between sites, see Figure 1. We flew around noon (earliest 10:00 latest 15:00) to have similar light conditions and avoid shadows on sunny days. We only flew on days with stable light conditions, such as blue skies (for study sites TOD, SAS) or continuous overcast (study site JAN), and wind speeds below 7 m/s. We used a Leica GS10 base station (Leica Geosystems) with an in-flight RTK (real time kinematic) link to the drone to obtain camera positions in real time and thus readily georeferenced images. We placed four ground control points (GCPs) in the area to validate spatial location accuracy, for which the maximum error was 5 cm.

We collected ground truth (GT) information, i.e., ground-cover class validation points, with the help of an accurate GPS system (Leica GS10 base station and rover) at 2 cm accuracy and real time correction. We aimed at collecting GT points spread throughout each of the study sites, with ca. 40 GT points per ground-cover class for each site, resulting in a total of 1782 GT points. We identified a total of 17 ground-cover classes, but not all classes were present at all sites. For each GT point, we noted the dominant ground-cover type in a ca. 7.5 cm radius around the point. Figure 2 illustrates the variation of the ground-cover classes. See Appendix A for detailed descriptions of each of the ground-cover classes.

Figure 2. Images of 16 ground-cover classes that were included in the initial classification scheme. **row 1**: *moss-graminoid, moss-equisetum, wetland, moss-brown-wetland*; **row 2**: *Carex subspathacea, heath-moss, heath-graminoid, dryas*; **row 3**: *cassiope, winter-damage-woody, biological crust, grubbing*; **row 4**: *winter-damage-moss, bare ground, gravel, water*. The class *snow* is not shown in this graphic. Detailed information for each class can be found in the Appendix A.

2.4. Data Preparation

Using Pix4D Mapper [78], we generated five orthomosaics with images from the Sequoia+ sensor (four optical bands and the normalized difference vegetation index (NDVI)). With the same software, we created a digital surface model (DSM) using the 3D point cloud from the images of the AeriaX RGB camera. Our AeriaX RGB camera switched from manual to automatic settings during several instances and could therefore not be radiometrically calibrated. As a result, we used RGB images only to generate the DSM. We used the R (version 4.0.0.) software for all further analyses [79], and scripts can be found in the Supplementary Materials. To obtain textural information from our orthomosaics, we calculated gray level co-occurence matrices (GLCMs), as conducted in Wang et al., 2015 [61], using the R package glcm [80]. We calculated seven types of GLCMs for the NDVI, green and red edge orthomosaics and using four different neighborhood sizes (0.3 m, 0.9 m, 1.5 m, 2 m) with equal offsets in all directions as we did not expect any specific directionality in our data. Larger neighborhood sizes for the GLCMs were not computationally feasible with our setup, as processing times per variable reached multiple days. We did not include the DSM in our analysis because we found that the absolute height above sea level is not a useful predictor for generalizing across sites as it is too specific for each study site. Instead, we used it to calculate several of the terrain variables that were tested and discussed in [81]: slope at 1 m, 5 m and 10 m resolution, vector ruggedness, dissection, several curvature functions and terrain ruggedness (R package: SpatialEco [82]), all of the latter at 0.2 m and 1 m resolution using a range of neighborhoods sizes (0.6 m–2.2 m for the 0.2 m resolution, with 0.4 m intervals, and 3 m–131 m for the 1 m resolution, with 10 m intervals). We chose the smaller resolutions to capture the local heterogeneity of the terrain, such as tussocks and grubbing craters, and the larger resolutions to describe more general terrain positioning at the study sites.

2.5. Variable Selection

To reduce the number of drone-imagery-derived variables for further analyses, we used two exclusion criteria. First, we visually investigated the potential of the terrain variables to distinguish ground-cover classes by plotting maps and generating boxplots. This led us to exclude the curvature functions and terrain ruggedness and only continue with slope, vector ruggedness and dissection. Both GLCMs and terrain functions are dependent on the analysis window size (neighborhoods). For each site, we explored the GLCMs and terrain variable neighborhoods in preliminary models to assess their predictive ability for at least one of the ground-cover classes. This way we reduced the number of GLCMs from 84 to 39 and the terrain variables from 61 to 21. More detailed information on variable selection considerations can be found in Table 1—(2) *Variable selection*.

2.6. Ground-Cover Class Selection

We used a research-driven class selection approach (see Table 1—(3) *Ground-cover class selection*). Since the final ground-cover classes were not pre-defined, we created a flowchart to guide our decision making on how to systematically test our ability to detect ground-cover classes (Figure 3a) at different levels of detail. As a starting point of selecting ground-cover classes, we created two lists of class detail (Figure 3b). Our scheme with *initial* classes included all the ground-cover classes that we identified as ecologically relevant to characterize the ground. We were mostly interested in detecting disturbance ground-cover classes, and our class selection is oriented towards that. We further created a *simplified* classes scheme that would fulfill our minimum requirements of detecting the disturbance-related classes (*bare ground*, *grubbing*, *winter damage*, *biological crust*) and merged all vegetated classes into one. After running the classification algorithm on both schemes and evaluating the results, we created a third, *optimized* classes list (Figure 3b). The *optimized* classes differ from initial classes in three ways: (i) *heath vegetation* is grouped into a single class, (ii) the *winter-damage-moss* class is excluded as it was a very localized phenomena and challenging to validate, and (iii) the *biological crust* and *winter damage classes* are grouped into a single (mixed) class.

2.7. Data Analysis

2.7.1. Disturbance Detection Based on NDVI

In an exploratory examination, we investigated our ability of using drone images and a GSD of approximately 10 cm to detect pink-footed goose grubbing, winter damage (i.e., grey areas of dead vegetation) and bare ground (open soil). We compared the NDVI value differences of the GT points in both disturbed and undisturbed areas within the same vegetation type: For dry-mesic habitats, we compared *cassiope* and *dryas* against *biological crust* and *winter damage*. For moist-mesic habitats, we compared *wetland* and *graminoid moss* against both *grubbing* and *bare ground*. We use box-and-whisker plots to show the range of NDVI values for each study site and ground-cover class.

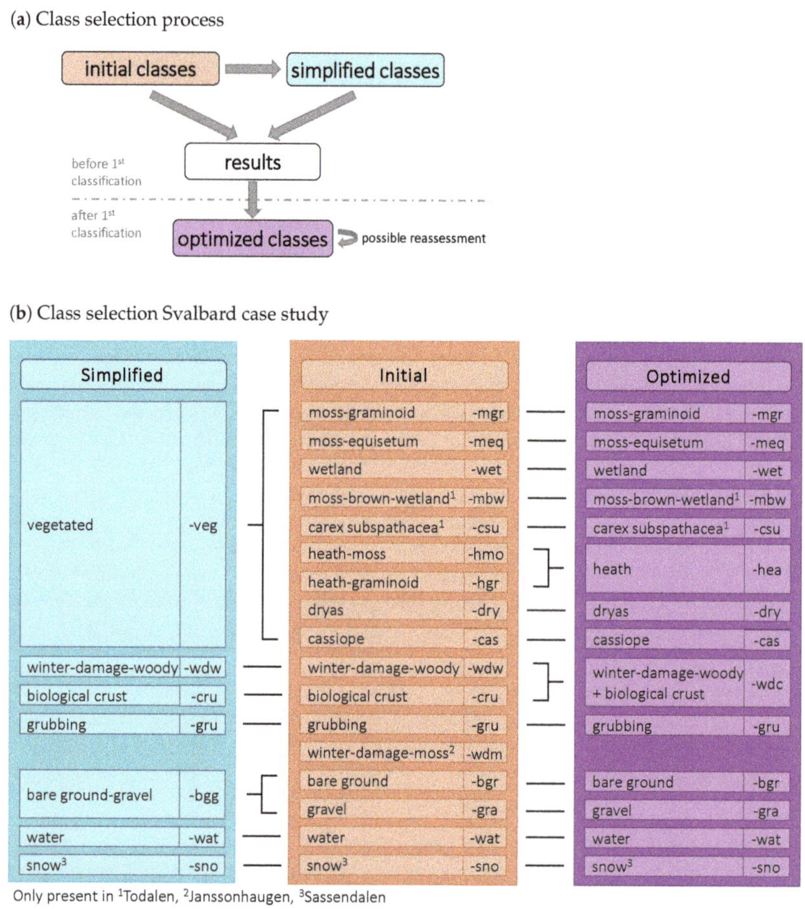

Figure 3. Ground-cover class selection schemes. (**a**) Stages to determine ground-cover classes balancing class detail and classification accuracy. The first stage includes testing two class schemes: one with the most possible detail and one with the minimum required class detail. In the second stage, the optimal classes are determined by merging the classes that improve the classification accuracy using the simplified class scheme as a guideline for meaningful merges. (**b**) The initial, simplified and optimized ground-cover classes and class abbreviations used in the case study to classify tundra vegetation on Svalbard.

2.7.2. Classifier Development and Validation

See Table 1 for considerations on (4) *Classifier development and validation*. We used a random forest (RF) classifier [83] to develop ground-cover classification models for each of the study sites. We selected the RF classifier as it has been shown to be successful for spatial vegetation data [84]. For each of the study sites, we generated six RF models using three levels of detail for output classes (*initial, simplified* and *optimized*) and two sets of input variables (*optical and terrain* and *optical* only). Note that the ground-cover classification schemes differed slightly between study sites as not all classes were present at each site (Figure 3b). We used our field ground truthing (GT) data as training data by extracting the pixel values for each layer that were within the 7.5 cm radius at each GT coordinate. For some classes (*snow, water, Carex subspathacea, dryas, cassiope*), we lacked sufficient GT points and therefore obtained additional training data from the drone imagery. To do this, we drew polygons of the classes with insufficient training data using the RGB orthomosaics and field recordings and added all pixels within those polygons to our GT dataset.

Each RF model was developed as follows (Figure 4): we first split the extracted GT data 70–30% (stratified random, i.e., random 70–30% split within ground-cover class) into a training dataset and a validation dataset. As each GT coordinate encompassed several pixels, we reduced autocorrelation by not using any of the pixels from the same GT point for both training and validation. We then trained the RF model with the training dataset using the "fit" function ("rminer" package [85]) with ntree = 500 and a default set of mtry = 8. We tested the classifier output on the independent validation dataset using the "predict" function in "rminer". To assess the robustness of each RF model, we repeated the process (i.e., the dataset split, model fitting and validation) thirty times. We used F1 scores to compare model performances because this measure captures both observation accuracy (recall, i.e., proportion of correctly classified pixels of a ground-cover class among all pixels belonging to that class) and prediction accuracy (precision, i.e., proportion of correctly classified pixels of a ground-cover class among all pixels classified to of that class) in a single score ($F1 = 2 \cdot \frac{precision \cdot recall}{precision + recall}$). F1 scores range from 0 to 100%, with values close to 100% representing high observation and high prediction accuracy. We calculated mean F1 scores for each ground-cover class and the mean macro-F1 score (arithmetic mean of all ground-cover class F1 scores) to summarize the results of the 30 cross-validation runs. We used confusion matrices to assess in more detail where class mixing occurred. In addition, we assessed the selection of predictor variables by analyzing variable importance in each of the RF models.

We created classified maps for each site with the *optical and terrain* option with the *optimized* classes. We then computed maps from these model outputs for each site using the predict function ("raster package" [86]) with parallel processing to speed up the process (ClusterR package [87]). We validated the classification by comparing the classified maps with the drone orthomosaics and hand-held pictures taken in the field.

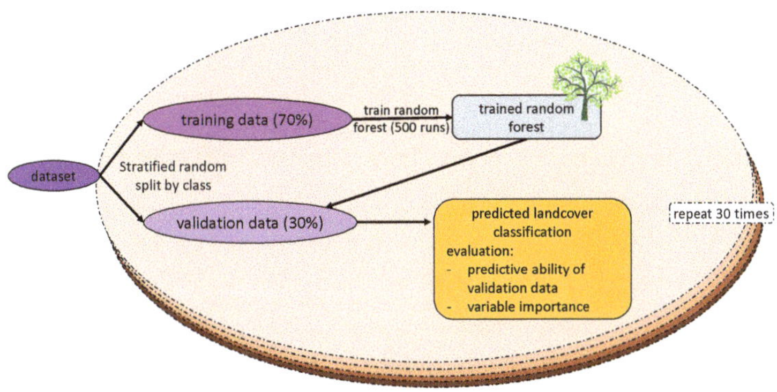

Figure 4. Data selection process, model validation and evaluation. The classifier was developed using 70% of the dataset as training data and 30% as validation data. The dataset split and classifier was run 30 times. Classifier performance mean and variance as well as variable importance values were calculated from the outputs.

2.7.3. Spatial Transferability

We tested how accurately a model, trained with the optimized classes developed for one study site, performed at the other two sites. First, we reduced the dataset to the ground-cover classes by excluding classes that were not present in all three sites. Then we ran the training on RF for each layer selection set (*optical and terrain* and *optical only*). Finally, we used these models to predict the ground-cover class for each pixel at the two other sites that were left out. We only tested the RF classifier transferability "as is" and did not adjust the RF trees by pruning or using transfer functions. See Table 1—(5) *Transferability of classifer* for more information. We used macro-F1 scores to compare the prediction outputs with one another. To gain a better understanding of the similarities between the classifiers, we compared their variable importance rankings using the mean decrease accuracy measure.

3. Results

3.1. Disturbance Detection Based on NDVI

Disturbed and associated undisturbed ground-cover types differed in their median normalized difference vegetation index (NDVI) values but had overlapping interquartile ranges. This was consistent for the three study sites (Figure 5). Most of the *wetland* GT points had NDVI values of 0.75 or higher, and the median of registered *moss graminoid* points was around 0.7. Most grubbing GT points had NDVI values that were lower than the undisturbed classes and had median values between 0.53 and 0.58. The *bare ground* class had the lowest NDVI, and none of the values overlapped with the undisturbed classes except for the outliers (Figure 5a). Among the drier ground-cover classes, *dryas* and *cassiope* had NDVI value medians between 0.59 and 0.75 across the three sites, whereas the classes *winter damage* and *biological crust* had median values between 0.30 and 0.48 (Figure 5b). The difference in NDVI values between the undisturbed and disturbed classes in the dry-mesic habitat was less pronounced in TOD compared to JAN and SAS.

3.2. Class and Layer Selection

Comparisons of the RF classifier performances between the three classification schemes and two sets of variable selections at each site (Figure 6) showed that the ground-cover classifications based on both the terrain and optical layers had higher macro-F1 scores than the ones using only optical layers. The magnitude of the difference depended on the study site and chosen classification scheme. The classifiers using terrain and optical variables showed that: (i) the optimized classification schemes performed similarly (macro-F1 score 82–83%) in all three sites, and (ii) sites differed in terms of performance of the simplified and initial classification schemes. In site TOD, the simplified classification scheme performed better than in the sites JAN and SAS. The optimized classifier improved (compared to the initial and simplified) macro-F1 scores most in TOD.

3.3. Variable Importance

The order of importance values (measured as the mean decrease accuracy for each variable in the RF) varied between the three study sites. Among the optical variables, NDVI-based gray level co-occurence matrices (GLCM) variables were important in all three sites. NDVI was the most important layer for SAS and JAN, but ranked 19th in TOD. The terrain variables that were most important were dissection (especially at large neighborhoods) for JAN and TOD and slope for SAS. Terrain variables that were calculated at 0.2 m resolution were among the least important for predicting ground-cover classes. Complete lists with all importance values can be found in the Supplementary Materials.

3.4. Ground-Cover Classification

Most ground-cover classes of the *optimized* classifier had F1 scores of 70% or higher, often over 80–90% (Table 2). The scores were generally highest for the classes *snow, Carex subspathacea, water* and *wetland.* The classes *heath* and *gravel* had the largest variation in F1 scores between study sites. There was no consistent bias in model-misclassification of the ground-cover classes in the three classifiers. Confusion matrices that show the ground-cover class distributions for each of the classifiers can be found in the Supplementary Materials. The two disturbance classes *grubbing* and *winter damage/crust* had higher observation accuracy (recall) than prediction accuracy (precision) in all three sites. Classification errors occurred mainly within the moist-mesic and the dry-mesic habitats: For example, in JAN, the most common example of misclassification was that *cassiope* (precision: 91.9%, recall: 72.1%) was classified as *heath* (precision: 67.1%, recall: 83.2%) or *winter damage/crust* (precision: 80.2%, recall: 87.1%). In TOD, *grubbing* (precision: 78.6%, recall: 78.3%) was most commonly falsely misclassfied (both directions) with the *wetland* (precision: 78.6%, recall: 83.0%) and *graminoid-moss* (precision: 88.8%, recall: 81.4%) classes.

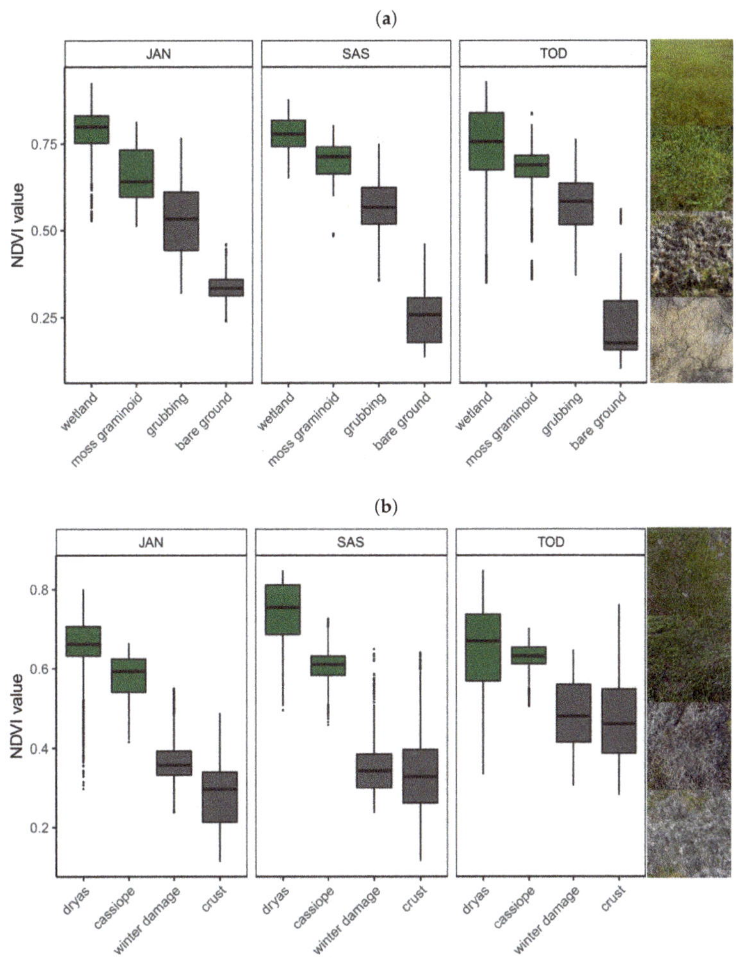

Figure 5. The distribution of NDVI values of selected ground-cover classes and their associated disturbances in the three study sites: The disturbed associates of classes *wetland* and *moss graminoid* are *grubbing* and *bare ground*, and the associated disturbed classes of *dryas* and *cassiope* are *winter damage* and *biological crust*. The boxes denote the interquartile ranges of NDVI values around the median for all registered ground-truthing pixels for each ground-cover class, and dots represent outliers. The order of the images (up-down) follows the x-axis from left to right for each plot. (**a**) Moist-mesic habitat; (**b**) Dry-mesic habitat.

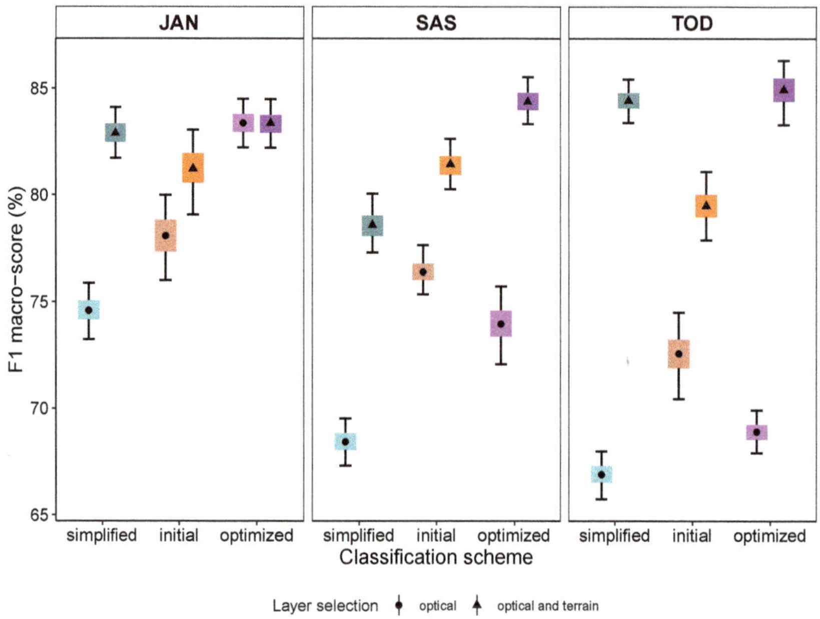

Figure 6. Comparison of site, classification scheme and layer selection. Black symbols represent mean macro-F1 scores after 30 runs, the colored boxes indicate confidence intervals of 0.5 and the bars represent the interval at 0.95.

Table 2. Macro-F1 scores of ground-cover classifications at the three study sites on Svalbard using the *optimized* classes and all variables (*optical and terrain*). Explanations to the class abbreviations can be found in Figure 3b.

	Class													
site	**mgr**	**meq**	**wet**	**mbw**	**csu**	**hea**	**dry**	**cas**	**wdc**	**gru**	**bgr**	**gra**	**wat**	**sno**
JAN	70.5	82.3	91.4	-	-	57.9	93.7	80.6	81.9	83.8	84.2	90.7	83.6	100
SAS	75.0	83.8	92.8	-	-	77.3	80.2	92.0	88.7	86.1	75.0	84.0	93.4	-
TOD	84.7	85.9	80.8	86.4	97.7	95.6	80.7	74.7	77.5	78.3	88.9	74.0	98.9	-

3.5. Visual Evaluation of Predicted Ground-Cover

Based on visual inspection, the classified maps (Figures 7–9) captured all ground-cover classes well. The *graminoid moss* tundra, the *wetlands* and the drier *Dryas* and *barren* classes were predicted as expected from our knowledge of the sites. Pink-footed goose grubbing was detected in *graminoid moss* tundra and in *wetland* vegetation (e.g., Figures 8b and 9b,d). *Winter damage/crust* was detected in dry ridges, within *Cassiope* belts and on top of polygonal moss tundra (e.g., Figures 8c and 9b). Based on our detailed in-field knowledge of the sites, there were site-specific misclassifications in "brown rugged terrain" (see discussion).

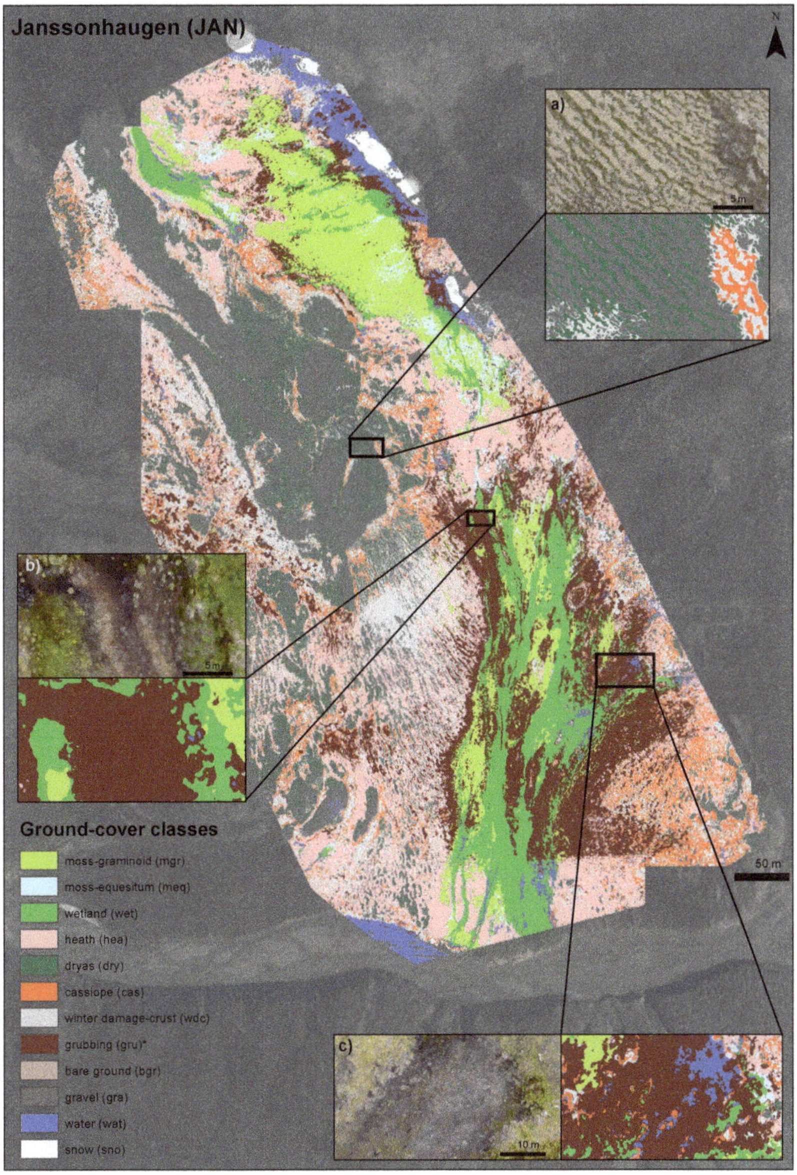

Figure 7. Classified map of Janssonhaugen (JAN). Zoomed areas highlight examples of classifications: (**a**) The classification of dry-mesic habitat with *dryas*, *cassiope*, *gravel* and *winter damage/crust*. (**b**) Wet, brown-colored mosses classified as *grubbing*. (**c**) Winter-damaged moss, classified as *grubbing* and *water*.

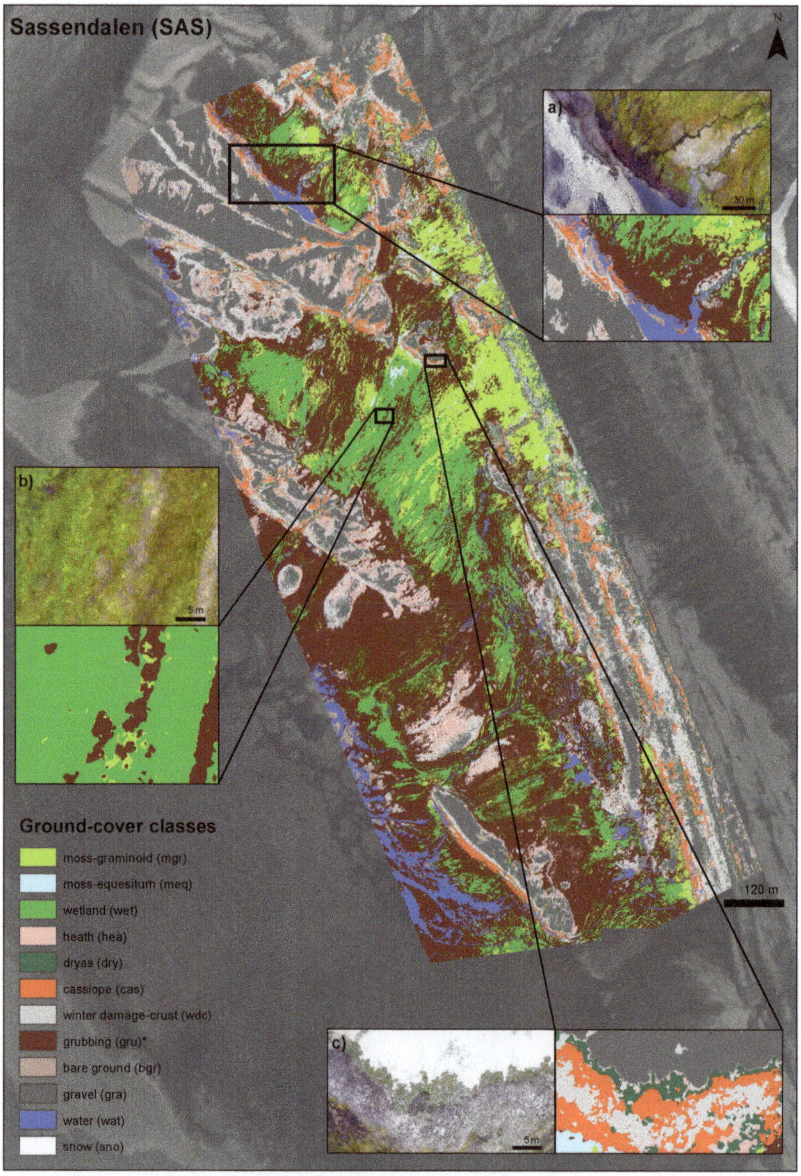

Figure 8. Classified map of Sassendalen (SAS). Zoomed areas highlight examples of classifications. (**a**) Dry and moist habitats: On the left, classification of dry-mesic habitat with *gravel* and a *cassiope belt*. On the right, moist habitat, classified as *wetland* and *moss-graminoid*, brown and rugged mosses classified as *grubbing*. (**b**) Grubbing detection in wetland area. (**c**) Zonation of dry-mesic ground-cover classes along a gravel ridge.

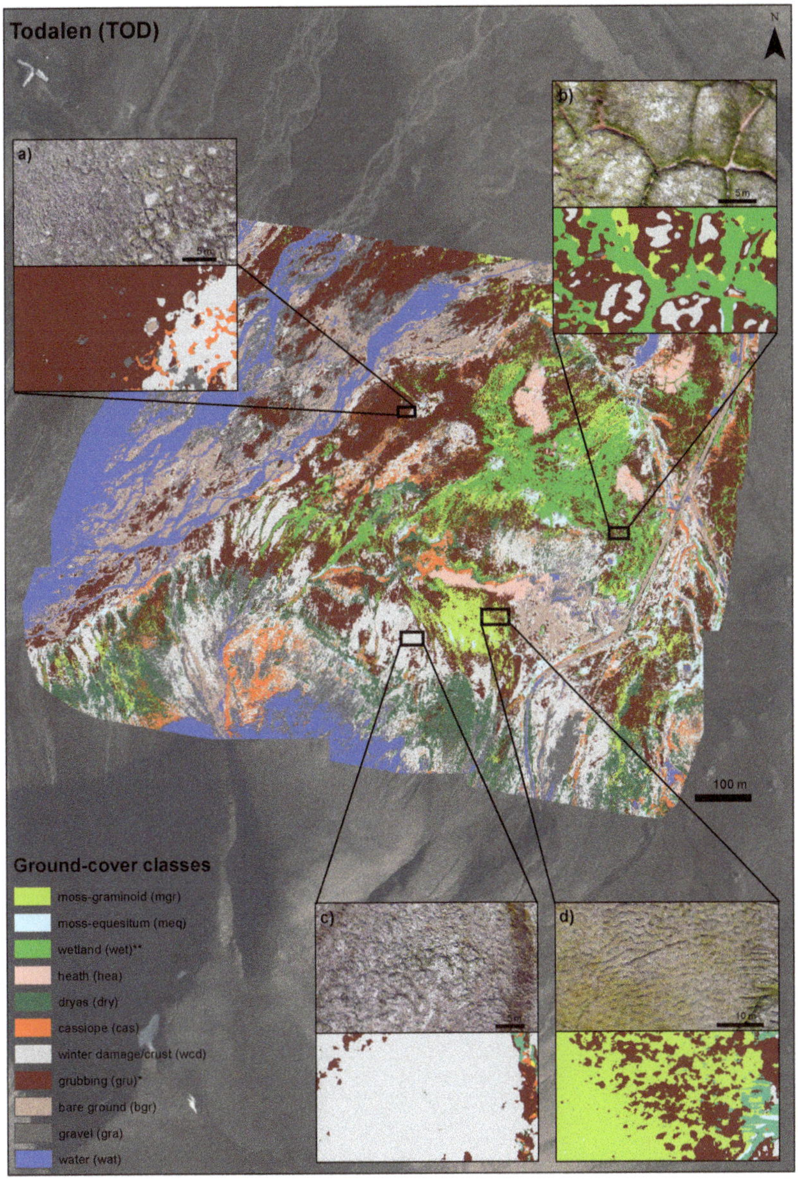

Figure 9. Classified map of Todalen (TOD). Zoomed areas highlight examples of classifications. (**a**) *Cassiope* classified as *grubbing*. (**b**) *Grubbing* and *winter damage/crust* on ice wedge polygons. (**c**) A dry-mesic habitat with *dryas*, *gravel* and *winter damage/crust* classified as *winter damage/crust*. (**d**) *Grubbing* in moss-graminoid habitat.

3.6. Spatial Transferability

In all cases, the models provided less accurate ground-cover classifications if trained in another site (transferred models (Figure 10). Transferability was generally highest between the sites JAN and SAS, and using only optical variables was sufficient. The highest macro-F1 score was found when using the model developed for SAS (*optical* variables only) to predict vegetation at JAN (macro-F1 score of 45.4%, Figure 10). Generally, transferability did not differ between using *optical and terrain* variables or using *optical* variables only. Ground-cover classes in TOD had the lowest prediction accuracy using models trained at the other study sites. Likewise, the model trained in TOD did not perform well in either of the other two study sites. The classifiers for JAN and SAS shared more of the top ten variable importance scores (see supplementary materials for details).

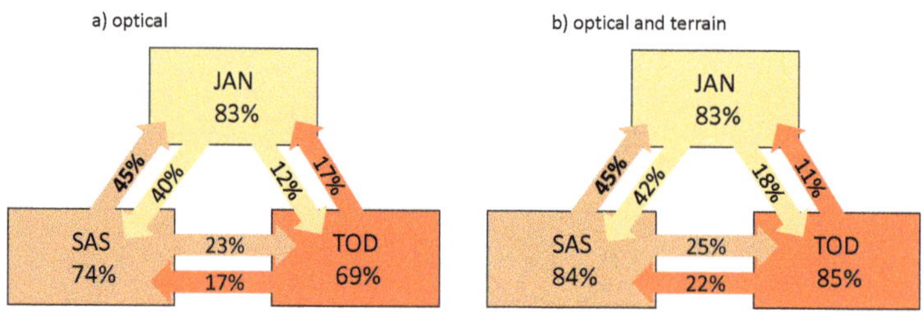

Figure 10. Classifier transferability between the three sites using the optimized classification scheme. Numbers inside the squares show the mean macro-F1 scores for in-site classifiers.The arrows indicate the direction and macro-F1 scores of using the classifier on the other sites. (**a**) Optical variables and (**b**) optical and terrain variables

4. Discussion

Our case study, using drone imagery in a high Arctic tundra landscape, showed that the classification of images can be reliably used to detect both vegetation types and small-scale disturbances from herbivory and winter weather events. The workflow we developed evaluates which combination of spatial variables and ground-cover classes yields random forest (RF) models with high predictive ability. Our suggested workflow emphasizes steps in drone image processing that have previously received less attention in ecological applications. The poor transferability of models between different sites highlights that RF models are spatially specific and degrade in accuracy when applied in different spatial contexts (sites). A point worth noting is that replicability is related to the method and that each of our RF models needs to be trained with site-specific data and care must be taken when using it in another context.

The lower normalized difference vegetation index (NDVI) values for the disturbed classes *winter damage/crust* and *grubbing* compared to their non-disturbed counterparts suggest that small-scale disturbances of tundra vegetation in Svalbard can be detected correctly using drone images. However, a range of NDVI values is commonly not exclusive to a certain vegetation class (see, e.g., Raynolds et al. [88]). This was also the case for the vegetation classes and their disturbed associates in our study that had partly overlapping ranges, especially in the site TOD. A classification approach is therefore needed to distinguish among the different classes, especially in cases where repeated NDVI measurements before and after the disturbance occurred are not available.

The F1 scores for the disturbed classes were generally high (77.5–88.7%), but reaching this classifier performance required merging *winter damage* and *biological crust* to one class. More work is thus needed to distinguish the different types of biological crust that have varied coloring from white to gray and dark brown/black from winter damage in the form of dead plants that make up a gray layer on the ground. Consistently higher recall than precision values for the disturbance classes *grubbing* and *winter damage/crust* suggest high detection rates coupled with slight but consistent overestimation of these classes. Field assessments supported these findings. We detected the following inconsistencies in the predicted map compared to knowledge from the field. We recognize that grubbing was overestimated in areas with brown-colored mosses where the surface of the tundra has small-scale ruggedness due to soil movements, surface runoffs or reindeer trampling. The overestimation of grubbing also occurs in brown-colored, senescent *cassiope* areas and in patches of muddy water (e.g., Figures 7b,c, 8a and 9a, Eischeid, Ravolainen, personal observation). *Winter damage/crust* can be overestimated in areas with small, interspersed patches of healthy vegetation (e.g., Figure 9c, Eischeid, Ravolainen, personal observation). A mountain side shadow in the southern side of TOD was classified as water. These examples illustrate the need for additional ground-truthing and further fine-tuning of classification of the tundra vegetation types, particularly related to small-scale disturbances.

The overall high classification accuracy with macro-F1 values of over 80% for all three sites suggests that the classification performed well, comparable or better than other drone-based classifications in the high Arctic, such as Fraser et al. [89] and Thomson et al. [48]. We found no consistent bias in the misclassification of the ground-cover classes. Those misclassifications that occurred were between classes that are ecologically similar (e.g., within dry or wet habitats). The high detection of the classes *moss-graminoid, moss-equisetum, wetland, Carex subspathacea* and *grubbing* suggests that drone classification of tundra vegetation can be used to detect vegetation classes important to monitor future state changes in Svalbard's tundra ecosystem [14]. Increases in horsetails (*Equisetum* spp.) and *Carex subspathacea* abundances were recently reported from a revisit study in central Spitsbergen [90], highlighting the importance of developing image-based methods to support field-surveys to better obtain area estimates of such changes.

Our workflow (Table 1) builds on recent publications that aimed at systematizing the collection of drone images for research in natural systems. Review articles by Aasen et al. [31], Assmann et al. [54] and Tmušic et al. [53] focused on improving reproducibility and quality of drone-based surveys, including spectral calibration, standardized data collection and general quality control. With our workflow, we expanded this set of available guidelines, focusing on variable and class selection in an ecological context of Tundra ecosystems that can help researchers make decisions on drone data processing and image analysis. With our case study, we show how the workflow helped us to make systematic choices in planning our methods, as well as testing the limitations of our data, thereby ensuring ecologically sound research that is suitable for long-term monitoring.

Variable selection can be a very time-consuming process due to the large number of variables that can be retrieved from the original high-resolution layers and requires an understanding of the studied ecosystem to pre-select suitable variables. Relevant neighborhood sizes are an aspect of this process adding to selection/weight given to terrain and optical layers. We adapted the approach of Fan and Wang et al. [61,62], who systematically pre-tested different neighborhood sizes to find the most relevant ones. Analyzing three study sites simultaneously complicated the variable selection process and increased the total number of variables used for our analysis. Common systematic approaches to reducing the number of variables often rely on reruns with lower numbers of variables to find the smallest error rates [91]. These approaches were, however, not suitable when evaluating three classifiers (three study sites) simultaneously, and we, therefore, kept all the variables after the pre-test we determined to be important in at least one of the three study sites. Similarly, we kept the feature set size (mtry = 8) constant for all sites and classifiers to reduce the possible number of model outputs and comparisons. Further work on optimizing the site-specific models would allow us to find the most important predictor variables and optimized feature set.

As part of our workflow, we compared the RF classifier outputs that were based on optical and terrain variables or optical variables only. Using both terrain and optical variables improved in-site classifications compared to using only optical variables. In other contexts, different combinations of variables can lead to the best classification. For instance, drone surveys of structural landscape changes, such as landslides or river morphology, have achieved good results using terrain information only [92,93]. The optimal set of variables will depend on plant community composition, the ground-cover classification detail needed and the types of sensors available [89,94–96]. We therefore recommend future drone-based ground-cover classification studies to include a wide range of variables and utilize our suggested selection procedures to find the most suitable ones for analysis.

Ground-cover classification maps have become a key resource in institutional, policy and law-making practice, but so far, there is a lack of conceptual frameworks and agreeable standards for map based monitoring [97]. Structuring ground-cover classes (i.e., in a hierarchical approach) can help with choosing appropriate class level detail and allow transferring across scales [97]. In our study, we found a structured selection approach, in line with adaptive monitoring goals [98], appropriate to find the optimal class level detail. It allowed us to focus on the ecological relevance of each of the ground-cover classes. We can therefore recommend our ground-cover class selection framework as a tool to create purpose-specific maps and find the balance between model accuracy and obtaining classes relevant for the monitoring goals.

Our results indicated low model transferability between study sites. Low regional transferability has also been reported in other RF classification studies [81,99]. Transferability can be improved by either reducing the number of classes [95], applying algorithm transfer functions, such as transformation matrices or tree pruning mechanism [69,70], or using different classifier types (e.g., [100,101]. Some deep learning approaches have recently been shown to provide classifications of drone and satellite data with reasonable transferability across space [102,103]. These approaches were outside the scope of our study as transferability was not the central goal. The right balance between a locally precise or transferable classifier will depend on the site-specific use-case, and a multi-approach framework may be needed to satisfy both local and regional monitoring requirements.

5. Conclusions

This study provides a planning workflow for generating vegetation cover maps with drones for ecological monitoring purposes in high Arctic tundra. Using random forest classifiers, we were able to successfully distinguish up to 15 different ground-cover classes, including two disturbance classes, goose grubbing and winter damage, which had lower NDVI values than their undisturbed counterparts. Although the models indicate a high predictive ability for the disturbance classes, field assessments have shown an overestimation of disturbances in parts of the moss tundra landscape. We show that a direct transfer of the models between study sites was not possible without further fine-tuning of methodologies. Future work might benefit from focusing separately on locally optimized maps and developing more generalized, transferable models. We have shown that it is possible to map the ground-cover classes that will likely be important to study state changes in Svalbard's tundra ecosystem. Future drone times-series are necessary to work on the detection of state transitions. We encourage closer interdisciplinary collaboration between experts of remote sensing, informatics and ecology to combine the knowledge-base and further improve the quality of map-based ecological monitoring.

Supplementary Materials: The following are available online at https://www.mdpi.com/article/10.3390/rs13214466/s1.

Author Contributions: The following authors contributed with the conceptualization, I.E., R.A.I., J.M., Å.Ø.P., E.M.S. and V.T.R.; methodology: I.E., J.J.A., F.P. and V.T.R.; investigation: I.E. and V.T.R.; formal analysis: I.E., F.P. and V.T.R.; visualization: I.E. and E.M.S.; writing—original draft preparation: I.E., E.M.S., Å.Ø.P. and V.T.R., writing—review and editing: I.E., E.M.S., J.J.A., R.A.I., J.M., Å.Ø.P., F.P., N.G.Y. and V.T.R.; funding acquisition: E.M.S., R.A.I., J.M., Å.Ø.P., N.G.Y. and V.T.R. All authors have read and agreed to the published version of the manuscript.

Funding: This research was funded by the Tromsø Research Foundation and made possible by a grant from the University of Aarhus, and the Norwegian Polar Institute provided personnel and field logistics.

Data Availability Statement: The data presented in this study are available on request from the corresponding author.

Acknowledgments: We thank Stein Tore Pedersen and the Norwegian Polar Institute for providing logistic support for our field work. We thank Ingrid G. Paulsen and Torgeir T. Blæsterdalen for support in the field. We thank the Governor of Svalbard and Hurtigruten for access to field accommodation. Anne Urset and Harald Faste Aas at the NPI mapping section provided valuable support with the GPS equipment. Furthermore, we thank Rolf Andersen at the Arctic University of Norway (UiT) for providing us processing services and support for using large datasets. This publication is a contribution from the Climate-Ecological Observatory for Arctic Tundra (COAT).

Conflicts of Interest: The authors declare no conflict of interest. The funders had no role in the design of the study; in the collection, analyses, or interpretation of data; in the writing of the manuscript; or in the decision to publish the results.

Appendix A. Ground-Cover Class Descriptions

The three sites for the drone flights were chosen so that they include ground-cover classes that are of high importance in tundra plant–herbivore interactions and have effects on plants from extreme winter weather events [14]. The meso-topography of mountainous tundra landscapes causes differentiation in vegetation types within relatively short distances (e.g., Mörsdorf et al. [104]), enabling study of dry, mesic, moist and wet ground-cover classes at the the same sites.

In the high Arctic Svalbard, the drier habitats of convex landscape forms are often dominated by *Dryas octopetala* heath [105] that is an available and important foraging habitat to the Svalbard reindeer and rock ptarmigan in the winter [106,107]. In a lower belt, below the ridges, in central parts of Spitsbergen, there is often *Cassiope tetragona* vegetation that is snow-covered during the winter. Winter extreme weather with rain-on-snow events have been shown to cause damage to these dwarf shrubs [20,21], although the area extent of such damage remains to be documented. Mountain sides and slightly sloping parts of the landscapes are often covered with a thick moss layer where herbaceous plants dominate alongside *Salix polaris*. In the lower parts of the landscapes, the moss tundra gradually changes to wetlands with cotton grasses and grasses adapted to wet conditions. The moss tundra and wetland parts of the landscapes are important summer foraging areas for all the vertebrate herbivores [14].

These landscapes encompass a selection of functionally important habitats with high potential for change due to climate change and the dynamic herbivore populations [14,58]. The expected trajectories of change differ between the dry, moist and wet habitats and between the disturbance types [14]. This highlights the importance of developing tools to quantify changes in areas of vegetation states and areas effected by the disturbances.

Vegetation types in Svalbard have been described in various sources earlier, including a vegetation map based on satellite data [72], plant sociological and classification studies [73,108] in Svalbard's flora [105] and in other research literature [22,30,109,110]. The previously defined vegetation state or type descriptions differ from each other depending on the purpose of the study. For our purpose of mapping especially the dwarf shrub vegetation types and the moss tundra, in combination with the most important disturbance to them (winter damage and herbivory by goose, respectively), we defined the following initial list of ground-cover classes so that it allowed us to work on vegetation and disturbances in combination.

(1a) Moss-graminoid

A common denominator for this class is a layer of mosses (typically 10–30 cm, or deeper) in moist terrain. Typical vascular plants are grasses in the genera *Poa*, *Festuca*, *Alopecurus* and *Dupontia*, the deciduous dwarf shrub *Salix polaris*, forbs in different genera such as *Ranunculus*, *Saxifraga*, *Micranthes* and *Pedicularis*, as well as some sedges (*Carex*). Moss flora is diverse, with dominant species from genera *Aulacomnium*, *Tomentypnum* and *Sanionia*.

(1b) Moss-equisetum

This class appears in habitats similar to the class moss-graminoid and in areas transitioning to wetland, but *Equisetum arvense* and *Equisetum scirpoides* dominate the vascular plant community.

(1c) Wetland

The wetlands in our study sites are located in slightly sloping areas and intermixed or downslope from the moist moss tundra. This class includes wetland dominated by mosses (such as *Drepanocladus* spp. and *Scorpidium* spp.) and graminoids such as *Eriophorum* spp. and *Dupontia fisheri*.

(1d) Moss-brown-wetland

This class is a specific type of wetland with less graminoids and mostly covered by wet, brown-colored mosses, giving the patches a potentially different spectral signature from 1c. This class was only mapped in TOD but also occurs in JAN and SAS, although less frequent there. These patches can be covered in water during wet years and dry out during drier years or seasons.

(2a) *Carex subspathacea*

This class occurs in habitats that otherwise would be classified as *wetland* but is characterized by a dense cover of *C. subspathacea*. It is one of the main arctic salt marsh plant species and has been documented to replace graminoids in areas of intensive pink-footed goose grubbing [90]. We only mapped this class in TOD, although it is also present in SAS and to a smaller extent in JAN.

(2b) Heath-moss

This class is typically found in slight depressions between ridges on the slopes and is characterized by a thin (approx. 1–5 cm) and drier moss layer often dominated by *Sanionia* spp. Vascular plants are usually sparse but include *Salix polaris*, *Saxifraga oppositifolia* or *Cerastium* spp.

(2c) Heath-graminoid

The heath-graminoid class is similar to the *heath-moss* class, with the difference that the moss layer is covered by graminoids, typically *Luzula confusa*, or in dry and convex terrain, such as *Carex* species.

(2d) Dryas

We assigned this class to ridges where open *Dryas octopetala* vegetation is found. The substrate in our *dryas* class is of gravel or thin silty soils on mainly alluvial deposits or terraces. *Dryas octopetala* occurs also in tussock tundra in Svalbard, but such *Dryas* areas were not common in the study sites.

(3a) Cassiope

We assigned this class if the dominant vegetation was *Cassiope tetragona*. This class is typically found downslope from the convex *Dryas* areas but above moss tundra and wetland areas where snow cover is stable in the winter.

(3b) Winter-damage-woody

This class is the disturbed counterpart to the *dryas* and *cassiope* classes. Periodic warm periods during winter result in sequential freezing that damages plant tissue, and coupled with rain, encapsulate plants with an ice coat. In addition, extreme winter warming can lead to desiccation when plants leave hibernation but cannot access water due to frozen soils [21]. Freezing temperatures after snow melt can damage plant tissues and decrease flowering success [20]. If the vegetation within a 15 cm circle was more than 80% damaged (gray), it was recorded for this class.

(3c) Biological crust

A layer of micro-organisms can cover bare soils, in which case, it is usually called biotic soil crust [111], but can also grow on damaged plants, in which case, it can be called biotic film. The biotic crust/film that is composed of a variety of bacteria, lichens and algae can grow over areas or that have been opened due to disturbances or over plants that were damaged. In our classification, we combined these two types of cryptogamic cover. The color of biotic crust can vary from almost white to gray or brown and black.

(3d) Grubbing

The grubbing class is the disturbed counterpart to the *moss tundra* and *wetland* classes. While grubbing, pink-footed geese often remove large amounts of moss in order to access the desired plant parts and opening up bare ground [27]. Grubbing can occur at different intensities. Our grubbing class describes areas where geese have removed enough moss to result in a continuous surface of dried moss interspersed with holes created by their beaks. We did not record singular holes (lighter intensity than our grubbing class). If grubbing has removed all vegetation and erosion processes have begun (higher intensity than our grubbing class), this would have been recorded as *bare ground* as it would not be possible to distinguish these patches from other bare areas that may have formed differently.

(4a) Winter-damage-moss

In one of our study sites, we recorded large patches of black/gray, previously moss-covered areas, probably disturbed by winter weather events. The most important process is not known but could include prolonged anaerobic conditions, freeze damage, perhaps combined with drought, but no apparent signs of fungal infections. This class was only recorded for JAN. These areas were black in color and different in appearance from the "winter-damage-woody" class.

(4b) Bare ground

This class encompasses all areas that are not covered with vegetation and have a soil substrate.

(4c) Gravel

This classes describes gravel substrates, pebbles and up over in size. It also includes rocks.

(4d) Water

The class *water* includes lakes, rivers, streams and surface run-offs. The water can either be clear or (in most cases) enriched with sediments and brown in color.

Snow

A small snow field was only present in JAN.

References

1. Bhatt, U.; Walker, D.; Raynolds, M.; Comiso, J.; Epstein, H.; Jia, G.; Gens, R.; Pinzon, J.; Tucker, C.; Tweedie, C.; et al. Circumpolar Arctic Tundra Vegetation Change Is Linked to Sea Ice Decline. *Earth Interact.* **2010**, *14*, 1–20. [CrossRef]
2. Meredith, M.; Sommerkorn, M.; Cassota, S.; Derksen, C.; Ekaykin, A.; Hollowed, A.; Kofinas, G.; Mackintosh, A.; Melbourne-Thomas, J.; Muelbert, M.M.C.; et al. Polar regions. In *The Ocean and Cryosphere in a Changing Climate. A Special Report of the Intergovernmental Panel on Climate Change*; Pörtner, H.O., Roberts, D., Masson-Delmotte, V., Zhai, P., Tignor, M., Poloczanska, E., Mintenbeck, K., Alegría, A., Nicolai, M., Okem, A., et al., Eds.; IPCC: Geneva, Switzerland, 2019; pp. 203–320.
3. Taylor, J.; Lawler, J.; Aronsson, M.; Barry, T.; Bjorkman, A.; Christensen, T.; Coulson, S.; Cuyler, C.; Ehrich, D.; Falk, K.; et al. Arctic terrestrial biodiversity status and trends: A synopsis of science supporting the CBMP State of Arctic Terrestrial Biodiversity Report. *Ambio* **2020**, *49*, 833–847. [CrossRef] [PubMed]
4. Aronsson, M.; Heiðmarsson, S.; Jóhannesdóttir, H.; Barry, T.; Braa, J.; Burns, C.; Coulson, S.; Cuyler, C.; Falk, K.; Helgason, H.; et al. State of the Arctic Terrestrial Biodiversity Report; Conservation of Arctic Flora and Fauna International Secretariat: Akureyri, Iceland, 2021.
5. AMAP. *Snow, Water, Ice and Permafrost. Summary for Policy-Makers. Arctic Monitoring and Assessment Programme (AMAP)*; AMAP: Oslo, Norway, 2017.
6. Box, J.; Colgan, W.; Christensen, T.R.; Schmidt, N.; Lund, M.; Parmentier, F.J.; Brown, R.; Bhatt, U.; Euskirchen, E.; Romanovsky, V.; et al. Key indicators of Arctic climate change: 1971–2017. *Environ. Res. Lett.* **2019**, *14*, 045010. [CrossRef]
7. Niittynen, P.; Heikkinen, R.K.; Aalto, J.; Guisan, A.; Kemppinen, J.; Luoto, M. Fine-scale tundra vegetation patterns are strongly related to winter thermal conditions. *Nat. Clim. Chang.* **2020**, *10*, 1143–1148. [CrossRef]
8. Wrona, F.J.; Johansson, M.; Culp, J.M.; Jenkins, A.; Mård, J.; Myers-Smith, I.H.; Prowse, T.D.; Vincent, W.F.; Wookey, P.A. Transitions in Arctic ecosystems: Ecological implications of a changing hydrological regime. *J. Geophys. Res. Biogeosci.* **2016**, *121*, 650–674. [CrossRef]
9. Jorgenson, M.; Shur, Y.; Pullman, E. Abrupt increase in permafrost degradation in Arctic Alaska. *Geophys. Res. Lett.* **2006**, *33*, L02503. [CrossRef]
10. Rennert, K.J.; Roe, G.; Putkonen, J.; Bitz, C.M. Soil Thermal and Ecological Impacts of Rain on Snow Events in the Circumpolar Arctic. *J. Clim.* **2009**, *22*, 2302–2315. [CrossRef]
11. Biskaborn, B.; Smith, S.; Noetzli, J.; Matthes, H.; Vieira, G.; Streletskiy, D.; Schoeneich, P.; Romanovsky, V.; Lewkowicz, A.; Abramov, A.; et al. Permafrost is warming at a global scale. *Nat. Commun.* **2019**, *10*, 1–11. [CrossRef]

12. Kerbes, R.H.; Kotanen, P.M.; Jefferies, R.L. Destruction of Wetland Habitats by Lesser Snow Geese: A Keystone Species on the West Coast of Hudson Bay. *J. Appl. Ecol.* **1990**, *27*, 242–258. [CrossRef]
13. Hansen, B.B.; Grøtan, V.; Aanes, R.; Sæther, B.E.; Stien, A.; Fuglei, E.; Ims, R.A.; Yoccoz, N.G.; Pedersen, Å.Ø. Climate Events Synchronize the Dynamics of a Resident Vertebrate Community in the High Arctic. *Science* **2013**, *339*, 313–315. [CrossRef]
14. Ravolainen, V.; Soininen, E.M.; Jónsdóttir, I.S.; Eischeid, I.; Forchhammer, M.; van der Wal, R.; Pedersen, Å.Ø. High Arctic ecosystem states: Conceptual models of vegetation change to guide long-term monitoring and research. *Ambio* **2020**, *49*, 666–677. [CrossRef]
15. Chapin, F.S., III; Woodwell, G.; Randerson, J.; Rastetter, E.; Lovett, G.; Baldocchi, D.; Clark, D.; Harmon, M.; Schimel, D.; Valentini, R.; et al. Reconciling Carbon-cycle Concepts, Terminology, and Methods. *Ecosystems* **2006**, *9*, 1041–1050. [CrossRef]
16. Scheffer, M.; Carpenter, S.; Lenton, T.; Bascompte, J.; Brock, W.; Dakos, V.; van de Koppel, J.; van de Leemput, I.; Levin, S.; Nes, E.; et al. Anticipating Critical Transitions. *Science* **2012**, *338*, 344–348. [CrossRef]
17. Jefferies, R.L.; Jano, A.P.; Abraham, K.F. A biotic agent promotes large-scale catastrophic change in the coastal marshes of Hudson Bay. *J. Ecol.* **2006**, *94*, 234–242. [CrossRef]
18. Van der Wal, R. Do herbivores cause habitat degradation or vegetation state transition? Evidence from the tundra. *Oikos* **2006**, *114*, 177–186. [CrossRef]
19. Peeters, B.; Pedersen, Å.; Loe, L.E.; Isaksen, K.; Veiberg, V.; Stien, A.; Kohler, J.; Gallet, J.C.; Aanes, R.; Hansen, B. Spatiotemporal patterns of rain-on-snow and basal ice in high Arctic Svalbard: Detection of a climate-cryosphere regime shift. *Environ. Res. Lett.* **2019**, *14*, 015002. [CrossRef]
20. Milner, J.M.; Varpe, Ø.; van der Wal, R.; Hansen, B.B. Experimental icing affects growth, mortality, and flowering in a high Arctic dwarf shrub. *Ecol. Evol.* **2016**, *6*, 2139–2148. [CrossRef] [PubMed]
21. Bjerke, J.W.; Treharne, R.; Vikhamar-Schuler, D.; Karlsen, S.R.; Ravolainen, V.; Bokhorst, S.; Phoenix, G.K.; Bochenek, Z.; Tømmervik, H. Understanding the drivers of extensive plant damage in boreal and Arctic ecosystems: Insights from field surveys in the aftermath of damage. *Sci. Total Environ.* **2017**, *599–600*, 1965–1976. [CrossRef]
22. Speed, J.D.; Woodin, S.; Tømmervik, H.; Tamstorf, M.; Wal, R. Predicting Habitat Utilization and Extent of Ecosystem Disturbance by an Increasing Herbivore Population. *Ecosystems* **2009**, *12*, 349–359. [CrossRef]
23. Speed, J.D.M.; Cooper, E.J.; Jónsdóttir, I.S.; Van Der Wal, R.; Woodin, S.J. Plant community properties predict vegetation resilience to herbivore disturbance in the Arctic. *J. Ecol.* **2010**, *98*, 1002–1013. [CrossRef]
24. Pedersen, Å.Ø.; Tombre, I.; Jepsen, J.U.; Eidesen, P.B.; Fuglei, E.; Stien, A. Spatial patterns of goose grubbing suggest elevated grubbing in dry habitats linked to early snowmelt. *Polar Res.* **2013**, *32*, 19719. [CrossRef]
25. Pedersen, A.S.; Speed, J.; Tombre, I. Prevalence of pink-footed goose grubbing in the arctic tundra increases with population expansion. *Polar Biol.* **2013**, *36*, 1569–1575. [CrossRef]
26. Madsen, J.; Williams, J.H.; Johnson, F.A.; Tombre, I.M.; Dereliev, S.; Kuijken, E. Implementation of the first adaptive management plan for a European migratory waterbird population: The case of the Svalbard pink-footed goose Anser brachyrhynchus. *Ambio* **2017**, *46*, 275–289. [CrossRef] [PubMed]
27. Fox, A.; Francis, I.; Bergersen, E. Diet and habitat use of Svalbard Pink-footed Geese Anser brachyrhynchus during arrival and pre-breeding periods in Adventdalen. *Ardea* **2006**, *94*, 691–699.
28. Van der Wal, R.; Sjögersten, S.; Woodin, S.J.; Cooper, E.J.; Jónsdóttir, I.S.; Kuijper, D.; Fox, T.A.D.; Huiskes, A.D. Spring feeding by pink-footed geese reduces carbon stocks and sink strength in tundra ecosystems. *Glob. Chang. Biol.* **2007**, *13*, 539–545. [CrossRef]
29. Speed, J.; Woodin, S.; Tømmervik, H.; van der Wal, R. Extrapolating herbivore-induced carbon loss across an arctic landscape. *Polar Biol.* **2010**, *33*, 789–797. [CrossRef]
30. Van der Wal, R.; Stien, A. High-arctic plants like it hot: A long-term investigation of between-year variability in plant biomass. *Ecology* **2014**, *95*, 3414–3427. [CrossRef]
31. Aasen, H.; Honkavaara, E.; Lucieer, A.; Zarco-Tejada, P.J. Quantitative Remote Sensing at Ultra-High Resolution with UAV Spectroscopy: A Review of Sensor Technology, Measurement Procedures, and Data Correction Workflows. *Remote Sens.* **2018**, *10*, 1091. [CrossRef]
32. Assmann, J.J.; Myers-Smith, I.H.; Kerby, J.T.; Cunliffe, A.M.; Daskalova, G.N. Drone data reveal heterogeneity in tundra greenness and phenology not captured by satellites. *Environ. Res. Lett.* **2020**, *15*, 125002. [CrossRef]
33. Lovitt, J.; Rahman, M.M.; McDermid, G.J. Assessing the Value of UAV Photogrammetry for Characterizing Terrain in Complex Peatlands. *Remote Sens.* **2017**, *9*, 715. [CrossRef]
34. Lousada, M.; Pina, P.; Vieira, G.; Bandeira, L.; Mora, C. Evaluation of the use of very high resolution aerial imagery for accurate ice-wedge polygon mapping (Adventdalen, Svalbard). *Sci. Total Environ.* **2018**, *615*, 1574–1583. [CrossRef]
35. Siewert, M.B.; Olofsson, J. Scale-dependency of Arctic ecosystem properties revealed by UAV. *Environ. Res. Lett.* **2020**, *15*, 094030. [CrossRef]
36. Harder, P.; Schirmer, M.; Pomeroy, J.; Helgason, W. Accuracy of snow depth estimation in mountain and prairie environments by an unmanned aerial vehicle. *Cryosphere* **2016**, *10*, 2559–2571. [CrossRef]
37. Pedersen, S.H.; Tamstorf, M.P.; Abermann, J.; Westergaard-Nielsen, A.; Lund, M.; Skov, K.; Sigsgaard, C.; Mylius, M.R.; Hansen, B.U.; Liston, G.E.; et al. Spatiotemporal Characteristics of Seasonal Snow Cover in Northeast Greenland from in Situ Observations. *Arct. Antarct. Alp. Res.* **2016**, *48*, 653–671. [CrossRef]

38. Cimoli, E.; Marcer, M.; Vandecrux, B.; Boggild, C.E.; Williams, G.; Simonsen, S.B. Application of Low-Cost UASs and Digital Photogrammetry for High-Resolution Snow Depth Mapping in the Arctic. *Remote Sens.* **2017**, *9*, 1144. [CrossRef]
39. Ewertowski, M.W.; Evans, D.J.A.; Roberts, D.H.; Tomczyk, A.M. Glacial geomorphology of the terrestrial margins of the tidewater glacier, Nordenskiöldbreen, Svalbard. *J. Maps* **2016**, *12*, 476–487. [CrossRef]
40. Tonkin, T.; Midgley, N.; Cook, S.; Graham, D. Ice-cored moraine degradation mapped and quantified using an unmanned aerial vehicle: A case study from a polythermal glacier in Svalbard. *Geomorphology* **2016**, *258*, 1–10. [CrossRef]
41. Phillips, E.; Everest, J.; Evans, D.J.; Finlayson, A.; Ewertowski, M.; Guild, A.; Jones, L. Concentrated, 'pulsed' axial glacier flow: Structural glaciological evidence from Kvíárjökull in SE Iceland. *Earth Surf. Process. Landforms* **2017**, *42*, 1901–1922. [CrossRef]
42. Nehyba, S.; Hanáček, M.; Engel, Z.; Stachoň, Z. Rise and fall of a small ice-dammed lake—Role of deglaciation processes and morphology. *Geomorphology* **2017**, *295*, 662–679. [CrossRef]
43. Jouvet, G.; Weidmann, Y.; Seguinot, J.; Funk, M.; Abe, T.; Sakakibara, D.; Seddik, H.; Sugiyama, S. Initiation of a major calving event on the Bowdoin Glacier captured by UAV photogrammetry. *Cryosphere* **2017**, *11*, 911–921. [CrossRef]
44. Jones, C.; Ryan, J.; Holt, T.; Hubbard, A. Structural glaciology of Isunguata Sermia, West Greenland. *J. Maps* **2018**, *14*, 517–527. [CrossRef]
45. Jouvet, G.; Weidmann, Y.; Kneib, M.; Detert, M.; Seguinot, J.; Sakakibara, D.; Sugiyama, S. Short-lived ice speed-up and plume water flow captured by a VTOL UAV give insights into subglacial hydrological system of Bowdoin Glacier. *Remote Sens. Environ.* **2018**, *217*, 389–399. [CrossRef]
46. Van der Sluijs, J.; Kokelj, S.V.; Fraser, R.H.; Tunnicliffe, J.; Lacelle, D. Permafrost Terrain Dynamics and Infrastructure Impacts Revealed by UAV Photogrammetry and Thermal Imaging. *Remote Sens.* **2018**, *10*, 1734. [CrossRef]
47. Cunliffe, A.M.; Tanski, G.; Radosavljevic, B.; Palmer, W.F.; Sachs, T.; Lantuit, H.; Kerby, J.T.; Myers-Smith, I.H. Rapid retreat of permafrost coastline observed with aerial drone photogrammetry. *Cryosphere* **2019**, *13*, 1513–1528. [CrossRef]
48. Thomson, E.; Spiegel, M.; Althuizen, I.; Bass, P.; Chen, S.; Chmurzynski, A.; Halbritter, A.; Henn, J.; Jónsdóttir, I.; Klanderud, K.; et al. Multiscale mapping of plant functional groups and plant traits in the High Arctic using field spectroscopy, UAV imagery and Sentinel-2A data. *Environ. Res. Lett.* **2021**, *16*, 055006. [CrossRef]
49. Turner, I.L.; Harley, M.D.; Drummond, C.D. UAVs for coastal surveying. *Coast. Eng.* **2016**, *114*, 19–24. [CrossRef]
50. Gillan, J.K.; Karl, J.W.; van Leeuwen, W.J.D. Integrating drone imagery with existing rangeland monitoring programs. *Environ. Monit. Assess.* **2020**, *192*, 1–20. [CrossRef]
51. Hoffmann, H.; Jensen, R.; Thomsen, A.; Nieto, H.; Rasmussen, J.; Friborg, T. Crop water stress maps for entire growing seasons from visible and thermal UAV imagery. *Biogeosci. Discuss.* **2016**, 13, 6545–6563. [CrossRef]
52. Tirado, S.B.; Hirsch, C.N.; Springer, N.M. UAV-based imaging platform for monitoring maize growth throughout development. *Plant Direct* **2020**, *4*, e00230. [CrossRef] [PubMed]
53. Tmušić, G.; Manfreda, S.; Aasen, H.; James, M.R.; Gonçalves, G.; Ben-Dor, E.; Brook, A.; Polinova, M.; Arranz, J.J.; Mészáros, J.; et al. Current Practices in UAS-based Environmental Monitoring. *Remote Sens.* **2020**, *12*, 1001. [CrossRef]
54. Assmann, J.J.; Kerby, J.T.; Cunliffe, A.M.; Myers-Smith, I.H. Vegetation monitoring using multispectral sensors—Best practices and lessons learned from high latitudes. *J. Unmanned Veh. Syst.* **2019**, *7*, 54–75. [CrossRef]
55. Arctic Long-Term Ecological Research Site. Available online: https://arc-lter.ecosystems.mbl.edu/ (accessed on 23 October 2020).
56. Bylot Island—Sirmilik National Park Long Term Monitoring Program. Available online: http://www.cen.ulaval.ca/bylot/en/bylothistory.php (accessed on 16 October 2020).
57. Zackenberg Ecological Research Operations. Available online: https://g-e-m.dk/gem-localities/zackenberg-research-station/ (accessed on 23 October 2020).
58. Ims, R.; Jepsen, J.U.; Stien, A.; Yoccoz, N.G. *Science plan for COAT: Climate-ecological Observatory for Arctic Tundra*; Fram Centre Report Series 1; Fram Centre: Tromsø, Norway, 2013.
59. High-Latitude Drone Ecology Network (HiLDEN). Available online: https://arcticdrones.org/ (accessed on 4 September 2020).
60. Zhou, Y.; Daakir, M.; Rupnik, E.; Pierrot-Deseilligny, M. A two-step approach for the correction of rolling shutter distortion in UAV photogrammetry. *ISPRS J. Photogramm. Remote Sens.* **2020**, *160*, 51–66. [CrossRef]
61. Wang, H.; Pu, R.; Zhu, Q.; Ren, L.; Zhang, Z. Mapping health levels of Robinia pseudoacacia forests in the Yellow River delta, China, using IKONOS and Landsat 8 OLI imagery. *Int. J. Remote Sens.* **2015**, *36*, 1114–1135. [CrossRef]
62. Fan, H. Land-cover mapping in the Nujiang Grand Canyon: Integrating spectral, textural, and topographic data in a random forest classifier. *Int. J. Remote Sens.* **2013**, *34*, 7545–7567. [CrossRef]
63. Weinmann, M.; Jutzi, B.; Hinz, S.; Mallet, C. Semantic point cloud interpretation based on optimal neighborhoods, relevant features and efficient classifiers. *ISPRS J. Photogramm. Remote Sens.* **2015**, *105*, 286–304. [CrossRef]
64. Mascaro, J.; Asner, G.P.; Knapp, D.E.; Kennedy-Bowdoin, T.; Martin, R.E.; Anderson, C.; Higgins, M.; Chadwick, K.D. A Tale of Two "Forests": Random Forest Machine Learning Aids Tropical Forest Carbon Mapping. *PLoS ONE* **2014**, *9*, e85993. [CrossRef] [PubMed]
65. Belgiu, M.; Drăguţ, L. Random forest in remote sensing: A review of applications and future directions. *ISPRS J. Photogramm. Remote Sens.* **2016**, *114*, 24–31. [CrossRef]
66. Colditz, R.R. An Evaluation of Different Training Sample Allocation Schemes for Discrete and Continuous Land Cover Classification Using Decision Tree-Based Algorithms. *Remote Sens.* **2015**, *7*, 9655–9681. [CrossRef]

67. Roberts, D.R.; Bahn, V.; Ciuti, S.; Boyce, M.S.; Elith, J.; Guillera-Arroita, G.; Hauenstein, S.; Lahoz-Monfort, J.J.; Schröder, B.; Thuiller, W.; et al. Cross-validation strategies for data with temporal, spatial, hierarchical, or phylogenetic structure. *Ecography* **2017**, *40*, 913–929. [CrossRef]
68. Hastie, T.; Tibshirani, R.; Friedman, J. *The Elements of Statistical Learning: Data Mining, Inference and Prediction*, 2nd ed.; Springer: Berlin/Heidelberg, Germany, 2009.
69. Segev, N.; Harel, M.; Mannor, S.; Crammer, K.; El-Yaniv, R. Learn on Source, Refine on Target:A Model Transfer Learning Framework with Random Forests. *IEEE Trans. Pattern Anal. Mach. Intell.* **2015**, *39*, 1811–1824. [CrossRef]
70. Sukhija, S.; Krishnan, N.C. Supervised heterogeneous feature transfer via random forests. *Artif. Intell.* **2019**, *268*, 30–53. [CrossRef]
71. Walker, D.; Raynolds, M.; Daniëls, F.; Einarsson, E.; Elvebakk, A.; Gould, W.; Katenin, A.; Kholod, S.; Markon, C.; Melnikov, E.; et al. The Circumpolar Arctic Vegetation Map. *J. Veg. Sci.* **2005**, *16*, 267–282. [CrossRef]
72. Johansen, B.E.; Karlsen, S.R.; Tømmervik, H. Vegetation mapping of Svalbard utilising Landsat TM/ETM data. *Polar Rec.* **2012**, *48*, 47–63. [CrossRef]
73. Elvebakk, A. A vegetation map of Svalbard on the scale 1:3.5 mill. *Phytocoenologia* **2005**, *35*, 951–967. [CrossRef]
74. Elvebakk, A. A survey of plant associations and alliances from Svalbard. *J. Veg. Sci.* **1994**, *5*, 791–802. [CrossRef]
75. Lawrimore, J.; Ray, R.; Applequist, S.; Korzeniewski, B.; Menne, M. Global Summary of the Month (GSOM), Version 1. In *NOAA National Centers for Environmental Information*. Available online: https://www.ncei.noaa.gov/access/metadata/landing-page/bin/iso?id=gov.noaa.ncdc:C00946 (accessed on 10 September 2021).
76. Migała, K.; Wojtuń, B.; Szymański, W.; Muskała, P. Soil moisture and temperature variation under different types of tundra vegetation during the growing season: A case study from the Fuglebekken catchment, SW Spitsbergen. *Catena* **2014**, *116*, 10–18. [CrossRef]
77. Cubero-Castan, M.; Schneider-Zapp, K.; Bellomo, M.; Shi, D.; Rehak, M.; Strecha, C. Assessment Of The Radiometric Accuracy In A Target Less Work Flow Using Pix4D Software. In Proceedings of the 2018 9th Workshop on Hyperspectral Image and Signal Processing: Evolution in Remote Sensing (WHISPERS), Amsterdam, The Netherlands, 23–26 September 2018; pp. 1–4. [CrossRef]
78. Pix4Dmapper. Available online: https://www.pix4d.com/product/pix4dmapper-photogrammetry-software (accessed on 1 November 2021).
79. R Core Team. *R: A Language and Environment for Statistical Computing*; R Foundation for Statistical Computing: Vienna, Austria, 2013.
80. Zvoleff, A. glcm: Calculate Textures from Grey-Level Co-Occurrence Matrices (GLCMs). R Package Version 1.6.5. Available online: https://cran.r-project.org/web/packages/glcm/glcm.pdf (accessed on 1 November 2021).
81. Maxwell, A.E.; Warner, T.A.; Strager, M.P. Predicting Palustrine Wetland Probability Using Random Forest Machine Learning and Digital Elevation Data-Derived Terrain Variables. *Photogramm. Eng. Remote Sens.* **2016**, *82*, 437–447. [CrossRef]
82. Evans, J.S. spatialEco: Spatial Analysis and Modelling Utilities. R Package Version 1.3-2. Available online: https://cran.r-project.org/web/packages/spatialEco/spatialEco.pdf (accessed on 1 November 2021).
83. Cutler, A.; Cutler, D.R.; Stevens, J.R. Random Forests. In *Ensemble Machine Learning: Methods and Applications*; Zhang, C., Ma, Y., Eds.; Springer: Boston, MA, USA, 2012; pp. 157–175. [CrossRef]
84. Karami, M.; Westergaard-Nielsen, A.; Normand, S.; Treier, U.A.; Elberling, B.; Hansen, B.U. A phenology-based approach to the classification of Arctic tundra ecosystems in Greenland. *ISPRS J. Photogramm. Remote Sens.* **2018**, *146*, 518–529. [CrossRef]
85. Cortez, P. rminer: Data Mining Classification and Regression Methods. R Package Version 1.4.6. Available online: https://cran.r-project.org/web/packages/spatialEco/spatialEco.pdf (accessed on 1 November 2021).
86. Hijmans, R.J. raster: Geographic Data Analysis and Modeling. R Package Version 3.3-13. Available online: https://cran.r-project.org/web/packages/raster/raster.pdf (accessed on 1 November 2021).
87. Mouselimis, L. ClusterR: Gaussian Mixture Models, K-Means, Mini-Batch-Kmeans, K-Medoids and Affinity Propagation Clustering. R Package Version 1.2.2. Available online: https://cran.r-project.org/web/packages/ClusterR/ClusterR.pdf (accessed on 1 November 2021).
88. Raynolds, M.K.; Comiso, J.C.; Walker, D.A.; Verbyla, D. Relationship between satellite-derived land surface temperatures, arctic vegetation types, and NDVI. *Remote Sens. Environ.* **2008**, *112*, 1884–1894.
89. Fraser, R.H.; Olthof, I.; Lantz, T.C.; Schmitt, C. UAV photogrammetry for mapping vegetation in the low-Arctic. *Arct. Sci.* **2016**, *2*, 79–102. [CrossRef]
90. Van der Wal, R.; Anderson, H.; Stien, A.; Loe, L.E.; Speed, J. Disturbance, Recovery and Tundra Vegetation Change Final Report project 17/92—to Svalbard Environmental Protection Fund. 2020. Available online: https://aura.abdn.ac.uk/bitstream/handle/2164/16573/SMF_Distrubance_recovery_veg_change.pdf;jsessionid=65802B34A907DB989FF65BD2D7FDB248?sequence=1 (accessed on 13 September 2021).
91. Hapfelmeier, A.; Ulm, K. A new variable selection approach using Random Forests. *Comput. Stat. Data Anal.* **2013**, *60*, 50–69. [CrossRef]
92. Rossi, G.; Tanteri, L.; Tofani, V.; Vannocci, P.; Moretti, S.; Casagli, N. Multitemporal UAV surveys for landslide mapping and characterization. *Landslides* **2018**, *15*, 1045–1052. [CrossRef]
93. Özcan, O.; Özcan, O. Multi-temporal UAV based repeat monitoring of rivers sensitive to flood. *J. Maps* **2021**, *17*, 163–170. [CrossRef]

94. Chen, J.; Yi, S.; Qin, Y.; Wang, X. Improving estimates of fractional vegetation cover based on UAV in alpine grassland on the Qinghai–Tibetan Plateau. *Int. J. Remote Sens.* **2016**, *37*, 1922–1936. [CrossRef]
95. Miranda, V.; Pina, P.; Heleno, S.; Vieira, G.; Mora, C.; Schaefer, C.E. Monitoring recent changes of vegetation in Fildes Peninsula (King George Island, Antarctica) through satellite imagery guided by UAV surveys. *Sci. Total Environ.* **2020**, *704*, 135295. [CrossRef] [PubMed]
96. Morgan, B.E.; Chipman, J.W.; Bolger, D.T.; Dietrich, J.T. Spatiotemporal Analysis of Vegetation Cover Change in a Large Ephemeral River: Multi-Sensor Fusion of Unmanned Aerial Vehicle (UAV) and Landsat Imagery. *Remote Sens.* **2021**, *13*, 51. [CrossRef]
97. Cullum, C.; Rogers, K.; Brierley, G.; Witkowski, E. Ecological classification and mapping for landscape management and science: Foundations for the description of patterns and processes. *Prog. Phys. Geogr.* **2015**, *40*, 38–65. [CrossRef]
98. Lindenmayer, D.B.; Likens, G.E. Adaptive monitoring: A new paradigm for long-term research and monitoring. *Trends Ecol. Evol.* **2009**, *24*, 482–486. [CrossRef] [PubMed]
99. Juel, A.; Groom, G.B.; Svenning, J.C.; Ejrnæs, R. Spatial application of Random Forest models for fine-scale coastal vegetation classification using object based analysis of aerial orthophoto and DEM data. *Int. J. Appl. Earth Obs. Geoinf.* **2015**, *42*, 106–114. [CrossRef]
100. Kalantar, B.; Mansor, S.B.; Sameen, M.I.; Pradhan, B.; Shafri, H.Z.M. Drone-based land-cover mapping using a fuzzy unordered rule induction algorithm integrated into object-based image analysis. *Int. J. Remote Sens.* **2017**, *38*, 2535–2556. [CrossRef]
101. Wessel, M.; Brandmeier, M.; Tiede, D. Evaluation of Different Machine Learning Algorithms for Scalable Classification of Tree Types and Tree Species Based on Sentinel-2 Data. *Remote Sens.* **2018**, *10*, 1419. [CrossRef]
102. Zhang, C.; Sargent, I.; Pan, X.; Li, H.; Gardiner, A.; Hare, J.; Atkinson, P.M. Joint Deep Learning for land cover and land use classification. *Remote Sens. Environ.* **2019**, *221*, 173–187. [CrossRef]
103. Tong, X.Y.; Xia, G.S.; Lu, Q.; Shen, H.; Li, S.; You, S.; Zhang, L. Land-cover classification with high-resolution remote sensing images using transferable deep models. *Remote Sens. Environ.* **2020**, *237*, 111322. [CrossRef]
104. Mörsdorf, M.A.; Ravolainen, V.T.; Yoccoz, N.G.; Thórhallsdóttir, T.E.; Jónsdóttir, I.S. Decades of Recovery From Sheep Grazing Reveal No Effects on Plant Diversity Patterns Within Icelandic Tundra Landscapes. *Front. Ecol. Evol.* **2021**, *8*, 502. [CrossRef]
105. Rønning, O.I. *Svalbards Flora*; Norsk Polarinstitutt: Tromsø, Norway, 1996.
106. Pedersen, Å.Ø.; Overrein, Ø.; Unander, S.; Fuglei, E. *Svalbard Rock Ptarmigan (Lagopus Mutus Hyperboreus): A Status Report*; Norwegian Polar Institute (Norsk Polarinstitutt): Tromsø, Norway, 2005.
107. Pedersen, Å.; Paulsen, I.; Albon, S.; Arntsen, G.L.; Hansen, B.; Langvatn, R.; Loe, L.E.; Le Moullec, M.; Overrein, Ø.; Peeters, B.; et al. *Svalbard Reindeer (Rangifer Tarandus Platyrhynchus): A Status Report*; Rapportserie—Norsk Polarinstitutt, Norwegian Polar Institute: Tromsø, Norway, 2019.
108. Vanderpuye, A.W.; Elvebakk, A.; Nilsen, L. Plant communities along environmental gradients of high-arctic mires in Sassendalen, Svalbard. *J. Veg. Sci.* **2002**, *13*, 875–884. [CrossRef]
109. Eurola, S.; Hakala, A. The bird cliff vegetation of Svalbard. *Aquil. Ser. Bot* **1977**, *15*, 1–18.
110. Jónsdóttir, I.S. Terrestrial ecosystems on Svalbard: Heterogeneity, complexity and fragility from an Arctic island perspective. In *Biology and Environment: Proceedings of the Royal Irish Academy (JSTOR)*; Royal Irish Academy: Dublin, Ireland, 2005; pp. 155–165.
111. Agnelli, A.; Corti, G.; Massaccesi, L.; Ventura, S.; D'Acqui, L.P. Impact of biological crusts on soil formation in polar ecosystems. *Geoderma* **2021**, *401*, 115340. [CrossRef]

remote sensing

MDPI

Article

Evaluation of Satellite-Derived Estimates of Lake Ice Cover Timing on Linnévatnet, Kapp Linné, Svalbard Using In-Situ Data

Samuel E. Tuttle [1,*], Steven R. Roof [2], Michael J. Retelle [3,4], Alan Werner [5], Grant E. Gunn [6] and Erin L. Bunting [7]

1 Department of Earth and Environmental Sciences, College of Arts and Sciences, Syracuse University, Syracuse, NY 13244, USA
2 School of Natural Science, Hampshire College, Amherst, MA 01002, USA; sroof@hampshire.edu
3 Department of Earth and Climate Sciences, Bates College, Lewiston, ME 04240, USA; mretelle@bates.edu
4 Arctic Geology Department, University Centre in Svalbard, N-9171 Svalbard, Norway
5 Department of Geology and Geography, Mount Holyoke College, South Hadley, MA 01075, USA; awerner@mtholyoke.edu
6 Department of Geography and Environmental Management, University of Waterloo, Waterloo, ON N2L 3G1, Canada; g2gunn@uwaterloo.ca
7 Department of Geography, Environment, and Spatial Sciences, College of Social Science, Michigan State University, East Lansing, MI 48824, USA; ebunting@msu.edu
* Correspondence: setuttle@syr.edu

Citation: Tuttle, S.E.; Roof, S.R.; Retelle, M.J.; Werner, A.; Gunn, G.E.; Bunting, E.L. Evaluation of Satellite-Derived Estimates of Lake Ice Cover Timing on Linnévatnet, Kapp Linné, Svalbard Using In-Situ Data. *Remote Sens.* **2022**, *14*, 1311. https://doi.org/10.3390/rs14061311

Academic Editors: Shridhar Jawak, Andreas Kääb, Veijo Pohjola, Hiroyuki Enomoto, Geir Moholdt, Kjell Arild Høgda, Małgorzata Błaszczyk, Bo N. Andersen, Ann Mari Fjæraa, Bartłomiej Luks, Roberto Salzano and Frode Dinessen

Received: 1 January 2022
Accepted: 3 March 2022
Published: 9 March 2022

Abstract: Arctic lakes are sensitive to climate change, and the timing and duration of ice presence and absence (i.e., ice phenology) on the lake surface can be used as a climate indicator. In this study of Linnévatnet, one of the largest lakes on Svalbard, we compare inferences of lake ice duration from satellite data with continuously monitored lake water temperature and photographs from automatic cameras. Visible surface reflectance data from the moderate-resolution imaging spectroradiometer (MODIS) were used to observe the change in the lake-wide mean surface reflectance of Linnévatnet from 2003–2019, and smoothing splines were applied to the to determine the date of summer ice-off (also called "break-up end"—BUE). Similarly, BUE and fall ice-on (or "freeze-up end"—FUE) were determined from lake-wide mean time series of Sentinel-1 microwave backscatter from 2014–2019. Overall, the ice timing dates identified from the satellite observations agree well with the in-situ observations (RMSE values of approximately 2–7 days for BUE and FUE, depending on the method and in-situ dataset), lending confidence to the accuracy of remote sensing of lake ice phenology in remote Arctic regions. Our observations of Linnévatnet indicate that BUE dates do not have a significant trend, while FUE dates have been occurring approximately 1.5 days later per year during the study period. These results support an overall decrease in annual duration of lake ice cover in this part of Svalbard.

Keywords: lake ice; remote sensing; Svalbard; MODIS; Sentinel-1; water temperature

1. Introduction

Arctic regions are highly responsive to climate change, with large climate shifts occurring over the past two decades [1–3]. It is important to document changes in arctic climate because they can instigate further effects of global consequence [4]. Arctic lakes are sensitive climate indicators because they reflect ambient environmental conditions and effectively integrate short-term variations in atmospheric conditions [5,6]. Past studies have shown that lake ice phenology is primarily driven by local air temperature [7–9], but also depends on a variety of secondary meteorologic, hydrologic, morphologic, and geographic factors [9–12]. The phenology of lake ice (i.e., timing of ice formation and break-up) influences a variety of lake properties, including water temperature distribution, stratification, water level, and ecology [13–16]. Regional changes in lake ice phenology can thus affect landscape energy and water balance (e.g., evapotranspiration), with downstream impacts on regional hydrology and riverine habitats (e.g., river flow regimes, sediment and nutrient fluxes, river ice), as well as hydroelectric production and high latitude transportation (e.g., ice roads) [13]. Numerous studies have documented widespread increases in lake water temperatures, decreased duration of seasonal lake ice cover, and large-scale ecosystem shifts [3,13,17–26]. Continuation of these trends are expected in the future [27–29], highlighting the importance of continued lake ice monitoring.

Lake ice phenology has been recorded for up to two centuries, mainly in populated regions [17,21,26,30–32]. Modeling efforts have extended estimates of lake ice phenology to lakes in remote regions [33,34], but are limited by model representativeness and a lack of validation data. In recent decades, satellites have greatly increased the capability of monitoring lake ice, especially in remote areas [18,23,35–42]. However, satellite observations are not always ground-truthed with detailed in-situ measurements or direct observations, especially in remote regions. It is important to understand the accuracy of remote sensing methods to observe lake ice phenology, so that we can better monitor and understand local and regional changes.

Svalbard's high-latitude location and its position at the end of the northern branch of the Gulf Stream (the West Spitsbergen Current) make it an ideal place to study changes in lake ice phenology in the High Arctic. The Svalbard archipelago and the adjacent Northern Barents Sea is one of the regions in the arctic that has experienced the greatest warming, with mean annual surface temperature increase of 3 to 5 °C over the past few decades [43]. The warm northward-flowing West Spitsbergen current, an extension of the North Atlantic Current and Gulf Stream, along with associated air masses, contribute to the Atlantic High Arctic maritime climate of the archipelago [44]. The recent intensified Atlantification of the Northern Barents Sea and the intrusion of warm Atlantic waters into the western fjords of Svalbard in recent decades [45] have caused far-reaching environmental impacts in both marine and terrestrial systems. Sea ice has become less extensive around Svalbard, and winter ice has not appeared in the west coast fjords since 2006 [46]. On land, the recent warming has accelerated glacier thinning and retreat [47,48], warming of permafrost [49], and lengthening of the annual hydrological cycle [50]. Warming temperatures, especially during the freezing season, and the shifting seasonality of temperature and precipitation have also been observed in Spitsbergen [51,52].

We compare lake ice timing estimates from satellite observations to a robust 16-year record of in-situ observations of water temperature and lake surface photographs from a lake in western Svalbard. The goal of the study is to confront satellite lake ice phenology estimates for an Arctic lake with high quality validation data in order to evaluate the accuracy and limitations of monitoring of lake ice dynamics using current remote sensing methods. The study area, datasets, and methods for determining lake ice timing are detailed in Section 2. Results of the analysis are presented in Section 3, including observations of lake water temperature and ice dynamics, and a comparison between in-situ and satellite methods. Finally, the findings are discussed in Section 4, including implications for satellite lake ice monitoring in the broader Arctic.

2. Materials and Methods

2.1. Study Area

Linnévatnet (or Lake Linné) is a glacial lake in the Linnédalen valley in the western portion of Nordenskiöld Land on the island of Spitsbergen in the Norwegian archipelago of Svalbard (Figure 1a). The lake sits 14 m above sea level and is approximately 4.5 km long and 1 km wide, with a long axis oriented at approximately N 30° W. Mountains border the lake, forming a U-shaped, elongate valley. A glacier-fed river (Linnéelva) flows into the lake at its southern tip, bringing glacial sediments. Outflow exits the lake at its northern end and continues northward to Russekeila bay, in Isfjorden, 2 km to the north. The lake is shallow in its southern reaches, no more than 15 m deep, and increases to up to 40 m depth in its center (Figure 1b). Svalbard is located above the Arctic Circle, so Linnévatnet experiences up to 24 h of sunlight in the summer and 24 h of darkness in the winter (i.e., the polar night). This extreme seasonality leads to freezing of the lake surface in the fall, with ice reaching up to 2 m thick, and melting of the lake ice in the summer.

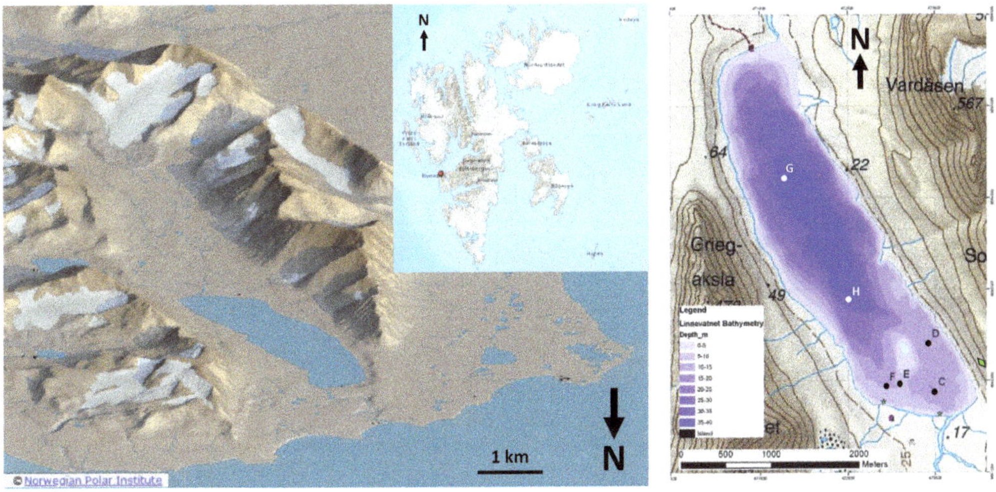

(**a**) (**b**)

Figure 1. (**a**) Three-dimensional digital elevation model (DEM) of Linnévatnet and the surrounding Linnédalen valley, along with inset map of Svalbard showing the location of Linnévatnet on the western coast of Spitsbergen (red dot). Images from TopoSvalbard, courtesy of the Norwegian Polar Institute. (**b**) Bathymetric map of Linnévatnet, showing the locations of water temperature profiles (indicated by white and black dots and labeled with capital letters). Water temperature from mooring G, located in the deepest part of the lake, was used in this analysis. The location of the automated cameras is shown in the green diamond near the southeast corner of the lake.

2.2. Data

2.2.1. In-Situ Observations

Linnévatnet has been studied for 18 years as part of a United States National Science Foundation (NSF) Research Experiences for Undergraduates (REU) program and the University Centre in Svalbard (UNIS). Observations collected within or nearby the lake include lake water temperature, meteorology, sediment influx, and automated photographs of the lake [53]. The observations of most interest for this study are lake water temperature measurements and automated photographs.

Beginning in 2003, lake water temperature profiles have been collected at 5 locations within Linnévatnet (C, D, E, F, and G in Figure 1). An additional temperature profile (H) was added in the center of the lake in 2009. Each profile consists of a string of Onset HOBO Water Temperature Pro sensors that extends from just below the water surface (typically 2–3 m depth) to the lakebed (up to 35 m). A float allows the top of the profile to migrate slightly as the lake water and ice moves, while a rock anchor keeps the base of each profile fixed on the lakebed. Each profile consists of 5–8 sensors at different depths, where the depths have varied slightly throughout the different years at the different profile locations. Each summer, the sensors are retrieved from the lake and the data are downloaded. Then, the sensors are replaced and redeployed until they can be recovered the following summer. For this analysis, we focus only on the deepest lake temperature profile (G). Profile G is located in the center of the lake, displays strong stratification with depth during the winter, and is also farthest from the lake inlet (and thus, farthest removed from any impact from the influx of spring meltwater). Thus, it is likely the most representative of lake-wide ice cover conditions.

Automated photographs of the lake surface have been collected since 2006 using a variety of digital SLR cameras and electronic timelapse controllers. Two different views of the lake have been collected from a location near the southeast corner of the lake (see Figure 2a). The "North" camera collects images facing northwest, along the long axis of the lake, while the "Plume" camera focuses on the southeast portion of the lake near the inlet stream (Figure 2b). Images have been taken at least twice per day at 10 a.m. and 4 p.m. local time (to avoid direct exposure and to optimize daily visibility) since 2006 for the Plume camera and 2009 for the North camera. In some years, images have been lost for part or all of the winter due to camera battery and timestamp issues, avalanches, or tampering from local wildlife (e.g., 2014, 2018, and 2019). The lake surface is clearly visible in the late spring through early fall (i.e., when sufficient sunlight is present) in these camera images, except for periods of inclement weather. This allows for the timing of lake ice formation and breakup to be directly observed. In some years, lake ice formation did not occur until late November, when daytime light levels were very low. In these cases, the lake ice formation was more difficult to identify, and thus there is less confidence in these estimates. In very low light conditions, snowfall events that accumulate on the ice surface are the only way to identify ice presence, because the white snow is much more distinguishable from open water than fresh ice.

(a)

(b)

Figure 2. (**a**) Example image from the "Plume" camera from 25 June 2016 at 11:30 local time, showing the southeast portion of the lake during ice break-up. Patchy lake ice (which appears in blue-green color in the image) is present on the lake surface, surrounded by a moat of open water along the shoreline. (**b**) Same as (**a**), but for the "North" camera, that looks northwest along the long axis of the lake. In this particular year, BUE (the first date when ice was no longer visible on the lake surface) was observed on June 30.

2.2.2. Satellite Observations

Observations from two satellite instruments were compared to the in-situ lake observations, in order to determine how well lake ice timing can be detected with widely used satellite technology. These include surface reflectance from the Moderate Resolution Imaging Spectroradiometer (MODIS) and microwave backscatter from Sentinel-1. The datasets for both satellites were accessed and processed using Google Earth Engine (GEE) [54].

MODIS is a passive instrument that measures upwelling visible and infrared radiation that has been reflected or emitted from the Earth. It is flown on both the National Aeronautics and Space Administration (NASA) Terra and Aqua satellites. The MODIS products used in this analysis were MOD09GA version 6 ("MOD09GA.006 Terra Surface Reflectance Daily Global 1 km and 500 m") and MYD09GA version 6 ("MOD09GA.006 Aqua Surface Reflectance Daily Global 1 km and 500 m"), which provide spectral bands 1–7 at daily temporal and 500 m spatial resolution [55,56]. We accessed the data from 1 January 2004 through 31 December 2019 via Google Earth Engine. After initial investigation, only band 3 (459–497 nm; blue light) was used in this analysis due to its high sensitivity to the presence of snow and ice vs. open water, and the redundancy of information in bands 1, 2, and 4.

Sentinel-1 is a C-band (5.405 GHz) synthetic aperture radar (SAR) instrument that emits microwave radiation and observes the properties of the radiation that returns to the instrument after interacting with the Earth. Sentinel-1 is flown on two European Space Agency (ESA) satellite platforms: Sentinel-1A and Sentinel-1B. The Sentinel-1 SAR GRD product on Google Earth Engine ("Sentinel-1 SAR GRD: C-band Synthetic Aperture Radar Ground Range Detected, log scaling") was used [57], which provides ground range detected (GRD) scenes that have been calibrated and ortho-corrected. The polarizations that were most frequently available at this latitude are dual-band horizontal transmit/horizontal receive (HH) and horizontal transmit/vertical receive (HV) at approximately 10 m spatial resolution, so these were the data selected for this analysis. We accessed the data from 1 January 2015 to 31 December 2019 via Google Earth Engine.

2.3. Methods

2.3.1. Lake Ice Timing from In-Situ Observations

Two key dates in the annual lake ice cycle were analyzed in this study: freeze-up end (FUE) (i.e., the date when the lake surface freezes and becomes entirely covered by ice for the rest of the winter) and break-up end (BUE) (i.e., the date when the last ice melts and the lake surface becomes entirely open water for the rest of the summer). For Linnévatnet, BUE took place in the summer (July), while FUE took place in the fall (October–November). These dates were identified in each year from the in-situ data, using two different methodologies. The first method consisted of manually examining the automated photographs to observe when the BUE and FUE occurred. This was carried out separately for each of the "Plume" and the "North" cameras. If the lake surface was not clearly visible in the camera images, the first date where ice was no longer observed was selected as the BUE, and the first date where the lake was found to be completely covered in ice was selected as the FUE.

The second method involved both manual and automated analysis of the lake water temperature profile time series. Because the lake water temperature was measured at multiple depths, we can observe the change in temperature with depth over time (see Figure 3). From these data, we observe periods of stratification (where the temperature at different depths diverges) and mixing (where the temperature is approximately the same at all depths). Thus, it is possible to infer when the lake water surface was free of ice, and thus could be mixed by wind action and water density changes. Conversely, when the lake is ice-covered, it is poorly mixed and cold, leading to stratification, with colder temperatures near the icy surface and warmer temperatures at greater depths. Winter water temperatures were always below 4 °C, which is the temperature at which water is most dense, so the warmer water at depth was denser than the colder water near the surface, favoring stability. These processes resulted in a general seasonal pattern marked by warmer, well-mixed water during periods of open water (i.e., summer) and colder, stratified water temperature during periods of ice cover (i.e., winter). Immediately after FUE, the water temperature at different depths diverged as the water stratified, while convergence was observed at BUE. We manually (i.e., visually) identify the transitions as indicators of ice cover appearance and disappearance in all years, using the time series of water temperature at different depths (hereafter the "WTemp Manual" method. These manual observations were aided by calculating a time series of the standard deviation of the water temperature across all depths, because areas of strong mixing resulted in a low standard deviation, while stratification resulted in a high standard deviation.

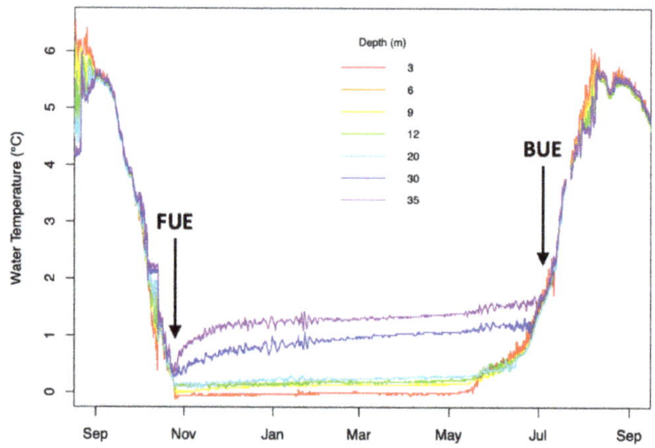

Figure 3. Example time series of lake water temperature measured at different depths from September 2009 to September 2010. During the summer period of open water (typically July–October/November), the lake is typically well-mixed, and the water temperature does not vary much with depth. However, once ice appears on the lake, it stratifies until the ice disappears, with colder temperatures near the surface and warmer temperatures near the bottom. The appearance (FUE) and disappearance (BUE) of ice is indicated with arrows.

We also automate the detection of FUE using the mean of the water temperature across all depths. During freeze-up, the mean water temperature (across all depths) reaches its annual minimum and then the mean water temperature increases significantly as the water stratifies, with water at 0 °C at the surface and slightly warmer with depth. Thus, an FUE estimate could easily be identified using this method (hereafter called the "WTemp AnnMin" method).

2.3.2. Lake Ice Phenology from MODIS

For simplicity in comparison to in-situ lake ice timing data, the MODIS surface reflectance from band 3 was averaged over the entire lake surface for each satellite overpass. This was accomplished by drawing a polygon in Google Earth Engine that matched the boundary of the lake, and then averaging the MODIS reflectance data within the polygon. Thus, each overpass was treated as a single value in order to create a time series of mean lake surface reflectance (Figure 4). The lake-wide mean MODIS surface reflectance time series also showed a seasonal pattern, with generally higher reflectance in the winter (when the lake surface was covered with snow and ice) and lower reflectance in the summer (when the lake surface consisted of less reflective open water). This difference allowed for identification of BUE as the end of the transition period from the high winter reflectance to lower summer reflectance.

Unfortunately, the MODIS reflectance product was not available after late September each year, presumably due to low levels of reflected light as the Arctic transitioned toward polar night. Freeze-up of Linnévatnet did not occur until October or later, so it was not possible to observe FUE using MODIS.

The frequencies of electromagnetic radiation observed by MODIS do not penetrate through clouds. Cloud cover added considerable noise to this seasonal reflectance pattern because clouds were present throughout the year and their reflectance was usually in between the low reflectance of open water and the high reflectance of snow and ice. Unfortunately, we found that the MODIS internal cloud algorithm flag was not completely effective at filtering out cloudy days, probably due to the difficulty of distinguishing clouds from lake ice and snow, as well as partial cloud cover. So, we manually filtered out cloudy MODIS observations by viewing the MODIS data on Google Earth Engine Explorer. When the outline of the lake (either ice-free or ice-covered) was visible, the data were identified as cloudless, while images where the lake was obscured were discarded. This resulted in a time series of cloud-free MODIS surface reflectance observations, but with many gaps due to missing data (~75–80%).

In order to provide the most accurate BUE and FUE dates possible, it was necessary to fill the gaps in the time series. We tested multiple gap-filling methods, including linear interpolation, logistic curves, kernel smoothers, and splines. Ultimately, we found that a cubic smoothing spline ("smooth.spline" in R) was effective at retaining the features of the MODIS time series without overfitting the data. Smoothing parameters were calculated using ordinary leave-one-out cross-validation.

We tested three automated methods (described below) to identify BUE from the smoothing spline fit to cloud-free, lake-wide mean surface reflectance (i.e., a smoothed, gap-filled representation of the time series). These included (1) identifying the first date at the bottom of the downward trend in the smoothing spline, where the first derivative was equal to zero (i.e., first location where the spline leveled out in the early summer; hereafter the "MODIS dSR" method); and (2) the date nearest to the bottom of the spline trend where the second derivative was equal to zero (i.e., point of maximum curvature; hereafter the "MODIS d^2SR" method). Additionally, we found (3) the earliest date in each year where the smoothing spline crossed below a reflectance threshold, calculated as the third quartile (i.e., 75th percentile) of the clear sky MODIS surface reflectance values during the summer (20 July–31 August) of the given year (hereafter, the "MODIS Summer Q_3" method). The mid-July through August period was chosen as it was a consistent annual period of open water on the lake, before some high surface reflectance values in the fall (presumably due to sun angle dynamics).

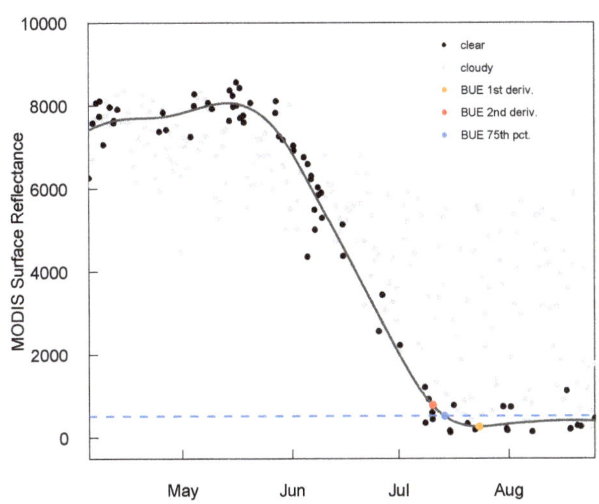

Figure 4. Example time series of lake-wide mean MODIS surface reflectance from year 2012. Clear observations are shown in black points, while cloudy observations are shown in gray points. A smoothing spline fit to the clear observations is shown in the solid line. BUE estimates from the first and second derivatives of the smoothing spline are shown in orange and red points, respectively. The blue dashed line shows the 75th percentile of summer (20 July–31 August) surface reflectance for clear observations, and the blue point shows estimated BUE based on when the spline curve first intersects this threshold.

2.3.3. Lake Ice Phenology from Sentinel-1

Similar to the processing of the MODIS data, Google Earth Engine was used to create a lake-wide average time series of Sentinel-1 backscatter, where the HH and HV polarizations (and incidence angle) were averaged separately over the entire lake surface in order to create a backscatter time series for each polarization.

For the Sentinel-1 backscatter, an angular dependence was observed. Higher backscatter values were observed at lower incidence angle (i.e., the angle between the radiation emitted from the instrument and the ground surface), which introduced variability into the time series that made it more difficult to interpret. So, a simple regression-based normalization was applied to the Sentinel-1 backscatter data in order to reduce angle-based noise in the data. First, we smoothed the backscatter times series using locally estimated scatterplot smoothing (LOESS), a local regression smoothing method, in order to approximate the seasonal evolution of lake backscatter properties. Then, the moving average was subtracted from the original backscatter times series to obtain a time series of residuals (which were partly a function of incidence angle). The incidence angle was regressed on the backscatter residuals using simple linear regression to obtain a statistical relationship between residual and incidence angle (R^2 = 0.63 for HH and 0.54 for HV). Then, this linear relationship was used to predict an expected backscatter residual time series based on the known incidence angles. The regression-predicted backscatter residual was subtracted from the observed backscatter residuals in order to remove the expected effect of incidence angle on backscatter. Finally, this difference was added to the moving average backscatter to obtain an incidence angle-corrected backscatter time series. The method was applied separately to the HH and HV polarizations. This correction method is imperfect, as a higher influence of backscatter was observed during periods of open water as compared to ice cover, but it did help to reduce noise in the time series and achieved the main purpose of increasing the interpretability of the backscatter time series. During the ice-covered time periods, the normalization of backscatter over incidence angle is appropriate for lake ice as the dominant scattering mechanism observed in shallow lake ice is single-bounce interactions occurring at the ice–water interface [58].

The incidence angle-corrected, lake-wide mean Sentinel-1 backscatter showed a seasonal pattern (Figure 5), but it was different from that of MODIS. Generally, the backscatter increased throughout the winter, followed by a sharp drop in the spring. This drop was followed by a gradual increase and then another sharp drop in the early summer. During the summer, the backscatter magnitude remained generally lower than the rest of the year (but also with high variability). Finally, the backscatter increased, sometimes gradually and sometimes abruptly, in the fall into winter. Based on previous studies (e.g., [36,59]), we interpret the first sharp drop in backscatter in the spring to indicate the first melting of the snow and lake ice (i.e., break-up start—BUS) which leads to absorption of the incident microwave signal, while the second drop (which was followed by low backscatter throughout the summer) indicates the disappearance of ice from the lake (BUE), resulting in a smooth water surface that reflects incident microwaves away from the sensor. Due to the difficulty in identifying the individual dates of these features (because of noise in the time series), we identified the BUE date manually from both HH and HV polarizations as the last date of the second drop in backscatter (hereafter, the "S1 HH Manual" and "S1 HV Manual" methods). FUE dates were identifiable by a decrease in the variability of the backscatter values over time in the HH polarization, likely due to more diffuse scattering from ice or snow on the lake surface, as compared with open water. However, this change in backscatter variability was not easily distinguishable in the HV polarization.

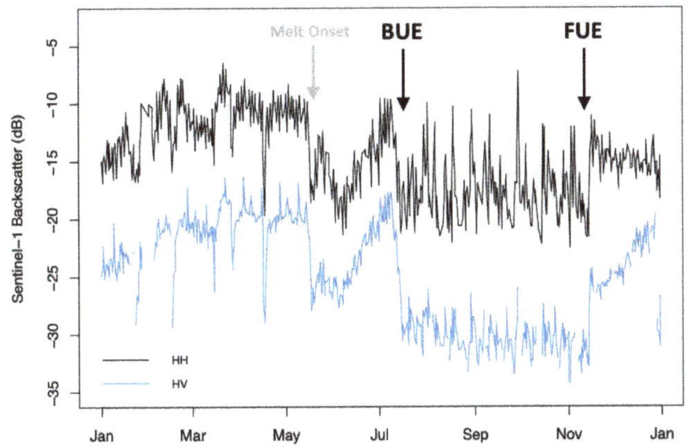

Figure 5. Example time series of lake-wide average Sentinel-1 backscatter from 2015. The black line shows the backscatter at HH polarization, while the blue line is for HV polarization. The sharp drop in backscatter in mid-May indicates the onset of melting of the snow and ice on the lake (melt onset), while the sharp drop in early July marks the disappearance of ice (BUE). In this particular year, a sharp increase in backscatter was observed in early November, which also marks a change in the variability of the backscatter time series. This is taken to be the date at which the lake froze over (FUE). However, in some years, this transition was gradual and only identifiable by the change in backscatter variability.

An automated algorithm to estimate freeze onset from Sentinel-1 was also used, which could be broadly applied in other regions to determine lake ice phenology. This method is illustrated with a flow diagram in Figure 6. The process is as follows: In Google Earth Engine, the Sentinel-1 A/B repository is called and filtered for the desired time periods (typically a one-year period (e.g., from 1 August 2017 to 30 July 2018) (Figure 6A). Filters are applied to ensure acquisition property consistency throughout the time series with respect to orbit path, imaging mode, and polarization. For the purpose of this study, the filters applied were that all images had to be acquired in Interferometric Wideswath (IW) mode using HH co-polarization. For Linnévatnet, Sentinel-1 acquisitions were acquired on the same orbit track, which did not necessitate the normalization of backscatter for multiple incidence angles.

Once the repository is queried, an image collection of all filtered images is produced with a 40 m spatial resolution. An image stack is then generated to allow for the tracking of backscatter over time (Figure 6C). Recent understanding of microwave interactions with freshwater ice has indicated the source of backscatter is surface roughness at the ice–water interface, allowing for the normalization of backscatter, following [60,61] (Figure 6D). A time series of backscatter is generated for each pixel, and a smoothing algorithm is applied. The smoothing algorithm extracts the mean value for pixels acquired within five days before or after the initial acquisition to capture variability of incidence angle, and orbit path without including repeat passes. The approach of averaging the most recent acquisitions is not used because some periods and regions in this study occasionally do not have dense enough coverage to include those acquisitions without incorporating repeat passes, which would limit the observation of change in backscatter due to thermodynamic shifts. A user-defined threshold is set to monitor all drops in the smoothed backscatter time series for each pixel, with the timing of any drops flagged and recorded (Figure 6E). For consistency from year to year, the date of the first backscatter drop greater than 5 dB—in this case for the smoothed time series—is extracted and is recorded as the timing of freeze- or melt-onset for that pixel within the lake boundary (Figure 6F). Below, we report the median freeze onset date for all 40 m pixels within the lake boundary, for comparison to other freeze-up estimates (hereafter, the "S1 Auto Median" method).

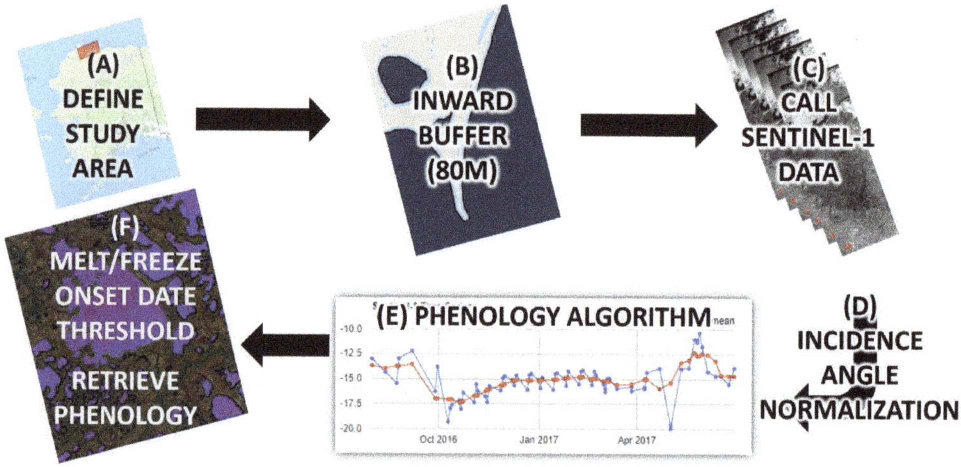

Figure 6. Flowchart of automated ice phenology retrieval using Sentinel-1 acquisitions within the Google Earth Engine coding environment. The steps within the processing chain are as follows: (**A**) a study area is defined using either a user-defined polygon or imported lake boundary. (**B**) An inward buffer of the polygon is generated at a distance of 80 m to avoid mixed pixel effects of backscatter from the lake edge. (**C**) Sentinel-1 acquisitions intersecting the boundary are called from the Sentinel-1 GRD GEE repository, where (**D**) an incidence angle normalization is applied as described by [60,61], and (**E**) a smooth algorithm is applied to the winter-time series of Sentinel-1 backscatter on a per pixel basis. (**F**) Once a drop in the smoothed backscatter is detected that exceeds a user-defined threshold, the date of the drop is recorded and flagged as freeze-onset, or melt-onset.

2.3.4. Agreement between Lake Ice Phenology Estimates

To assess the viability of the satellite-derived BUE estimates, we calculated agreement statistics between the satellite-derived BUE estimates and the in-situ BUE estimates. These statistics included bias, mean absolute error (MAE), and root mean squared error (RMSE). Agreement statistics were computed for each different combination of satellite and in-situ methods, treating the BUE values from each individual year as a data pair. We followed the same procedure to calculate agreement statistics for FUE. Additionally, we calculated agreement statistics between the two in-situ methods (camera images and water temperature) for each of BUE and FUE.

3. Results

3.1. Lake Water Temperature Dynamics

Considerable annual variability was observed in both mean lake water temperature throughout the summer and in the degree of water temperature stratification during the winter. However, a general increasing trend in lake temperature is evident in Figure 7, with warmer years clustered in the later portion of the time series. The mean water temperature from June–November (DOY 152–334) showed a significant increasing trend of 0.06 °C per year, according to a Theil–Sen line fit to the data (Mann–Kendall (M-K) trend test: $p = 0.008$). Additionally, the lake stayed warm for a longer period each year. The latest day with a mean temperature above 4 °C was found for each year, and significant increasing trend of 1 day per year was observed (Theil–Sen line; M-K test: $p = 0.012$). Similar significant trends were found for 3 °C and 2 °C. These findings are consistent with extensive literature that documents rapid Arctic warming in the past two decades [26,62].

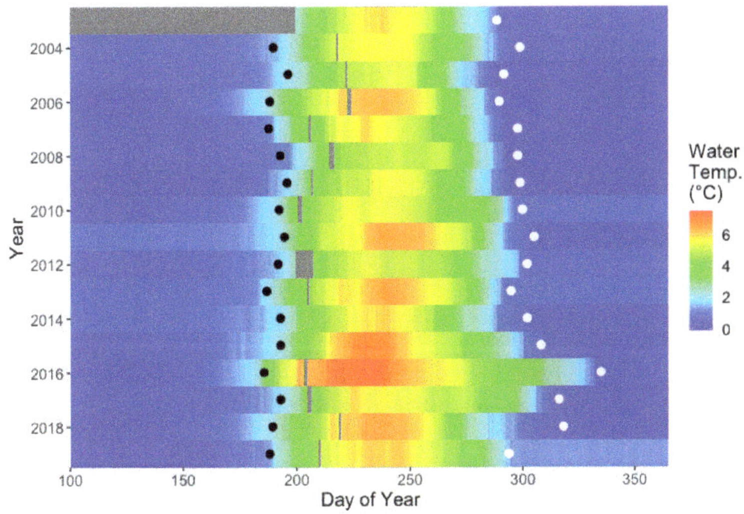

Figure 7. Daily mean water column temperature for mooring G in Linnévatnet as a function of year (y-axis) and day-of-year (x-axis). Blue indicates colder temperatures, while red indicates warmer temperatures. Gray indicates missing data. Estimates of BUE and FUE dates are shown for each year (black and white dots, respectively), based on lake water temperature dynamics.

3.2. *In-Situ Observations of Lake Ice Timing*

BUE and FUE dates from the automated photographs of the lake surface, and inferred from lake water temperature, showed that BUE did not change significantly from 2004 to 2019 (M-K trend test on water temperature BUE estimates: $p = 0.30$) (Figure 8). While Svalbard lies far off the northern European coast, this is consistent with findings of little change in BUE in Northern Europe [23]. In contrast, the FUE date was observed to occur much later in recent years as compared with the mid-2000s, and there is much more year-to-year variability. The mean FUE from 2003–2007 was day-of-year (DOY) 294, while the mean FUE from 2015–2019 was DOY 314, based on lake water temperature FUE estimates. This corresponded to an increase of 1.5 days per year over the period 2003–2019, according to a Theil–Sen line fit to the data (M-K test: $p = 0.002$). These findings are consistent with a rise in winter air temperatures in Svalbard [52]. Mainly due to the later freeze-up, there was an increase in the annual period of open water of approximately 1.4 days per year from 2003 to 2019 (Theil–Sen line; M-K test: $p = 0.01$). While the 16-year period of this study is less than the commonly accepted 30-year period to define climate normals, our findings are consistent with trends in the lake water temperature and with extensive observations of rapid Arctic warming [26,62].

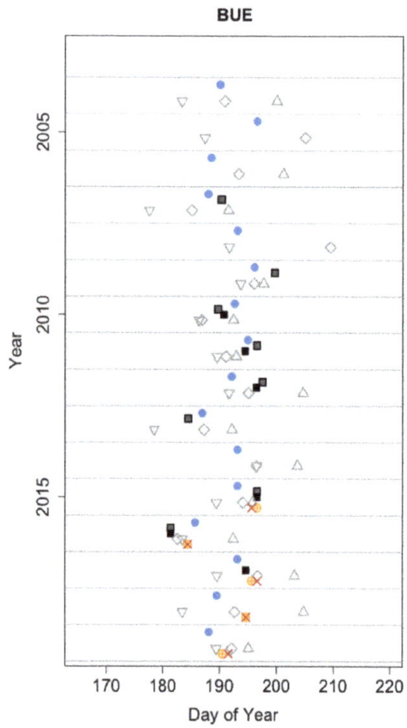

(a)

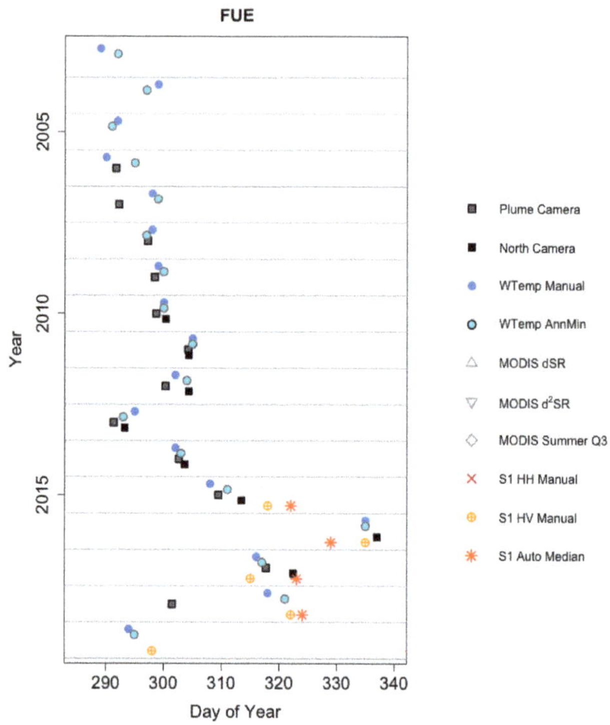

(b)

Figure 8. (**a**) Timing of BUE (x-axis) from the various in-situ and remote sensing methods in this analysis for each year in the study period (y-axis). (**b**) Same as (**a**), but for FUE. The dark, square markers indicate dates estimated from camera images, blue circular markers are estimates from water temperature dynamics, light gray markers are estimates from MODIS, and orange and red markers are estimates from Sentinel-1. The symbols are slightly offset on the vertical axis to reduce overlap. A trend toward later FUE is visible in (**b**), while no BUE trend is observed in (**a**).

The agreement in ice timing between the two in-situ methods (photographs and lake water temperature profiles) is comparable for FUE dates and BUE dates, but depends on which view of the lake is considered. The RMSE for FUE, identified manually from water temperature dynamics ("WTemp Manual" method), is 2.6 and 5.3 days for the North and Plume cameras, respectively, while the RMSE for BUE dates is 3.0 and 3.5 days, respectively (see Tables 1 and 2). FUE estimates from the annual minimum in water column temperature ("WTemp AnnMin" method) resulted in a RMSE of 2.3 days with the North camera, but 6.2 days with the Plume camera. This difference can likely be attributed to the fact that the North camera view includes a much broader view of the lake surface and also contains mooring G (which is the source of the water temperatures in this analysis), while the Plume camera observes a very small portion of the lake that also does not contain mooring G. It is somewhat surprising that BUE can be identified with comparable error to FUE, mainly because it can be difficult to identify the BUE from the lake water temperature profiles due to gradual mixing of the lake in the spring and early summer, while it is straightforward to observe ice disappearance from the photographs. For FUE, both methods have highly identifiable transitions, and thus good agreement, except in years when FUE occurs late in the fall. In those cases, light levels are low, and it can be hard to distinguish open water from ice cover in the photographs, so we sometimes relied on snow accumulation on the lake surface to indicate the presence of ice. This may have introduced error into FUE camera estimates. The FUE transition is very identifiable from the lake water temperatures because it is marked by the beginning of winter-long stratification of water temperature within the lake. This is evident in the good agreement between the manual FUE estimates ("WTemp Manual"), which used the stratification dynamics to identify FUE, and the FUE estimates from the annual minimum in water column temperature ("WTemp AnnMin"; MAE = 1.6 days, RMSE = 2.0 days; see Table 4).

Table 1. Agreement between BUE dates from the camera images and other BUE estimation methods.

Data	Method [1]	North Camera (n = 6)			Plume Camera (n = 8)		
		Bias (Days)	**MAE (Days)**	**RMSE (Days)**	**Bias (Days)**	**MAE (Days)**	**RMSE (Days)**
In-situ water temp.	Manual	−0.5	2.64	3.0	−0.9	3.3	3.5
MODIS	dSR	4.5	5.3	6.7	2.9	4.5	5.6
	d^2SR	−3.9	4.6	4.9	−5.7	6.3	6.9
	Summer Q_3	−1.3	2.4	2.5	−2.3	3.2	3.5
Sentinel-1 [2]	HH Manual	1.4	2.0	2.2	1.0	2.0	2.2
	HV Manual	1.4	1.4	1.8	1.5	1.5	2.1

[1] The various BUE identification methods are described in Sections 2.3.1–2.3.3. [2] The statistics for Sentinel-1 are based on very few data pairs (n = 3 for North camera, n = 2 for Plume camera) due to loss of camera images in 2017–2019 (see Figure 8a).

Table 2. Agreement between FUE dates from the camera images and other FUE estimation methods.

Data	Method [1]	North Camera (n = 8)			Plume Camera (n = 12)		
		Bias (Days)	**MAE (Days)**	**RMSE (Days)**	**Bias (Days)**	**MAE (Days)**	**RMSE (Days)**
In-situ water temp.	Manual	−2.0	2.5	3.3	2.1	3.1	5.3
	AnnMin	−1.3	1.5	2.3	3.3	3.5	6.2
Sentinel-1 [2]	HV Manual	−1.6	4.7	5.2	8.8	10.6	12.9
	Auto Median	0.4	5.7	6.7	13.5	13.5	15.2

[1] The various FUE identification methods are described in Sections 2.3.1–2.3.3. [2] The statistics for Sentinel-1 are based on very few data pairs (n = 3 for each camera) due to loss of camera images in 2016, 2018, and 2019 (see Figure 8b).

3.3. Satellite vs. In-Situ Lake Ice Phenology

The agreement between the different satellite methods and the in-situ methods varies by sensor, method, and transition (BUE vs. FUE). Agreement statistics are provided in Tables 1–4.

For BUE, Sentinel-1 outperformed MODIS in terms of agreement with the in-situ observations, with MAE of 1.4–3.1 days and RMSE of 1.8–3.4 days among the BUE estimation method and in-situ combinations (i.e., cameras and water temperature). For MODIS, BUE agreement ranged from MAE of 2.5–8.7 days and RMSE ranged from 2.5 to 11.4 days among the BUE estimation methods and cameras. However, the Sentinel-1 statistics are based on a very small sample size (n = 5 for comparison to water temperature BUE dates, and n = 3 for North camera and n = 2 for Plume camera due to loss of camera images in 2017–2019). Of the MODIS methods, the "Summer Q3" method (which determined BUE based on when the smoothing spline fit to the surface reflectance data crossed below a threshold equal to the 75th percentile of open-water surface reflectance) showed the best agreement, with MAE of 2.4–4.0 days and RMSE of 2.5–5.5 days among the different in-situ comparisons.

The agreement between the Sentinel-1 and the in-situ FUE estimates was worse than for BUE, with MAE of 3.8–13.6 days and RMSE of 5.2–23.3 days (Tables 2 and 4). However, some of this disagreement can be contributed to the difficulty in identifying the FUE transition in the camera images in recent years due to later freeze-up, as well as small sample size for the comparison (n = 3 for both cameras due to loss of images in 2016, 2018, and 2019). The agreement with manual FUE estimates from water temperature dynamics, which did not suffer from those limitations, was better than with either camera (MAE = 3.8 days; RMSE = 5.2 days for manual HV FUE estimates). The larger disagreement between the Sentinel-1 "Auto Median" method and the water temperature FUE estimates is largely driven by a large difference in 2019. Without this, the MAE and RMSE decrease to 8.3 and 8.9 days, respectively. FUE could not be calculated from MODIS surface reflectance due to the low light levels during fall freeze-up.

Table 3. Agreement between manual BUE dates from water temperature dynamics (the "WTemp Manual" method) and other BUE estimation methods (n = 16).

Data	Method [1]	Bias (Days)	MAE (Days)	RMSE (Days)
MODIS	dSR [2]	8.4	8.7	11.4
	d^2SR [3]	−4.0	4.7	5.5
	Summer Q_3	2.0	4.0	5.5
Sentinel-1 [4]	HH Manual	2.6	3.1	3.4
	HV Manual	2.4	2.9	3.2

[1] The various BUE identification methods are described in Sections 2.3.1–2.3.3. [2] For the year 2005, the smoothing spline fit to the MODIS surface reflectance did not reach a slope of zero as it declined through the melt period, so a BUE date could not be calculated. Thus, n = 15 for this method. [3] For the year 2006, the second derivative of the smoothing spline was not calculated. Thus, n = 15 for this method. [4] The statistics for Sentinel-1 are based on n = 5 data pairs (see Figure 8a).

Table 4. Agreement between manual FUE dates from water temperature dynamics (the "WTemp Manual" method) and other FUE estimation methods (n = 16).

Data	Method [1]	Bias (Days)	MAE (Days)	RMSE (Days)
In-situ water temp.	AnnMin	0.9	1.6	2.0
Sentinel-1 [2]	HV Manual	3.4	3.8	5.2
	Auto Median	14.0	16.4	23.3

[1] The various FUE identification methods are described in Sections 2.3.1–2.3.3. [2] The statistics for Sentinel-1 are based on n = 5 data pairs (see Figure 8b).

4. Discussion

4.1. Considerations for Determining Lake Ice Timing

In this analysis, we identified a single date each year where the lake became completely ice-free in the spring (BUE) and a single date where the lake became completely ice-covered in the fall (FUE), remaining ice-covered for the rest of the winter. However, it is important to note that the break-up of ice in the spring and freeze-up in the fall are not instantaneous processes. A recent synthesis article [2] noted this limitation and highlighted that moving from binary to continuous monitoring of lake ice is a path to advance lake ice research through interdisciplinary collaboration. The process of spring melting of lake ice (and the snow that may cover it) can take over a month. This is evident in both the Sentinel-1 time series, in which a date of first spring melting can often be identified (see Figure 5), as well as the MODIS surface reflectance time series, which shows a decrease in surface reflectance that lasts a month or more (Figure 4). Additionally, the period over which ice disappears from the lake surface to reveal open water can last over a week, with fragmented ice cover being blown back and forth over the lake surface by the wind. This gradual ice breakup period averaged 7–8 days for the years with automatic camera images of the lake. For this analysis, we have chosen the most identifiable events in the lake ice break-up and freeze-up processes (BUE and FUE, respectively), but it is important to consider that distilling gradual processes into individual points in time (or binary: ice vs. no ice) may introduce some amount of uncertainty. This uncertainty may be reflected in the level of agreement between the satellite and in-situ ice timing estimates.

4.2. Effect of Clouds on MODIS BUE Dates

In order to interpret the level of agreement between the BUE dates from MODIS and the BUE dates from in-situ observations, we conducted an additional simulation. The goal of the simulation was to evaluate the effect of data loss due to cloud cover on MODIS BUE estimates, because the lake surface was not visible on cloudy days and the MODIS data on those days were discarded.

First, all MODIS surface reflectance observations (both MOD09GA and MYD09GA) of the lake on cloudless days for the entire 16-year time series were plotted as a function of decimal day of year (April–August), with the Terra and Aqua observations plotted according to their overpass times (10:30 and 13:30 local time, respectively). Then, an idealized MODIS seasonal pattern was generated by fitting a 10th degree polynomial function to the MODIS observations from winter until the annual mean BUE date (7 July), and all days after 7 July were assigned a constant value equal to the mean reflectance for the remainder of the summer. The result was an idealized time series of alternating Terra and Aqua MODIS observations (10:30 and 13:30 local time, respectively) for Linnévatnet for each day of the year from 1 April–31 August, as shown in Figure 9a.

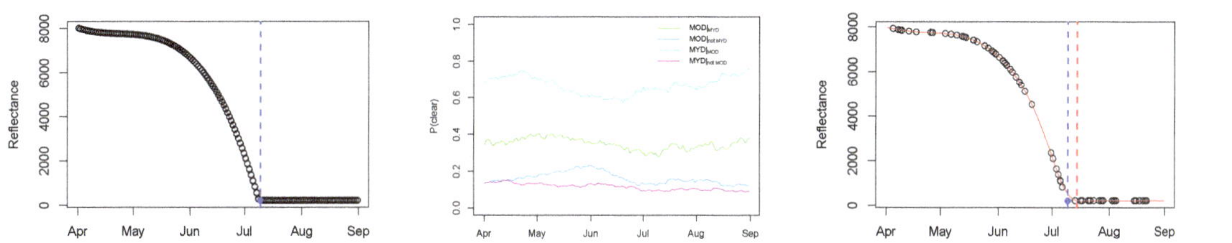

(**a**) (**b**) (**c**)

Figure 9. (**a**) Idealized seasonal time series of MODIS surface reflectance for Linnévatnet, based on the data from 2004 to 2019. The blue dot shows the surface reflectance at the BUE date, while the blue dashed line indicated the BUE date; (**b**) Markov transition matrix, as a function of day of year. MOD stands for the MODIS instrument on the Terra satellite, and MYD stands for the Aqua satellite. The four lines show the transition probabilities between clear and cloudy conditions (i.e., probability of a clear observation) for the two MODIS overpasses as a function of day of year, calculated using a 31-day window around the day-of-year across all years. For instance, $MOD|_{MYD}$ means the probability that the Terra observation (10:30 local time) was clear, given that the previous Aqua observation (13:30 local time on the previous day) was also clear. Similarly, $MOD|_{not\ MYD}$ is the probability that the Terra observation was clear, given that the previous Aqua observation was cloudy. (**c**) Same as (**a**), but with observations missing based on one realization of a Markov chain model of clear and cloudy conditions, where cloudy observations are removed. The red solid line is a smoothing spline fit to the remaining data, and the red dashed line is the inferred BUE date from the smoothing spline. The difference between the blue and red dashed lines represents the amount of error in the MODIS BUE estimate introduced by data loss due to cloudy conditions.

Next, the cloud cover for the lake was modeled as a Markov process. For each day of the year (in a 31-day moving window across all years), the probability of observing a clear (i.e., not cloudy) day was calculated, dependent on (1) whether the previous observation was cloudy or clear, and (2) whether the previous observation was from the Terra or Aqua MODIS instrument. This second condition was deemed necessary because the Aqua instrument observes the lake only 3 h after the Terra instrument, but there are 21 h between Aqua and Terra observations. The average probability of a clear day from April–August was 0.21 for Terra and 0.23 for Aqua, reinforcing the idea that clouds are very common in this Arctic maritime location. The probability of observing a clear day was also highly dependent on whether the previous observation was cloudy or clear (see Figure 9b), indicating a high degree of persistence in cloud conditions.

This Markov transition matrix was then used to randomly generate time series of cloudy vs. clear observations, which was applied to the idealized MODIS time series. For each cloudy observation, the value from the idealized times series was replaced with a missing value, resulting in a time series where the data were randomly knocked out due to simulated clouds (see Figure 9c). This process was repeated 1000 times, generating 1000 simulated times series of cloud-impacted MODIS observations. Then, a smoothing spline was independently fit to each simulated time series, using the same method that was used for the MODIS observations (Section 2.3.2), and BUE dates were calculated according to the MODIS "dSR" and "Summer Q_3" methods (Section 2.3.2). These BUE dates were then averaged across all 1000 simulations of data loss due to cloud cover. Finally, the simulated BUE dates were compared to the "known" BUE date from the simulated time series (July 7 at 16:00 local time), in order to evaluate the uncertainty introduced into the MODIS BUE identification method due to cloud cover.

The RMSE between the "known" BUE date and the mean BUE date from the simulations was 3.0 days for the "Summer Q_3" method, and 4.4 days for the "dSR" method. The MAE for the two methods were 1.8 and 3.3 days, respectively. These simulation results show that the error in estimated BUE dates due to data loss from cloud cover is of similar magnitude to the error between observed BUE dates from the in-situ data and estimated BUE dates from MODIS (Tables 1 and 3). Thus, it is reasonable to assume that a large majority of the error in the MODIS-derived BUE dates is due to loss of data due to cloud cover, rather than flaws in the methodology or MODIS observations. This lends confidence to the ability of MODIS to accurately observe lake ice conditions in clear conditions. The accuracy of the BUE dates from MODIS will thus be primarily dependent on the frequency of cloud cover at a given location (where fewer clouds will result in more accurate BUE dates), as well as the accuracy of cloud masking over ice and snow surfaces.

4.3. *Lake Ice Phenology from MODIS vs. Sentinel-1*

The results of this analysis showed that both MODIS and Sentinel-1 were reasonably accurate in determining the date of ice disappearance (i.e., BUE date) for a large Arctic lake. These findings lend confidence to remote sensing of lake ice phenology, consistent with previous studies that used ground-based observations for validation [12,19,23,39,40]. Sentinel-1 was slightly more accurate in determining BUE (Table 1). This difference can be attributed to the fact that electromagnetic radiation penetrates through clouds at the microwave frequency of Sentinel-1, but does not penetrates through clouds at the optical frequencies of MODIS. Thus, Sentinel-1 did not suffer loss of data due to cloud cover, while nearly 80% of MODIS observations were discarded due to clouds. A large portion of the error in Sentinel-1 BUE estimates can likely be attributed to differing scattering properties at different view angles. A simple method was used in this manuscript to reduce these errors, but a more sophisticated method may improve the accuracy of BUE date identification from Sentinel-1.

FUE was more difficult to identify than BUE using satellite observations. MODIS surface reflectance observations were not available during FUE due to low light levels during the polar night. Based on in-situ observations, most FUE dates occurred in late October, with no FUE dates before October 16th, while MODIS reflectance decreased to extremely low values after approximately October 10th each year and thus were not interpretable. Sentinel-1 observations were available year-round, but the fall freeze-up was less identifiable in the backscatter time series than for spring break-up. This led to a large amount of uncertainty in FUE dates (RMSE of at least 5.2 days). This uncertainty may partially be due to the gradual nature of freeze-up, punctuated freezing interrupted by melting (and thus mixed, snow, ice, and liquid water), or deformation features [35,36]. The greater uncertainty in satellite FUE estimates is unfortunate because the fall freeze-up is when the most long-term change is observed for Linnévatnet. The Arctic is a remote region, so satellite observations are important for tracking indicators of climate change. Numerous ground-based studies have documented later freeze-up and earlier break-up of lake ice across northern latitudes, and remote sensing has corroborated trends in earlier break-up (e.g., [20,23]). However, the slightly larger uncertainty in Sentinel-1 FUE estimates, combined with the relatively short Sentinel-1 record, may make it difficult to confidently identify trends in FUE. Additionally, trends in FUE for some high Arctic lakes may have gone unobserved by optical satellites over the past two decades due to low light levels in the fall [23].

Despite the complication of low light levels in the fall at high latitudes, the identification of ice phenology from both the Sentinel-1 and MODIS benefitted from the high latitude of the study site combined with the polar orbits of the satellites, as well as the fact that each instrument is flown on two satellite platforms (Sentinel-1A and Sentinel-1B, and Terra and Aqua, respectively). These factors resulted in high frequency observations of the study area (1 day or less for Sentinel-1 and at least twice daily for MODIS). However, the frequency of overpasses for both Sentinel-1 and MODIS will decrease at lower latitudes. This effect is most pronounced for Sentinel-1 because MODIS has a much wider observation swath. While Sentinel-1 outperformed MODIS for Linnévatnet, it is possible (or even likely) that MODIS will be more useful for determining ice timing at lower latitudes, where Sentinel-1 overpass frequency decreases significantly (every 6 days over much of the continents outside of Europe) and the amount of sunlight received in the fall increases (making identification of FUE possible). Similarly, Sentinel-1 overpasses are less frequent outside of the European Arctic, with revisit frequency decreasing to 1–4 days. Additionally, Sentinel-1 operates at different polarizations over the Arctic (HH and HV) than over much of the rest of the Earth's surface (VV and VH), which could affect the detectability of the features in the backscatter time series that were used infer lake ice phenology in this analysis. Lastly, MODIS benefits from a longer history of observations as compared to Sentinel-1 (year 2000 to present for MODIS vs. 2014 to present for Sentinel-1), so MODIS is most useful for evaluating longer-term trends in climate indicators.

In summary, this case study suggests that optical and SAR satellite data can be confidently used to monitor lake ice changes in the High Arctic, and at regional scales across the Arctic, with some sensor-specific considerations. However, lake-specific conditions (e.g., size, depth, orientation, and hydrology) may influence individual lake melt-out and freeze-up processes and the ability to detect ice phenology from satellite imagery.

Author Contributions: Conceptualization, S.E.T., S.R.R. and A.W.; methodology, S.E.T., S.R.R. and A.W.; software, S.E.T., G.E.G. and E.L.B.; validation, S.E.T.; formal analysis, S.E.T. and G.E.G.; investigation, S.E.T., G.E.G., S.R.R., M.J.R. and A.W.; resources, S.R.R., M.J.R. and A.W.; data curation, S.E.T., S.R.R., M.J.R. and A.W.; writing—original draft preparation, S.E.T.; writing—review and editing, S.E.T., S.R.R., M.J.R. and A.W.; visualization, S.E.T.; supervision, S.E.T., S.R.R., M.J.R. and A.W.; project administration, S.E.T., S.R.R., M.J.R. and A.W.; funding acquisition, S.R.R., M.J.R. and A.W. All authors have read and agreed to the published version of the manuscript.

Funding: This research received funding from the U.S. National Science Foundation from 2003 to 2012 through REU site awards 0244097 and 0649006.

Institutional Review Board Statement: Not applicable.

Informed Consent Statement: Not applicable.

Data Availability Statement: The MODIS and Sentinel-1 data used in this study are freely available from Google Earth Engine (MOD09GA: https://developers.google.com/earth-engine/datasets/catalog/MODIS_006_MOD09GA, accessed 27 December 2021; MYD09GA: https://developers.google.com/earth-engine/datasets/catalog/MODIS_006_MYD09GA, accessed 27 December 2021; Sentinel-1 SAR GRD: https://developers.google.com/earth-engine/datasets/catalog/COPERNICUS_S1_GRD, accessed 30 December 2021). The water temperature data and camera images presented in this study are available on request from the corresponding author.

Acknowledgments: The authors acknowledge contributions of fieldwork, data analysis, and inspiration from the many students that have participated in Svalbard summer field research from 2003 to 2021. Hampshire College students Katy Hofmeister and Amber Bonarrigo completed initial studies that helped establish our lake ice timing method using in-situ temperature data. We also acknowledge the contribution of Mount Holyoke College student Jin Cao for her work on initial studies that helped to establish the methods used to analyze the satellite data. Finally, we acknowledge personnel and logistical support from the University Centre in Svalbard from 2013 to 2019.

Conflicts of Interest: The authors declare no conflict of interest. The funders had no role in the design of the study; in the collection, analyses, or interpretation of data; in the writing of the manuscript, or in the decision to publish the results.

References

1. Serreze, M.C.; Barry, R.G. Processes and impacts of Arctic amplification: A research synthesis. *Global Planet. Change* **2011**, *77*, 85–96. [CrossRef]
2. Sharma, S.; Meyer, M.F.; Culpepper, J.; Yang, X.; Hampton, S.; Berger, S.A.; Brousil, M.R.; Fradkin, S.C.; Higgins, S.N.; Jankowski, K.J.; et al. Integrating perspectives to understand lake ice dynamics in a changing world. *J. Geophys. Res.-Biogeo.* **2020**, *125*, e2020JG005799. [CrossRef]
3. Sharma, S.; Richardson, D.C.; Woolway, R.I.; Imrit, M.A.; Bouffard, D.; Blagrave, K.; Daly, J.; Filazzola, A.; Granin, N.; Korhonen, J.; et al. Loss of ice cover, shifting phenology, and more extreme events in Northern Hemisphere lakes. *J. Geophys. Res.-Biogeo.* **2021**, *126*, e2021JG006348. [CrossRef]
4. Goosse, H.; Kay, J.E.; Armour, K.C.; Bodas-Salcedo, A.; Chepfer, H.; Docquier, D.; Jonko, A.; Kushner, P.J.; Lecomte, O.; Massonnet, F.; et al. Quantifying climate feedbacks in polar regions. *Nat. Commun.* **2018**, *9*, 1919. [CrossRef]
5. Williamson, C.E.; Dodds, W.; Kratz, T.K.; Palmer, M.A. Lakes and streams as sentinels of environmental change in terrestrial and atmospheric processes. *Front. Ecol. Environ.* **2008**, *6*, 247–254. [CrossRef]
6. Adrian, R.; O'Reilly, C.M.; Zagarese, H.; Baines, S.B.; Hessen, D.O.; Keller, W.; Livingstone, D.M.; Sommaruga, R.; Straile, D.; van Donk, E.; et al. Lakes as sentinels of climate change. *Limnol. Oceanogr.* **2009**, *54*, 2283–2297. [CrossRef] [PubMed]
7. Weyhenmeyer, G.A.; Meili, M.; Livingstone, D.M. Nonlinear temperature response of lake ice breakup. *Geophys. Res. Lett.* **2004**, *31*, L07203. [CrossRef]
8. Walsh, J.E.; Anisimov, O.; Hagen, J.O.; Jakobsson, T.; Oerlemans, J.; Prowse, T.D.; Romanovsky, V.; Savelieva, N.; Serreze, M.; Shiklomanov, A.; et al. Cryosphere and hydrology. In *Arctic Climate Impact Assessment. Scientific Report*; Arris, L., Ed.; Cambridge University Press: Cambridge, UK, 2005; pp. 183–242.
9. Higgins, S.N.; Desjardins, C.M.; Drouin, H.; Hrenchuk, L.E.; Van der Sanden, J.J. The role of climate and lake size in regulating the ice phenology of boreal lakes. *J. Geophys. Res.-Biogeo.* **2021**, *126*, e2020JG005898. [CrossRef]
10. Liston, G.E.; Hall, D.K. An energy-balance model of lake-ice evolution. *J. Glaciol.* **1995**, *41*, 373–382. [CrossRef]
11. Brown, L.C.; Duguay, C.R. The response and role of ice cover in lake-climate interactions. *Prog. Phys. Geog.* **2010**, *34*, 671–704. [CrossRef]
12. Arp, C.D.; Jones, B.M.; Grosse, G. Recent lake ice-out phenology within and among lake districts of Alaska, USA. *Limnol. Oceanogr.* **2013**, *58*, 2013–2028. [CrossRef]
13. Prowse, T.; Alfredsen, K.; Beltaos, S.; Bonsal, B.R.; Bowden, W.B.; Duguay, C.R.; Korhola, A.; McNamara, J.; Vincent, W.F.; Vuglinsky, V.; et al. Effects of changes in arctic lake and river ice. *Ambio* **2011**, *40*, 63–74. [CrossRef]
14. Dibike, Y.; Prowse, T.; Saloranta, T.; Ahmed, R. Response of Northern Hemisphere lake-ice cover and lake-water thermal structure patterns to a changing climate. *Hydrol. Process.* **2011**, *25*, 2942–2953. [CrossRef]
15. Hampton, S.E.; Galloway, A.W.; Powers, S.M.; Ozersky, T.; Woo, K.H.; Batt, R.D.; Labou, S.G.; O'Reilly, C.M.; Sharma, S.; Lottig, N.R.; et al. Ecology under lake ice. *Ecol. Lett.* **2017**, *20*, 98–111. [CrossRef] [PubMed]
16. Dugan, H.A. A Comparison of ecological memory of lake ice-off in eight north-temperate lakes. *J. Geophys. Res.-Biogeo.* **2021**, *126*, e2020JG006232. [CrossRef]
17. Magnuson, J.J.; Robertson, D.M.; Benson, B.J.; Wynne, R.H.; Livingstone, D.M.; Arai, T.; Assel, R.A.; Barry, R.G.; Card, V.; Kuusisto, E.; et al. Historical trends in lake and river ice cover in the Northern Hemisphere. *Science* **2000**, *289*, 1743. [CrossRef] [PubMed]
18. Duguay, C.R.; Prowse, T.D.; Bonsal, B.R.; Brown, R.D.; Lacroix, M.P.; Ménard, P. Recent trends in Canadian lake ice cover. *Hydrol. Process.* **2006**, *20*, 781–801. [CrossRef]
19. Latifovic, R.; Pouliot, D. Analysis of climate change impacts on lake ice phenology in Canada using the historical satellite data record. *Remote Sens. Environ.* **2007**, *106*, 492–507. [CrossRef]
20. Prowse, T.; Alfredsen, K.; Beltaos, S.; Bonsal, B.; Duguay, C.; Korhola, A.; McNamara, J.; Pienitz, R.; Vincent, W.F.; Vuglinsky, V.; et al. Past and future changes in Arctic lake and river ice. *Ambio* **2011**, *40*, 53–62. [CrossRef]
21. Benson, B.J.; Magnuson, J.J.; Jensen, O.P.; Card, V.M.; Hodgkins, G.; Korhonen, J. Extreme events, trends, and variability in Northern Hemisphere lake-ice phenology (1855–2005). *Clim. Change* **2012**, *112*, 299–323. [CrossRef]

22. O'Reilly, C.M.; Sharma, S.; Gray, D.K.; Hampton, S.E.; Read, J.S.; Rowley, R.J.; Schneider, P.; Lenters, J.D.; McIntyre, P.B.; Kraemer, B.M.; et al. Rapid and highly variable warming of lake surface waters around the globe. *Geophys. Res. Lett.* **2015**, *42*, 10–773. [CrossRef]
23. Šmejkalová, T.; Edwards, M.E.; Dash, J. Arctic lakes show strong decadal trend in earlier spring ice-out. *Sci. Rep.* **2016**, *6*, 38449. [CrossRef] [PubMed]
24. Griffiths, K.; Michelutti, N.; Sugar, M.; Douglas, M.S.; Smol, J.P. Ice-cover is the principal driver of ecological change in High Arctic lakes and ponds. *PLoS ONE* **2017**, *12*, e0172989. [CrossRef] [PubMed]
25. Lehnherr, I.; Louis, V.L.S.; Sharp, M.; Gardner, A.S.; Smol, J.P.; Schiff, S.L.; Muir, D.C.G.; Mortimer, C.A.; Michelutti, N.; Tarnocai, C.; et al. The world's largest High Arctic lake responds rapidly to climate warming. *Nat. Comm.* **2018**, *9*, 1290. [CrossRef] [PubMed]
26. Newton, A.M.W.; Mullan, D.J. Climate change and Northern Hemisphere lake and river ice phenology from 1931–2005. *Cryosphere* **2021**, *15*, 2211–2234. [CrossRef]
27. Post, E.; Alley, R.B.; Christensen, T.R.; Macias-Fauria, M.; Forbes, B.C.; Gooseff, M.N.; Iler, A.; Kerby, J.T.; Laidre, K.L.; Mann, M.E.; et al. The polar regions in a 2 C warmer world. *Sci. Adv.* **2019**, *5*, eaaw9883. [CrossRef]
28. Brown, L.C.; Duguay, C.R. The fate of lake ice in the North American Arctic. *Cryosphere* **2011**, *5*, 869–892. [CrossRef]
29. Sharma, S.; Blagrave, K.; Magnuson, J.J.; O'Reilly, C.M.; Oliver, S.; Batt, R.D.; Straile, D.; Weyhenmeyer, G.A.; Winslow, L.; Woolway, R.I. Widespread loss of lake ice around the Northern Hemisphere in a warming world. *Nat. Clim. Change* **2019**, *9*, 227–231. [CrossRef]
30. Korhonen, J. Long-term changes in lake ice cover in Finland. *Hydrol. Res.* **2006**, *37*, 347–363. [CrossRef]
31. Robertson, D.M.; Ragotzkie, R.A.; Magnuson, J.J. Lake ice records used to detect historical and future climatic changes. *Clim. Change* **1992**, *21*, 407–427. [CrossRef]
32. Hodgkins, G.A.; James, I.C.; Huntington, T.G. Historical changes in lake ice-out dates as indicators of climate change in New England, 1850–2000. *Int. J. Climatol.* **2002**, *22*, 1819–1827. [CrossRef]
33. Walsh, S.E.; Vavrus, S.J.; Foley, J.A.; Fisher, V.A.; Wynne, R.H.; Lenters, J.D. Global patterns of lake ice phenology and climate: Model simulations and observations. *J. Geophys. Res.-Atmos.* **1998**, *103*, 28825–28837. [CrossRef]
34. Duguay, C.R.; Flato, G.M.; Jeffries, M.O.; Ménard, P.; Morris, K.; Rouse, W.R. Ice-cover variability on shallow lakes at high latitudes: Model simulations and observations. *Hydrol. Process.* **2003**, *17*, 3465–3483. [CrossRef]
35. Duguay, C.R.; Pultz, T.J.; Lafleur, P.M.; Drai, D. RADARSAT backscatter characteristics of ice growing on shallow sub-arctic lakes, Churchill, Manitoba, Canada. *Hydrol. Process.* **2002**, *16*, 1631–1644. [CrossRef]
36. Howell, S.E.; Brown, L.C.; Kang, K.K.; Duguay, C.R. Variability in ice phenology on Great Bear Lake and Great Slave Lake, Northwest Territories, Canada, from SeaWinds/QuikSCAT: 2000–2006. *Remote Sens. Environ.* **2009**, *113*, 816–834. [CrossRef]
37. Cook, T.L.; Bradley, R.S. An analysis of past and future changes in the ice cover of two High-Arctic lakes based on synthetic aperture radar (SAR) and Landsat imagery. *Arct. Antarct. Alp. Res.* **2010**, *42*, 9–18. [CrossRef]
38. Murfitt, J.; Brown, L.C. Lake ice and temperature trends for Ontario and Manitoba: 2001 to 2014. *Hydrol. Process.* **2017**, *31*, 3596–3609. [CrossRef]
39. Zhang, S.; Pavelsky, T.M. Remote sensing of lake ice phenology across a range of lakes sizes, ME, USA. *Remote Sens.* **2019**, *11*, 1718. [CrossRef]
40. Zhang, S.; Pavelsky, T.M.; Arp, C.D.; Yang, X. Remote sensing of lake ice phenology in Alaska. *Environ. Res. Lett.* **2021**, *16*, 064007. [CrossRef]
41. Yang, X.; Pavelsky, T.M.; Bendezu, L.P.; Zhang, S. Simple Method to Extract Lake Ice Condition from Landsat Images. *IEEE Trans. Geosci. Remote* **2021**, *60*, 1–10. [CrossRef]
42. Liu, C.; Huang, H.; Hui, F.; Zhang, Z.; Cheng, X. Fine-Resolution Mapping of Pan-Arctic Lake Ice-Off Phenology Based on Dense Sentinel-2 Time Series Data. *Remote Sens.* **2021**, *13*, 2742. [CrossRef]
43. Hanssen-Bauer, I.; Førland, E.J.; Hisdal, H.; Mayer, S.; Sandø, A.B.; Sorteberg, A. *Climate in Svalbard 2100. A Knowledge Base for Climate Adaptation*; NCCS Report no 1/2019; Norwegian Centre of Climate Services (NCCS) for Norwegian Environment Agency: Trondhjem, Norway, 2019; 208p.
44. Eckerstorfer, M.; Christiansen, H.H. The "High Arctic maritime snow climate" in central Svalbard. *Arct. Antarct. Alp. Res.* **2011**, *43*, 11–21. [CrossRef]
45. Barton, B.I.; Lenn, Y.D.; Lique, C. Observed Atlantification of the Barents Sea causes the polar front to limit the expan-sion of winter sea ice. *J. Phys. Oceanogr.* **2018**, *48*, 1849–1866. [CrossRef]
46. Nilsen, F.; Skogseth, R.; Vaardal-Lunde, J.; Inall, M. A simple shelf circulation model: Intrusion of Atlantic water on the West Spitsbergen shelf. *J. Phys. Oceanogr.* **2016**, *46*, 1209–1230. [CrossRef]
47. Kohler, J.; James, T.D.; Murray, T.; Nuth, C.; Brandt, O.; Barrand, N.E.; Aas, H.F.; Luckman, A. Acceleration in thinning rate on western Svalbard glaciers. *Geophys. Res. Lett.* **2007**, *34*, L18502. [CrossRef]
48. Schuler, T.V.; Kohler, J.; Elagina, N.; Hagen, J.O.; Hodson, A.; Jania, J.; Kääb, A.M.; Luks, B.; Małecki, J.; Moholdt, G.; et al. Reconciling Svalbard glacier mass balance. *Front. Earth Sci.* **2020**, *8*, 156. [CrossRef]
49. Christiansen, H.H.; Gilbert, G.L.; Demidov, N.; Guglielmin, M.; Isaksen, K.; Osuch, M.; Boike, J. Permafrost thermal snapshot and active-layer thickness in Svalbard 2016–2017. *First SIOS SESS Rep.* **2019**, 27–47. Available online: https://sios-svalbard.org/SESS_Issue1 (accessed on 31 December 2021).

50. Nowak, A.; Hodson, A. Hydrological response of a High-Arctic catchment to changing climate over the past 35 years: A case study of Bayelva watershed, Svalbard. *Polar Res.* **2013**, *32*, 19691. [CrossRef]
51. Førland, E.J.; Benestad, R.; Hanssen-Bauer, I.; Haugen, J.E.; Skaugen, T.E. Temperature and precipitation development at Svalbard 1900–2100. *Adv. Meteorol.* **2011**, *2011*, 893790. [CrossRef]
52. Osuch, M.; Wawrzyniak, T. Inter-and intra-annual changes in air temperature and precipitation in western Spitsbergen. *Int. J. Climatol.* **2017**, *37*, 3082–3097. [CrossRef]
53. Retelle, M.; Christiansen, H.; Hodson, A.; Nikulina, A.; Osuch, M.; Poleshuk, K.; Romashova, K.; Roof, S.; Rouyet, L.; Strand, S.M.; et al. Environmental Monitoring in the Kapp Linne-Gronfjorden Region (KLEO). The State of Environmental Science in Svalbard. 2019. Available online: https://par.nsf.gov/biblio/10140004-environmental-monitoring-kapp-linne-gronfjorden-region-kleo (accessed on 31 December 2021).
54. Gorelick, N.; Hancher, M.; Dixon, M.; Ilyushchenko, S.; Thau, D.; Moore, R. Google Earth Engine: Planetary-scale geospatial analysis for everyone. *Remote Sens. Environ.* **2017**, *202*, 18–127. [CrossRef]
55. Vermote, E.; Wolfe, R. *MOD09GA MODIS/Terra Surface Reflectance Daily L2G Global 1kmand 500m SIN Grid V006*; NASA EOSDIS Land Processes DAAC: Sioux Falls, SD, USA, 2015. [CrossRef]
56. Vermote, E.; Wolfe, R. *MYD09GA MODIS/Aqua Surface Reflectance Daily L2G Global 1kmand 500m SIN Grid V006*; NASA EOSDIS Land Processes DAAC: Sioux Falls, SD, USA, 2015. [CrossRef]
57. Copernicus Sentinel Data. Sentinel-1A/B Synthetic Aperture Radar (SAR) Ground Range Detected (GRD). Available online: https://developers.google.com/earth-engine/datasets/catalog/COPERNICUS_S1_GRD (accessed on 31 December 2021).
58. Gunn, G.E.; Duguay, C.R.; Atwood, D.K.; King, J.; Toose, P. Observing scattering mechanisms of bubbled freshwater lake ice using polarimetric RADARSAT-2 (C-Band) and UW-Scat (X-and Ku-Bands). *IEEE Trans. Geosci. Remote* **2018**, *56*, 2887–2903. [CrossRef]
59. Howell, S.E.; Yackel, J.J.; De Abreu, R.; Geldsetzer, T.; Breneman, C. On the utility of SeaWinds/QuikSCAT data for the estimation of the thermodynamic state of first-year sea ice. *IEEE Trans. Geosci. Remote* **2005**, *43*, 1338–1350. [CrossRef]
60. Atwood, D.K.; Gunn, G.E.; Roussi, C.; Wu, J.; Duguay, C.; Sarabandi, K. Microwave backscatter from Arctic lake ice and polarimetric implications. *IEEE Trans. Geosci. Remote* **2015**, *53*, 5972–5982. [CrossRef]
61. Mladenova, I.E.; Jackson, T.J.; Bindlish, R.; Hensley, S. Incidence angle normalization of radar backscatter data. *IEEE Trans. Geosci. Remote* **2015**, *51*, 1791–1804. [CrossRef]
62. Nordli, Ø.; Wyszyński, P.; Gjelten, H.; Isaksen, K.; Łupikasza, E.; Niedźwiedź, T.; Przybylak, R. Revisiting the extended Svalbard Airport monthly temperature series, and the compiled corresponding daily series 1898–2018. *Polar Res.* **2020**, *39*, 3614. [CrossRef]

remote sensing

Article

The Collection of Hyperspectral Measurements on Snow and Ice Covers in Polar Regions (SISpec 2.0)

Rosamaria Salvatori [1,†], Roberto Salzano [2,*,†], Mauro Valt [3], Riccardo Cerrato [2] and Stefano Ghergo [4]

1 CNR—Institute of Polar Sciences, 00010 Rome, Italy; rosamaria.salvatori@cnr.it
2 CNR—Institute of Atmospheric Pollution Research, 50019 Florence, Italy; riccardo.cerrato@iia.cnr.it
3 ARPAV—Avalanche Center, 32020 Belluno, Italy; mauro.valt@arpa.veneto.it
4 CNR—Institute of Water Research, 00010 Rome, Italy; stefano.ghergo@irsa.cnr.it
* Correspondence: roberto.salzano@cnr.it
† These authors contributed equally to this work.

Abstract: The data value of hyperspectral measurements on ice and snow cover is strongly impacted by the availability of data services, where spectral libraries are integrated to detailed descriptions of the observed surface cover. For snow and ice cover, we present an updated version of the Snow/Ice Spectral Archive (SISpec 2.0), which has been integrated into a web portal characterized by different functionalities. The adopted metadata scheme features basic geographic data, information about the acquisition setup, and parameters describing the different surface types. While the implementation of the IACS Classification of Seasonal Snow on the Ground is the core component for snow cover, ice cover is approached using different parameters associated with its surface roughness and location. The web portal is not only a visualization tool, but also supports interoperability functionalities, providing data in the NetCDF file format. The availability of these functionalities sets the foundation for sharing a novel platform with the community and is an interesting tool for calibrating and validating data and models.

Keywords: snow cover; ice cover; Arctic; Antarctic; spectral reflectance; hyperspectral data

Citation: Salvatori, R.; Salzano, R.; Valt, M.; Cerrato, R.; Ghergo, S. The Collection of Hyperspectral Measurements on Snow and Ice Covers in Polar Regions (SISpec 2.0). *Remote Sens.* **2022**, *14*, 2213. https://doi.org/10.3390/rs14092213

Academic Editor: Gareth Rees

Received: 8 April 2022
Accepted: 30 April 2022
Published: 5 May 2022

Publisher's Note: MDPI stays neutral with regard to jurisdictional claims in published maps and institutional affiliations.

1. Introduction

The cryosphere is a complex domain where surface snow and ice cover play a key role in the climate change framework. The spatial and temporal monitoring of these components is a critical task that requires both in-situ and remotely sensed observations. The description of snow and ice cover, defined as an aggregation of ice crystals with different sizes and shapes [1], can be approached using microphysical observations and optical measurements combining in-situ and remotely sensed data. Moreover, the spectral behavior of the surface is controlled by further factors associated with the microphysical conditions of the considered layer (snow grain shape and size), including the surface roughness and the chemical composition of the most superficial layer. The snow surface response in the visible wavelength domain is mainly affected by light-absorbing impurities (e.g., lithogenic dust, algae, soot) [2], whereas the snow properties in the short-wave infrared range (1400–2500 nm) are widely influenced by the size and shape of the ice crystals [3]. The spectral behavior in the visible range of compact ice and snow surfaces is similar, but in the short-wave infrared ranges, the former absorbs the incident radiation almost completely [1]. The calibration and validation of the satellite data of glacial and periglacial environments are therefore widely dependent on the knowledge of the optical behavior of snow and ice. From this point of view, the ground-based data obtained by field spectroradiometers represent an ideal and fundamental data source. Hyperspectral observations are an extremely interesting feature, especially when they are obtained during field surveys located in remote areas. The need for hyperspectral measurements on snow cover, obtained by different experimental setups [4], will be further evidenced by the deployment of hyperspectral sensors on satellite platforms in the near future (e.g., PRISMA, CHIME, EnMAP). The combination of traditional manned observations [5,6] and autonomous acquisitions [7–11] will support such a demand, but the data value will only increase if observations are described in detail, especially regarding the acquisition setup [4], snow microphysics [12–15], and surface conditions [16].

The first pillar of the data value is data interoperability and, consequently, the data sharing of the described items using a standardized metadata profile. The availability of a snow-related data model [17], defined considering the first version of the Snow and Ice Spectral Library (SISpec 1.0) published by [18,19], matches this requirement. In fact, specific extensions have been combined in order to describe the three abovementioned information components (Base, Acquisition, and Domain). The paradigm shift from the crystal size-oriented classification to a more exhaustive classification that takes into account a complex mixture of crystal shapes and sizes is supported by the proposed scheme. The change of the data model allows the possibility of describing the snow surface as an ensemble of crystal types, sizes, shapes and their genesis, permitting the potential classification of the crystals into more than 30 classes, as indicated by the International Classification of Seasonal Snow on the Ground [20]. The snow crystallography is not the only feature described by the metadata profile, but it also allows a harmonization of microphysical properties of the surface (e.g., hardness and roughness), which were observed following the international guidelines. The encoding specifications drive the interoperability action through the preparation of NetCDF file formats compliant with INSPIRE and ISO guidelines. This background is a primary component for preparing a specific data service capable of showing, querying, and sharing its content with the community, or to the already available spectral archives (SA), spectral libraries (SL), and spectral information systems (SIS). The availability of the SISpec metadata profile also facilitates the connection between domain-specific collections and general-purpose systems (SA, SL, or SIS), improving machine-to-machine interactions with dedicated tools.

Following a survey about hyperspectral collections of snow surfaces, the available datasets could be grouped in domain-specific and general-purpose collections. The first group includes the SPECLIB SL [21], the ECOsystem Spaceborne Thermal Radiometer Experiment on Space Station (ECOSTRESS) SL [22], the SPECCHIO SIS [23], and the LUCAS database with its SL [24]. The second group includes two soil-oriented collections, namely, the INTA SL [25] and the Global vis-NIR SL [26], and one snow-related SISpec SL [18]. The interoperability issue is, of course, a major aspect, since some collections, like ECOSTRESS and SPECLIB, consist of text files reporting the single measurements (spectra) associated with a limited amount (about 20) of unstandardized metadata. Nevertheless, SPECCHIO and INTA are characterized by a more detailed and standardized data model with more than 50 attributes. The Global vis-NIR SL and LUCAS SL are dedicated to soils, and their structure has specific attributes (8 and more than 30, respectively) for chemistry and physics that are not suitable for hyperspectral measurements on snow surfaces. Regarding snow observation databases, ECOSTRESS, SPECLIB, INTA, and SPECCHIO provide few spectra with a very coarse description, which can be difficult to utilize in certain studies. SISpec 1.0 [18] is the only fully snow-oriented collection where snow measurements are fully described in compliance with the international classification [20]. This library was released with physical support and this limitation required an upgrade to SISpec 2.0, with the creation of a web service that must be user-friendly, flexible, and interoperable. The aim of this paper is to present such a web service and associated database, which can promote the sharing of available measurements and the harvesting of novel observations in snow-covered areas.

2. Methods

The presented spectral library contains data collected during field campaigns in polar regions between 1998 and 2011. The ability to analyze flat surfaces is a prerogative that makes polar areas the most suitable for collecting snow and ice spectral measurements, because the spectral response of these surfaces is not appreciably affected by the presence of nearby reliefs and vegetation or by light-absorbing impurities of anthropogenic origin. The field data were collected during cold and clear sky days (with air temperatures always below 0 °C), to avoid bias due to the snow cover melting. All of the selected sites were located in wide and plain areas in order to minimize slope and adjacent pixel effects, and to be clearly identifiable at Landsat satellite data resolution [18]. The collection includes more than 250 spectra obtained using a standard procedure both for the acquisition of reflectance measurements and for the manual snow observation [20].

2.1. Regions of Interest

The dataset includes 152 hyperspectral measurements acquired in the Arctic, where radiometric and snow/ice data were collected from 26 different measurement sites located in the Brøggerhalvøya (Brøgger Peninsula), up to 160 km from the International Research Station of Ny-Ålesund (Figure 1a). The field campaigns were carried out in six spring seasons (1998, 2000, 2001, 2003, 2010, and 2011). The selected region is characterized by flat snow/ice surfaces (mainly tundra snow and various ice covers), large enough to be detected on satellite images and far enough from dense human settlements, in order to provide relatively uncontaminated snow spectral signatures. The peninsula is characterized by a relatively large variability of snow conditions in the framework of relatively short distances due to the different exposure of the coastline towards local-scale meteorology [27,28].

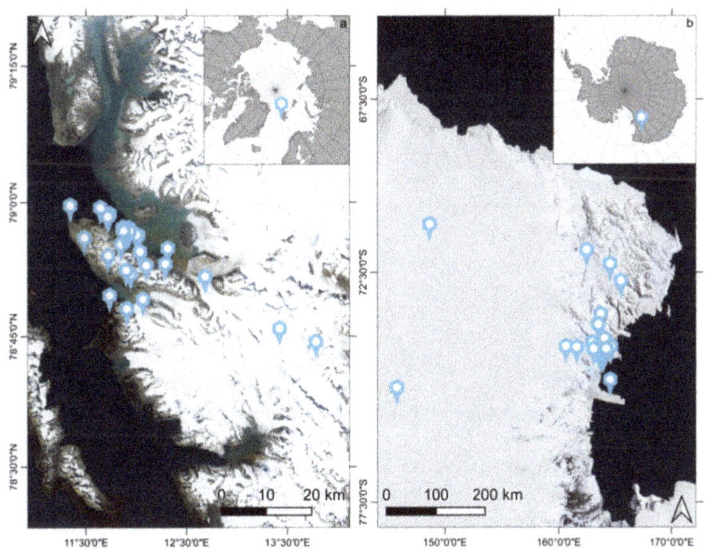

Figure 1. Localization of the available measurement sites in the Arctic region (**a**) and in the Antarctic continent (**b**).

The 105 spectral snow reflectance observations in Antarctica were acquired from 25 measure sites at Terra Nova Bay (Victoria Land—Ross Sea Region) in the Austral summers of 1996, 1998, and 2002. The location of the investigated sites is shown in Figure 1b. A large variety of glacial landforms and snow/ice types were detected due to the intense glacial activity and to the major difference in elevation between areas next to the coast and areas on the large inner plateau.

2.2. Hyperspectral Measurements

The snow and ice hyperspectral measurements were acquired by different field spectroradiometers: Fieldspec FR and FR3 (Analytical Spectral Devices Inc., Boulder, CO, USA), covering the wavelength range between 350 and 2500 nm. Measurements were acquired using a hemispherical–conical setup (Figure 2), where the calculated reflectance factor was estimated as the ratio of incident solar radiation reflected from the snow target and the incident radiation reflected from a white reference Spectralon (about 30 cm × 30 cm), known as a Lambertian reflector. The acquisition protocol was standardized, with the sensor optic fiber on a tripod 50 cm above the surface target in a nadir position. The radiometer was used with the bare optics that correspond to a field of view of 25°, thus covering ground areas of about 23 cm in diameter. All of the measurements were acquired by positioning the target towards the sun and minimizing interferences on the surface from the operator and the tripod. The absolute reflectance was obtained by multiplying the reflectance factor by the calibration coefficient of the reference panel. Possible sources of errors or noise in field spectroradiometry could be operational due to the incorrect viewing geometry during the data acquisition, or due to internal (e.g., random noise produced by electronic components of the instrument) or external (e.g., the atmospheric water vapor absorption, the low atmospheric irradiance at wavelengths of 1400 nm and beyond 1700 nm) factors causing a low signal-to-noise (S/N) ratio. To avoid incorrect or anomalous reflectance values and/or patterns, especially in the visible and infrared wavelengths, a correct orientation of the spectroradiometer over the calibration panel and the snow surface target is necessary. The optimization of the S/N ratio is a major issue in field measurements; the selected compromise was found using a sample average from 10 to 50 acquisitions. For every target, a statistically meaningful sample of the surface was obtained by acquiring 20 to 30 spectral curves; the evaluation of the signal's stability and of the surface heterogeneity contribute to defining the final number of acquisitions. This procedure increases the spectral characterization of the snow target—the higher the number of spectral curves, the smaller the random errors. No spectral polishing algorithm was applied to the published data. Values in the wavelength range around 1400, 1940, and 2400 nm are affected by the presence of atmospheric water vapor [29].

Figure 2. Field hyperspectral observation setup.

2.3. *Surface Characterization*

The description of each snow/ice target was performed following a protocol based on the preliminary identification of the most representative target of the entire area (about 100 × 100 m) in terms of surface type: ice and snow cover. The surface roughness is a major feature in both cases, and the selected protocol for such a description was based on acquiring a visual overview of the surface characteristics with a terrestrial image as well as to measure the roughness in terms of the length and height of the identified geometric elements (e.g., ripples, furrows). The different pattern and size of the surface elements may affect the spectral response due to shadowing effects and backscattering. Surface roughness (furrow distance and depth) was also measured in mm, in accordance with [20].

2.3.1. Snow Cover Microphysics

The description of the microphysical characteristics of the snow was carried out by determining the shape and size of the snow grains based on the observation of the individual grains, with a magnifying glass, on a graduated crystals card, as indicated by [20]. Considering that the surface of the snow is often made up of a mixture of grains of different sizes and shapes, the three most abundant grain types were reported for each target surface. Snow observations were performed on the selected target area immediately after the spectral measurement. A conventional survey of the first 20 cm depth of the snow surfaces was carried out at all measurement sites in order to measure the hardness, density, and liquid water content of the snow cover. Hardness measurements were performed using the hand test [30]; the snow density of the first 10 cm (from the surface) was measured using a core drill with a volume of 10^{-4} m^3 and a properly calibrated steelyard scale. The liquid water content was estimated with a SnowFork (Toikka Engineering Ltd., Espoo, Finland). The temperature of the air and snow (10 cm below the surface) was measured at each site immediately after the spectra were acquired. Snow and air temperature measurements were carried out using a digital contact thermometer.

2.3.2. Ice Cover Classification

The description of the ice cover followed the principles defined in [31], where the major cryospheric components that complete the framework with the snow cover are glaciers and ice caps, freshwater ice, and sea ice. The ice cover classification in SISpec is limited to the geographical framework of the considered site: sea ice, glacier ice (the surface of the glacier not covered by snow), and lake or river ice. According to the International Classification of Seasonal Snow on the Ground [20], codes with two uppercase letters are selected for these iced surfaces (IS) and two lowercase letters: we use ISsi for sea ice, ISgl for glacier ice, and ISlr for lake or river ice.

2.4. Other Info

Environmental parameters (e.g., sky conditions, air temperature, and humidity), the GPS coordinates, the elevation of the observation site, and the date and time of the spectral measurement were also acquired for each measurement site. The date and time data were then used to calculate the elevation and azimuth of the sun. When possible, pictures of the target and panoramic photos of the measuring site were acquired.

2.5. SISpec Spectral Information System

The SISpec database and management procedures (Figure 3) are designed to facilitate the storage, organization, and elaboration of the spectral data and of the supplementary observations collected during field measurements. The SISpec web service (https://niveos.cnr.it/SISpec, accessed on 2 May 2022) supports specific tasks that steer the implementation of the data management/data output tool system, enhancing the following actions: (i) allows local operators and visiting users constant access to basic information and processed data; (ii) centralizes data management activities; and (iii) standardizes the most-used visualization and processing procedures. The relational database management system (RDBMS) used for this project is MySQL (https://www.mysql.com/ accessed on 2 May 2022) and the server-side procedures are written using PHP (https://www.php.net accessed on 2 May 2022) as a programming language. To facilitate client-side interactions with users, the procedures contain code in JavaScript. The relational structure of the database was designed considering the SISpec web system purposes and the different types of data collected during the field campaigns. The SISpec core component (left side of Figure 3) supports the hyperspectral data (reflectance values as well as wavelength ranges) with the information about the observation site (metadata about the geographic location), the primary surface characterization (description and metrics), and the environmental conditions (using categories and numerical variables) during the acquisition of the data and photographic material regarding the target. Additional information is present in the nivological component (right side of Figure 3) for the snow cover, and the attributes are organized following the international convention available in the literature [20]. The most relevant information of the nivological component is stored as coded fields that need support tables to supply a more user-friendly appearance for the output forms. The code conventions used in these support tables are reported in [20]. Reflectance spectra in the database are stored as pointers to external files as well as the nivological profiles, the target photos, and the metainformation (in NetCDF format) files. The whole system is supported by predefined standardized queries, which facilitate the navigation and visualization of the available dataset. These functionalities are based on code lists developed using specific domain knowledge, such as the snow classification, surface classification, satellite band specifications, and geographic toponyms. The SISpec web system is equipped with an authentication procedure currently only used for database table maintenance, but this can be easily upgraded for the implementation of new features of the system such as new contributions from other users or the direct download of selected data.

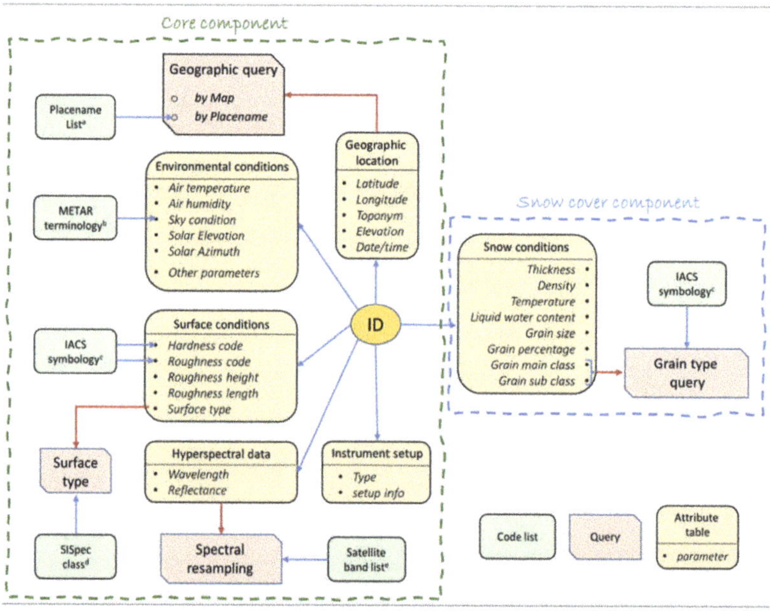

Figure 3. Database structure and relational tables breakdown: the core and nivological components are characterized by attribute tables, functionalities, and code lists. [a] is the placename list defined in [32], [b] refers to the METAR terminology reported in [20], [c] refers to the IACS classification [20], [d] are the classes defined in this paper combining [20,31], [e] are the bands declared by each satellite platform [33–36].

3. Results

The SISpec collection presented in this study includes the dataset published in the first version [18] and has been updated with newer observations acquired in recent years. While the accessibility to the collection is approached by developing a specific web service, the interoperability is handled by preparing data following the SISpec metadata profile defined by [17].

3.1. SISpec Web Portal

The published web service is based on a user interface that supports different functionalities, aimed at data querying and visualization. The collection can be queried using two different strategies (Figure 4) based on geographic location and snow microphysics, respectively.

Figure 4. The search features of the SISpec web portal.

The geographic search is currently limited to the two regions of interest (Svalbard and Antarctica) by constraining the searchable geographic extent. The search functionality focusing on snow microphysics is currently limited to the snow grain type, defined by [20]. Both search queries produce a standardized output (Figure 5), which supplies the list of all records satisfying the search parameters.

The left panel of the search output is the summary of the results obtained, reporting the file ID, the geographic location, the acquisition time, the grain type symbology, and the snow layer thickness. Each record can be further displayed interactively with a single or multiple selection. The full data information can only be displayed for each measurement, and the right-side panel reports the complete metadata description of the selected observation in terms of geographic location, nivological properties, environmental conditions, and acquisition setup. The output list of a query provides the functionality to plot each of the reported records with a specific panel displaying the reflectance spectrum (Figure 6).

Such a representation also supports the multiple selection of up to six spectra, which can be displayed as separate curves in a collated plot, as an averaged spectrum, or as a plot grid with single records. The proposed web interface is therefore aimed at supporting the analysis and interpretation of satellite images. This final objective implies the availability of a specific tool focused on resampling selected spectra in the corresponding intervals of the bands of different satellite sensors. It is possible to visualize the curves not only in the satellite bands that have the same spatial resolution, but also to select the option that allows the visualization of the resampled value in all of the available bands. This latter opportunity allows users to support many different applications to study snow cover. In addition, for the multiple selection option, the satellite bands resampled output is available (Figure 7). The sensor portfolio included in the presented tool includes the band list of instruments deployed on different platforms: Landsat 7 [33]; Landsat 8 [34]; Sentinel-2 [35]; and Terra/Aqua [36]. The band resampling was developed by calculating the resampled values using the specific full width half maximum (FWHM) of each sensor band.

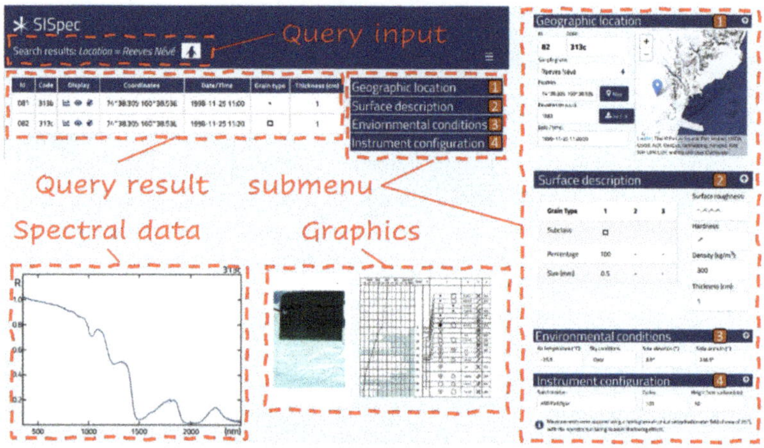

Figure 5. Results of the query by location and available information of the Reeves Névé site. Information sections rearranged for clarity.

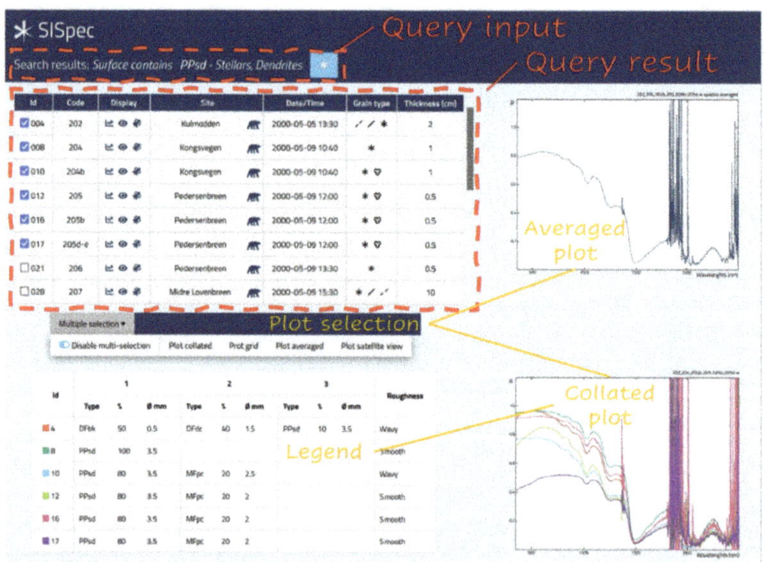

Figure 6. Results of the query by grain type PPsd with spectrum visualization and multiple selection. Information sections rearranged for clarity.

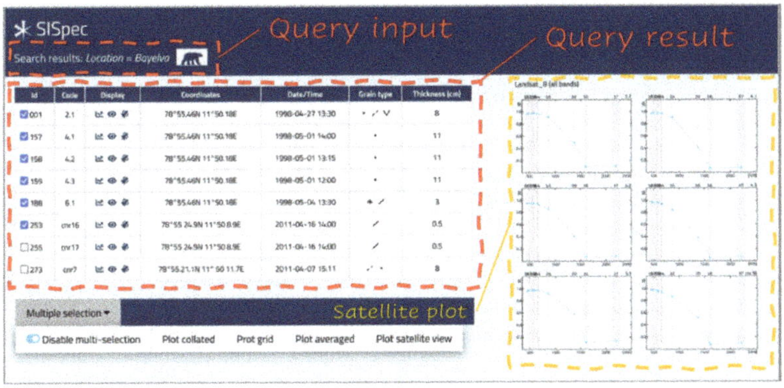

Figure 7. Results of the query by location and spectrum for the Bayelva sites. Example of band resampling considering the sensors deployed on the Landsat 8 platform. Information sections rearranged for clarity.

3.2. SISpec Interoperability

The interoperability issue of the data value chain was approached considering the latest standards and the available convention to prepare sharable and reusable datasets. The preparation of datasets in the NetCDF file format, following the encoding technical specification prepared by [17], provided a solution for having a formal and shared standardization aimed at producing well-documented and sound metadata for hyperspectral measurements of the snow/ice surfaces. The dataset was prepared with single files for each measurement that can be downloaded using the dedicated functionalities. The prepared data are fully described following the ISO and INSPIRE guidelines. They are ready for machine-to-machine interactions and can be easily read by different open tools. Metainformation and data for each measurement, in NetCDF format, are downloadable through the field acquisition form, which provides a map of the sampling site.

4. Discussion

4.1. Usefulness of the SISpec Archive

The available records presented in the SISpec web portal include observations acquired both in the Arctic and Antarctic regions, 152 and 105, respectively. The archive provides hyperspectral measurements obtained on snow (209) and ice (48) covers, which are described in detail from the acquisition point of view, as well as in terms of surface and nivological conditions. The observations are representative of a large variety of environmental conditions, considering that the air temperatures covered a range from −21 to 3 °C in Svalbard, and from −34 to −2 °C in Antarctica. The study sites included 66% locations in coastal areas (i.e., below 100 m a.s.l.), 17% low elevated lands between 100 and 500 m a.s.l., 6% medium elevated areas between 500 and 1000 m a.s.l., and 11% high elevated areas above 1000 m a.s.l. The sky condition is additional metadata information that has a double impact on both the snow metamorphism and the reflectance estimation. The primary target of the considered field campaigns was to maximize measurements obtained under clear or white sky conditions with 47% and 20% of the available records, respectively. The key parameter for discriminating between ice and snow cover was the hardness, but the description of snow surfaces was completed by snow densities (with an average of 253 ± 107 kg m^{-3}) and a roughness of almost smooth in more than 50% of observations (Figure 8). Furthermore, measurements showed conditions where the average snow temperature in Svalbard was -6.2 ± 4.3 °C and in Antarctica was -14.6 ± 8.7 °C. Therefore, the snow surfaces were described using the IACS Classification of Seasonal Snow on the Ground, and interesting features were obtained about the snow microphysics. The observed surfaces were, in fact, characterized by single shape classes in 33% of cases, by two different shape types in 51% of occurrences, and by three components in the rest of the available measurements. The median size of the observed snow grains varied from recently formed crystals (PP, DF, and SH) at 1.2 mm, to surface modified forms (FC and RG) at 0.4 mm, and to melt forms (MF) at 1.65 mm.

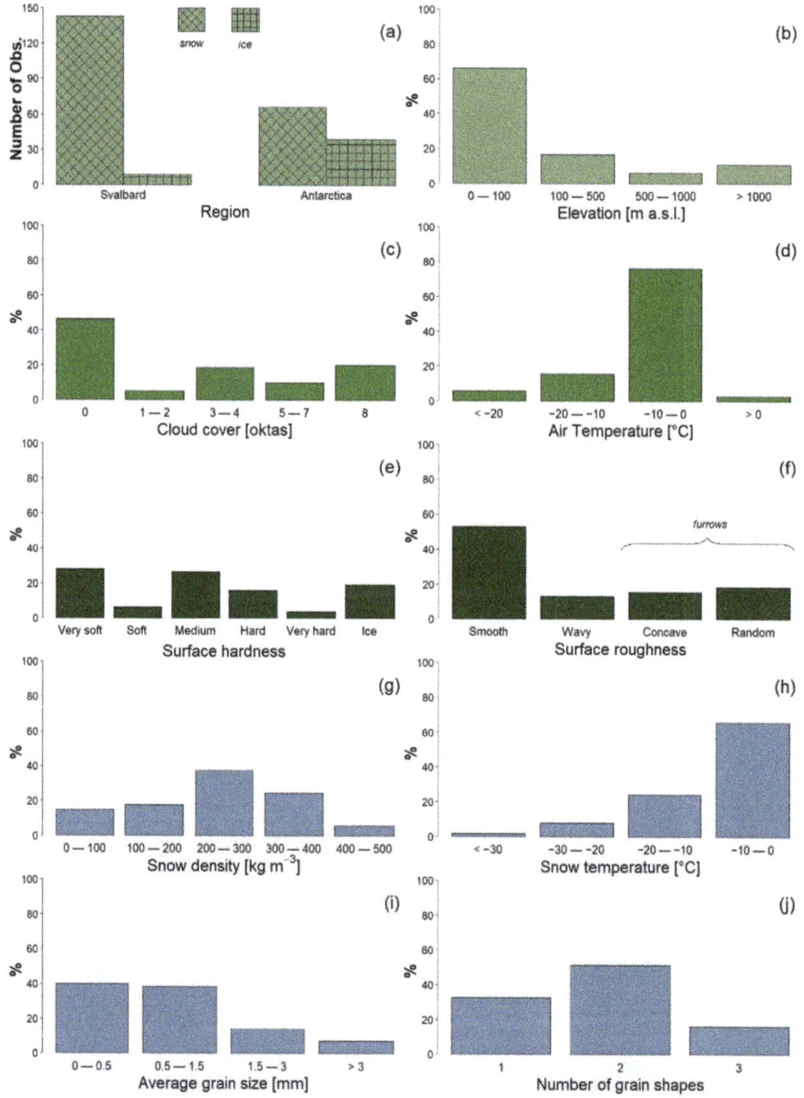

Figure 8. Summary plot of observations: (**a**) region of interest, (**b**) site elevation, (**c**) cloud cover during the acquisition, (**d**) air temperature, (**e**) surface hardness, (**f**) surface roughness, (**g**) snow density, (**h**) snow temperature, (**i**) average snow grain size, and (**j**) number of identified snow crystal forms.

All of these data are useful metrics that support the processing of hyperspectral measurements (Figure 9). The description of snow and ice cover is a major scope of the support provided by the SISpec web portal and it can be approached considering preliminary criteria aimed at discriminating different surface types. The SISpec archive provides continuous variability between different surface types that are significantly discriminated between snow (80%)- and ice (20%)-covered end members. The separation between these surface types in terms of reflectance increases from the visible range to the short-wave infrared wavelength domain (above 1100 nm). This behavior is, of course, consistent with the literature evidence [1], but the management of different metadata patterns associated with the two different surface types is a primary requirement. The analysis of snow surfaces considering the IACS classification [20] is a major challenge of the SISpec web portal, since it is possible to combine and share a range of information impacting on the final spectral behavior of the observed surface. The potentialities offered by the web portal include the sharing and visualization of data, described as complex mixtures of different snow grain shapes, that will enhance studies concerning radiative transfer models [2,12,15] and novel sensing capabilities of the snow cover [37–39].

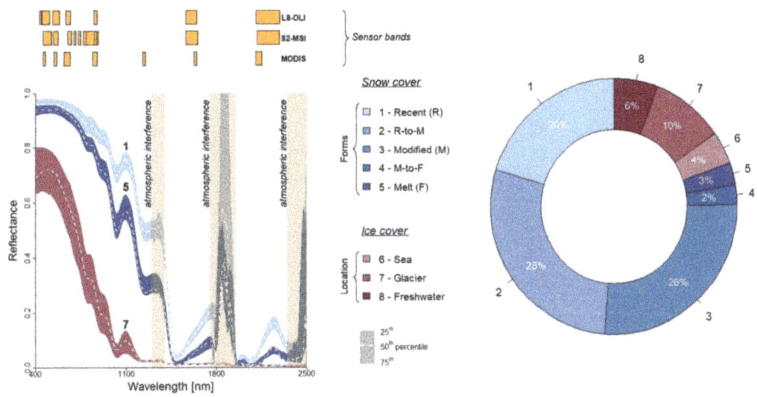

Figure 9. Aggregation of hyperspectral measurements of snow and ice cover (**left**) and statistics on the occurrence of different snow forms and ice locations (**right**). Snow cover is aggregated considering the occurrence of different snow grain types.

The proposed approach is based on aggregating the different snow forms, considering the texture modification from recently deposited or precipitated crystals to advanced melting and refreezing of forms. While precipitated particles (PP), decomposed and fragmented grains (DF), and surface hoars (SH) are included in the recent forms, rounded grains (RG) and faceted crystals (FC) occur in the modified forms, produced by both equilibrium and kinetic growth metamorphism. Furthermore, the modified forms are dominated by equilibrium gradient grains (RG), since the environmental conditions observed during the fieldwork favor the occurrence of those forms on the surface. The occurrence of melt forms (MF) represents the final stage before the transformation in perpetual snow and ice formation. It is, of course, not possible to have pure terms, and mixed elements must be considered. Five classes are presented from recently formed snow grains, to modified forms, and finally to refrozen melt snow, where the spectral features are gradually lower in the whole spectrum range. Furthermore, the decrease is more intense in the short-wave infrared, as expected from the lengthening of the mean free path through the ice in relation to the decrease of the specific surface area already observed by [3]. The analysis of iced surfaces is, on the other hand, not mature enough, since the number of available observations is limited. From this perspective, the SISpec web portal only supports a geographic classification based on discriminating ice formations (IF) in the snowpack from iced surfaces associated with sea ice (ISsi), with freshwater entities such as lake and rivers (ISlr) and glacier areas (ISgl).

The flexibility of the SISpec archive is based on the ability to describe complex mixtures of several snow grain types instead of using sharp separation between single elements considering just size, shape, or density. The detailed description offered by the IACS classification [20] is, in fact, an added value that must be considered during the intercomparison between ground-based and satellite observations. It is, therefore, limiting to discriminate between wet or dry snow cover, coarse or small-sized snow crystals, and new or old snow. SISpec allows a full description of complex surface layers, where different and heterogeneous snow grains could represent the deposition of new fallen snow onto intensively melted snow cover types. The opportunity to describe mixed surfaces where the snow cover is momentarily deposited on ice covers is also worthy of note. Finally, the aim of the SISpec functionalities is to support the full description offered by the IACS Classification of Seasonal Snow on the Ground, and to match the requirements of hyperspectral measurements from different disciplines and communities.

4.2. SISpec Collection in the Spectral Archive Framework

The integration of the international classification of snow [20] into a web service represents a novel tool for visualizing and sharing spectroradiometric observations of snow and ice surfaces. The available services, such as the ECOSTRESS SL, include snow measurements in the water type, distinguishing between granular snow (coarse, medium, and fine) and frost. The size distribution of the snow cover is, unfortunately, a limiting approach since the grain shape and composition, as well as the surface roughness, are driving factors of the final spectral behavior of the surface covered by snow and ice. This is the reason why the key novel aspect of the SISpec web portal is the description of several crystal types, limited to three different classes, where dimensions and crystal genesis define more than 30 different classes. The description of additional microphysical properties (hardness and roughness, for example), estimated with internationally recognized methods, is an additional key feature of the presented data portal. Having a fully described dataset is crucial in terms of gathering information on the acquisition setup, the surface condition, and the microphysical framework. It is a powerful tool featuring further advances on interpreting the optical behavior of snow and ice cover.

The flexibility of the SISpec data service supports different interactions between the offered dataset of the available snow/ice standard classifications. While the IACS classification supports avalanche-oriented communities, the available metadata and data format include the ability, among others, to revisit the snow classification using different perspectives: e.g., microphysical or meteoclimatic. The integration of the web portal to the NetCDF file format represents an ideal solution for human and machine-based interactions. The proposed strategy increases the data value of the SISpec dataset, offering the perspective to facilitate the interaction between smaller databases and wider services useful for different communities. The integration of the SISpec data model into collections of spectral data supported by other databases will be possible, developing a middleware aimed at translating the SISpec schema into selected wide-audience services. The opposite flow direction is unfortunately limited by the occurrence of snow measurements in other services (SPECLIB, ECOSTRESS, and SPECCHIO). While it is possible to ensure a complete overlap between the considered collections and SISpec, the opposite direction is limited by the available metadata components. The match between different attributes included in the SISpec scheme and the other collections is the only requirement for the translation tool. On the opposite side, the to-SISpec process is more critical due to missing snow information, and gaps in terms of mandatory SISpec attributes. The SISpec web portal represents, in conclusion, an ideal and flexible solution for the cryosphere community to share data, and the web portal is potentially open to include datasets acquired from different locations, being compliant with FAIR principles and INSPIRE guidelines.

5. Conclusions

The presented version of the Snow/Ice Spectral Archive (SISpec 2.0) is an innovative web portal where hyperspectral measurements on ice and snow cover are available in terms of accessibility and interoperability. The revisited spectral library is now integrated with detailed descriptions of the observed surfaces, and the adopted metadata scheme includes information about basic geographic features, the acquisition setup, and most of all, about the parameters describing the different surface types. While snow cover is fully described following the IACS Classification of Seasonal Snow on the Ground, the approach to ice cover involves applying different parameters regarding surface roughness and location. The web portal is designed as a visualization tool, but it also supports different interoperability functionalities. The use of a standardized NetCDF file format allows the presented archive to connect to other platforms and to prepare the ground for investigating the relationship between the optical classification of the surface and the macro-/microphysical conditions of the observed cover. The availability of these functionalities sets the stage for sharing a novel platform with the community, creating an interesting tool for calibrating and validating data and models.

Author Contributions: All authors are intellectually responsible for the conducted research, work design, and manuscript preparation. Conceptualization, R.S. (Rosamaria Salvatori), S.G. and R.S. (Roberto Salzano); methodology, R.S. (Rosamaria Salvatori), M.V. and S.G.; software, R.S. (Rosamaria Salvatori) and S.G.; validation, R.S. (Rosamaria Salvatori), R.S. (Roberto Salzano) and M.V.; data curation, R.S. (Rosamaria Salvatori), M.V. and R.S. (Roberto Salzano); writing—original draft preparation, R.S. (Rosamaria Salvatori) and R.S. (Roberto Salzano); writing—review and editing, R.S. (Rosamaria Salvatori), R.S. (Roberto Salzano), R.C., M.V. and S.G.; visualization, R.S. (Rosamaria Salvatori), R.S. (Roberto Salzano) and R.C. All authors have read and agreed to the published version of the manuscript.

Funding: The development of the SISpec web portal is supported by the CRASI project (PNRA18-00131) and by the EcoClimate project (PRA21-19) in the framework of the reusing strategy of past datasets and the data management and harmonization of future novel datasets.

Data Availability Statement: The dataset is available for research purposes directly on the website after user registration.

Acknowledgments: The authors are extremely grateful to the staff of the ARPAV-Avalanche Center of Arabba (Renato Zasso, Anselmo Cagnati) and the CNR staff involved in the scientific support (Ruggero Casacchia) and the outstanding logistical support (Roberto Sparapani, Paolo Plini, Alessandro Mei) received during the field campaigns. We would like to acknowledge Giorgia Ghergo for the graphic project of the website and Massimiliano Olivieri for the server configuration.

Conflicts of Interest: The authors declare no conflict of interest.

References

1. Rees, W.G. *Remote Sensing of Snow and Ice*; CRC Press—Taylor & Francis Group: Boca Raton, FL, USA, 2006; pp. 1–285.
2. Warren, S.G. Optical properties of ice and snow. *Philos. Trans. R. Soc. London. Ser. A Math. Phys. Eng. Sci.* **2019**, *377*, 20180161. [CrossRef] [PubMed]
3. Dominé, F.; Salvatori, R.; Legagneux, L.; Salzano, R.; Fily, M.; Casacchia, R. Correlation between the specific surface area and the short-wave infrared (SWIR) reflectance of snow. *Cold Reg. Sci. Technol.* **2006**, *46*, 60–68. [CrossRef]
4. Schaepman-Strub, G.; Schaepman, M.E.; Painter, T.H.; Dangel, S.; Martonchik, J.V. Reflectance quantities in optical remote sensing—Definitions and case studies. *Remote Sens. Environ.* **2006**, *103*, 27–42. [CrossRef]
5. Carmagnola, C.M.; Domine, F.; Dumont, M.; Wright, P.; Strellis, B.; Bergin, M.; Dibb, J.; Picard, G.; Libois, Q.; Arnaud, L.; et al. Snow spectral albedo at Summit, Greenland: Measurements and numerical simulations based on physical and chemical properties of the snowpack. *Cryosphere* **2013**, *7*, 1139–1160. [CrossRef]
6. Painter, T.H.; Dozier, J. The effect of anisotropic reflectance on imaging spectroscopy of snow properties. *Remote Sens. Environ.* **2004**, *89*, 409–422. [CrossRef]
7. Dumont, M.; Arnaud, L.; Picard, G.; Libois, Q.; Lejeune, Y.; Nabat, P.; Voisin, D.; Morin, S. In situ continuous visible and near-infrared spectroscopy of an alpine snowpack. *Cryosphere* **2017**, *11*, 1091–1110. [CrossRef]
8. Picard, G.; Libois, Q.; Arnaud, L.; Verin, G.; Dumont, M. Development and calibration of an automatic spectral albedometer to estimate near-surface snow SSA time series. *Cryosphere* **2016**, *10*, 1297–1316. [CrossRef]
9. Kokhanovsky, A.; Di Mauro, B.; Garzonio, R.; Colombo, R. Retrieval of Dust Properties From Spectral Snow Reflectance Measurements. *Front. Environ. Sci.* **2021**, *9*, 644551. [CrossRef]
10. Salzano, R.; Lanconelli, C.; Salvatori, R.; Esposito, G.; Vitale, V. Continuous monitoring of spectral reflectance of snowed surfaces in Ny-Ålesund. *Rend. Lincei* **2016**, *27*, 137–149. [CrossRef]
11. Salzano, R.; Lanconelli, C.; Esposito, G.; Giusto, M.; Montagnoli, M.; Salvatori, R. On the Seasonality of the Snow Optical Behaviour at Ny Ålesund (Svalbard Islands, Norway). *Geosciences* **2021**, *11*, 112. [CrossRef]
12. Kokhanovsky, A. Light penetration in snow layers. *J. Quant. Spectrosc. Radiat. Transf.* **2022**, *278*, 108040. [CrossRef]
13. Horton, S.; Jamieson, B. Spectral measurements of surface hoar crystals. *J. Glaciol.* **2017**, *63*, 477–486. [CrossRef]

14. He, C.; Liou, K.-N.; Takano, Y.; Yang, P.; Qi, L.; Chen, F. Impact of grain shape and multiple black carbon internal mixing on snow albedo: Parameterization and radiative effect analysis. *J. Geophys. Res. Atmos.* **2018**, *123*, 1253–1268. [CrossRef]
15. Saito, M.; Yang, P.; Loeb, N.G.; Kato, S. A Novel Parameterization of Snow Albedo Based on a Two-Layer Snow Model with a Mixture of Grain Habits. *J. Atmos. Sci.* **2019**, *76*, 1419–1436. [CrossRef]
16. Larue, F.; Picard, G.; Arnaud, L.; Ollivier, I.; Delcourt, C.; Lamare, M.; Tuzet, F.; Revuelto, J.; Dumont, M. Snow albedo sensitivity to macroscopic surface roughness using a new ray-tracing model. *Cryosphere* **2020**, *14*, 1651–1672. [CrossRef]
17. Di Franco, S.; Salzano, R.; Boldrini, E.; Salvatori, R. Increasing the interoperability of snow/ice hyperspectral observations. *Comput. Geosci.* **2022**, *162*, 105076. [CrossRef]
18. Casacchia, R.; Salvatori, R.; Cagnati, A.; Valt, M.; Ghergo, S. Field reflectance of snow/ice covers at Terra Nova Bay, Antarctica. *Int. J. Remote Sens.* **2002**, *23*, 4563–4667. [CrossRef]
19. Casacchia, R.; Lauta, F.; Salvatori, R.; Cagnati, A.; Valt, M.; Ørbæk, J.B. Radiometric investigation of different snow covers in Svalbard. *Polar Res.* **2001**, *20*, 13–22. [CrossRef]
20. Fierz, C.; Armstrong, R.L.; Durand, Y.; Etchevers, P.; Greene, E.; McClung, D.M.; Nishimura, K.; Satyawali, P.K.; Sokratov, S.A. *The International Classification for Seasonal Snow on the Ground*; IHP-VII Technical Documents in Hydrology N°83, IACS Contribution N°1; UNESCO-IHP: Paris, France, 2009; pp. 1–81.
21. Kokaly, R.F.; Clark, R.N.; Swayze, G.A.; Livo, K.E.; Hoefen, T.M.; Pearson, N.C.; Wise, R.A.; Benzel, W.M.; Lowers, H.A.; Driscoll, R.L.; et al. *USGS Spectral Library Version 7*; USGS Data Series; U.S. Geological Survey: Reston, WV, USA, 2017; Volume 1035, pp. 1–61.
22. Meerdink, S.K.; Hook, S.J.; Roberts, D.A.; Abbott, E.A. The ECOSTRESS spectral library version 1.0. *Remote. Sens. Environ.* **2019**, *230*, 111196. [CrossRef]
23. Hueni, A.; Nieke, J.; Schopfer, J.; Kneubühler, M.; Itten, K.I. The spectral database SPECCHIO for improved long-term usability and data sharing. *Comput. Geosci.* **2009**, *35*, 557–565. [CrossRef]
24. Orgiazzi, A.; Ballabio, C.; Panagos, P.; Jones, A.; Fernández-Ugalde, O. LUCAS Soil, the largest expandable soil dataset for Europe: A review. *Eur. J. Soil Sci.* **2018**, *69*, 140–153. [CrossRef]
25. Jiménez, M.; González, M.; Amaro, A.; Fernández-Renau, A. Field Spectroscopy Metadata System Based on ISO and OGC Standards. *ISPRS Int. J. Geo-Inf.* **2014**, *3*, 1003–1022. [CrossRef]
26. Viscarra Rossel, R.A.; Behrens, T.; Ben-Dor, E.; Brown, D.J.; Demattê, J.A.M.; Shepherd, K.D.; Shi, Z.; Stenberg, B.; Stevens, A.; Adamchuk, V.; et al. A global spectral library to characterize the world's soil. *Earth Sci. Rev.* **2016**, *155*, 198–230. [CrossRef]
27. Esau, I.; Repina, I. Wind Climate in Kongsfjorden, Svalbard, and Attribution of Leading Wind Driving Mechanisms through Turbulence-Resolving Simulations. *Adv. Meteorol.* **2012**, *2012*, 568454. [CrossRef]
28. Valt, M.; Salvatori, R. Snowpack characteristics of Brøggerhalvøya, Svalbard Islands. *Rend. Lincei* **2016**, *27*, 129–136. [CrossRef]
29. Jensen, J.R. *Introductory Digital Image Processing: A Remote Sensing Perspective*, 4th ed.; Pearson Education: Glenview, IL, USA, 2015; pp. 1–623.
30. Höller, P.; Fromm, R. Quantification of the hand hardness test. *Ann. Glaciol.* **2010**, *51*, 39–44. [CrossRef]
31. Lemke, P.; Ren, J.; Alley, R.B.; Allison, I.; Carrasco, J.; Flato, G.; Fujii, Y.; Kaser, G.; Mote, P.; Thomas, R.H.; et al. Observations: Changes in Snow, Ice and Frozen Ground. In *Climate Change 2007: The Physical Science Basis. Contribution of Working Group I to the Fourth Assessment Report of the Intergovernmental Panel on Climate Change*; Solomon, S., Qin, D., Manning, M., Chen, Z., Marquis, M., Averyt, K.B., Tignor, M., Miller, H.L., Eds.; Cambridge University Press: Cambridge, UK, 2007; pp. 339–383.
32. Norwegian Polar Institute. *Stedsnavn Svalbard (S0 Stedsnavn) [Data Set]*; Norwegian Polar Institute: Tromsø, Norway, 2014. [CrossRef]
33. Selkowitz, D.J.; Forster, R.R. Automated mapping of persistent ice and snow cover across the western U.S. with Landsat. *ISPRS J. Photogramm. Remote Sens.* **2016**, *117*, 126–140. [CrossRef]
34. Rittger, K.; Bormann, K.J.; Blair, E.H.; Dozier, J.; Painter, T.H. Evaluation of VIIRS and MODIS Snow Cover Fraction in High-Mountain Asia Using Landsat 8 OLI. *Front. Remote Sens.* **2021**, *2*, 2673–6187. [CrossRef]
35. Gascoin, S.; Grizonnet, M.; Bouchet, M.; Salgues, G.; Hagolle, O. Theia Snow collection: High-resolution operational snow cover maps from Sentinel-2 and Landsat-8 data. *Earth Syst. Sci. Data* **2019**, *11*, 493–514. [CrossRef]
36. Hall, D.K.; Riggs, G.A.; Salomonson, V.V.; Di Girolamo, N.E.; Bayr, K.J. Modis snow-cover products. *Remote Sens. Environ.* **2002**, *83*, 181–194. [CrossRef]
37. El Oufir, M.K.; Chokmani, K.; El Alem, A.; Bernier, M. Estimating Snowpack Density from Near-Infrared Spectral Reflectance Using a Hybrid Model. *Remote Sens.* **2021**, *13*, 4089. [CrossRef]
38. Langlois, A.; Royer, A.; Montpetit, B.; Roy, A.; Durocher, M. Presenting Snow Grain Size and Shape Distributions in Northern Canada Using a New Photographic Device Allowing 2D and 3D Representation of Snow Grains. *Front. Earth Sci.* **2020**, *7*, 347. [CrossRef]
39. Bohn, N.; Di Mauro, B.; Colombo, R.; Thompson, D.R.; Susiluoto, J.; Carmon, N.; Turmon, M.J.; Guanter, L. Glacier ice surface properties in South-West Greenland Ice Sheet: First estimates from PRISMA imaging spectroscopy data. *J. Geophys. Res. Biogeosci.* **2022**, *127*. [CrossRef]

remote sensing MDPI

Article

Spatio-Temporal Variability of Suspended Particulate Matter in a High-Arctic Estuary (Adventfjorden, Svalbard) Using Sentinel-2 Time-Series

Daniela M. R. Walch [1,2,3,*], Rakesh K. Singh [3], Janne E. Søreide [1], Hugues Lantuit [2,4] and Amanda Poste [5,6]

1 Arctic Biology Department, The University Centre in Svalbard (UNIS), P.O. Box 156, 9171 Longyearbyen, Norway; janne.soreide@unis.no
2 Insititut für Umweltwissenschaften und Geografie, Universität Potsdam, 14461 Potsdam, Germany; hugues.lantuit@awi.de
3 Département de Biologie, Chimie et Géographie, Université du Québec à Rimouski, Rimouski, QC G5L 3A1, Canada; rakeshkstp@gmail.com
4 Permafrost Research Section, Alfred Wegener Institute for Polar and Marine Research, 14473 Potsdam, Germany
5 Norwegian Institute for Water Research (NIVA), Fram Centre for High North Research, Hjalmar Johansens Gate 14, 9007 Tromsø, Norway; amanda.poste@niva.no
6 Department of Arctic and Marine Biology, UiT: The Arctic University of Norway, P.O. Box 6050, 9037 Tromsø, Norway
* Correspondence: wald0001@uqar.ca

Abstract: Arctic coasts, which feature land-ocean transport of freshwater, sediments, and other terrestrial material, are impacted by climate change, including increased temperatures, melting glaciers, changes in precipitation and runoff. These trends are assumed to affect productivity in fjordic estuaries. However, the spatial extent and temporal variation of the freshwater-driven darkening of fjords remain unresolved. The present study illustrates the spatio-temporal variability of suspended particulate matter (SPM) in the Adventfjorden estuary, Svalbard, using in-situ field campaigns and ocean colour remote sensing (OCRS) via high-resolution Sentinel-2 imagery. To compute SPM concentration ($C_{SPM_{sat}}$), a semi-analytical algorithm was regionally calibrated using local in-situ data, which improved the accuracy of satellite-derived SPM concentration by ~20% (MRD). Analysis of SPM concentration for two consecutive years (2019, 2020) revealed strong seasonality of SPM in Adventfjorden. Highest estimated SPM concentrations and river plume extent (% of fjord with $C_{SPM_{sat}} > 30$ mg L^{-1}) occurred during June, July, and August. Concurrently, we observed a strong relationship between river plume extent and average air temperature over the 24 h prior to the observation ($R^2 = 0.69$). Considering predicted changes to environmental conditions in the Arctic region, this study highlights the importance of the rapidly changing environmental parameters and the significance of remote sensing in analysing fluxes in light attenuating particles, especially in the coastal Arctic Ocean.

Keywords: ocean colour; coastal darkening; SPM; sediment plumes; Arctic coast; remote sensing; regional tuning; coastal ecosystems; land-ocean-interaction; riverine inputs

Citation: Walch, D.M.R.; Singh, R.K.; Søreide, J.E.; Lantuit, H.; Poste, A. Spatio-Temporal Variability of Suspended Particulate Matter in a High-Arctic Estuary (Adventfjorden, Svalbard) Using Sentinel-2 Time-Series. *Remote Sens.* **2022**, *14*, 3123. https://doi.org/10.3390/rs14133123

Academic Editor: Dino Ienco

Received: 22 April 2022
Accepted: 18 June 2022
Published: 29 June 2022

1. Introduction

Changes to the Arctic have been well-documented, including increasing annual temperature and precipitation as well as distinct changes in seasonality [1–3]. These changes can be expected to affect catchment processes in Arctic river systems and impact adjacent coastal ecosystems through an increase in melting and drainage of glaciers during the ablation season [4], permafrost thaw [5,6] and more general shifts in the local water budget by extreme weather events such as heavy rainfall events [1].

Estuaries and fjords, as a specific type of coast, are characterised by strong physico-chemical gradients from land to sea [7]. These gradients vary corresponding to the distance from land and river outlets as well as coastal geomorphology. Riverine inputs, especially to these fjord-type estuaries [8,9], demonstrate strong variability in levels of particle load and distribution over a hydrological year [10] and have the potential to support primary production through delivery of terrestrial nutrients [11,12]. Environmental conditions in these transition zones are, thus, driven by tides, wind, as well as fluxes of freshwater, nutrients, and organic matter [12–15].

Water quality is a crucial assessment in these "Aquatic Critical Zones" [16], commonly through the characterisation of concentrations of chlorophyll-a (Chl-a), chromophoric dissolved organic matter (CDOM), turbidity and suspended particulate matter (SPM). The latter is the key driver of light attenuation in surface waters [17] as it increases scattering of incoming light [18]. Thus, high turbidity in sediment-laden freshwater plumes and comparably high amounts of CDOM may limit light availability for coastal phytoplankton and benthic primary producers [12,19,20]. Coastal darkening [21,22] refers to this ongoing change through increased light attenuation from terrestrial inputs in coastal ecosystems, which results in decreased benthic light availability. SPM may consist of organic and inorganic material [23]. In highly sedimentary environments, such as fjordic estuaries that are characterised by glaciogenic sediment delivery during the ablation season [24], SPM can be used as a proxy for estimation of sediment fluxes in these transition zones [25–27].

Polar warming, driving increased glacial river runoff [28], as well as an increase in precipitation [1] and coastal erosion may lead to massive sediment plumes along Arctic coastlines [29] such as in the archipelago of Svalbard. These plumes vary rapidly on spatio-temporal scales and hence, estimating the extent by traditional in-situ measurements is challenging. Consequently, ocean colour remote sensing (OCRS) for water quality monitoring has become more and more important [30] and analysis of satellite imagery for water quality monitoring at lower spatial resolution for off-shore applications and in lower latitudes is now quite common. The use of optical satellite imagery at high to medium resolution has, however, only seldomly been applied to the narrow and optically complex environments of polar fjords and estuaries [31,32], but is likely to improve our understanding of dynamic changes that occur over seasonal and shorter time scales (e.g., in response to extreme weather events or tidal cycles) and/or over a range of spatial scales (very local effects and regional patterns). Additionally, in polar regions, and even more so along the vast coasts that host areas of enhanced productivity, sufficient ground-truthing datasets for the validation of arctic tuned OCRS algorithms and consistent analysis remain scarce. With growing applications of remote observations in the polar regions from multispectral imagery [31,33–36], datasets of matchup are of increasing importance.

One of the greatest challenges associated with OCRS at higher latitudes is the dependency on favourable weather conditions: fjords and other nearshore environments are prone to higher levels of air moisture, often resulting in haze and clouds [37]. This is especially true during the high Arctic summer [38,39]. Polar-orbiting satellite missions overcome these limitations by featuring a higher temporal resolution with multiple overpasses per day during the summer season. In particular, the Sentinel-2 mission (ESA Copernicus), with its twin-satellites (S2A and S2B) equipped with MultiSpectral Imager (MSI) sensors, is providing high resolution and high frequency monitoring of Arctic regions, allowing for more cloud-free images over the course of a season. They operate with a primary focus on land imaging since 2015 and 2017 for S2A and S2B, respectively, and record 13 bands between violet and near-infrared at high to medium spatial resolution (band dependent between 10–60 m).

OCRS algorithms for turbidity [40] and SPM [41] in coastal areas have become a new asset. A wide range of ratio-based [26] and semi-analytical approaches for retrieving water quality parameters such as SPM from optical satellite imagery exists. An in-depth discussion of the performances can be found elsewhere (e.g., [42,43]) and would go beyond the scope of this introduction. Briefly, band-ratio algorithms rely on direct empirical relationships between in-situ concentrations of SPM (C_{SPM}) and recorded reflectance [26,44] and are, as such, data-driven [43] They can be distinguished from analytical or semi-analytical algorithms [41] that build (at least in part) on physical assumptions of the interaction between light and the parameter in question [43].

Studies emphasise the use of switching or algorithms to avoid saturation of the algorithm at use of a single-band [42]. Classically, reflectance in the red (for instance at 665 nm in the case of MSI) would in this case be used up to a threshold value of C_{SPM}, while for very high C_{SPM} longer wavelengths in the near infrared (at 865 nm), seem to model these concentrations more accurately. For highly productive fjordic areas, we have, however, to consider the possibility of misclassification of elevated signals in the near infrared—due to increased chlorophyll concentrations—as SPM [45]. Therefore, for this study, a single-band approach was chosen based on solely the red channel (at 665 nm). Therefore, a single-band approach based on the red channel was chosen to model C_{SPM} in the present study.

AOPs depend upon many region-specific chemical and physical factors (given by the IOPs) as well as the light field, and algorithms developed for the open ocean or other coastal regions may not be applicable to Arctic coasts. Klein et al. [35] therefore recently used in-situ measurements to regionally calibrate a turbidity model [40] for the Arctic nearshore zone of Herschel Island, Canada. Regional tuning of the existing generic algorithms for SPM [41,46,47] offers the possibility to build a strong RS-based dataset with high temporal and spatial resolution. Although the matchup cal/val dataset established for this study encompasses multiple days of measurements, it is not suitable for empirical single-band or band-ratio approaches [46] as they gain accuracy with the number of observations used as input. We therefore used the calibration dataset and statistical tools to regionally calibrate the semi-analytical single-band SPM algorithm, first introduced by Nechad et al. [41] (here named NeCal). We call this calibrated model 'AdvFCal' for this specific study, which enabled the observation of the seasonal evolution of modelled C_{SPM} and plume extent in Adventfjorden as well as the statistical evaluation of a set of potential environmental drivers.

With our study, the general goal was to enhance our understanding of the physicochemical conditions in highly dynamic coastal ecosystems, specifically the estuary of Adventfjorden, by following the objectives to (1) calibrate and validate (cal/val) the OCRS SPM algorithm and assess its performance, (2) analyse the spatial and temporal variability of C_{SPM} based on in-situ and Sentinel-2 data, and finally (3) couple time-series data of RS-derived SPM ($C_{SPM_{sat}}$) with additional meteorological and hydrological datasets to explore existing relationships with environmental drivers.

2. Materials and Methods

2.1. Study Site

Adventfjorden is one of several inner fjord arms in Isfjorden, a fjord system on the west coast of Spitsbergen in the archipelago of Svalbard, Norway [48]. The fjord is 8.3 km long and 3.4 km wide and has a northwest orientation within 78°13′ and 78°17′N and 15°25′ and 15°46′E [49] (Figure 1). A recent report highlighted the climate changes to this region, such as the ongoing rise in average annual temperature and the seasonality in precipitation [1]. The largest settlement of Svalbard, Longyearbyen, lies at the southwest shore of the fjord. Situated near the University Centre in Svalbard (UNIS), Adventfjorden offers high accessibility and short transport and processing time for field samples.

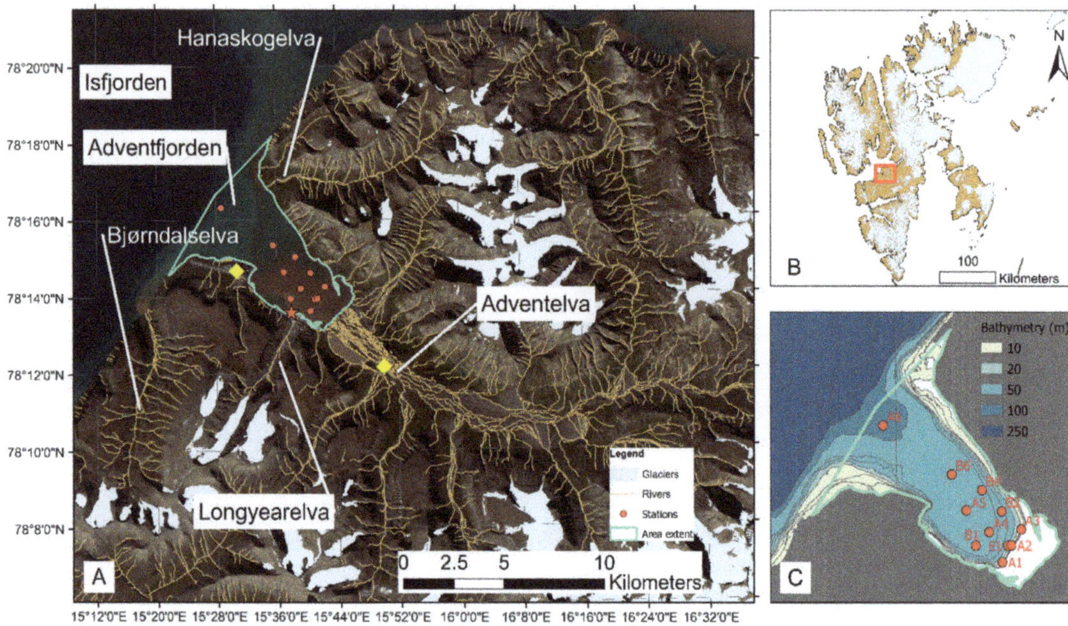

Figure 1. (**A**) Map of the Adventfjorden study area and the catchment of Adventelva and Longyearelva. Longyearbyen is marked with a red star. Yellow squares mark monitoring stations for weather (at Longyearbyen Airport) and river-hydrology (monitoring station operated by NIVA at Aventelva), (**B**) Overview of the study region in Svalbard archipelago, the highlighted outline indicating the area of interest (AOI) shown in (**A**,**C**) Station map of Adventfjorden for 2019 and 2020 field campaigns within the area extent defined for Adventfjorden for the purpose of this study.(map was created based on S-100 vector data provided by Norwegian Polar Institute [50]; bathymetry data from the Norwegian Mapping Authority [51]; contains modified Copernicus Sentinel-2 data (2020, Sentinel-2 B image 27 July 2020) processed by Sentinel-Hub (https://scihub.copernicus.eu/, accessed: 28 June 2021).

Adventdalen is a broad U-shaped valley with a catchment area of more than 500 km^2 [2,52]. The main discharging river is the strongly braided river Adventelva. A recent report indicate an increase in mean annual freshwater fluxes in the Adventelva river over the last two decades, with discharge reaching as high as 376,143 $\times 10^3$ m^3 per year (averaged over the years 2011–2019) [2]. The fjord is fed by several land-terminating glaciers [52,53] and receives discharge from the rivers Adventelva and Longyearelva, as well as several smaller river systems [48]. Thus, glaciogenic particles being mobilised throughout the ablation season dominate the suspended sediments in the runoff to the fjord [52,54].

Generally, fjords are areas of sediment deposition, especially in close proximity to river outlets [8,24], while transport to the outer parts of the fjord is limited by grain size of the transported sediments [49]. Riverine transport of terrigenous material to Adventfjorden is limited to approximately four months of the year, when the rivers and tributaries are flowing [55]. The fjord features seasonal ice-cover [55,56], but has remained without a solid sea-ice cover for several years [57], thus it is commonly regarded as ice-free. It should be noted that the studied year 2020 was an exception, when extensive ice-cover was observed in the fjord.

2.2. In-Situ Measurements

Field sampling and collection of in-situ data was carried out on 10 occasions in 2019 and 2020. Field work was carried out on small open boats and consisted of the collection of surface water samples from a station grid in Adventfjorden (n = 11 stations) that was designed to cover the main plume area. To capture the variability of values for SPM in the fjord, a transect from inner to outer fjord (stations A2, A4, A5, B6, A6, see Figure 1) was prioritised in the sampling campaigns, as was one cross-section close to the tidal flat (A1–A3) when sampling time was limited. Sampling days were planned to coincide with Sentinel-2 (S2A and S2B) overpasses and (where possible) clear skies, allowing for the in-situ samples to be used as a ground truth dataset to match-up with corresponding cloudless satellite images. The duration of field campaigns varied between two to five hours depending on the number of measurements and water samples taken. This resulted in varying differences between the in-situ sampling and the satellite passing, which occurred between 12:00 and 13:30 (UTC) on average.

Water samples were collected from the uppermost 10–20 cm of surface water with a clean stainless-steel bucket, pre-screened through 200 µm mesh to remove larger zooplankton and transferred into a well-rinsed opaque 5 L carboy. Samples were kept cool in a dark transport cooler box until further processing at the University Centre in Svalbard (UNIS). Secchi depth Z_{SD} was recorded on every station using a standardised secchi disk and a measuring tape lowered from the shadow side of the vessel. A handheld turbidity meter (TN-100 model, Thermo Scientific Eutech) was used to record T in NTU (Nephelometric Turbidity Unit) from well-mixed samples immediately after collection. In this study, measurements of turbidity were taken as triplicates at each station in 2019 only since the instrument was not available for use during the field operations in 2020. Averaged values of turbidity per station were used for further analysis.

C_{SPM} in the surface water of Adventfjorden was determined for n = 96 surface samples, over the course of the two ablation seasons in 2019 and 2020, respectively (see Supplementary Table S1). SPM concentration is calculated from laboratory filtration of water samples through filters with commonly 0.7 µm pore size [58] via gravimetric methods (e.g., [23]). To determine the dry weight of SPM (hereon SPM_{insitu}), the obtained water samples were filtered through pre-combusted, pre-weighed 47 mm GF/F glass microfibre filters (Whatman®) with a nominal pore size of 0.7 µm [23], following protocols of previous work in the FreshFate project (see [12]). Filtration of samples took place promptly, with the time between field sampling and processing rarely exceeding four hours. Maximum filtration volumes were chosen according to total available sample volume and the present water clarity at the station. The final filtration volume was noted, the filters were transferred into petri dishes (FALCON®, 50 mm × 9 mm) and stored horizontally in the dark at −20 °C until further analysis (within two months of collection). Filters were dried at 104 °C for 60 min, then cooled in a desiccator. This drying process was repeated until the weight was stable (mass loss of less than 4% or 0.04 mg between weighing). The final concentration of $C_{SPM_{insitu}}$ (mg L^{-1}) was determined from the difference in weight of the filter before and after filtration (in mg) divided by the filtration volume (in L).

2.3. Sentinel-2 Satellite Imagery

In aquatic applications of Earth Observation (EO), airborne and satellite-based remote sensing (RS) are used to assess the magnitude and variability of optically active constituents (OACs), such as SPM, present in the water body. The Inherent Optical Properties (IOPs) of these result in different spectral patterns due to selective absorption (a) and scattering (b) in the backward direction (b_b) [59]. Ocean Colour Remote Sensing (OCRS) is, thus, used to estimate the concentration of the OACSs with a high-degree of accuracy [60,61] by building on inverse relationships between IOP, which are independent on the geometry of the incident and reflected light, and apparent optical properties (AOP, i.e., the water's colour) recorded as remote sensing reflectance R_{rs}. The water-leaving reflectance ρ_w, which is more commonly used in OCRS, is calculated from R_{rs},

$$\rho_w(\lambda) = \pi \times R_{rs}(\lambda) \tag{1}$$

For this study, available Copernicus Sentinel 2019 and 2020 data, that is S2A and S2B Level 1C (L1C) images were downloaded from the Copernicus Open Access Hub (European Space Agency, ESA) consisting of calibrated top-of-the-atmosphere (TOA) reflectance values. Images throughout the sensing period 01 March 2019 to 30 September 2019 and 01 March 2020 to 30 September 2020 that fit the criteria were further investigated as potential data based on the relative cloud cover present in the image at the time of the observation. The threshold for the selection was set to 50% cloud cover to account for the potential valid observations over the area of interest (AOI), Adventfjorden.

The TOA reflectance from the L1C images was corrected for the influence of atmospheric scattering, which includes scattering by air molecules and aerosols, along with the surface scattering in the form of, for instance, sun-glint [62,63]. This correction of TOA reflectance to calculate bottom of atmosphere (BOA) reflectance, called atmospheric correction (AC), is a crucial part of OCRS. Over optically complex coastal and inland water surfaces, AC is a challenging process and may lead to large differences between satellite estimates and in-situ conditions [62]. The remote sensing reflectance (R_{rs}) was, therefore, determined from the L1C products by the ocean colour processing software SeaDAS v7.5.3 (SeaWiFS Data Analysis System, NASA) with a modified aerosol correction method adapted for optically complex and turbid water masses, called SSP aerosol correction algorithm [62], which computed aerosol radiance in the coastal waters of a highly sedimentary estuary with exceptional accuracy. Therefore, SSP is used in the present study. Due to lacking validation reflectance data, we were not able to perform the validation of the here-chosen AC for Adventfjorden. The application of this AC workflow has been, however, recently evaluated for very turbid waters of the St. Lawrence estuary and James Bay, Canada [64] and showed good results for the red band of the MSI sensor onboard Sentinel-2. Using this adapted AC within the framework of the SeaDAS command line allowed for the automated processing of all available images.

An initial number of 181 images had to be reduced to 116 based on requirements for cloud presence over the area of interest. The images were cropped to the area of interest. The land and clouds were masked, leaving the area of Adventfjorden, from the tidal flat to the farthest station A6 for time-series and statistical analysis.

2.4. Calibration and Validation of the SPM Algorithm

The semi-analytical single-band SPM algorithm introduced by [41] assumes that the IOPs of suspended particles are directly proportional to C_{SPM}, with negligible influence from other (non-particulate) constituents. By applying these assumptions, C_{SPM} may be calculated as [41]

$$C_{SPM} = \frac{A^{\rho}\rho_w(665)}{1 - \rho_w(665)/C^{\rho}} + B^{\rho} \tag{2}$$

where ρ_w is the water-leaving reflectance, and A^{ρ} $\left(\text{g m}^{-3}\right)$ and C^{ρ} are coefficients representing IOPs of the water column (i.e., algal and non-algal particles). B^{ρ} is an offset to avoid underestimation by the model at low SPM concentrations [41,42]. This generic algorithm was originally calibrated based on data from the southern North Sea [41] and should therefore be re-calibrated to suit the regional characteristics of waters in Arctic Adventfjorden, which is impacted by high sediment loads from land-terminating glaciers in the adjacent catchment.

Given the characteristic signature of absorption, the choice of wavelength (and the corresponding band in the sensor) can influence the accuracy of $\rho_w(\lambda)$ estimates significantly [42,46,65]. This study used ρ_w at the red band of the Sentinel-2 MSI sensors (λ = 665 nm) to derive remotely sensed values SPM (here on, C_{SPMsat}). The red band was chosen as Chl-a exhibits an absorption maximum in the red band while the reflectance of sediments is significantly high, which will reduce potential misclassification of phytoplankton particles as SPM for the purpose of this study.

Non-linear least squared analysis was then performed on a calibration dataset for $C_{SPM_{insitu}}$ against the pixel value for the reflectance $\rho_w(665)$ in Python v3.7.9 using the open-source SciPy package [66] and averaging 100 rounds of random subsamples. Bounds for the three coefficients (A, B and C in Equation (2)) were chosen in accordance with the suggestions in Nechad et al. [41] for the chosen band wavelength of 665 nm. The calculated $C_{SPM_{sat}}$ data was then tested against $C_{SPM_{insitu}}$ measurements from the field campaign. The performance of the calibration was assessed with the standard measurement of the normalised root-mean-square difference (RMSD), the mean relative difference (MRD), and bias, which are used in performance testing of modelled values in OCRS [62,67]. These were calculated as

$$MRD = \frac{1}{n}\sum \frac{C_{SPM_{sat}} - C_{SPM_{insitu}}}{C_{SPM_{insitu}}} \times 100 \quad (3)$$

$$RMSD = \sqrt{\frac{1}{n}\sum \left(C_{SPM_{insitu}} - C_{SPM_{sat}}\right)^2} \quad (4)$$

$$Bias = \frac{1}{n}\sum \left(C_{SPM_{insitu}} - C_{SPM_{sat}}\right) \quad (5)$$

where n is the number of observations.

2.5. Environmental Datasets

Weather data, recorded at the nearest permanently operated weather station (Svalbard airport [68]) was downloaded from the Norwegian MET service homepage [69]. The values were processed and used as background data for regression analysis, i.e., environmental forcing of freshwater-derived turbid river plumes in Adventfjorden. Daily averages of air temperature (°C) and wind speed (ms^{-1}) as well as sums of precipitation are used to describe the meteorology present in Adventfjorden over the course of the studied period. Tidal data for the semi-diurnal tidal regime of Adventfjorden (~104 cm tidal range) was downloaded for Longyearbyen and Adventfjorden as recorded by Norwegian Mapping Authorities, Hydrographic Service [70]. The tide tables are openly accessible. In-situ sensor-based monitoring platforms are operated by the Norwegian Institute for Water Research (NIVA, PI: Amanda Poste) in the Adventelva river. The river mooring is typically deployed in mid-June and retrieved in late September, and provides continuous measurements of temperature, conductivity, and pH (2017–2020); turbidity (2019–2020); and water level (2017–2019). Here, values for turbidity and water level were used.

2.6. Time-Series Analysis and Environmental Statistics

A workflow in R 4.1.0 [71], was used for further processing based on the packages "raster" [72] and "ncdf4" [73]. The regionally adapted coefficients for the OCRS Nechad et al. [41] SPM algorithm were applied using $\rho_w(665)$ values from L2 processed images to retrieve $C_{SPM_{sat}}$, with a maximum 500 mg L^{-1} $C_{SPM_{sat}}$ defined as valid observations per pixel. Projected image stacks for 2019–2020 of satellite images were created. Additional bands were calculated, including statistical information per temporally binned pixel. More specifically minimum (min), maximum (max), standard deviation (SD), coefficient of variation (CV), mean, median, and finally the number of valid observations. The images with zero valid observations were discarded. Pixel values of $C_{SPM_{sat}}$ were extracted from all images. Based on this dataset, elementary statistics were applied to all images (i.e., minimum, maximum, mean, median, interquartile range (IQR)) based on the total number of available pixels per image.

For the analysis of the spatial extent of very turbid sediment-laden freshwater plumes in Adventfjorden, that we call the river plume, a threshold value for $C_{SPM_{sat}}$ of 30 mg L^{-1} was identified from processed images and modelled $C_{SPM_{sat}}$ and is furthermore based on

Secchi depth (Z_{SD}) being less than 1 m for most occasions where $C_{SPM_{insitu}}$ observations reached or exceeded 30 mg L^{-1}. In a comparable study on OCRS of sediment plumes, [74] defined 10 mg L^{-1} as a threshold. However, in the highly sedimentary systems of Svalbard this value might be easily exceeded, hence, a higher value was chosen to differentiate between distinct river plumes and assess its seasonal variability. This threshold value functions as a mask to the time-series data, and the relative area of the fjord covered with the river plume was estimated with the following equation.

$$\text{Pixel}_{\text{plume}}[\%] = \frac{N_{>30}}{N} \times 100 \quad (6)$$

Here, $N_{>30}$ depicts the number of pixels with $C_{SPM_{sat}}$ greater than 30 mg L^{-1} and N is the total number of pixels in the fjord and the catchment area.

June, July, and August (JJA) satellite data were selected as the months with the most available supplementary data. Only images with more than 50% valid pixel observations were chosen for correlation analysis. For 2019, the year with available data from environmental variables, a linear model was developed for predictors of median $C_{SPM_{sat}}$ values and plume extent. Statistical analysis was performed on averaged values for satellite imagery (i.e., median values for $C_{SPM_{sat}}$ and river plume extent) and meteorological-environmental data from the weather- and river-monitoring stations. Therefore, the supplementary dataset based on the weather and environmental data was merged to the summary statistics per RS image based on the averaged values of environmental data over 24 h prior to the satellite observation. The following environmental predictors were tested against the log-transformed median values for $C_{SPM_{sat}}$ and river plumes extent for the datapoints in the JJA subset:

- Averaged air temperature (°C);
- Precipitation sums (mm);
- Averaged water level at the river station (m);
- Averaged turbidity T (NTU) at the river station.

Here, a combination of correlation matrix and regression analysis is used to determine relationships of environmental parameters and the evolution of SPM (concentration and river plume extent) in the fjord. All statistical analyses on extracted $C_{SPM_{sat}}$ and additional environmental datasets were performed in R 4.1.0 [71]. The functions "cor" (R basic package), and packages "corrplot" [75], and "reshape2" [76] were used for correlation analyses and to create elementary matrices, while "ggpmisc" [77] was used for the regressions and statistical evaluation of potential environmental predictors of SPM concentrations and plume extent in Adventfjorden. Both steps in the analysis were based on only complete observations. The packages "ggplot2" [78] and "RColorBrewer" [79] were used for arranging the data and visualization.

3. Results

3.1. Meteorological and Environmental Conditions in 2019 and 2020

Meteorologically, the year 2019 was colder and wetter than 2020, with average temperature and sum precipitation being −3.43 °C and 167 mm in 2019 and −3.38 °C and 137.9 mm in 2020, respectively (an overview of the meteorological conditions can be found in the Supplementary Figure S4). The average air temperature in JJA for 2019 was 6.46 °C, while in JJA 2020 it rose by 11.5% to an average of 7.2 °C. July 2020 also experienced an extreme heat event, with maximum air temperatures reaching higher than 20 °C for four consecutive days (25–29 July 2020). For both spring and autumn, precipitation and snow cover were strikingly higher 2019 compared to 2020. The lower air temperatures in winter and spring 2020 resulted in an exceptional sea-ice cover in the innermost parts of Adventfjorden, which was visible on satellite images until mid-May, whereas in 2019, sea ice was not observed in Adventfjorden (as is common for this fjord). Wind is a considerable factor in surface distribution of suspended and dissolved matter. Easterly and south-easterly winds in Adventfjorden are predominant throughout the year and only disturbed by a period of westerly and south-westerly winds during the summer months (JJA).

3.2. In-Situ Measurements

SPM values presented an overall high variability in Adventfjorden in 2019 and 2020 ablation seasons (Table 1). In 2019, highest SPM values were found in early August, while in 2020, the values for SPM_{insitu} were particularly high in both June and July. It should be noted that the timing of the sampling in 2019 and 2020 is different, which will lead to differences in the measurements and subsequently monthly averaged values of particulate matter load in the fjord surface water.

Table 1. Summary of performed in-situ sampling during the 2019 and 2020 summer campaigns. Averaged values (± SD) are given for suspended particulate matter (C_{SPM}), turbidity (*T*), and secchi depth (Z_{SD}) at *n* stations sampled. In 2020, turbidity was not measured.

Date	Platform	C_{SPM} [mg L^{-1}]	*T* [NTU]	Z_{SD} [m]	*n*
14 June 2019	Basecamp	27.32 ± 7.5	19.50 ± 16.8	1.44 ± 1.32	11
17 June 2019	UNIS Kolga	25.73 ± 12.3	12.60 ± 11.98	1.01 ± 0.8	8
06 August 2019	UNIS Kolga	57.80 ± 24.13	40.53 ± 35.29	-	10
07 August 2019	UNIS Kolga	68.22 ± 56.34	19.30 ± 20.58	1.81 ± 1.72	8
12 June 2020	UNIS Polaris	115.45 ± 131.37	-	0.90 ± 1.31	11
17 July 2020	UNIS Polaris	70.41 ± 75.16	-	0.93 ± 1.03	11
30 July 2020	UNIS Polaris	156.68 ± 108.61	-	0.64 ± 0.75	9
26 August 2020	UNIS Polaris	28.01 ± 9.11	-	3.80 ± 2.04	10
22 September 2020	UNIS Kolga	7.32 ± 3.44	-	5.20 ± 1.59	11

The $C_{SPM_{insitu}}$ in Adventfjorden ranged from 4.7 to 152.36 mg L^{-1} and 3.57 to 434.95 mg L^{-1} in the studied years 2019 and 2020, respectively. In both years, the highest values for $C_{SPM_{insitu}}$ were found in the innermost stations of the fjord, with values ranging from 16.7 to 152.4 mg L^{-1} for the stations E1 and A1, A2, and A3, which were located within 1 km distance to the tidal flat of the estuary. The entry station of the river, E1, shows comparable values to the inner transect stations. The values for SPM_{insitu} generally decreased towards outer Adventfjorden. The lowest values for SPM were, thus, found at A6, the entry point to Isfjorden, with minimum of 4.7 mg L^{-1} $C_{SPM_{insitu}}$ over the course of both years. Turbidity was measured in the 2019 field campaigns only and followed largely the values of $C_{SPM_{insitu}}$. Averaged values and SD increased noticeably in August, illustrating the high variability seen in the values for C_{SPM}. Secchi depth Z_{SD} was measured as a solid indicator of photosynthetically available radiation within the water column. The negative non-linear relationship of Z_{SD} with $C_{SPM_{insitu}}$ is shown in the Supplementary Figure S3. Z_{SD} were shallowest during the June and July sampling days in 2020, with values as low as 0.05 m for the innermost station (A1) on 12 June 2020.

3.3. SPM Algorithm Calibration and Validation

Four match-up days of in-situ field measurements and clear-weather satellite images were achieved during the study period of 2019 and 2020 (Table 2, $n = 60$). However, only three dates were used: 14 June 2019, 06 August 2019, and 17 July 2020. The sampling day on the 30 July 2020 was classified as an outlier and removed due to the unique characteristics of the very high SPM load and turbidity likely caused by the extreme heatwave event (see above; an overview of the meteorological conditions is given in the Supplementary Figure S4). Quality assessment of the dataset revealed unusable data from two stations sampled on 06 August 2019. Subsequently, the remaining dataset ($n = 46$) was split based on values for $\rho_w(665)$ and $n = 23$ and $n = 23$ of $C_{SPM_{insitu}}$ paired with contemporaneous $\rho_w(665)$ values were used in the calibration/validation (cal/val) process, respectively.

Table 2. Overview of timing of sampling in Adventfjorden that coincided with the acquisition of the Sentinel-2 satellite imagery. Start and End time for field campaigns was taken from CTD casts and corrected for local time. The time (UTC) is given for every matchup date. Note that the cal/val data obtained on 30 July 2020 (in italic) was later excluded from the calibration.

Matchup Date	In-Situ Timing (UTC)		Sentinel-2 Acquisition (UTC)	
	Start	End	S2A	S2B
14 June 2019	10:48	13:18	12:57	12:06
06 August 2019	11:49	14:17	13:07	12:16
17 July 2020	11:14	13:46	–	12:37
30 July 2020	*11:03*	*14:23*	*11:58*	*12:47*

The results of the non-linear least squared regression analysis are presented in Table 3 (and visually in Figure 2) and include the results of the accuracy assessment based on the validation dataset. The A coefficient, essentially reflecting the backscattering characteristics of algal and non-algal particles was found to be higher and the C coefficient lower but within reasonable margins of the original calibration in Nechad et al. [41]. AdvFCal performed better on the validation dataset with MRD 47.5% and 29.1% for the Nechad- (NeCal) and the Adventfjorden calibration (AdvFCal), respectively. The calibrated algorithm also shows better values for RMSD and bias.

Table 3. Coefficients for the SPM-algorithm in the original (NeCal) [41] and the adapted regional (AdvFCal) calibration. Measures of accuracy are given with mean relative difference (MRD), root mean square difference (RMSD), bias, and the coefficient of determination (R^2).

	Coefficients for $\rho_w(665)$	MRD	RMSD	Bias	R^2
NeCal	A: 355.85, B: 1.74, C: 0.1728	47.5%	23.3%	−17.87	0.55
AdvFCal	A: 523.78, B: 1.97, C: 0.158	29.1%	15.9%	−7.72	0.55

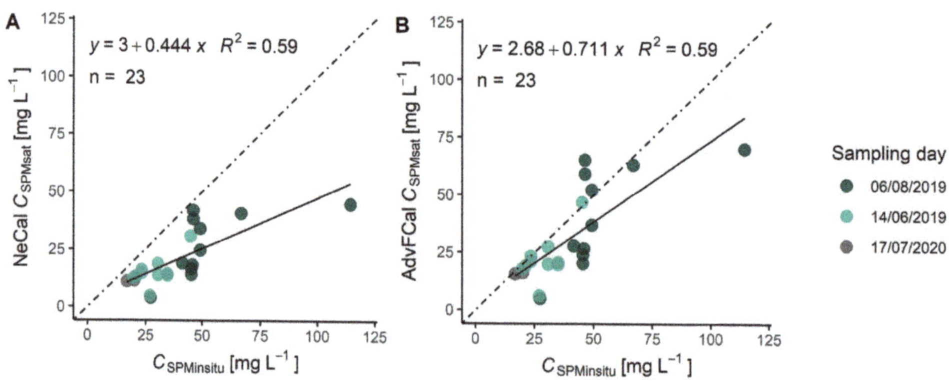

Figure 2. Validation comparison of (**A**) the original SPM algorithm calibration (NeCal) and (**B**) the regionally tuned algorithm (AdvFCal) based on the match-up dataset from three sampling days, shown by the different colours. The dashed line illustrates a 1:1 relationship between modelled and in-situ values.

3.4. Sensitivity Analysis of AdvFCal

The visual validation (Figure 2) shows the underestimation of in-situ values in the NeCal calibration. While the retuned AdvFCal performs better, both calibrations show outliers for two match-up data points. Therefore, the sensitivity in the training data to different parameters was analysed. The sensitivity analysis on the cal/val dataset showed marginal correlation between residuals of predicted and observed C_{SPM} values and the time-difference between the in-situ measurements and the satellite observation (Supplementary Figure S1). Additionally, we explored potential sensitivity to different concentrations of C_{SPM} and how it might affect the algorithm's performance. The use of different window sizes was explored (1 × 1, 3 × 3, 5 × 5 pixel) and showed negligible effects on residuals in the matchup dataset. Considering the very local processes in the estuary and the potential effects of pixel-mixing at the land-ocean interface, a high spatial resolution is needed [80]. We therefore processed the data at its native resolution of ~20 m.

3.5. Time-Series Analysis

A total of 116 Sentinel-2 (MSI) images were found to fit the criteria for the performed time-series analysis. The area of the polygon created as bounds for the satellite estimates over our AOI was ~29 km^2 (see Figure 1), consisting of more than 200.000 potential pixel estimates. In total, more than 15 × 10^6 valid pixel estimates could be yielded from the time-series dataset. For 2020, land-fast ice was still present in the inner parts of Adventfjorden, thus, limiting the number of valid retrievals in spring 2020 spatially to the mid-fjord and outer fjord regions and temporarily to the time after the ice-breakup (around mid-May). Maps showing the valid observations per pixel can be found in the Supplementary Figure S2. The spatial resolution after processing was found to be ~20–27 m due to the specifications in the SeaDAS l2gen module.

The river plume extent exhibited a high degree of seasonal variability in both 2019 and 2020 (Figure 3). In both years the river plume extent remained well below 10% of the fjord area for most parts of the observation period, however, a strong increase in the plume extent was observed in mid-June, with values remaining elevated until late August both in 2019 and 2020. In 2019, the strong influence of the river plume began in mid-June (16 June 2019), when it increased from 5% to over 20% of the fjord over a two-day period. This increase also coincided with a 17 cm increase in river water level, observed at the river sensor station (marked in Figure 1). The maximum plume extent was 34.8% of the fjord area, on 09 July 2019, while the last observed day with significant plume extent in Adventfjorden was 06 August 2019. In 2020, the plume also developed toward mid-June, reaching 5.9% on the 17 June 2020. The maximum plume extent of 40.5% was observed between 26–27 July 2020, following the extreme heat event with very high (>20 °C) air temperatures for four consecutive days. Overall, the turbid plume in Adventfjorden does not exceed ~35% to ~41% of the total fjord area for 2019 and 2020, respectively. Noteworthily, the sampling days considered for the cal/val dataset coincided with the onset and offset of the plume influence in 2019, while the sampling campaigns were conducted during the time of enhanced riverine influence.

Variability between stations and years, observed in the in-situ measurements are reflected in the RS dataset. The RS data shows higher variability in the inner fjord stations for 2020 in contrast to 2019 (Figure 4). Generally, it is the inner fjord stations close to the Adventelva tidal flat that experience largest variability but also highest median $C_{SPM_{sat}}$ in both years. Reduced valid observations in early 2020 due to exceptional ice cover in Adventfjorden may lead to higher median values.

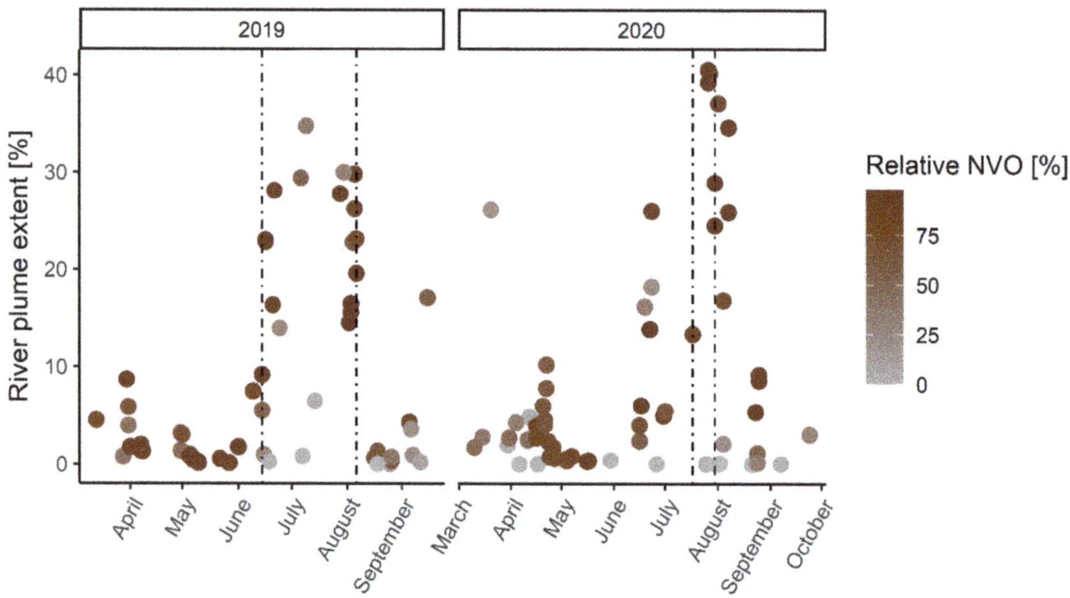

Figure 3. Relative area of Adventfjorden classified as river plume (i.e., covered by $C_{SPM_{sat}}$ values of 30 to 500 mg L^{-1}). Cloud cover as estimated from the relative number of valid observations (NVO) per day is shown as the gradient of colour. Images in JJA with extensive (>50%) cloud cover, i.e., invalid observations, were excluded from the environmental statistics later on. Dashed lines represent the in-situ matchup days considered for the cal/val dataset.

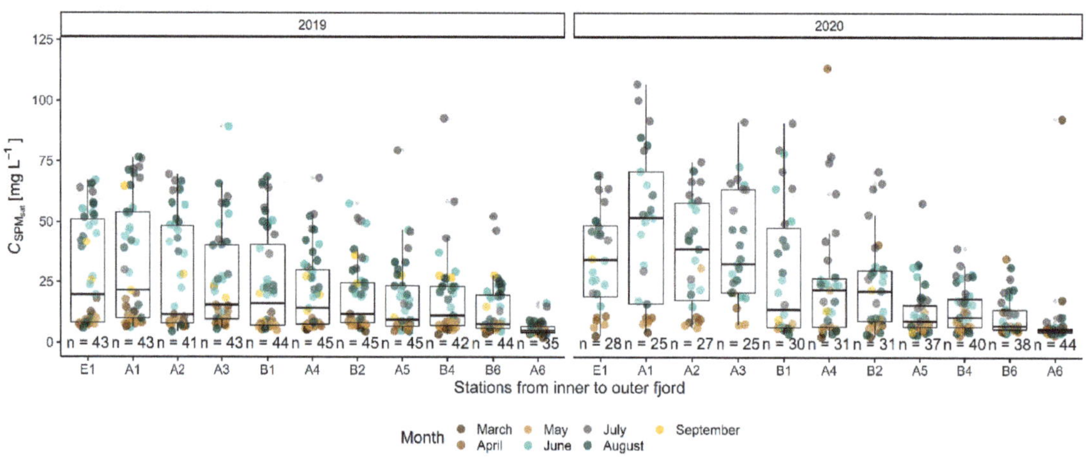

Figure 4. Box and whiskers plot with underlying point observations for $C_{SPM_{sat}}$ at the sampling stations in Adventfjorden in 2019 and 2020 (for FreshFate station grid see map in Figure 1). NVO (n) for each station are indicated below each boxplot. The stations are sorted from inner to outer fjord stations to illustrate spatial variability.

For the summer months (JJA), monthly composite pictures of $C_{SPM_{sat}}$ (calculated from averaged mean-values of $C_{SPM_{sat}}$) reveal strong differences in concentration and distribution of $C_{SPM_{sat}}$ between months (Figure 5). The maps show that the sediment-laden water masses do not penetrate far out into the fjord in June, while for July the strong influence of SPM to nearly the whole surface area of Adventfjorden becomes apparent. $C_{SPM_{sat}}$ decreases significantly in August, leading to a characteristic threefold dynamic of SPM influence to Adventfjorden with strong gradients during the summer shoulder months June and August. Maps of binned monthly satellite images for JJA of coefficient of variation (CV) per pixel were chosen as an indicator of the spatial variability over time. Variability "hotspots" were mostly located in the very nearshore environments in Adventfjorden but are especially pronounced in the direct proximity of river outlets, such as Longyearelva or Hanaskogelva. Surprisingly, the CV values per pixel were less high in the tidal flat of Adventelva, indicating less pronounced changes over the monitoring period. Other areas of stronger variability appear to be the outlet in Hanaskogdalen on the northern side of Adventfjorden as well as the southwestern coast of Adventfjorden, where the data suggests influences from dispersed plumes from the close-by Bjørndalselva. Variability is generally higher in the middle parts of Adventfjorden in July compared to the other months.

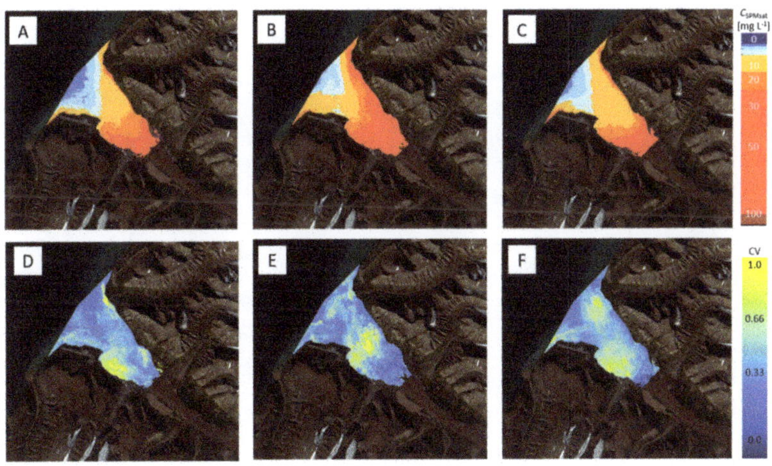

Figure 5. Composite images of mean values for $C_{SPM_{sat}}$ (upper panels) and coefficient of variation (CV, lower panel) per pixel during the months of June (**A**,**D**), July (**B**,**E**), and August (**C**,**F**). Composite are based on both 2019 and 2020 satellite imagery (background corresponding to Sentinel-2 image in overview map of this study). Contains modified Copernicus Sentinel-2 data (2020, Sentinel-2 B image 27 July 2020) processed by Sentinel-Hub (https://scihub.copernicus.eu/, accessed on 28 June 2021).

3.6. Environmental Statistics

Pearson correlation analysis was performed on the JJA 2019 subset to the dataset (n = 21). The strongest relationship defined by the Pearson correlation coefficient r was found between the median $logC_{SPM_{sat}}$ values and the river water level over the previous 24 h (r = 0.89, $p < 0.005$) or the river plume extent ($C_{SPM_{sat}}$>30 mg L^{-1}) and the air temperature during the previous 24 h (r = 0.78, $p < 0.005$). Since air temperature data was available for the analysis of the 2020 dataset as well, we can show that the relationship persisted over the whole dataset, increasing the correlation to r = 0.83. An overview is given in the correlation plot in the Supplementary Figure S5. Linear models for the above-mentioned predictors of median $C_{SPM_{sat}}$ and plume extent confirmed the high explanatory power of the tested environmental parameters with R^2 (coefficient of determination) being 0.69 for the relationship of averaged air temperature and the relative cover of the fjord by plume (Figure 6).

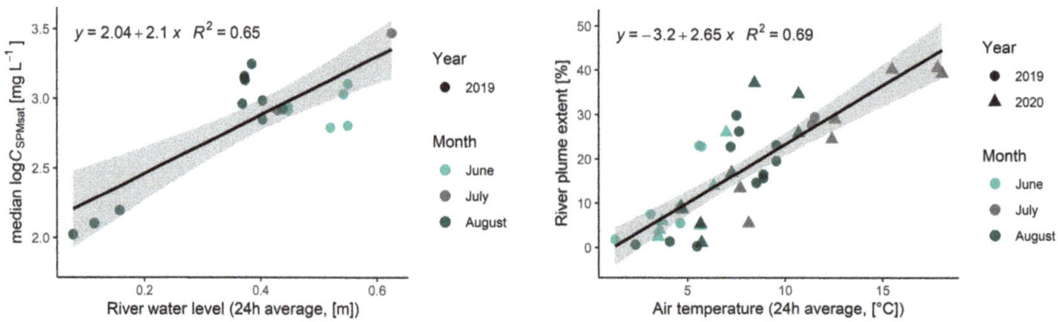

Figure 6. Linear model showing the analysed relationship of the satellite derived river plume extent ($C_{SPM_{sat}}$> 30 mg L^{-1}) and air temperature (°C) as well as the river water level and median values of $logC_{SPM}$. The equation and R^2, the coefficient of determination, are given. Colours and shapes represent the month and year of the satellite acquisition, respectively.

4. Discussion

This study proposed a regionally calibrated algorithm (AdvFCal) based on the well-established NeCal by Nechad et al. [41]. We show that regional or site-specific tuning is crucial, especially in complex nearshore environments [35], as it enhanced the accuracy of satellite-derived time-series analysis of $C_{SPM_{sat}}$ by ~20% for the targeted study region. The yielded RS dataset covered two complete summer seasons in 2019 and 2020 and managed in a satisfactory way to detect the onset and length of the runoff season from satellite imagery based on the distinct change in the freshwater-driven $C_{SPM_{sat}}$. With the ~20 m spatial resolution achieved in our workflow on S2 data, trends in riverine inputs to the dynamic and complex nearshore zones of inaccessible Arctic fjords can be explored. Especially during the summer months, the extensive time-series dataset allows a near-daily monitoring of river plume evolution and has the potential to monitor the long-term trends, such as coastal darkening and related decrease of benthic light availability [22]. It is reasonable to believe that the calibration holds for comparable catchments in polar regions, dominated by glaciogenic sediments. Indeed, a comparable study showed a solid reliability for the transfer of a regionally tuned turbidity algorithm [35], which may indicate the broader applicability of the calibration AdvFCal to a wider spectrum of Arctic coasts.

4.1. Surface Water SPM Exhibits High Variability in Space and Time

The particle load in the surface water in Adventfjorden showed considerable spatial differences within the in-situ data, and also within the spatially more-comprehensive RS dataset (see also Figures 3–5). The concentration and distribution of SPM varied strongly during the studied period with persistently high values found in the summer months of June, July, and August of 2019 and 2020, respectively. The estimated values for C_{SPM} (both in-situ and from OCRS) are consistent with earlier studies from Zajączkowski [49] and Weslawski [48], with 112 mg L^{-1} and 200–500 mg L^{-1}, respectively. The range of values for SPM was naturally found to be particularly high in the inner fjord stations, in accordance with previous in-situ studies [24,48,56,81]. The spatial distribution of sediment-laden water masses seems to be most commonly observed as a deflection to the right of the river mouth, following the Coriolis force [8,15].

We were able to extract the coefficient of variation (CV) per pixel to describe the most variable locations of surface C_{SPM} in Adventfjorden (Figure 5). We see CV increasing during the summer months (JJA), especially close to river outlets from catchments with steeper terrain. The river outlet of Longyearelva is, in our dataset, a striking source of variability compared to the wider Adventelva river outlet. The longer passage of discharged glacial meltwater through the valley of Adventdalen might offer a possible explanation, as particles will have the time to settle and accumulate in river channels [82]. In contrast, glaciers in

Longyeardalen [83] as well as the steeper catchment terrain may support higher discharge velocities with subsequent effects on SPM concentration and variability. In our data, the considerably higher water levels in 2020 measured at the river monitoring station may be relatable to Longyearelva and explain the generally higher values of SPM load during JJA 2020. Areas of high variability will favour a community structure that is adapted to a wider range of physico-chemical properties. Identification of hotspots of variability through binned satellite imagery might deliver a crucial metric for ecosystem studies targeting community structure or shifts in productivity.

A study on subglacial discharge in Raudvika in Kongsfjorden, Svalbard, found C_{SPM} ranging as high as 600 mg L^{-1} at the glacier front using Landsat-8 imagery and a band-ratio algorithm [84]. The direct discharge of turbid subglacial meltwater enriched with inorganic particles may, however, only compare to a certain degree to the retention potential of extensive river valleys that drain a comparably large area that is under various influences. This may also be evident in the differences in suspended particle load in subpolar fjord systems such as the Norwegian Trondheimsfjord and Sognefjord, where values for suspended matter were substantially lower (maximal C_{SPM} 1.40 and 2.23 mg L^{-1}, respectively) [10]. Here, not only catchment size but also the presence of vegetation in the river valley may be a reasonable explanation for the observed differences, as longer retention times and disturbance by vegetation may prevent sediments from mobilising to the same degree and cause sediments and particles to settle more readily.

Arctic coastal zones and fjords off Svalbard generally exhibit comparably lower C_{SPM}. In the case of several studies in Greenland [19,28,85], C_{SPM} estimated mainly from in-situ measurements were typically a fractional amount (~10%) of the values observed in Adventfjorden. When comparing these values, it is, however, important to acknowledge the differences in study design as well as the characteristics of the fjord or estuary studied. The length of most of the other Arctic study areas encompassed longer distances, leading to more sediment loss over longer distances from the river mouth or glacier front, which subsequently might explain lower median values. In the case of the Uummannaq fjord and Vaigat-Disko bay [19], the length of both fjords is tenfold that of Adventfjorden.

4.2. Temperature Drives Mobilisation and Transport of SPM to Arctic Fjords

The values of averaged air temperature 24 h prior to the observation showed a strong (r = 0.83, R^2 = 0.69) correlation with the river plume extent and a less pronounced, yet significant relationship, with the median of log-transformed $C_{SPM_{sat}}$ values (r = 0.59). This might illustrate how dilution of particles during times of higher discharge can foster a larger extent of the plume itself while C_{SPM} might not grow proportionally. Mobilisation of particles from subglacial discharge and erosion processes in the catchment are generally tied to discharged volume and velocity of water flow in the catchment [54].

The here-presented satellite-based dataset could not support a strong relationship of either median values of $C_{SPM_{sat}}$ or the river plume extent ($C_{SPM_{sat}}$ > 30 mg L^{-1}) to the precipitation values in the previous 24 h—a relationship discussed in other publications (e.g., [86]). However, continuous rainfall over longer periods of time as well as extreme rainfall events will sustain higher water levels in the rivers and have subsequent effects on particle fluxes from mobilised sediments. In fact, air temperature was reported to be the more significant predictor in the first half of the runoff season [86], while during the second half, the effects of rainfall events have been found to be more pronounced in an upstream transect in Adventelva [82]. From lack of valid RS observations in late August and September, paired with our own on-ground observations of the occurrence of a substantial river plume after rainfall events, we conclude that the RS-based analysis might underestimate the importance of precipitation-driven riverine impacts on Adventfjorden during late summer and early autumn.

The results are evidence that supports the mobilisation potential with increasing temperatures (both during the ablation season as well as over longer time scales) but it should be noted that these results cannot exclude the influence of additional and/or connected environmental drivers. The generalization of presented results on other fjords should be critically reviewed as strong regionality and complex interplay of influencing factors may apply. However, predicting ecologically important endpoints such as river plume extent and SPM concentration in productive systems, such as fjords, from elementary environmental data (i.e., air temperature) highlights the exceptional value of combinations of OCRS in combination with extensive time-series data on environmental conditions.

4.3. Ecological Implications under Climate Change

Warming temperatures will affect the length of the ablation season in high latitude and Arctic regions, leading to enhanced melt of snow and ice. This might in turn fuel longer runoff seasons and subsequent mobilisation of sediments. In fact, recent studies suggest a general retreat for most glaciers and negative mass balances on the Svalbard archipelago: simulations based on Representative Concentration Pathways (RCP) 4.5 and 8.5 used by van Pelt et al. [4] found highest glacial retreat probability for the southern part of Spitsbergen, due to increase in runoff from both glaciers and land. Nowak et al. [2] come to a comparable conclusion, while acknowledging that "peak water" (i.e., a levelling-off in runoff values [87]) is likely to happen if catchment potential is exhausted. Nonetheless, major changes to the glacial landscape in Svalbard are happening, causing the retreat of previously marine-terminating glacier fronts to move further inland [88] with implications for the hydrological and sedimentary characteristics in the transition zones from land to sea. Indeed, primary productivity is significantly different for marine-terminating and land-terminating glaciers [89,90]. While the upwelling by subglacial discharge at the marine terminus delivers essential nutrients to the surface waters [89,91,92], non-glaciated fjords will feature strong freshwater-driven stratification and enhanced turbidity, restraining the majority of primary productivity and subsequent carbon fixation to surface waters [85]. We therefore propose that Adventfjorden may represent a model environment of "post-glaciation" (i.e., deglaciated) estuaries beyond the limits of the Svalbard archipelago.

The strong relationship observed between $C_{SPM_{sat}}$ and Z_{SD} (Supplementary Figure S3) highlights the role of turbid river plumes in driving high light attenuation and thereby reducing coastal light availability [17]. The presence of turbid freshwater plumes negatively impacts the productivity of primary producers in the nearshore zone by reducing the light availability in the water column. In these deglaciated and (mostly) ice-free Arctic fjords, light attenuation driven by high concentrations of suspended matter become a considerable force in structuring the phenology of primary production and carbon uptake [17,19]. High values of SPM (higher than 30 mg L^{-1}) in this study have been found to be limiting the illumination by incident light to the uppermost 50 cm in most cases (based on the Z_{SD} and $C_{SPM_{insitu}}$ dataset). The erosion potential of the catchment was not considered in this study but could allow the evaluation of C_{SPM}, which would in turn lead to a better understanding of the vulnerability of these coastal environments due to enhanced particle transport in glaciated catchments and turbidity fluxes in the productive nearshore zone. Further investigation, drawing on available and additional data on SPM, Chl-a, CDOM and other OACs may be necessary to fully understand the dynamics. OCRS techniques may provide a larger spatial understanding of optical water properties driven by these parameters. The application of regionally adapted OCRS algorithms, especially in remote but productive Arctic coastal environments, has the potential to detect changes to the light availability and may offer the opportunity to develop an optical classification of the water column [93]. This study has targeted the prevailing concentrations of suspended particles, however, dissolved organic and inorganic matter (CDOM) can contribute largely to the absorption and benthic light availability [10]. There is a large potential in resolving the complexity in the chemical physicochemical composition of optically active constituents to describe the variability of coastal marine ecosystems.

4.4. Potential for Remote and Satellite Observations of SPM in Arctic Fjord Estuaries

The use of OCRS fills gaps in our understanding of variability of ecologically relevant water quality parameters, such as SPM, which has been targeted in this study. The hereby proposed use of $C_{SPM_{sat}}$ analysis from high resolution satellite data can provide a necessary metric to model changes to the water level by providing a strong relationship between the satellite detected median $C_{SPM_{sat}}$ over Adventfjorden and river water level in Adventelva. In contrast to in-situ-based studies, this OCRS-based approach can resolve SPM variability at unprecedented frequency and spatial scale. Regional tuning of an existing algorithm was necessary to enhance the quality of satellite-derived C_{SPM}. Certain limitations of this approach became, however, apparent and should be addressed in further studies.

In our study, we found the performance for the regional calibration AdvFCal based on the Nechad et al. [41] algorithm to not fully detect the highest values of SPM in the nearshore zone. The lack of calibration and ground-truthing data for outer fjord stations with lower concentrations of SPM was most likely responsible for the larger offset values in the regional calibrated AdvFCal algorithm in this study, which caused a general overestimation of SPM values in clearer outer fjord waters. Furthermore, the generality of SPM detection through OCRS might be constrained by a limited range of values that can be modelled by semi-analytical algorithms with certainty [42]. In this study, a single-band approach delivered satisfactory results. However, the quantification of pixel-values, especially for high to extremely high C_{SPM} can likely be improved using switching algorithms [47,94]. Especially in regions with higher fluxes in allochthonous sediments and less primary productivity, the accuracy in otherwise underestimated C_{SPM} values > 60 mg L^{-1} could be enhanced. For the purpose of detecting a threshold value as used in this study based on ecologically meaningful 30 mg L^{-1} C_{SPM}, this approach allowed for the very detailed discussion of the variability of turbid plumes in the Arctic fjord of Adventfjorden within margins of acceptable uncertainty of less than 30% [65,95].

The IOPs of SPM may not always correspond well to the signal that is recorded by the satellite. In fact, the optical properties of SPM change with the particle mass, size, and density [96]. In the Adventfjorden system, there are also likely to be strong seasonal changes [12,83] which influence SPM composition and optical properties (e.g., with higher inorganic content during periods with high contributions of glacial-meltwater [32], and higher organic content during later summer rainfall events when the permafrost active layer is deeper [12,97]). We therefore suggest that monthly or seasonal calibrations to model C_{SPM} could provide better predictive power, as proposed by [98] for the Baltic Sea region.

We propose the use of the system of the Sentinel-2 mission for monitoring purposes at high to medium resolution. However, there are obvious limitations to optical satellite sensors regarding cloud cover or spatial resolution. Drone-based surveying of complex nearshore waters should therefore be considered [34]. They can complement satellite observations while being currently limited in their spatial coverage and revisit time. In the case of monitoring of Arctic nearshore waters in Svalbard, the recently launched hyperspectral sensor on board the Dornier small-wing plan stationed in Longyearbyen may prove a necessary and useful addition, especially during the cloudier summer season [99]. These high-resolution sensors may, however, be more adequate for targeted surveying while continuous monitoring can build on the temporal resolution of satellite data.

5. Conclusions

This study explored the fluxes of SPM into an Arctic fjord-type estuary over the course of two runoff seasons in 2019 and 2020 on a spatially and temporally improved dataset by using in-situ observations to calibrate a generic semi-analytical algorithm from Nechad et al. [41]. We located variability "hotspots" for SPM at the more channelled riverine entry points, of, e.g., Longyearelva, to the fjord, while values were more stable in the tidal flat of Adventelva, indicating the effects of catchment topography on the velocity of the runoff throughout the season and the potential for particle mobilisation. Combining the time-series data with environmental driver data, we identified ambient air temperature (°C) as a predictor to the elevated C_{SPM} and its spatial coverage in the fjord, supporting the well-reported effects of temperature on glacial discharge and the mobilisation of terrestrial particulate matter. This highlights the explanatory power of OCRS combined with environmental data in these dynamic environments.

The recorded and ongoing changes to temperature in the Arctic will likely change the phenology and variability of light attenuating SPM in near-shore regions and fjords. This study suggests that the Adventfjorden estuary can function as a valid model area for regions that are presently governed by marine terminating glaciers and predominantly glaciated catchments. Fine-tuning of the OCRS algorithms by in-situ measurements led to acceptable estimates of spatio-temporal variability of SPM in the highly dynamic Adventfjorden estuary and, thus, closes knowledge gaps related to more snapshot-like in-situ observations of optically active water constituents, specifically for SPM. This is a substantial improvement to previous study designs in the area that were either based on in-situ observations or could not provide a comparable resolution of RS based estimation. To improve the accuracy even more, longer time series of cal/val and matchup datasets may be required, covering clearer water in the outer fjord stations to complex turbid waters in the inner parts of the fjord equally. With the seasonality in runoff and SPM concentration in the surface waters of fjordic estuaries, such as the here-studied Adventfjorden, serious consideration should be given to seasonally tuned calibrations. The here-presented results highlight the use of remote sensing data for the monitoring purposes of complex Arctic coasts.

Supplementary Materials: The following are available online at https://www.mdpi.com/article/10.3390/rs14133123/s1, Table S1: Overview of FreshFate Campaigns relevant to this study, Figure S1: Sensitivity Analysis of matchup cal/val dataset. Figure S2: Number of valid observations (NVO [count]) for the studied years 2019 (left panel) and 2020 (right panel) over Adventfjorden. Figure S3: Relationship of in-situ Secchi depth and SPM, Figure S4: Meterological data overview (courtesy METNorway, seklima.no), Figure S5: Pearson correlation matrix for environmental statistics, Table S2: List of Sentinel-2 L1C data used in the time-series analysis.

Author Contributions: Conceptualization, A.P., J.E.S. and D.M.R.W.; methodology, A.P. and R.K.S.; software, R.K.S.; validation, D.M.R.W.; formal analysis, D.M.R.W., R.K.S. and A.P.; investigation, D.M.R.W., A.P., J.E.S. and R.K.S.; data curation, D.M.R.W.; writing—original draft preparation, D.M.R.W.; visualization, D.M.R.W.; supervision, A.P., J.E.S. and H.L.; project administration, A.P. and J.E.S.; funding acquisition, A.P. and D.M.R.W. All authors contributed to the reviewing process of the submitted manuscript. All authors have read and agreed to the published version of the manuscript.

Funding: This work was supported by the Research Council of Norway ('TerrACE' -project; project number 268458), and the Fram Center for High North Research "Fjord and Coast" -flagship ('FreshFate'-project; project number 132019). This research was also part of the project ACCES, funded through the 2017–2018 Belmont Forum and BiodivERsA joint call for research proposals, under the BiodivScen ERA-Net COFUND programme, and with the funding organisations Research Council of Norway (project nr. 296836) and Fonds de recherche du Québec. H. Lantuit received support from the Nunataryuk project, funded by the European Union's Horizon 2020 Research and Innovation Program under grant agreement No. 773421. Fieldwork in 2020 was funded by the Svalbard Science Forum's Arctic Field Grant (RIS-ID: 11386). In addition, strategic funding for field investigations and cal/val work were received from Svalbard Integrated Arctic Earth Observing System (SIOS), Norway, (Research Infrastructure access call). The article processing charges (APC) were funded by SIOS through the ECR presentation award (SIOS Online Conference, 2020).

Data Availability Statement: Contains modified Copernicus Sentinel-2 data (2019–2020). Sentinel-2 images used for this study are freely accessible via the Copernicus SciHub https://scihub.copernicus.eu/ (last accessed 30 March 2022). The cal/val matchup datasets presented in this study are available upon request from the corresponding author. Restrictions apply to the availability of river monitoring data, which was obtained from NIVA (PI Amanda Poste) and can be made available upon request.

Acknowledgments: We especially thank Maeve McGovern for her involvement in the field campaigns and the processing of the SPM and Turbidity samples and data for the Adventfjorden project in 2019. Uta Brandt is acknowledged for the used NIVA river monitoring data. We thank Liv Sletten, Anne Deininger, Cathrine Gundersen, Sarah Nelson, Tobias Vonnahme, Ulrike Dietrich, Vanessa

Pitusi, and Sine Sara for their assistance in field and lab work. UNIS Svalbard is acknowledged for the provision of laboratories. We are grateful for the support by UNIS logistics during field operations in Adventfjorden. We would like to thank the Ocean Biology Processing Group (NASA) for developing and maintaining SeaDAS v7.5.3 (seadas.gsfc.nasa.gov, accessed on 22 April 2022) used to process the MSI Level-1C data used in this study. European Space Agency (ESA) is thanked for the freely available Sentinel-2 imagery.

Conflicts of Interest: The authors declare no conflict of interest. The funders had no role in the design of the study; in the collection, analyses, or interpretation of data; in the writing of the manuscript, or in the decision to publish the results.

References

1. Hanssen-Bauer, I.; Førland, E.; Hisdal, H.; Mayer, S.; Sandø, A.B.; Sorteberg, A. (Eds.) *Climate in Svalbard 2100: A Knowledge Base for Climate Adaptation*; NCCS Report No. 1/2019; NCCS: Albuquerque, NM, USA, 2019.
2. Nowak, A.; Hodgkins, R.; Nikulina, A.; Osuch, M.; Wawrzyniak, T.; Kavan, J.; Łepkowska, E.; Majerska, M.; Romashova, K.; Vasilevich, I.; et al. *From Land to Fjords: The Review of Svalbard Hydrology from 1970 to 2019 (SvalHydro)*; Loughborough University: Loughborough, UK, 2021.
3. IPCC. *IPCC Special Report on the Ocean and Cryosphere in a Changing Climate*; Pörtner, H.-O., Roberts, D.C., Masson-Delmotte, V., Zhai, P., Tignor, M., Poloczanska, E., Mintenbeck, K., Alegría, A., Nicolai, M., Okem, A., et al., Eds.; Summary for policy makers; IPCC: Geneva, Switzerland, 2019.
4. van Pelt, W.J.J.; Schuler, T.V.; Pohjola, V.A.; Pettersson, R. Accelerating future mass loss of Svalbard glaciers from a multi-model ensemble. *J. Glaciol.* **2021**, *67*, 485–499. [CrossRef]
5. Lantuit, H.; Overduin, P.P.; Couture, N.; Wetterich, S.; Aré, F.; Atkinson, D.; Brown, J.; Cherkashov, G.; Drozdov, D.; Forbes, D.L.; et al. The Arctic Coastal Dynamics Database: A New Classification Scheme and Statistics on Arctic Permafrost Coastlines. *Estuaries Coasts* **2012**, *35*, 383–400. [CrossRef]
6. Vonk, J.E.; Tank, S.E.; Bowden, W.B.; Laurion, I.; Vincent, W.F.; Alekseychik, P.; Amyot, M.; Billet, M.F.; Canário, J.; Cory, R.M.; et al. Reviews and syntheses: Effects of permafrost thaw on Arctic aquatic ecosystems. *Biogeosciences* **2015**, *12*, 7129–7167. [CrossRef]
7. Cottier, F.R.; Nilsen, F.; Skogseth, R.; Tverberg, V.; Skarðhamar, J.; Svendsen, H. Arctic fjords: A review of the oceanographic environment and dominant physical processes. *Geol. Soc. Lond. Spec. Publ.* **2010**, *344*, 35–50. [CrossRef]
8. Syvitski, J.P.M.; Burrell, D.C.; Skei, J.M. *Fjords: Processes and Products*; Springer: New York, NY, USA; London, UK, 1987; ISBN 0-387-96342-1.
9. Lund-Hansen, L.C.; Andersen, T.J.; Nielsen, M.H.; Pejrup, M. Suspended Matter, Chl-a, CDOM, Grain Sizes, and Optical Properties in the Arctic Fjord-Type Estuary, Kangerlussuaq, West Greenland during Summer. *Estuaries Coasts* **2010**, *33*, 1442–1451. [CrossRef]
10. Mascarenhas, V.J.; Voß, D.; Wollschlaeger, J.; Zielinski, O. Fjord light regime: Bio-optical variability, absorption budget, and hyperspectral light availability in Sognefjord and Trondheimsfjord, Norway. *J. Geophys. Res. Oceans* **2017**, *122*, 3828–3847. [CrossRef]
11. Terhaar, J.; Lauerwald, R.; Regnier, P.; Gruber, N.; Bopp, L. Around one third of current Arctic Ocean primary production sustained by rivers and coastal erosion. *Nat. Commun.* **2021**, *12*, 169. [CrossRef]
12. McGovern; MPavlov, A.K.; Deininger, A.; Granskog, M.A.; Leu, E.; Søreide, J.E.; Poste, A.E. Terrestrial Inputs Drive Seasonality in Organic Matter and Nutrient Biogeochemistry in a High Arctic Fjord System (Isfjorden, Svalbard). *Front. Mar. Sci.* **2020**, *7*. [CrossRef]
13. Wolanski, E. *Estuarine Ecohydrology*, 1st ed.; Elsevier: Amsterdam, The Netherlands, 2007; ISBN 978-0-444-530660.
14. Mann, K.H. *Ecology of Coastal Waters: With Implications for Management*, 2nd ed.; Blackwell Science: Malden, MA, USA, 2000; ISBN 0-86542-550-7.
15. Forwick, M.; Vorren, T.O.; Hald, M.; Korsun, S.; Roh, Y.; Vogt, C.; Yoo, K.-C. Spatial and temporal influence of glaciers and rivers on the sedimentary environment in Sassenfjorden and Tempelfjorden, Spitsbergen. *Geol. Soc. Lond. Spec. Publ.* **2010**, *344*, 163–193. [CrossRef]
16. Bianchi, T.S.; Arndt, S.; Austin, W.E.; Benn, D.I.; Bertrand, S.; Cui, X.; Faust, J.C.; Koziorowska-Makuch, K.; Moy, C.M.; Savage, C.; et al. Fjords as Aquatic Critical Zones (ACZs). *Earth-Sci. Rev.* **2020**, *203*, 103145. [CrossRef]
17. Pavlov, A.K. The Underwater Light Climate in Kongsfjorden and Its Ecological Implications. In *The Ecosystem of Kongsfjorden, Svalbard*; Hop, H., Wiencke, C., Eds.; Springer: Cham, Switzerland, 2019; ISBN 978-3-319-46425-1.
18. Kirk, J.T.O. *Light and Photosynthesis in Aquatic Ecosystems*; Cambridge University Press: Cambridge, UK, 2010; ISBN 9781139168212.
19. Holinde, L.; Zielinski, O. Bio-optical characterization and light availability parameterization in Uummannaq Fjord and Vaigat–Disko Bay (West Greenland). *Ocean Sci.* **2016**, *12*, 117–128. [CrossRef]
20. Frigstad, H.; Kaste, Ø.; Deininger, A.; Kvalsund, K.; Christensen, G.; Bellerby, R.G.J.; Sørensen, K.; Norli, M.; King, A.L. Influence of Riverine Input on Norwegian Coastal Systems. *Front. Mar. Sci.* **2020**, *7*. [CrossRef]
21. Aksnes, D.L.; Dupont, N.; Staby, A.; Fiksen, Ø.; Kaartvedt, S.; Aure, J. Coastal water darkening and implications for mesopelagic regime shifts in Norwegian fjords. *Mar. Ecol. Prog. Ser.* **2009**, *387*, 39–49. [CrossRef]

22. Konik, M.; Darecki, M.; Pavlov, A.K.; Sagan, S.; Kowalczuk, P. Darkening of the Svalbard Fjords Waters Observed With Satellite Ocean Color Imagery in 1997–2019. *Front. Mar. Sci.* **2021**, *8*. [CrossRef]
23. Neukermans, G.; Ruddick, K.; Loisel, H.; Roose, P. Optimization and quality control of suspended particulate matter concentration measurement using turbidity measurements. *Limnol. Oceanogr. Methods* **2012**, *10*, 1011–1023. [CrossRef]
24. Elverhøi, A.; Lønne, Ø.; Seland, R. Glaciomarine sedimentation in a modern fjord environment, Spitsbergen. *Polar Res.* **1983**, *1*, 127–149. [CrossRef]
25. IOCCG. *Earth Observations in Support of Global Water Quality Monitoring*; Greb, S., Dekker, A., Binding, C., Eds.; IOCCG Report Series, No. 17; IOCCG: Dartmouth, NS, Canada, 2018.
26. Doxaran, D.; Ehn, J.; Bélanger, S.; Matsuoka, A.; Hooker, S.; Babin, M. Optical characterisation of suspended particles in the Mackenzie River plume (Canadian Arctic Ocean) and implications for ocean colour remote sensing. *Biogeosciences* **2012**, *9*, 3213–3229. [CrossRef]
27. Ehn, J.K.; Reynolds, R.A.; Stramski, D.; Doxaran, D.; Lansard, B.; Babin, M. Patterns of suspended particulate matter across the continental margin in the Canadian Beaufort Sea during summer. *Biogeosciences* **2019**, *16*, 1583–1605. [CrossRef]
28. McGrath, D.; Steffen, K.; Overeem, I.; Mernild, S.H.; Hasholt, B.; van den Broeke, M. Sediment plumes as a proxy for local ice-sheet runoff in Kangerlussuaq Fjord, West Greenland. *J. Glaciol.* **2010**, *56*, 813–821. [CrossRef]
29. Klein, K.P.; Lantuit, H.; Heim, B.; Fell, F.; Doxaran, D.; Irrgang, A.M. Long-Term High-Resolution Sediment and Sea Surface Temperature Spatial Patterns in Arctic Nearshore Waters Retrieved Using 30-Year Landsat Archive Imagery. *Remote Sens.* **2019**, *11*, 2791. [CrossRef]
30. Groom, S.; Sathyendranath, S.; Ban, Y.; Bernard, S.; Brewin, R.; Brotas, V.; Brockmann, C.; Chauhan, P.; Choi, J.; Chuprin, A.; et al. Satellite Ocean Colour: Current Status and Future Perspective. *Front. Mar. Sci.* **2019**, *6*, 28. [CrossRef]
31. Neder, C.; Fofonova, V.; Androsov, A.; Kuznetsov, I.; Abele, D.; Falk, U.; Schloss, I.R.; Sahade, R.; Jerosch, K. Modelling suspended particulate matter dynamics at an Antarctic fjord impacted by glacier melt. *J. Mar. Syst.* **2022**, *231*, 103734. [CrossRef]
32. Moskalik, M.; Ćwiąkała, J.; Szczuciński, W.; Dominiczak, A.; Głowacki, O.; Wojtysiak, K.; Zagórski, P. Spatiotemporal changes in the concentration and composition of suspended particulate matter in front of Hansbreen, a tidewater glacier in Svalbard. *Oceanologia* **2018**, *60*, 446–463. [CrossRef]
33. Melton, S.M.; Alley, R.B.; Anandakrishnan, S.; Parizek, B.R.; Shahin, M.G.; Stearns, L.A.; LeWinter, A.L.; Finnegan, D.C. Meltwater drainage and iceberg calving observed in high-spatiotemporal resolution at Helheim Glacier, Greenland. *J. Glaciol.* **2022**, 1–17. [CrossRef]
34. Wójcik, K.A.; Bialik, R.J.; Osińska, M.; Figielski, M. Investigation of Sediment-Rich Glacial Meltwater Plumes Using a High-Resolution Multispectral Sensor Mounted on an Unmanned Aerial Vehicle. *Water* **2019**, *11*, 2405. [CrossRef]
35. Klein, K.P.; Lantuit, H.; Heim, B.; Doxaran, D.; Juhls, B.; Nitze, I.; Walch, D.; Poste, A.; Søreide, J.E. The Arctic Nearshore Turbidity Algorithm (ANTA) - A multi sensor turbidity algorithm for Arctic nearshore environments. *Sci. Remote Sens.* **2021**, *4*, 100036. [CrossRef]
36. Petit, T.; Hamre, B.; Sandven, H.; Röttgers, R.; Kowalczuk, P.; Zablocka, M.; Granskog, M.A. Inherent optical properties of dissolved and particulate matter in an Arctic fjord (Storfjorden, Svalbard) in early summer. *Ocean Sci.* **2022**, *18*, 455–468. [CrossRef]
37. Slonecker, E.T.; Jones, D.K.; Pellerin, B.A. The new Landsat 8 potential for remote sensing of colored dissolved organic matter (CDOM). *Mar. Pollut. Bull.* **2016**, *107*, 518–527. [CrossRef]
38. IOCCG. *Ocean Colour Remote Sensing in Polar Seas*; Babin, M., Arrigo, K., Bélanger, S., Forget, M.-H., Eds.; IOCCG Report Series, No. 16; IOCCG: Dartmouth, NS, Canada, 2015.
39. Bélanger, S.; Babin, M.; Tremblay, J.-É. Increasing cloudiness in Arctic damps the increase in phytoplankton primary production due to sea ice receding. *Biogeosciences* **2013**, *10*, 4087–4101. [CrossRef]
40. Dogliotti, A.I.; Ruddick, K.G.; Nechad, B.; Doxaran, D.; Knaeps, E. A single algorithm to retrieve turbidity from remotely-sensed data in all coastal and estuarine waters. *Remote Sens. Environ.* **2015**, *156*, 157–168. [CrossRef]
41. Nechad, B.; Ruddick, K.G.; Park, Y. Calibration and validation of a generic multisensor algorithm for mapping of total suspended matter in turbid waters. *Remote Sens. Environ.* **2010**, *114*, 854–866. [CrossRef]
42. Han, B.; Loisel, H.; Vantrepotte, V.; Mériaux, X.; Bryère, P.; Ouillon, S.; Dessailly, D.; Xing, Q.; Zhu, J. Development of a Semi-Analytical Algorithm for the Retrieval of Suspended Particulate Matter from Remote Sensing over Clear to Very Turbid Waters. *Remote Sens.* **2016**, *8*, 211. [CrossRef]
43. IOCCG. *Synergy between Ocean Colour and Biogeochemical/Ecosystem Models*; Dutkiewicz, S., Ed.; IOCCG Report Series, No. 19; IOCCG: Dartmouth, NS, Canada, 2020.
44. Matthews, M.W. A current review of empirical procedures of remote sensing in inland and near-coastal transitional waters. *Int. J. Remote Sens.* **2011**, *32*, 6855–6899. [CrossRef]
45. Shanmugam, P.; He, X.; Singh, R.K.; Varunan, T. A modern robust approach to remotely estimate chlorophyll in coastal and inland zones. *Adv. Space Res.* **2018**, *61*, 2491–2509. [CrossRef]
46. Novoa, S.; Doxaran, D.; Ody, A.; Vanhellemont, Q.; Lafon, V.; Lubac, B.; Gernez, P. Atmospheric Corrections and Multi-Conditional Algorithm for Multi-Sensor Remote Sensing of Suspended Particulate Matter in Low-to-High Turbidity Levels Coastal Waters. *Remote Sens.* **2017**, *9*, 61. [CrossRef]

47. Ody, A.; Doxaran, D.; Verney, R.; Bourrin, F.; Morin, G.P.; Pairaud, I.; Gangloff, A. Ocean Color Remote Sensing of Suspended Sediments along a Continuum from Rivers to River Plumes: Concentration, Transport, Fluxes and Dynamics. *Remote Sens.* **2022**, *14*, 2026. [CrossRef]
48. Weslawski, J. Influence of salinity and suspended matter on benthos of an Arctic tidal flat. *ICES J. Mar. Sci.* **1999**, *56*, 194–202. [CrossRef]
49. Zajączkowski, M. Sediment supply and fluxes in glacial and outwash fjords, Kongsfjorden and Adventfjorden, Svalbard. *Pol. Polar Res.* **2008**, *29*, 59–72.
50. Norwegian Polar Institute. *Kartdata Svalbard 1:100 000 (S100 Kartdata)/Map Data*; Norwegian Polar Institute: Tromsø, Norway, 2014. [CrossRef]
51. Kartverket—Norwegian Mapping Authority. Sjøkart—Dybdedata. Available online: https://kartkatalog.geonorge.no/metadata/kartverket/dybdedata/2751aacf-5472-4850-a208-3532a51c529a (accessed on 30 January 2022).
52. Zajączkowski, M.; Szczuciński, W.; Bojanowski, R. Recent changes in sediment accumulation rates in Adventfjorden, Svalbard. *Oceanologia* **2004**, *46*, 217–231.
53. Hagen, J.O.; Lisestøl, O.; Erik, R.; Jørgensen, T. *Glacier Atlas of Svalbard and Jan Mayen*; Meddelelser No. 129; Norsk Polarinstitutt: Oslo, Norway, 1993; Available online: https://brage.npolar.no/npolar-xmlui/handle/11250/173065 (accessed on 18 April 2022).
54. Bogen, J.; Bønsnes, T.E. Erosion and sediment transport in High Arctic rivers, Svalbard. *Polar Res.* **2003**, *22*, 175–189. [CrossRef]
55. Węsławski, J.M. *Adventfjorden: Arctic Sea in the Backyard*; Institute of Oceanology PAS: Sopot, Poland, 2011.
56. Zajączkowski, M.; Nygård, H.; Hegseth, E.N.; Berge, J. Vertical flux of particulate matter in an Arctic fjord: The case of lack of the sea-ice cover in Adventfjorden 2006–2007. *Polar Biol.* **2010**, *33*, 223–239. [CrossRef]
57. Wiedmann, I.; Reigstad, M.; Marquardt, M.; Vader, A.; Gabrielsen, T.M. Seasonality of vertical flux and sinking particle characteristics in an ice-free high arctic fjord—Different from subarctic fjords? *J. Mar. Syst.* **2016**, *154*, 192–205. [CrossRef]
58. *ISO 11923:1997*; Water Quality—Determination of Suspended Solids by Filtration through Glass-Fibre Filters. ISO: Geneva, Switzerland, 1997. Available online: https://www.iso.org/standard/20654.html (accessed on 21 June 2021).
59. Babin, M.; Morel, A.; Fournier-Sicre, V.; Fell, F.; Stramski, D. Light scattering properties of marine particles in coastal and open ocean waters as related to the particle mass concentration. *Limnol. Oceanogr.* **2003**, *48*, 843–859. [CrossRef]
60. IOCCG. *Minimum Requirements for an Operational Ocean-Colour Sensor for the Open Ocean*; Morel, A., Ed.; Reports of the International Ocean-Colour Coordinating Group, No. 13; IOCCG: Dartmouth, NS, Canada, 1998.
61. IOCCG. *Mission Requirements for Future Ocean-Colour Sensors*; McClain, C.R., Meister, G., Eds.; Reports of the International Ocean-Colour Coordinating Group, No. 13; IOCCG: Dartmouth, NS, Canada, 2012; Available online: http://www.ioccg.org/reports/IOCCG_Report13.pdf (accessed on 23 June 2021).
62. Singh, R.K.; Shanmugam, P.; He, X.; Schroeder, T. UV-NIR approach with non-zero water-leaving radiance approximation for atmospheric correction of satellite imagery in inland and coastal zones. *Opt. Express* **2019**, *27*, A1118–A1145. [CrossRef]
63. Mobley, C.D.; Werdell, J.; Franz, B.; Ahmad, Z.; Bailey, S. Atmospheric correction for satellite ocean color radiometry. In *A Tutorial and Documentation NASA Ocean Biology Processing Group (Issue June)*; Goddart Space Flight Center: Greenbelt, MD, USA, 2016.
64. Mabit, R.; Araújo, C.A.S.; Singh, R.K.; Bélanger, S. Empirical Remote Sensing Algorithms to Retrieve SPM and CDOM in Québec Coastal Waters. *Front. Remote Sens.* **2022**, *3*, 834908. [CrossRef]
65. IOCCG. *Remote Sensing of Ocean Colour in Coastal, and Other Optically-Complex, Waters*; IOCCG: Dartmouth, NS, Canada, 2000.
66. Virtanen, P.; Gommers, R.; Oliphant, T.E.; Haberland, M.; Reddy, T.; Cournapeau, D.; Burovski, E.; Peterson, P.; Weckesser, W.; Bright, J.; et al. SciPy 1.0: Fundamental algorithms for scientific computing in Python. *Nat. Methods* **2020**, *17*, 261–272. [CrossRef]
67. Seegers, B.N.; Stumpf, R.P.; Schaeffer, B.A.; Loftin, K.A.; Werdell, P.J. Performance metrics for the assessment of satellite data products: An ocean color case study. *Opt. Express* **2018**, *26*, 7404–7422. [CrossRef]
68. Nordli, Ø. The Svalbard Airport Temperature Series. *Bull. Geography. Phys. Geogr. Ser.* **2010**, *3*, 5–25. [CrossRef]
69. Meteorologisk Institutt. Norsk Klimaservicesenter. Seklima. Observasjoner og værstatistikk. Available online: https://seklima.met.no/observations/ (accessed on 20 January 2022).
70. Kartverket—Norwegian Mapping Authority. Tidal and Water Level Data. Available online: https://www.kartverket.no/en/at-sea/se-havniva (accessed on 20 January 2022).
71. R Core Team. *R: A Language and Environment for Statistical Computing*; R Core Team: Vienna, Austria, 2021.
72. Hijmans, R.J. Raster: Geographic Data Analysis and Modeling. R package version 3.5-11. 2021. Available online: http://CRAN.R-project.org/package=raster (accessed on 15 June 2022).
73. Pierce, D. ncdf4: Interface to Unidata netCDF (Version 4 or Earlier) Format Data Files: R Package Version 1.18. 2021. Available online: https://CRAN.R-project.org/package=ncdf4 (accessed on 15 June 2022).
74. Ody, A.; Doxaran, D.; Vanhellemont, Q.; Nechad, B.; Novoa, S.; Many, G.; Bourrin, F.; Verney, R.; Pairaud, I.; Gentili, B. Potential of High Spatial and Temporal Ocean Color Satellite Data to Study the Dynamics of Suspended Particles in a Micro-Tidal River Plume. *Remote Sens.* **2016**, *8*, 245. [CrossRef]
75. Wei, T.; Simko, V. R package 'corrplot': Visualization of a Correlation Matrix. R paclage version 0.92. 2021. Available online: https://github.com/taiyun/corrplot (accessed on 23 June 2022).
76. Wickham, H. Reshaping Data with the reshape Package. *J. Stat. Softw.* **2007**, *21*, 1–20. [CrossRef]
77. Aphalo, P.J. ggpmisc: Miscellaneous Extensions to 'ggplot2'. 2021. Available online: https://CRAN.R-project.org/package=ggpmisc (accessed on 15 June 2022).

78. Wickham, H. *ggplot2: Elegant Graphics for Data Analysis*; Springer: New York, NY, USA, 2016; ISBN 978-3-319-24277-4.
79. Neuwirth, E. RColorBrewer: ColorBrewer Palettes. R package version 1.1-2. 2014. Available online: https://CRAN.R-project.org/package=RColorBrewer (accessed on 15 June 2022).
80. Bailey, S.W.; Werdell, P.J. A multi-sensor approach for the on-orbit validation of ocean color satellite data products. *Remote Sens. Environ.* **2006**, *102*, 12–23. [CrossRef]
81. Zajączkowski, M.; Włodarska-Kowalczuk, M. Dynamic sedimentary environments of an Arctic glacier-fed river estuary (Adventfjorden, Svalbard). I. Flux, deposition, and sediment dynamics. *Estuar. Coast. Shelf Sci.* **2007**, *74*, 285–296. [CrossRef]
82. Rothenburg, M.A.J. The Advent River system, Central Svalbard: A High Temporal Resolution Analysis of Sediment Flux from a Dynamic Arctic River-Mouth. Master's Thesis, Vrije Universiteit, Amsterdam, The Netherlands, 2019.
83. Yde, J.C.; Riger-Kusk, M.; Christiansen, H.H.; Tvis Knudsen, N.; Humlum, O. Hydrochemical characteristics of bulk meltwater from an entire ablation season, Longyearbreen, Svalbard. *J. Glaciol.* **2008**, *54*, 259–272. [CrossRef]
84. Urbanski, J.A.; Stempniewicz, L.; Węsławski, J.M.; Dragańska-Deja, K.; Wochna, A.; Goc, M.; Iliszko, L. Subglacial discharges create fluctuating foraging hotspots for sea birds in tidewater glacier bays. *Sci. Rep.* **2017**, *7*, 1–12. [CrossRef] [PubMed]
85. Holding, J.M.; Markager, S.; Juul-Pedersen, T.; Paulsen, M.L.; Møller, E.F.; Meire, L.; Sejr, M.K. Seasonal and spatial patterns of primary production in a high-latitude fjord affected by Greenland Ice Sheet run-off. *Biogeosciences* **2019**, *16*, 3777–3792. [CrossRef]
86. Nowak, A.; Hodson, A. Hydrological response of a High-Arctic catchment to changing climate over the past 35 years: A case study of Bayelva watershed, Svalbard. *Polar Res.* **2013**, *32*, 19691. [CrossRef]
87. Huss, M.; Hock, R. Global-scale hydrological response to future glacier mass loss. *Nature Clim. Chang.* **2018**, *8*, 135–140. [CrossRef]
88. Jørgensen, B.B.; Laufer, K.; Michaud, A.B.; Wehrmann, L.M. Biogeochemistry and microbiology of high Arctic marine sediment ecosystems—Case study of Svalbard fjords. *Limnol. Oceanogr.* **2021**, *66*, S273–S292. [CrossRef]
89. Meire, L.; Mortensen, J.; Meire, P.; Juul-Pedersen, T.; Sejr, M.K.; Rysgaard, S.; Nygaard, R.; Huybrechts, P.; Meysman, F.J.R. Marine-terminating glaciers sustain high productivity in Greenland fjords. *Glob. Chang. Biol.* **2017**, *23*, 5344–5357. [CrossRef]
90. Hopwood, M.J.; Carroll, D.; Dunse, T.; Hodson, A.; Holding, J.M.; Iriarte, J.L.; Ribeiro, S.; Achterberg, E.P.; Cantoni, C.; Carlson, D.F.; et al. Review article: How does glacier discharge affect marine biogeochemistry and primary production in the Arctic? *Cryosphere* **2020**, *14*, 1347–1383. [CrossRef]
91. Bhatia, M.P.; Waterman, S.; Burgess, D.O.; Williams, P.L.; Bundy, R.M.; Mellett, T.; Roberts, M.; Bertrand, E.M. Glaciers and Nutrients in the Canadian Arctic Archipelago Marine System. *Global Biogeochem. Cycles* **2021**, *35*. [CrossRef]
92. Halbach, L.; Vihtakari, M.; Duarte, P.; Everett, A.; Granskog, M.A.; Hop, H.; Kauko, H.M.; Kristiansen, S.; Myhre, P.I.; Pavlov, A.K.; et al. Tidewater Glaciers and Bedrock Characteristics Control the Phytoplankton Growth Environment in a Fjord in the Arctic. *Front. Mar. Sci.* **2019**, *6*, 254. [CrossRef]
93. Mélin, F.; Vantrepotte, V. How optically diverse is the coastal ocean? *Remote Sens. Environ.* **2015**, *160*, 235–251. [CrossRef]
94. Yu, X.; Lee, Z.; Shen, F.; Wang, M.; Wei, J.; Jiang, L.; Shang, Z. An empirical algorithm to seamlessly retrieve the concentration of suspended particulate matter from water color across ocean to turbid river mouths. *Remote Sens. Environ.* **2019**, *235*, 111491. [CrossRef]
95. Blondeau-Patissier, D.; Gower, J.F.; Dekker, A.G.; Phinn, S.R.; Brando, V.E. A review of ocean color remote sensing methods and statistical techniques for the detection, mapping and analysis of phytoplankton blooms in coastal and open oceans. *Prog. Oceanogr.* **2014**, *123*, 123–144. [CrossRef]
96. Neukermans, G.; Loisel, H.; Mériaux, X.; Astoreca, R.; McKee, D. In situ variability of mass-specific beam attenuation and backscattering of marine particles with respect to particle size, density, and composition. *Limnol. Oceanogr.* **2012**, *57*, 124–144. [CrossRef]
97. Christiansen, H.H.; French, H.M.; Humlum, O. Permafrost in the Gruve-7 mine, Adventdalen, Svalbard. *Nor. Geogr. Tidsskr.-Nor. J. Geogr.* **2005**, *59*, 109–115. [CrossRef]
98. Ligi, M.; Kutser, T.; Kallio, K.; Attila, J.; Koponen, S.; Paavel, B.; Soomets, T.; Reinart, A. Testing the performance of empirical remote sensing algorithms in the Baltic Sea waters with modelled and in situ reflectance data. *Oceanologia* **2017**, *59*, 57–68. [CrossRef]
99. Jawak, S.D.; Andersen, B.N.; Pohjola, V.A.; Godøy, Ø.; Hübner, C.; Jennings, I.; Ignatiuk, D.; Holmén, K.; Sivertsen, A.; Hann, R.; et al. SIOS's Earth Observation (EO), Remote Sensing (RS), and Operational Activities in Response to COVID-19. *Remote Sens.* **2021**, *13*, 712. [CrossRef]

remote sensing

MDPI

Article

Eddies in the Marginal Ice Zone of Fram Strait and Svalbard from Spaceborne SAR Observations in Winter

Igor E. Kozlov [1,*] and Oksana A. Atadzhanova [1,2]

1 Marine Hydrophysical Institute of RAS, 299011 Sevastopol, Russia; atadzhanova.oa@spb.ocean.ru
2 Shirshov Institute of Oceanology RAS, 117997 Moscow, Russia
* Correspondence: ik@mhi-ras.ru

Abstract: Here we investigate the intensity of eddy generation and their properties in the marginal ice zone (MIZ) regions of Fram Strait and around Svalbard using spaceborne synthetic aperture radar (SAR) data from Envisat ASAR and Sentinel-1 in winter 2007 and 2018. Analysis of 2039 SAR images allowed identifying 4619 eddy signatures. The number of eddies detected per image per kilometer of MIZ length is similar for both years. Submesoscale and small mesoscale eddies dominate with cyclones detected twice more frequently than anticyclones. Eddy diameters range from 1 to 68 km with mean values of 6 km and 12 km over shallow and deep water, respectively. Mean eddy size grows with increasing ice concentration in the MIZ, yet most eddies are detected at the ice edge and where the ice concentration is below 20%. The fraction of sea ice trapped in cyclones (53%) is slightly higher than that in anticyclones (48%). The amount of sea ice trapped by a single 'mean' eddy is about 40 km^2, while the average horizontal retreat of the ice edge due to eddy-induced ice melt is about 0.2–0.5 km·d^{-1} ± 0.02 km·d^{-1}. Relation of eddy occurrence to background currents and winds is also discussed.

Keywords: ocean eddies; marginal ice zone; sea ice; SAR imaging; Fram Strait; Svalbard; Greenland Sea; Hopen Island; Arctic Ocean

Citation: Kozlov, I.E.; Atadzhanova, O.A. Eddies in the Marginal Ice Zone of Fram Strait and Svalbard from Spaceborne SAR Observations in Winter. *Remote Sens.* **2022**, *14*, 134. https://doi.org/10.3390/rs14010134

Academic Editor: Shridhar D. Jawak

Received: 24 October 2021
Accepted: 25 December 2021
Published: 29 December 2021

1. Introduction

Eddies forming at the ice edge and within marginal ice zones (MIZ) are a common dynamic feature of the ice edge evolution under varying winds and ocean currents in polar oceans. They are known to be important for sea ice deformation, horizontal transport and melting. Previous studies show that MIZ eddies are common in polar ocean regions, such as Fram Strait, and discuss the broad variety of their generation mechanisms from available in situ, aircraft and satellite observations [1–4].

Recent studies also show that polar MIZs are rich in submesoscale flows with high Rossby numbers and strong ageostrophic effects [5–10]. Submesoscale ocean variability induces large vertical velocities bringing warm subsurface waters into the mixed layer and results in pronounced ocean-sea ice heat fluxes localized over cyclonic eddies and filaments reaching about 100 W m^{-2} [8]. It also leads to enhanced mixing of water masses over short horizontal scales, so impacting the sea ice and biological structures within the MIZ [10].

Available eddy observations in the Arctic Ocean are still very sparse, and hence high-resolution hydrodynamic models are used to fill these gaps [11–15]. Though current state-of-the-art 1 km and finer resolution models potentially resolve small eddies in the MIZ and around sea ice leads [13], a realistic forecasting of meso- and submeso-scale MIZ dynamics is still challenging and needs a good observational basis for model improvement and validation. In this sense, satellite remote sensing using spaceborne synthetic aperture radar (SAR) is a good and effective source of information to explore and better understand the eddy field in the Arctic Ocean, including the ice-covered regions [1,3,7,10,16–19]. What is especially good about SAR is that it can map MIZ dynamics, including eddy motions, under broad variety of wind speeds including high wind conditions [1,10].

Historically, the eddies forming in the Fram Strait (FS) MIZ obtained a lot of attention during a set of specialized experimental programs such as MIZEX'83, '84 that used a variety of observational techniques including in situ, satellite and aerial remote sensing (e.g., [1–3]). Here MIZ eddies occur in the region where two very different water masses co-exist, namely the cold Polar Water (PW) going with the East Greenland Current (EGC) southward and the warm Atlantic Water (AW) carried by the West Spitsbergen Current (WSC) northward. The latter splits into two recirculation branches interacting with the cold EGC, and Svalbard and Yermak branches entering the Arctic Ocean (Figure 1). Such a close occurrence of the cold PW, the warm AW with very high levels of eddy kinetic energy [13,14,19], and the year-round MIZ creates very favorable conditions for intensive eddy generation [1,7].

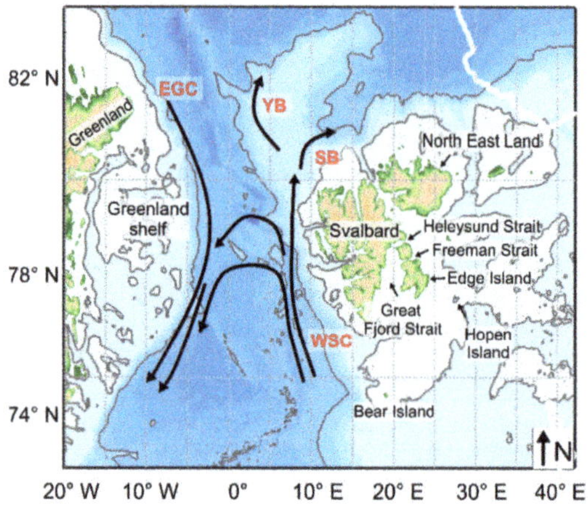

Figure 1. Bathymetry map of study site showing schematic of main currents in Fram Strait and near Svalbard. Arrows indicate the directions of currents: WSC—West Spitsbergen Current, EGC—East Greenland Current, SB—Svalbard Branch, YB—Yermak Branch. Gray lines mark 200 and 2000 m isobaths.

Though previous studies in FS gained much information about eddy generation mechanisms, their spatial and kinematic properties and influence on sea ice, the obtained results were still episodic and resolving only a limited number of eddies. The only work reporting a detailed statistics of MIZ eddies in FS and the Greenland Sea using Envisat ASAR data of 2008 and 2009 is that of Bondevik [18]. According to [18], the maximum occurrence of MIZ eddies takes place in summer when the ice extent is near its minimum, while generation and surface manifestation of eddies is favored at relatively low winds of predominant northern direction. In that work, the regions north and northeast of Svalbard, as well as coastal regions around it were not considered.

The aim of this study is to analyze the intensity of eddy generation and document their properties over the region of 74°–83° N and 20° W–40° E (Figure 1), while also assessing their influence on ice trapping and ice edge retreat, and discussing relation to background currents and winds using historical Envisat ASAR and contemporary Sentinel-1 SAR observations in winter season.

2. Materials and Methods

To investigate eddy signatures in FS and around Svalbard we analyze SAR C-band images acquired by ENVISAT Advanced SAR (ASAR) in winter 2007 (December 2006–March 2007) and Sentinel-1 A/B in winter 2018 (January–April 2018). These two winter seasons were selected primarily due to better data availability without much consideration about the sea ice and hydrographic conditions that were somewhat unusual in the study region, as discussed below.

For the analysis we use 212 ASAR images acquired in wide swath mode and image mode with spatial resolution of 150 m and 30 m, respectively. For Sentinel-1 A/B, we use 1827 images taken in interferometric wide swath and extra-wide swath modes with spatial resolution of 20 m and 90 m, respectively. For both data sets we use all publicly available data for the considered time periods. Envisat ASAR data were obtained from the European Space Agency rolling archive, while Sentinel-1 A/B data were obtained from Copernicus Open Access Hub (https://scihub.copernicus.eu (accessed on 1 July 2021). A summary of the data used, their quantity and time periods are given in Table 1.

Table 1. Number of spaceborne SAR images used in the analysis during winter 2007 and 2018.

Year	Sensor	December	January	February	March	April	Total
2007	ASAR	13	69	52	78	-	212
2018	S-1 A/B	-	500	418	383	526	1827

As seen from Table 1, the total number of SAR images analyzed in winter 2018 is nearly 9 times larger than that in 2007. The number of images also varies from month to month in both years, but not significantly. Figure 2 shows the spatial coverage of the study site by SAR data in 2007 and 2018. The maximal density of SAR observations in 2007 is about 70–90 scenes per unit area over the northern Greenland Sea, while for the rest of the region the mean value is about 40 images per unit area. In 2018, the overall density of SAR observations is much higher than in 2007, yet somewhat less homogeneous. The highest image density is found over the northern FS and north of Svalbard where it exceeds 200 images per unit area, in other regions it varies from 30 to 150 images per unit area. Such a big difference in spatial SAR data coverage between 2007 and 2018 is primarily linked to availability of two Sentinel-1 sensors and their more frequent sampling compared to Envisat ASAR. In addition, also note that ASAR has a very poor coverage north of 80° N and east of 10° E which is critical for eddy observations in this important region.

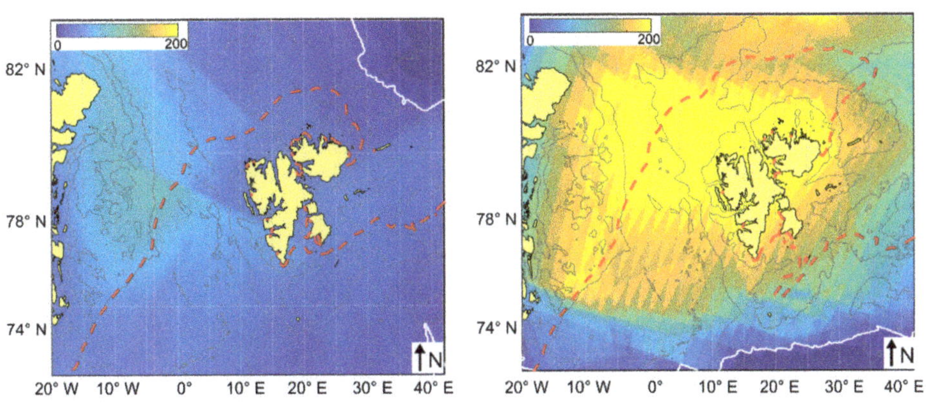

Figure 2. Spatial coverage of the study site by SAR data during: (**a**) December 2006–March 2007; (**b**) January–April 2018. The number of available SAR images is shown in color. The white line denotes the boundary of a region where the number of SAR images is below 10. Overlaid grey lines are the 200- and 2000-m isobaths taken from IBCAO v.3.0. Red dashed lines show the season-mean position of ice-water boundary for particular winter season.

It is also important to note that the season-mean ice margin length and position are different between 2007 and 2018 (Figure 2). In particular, the ice margin is found much further north of Svalbard and there is a much larger ice field southeast of archipelago in winter 2018. As a result, the mean MIZ length is ~3430 km in winter 2007 and ~4160 km in winter 2018, i.e., 1.2 times longer in 2018 than in 2007.

Under light to moderate winds the morphology of the marginal ice zone reflects the underlying ocean circulation, including eddy features [3]. Eddies in the MIZ and at the ice edge are visible in spaceborne SAR images owing to spatial redistribution of drifting ice fields tending to accumulate in the surface current convergence zones associated with spiraling motion of eddy-induced currents [1].

SAR manifestations of MIZ eddies depend on background wind and thickness of drifting ice. In this case thinner ice (e.g., newly forming or melting ice) would be seen in SAR image as dark patterns (low backscatter), while thicker ice fields—as bright patterns (high backscatter). This is well illustrated in Envisat ASAR image acquired on 23 February 2007 over the 50–60 km wide MIZ in Fram Strait (Figure 3). As seen, very dark eddying patterns are seen further off the pack ice that has nearly uniform gray background in the upper left part of the image. Closer to the pack ice, eddy signatures have higher backscatter due to entrainment of more and thicker ice floes in their centers. The distance between the centers of adjacent eddies range from 10 km to 50 km, 30 km on average. Note also numerous bright elongated filaments that are formed due to enhanced ice accumulation in the surface current convergence zones and outline eddy boundaries. The visual inspection of such eddy-induced regions of enhanced/decreased radar backscatter shows the limits of the areas with positive or negative vorticity linked to eddy structures [7].

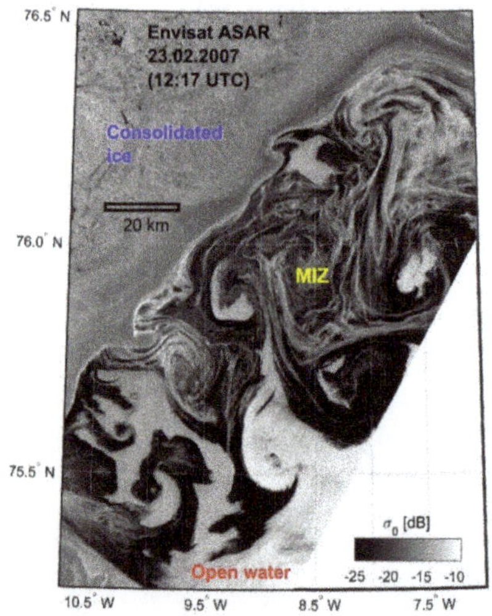

Figure 3. Map of normalized radar cross section showing an example of MIZ eddy manifestations in Envisat ASAR image acquired on 23 February 2007 at 12:17 UTC over Fram Strait. Notions given in different colors show position of consolidated ice (blue), MIZ (yellow) and open water (red).

Our method for eddy detection in SAR data is based on visual identification of their surface signatures frequently used in similar studies [18,20–22]. Earlier, it was successfully applied to quantify eddy properties during summer periods in the Western Arctic [23], northern Greenland Sea and Fram Strait [24,25], Barents, Kara and White seas [16]. In order to obtain a coherent pan-Arctic climatology of eddy properties, here we use the same visual detection of eddies in SAR data.

Following the methodology described in [23], every SAR image was visually inspected at full resolution in search for distinct eddy signatures and identification of their boundaries.

Then by looking at eddy boundaries outlined due to accumulation of floating ice, their location and vorticity sign (cyclone/anticyclone) were defined manually. The mean of the two quasi-perpendicular sections across the eddy center was taken as a mean eddy diameter.

Analysis of ASAR data was performed using Matlab-based software suggested in [26], while the analysis of Sentinel-1 data was performed using the open-source ESA SNAP software (http://step.esa.int/main/toolboxes/snap (accessed on 14 February 2021). The entire procedure of data pre-processing and eddy detection was similar for all the data. The original SAR data were calibrated to normalized radar cross-section units and smoothed to reduce the speckle noise using Lee filter [27].

The background sea ice concentration corresponding to each eddy center was obtained from daily AMSR-E and its successor AMSR-2 sea ice concentration maps produced by the University of Bremen [28], while a detailed estimate of the amount of sea ice entrained into the eddies was derived directly from SAR images. Depth values corresponding to eddy center coordinates were retrieved from IBCAO Version 3.0 [29]. To analyze the relation of eddy generation intensity to background winds and ocean currents we used ERA Interim Reanalysis 10 m winds and CMEMS GLORYS12V1 reanalysis data for January–April 2018, respectively. For ocean currents' data we used the surface layer at 0.5 m depth.

3. Results

3.1. Spatio-Temporal Statistics of MIZ Eddies

Analysis of spaceborne SAR data for two winter periods of 2007 and 2018 allowed identifying 4619 eddy signatures in the MIZ and at the ice edge with 399 eddy signatures detected in December 2006–March 2007, and 4220—in January–April 2018. Quantitatively, two Sentinel-1 sensors captured 10.5 times more eddies compared to Envisat ASAR, or 1.2 times more eddies when normalized on the number of available SAR images. However, when accounting for the 1.2 times difference in total MIZ length between 2007 and 2018, the average number of eddies detected per one image per kilometer of MIZ length is similar in 2007 and 2018 and does not depend much on the sensor used.

Apart of MIZ eddies, open-ocean eddies were also detected in the data, however, their number was surprisingly small, only 323 eddy signatures. In contrast, much more open-ocean eddies (1609 eddies) were detected in July–September 2007 [24,25]. Such a pronounced difference in open-ocean eddy detections between winter and summer seasons could be related to the less pronounced vertical stratification and the higher near-surface winds during winter months that limits eddy identification in SAR images [18]. Hence, we did not consider open-ocean eddies in the analysis. Table 2 provides a summary of eddy detection results in MIZ obtained upon the processing of all SAR data.

Table 2. Summary of the MIZ eddy detection in spaceborne SAR data in winter seasons of 2006/2007 and 2018.

Month	Number of Eddies						Mean Diameter (STD), km					
	C		AC		SUM		C		AC		SUM	
	2007	2018	2007	2018	2007	2018	2007	2018	2007	2018	2007	2018
December	31	-	7	-	38	-	11.2	-	12.3	-	11.4	-
January	56	752	13	410	69	1162	8	7.5	9.4	9.3	8.3	8.1
February	83	848	40	432	123	1280	11.4	7.1	14	9.5	12.2	7.9
March	109	767	60	424	169	1191	9.7	8.1	12.8	8.2	10.8	8.1
April	-	430	-	157	-	587	-	9.3	-	11.3	-	9.9
Total	**279**	**2797**	**120**	**1423**	**399**	**4220**	**10 (8.1)**	**7.8 (7.2)**	**12.8 (10.8)**	**9.2 (8.5)**	**10.9 (9.1)**	**8.3 (7.7)**

What is immediately apparent, the number of cyclones is twice larger than that of anticyclones (AC), i.e., 67% of C versus 33% of AC, similar for 2007 and 2018. This exactly matches the results of the summer-time observations of MIZ eddies in the Western Arctic [24], and is slightly less than was obtained in the Greenland Sea [18].

In terms of monthly variability, both records show the highest number of eddies in February and March (Table 2). This is also confirmed by the number of eddies detected per given month normalized by the SAR coverage during that month (Figure 4a) and shows similar high values in February and March, a bit lower value for January and a minimum in April.

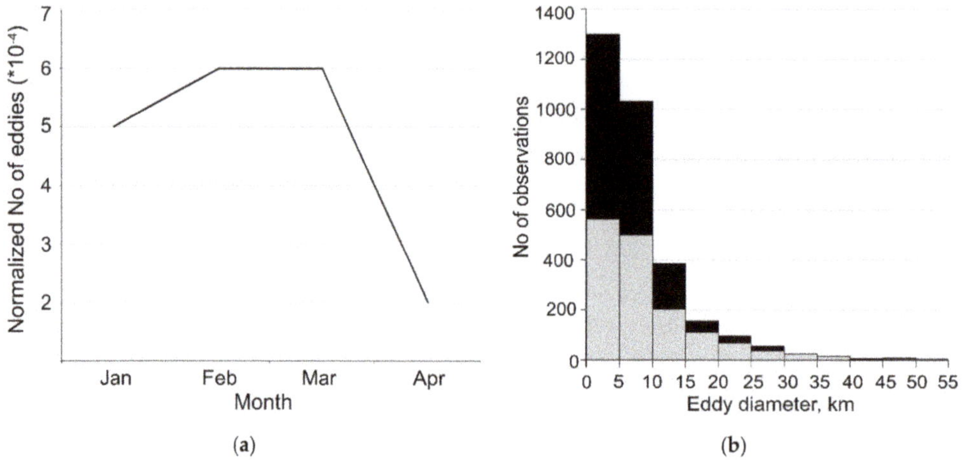

Figure 4. (**a**) Monthly variability of the normalized number of MIZ eddies detected in winter 2018; (**b**) histogram distributions of the number of eddies as a function of eddy diameter. Black and grey colors correspond to cyclones and anticyclones, respectively.

Table 2 also shows mean values of eddy diameters in 2007 and 2018 and their standard deviations (STDs). The minimum (maximum) value of eddy diameters registered in 2007 is about 1.3 km (66.8 km). In 2018, the minimum diameter value, ~0.6 km, is lower than in 2007 due to higher spatial resolution of Sentinel-1 data, while the maximum is nearly the same, 67.6 km. Large intervals between minimum and maximum values of eddy diameters in 2007 and 2018 explain high STDs in the record.

Figure 4b shows a histogram distribution of eddy diameters for eddies registered in winter 2018. Both cyclones and anticyclones have a peak in the range of diameters of 1–10 km. It is also apparent that at scales of 1–15 km cyclones strongly dominate over anticyclones. However, in the range of 15–30 km this difference is gradually vanishing, and for diameter values above 30 km anticyclones start to dominate slightly. This appears to be a characteristic feature previously documented for open-ocean eddies detected in SAR images elsewhere [22,30].

Figure 5 shows a spatial distribution of MIZ eddies detected in 2007 and 2018, and their probability calculated on a 35 km $\times$ 35 km grid as the ratio between the total number of eddies encountered within a given grid cell and the number of SAR looks of that cell. The probability color bars in Figure 5c,d have a maximum at 0.2 which means that at the particular grid cell at least one eddy is registered in every fifth SAR image. Note that at some locations the maximal probability values exceed this upper limit.

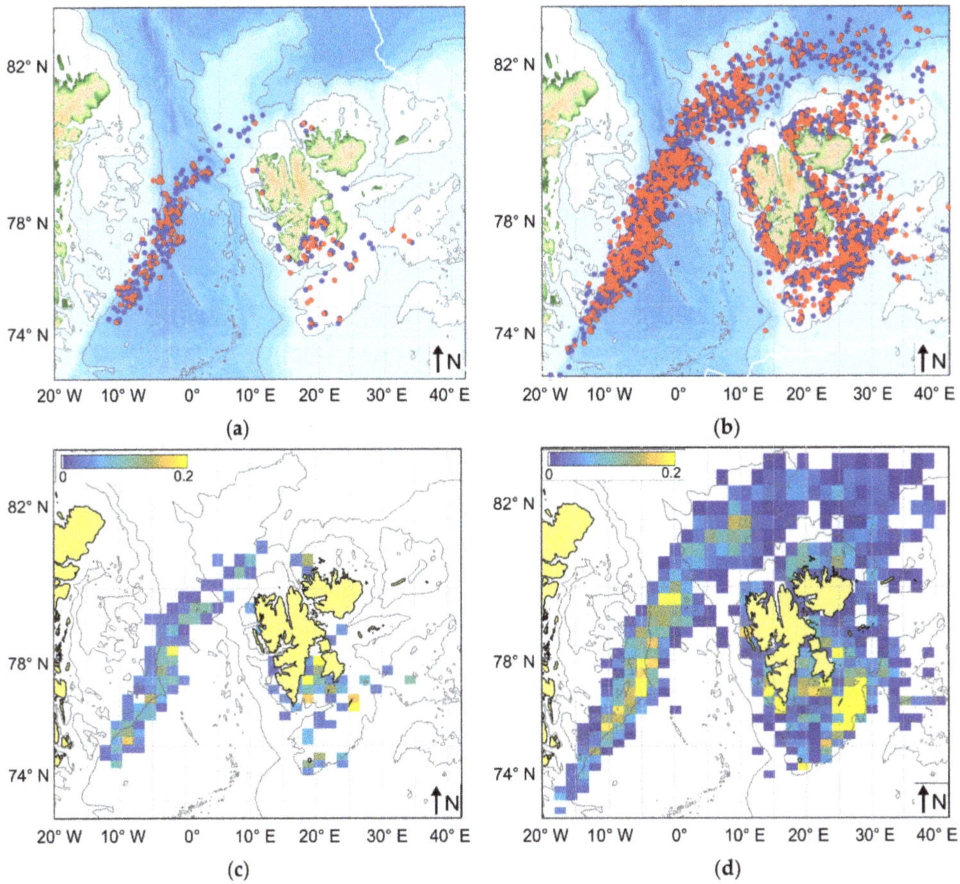

Figure 5. Locations (**a**,**b**) and probability (**c**,**d**) of the MIZ eddies identified in spaceborne SAR data during winter of 2007 (**a**,**c**) and 2018 (**b**,**d**). Blue (red) circles in (**a**,**b**) mark cyclonic (anticyclonic) eddies. The thick white line in (**a**,**b**) denotes the boundary of a region with the number of SAR images below 10. Grey lines are the 500- and 2000-m isobaths.

In 2007, the eddies were identified across the narrow band (hereinafter, Fram band) extending over 75°–80° N and 10° W–4° E (i.e., the region of maximum SAR coverage in 2007), in Great Fjord Strait and over the Spitsbergen Bank (Figure 5a).

The results for winter 2018 confirm and expand the above results for 2007. A very dense and 200–250 km wide band of eddy detections, resulting from the gradual week-to-week variability of the ice edge location, now expands all along the Greenland Sea shelf break and slope, Fram Strait, southern Yermak Plateau (YP) and further east up to 40° E over the deep water north of Svalbard (Figure 5b). These regions are also characterized by the highest probability of eddies of 0.2 and higher (Figure 5c,d).

Many eddies are also registered over the Spitsbergen Bank and in the coastal regions around Svalbard, in many fjords and straits, especially in the southern part. The highest probability of eddies is found east of Edge Island, around Hopen, Prince Karl Land and Bear Islands, and near the southern tip of Spitsbergen. The most pronounced hot spot of MIZ eddy generation is the region over the Spitsbergen Bank and near Hopen Island. The barotropic tidal currents are strong here [31,32], reaching 20 cm/s, and the eddies are generated due to instability of sheared current when the flow passes the orographic obstacle [33]. High eddy probability is also found in Great Fjord Strait, where frequent eddy generation was also reported in summer [25], and could be related to interaction of cold PW entering through Heleisund and Freeman straits and warm AW coming with the WSC.

From the comparison of two data sources one can clearly see that the ASAR-based eddy observations in winter 2007 partly lack some important regions of frequent eddy generation observed by Sentinel-1 in winter 2018, e.g., north of Svalbard and in many coastal areas around it. Of course, the observed difference results not only from a much wider coverage of the study site by Sentinel-1 data, but is also related to rather different sea ice extent and ice margin location in 2007 and 2018 (see Figure 2).

Figure 6 shows a spatial distribution of mean and maximum values of eddy diameters on the 35 × 35 grid for 2007 and 2018. Comparison of mean values for 2007 and 2018 (also shown in Table 2), where they overlap, suggests that higher mean values were observed in 2007. This is particularly pronounced at 76–79° N where mean eddy diameters of ~20–25 km are observed in 2007 versus ~12–20 km in 2018.

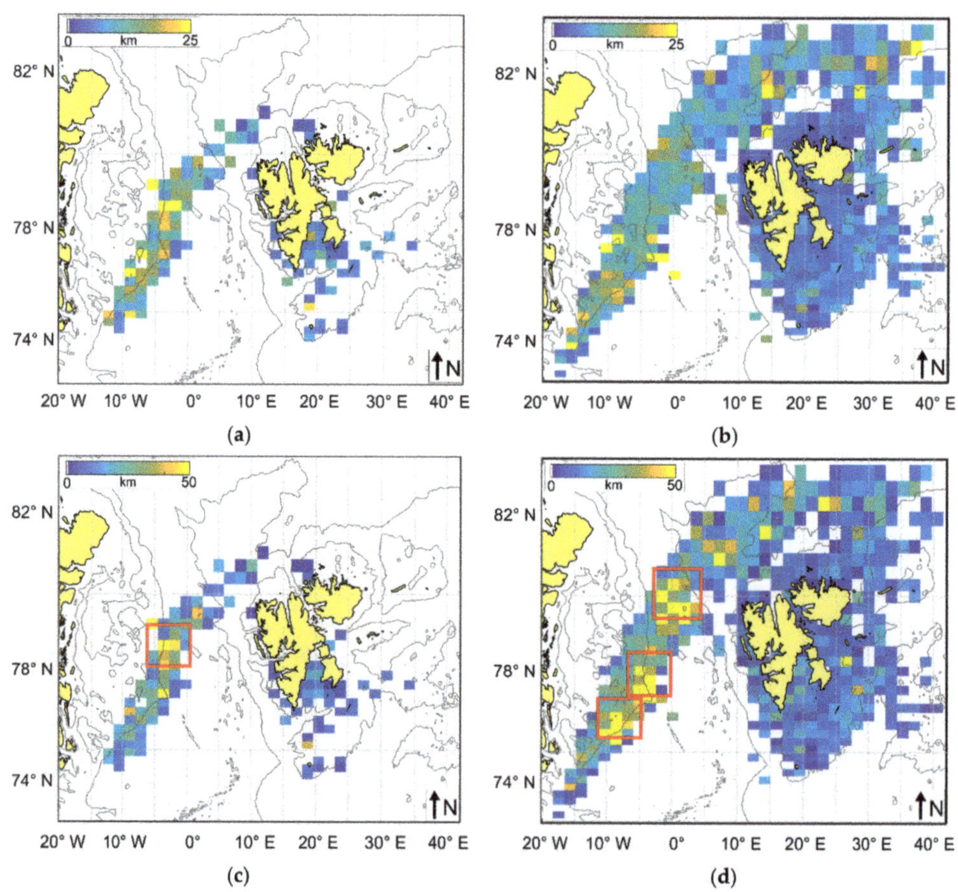

Figure 6. Spatial distribution of the mean (**a**,**b**) and the maximum (**c**,**d**) diameters of the MIZ eddies in 2006/2007 (**a**,**c**) and 2018 (**b**,**d**) in kilometers. Overlaid grey lines are the 200- and 2000-m isobaths.

However, it is not the case for the maximum eddy diameter values (Figure 6c,d). Where observational data of 2007 and 2018 overlap, the data of 2018 have, in general, higher values. Moreover, there are at least three hot spots in Fram eddy band (75°–80° N) where eddy diameters are of 40–60 km, while only one is seen in 2007 (marked by red boxes in Figure 6c,d). There are also other hot spots of high eddy diameters—over the southeastern YP and further eastward.

Figure 6 also clearly shows the overall difference in eddy diameters between the deep and the shallow regions with larger eddies usually found over the deep water. This is detailed in Figure 7a showing a distinct increase in eddy diameters from about 6–7 km over <500 m depths to maximum values of 11–15 km over 2000–3000 m depths, and another pronounced decrease for deeper regions.

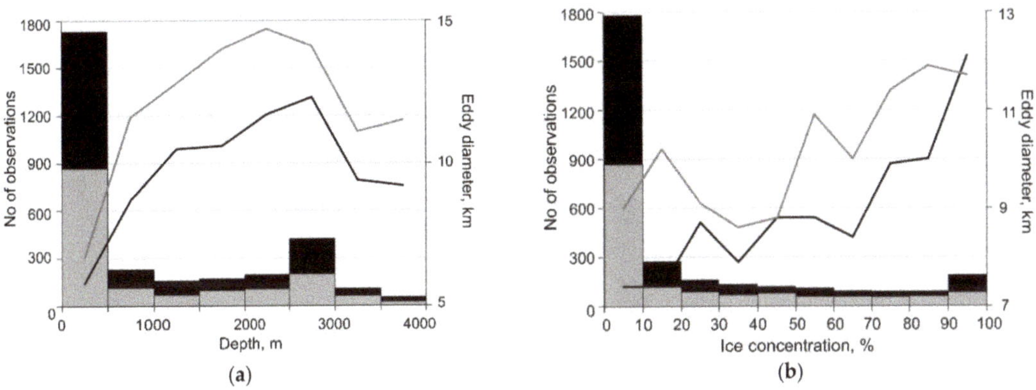

Figure 7. Histogram distributions of the number of eddies and their mean diameters observed in winter 2007 and 2018 as a function of: (**a**) depth corresponding to eddy locations; (**b**) sea ice concentration. Black and grey colors correspond to cyclones and anticyclones, respectively.

According to [34], the first mode Rossby radius is about 5–8 km over the deep-water regions of the Greenland Sea, Fram Strait and north of Svalbard in winter, and about 0.5–3 km in the shallow coastal regions around Svalbard. In our case, the observed mean eddy radii are about 3 km in shallow water and 5–7.5 km in deep water, suggesting that submesoscale and small mesoscale eddies dominate in the record. Indeed, the portion of submesoscale eddies for shallow water (i.e., eddies with radii <3 km) is 69% out of 2121 eddies, and 57% out of 2168 eddies found over deep water (i.e., eddies with radii <5 km).

Note also that anticyclones (gray curve in Figure 7a) are always 15–30% larger in diameter than cyclones (black curve in Figure 7a), which is also well seen from Table 2 for all months and years. This, again, appears to be a common feature in the Arctic Ocean [16,23] and elsewhere [22,33,35].

Figure 7b provides a general information about the background ice concentration at the locations of MIZ eddy encounters in 2007 and 2018. Clearly, most eddies were detected at the ice edge and in the regions with the low ice concentration (<20%). Once the ice concentration rises above 10%, the number of detected eddies rapidly drops down with the minimum found around 60–90% of the ice concentration. Interestingly, a small rise of eddy detections is also seen over the pack ice with the ice concentrations of 90–100% which makes it different from the results obtained in the Western Arctic in summer [23]. Another remarkable feature is that the size of detected eddies apparently rises with increasing ice concentration, especially prominent for cyclonic eddies (Figure 7b). Plausibly, the increase of mean floe size with increasing ice concentration in the MIZ limits the detection of small eddies under pack ice compared to low ice concentrations.

3.2. *Ice Trapping and Melting by Eddies*

Figure 8 further shows a histogram distribution of sea ice fraction inside detected eddies derived from SAR images. This property was counted only for eddies detected in January 2018 (1162 eddies), presuming that it should be relevant for the entire record. It describes the amount of sea ice trapped inside every single eddy. The ice fraction in eddies was visually estimated using five classes from 0% to 100% with a 20% step in sea ice concentration. For this, every eddy was zoomed at full resolution, and then a curve following the eddy boundaries and two perpendicular lines denoting the eddy diameter were put on. A relative amount of ice within the eddy was then evaluated visually and assigned to one of five classes with mean ice concentration values of 10%, 30%, 50%, 70% and 90%.

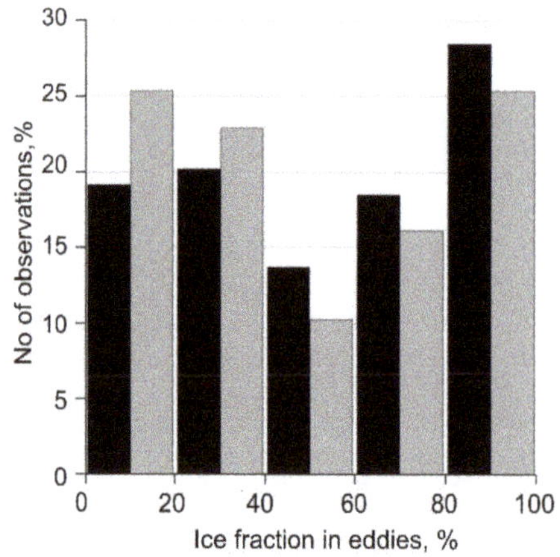

Figure 8. Histogram distribution of sea ice fraction in eddies detected in Sentinel-1 SAR images in January 2018. Black (gray) bars indicate cyclonic (anticyclonic) eddies.

Bearing in mind that cyclones were twice more frequent than anticyclones, the number of observations for cyclones (anticyclones) in Figure 8 is given relative to the total number of cyclones (anticyclones), not the total number of all eddies, so that their histogram distributions would be easily comparable. As seen from Figure 8, the distribution for cyclones (black bars) has peaks in two classes of 20–40% and 80–100% with higher number of observations in the latter. For anticyclones (gray bars), two equal peaks are in 0–20% and 80–100% classes with an overall higher portion of observations for ice fraction values below 50%. Indeed, the mean value of ice fraction for anticyclones is slightly lower (48%) than that for cyclones (53%). Hence, below we take the mean ice fraction in eddies equal to 50%.

Bondevik [18] used the same value of mean ice fraction of 50% and mean eddy radius of 30 km to quantify the area of sea ice trapped by a single eddy and associated integral eddy-induced retreat of the ice edge. Our results suggest a much smaller range of mean eddy radii, 3–7 km. Following [1,18] and taking the mean eddy radius $r = 5$ km, and the mean ice fraction $f = 0.5$ (i.e., 50%) results in trapping of $S_{ice} = \pi r^2 f = 3.14 \times 5^2 \times 0.5 \sim 40$ km^2 of sea ice by a single 'mean' eddy, i.e., about 40 times less than estimated in [18].

In general, MIZ eddies can persist from couple of days till several weeks [1,7]. Assuming that each eddy was detected in SAR data at least twice (but could be more frequent), the average number of eddies per winter month equals to ~500 eddies. As a result, a total area of sea ice trapped by all eddies per month equals to ~2×10^4 km^2.

Using the obtained information one can estimate an average horizontal retreat of the ice edge due to MIZ eddies following a simple formula suggested in [1] and applied in [18] in the form:

$$A = \left(\omega f \pi r^2\right)/(lh) \tag{1}$$

where $\omega = 0.38$ m·d^{-1} is the mean bottom ablation [36], $f = 0.5$, r has range of 3–6 km for shallow and deep water; l is the mean distance between the centers of adjacent eddies taken as 15 km for shallow water and 30 km for deep water, respectively; and $h = 1.5$ m is the mean ice thickness [1,18]. When using Equation (1), we assume that all parameters are constant apart of r that has error $\sigma_r \approx 0.1$ km (i.e., equal to the spatial resolution of the data that is 2 × pixel size ≈ 90 m). To calculate the error propagation, Equation (1) can be presented in the form $A = br^2$, where $b = (\omega f \pi)/(lh)$ is a constant value. The relative error is then $\sigma_A/A = 2\sigma_r/r$, and the error of $A \approx 0.02$ km·d^{-1}. In this case, the average horizontal retreat perpendicular to the ice band equals to 0.2–0.5 km·d^{-1} ± 0.02 km·d^{-1}.

Having spatial fields of some parameters (i.e., f and r) used in Equation (1) on the 35 × 35 grid, it is a natural step to obtain a spatial distribution of A first for January 2018 and then for the entire winter 2018. To calculate A for January 2018 we use ice fraction in eddies, f (Figure 9a), and eddy radii, r (Figure 9b), on the 35 × 35 grid derived from SAR observations, the mean distance between the centers of adjacent eddies equal to 15 km (depth < 300 m) and 30 km (depth > 300 m), and other parameters as constants described above.

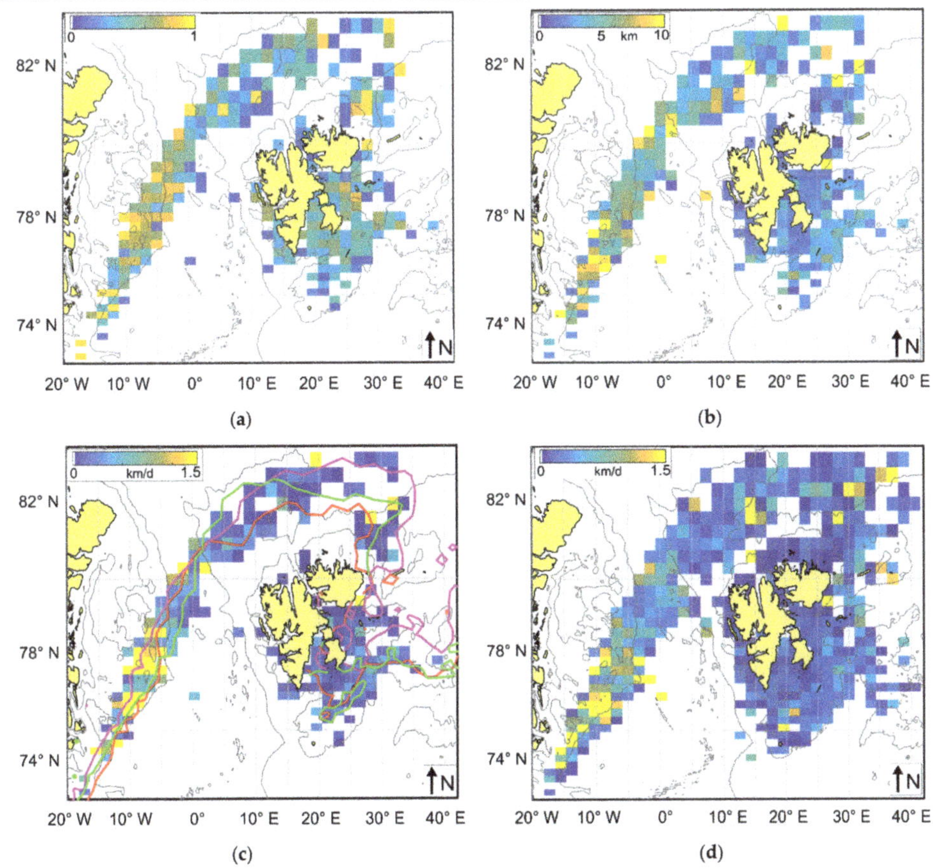

Figure 9. Spatial distributions of: (**a**) ice fraction in eddies and (**b**) eddy radii in January 2018; the average horizontal retreat of the ice edge due to MIZ eddies for (**c**) January 2018 calculated using the data from (**a**,**b**); (**d**) for the entire winter 2018, when only r variations in space (shown in Figure 6b) were considered. Red, pink and green lines in (**c**) show the ice edge location from AMSR-2 data in the beginning (5th), middle (15th) and end (25th) of January 2018.

Note that ice fraction in eddies (Figure 9a), in general, has higher values over the Fram eddy band, 0.5–0.8 (or 50–80%), and a similar range of values over the northern band and the Spitsbergen Bank near Hopen, 0.2–0.6 (20–60%). For the entire winter of 2018 (Figure 9d), only r is taken on the 35 × 35 grid ($r = d/2$ with the spatial distribution of d shown on Figure 6b), and the distance between adjacent eddies varies depending on depth, while other parameters are constant (including $f = 0.5$).

According to calculations, in January 2018, high values of horizontal ice retreat, up to 1.5–2 km·d^{-1}, are seen over the southern part of the Fram eddy band, between 74–79° N, and over the slope regions northeast of Svalbard and YP (Figure 9c). These regions are characterized by relatively high values of eddy radii and ice fraction inside eddies as compared to adjacent regions (Figure 9a,b). Minimal A values about 0.1–0.2 km·d^{-1} are seen in the coastal regions of Svalbard and over the central and eastern YP, while moderate values (0.3–1 km·d^{-1}) are found over the northern eddy band (southern flanks of YP) and northeast of Hopen Island.

The color lines in Figure 9c show the ice edge location taken from AMSR-2 data in the beginning (5th), middle (15th) and end (25th) of January 2018. As seen, during the month the ice edge boundary has moved either westward (for the Fram eddy band) or northward (for the northern band and over the Spitsbergen Bank), meaning a decrease in the actual MIZ area that could be at least partially attributed to the eddy-induced ice melt. The strongest decrease of MIZ area and shift of the ice edge location are seen during the first part of January (compare the location of red and pink lines), confirming that MIZ possess a strong week-by-week spatial variability. While the ice edge has partially restored back in some locations during the second half of January 2018 (compare red and green lines), the resulting area of MIZ has obviously decreased during the month. However, the amplitude of the ice edge movement does not entirely correlate with the plotted A values, though in some locations the correspondence is still observed.

The distribution of the horizontal ice edge retreat for the entire winter 2018 (Figure 9d) has many similar patterns to its distribution in January 2018 with higher A values observed over the southern part of the Fram eddy band and at some certain locations north, east and south of Svalbard. The spatial distribution of A values in Figure 9d is proportional to eddy size distribution shown in Figure 6b and basically suggests that larger eddies would have a stronger impact on the ice edge retreat in the MIZ. Due to longer time period and rather complex MIZ variability, here we do not attempt to make a direct comparison of A values plotted in Figure 9d with the ice edge locations throughout the 4-month period. Obviously, such a comparison is a very challenging task and would need a lot of additional information about the coupled ocean–ice–atmosphere state in the study region. Nevertheless, the presented methodology allows one to make a kind of sensitivity tests and a bulk estimate of eddy influence on the ice edge retreat in the MIZ.

3.3. *Relation to Boundary Currents and Winds*

As seen from Figures 1 and 5, bands of MIZ eddies are often found along or in close proximity to the main boundary currents of the region, i.e., the EGC, the WSC and its Svalbard and Yermak Branches, that follow topography and, in some locations, coincide with the ice edge boundary. In such case it seems reasonable to expect that intensity of boundary currents should have an impact on eddy generation at the ice edge and in the MIZ [1] as the eddy kinetic energy is highest along main currents over topography slopes [13]. Wind speed and direction are also key factors defining the movement of ice floes in the MIZ and resulting eddy generation [17,18,37].

To check the spatial correlation between eddies, currents and winds we have plotted monthly-mean distributions of these properties for January–April 2018 (Figure 10). The numbers of eddy detections between January and March are very similar (Table 2), but their probability maps have certain spatial differences (Figure 10, left column). The Fram eddy band between 74°–80° N is the most stable one and directly correlates with the position of main jet of the EGC directed southward (Figure 10a,b).

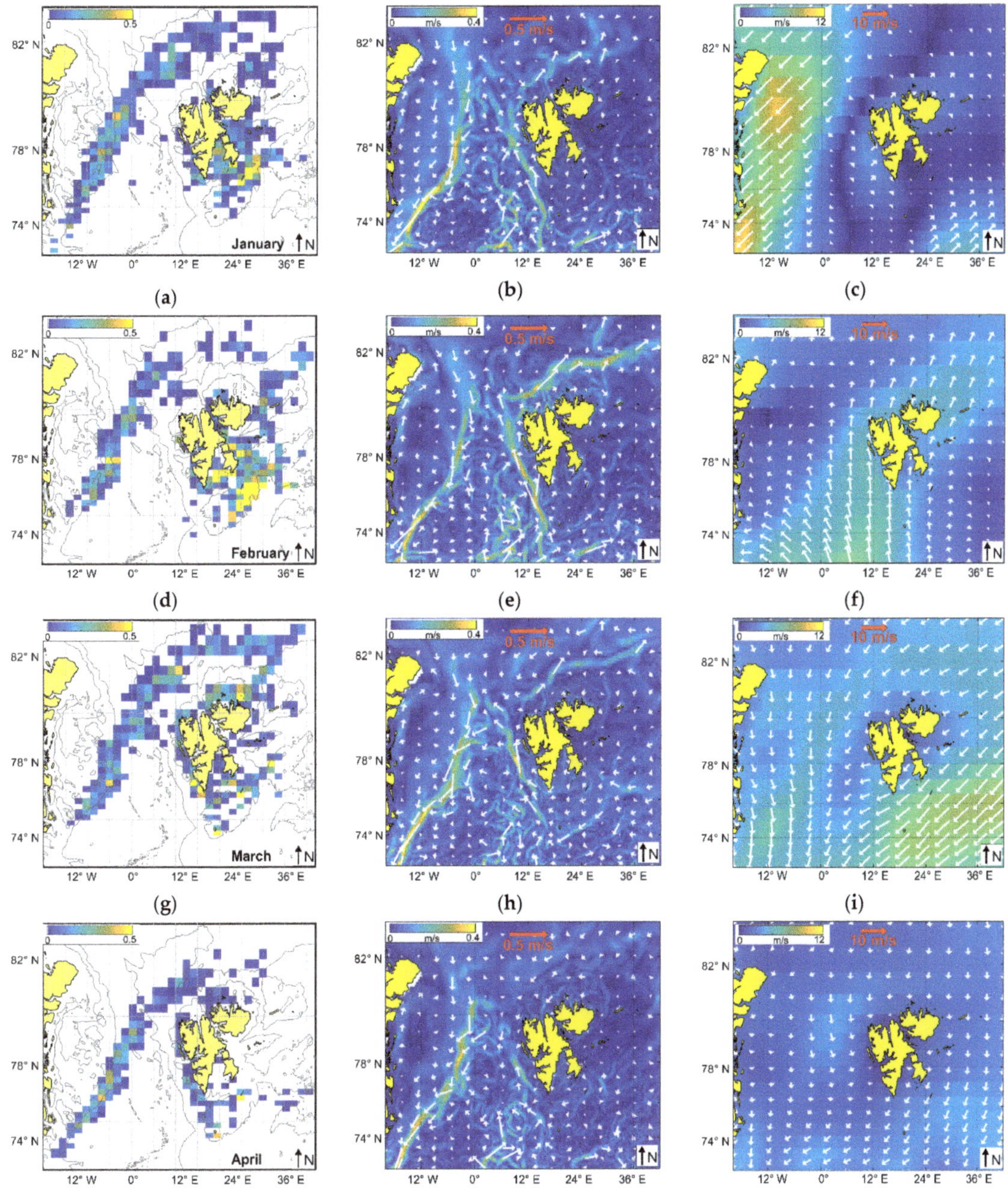

(a) (b) (c)

(d) (e) (f)

(g) (h) (i)

(j) (k) (l)

Figure 10. Monthly-mean spatial fields of eddy probability in the MIZ (**a**,**d**,**g**,**j**), surface current speed [m/s] and direction from GLORYS12V1 reanalysis (**b**,**e**,**h**,**k**), and 10 m wind speed [m/s] and direction from ERA Interim Reanalysis (**c**,**f**,**i**,**l**) for January–April 2018. Red arrows show velocity scales in the middle and right columns. Overlaid grey lines in left column are the 200- and 2000-m isobaths.

According to the GLORYS12V1 reanalysis data, the EGC intensity is highest in January, reaching 0.3–0.4 m/s (Figure 10b). At this time the Fram eddy band has high probability values and across-band width (Figure 10a). A monthly-mean wind field map shows a presence of northeasterly winds of 7–9 m/s over the Greenland Sea and western FS, and low winds of changing direction around Svalbard (Figure 10c). As seen from Figure 10a,c, high eddy probabilities over the Fram band are observed under moderate northeasterly winds that are favorable for eddy development at the ice edge in this region [18].

In turn, the eddy band north of 80° N (at 0–40° E, hereinafter, northern band) have largest area but moderate probability values that also correspond to moderate intensity of background currents (Figure 10b). As seen, the eddy activity is higher in the western part of the band influenced by northeasterly winds that blow parallel to the ice edge causing a westward movement of sea ice and upwelling along the ice edge. Lastly, rather intense eddy generation is observed near Hopen Island, where both currents and winds seem to be rather weak.

In February 2018, when the maximum number of eddies is registered, the WSC Svalbard Branch is intensified, but the eddy band is patchy and locates north of the main current. We do not see more eddies forming in this region. Instead, eddy probability has a maximum southeast from Hopen Island, where the currents are slightly more intense compared to January (Figure 10e). The wind field map shows a pattern of enhanced southerly winds that appear to favor eddy formation in FS and near Hopen.

In March 2018, the northern band has high probability values with the highest one registered over the southeastern slope of YP (Figure 10g), where currents are rather intense (Figure 10h). The wind speed is also relatively high and of favorable northern–northeastern direction. Further hot spots of eddy generation/observations are forming just north of North East Land (see Figure 1) and along the southwestern coast of Spitsbergen. Enhanced currents and 7–9 m/s northeasterly winds are seen over Hopen Island, but no apparent rise in eddy probabilities is seen in this case.

In April 2018, the overall intensities of the boundary currents and winds are lowest that, in general, correlates quite well with the smallest number of detected eddies and the lowest eddy probabilities apart of the Fram band (74–80° N) where the intensities of currents and eddies are still quite pronounced. The northern band shifts closer to Svalbard with no peaks in eddy probabilities, in agreement with weak currents in this region.

To sum up a general correspondence of eddies to background currents and winds, Figure 11 shows histogram distributions of the number of eddies observed in winter 2018 as a function of background current velocity and wind speed. Though there is a certain portion of eddies observed over the regions with enhanced currents of 0.2–0.4 m/s, most of them were observed in the regions with modest current velocity below 0.1 m/s (Figure 11a). As for the wind speed, most of the detected eddies were observed over low winds of 1–4 m/s, while about 15% of eddy observations were made under moderate winds of 4–8 m/s (Figure 11b).

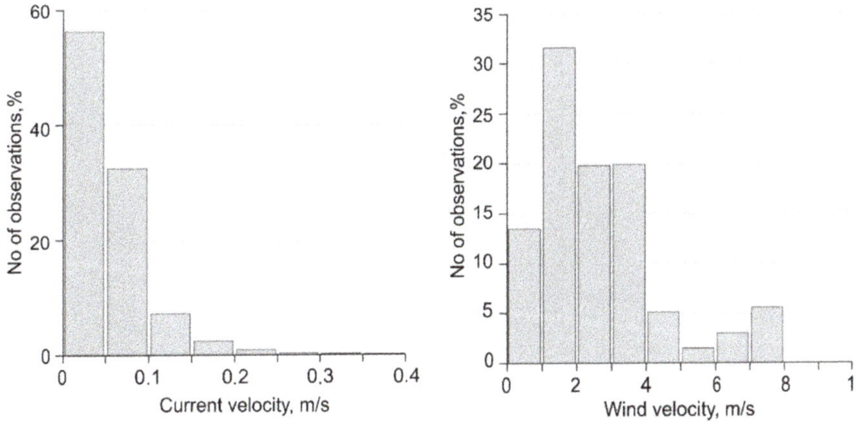

(a) (b)

Figure 11. Histogram distributions of the number of eddies observed in winter 2018 as a function of (**a**) current and (**b**) wind velocities [m/s] around the locations of detected eddies derived from GLORYS12V1 model and ERA Interim Reanalysis.

4. Discussion

Here we should make a remark about the reported eddy numbers, their changes from month to month, and the spatial maps of eddy probability. The typical lifetime of eddies generated at the ice edge or inside the MIZ could be from couple of days to couple of weeks. In such case almost every eddy (apart of very short-lived ones) in the record would be repeatedly identified in SAR images. Surely, the most accurate counting of MIZ eddies would need a precise tracking of every individual eddy whose surface/SAR signatures may strongly evolve throughout its lifetime. This is a difficult and a time-consuming task. To simplify it, we have applied a simple normalization of the total number of detected eddies by the SAR coverage. This allowed to answer two simple, yet very important questions—when and where the number of detected eddies, assumed to be proportional to their generation intensity, was highest over the study region during these two particular winter seasons.

As already mentioned above, the winter seasons of 2007 and 2018 were selected arbitrary. However, both these years were characterized by anomalous states in regard of sea ice and hydrographic conditions, e.g., the year of 2007 was known as one of the years with record minimum ice extent in the Arctic Ocean, while the year of 2018 had an anomalously low sea ice outflow through Fram Strait [38,39] and rather long MIZ. Without going into details, it is obvious that the results from these two years might not apply in all years. Specifically, the zonal position/extent of the main eddy bands, i.e., the Fram band, the northern band and near Hopen are different in 2007 and 2018, and might differ with those in other years.

The obtained results clearly show that submesoscale and small mesoscale eddies dominate in the record, while the overall range of eddy diameters spans from 1 km to 68 km. Both cyclones and anticyclones have a peak in the range of diameters of 1–10 km. At diameter scales of 1–15 km, cyclones strongly dominate over anticyclones, in agreement with historical field observations [1] and model results [8]. However, in the range of 15–30 km this difference is gradually vanishing, and for diameter values above 30 km we observe an approximate parity between them, in agreement with satellite altimetry results [24,40], with a slight dominance of anticyclones. Such a transitioning from dominating cyclones to anticyclones happens when the eddy size exceeds the Rossby deformation radius, i.e., when mesoscale eddies start to dominate over submesoscale, as earlier reported for open-ocean eddies in [22,30]. However, to our knowledge, such a peculiarity has never been reported before for the MIZ eddies.

Another notion should be completed regarding the specifics of SAR imaging of eddy features of certain spatial scales. For the ice-free regions, SAR observations are known to have a certain bias toward relatively small, predominantly submesoscale eddies often having pronounced surface patterns due to more intense convergence zones at the surface as compared to mesoscale eddies. Such signatures strongly depend on background winds and are well-detectable under low to moderate wind conditions [41]. In this case spatially varying oceanic and atmospheric conditions add complexity for identifying large eddy structures over open ocean. However, within MIZ and at the ice edge, where floating ice serves as a tracer, the visibility of eddy signatures in SAR is not limited by winds what allows the detection of large eddies as well (see examples e.g., in [1,4,7,23]). Therefore, we presume that our record should not have a pronounced bias toward small-scale eddies.

However, the observed dominance of cyclones at submesoscales might have a certain explanation that is linked to ice trapping by eddies. Using physical arguments, the model study [8] shows that the sea ice accumulates preferentially in submesoscale cyclones where the frictional Ekman pumping generates strong convergence, while anticyclones repel the sea ice. This fact is also supported by our results, though the observed difference in the amount of ice trapped by cyclones and anticyclones is not large. As a result, cyclones appear to have more pronounced ice-related surface signatures and, hence, are more easily identified in SAR data. Yet, the method of visual assessment of ice fraction in eddies used here is not very precise and can have certain errors. Therefore, future studies should address this question using automated machine learning methods (e.g., [42,43]) for sea ice detection and masking inside eddy patterns.

Here we also attempted to give a first-order estimate of the eddy-induced horizontal sea ice retreat using observed values of eddy radii and amount of sea ice trapped in the eddies, and empirical mean values of ice bottom ablation and ice thickness. The obtained average horizontal ice retreat is about 0.2–0.5 $km \cdot d^{-1} \pm 0.02\ km \cdot d^{-1}$. These values are several times smaller than those obtained by Johannessen et al. [1] and Bondevik [18] which is mainly due to the lower mean eddy radii observed in our data and used for calculations. In our calculations, we discarded the effect of lateral ablation though it might be important for relatively small ice floes dominating in the MIZ [1]. Moreover, we presumed the equal role of cyclones and anticyclones in sea ice melt, while the latter are thought to be more efficient in ice melting due to Ekman upwelling caused by 'Eddy Ice Pumping' mechanism prominent within the compact ice zone [44].

The spatial patterns of the eddy-induced horizontal sea ice retreat derived from SAR data suggest a pronounced decrease in MIZ area and a shift in the edge location that does not contradict the observations. Though the obtained values have many uncertainties that should be carefully examined in a more thorough study, presented methodology allows to obtain first-order bulk estimates of eddy influence on sea ice melt in the MIZ.

The analysis of the spatial correlation between eddies, currents and winds shows that the intensity of eddy generation/observations and their detectability in the MIZ, and the width of eddy bands correlate with the intensity of northern and northeasterly winds, as also previously shown in [18]. In some regions, e.g., along the Greenland Sea shelf break, in Fram Strait and over the Spitsbergen Bank the probability values of eddy occurrence in the MIZ seem to correlate with stronger boundary currents, while north of Svalbard and over Yermak Plateau higher probability values are observed under low/moderate currents and winds.

5. Conclusions

In this work, we have attempted to document the generation/occurrence hot spots and properties of MIZ eddies in FS and around Svalbard from the analysis of distinct quasi-circular features seen in spaceborne SAR images. We believe that these features in the sea ice distribution are most likely caused by mesoscale and submesoscale eddies in the ocean below, as numerous previous studies confirm. However, there may be plenty more eddies further away from the ice or into the denser pack ice that do not lead to similar signatures well-detectable in SAR data. There can also be eddies that are mostly or entirely subsurface and, therefore, do not have a strong surface expression. Additionally, there may be non-circular features such as narrow filaments that are not considered here. In this sense our SAR-based results present only a lower limit of the real eddy activity occurring in the MIZ.

Open-ocean eddies were also detected in the used SAR data; however, their number was surprisingly small as compared to the summer season when an order of magnitude more eddies were detected (e.g., [16,23–25]). We link this to the less pronounced vertical stratification and the relatively high winds in winter season limiting eddy manifestation in SAR images. Therefore, our current study does not allow us to conclude whether the eddy generation intensity over the open-ocean regions is higher in winter as compared to summer, as models suggest (e.g., [11,12,14]).

Analysis of SAR data from historical Envisat ASAR and contemporary Sentinel-1 SAR-C missions clearly shows a better performance of the latter in terms of spatial coverage of the study site that allowed a broader identification of eddy generation/occurrence hot spots in the marginal ice zone. Two Sentinel-1 sensors allowed to identify 1.2 times more eddies than ASAR when normalized on the number of available SAR images. However, when accounting for 1.2 times longer MIZ in winter 2018 compared to winter 2007, the average number of eddies detected per one image per kilometer of MIZ length is similar in both years and does not depend much on the sensor used.

Almost twice better spatial resolution of Sentinel-1 means that it may better resolve spatial patterns of fine- and small-scale dynamic features in the MIZ. The orbital design of two Sentinel-1 sensors allows sequential mapping of MIZ eddies on a daily regular basis which is critical to assess their dynamic properties and evolution [7], that was previously available only from altimetry observations of large mesoscale open-ocean eddies in the Arctic Ocean [24,40]. However, in this study, we do not consider kinematic properties of MIZ eddies which is a subject of future work.

Author Contributions: Conceptualization, I.E.K.; methodology, I.E.K. and O.A.A.; software, I.E.K.; formal analysis, I.E.K. and O.A.A.; investigation, I.E.K. and O.A.A.; data curation, O.A.A.; writing—original draft preparation, I.E.K.; writing—review and editing, I.E.K. and O.A.A.; visualization, O.A.A.; supervision, I.E.K.; project administration, I.E.K.; funding acquisition, I.E.K. All authors have read and agreed to the published version of the manuscript.

Funding: The analysis of eddy generation intensity, their spatio-temporal properties and relation to currents and winds were supported by the Russian Science Foundation Grant #21-77-10052. The analysis of sea ice conditions, ice trapping and melting by eddies was supported by the Russian Science Foundation Grant #21-17-00278. Data acquisition, processing and eddy identification in SAR data were supported by the Ministry of Science and Higher Education of the Russian Federation state assignments 075-00429-21-03 (MHI RAS) and 0128-2021-0014 (IO RAS).

Institutional Review Board Statement: Not applicable.

Data Availability Statement: Envisat ASAR images used in this work were available from European Space Agency within CAT-1 Project C1F.29721. The Envisat ASAR data can be ordered via https://earth.esa.int/web/guest/data-access/how-to-access-esa-data (accessed on 15 January 2019). Sentinel-1 data can be freely accessed from the Copernicus Open Access Hub at https://scihub.copernicus.eu (accessed on 1 July 2021). The University of Bremen sea ice concentration data can be accessed from https://seaice.uni-bremen.de/sea-ice-concentration/amsre-amsr2/ (accessed on 15 August 2021). Data on the wind velocity at 10 m height was obtained from the Era-5 reanalysis from Copernicus Climate Data Store https://cds.climate.copernicus.eu/ (accessed on 5 July 2021). Reanalysis ocean current data CMEMS GLORYS12V1 (product identifier GLOBAL_REANALYSIS_PHY_001_030) can be accessed from https://resources.marine.copernicus.eu/ (accessed on 10 July 2021).

Conflicts of Interest: The authors declare no conflict of interest. The funders had no role in the design of the study; in the collection, analyses, or interpretation of data; in the writing of the manuscript, or in the decision to publish the results.

References

1. Johannessen, J.A.; Johannessen, O.M.; Svendsen, E.; Shuchman, R.; Manley, T.; Campbell, W.J.; Josberger, E.G.; Sandven, S.; Gascard, J.C.; Olaussen, T.; et al. Mesoscale eddies in the Fram Strait marginal ice zone during the 1983 and 1984 Marginal Ice Zone Experiments. *J. Geophys. Res. Oceans* **1987**, *92*, 6754–6772. [CrossRef]
2. Johannessen, O.M.; Johannessen, J.A.; Svendsen, E.; Shuchman, R.A.; Campbell, W.J.; Josberger, E. Ice-edge eddies in the Fram Strait marginal ice zone. *Science* **1987**, *236*, 427. [CrossRef]
3. Shuchman, R.A.; Johannessen, O.M.; Campbell, W.J.; Lannelongue, N.; Burns, B.A.; Josberger, E.G.; Manley, T. Remote sensing of the Fram Strait marginal ice zone. *Science* **1987**, *236*, 427–439. [CrossRef]
4. Wadhams, P.; Squire, V.A. An ice-water vortex at the edge of the East Greenland Current. *J. Geophys. Res.* **1983**, *88*, 2770–2780. [CrossRef]
5. Biddle, L.C.; Swart, S. The observed seasonal cycle of submesoscale processes in the Antarctic marginal ice zone. *J. Geophys. Res. Oceans* **2020**, *125*, e2019JC015587. [CrossRef]
6. Brenner, S.; Rainville, L.; Thomson, J.; Lee, C. The evolution of a shallow front in the Arctic marginal ice zone. *Elem. Sci. Anth.* **2020**, *8*, 17. [CrossRef]
7. Kozlov, I.E.; Plotnikov, E.V.; Manucharyan, G.E. Brief Communication: Mesoscale and submesoscale dynamics in the marginal ice zone from sequential synthetic aperture radar observations. *Cryosphere* **2020**, *14*, 2941–2947. [CrossRef]
8. Manucharyan, G.E.; Thompson, A.F. Submesoscale sea ice–ocean interactions in marginal ice zones. *J. Geophys. Res. Oceans* **2017**, *122*, 9455–9475. [CrossRef]
9. Swart, S.; du Plessis, M.D.; Thompson, A.F.; Biddle, L.C.; Giddy, I.; Linders, T.; Mohrmann, M.; Nicholson, S.-A. Submesoscale fronts in the Antarctic marginal ice zone and their response to wind forcing. *Geophys. Res. Lett.* **2020**, *47*, e2019GL086649. [CrossRef]
10. von Appen, W.J.; Wekerle, C.; Hehemann, L.; Schourup-Kristensen, V.; Konrad, C.; Iversen, M.H. Observations of a submesoscale cyclonic filament in the marginal ice zone. *Geophys. Res. Lett.* **2018**, *45*, 6141–6149. [CrossRef]
11. Crews, L.; Sundfjord, A.; Albretsen, J.; Hattermann, T. Mesoscale eddy activity and transport in the Atlantic Water inflow region north of Svalbard. *J. Geophys. Res. Oceans* **2018**, *123*, 201–215. [CrossRef]
12. Hattermann, T.; Isachsen, P.E.; von Appen, W.-J.; Albretsen, J.; Sundfjord, A. Eddy-driven recirculation of Atlantic Water in Fram Strait. *Geophys. Res. Lett.* **2016**, *43*, 3406–3414. [CrossRef]
13. Wang, Q.; Koldunov, N.V.; Danilov, S.; Sidorenko, D.; Wekerle, C.; Scholz, P.; Bashmachnikov, I.L.; Jung, T. Eddy kinetic energy in the Arctic Ocean from a global simulation with a 1-km Arctic. *Geophys. Res. Lett.* **2020**, *47*, e2020GL088550. [CrossRef]
14. Wekerle, C.; Wang, Q.; Von Appen, W.-J.; Danilov, S.; Schourup-Kristensen, V.; Jung, T. Eddy-resolving simulation of the Atlantic water circulation in the Fram Strait with focus on the seasonal cycle. *J. Geophys. Res. Oceans* **2017**, *122*, 8385–8405. [CrossRef]
15. Wekerle, C.; Hattermann, T.; Wang, Q.; Crews, L.; von Appen, W.-J.; Danilov, S. Properties and dynamics of mesoscale eddies in Fram Strait from a comparison between two high-resolution ocean–sea ice models. *Ocean Sci.* **2020**, *16*, 1225–1246. [CrossRef]
16. Atadzhanova, O.A.; Zimin, A.V.; Romanenkov, D.A.; Kozlov, I.E. Satellite radar observations of small eddies in the White, Barents and Kara Seas. *Phys. Oceanogr.* **2017**, *2*, 75–83. [CrossRef]
17. Cole, S.T.; Toole, J.M.; Lele, R.; Timmermans, M.; Gallaher, S.G.; Stanton, T.P.; Shaw, W.J.; Hwang, B.; Maksym, T.; Wilkinson, J.P.; et al. Ice and ocean velocity in the Arctic marginal ice zone: Ice roughness and momentum transfer. *Elem. Sci. Anth.* **2017**, *5*, 55. [CrossRef]
18. Bondevik, E. Studies of Eddies in the Marginal Ice Zone along the East Greenland Current Using Spaceborne Synthetic Aperture Radar (SAR). Master's Thesis, The University of Bergen, Bergen, Norway, 2011; 95p.
19. von Appen, W.J.V.; Schauer, U.; Hattermann, T.; Beszczynska-Möller, A. Seasonal cycle of mesoscale instability of the West Spitsbergen Current. *J. Phys. Oceanogr.* **2016**, *46*, 1231–1254. [CrossRef]
20. Dokken, S.T.; Wahl, T. *Observations of Spiral Eddies along the Norwegian Coast in ERS SAR Image*; Rep. 96/01463; Norwegian Defence Research Establishment (NDRE): Kjeller, Norway, 1996; 29p.
21. Karimova, S.S. Spiral eddies in the Baltic, Black and Caspian seas as seen by satellite radar data. *Adv. Space Res.* **2012**, *50*, 1107–1124. [CrossRef]
22. Stuhlmacher, A.; Gade, M. Statistical Analyses of Eddies in the Western Mediterranean Sea based on Synthetic Aperture Radar Imagery. *Remote Sens. Environ.* **2020**, *250*, 112023. [CrossRef]
23. Kozlov, I.E.; Artamonova, A.V.; Manucharyan, G.E.; Kubryakov, A.A. Eddies in the Western Arctic Ocean from spaceborne SAR observations over open ocean and marginal ice zones. *J. Geophys. Res. Oceans* **2019**, *124*, 6601–6616. [CrossRef]
24. Bashmachnikov, I.L.; Kozlov, I.E.; Petrenko, L.A.; Glock, N.I.; Wekerle, C. Eddies in the North Greenland Sea and Fram Strait from satellite altimetry, SAR and high-resolution model data. *J. Geophys. Res. Oceans* **2020**, *125*, e2019JC015832. [CrossRef]
25. Petrenko, L.A.; Kozlov, I.E. Properties of eddies near Svalbard and in Fram Strait from spaceborne SAR observations in summer. *Sovr. Probl. DZZ Kosm.* **2020**, *17*, 167–177. [CrossRef]

26. Kozlov, I.E.; Kudryavtsev, V.N.; Zubkova, E.V.; Zimin, A.V.; Chapron, B. Characteristics of short-period internal waves in the Kara Sea inferred from satellite SAR data. *Izv. Atmos. Ocean. Phys.* **2015**, *58*, 1073–1087. [CrossRef]
27. Lee, J.-S. Digital image smoothing and the sigma filter. *Comput. Gr. Image Process.* **1983**, *24*, 255–269. [CrossRef]
28. Spreen, G.; Kaleschke, L.; Heygster, G. Sea ice remote sensing using AMSR-E 89 GHz channels. *J. Geophys. Res. Oceans* **2008**, *113*, 1–14. [CrossRef]
29. Jakobsson, M.; Mayer, L.; Coakley, B.; Dowdeswell, J.A.; Forbes, B.; Fridman, S.; Hodnesdal, H.; Noormets, R.; Pedersen, R.; Rebesco, M. The International Bathymetric Chart of the Arctic Ocean (IBCAO) Version 3.0. *Geophys. Res. Lett.* **2012**, *39*, 1–6. [CrossRef]
30. Karimova, S. Observations of asymmetric turbulent stirring in inner and marginal seas using satellite imagery. *Intern. J. Remote Sens.* **2017**, *38*, 1642–1664. [CrossRef]
31. Fer, I.; Drinkwater, K. Mixing in the Barents Sea polar front near hopen in spring. *J. Mar. Syst.* **2014**, *130*, 206–218. [CrossRef]
32. Marchenko, A.V.; Morozov, E.G.; Kozlov, I.E.; Frey, D.I. High-amplitude internal waves southeast of Spitsbergen. *Cont. Shelf Res.* **2021**, *227*, 104523. [CrossRef]
33. Zatsepin, A.; Kubryakov, A.; Aleskerova, A.; Elkin, D.; Kukleva, O. Physical mechanisms of submesoscale eddies generation: Evidences from laboratory modeling and satellite data in the Black Sea. *Ocean Dyn.* **2019**, *69*, 253–266. [CrossRef]
34. Nurser, A.J.G.; Bacon, S. The Rossby radius in the Arctic Ocean. *Ocean Sci.* **2014**, *10*, 967–975. [CrossRef]
35. Kubryakov, A.; Stanichny, S.V. Mesoscale eddies in the Black Sea from satellite altimetry data. *Oceanology* **2015**, *55*, 56–67. [CrossRef]
36. Josberger, E.G. Bottom ablation and heat transfer coefficients from the 1983 Marginal Ice Zone Experiments. *J. Geophys. Res.* **1987**, *92*, 7012–7016. [CrossRef]
37. McPhee, M.G.; Maykut, G.A.; Morison, J.H. Dynamics and thermodynamics of the ice/upper ocean system in the marginal ice zone of the Greenland Sea. *J. Geophys. Res.* **1987**, *92*, 701–77031. [CrossRef]
38. Mayot, N.; Matrai, P.A.; Arjona, A.; Bélanger, S.; Marchese, C.; Jaegler, T.; Ardyna, M.; Steele, M. Springtime export of Arctic sea ice influences phytoplankton production in the Greenland Sea. *J. Geophys. Res. Oceans* **2020**, *125*, e2019JC015799. [CrossRef]
39. Sumata, H.; de Steur, L.; Gerland, S.; Divine, D.; Pavlova, O. Unprecedented decline of Arctic sea ice outflow in 2018. *Res. Square* **2021**, 1–34. [CrossRef]
40. Kubryakov, A.A.; Kozlov, I.E.; Manucharyan, G.E. Large mesoscale eddies in the Western Arctic Ocean from satellite altimetry measurements. *J. Geophys. Res. Oceans* **2021**, *126*, e2020JC016670. [CrossRef]
41. Karimova, S.S.; Gade, M. Improved statistics of sub-mesoscale eddies in the Baltic Sea retrieved from SAR imagery. *Int. J. Rem. Sens.* **2016**, *37*, 2394–2414. [CrossRef]
42. Boulze, H.; Korosov, A.; Brajard, J. Classification of Sea Ice Types in Sentinel-1 SAR Data Using Convolutional Neural Networks. *Remote Sens.* **2020**, *12*, 2165. [CrossRef]
43. Park, J.-W.; Korosov, A.A.; Babiker, M.; Won, J.-S.; Hansen, M.W.; Kim, H.-C. Classification of sea ice types in Sentinel-1 synthetic aperture radar images. *Cryosphere* **2020**, *14*, 2629–2645. [CrossRef]
44. Gupta, M.; Marshall, J.; Song, H.; Campin, J.-M.; Meneghello, G. Sea-ice melt driven by ice-ocean stresses on the mesoscale. *J. Geophys. Res. Oceans* **2020**, *125*, e2020JC016404. [CrossRef]

remote sensing

MDPI

Article

Does the Intra-Arctic Modification of Long-Range Transported Aerosol Affect the Local Radiative Budget? (A Case Study)

Konstantina Nakoudi [1,2,*], Christoph Ritter [1], Christine Böckmann [3], Daniel Kunkel [4], Oliver Eppers [4,5], Vladimir Rozanov [6], Linlu Mei [6], Vasileios Pefanis [6,7], Evelyn Jäkel [8], Andreas Herber [7], Marion Maturilli [1] and Roland Neuber [1]

1 Alfred Wegener Institute, Helmholtz Centre for Polar and Marine Research, Telegrafenberg A45, 14473 Potsdam, Germany; Christoph.Ritter@awi.de (C.R.); Marion.Maturilli@awi.de (M.M.); roland.neuber@awi.de (R.N.)

2 Institute of Physics and Astronomy, University of Potsdam, Karl-Liebknecht 24/25, 14476 Potsdam, Germany

3 Institute of Mathematics, University of Potsdam, Karl-Liebknecht 24/25, 14476 Potsdam, Germany; bockmann@uni-potsdam.de

4 Institute for Atmospheric Physics, Johannes Gutenberg University, Johann-Joachim-Becherweg 21, 55128 Mainz, Germany; dkunkel@uni-mainz.de (D.K.); oleppers@uni-mainz.de (O.E.)

5 Particle Chemistry Department, Max Planck Institute for Chemistry, Hahn-Meitner-Weg 1, 55128 Mainz, Germany

6 Institute for Environmental Physics, University of Bremen, Otto-Hahn-Allee 1, 28359 Bremen, Germany; rozanov@iup.physik.uni-bremen.de (V.R.); mei@iup.physik.uni-bremen.de (L.M.); Vasileios.Pefanis@awi.de (V.P.)

7 Alfred Wegener Institute, Helmholtz Centre for Polar and Marine Research, Am Handelshafen 12, 27570 Bremerhaven, Germany; Andreas.Herber@awi.de

8 Leipzig Institute for Meteorology, University of Leipzig, Stephanstr. 3, 04103 Leipzig, Germany; evi.jaekel@uni-leipzig.de

* Correspondence: konstantina.nakoudi@awi.de

Received: 25 May 2020; Accepted: 29 June 2020; Published: 1 July 2020

Abstract: The impact of aerosol spatio-temporal variability on the Arctic radiative budget is not fully constrained. This case study focuses on the intra-Arctic modification of long-range transported aerosol and its direct aerosol radiative effect (ARE). Different types of air-borne and ground-based remote sensing observations (from Lidar and sun-photometer) revealed a high tropospheric aerosol transport episode over two parts of the European Arctic in April 2018. By incorporating the derived aerosol optical and microphysical properties into a radiative transfer model, we assessed the ARE over the two locations. Our study displayed that even in neighboring Arctic upper tropospheric levels, aged aerosol was transformed due to the interplay of removal processes (nucleation scavenging and dry deposition) and alteration of the aerosol source regions (northeast Asia and north Europe). Along the intra-Arctic transport, the coarse aerosol mode was depleted and the visible wavelength Lidar ratio (LR) increased significantly (from 15 to 64–82 sr). However, the aerosol modifications were not reflected on the ARE. More specifically, the short-wave (SW) atmospheric column ARE amounted to +4.4 - +4.9 W m^{-2} over the ice-covered Fram Strait and +4.5 W m^{-2} over the snow-covered Ny-Ålesund. Over both locations, top-of-atmosphere (TOA) warming was accompanied by surface cooling. These similarities can be attributed to the predominant accumulation mode, which drives the SW radiative budget, as well as to the similar layer altitude, solar geometry, and surface albedo conditions over both locations. However, in the context of retreating sea ice, the ARE may change even along individual transport episodes due to the ice albedo feedback.

Keywords: Arctic aerosol; aerosol transport; aged aerosol; aerosol modification; aerosol optical properties; aerosol microphysical properties; aerosol remote sensing; microphysical inversion; aerosol radiative effect; Arctic radiative budget

1. Introduction

As anthropogenic climate change drives major modifications in the Arctic environment, Arctic aerosol and its related feedback mechanisms are inextricably affected. Aerosol is involved in aerosol–cloud and aerosol–radiation interactions, which highly depend on the aerosol species, solar zenith angle, and surface albedo [1–3]. In the Arctic, the annual mean aerosol radiative forcing at the top-of-atmosphere (TOA) is estimated at −0.12 W m^{-2} with respect to pre-industrial emissions of anthropogenic aerosol and precursors [3]. Over the past century, the aerosol-induced cooling has offset the greenhouse gas-induced warming by 1.3 to 2.2 °C [4]. However, the aerosol radiative forcing in the Arctic region still entails high uncertainties [1–3]. Therefore, a better understanding of the aerosol-related physical processes is crucial in the changing Arctic environment, where the most rapid near-surface temperature increase is taking place, known as "Arctic amplification" [5,6].

Arctic aerosol properties exhibit a pronounced seasonal cycle due to variation in sources, transport pathways, and removal processes [7–10]. On a pan-Arctic scale, the accumulation mode concentration increases through winter and peaks in April, with median values between 100 and 200 cm^{-3} [11–13]. This seasonality is reflected upon the aerosol optical properties, with a late winter–spring scattering coefficient maximization (median values even higher than 10 Mm^{-1}) [10]. While in winter, maximal extinction occurs in the lower Arctic troposphere, in spring, there is a progressive shift towards the middle and upper troposphere [13–15]. The extinction enhancement in higher altitudes ("Arctic haze") is associated with the isentropic poleward transport of polluted mid-latitude air masses [16–18]. "Arctic haze" comprises mostly aged accumulation particles of sulfate composition. However, chemical analysis has illustrated the presence of nitrate, chloride, sea salt, ammonium, dust, and carbonaceous compounds [14,19–22]. Studies have also demonstrated contribution from biomass-burning sources [12,23–27].

Ny-Ålesund (78.9°N, 11.9°E), Spitsbergen island, Svalbard is an Arctic research site embedded in a complex orography of fjord and mountain ranges. The Svalbard region is the hotspot of winter warming with up to 2 °C per decade near-surface temperature increase in the past two decades [28] and a 5–20% decadal decrease in fjord sea ice coverage [29]. One quarter of this warming is attributed to the amplified advection of warm and moist air-masses over the mid-latitude Atlantic Ocean, which is facilitated by changes in large-scale circulation, such as atmospheric blocking patterns [30,31]. In the context of enhanced local humidity fluxes from the increasingly ice-free Arctic Ocean, the springtime AOD over Ny-Ålesund has diminished over the past two decades [32].

Despite the overall distinct annual patterns, the seasonal cycle of aerosol properties presents consistent discrepancies within the Arctic. Synergistic ground-based and air-borne aerosol observations can provide insight into intra-Arctic aerosol modification and improve its representation into chemical transport and climate models. Lidar systems are key observational tools of aerosol, with the asset of delivering vertically resolved measurements on high spatial and temporal scales [33]. Moreover, the Raman Lidar technique delivers a suite of optical parameters from which the aerosol microphysical properties can be inferred [34–37], allowing the evaluation of aerosol radiative effects.

Our study focused on a persistent aerosol transport episode, which was synergistically identified by remote sensing systems in remarkably high tropospheric altitudes. The elevated aerosol layers were revealed over two parts of the European Arctic, namely Fram Strait and Ny-Ålesund, three days apart in April 2018 (Figures 1 and 2). Our primary target was to investigate the intra-Arctic modification of the aerosol optical and microphysical properties. By incorporating the derived aerosol properties into a radiative transfer model, we aimed to assess the direct aerosol radiative effect (ARE) and quantify the related uncertainties.

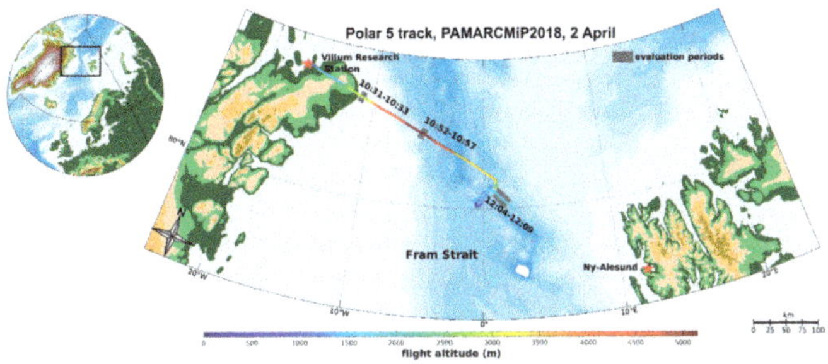

Figure 1. Flight track of Polar 5 during the outbound leg of the research flight over Fram on 2 April. The flight altitude is color-indicated. Polar 5 took off from Villum Research Station, Greenland, and performed a transect flight over Fram Strait. AMALi (10:31–10:33 and 12:04–12:09) and sun-photometer (10:52–10:57) observations were exploited synergistically. The aerosol properties evaluation periods are indicated with grey shading.

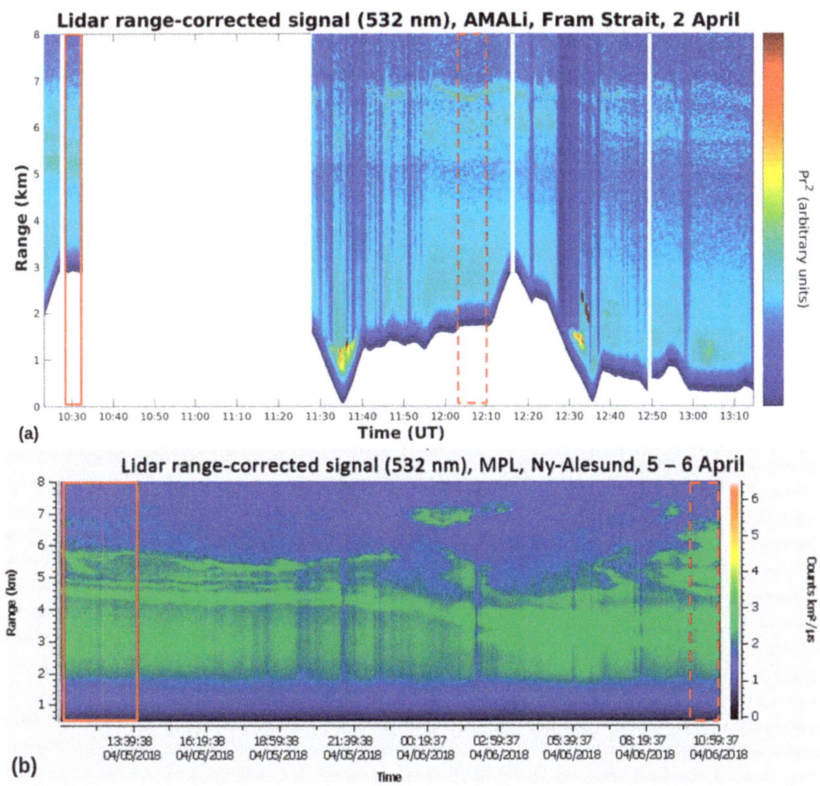

Figure 2. Persistent high tropospheric aerosol layers observed over Fram Strait and Ny-Alesund, Svalbard, by AMALi and MPL systems, respectively. The Lidar range-corrected signal over Fram Strait (**a**) and Ny-Ålesund (**b**) is shown. The evaluation periods of aerosol optical and microphysical properties are indicated with solid rectangles. Dashed rectangles denote periods where only the optical properties were evaluated, due to the lack of air-borne photometer and Raman Lidar observations at Fram Strait and Ny-Ålesund, respectively. Thick clouds over Ny-Ålesund did not allow for Raman Lidar observations in the interim days.

2. Methods: Instrumentation and Modeling Tools

In this study, we utilized active and passive remote sensing observations over two parts of the European Arctic in order to retrieve the optical and microphysical properties of long-range transported aerosol. The ensemble of inverted aerosol properties was incorporated into a radiative transfer model to assess the ARE and its related uncertainties. In parallel, the simulated irradiances were evaluated with ground-based and air-borne observations.

2.1. Lidar Systems

The elastic air-borne mobile aerosol Lidar (AMALi) is installed on board of the research aircraft Polar 5 [38]. During the polar air-borne measurements and Arctic Regional Climate Model Simulation Project 2018 (PAMARCMiP2018), AMALi was operated in zenith configuration during 9 flights (25 March–4 April for approximately 32 h in total) over the region of Fram Strait (84, 17 NW to 79, 20 NE). AMALi is equipped with an Nd:YAG laser that emits pulses at 15 Hz. A 10-cm off-axis telescope mirror of 3.1 mrad field of view (FOV) collects the backscattered light. Photomultiplier tubes (PMTs) are used for the detection of elastically backscattered photons at 532 and 355 nm. Additionally, the cross-polarized component is recorded at 532 nm, allowing for the determination of the aerosol linear depolarization ratio, hereafter mentioned as the aerosol depolarization ratio.

AMALi signal returns were acquired with a 1 s—-7.5 m resolution. Signals were vertically averaged every 30 m and smoothed with a 90-m window, while backscatter profiles were smoothed with a 150-m window. The altitude of the observed features was corrected with respect to the aircraft movements, with the pitch and roll angles changing by 2° and 3.5° during the evaluation windows. For this study, we extracted two observation periods of a 2- and 5-min duration (as indicated in Figures 1 and 2a), which were limited by the changing flight altitude. The signal-to-noise ratio (SNR) changed considerably due to the variable flight altitude (0.3–2.9 km) during AMALi observations, which limited the temporal averaging and led to higher uncertainties in the air-borne-derived optical properties compared to the ground-based ones.

The multi-wavelength Koldewey Aerosol Raman Lidar (KARL) is operated in the Alfred Wegener Institute Emille Victor (AWIPEV) research base, Ny-Ålesund, Svalbard. KARL is a "3β+2α+2δ+2wv" system equipped with an Nd:YAG laser emitting at 1064, 532, and 355 nm after frequency doubling and tripling, respectively [39]. The pulse repetition rate of the system is 50 Hz, emitting approximately 10 W in each wavelength. The receiver consists of a parabolic 70 cm diameter Newtonian telescope with a 1.75 mrad FOV and an aperture with an adaptable position and pinhole diameter. PMTs are used for the detection of backscattered photons in all channels, except for 1064 nm, where avalanche photodiodes (APDs) are used. KARL signal returns were obtained with a 1.5 min-7.5 m resolution. In order to reduce signal noise, we averaged the profiles for approximately 2.5 h and applied vertical smoothing of 157.5 m in the backscatter retrieval and 225 m in extinction retrieval, following the effective resolution concept [40].

We applied the Klett–Fernald method to the elastic AMALi and KARL signals in order to retrieve the backscatter profiles [41,42]. This method requires a priori information on the vertical profile of the extinction-to-backscatter ratio (Lidar ratio or LR) and a reference value for the backscatter ratio (total backscatter over molecular backscatter). Here, we assumed uncertainties of 10 sr and 3% for the LR and backscatter ratio, respectively. The positioning of the calibration window introduces additional uncertainty. Errors are also introduced due to statistical signal noise and the procedures of temporal averaging and vertical smoothing. We obtained extinction profiles from KARL by utilizing the Raman method [43]. Since Raman scattering cross-sections are orders of magnitude lower than Rayleigh cross-sections and our observations were performed during the Polar day period, we derived layer-mean values of extinction and LR. Aerosol optical properties are presented in Tables 1 and 2.

Table 1. Aerosol optical properties (mean ± one standard deviation) over two parts of Fram Strait and Ny-Ålesund retrieved from AMALi and KARL, respectively. For details on the retrieval procedure, see Section 2.1. The aerosol backscatter coefficient β, backscatter-related Angström exponent A_β, and aerosol depolarization ratio δ are presented.

Aerosol Optical Properties					
Aerosol Properties	**Fram Strait (AMALi)**		**Ny-Ålesund (KARL)**		
	2 April		5 April		6 April
	10:31–10:33	12:04–12:09	11:00–13:47		9:51–10:41
aerosol layer altitude (km)	5.15–6.8	5.3–6.9	4.3–4.9	4.9–5.75	4.6–5
β^{aer}_{355} ($Mm^{-1}sr^{-1}$)	1.3 ± 0.3	1.7 ± 0.5	0.5 ± 0.1	0.6 ± 0.1	1 ± 0.3
β^{aer}_{532} ($Mm^{-1}sr^{-1}$)	0.9 ± 0.1	1 ± 0.2	0.2 ± 0.05	0.3 ± 0.06	0.4 ± 0.2
β^{aer}_{1064} ($Mm^{-1}sr^{-1}$)	-	-	0.1 ± 0.03	0.1 ± 0.03	0.2 ± 0.1
$Å_{\beta355/\beta532}$	0.8 ± 0.6	1.3 ± 0.4	1.8 ± 0.2	2.2 ± 0.1	2.4 ± 0.3
$Å_{\beta532/\beta1064}$	-	-	1 ± 0.1	1.1 ± 0.04	1.2 ± 0.3
δ^{aer}_{355} (%)	-	-	3.8 ± 0.6	2.9 ± 0.3	3.9 ± 0.8
δ^{aer}_{532} (%)	1.3 ± 0.2	5.5 ± 1	3.2 ± 0.3	3 ± 0.2	3.4 ± 0.4

Table 2. Aerosol layer mean optical properties over the Greenland side of Fram Strait retrieved from the synergy of sun-photometer and KARL and over Ny-Ålesund as derived from KARL. The aerosol extinction coefficient α, Lidar ratio LR, and extinction-related Angström exponent A_α are presented. Values estimated by Mie theory are given in italic. For details, see Sections 2.1 and 2.2. The extinction over Fram Strait was estimated by Equation (2).

	Aerosol Layer-mean Optical Properties		
Aerosol Properties	**Fram Strait (Sun-photometer and KARL)**	**Ny-Ålesund (KARL)**	
	2 April	5 April	
	10:52–10:57	11:00–13:47	
aerosol layer altitude (km)	5.15–6.8	4.3–4.9	4.9–5.75
α^{aer}_{355} (Mm^{-1})	-	18 ± 7	30 ± 8
α^{aer}_{496} (Mm^{-1})	20 ± 2	-	-
α^{aer}_{532} (Mm^{-1})	*21 ± 2*	20 ± 5	16 ± 6
α^{aer}_{675} (Mm^{-1})	27 ± 1	*12 ± 2*	*12 ± 2*
α^{aer}_{779} (Mm^{-1})	28 ± 1	*10 ± 2*	*9 ± 2*
α^{aer}_{861} (Mm^{-1})	25 ± 1	*9 ± 2*	*7 ± 2*
α^{aer}_{1026} (Mm^{-1})	20 ± 1	*7 ± 2*	*4 ± 2*
LR_{355} (sr)	-	35 ± 15	48 ± 4
LR_{532} (sr)	*15 ± 3*	82 ± 25	64 ± 37
$Å_{\alpha355/\alpha532}$	-	−0.3 ± 0.8	1.5 ± 1.8
$Å_{\alpha496\text{-}\alpha1026}$	−0.09	-	-

2.2. *Sun-Photometers and Extinction Estimation at Fram Strait*

The air-borne Sun-photometer with an active tracking system (SPTA) was operated under a quartz dome on the top of Polar 5 during PAMARCMiP2018. The aerosol optical depth (AOD) was provided at 10 wavelengths between 396 and 1026 nm every 30 sec. Unfortunately, the 369–414 nm channels presented calibration issues and, thus, they were not used for further evaluation. Moreover, we did not use the 946-nm channel, which is dedicated to water vapor (WV) absorption. On 2 April, the AOD was measured at different flight levels between 0.04 and 5.15 km (Figure 3). We performed 5-min averaging in the AOD in order to reduce statistical noise. The aerosol extinction coefficient was estimated by consecutive AOD pairs within altitude layers of 0.5–2.2 km and temporal periods of 5–40 min, within which the atmospheric conditions were assumed to be horizontally homogeneous (similar AOD at similar altitudes, Figure 3).

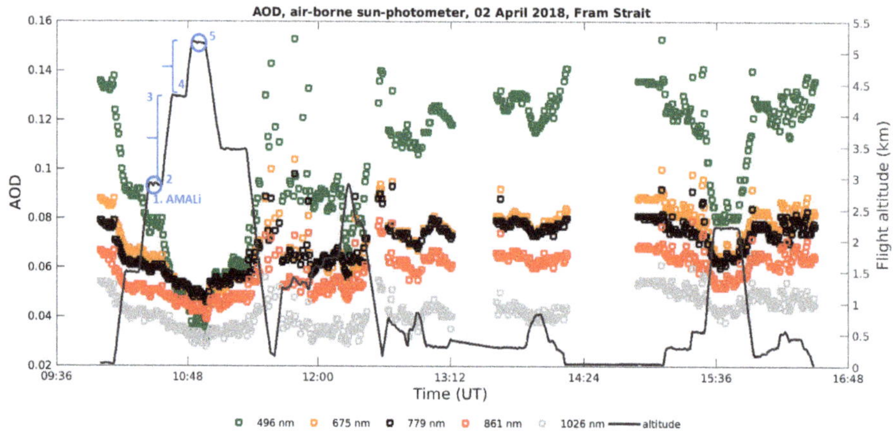

Figure 3. Evolution of aerosol optical depth (AOD) over Fram Strait obtained from the air-borne sun-photometer. A persistent aerosol layer was observed at 5.15 km. Selected measurement points (1–5) were used for estimating the aerosol extinction coefficient of the layer. These observations were combined with Raman Lidar measurements. For details, see Section 2.2. Note the changing flight altitude during the measurements.

The AOD of the elevated aerosol layer (5.15–6.8 km), observed close to the Greenland side of Fram Strait (Figures 1 and 2a), was estimated by combining photometer and Lidar observations. More specifically, we used the photometer AOD at five different measurement points below the layer, between the highest photometer observation (number 5, 10:52–10:57 UT; flight level 5.15 km) and the closest AMALi observation (number 1, 10:31–10:33 UT; flight level 2.8 km). We also used the AOD at three intermediate flight levels (numbers 2, 3, and 4 corresponding to 3, 4.3, and 4.4 km) so as to better constrain the estimated extinction. The AOD was calculated according to Equation (1), with the selected measurement points enumerated in Figure 3. Horizontal homogeneity was assumed due to the short spatial (maximum 100 km) and temporal differences (maximum 25 min) among the measurement points, as in the study of Mei et al. [44] that focused on satellite aerosol remote sensing. Above the aerosol layer, the AOD was derived from KARL. The ground-based Lidar signals were preferred over the air-borne ones due to their higher SNR. The aerosol load in the stratosphere was invariable between the air-borne and ground-based observations (three days and a maximum of 500 km apart). The AOD was the product of the integrated backscatter with the LR. The KARL AOD was interpolated to the photometer wavelengths using an appropriate Angström exponent (1.8 as fitted to the backscatter at 355, 532, and 1064 nm). Following the same steps, we estimated the stratospheric AOD. Finally, the extinction of the elevated aerosol layer was estimated by Equation (2):

$$AOD_{5.15-6.8km} = AOD_{2.8km}^{sun-photometer} - AOD_{3-4.3km}^{sun-photometer} - AOD_{4.4-5.15km}^{sun-photometer} \\ -AOD_{6.8km-tropopause}^{KARL} - AOD_{stratosphere}^{KARL} \tag{1}$$

$$a_{5.15-6.8km} = \frac{AOD_{5.15-6.8km}}{Layer geometrical depth}. \tag{2}$$

Over Ny-Ålesund the AOD was measured by an SP1a type sun-photometer in 10 wavelengths between 369 and 1023 nm. A cloud screening procedure that relies on the short-term AOD variability was applied to the data [45]. The total AOD uncertainty amounted to 0.01 [32]. In this study, we only utilized the wavelengths, which are spectrally closer to the air-borne photometer channels, except for the 1023 nm that was affected by calibration issues.

2.3. Aerosol Microphysical Properties Inversion Schemes

We incorporated the set of optical parameters derived for Fram Strait and Ny-Ålesund into inversion algorithms so as to retrieve the aerosol microphysical properties. In the inversion schemes, the extinction and backscatter coefficients were used as input for the derivation of the complex refractive index and aerosol volume size distribution using Mie theory. For each combination of the real and imaginary refractive index, the best solution was identified. More specifically, the least residuum between the backward (input) and forward calculated optical parameters was selected. The standard deviation of the best solutions represents the uncertainties of the complex refractive index. Statistical noise and systematic uncertainties of the input optical parameters together with mathematical approximation errors contribute additively to the total uncertainties in the microphysical parameters.

For Fram Strait, we utilized the AMALi-derived backscatter coefficient (355 and 532 nm) together with the sun-photometer-derived extinction (496, 675, 779, 861, and 1026 nm). We employed collocation discretization and an iterative Padé-regularization method with a fixed number of 30 iterations [35,36]. For Ny-Ålesund, the KARL-derived aerosol backscatter (355, 532 and 1064 nm) and extinction coefficients (355 and 532 nm) were inverted to the particle microphysics employing a truncated singular value decomposition regularization algorithm [34]. More details about the Lidar inversion algorithms, which have been developed at the University of Potsdam, can be found in [34,37,46].

The total number (n_t), surface-area (s_t), and volume (v_t) concentration, as well as the effective radius (r_{eff}) were derived for the fine and coarse aerosol modes by performing log-normal fits to the inverted volume distribution. Fitting errors amounted to 10% for the total concentrations and 5% for the r_{eff} estimation. The single-scattering albedo (SSA) and the asymmetry parameter were calculated from the fitted number size distributions utilizing Mie theory. The corresponding SSA and asymmetry parameter uncertainties were estimated by sensitivities based on different complex refractive index scenarios.

2.4. Trajectory Calculations for Air-Mass History

In order to assay the origin of the observed transport event and link the evolution of aerosol properties with intruding air masses, we performed 10-day LAGRANTO backward trajectories over Fram Strait and Ny-Ålesund [47,48]. A significant number of trajectories ending over Fram Strait and Ny-Ålesund originated from the lower troposphere, providing an indication for the surface aerosol sources. Due to the complex orography of Spitsbergen island, with the highest peak at 1.7 km, and the challenging to parameterize stable boundary layer, we are inclined to have higher confidence in higher altitude trajectories.

We used ECMWF analysis data to calculate trajectories with LAGRANTO, which utilizes the three-dimensional wind field to calculate kinematic trajectories. The ECMWF data for this study had a horizontal grid spacing of 0.5° x 0.5° with 137 hybrid sigma pressure levels in total. Analysis fields were available every six hours. For the flight locations over Fram Strait, we initialized the trajectories in a small region around the flight track. Trajectories from Ny-Ålesund were initialized every 0.05° in the horizontal within a 0.5° × 0.5° box centered at Ny-Ålesund. Trajectory information was then available every hour. Ancillary meteorological parameters, such as pressure and relative humidity (RH), were calculated along the trajectories. The spatial variability of the trajectories was taken into account by using a box, instead of a single point, for initializing our calculations. Moreover, the influence of atmospheric variability was considered by analyzing coherent bundles of trajectories instead of single trajectories.

2.5. Radiation and Meteorological Observations

We utilized air-borne and ground-based irradiance observations for evaluating the modeled irradiances at the surface and different flight altitudes. Radiation sensors were installed on the top and bottom of the Polar 5 fuselage during PAMARCMiP2018. More specifically, a CMP22 Pyranometer by Kipp&Zonen was measuring broadband (0.2–3.6 μm) global and reflected solar irradiances. The measurement repetition rate was 20 Hz, with all the irradiances being deconvoluted for their inertia time [49]. The overall uncertainty was lower than 10 W m^{-2} for all radiation quantities. Basic meteorological observations were also performed on board of the Polar 5 aircraft [50]. WV concentration was measured by the LI-7200 closed CO_2/H_2O analyzer [51].

At Ny-Ålesund, radiation measurements are performed within the frame of the Baseline Surface Radiation Network (BSRN) since 1992 [52]. Short-wave (SW) broadband radiation in the range 0.2 to 3.6 μm is measured with CMP22 pyranometers by Kipp&Zonen, installed in an up- and downward orientation for global and reflected radiation, respectively. In addition, the diffuse radiation is obtained by ball-shaded similar instrumentation. The basic surface radiation and meteorological measurements with a one-minute resolution applied in our study are available via the Pangaea data repository [53]. Vertical profiles of the temperature, RH, wind speed, and direction in the upper atmosphere are applied from daily radiosonde ascents performed at the AWIPEV station [54].

2.6. Radiative Transfer Simulations

In order to quantify the altitude-dependent direct ARE and its uncertainties, we performed simulations with the radiative transfer model SCIATRAN [55], using as input the retrieved aerosol optical and microphysical properties. For the evaluation of the simulated irradiances, the latter were integrated within the spectral range of the BSRN and air-borne pyranometers.

We performed simulations from the local surface up to 40 km. The vertical resolution varied from 50 m to 1 km (above 10 km). The radiative transfer equation was solved by means of the scalar discrete ordinate technique, while a plane-parallel atmosphere approximation was used. The absorption contribution of line absorbers was computed in line-by-line mode of 0.01-nm spectral resolution, based on the spectroscopic parameters of the HITRAN 2008 database [56]. The altitude-dependent ARE was derived from the difference of SW net irradiances between the respective aerosol and the off-aerosol scenario, in order to distinguish it from the total atmospheric effect. The difference between the TOA and surface ARE yielded the overall atmospheric column ARE.

The aerosol-related input comprised the spectrally resolved aerosol extinction coefficient (Table 2), the SSA, as well as the asymmetry factor calculated according to the Henyey–Greenstein parameterization (Table 3). The aerosol optical and microphysical properties were extrapolated to the model's spectral range (from 355–1026 nm) by forward Mie theory. More specifically, using the aerosol refractive index and number size distribution as input, we were able to calculate the aerosol extinction coefficient at any wavelength. The surface albedo over Fram Strait and Ny-Ålesund was obtained from air-borne and BSRN pyranometer observations, respectively. A surface albedo of approximately 0.7 was found for both locations, revealing the presence of sea ice at Fram Strait and snow-covered tundra in Ny-Ålesund.

For Fram Strait, we performed four simulation scenarios: Aerosol and off-aerosol for the aircraft positions during the AMALi and sun-photometer evaluation periods, as indicated in Figure 1. Thermodynamic profiles were constructed as a combination of appropriately selected air-borne meteorological observations below 5 km (pressure and temperature from aircraft meteorological observations, WV mixing ratio from LI-7200 instrument) and radiosonde ascents from Ny-Ålesund aloft (approximately 500 km distance from flight operations). Between 5 and 8 km, a hybrid air-borne – radiosonde profile was constructed, which gradually shifted from the highest air-borne towards the radiosonde observations.

For Ny-Ålesund, simulations were performed for the period of available Raman Lidar observations (5 April, 11:00–13:30). The simulations comprised four different scenarios: Off-aerosol, with two aerosol layers (4.3–5.75 km), with the lower layer (4.3–4.9 km) and finally with the upper layer (4.9–5.75 km). Ancillary meteorological information on the temperature at the surface, 2 and 10 m, as well as surface pressure, were also incorporated into the model. Pressure, temperature, WV mixing ratio, and O_3 concentration profiles were taken from a collocated ozonesonde ascent (5 April, AWIPEV). For both Arctic locations, further trace gas profiles were obtained from climatological profiles for the given month and latitude [57].

Table 3. Aerosol microphysical properties over the Greenland side of Fram Strait and Ny-Ålesund. The complex refractive index ri, the single-scattering albedo SSA, and the asymmetry parameter g are given. Moreover, the effective radius r_{eff} as well as the number n_t, surface s_t, and volume v_t concentrations are presented for the fine and coarse aerosol modes. For details on the retrieval schemes, see Section 2.3.

Aerosol Microphysical Properties						
Aerosol Properties	**Fram Strait (AMALi and Sun-Photometer)**		**Ny-Ålesund (KARL)**			
	2 April		**5 April**			
	10:31–10:57		11:00–13:47			
aerosol layer altitude (km)	5.15–6.8		4.3–4.9		4.9–5.75	
	Refractive Index ri (mean ± one standard deviation)					
ri	1.5 + 0.008 i ± 0.02 + 0.006 i		1.54 + 0.019 i ± 0.04 + 0.01 i		1.49 + 0.007 i ± 0.02 + 0.004 i	
	Single-Scattering Albedo SSA (mean ± uncertainty)					
SSA_{355}	0.84 ± 0.1		0.88 ± 0.06		0.96 ± 0.02	
SSA_{532}	0.91 ± 0.06		0.9 ± 0.05		0.96 ± 0.02	
	Asymmetry parameter g (mean ± uncertainty)					
g_{355}	0.69 ± 0.04		0.71 ± 0.04		0.72 ± 0.02	
g_{532}	0.7 ± 0.05		0.69 ± 0.03		0.68 ± 0.01	
	Effective radius r_{eff}, number n_t, surface s_t and volume v_t concentration					
	fine	coarse	fine	coarse	fine	coarse
r_{eff} (μm)	0.49	1.26	0.26	1.43	0.19	-
n_t (cm^{-3})	8.78	0.24	65.2	0.005	132.9	-
s_t (μm^2 cm^{-3})	24.09	4.68	21.6	0.13	36.4	-
v_t (μm^3 cm^{-3})	3.9	1.97	1.9	0.06	2.3	-

3. Aerosol Observations over the European Arctic and Meteorological Conditions

Air-borne observations were performed within the frame of PAMARCMiP2018 (March-April), Villum Research Station, Greenland (81.6°N, 16.7°W). The PAMARCMiP2018 period was characterized by a high-pressure system over the North Pole, while some weak lows occurred over northeast Greenland. On 2 April, a well-defined aerosol layer was identified at a remarkably high tropospheric altitude over Fram Strait by AMALi and sun-photometer (Figures 1 and 2a). Three days later, a geometrically similar aerosol layer was observed by KARL over Ny-Ålesund (Figure 2b), at a distance of 200–600 km from the flight operations. In the interim of Fram Strait and Ny-Ålesund on 3 April, clouds and aerosol remnants were captured in the upper troposphere by AMALi (Figure S1). At that day, Raman Lidar observations over Ny-Ålesund were not allowable due to low-level clouds. However, the micro-pulse Lidar (MPL, technical description given in Shibata et al. [18]) provided the geometrical evolution of the aerosol layers. It should be noted that the MPL optical products cannot be utilized for the derivation of aerosol microphysical properties due to the lack of Raman channels. Therefore, the optical and microphysical aerosol properties over Ny-Ålesund were investigated within periods of 5 and 6 April as indicated in Figure 2b. On the days when the aerosol layer was observed over Ny-Ålesund, the synoptic situation was characterized by the passing of a weak low-pressure system. According to radiosonde ascents from Ny-Ålesund, moderate N-NW winds (3–7 m/s) prevailed, while wind shear and RH were enhanced (RH of 47– 58% on 5 April and 55–65% on 6 April) at the aerosol layer atmospheric levels.

4. Results

4.1. Episode Evolution over Ny-Ålesund and Connection to Air Masses

The geometrical evolution of the transported aerosol plume over Ny-Ålesund along with the related air mass trajectories are presented in Figure 4. The columnar AOD over Ny-Ålesund and Fram Strait are demonstrated in Figure 5. Before the layer arrival (30–31 March), air masses originated from central Eurasia (4–7 km, Figure 4b). The daily mean AOD_{500} was relatively low (0.04 ± 0.01) with respect to background values (AOD_{500} = 0.06 from Yamanouchi et al. [58]). The influence of northeast Asian air masses is discernible for 2 April, with aerosol features appearing entangled with clouds. Due to persistent low-level clouds (2–4 April), Raman Lidar and sun-photometer observations were unfortunately not allowable during those days.

On 5 April, the layer displayed a well-defined structure. The AOD_{500} followed a moderate increasing tendency with persisting northeast Asian and emerging north European air masses (Figure 4d). On 6 April, the layer was vertically perturbed (Figure 4c,6 April 0-8 UT), possibly due to the passing of the weak low-pressure system. The AOD_{500} fluctuated and maximized (AOD_{500} = 0.13) some hours later. The layer gradually diluted (7 April), with the AOD approaching background levels. Re-emerging low-level clouds did not allow us to further investigate the event evolution, with intermittent aerosol features and mid-level clouds mostly observed (8 April). Overall, the AOD over Ny-Ålesund was clearly lower compared to Fram Strait. In Sections 4.3 and 4.4, we will investigate in more detail the modification of aerosol optical and microphysical properties within the elevated aerosol plume.

4.2. Origin of Observed Aerosol Over Fram Strait and Ny-Ålesund

Prior to investigating the modification of aerosol optical and microphysical properties, we assessed whether the observed aerosol plumes over Fram Strait and Ny-Ålesund originated from the same region or regions comprising similar aerosol sources. The atmospheric pressure along the LAGRANTO backward trajectories as well as the air mass transport time are displayed in Figure 6.

At Fram Strait, the layer close to Greenland (Figure 6a) was a receptor of central and northeast Asian air masses, which were at least nine days old (time since contact with main aerosol source region, Figure 6c). At Ny-Ålesund, the lower part of the plume originated from north Europe (Figure 6d), while the upper part was also affected by northeast Asian air masses (Figure 6e). On their arrival at Ny-Ålesund, the Asian air masses were at least nine days old, while the north European ones had last contact with the main aerosol sources four days ago (Figure 6f). The latter air masses mixed over Fram Strait (500–600 hPa) in the earlier days (Figure 6f).

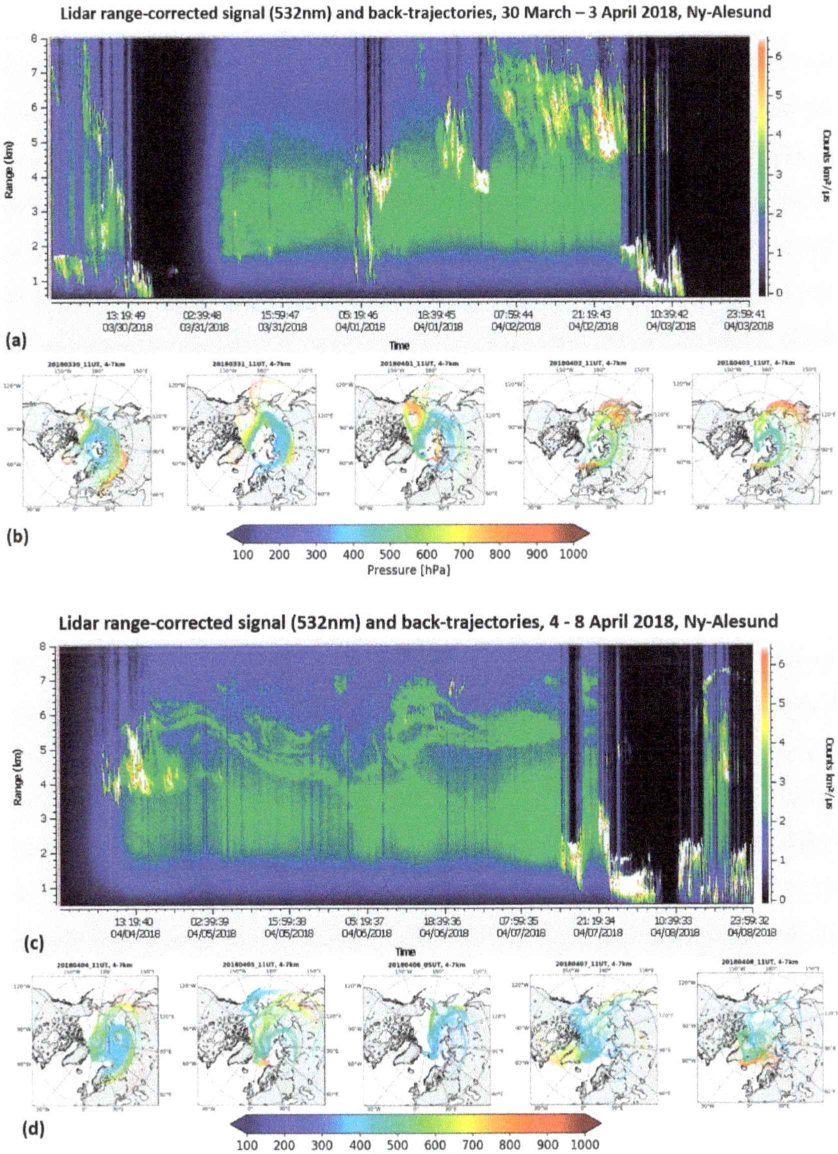

Figure 4. Evolution of the long-range transport episode over Ny-Ålesund. The MPL range-corrected signal is displayed (**a** and **c**) and LAGRANTO backward trajectories (**b** and **d**) with arrival height at 4–7 km. Increased aerosol burden air masses were funneled from northeast Asia and north Europe (2–5 April). Persistent elevated aerosol layers (approximately at 4–7 km) were observed on 5–7 April.

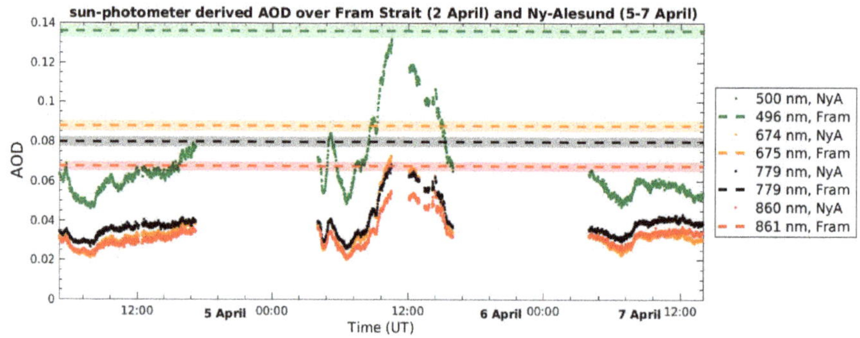

Figure 5. Evolution of AOD over Ny-Ålesund derived from sun-photometer, after cloud screening. For comparison, the AOD over Fram Strait as derived from the air-borne sun-photometer is given (average ± one standard deviation). Overall, the AOD over Ny-Ålesund was significantly lower than Fram Strait, reflecting the effect of removal processes and modified aerosol source efficiency.

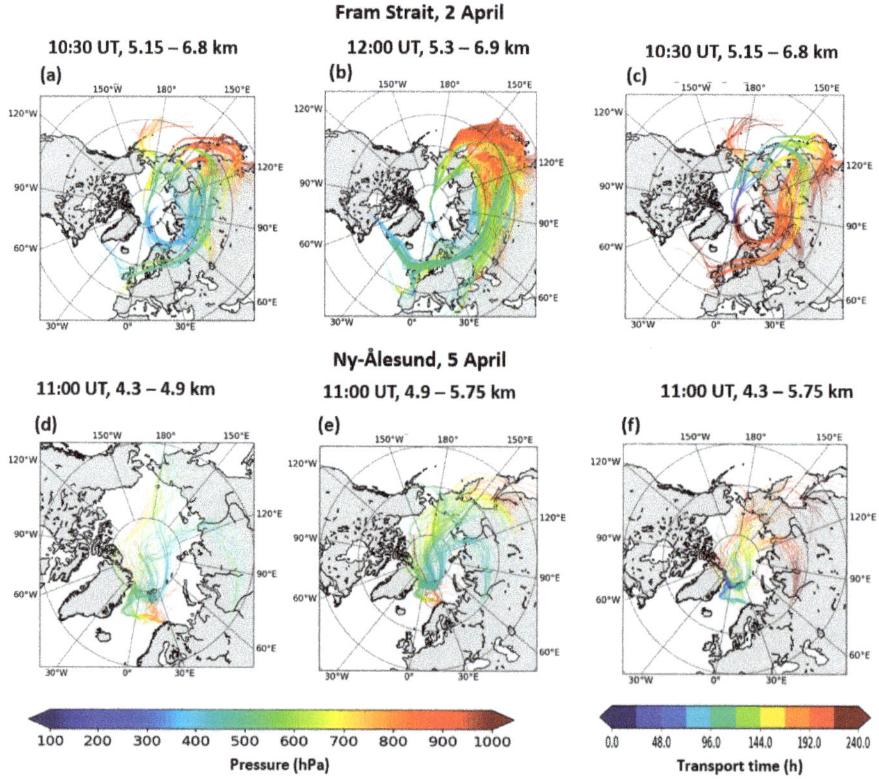

Figure 6. Backward trajectories ending at the aerosol layer altitude sectors over Fram Strait (**a–c**) and Ny-Ålesund (**d–f**). Atmospheric pressure (**a**, **b**, **d**, and **e**) and transport time (**c** and **f**) of the trajectories are color indicated. The transport time corresponds to air masses arriving at the Greenland side of Fram Strait as well as to both aerosol layer altitude sectors at Ny-Alesund (**c** and **f**). Aged Asian air masses affected Fram Strait. Ny-Ålesund was influenced by Asian and fresher air masses from north Europe. The latter air masses mixed over Fram Strait two to three days earlier (**f**).

4.3. Modification of Aerosol Optical Properties

The optical properties of the long-range transported aerosol over Fram Strait and Ny-Ålesund are summarized in Tables 1 and 2, for the evaluation periods indicated in Figures 1 and 2. The vertical profiles of the aerosol backscatter coefficient (proportional to the aerosol number concentration), aerosol depolarization ratio (rising with aerosol non-sphericity), and backscatter-related Angström exponent (inversely proportional to aerosol radius) are demonstrated in Figure 7. The aerosol extinction coefficient and extinction-related Angström exponent as derived by sun-photometer and Lidar are presented in Figure 8.

A shift towards smaller particles was observed, mainly between Fram Strait and Ny-Ålesund, as well as between 5 and 6 April over Ny-Ålesund. Over Fram Strait, the layers were characterized by low but variable aerosol depolarization ratios (1.1–5.5%). This indicates nearly spherical particles as well as connotes the presence of different aerosol mixtures. The lower depolarization close to Greenland (Figure 7b, black line) more likely represents a mixture of aerosol from industrial pollution and biomass burning (Figure S2, sparse fire hotspots and smoke plume observed over Jilin province, north China, on 22 March 2018). Nevertheless, higher depolarization in the mid-Fram Strait (Figure 7b, gray line) is more likely linked to an additional non-spherical aerosol component, such as dust from Gobi desert (supported by backward trajectories, Figure 6b).

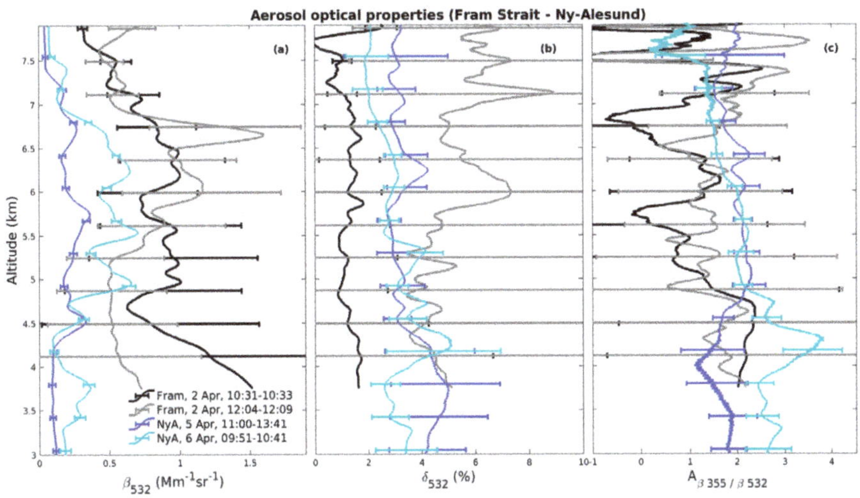

Figure 7. Aerosol optical properties over Fram Strait (Fram) and Ny-Ålesund (NyA) derived from the air-borne AMALi and ground-based KARL systems, respectively. The aerosol backscatter coefficient β, aerosol depolarization ratio δ, and backscatter-related Angström exponent A_β are presented. Error bars represent retrieval uncertainties (for details, see Section 2.1). Air-borne profiles exhibit higher uncertainties due to flight altitude changes that constrained the temporal averaging. The aerosol load was higher over Fram Strait (higher β, **a**). For both locations, particles were nearly spherical (low δ, **b**). A shift towards smaller particles was observed over Ny-Ålesund (higher A_β, **c**).

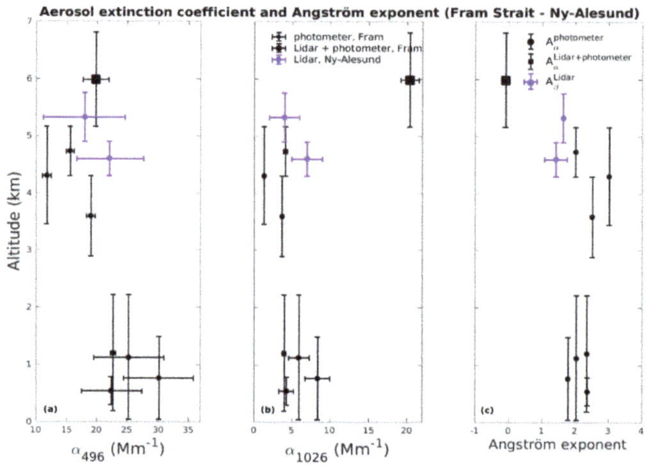

Figure 8. Aerosol extinction coefficient α (**a–b**) over Fram Strait (2 April) derived from the air-borne photometer (black dots) and from its synergy with Raman Lidar (black squares) as well as over Ny-Ålesund (5 April) from Raman Lidar (blue dots). The corresponding extinction-related Angström exponent is also demonstrated (**c**). Horizontal error bars indicate the uncertainties according to error propagation. Vertical error bars indicate the altitude sectors where the α was derived (for details, see Section 2.2). The α over Fram Strait was calculated by Equation (2). At the visible spectrum, α was similar over both locations (**a**). However, at near-infrared, α diminished significantly over Ny-Ålesund (**b**).

For Ny-Ålesund, a mixture of industrial pollution and biomass-burning aerosol is more likely for the upper part of the plume. The lower layer is more likely to comprise industrial pollution particles as it is affected by north European intrusions. The depolarization (approximately 3%) indicates more spherical particles compared to mid-Fram Strait but less spherical than the Greenland regime. The transition to more spherical aerosol could be attributed to the depletion of non-spherical components. In general, particles with a higher surface roughness are more likely to undergo scavenging due to their higher heterogeneous nucleation efficacy [59]. Close to Greenland, the LR_{532} was significantly lower (15 sr) compared to Ny-Ålesund (64–82 sr). We will discuss in more detail the modification of light scattering properties in Section 5.

4.4. Modification of Aerosol Microphysical Properties

The inverted aerosol microphysical properties for Fram Strait and Ny-Ålesund are summarized in Table 3, while the corresponding volume size distributions are illustrated in Figure 9. The microphysical retrieval periods are denoted in Figure 2. The low imaginary refractive index at the Greenland side of Fram Strait suggests the domination of scattering particles. The same holds true for the upper layer over Ny-Ålesund, while weakly absorbing particles dominated in the lower one, respectively. It should be noted that we considered the refractive index as wavelength independent, which is a common assumption. However, the wavelength-dependent SSA indicates the presence of relatively more absorbing small particles over Fram Strait (SSA_{355} = 0.84) as well as in the lower Ny-Ålesund plume (SSA_{355} = 0.88). The asymmetry parameter was similar over Fram Strait and Ny-Ålesund (approximately 0.7), indicating relatively large particles. Moreover, the asymmetry parameter was higher than the monthly average of different Arctic sites, which mainly lies between 0.55 and 0.65 (Figure 7 from Schmeisser et al. [10]), with such high values usually being pertinent to long-range transport events [10].

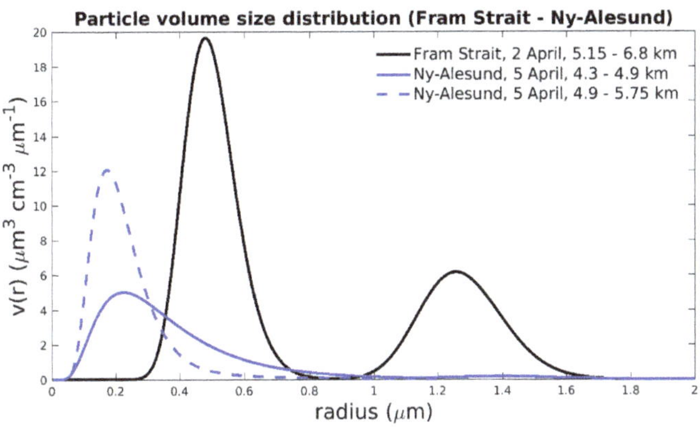

Figure 9. Particle volume size distributions over Fram Strait and Ny-Ålesund retrieved by regularization techniques (for details, see Section 2.3). A clear diminishing of the coarse aerosol mode was revealed for Ny-Ålesund.

The aerosol size distribution was clearly modified along the intra-Arctic transport (Figure 9). A bi-modal distribution was derived for Fram Strait (fine r_{eff} = 0.49 μm, coarse r_{eff} = 1.26 μm), whereas the coarse mode was strongly diminished over Ny-Ålesund. A weak bi-modal distribution (fine r_{eff} = 0.26 μm, coarse r_{eff} = 1.43 μm) emerged in the lower Ny-Ålesund layer, while the fine mode (r_{eff} = 0.19 μm) was predominant in the upper one layer. This transition could be related to the modified pollution pathways, as presented in Section 4.2, as well as to aerosol removal processes. The latter processes will be further discussed in Section 5.

4.5. Aerosol Radiative Effect (ARE)

We assessed the direct effect of aerosol on the local radiative budget by incorporating the retrieved aerosol properties into the radiative transfer model SCIATRAN. The vertical distribution of simulated SW downward, upward and net irradiances, as well as the ARE are depicted in Figure 10. For evaluation, the observed irradiances at the surface and different flight levels are displayed.

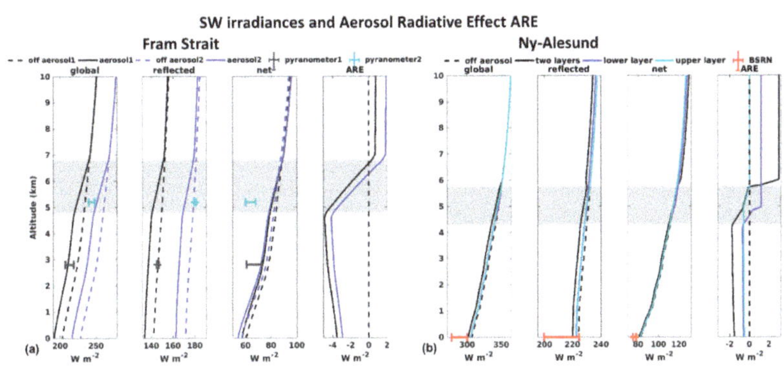

Figure 10. Altitude-dependent SW simulated irradiances and aerosol radiative effect (ARE) for AMALi (1) and sun-photometer (2) evaluation periods over Fram Strait (**a**) and over Ny-Ålesund (**b**). Observed irradiances are given for comparison (average ± one standard deviation). The aerosol layer altitude is indicated with grey shading. Over both locations, TOA warming was accompanied by surface cooling. Note the different scales.

The aerosol layer attenuated the down-welling irradiance through scattering and absorption. At Fram Strait, the surface up-welling irradiance was diminished to a less pronounced degree, in terms of the absolute magnitude compared to the downward component, and, thus, a "shadowing" effect (−3.5 to −2.9 W m^{-2}) was produced. The less intense depletion of upward surface irradiances connotes that their main driver is the surface albedo. Contrarily, the upward component above the layer was diminished more intensely, compared to the downward component, producing a warming effect at TOA (+0.9 to +2 W m^{-2}). Finally, +4.4 to +4.9 W m^{-2} were retained in the atmosphere, with the majority absorbed within the aerosol layer. Compared to air-borne radiation observations, the simulated irradiances presented a maximum bias of 3%, 8%, and 13 % in the SW down-welling, up-welling, and net components, respectively. These differences can be attributed to uncertainties in the thermodynamic and trace gas profiles, as well as the high solar zenith angles (larger than 75^0), which could introduce additional errors in the radiative transfer calculations, especially under the plane-parallel assumption.

At Ny-Ålesund, the TOA was characterized by positive ARE (two-layers +3 W m^{-2}, lower layer +1.15 W m^{-2}, upper layer +0.1 W m^{-2}). More pronounced warming was produced within the atmospheric column (two-layers +4.5 W m^{-2}, lower layer +1.8 W m^{-2}, upper layer +0.6 W m^{-2}), which led to a cooling effect at the surface (two-layers −1.5 W m^{-2}, lower layer −0.6 W m^{-2}, upper layer −0.5 W m^{-2}). In comparison to BSRN observations, the simulated full atmospheric scenario (two-layers) irradiances exhibited agreement with the range of uncertainty. It is worth noticing that the ARE of the two layers did not act cumulatively, indicating the complexity of aerosol radiation interactions. The positive TOA ARE revealed at both Arctic locations implies a contribution from black carbon. On an annual basis, fossil fuel-related black carbon drives a positive TOA forcing (+0.19 W m^{-2}) with respect to pre-industrial emissions, albeit the black carbon AOD is low compared to other aerosol species [3].

4.6. ARE Uncertainties and Comparison to WV-Related Radiative Effect

To investigate the effect of aerosol uncertainties on the simulated ARE, we performed a set of sensitivity tests. More specifically, we repeated the full atmospheric scenario over Ny-Ålesund, but we perturbed the aerosol input by its highest related uncertainties, as they are presented in Tables 2 and 3. We found out that a 30% increase in the aerosol extinction produced more pronounced warming at TOA (by +0.9 W m^{-2}) and atmosphere (by +1.3 W m^{-2}) and, thus, moderately enhanced cooling at the surface (by -0.4 W m^{-2}). Accordingly, a 10% decrease in the SSA resulted in even more pronounced TOA and atmospheric warming (by +1.7 and +2.1 W m^{-2}) but similar surface cooling (by −0.4 W m^{-2}) as the extinction perturbation. Finally, an 8% decline in the asymmetry parameter led to equal magnitude warming and cooling at TOA and the surface (by −0.1 W m^{-2}), respectively, and, thus, the atmospheric radiative budget was not modified. According to error propagation, the maximum cumulative uncertainty amounted to 2.7 W m^{-2} at TOA, 0.9 W m^{-2} at the surface, and 3.4 W m^{-2} within the atmospheric column. Consequently, the sign of the obtained ARE is not affected given the range of aerosol related uncertainties.

Finally, we investigated the effect of the WV concentration on the irradiances in order to assess its radiative effect comparatively to the aerosol effect. More specifically, we repeated the full atmospheric scenario over Ny-Ålesund, but we reduced the WV mixing ratio by 30%. Our sensitivity analysis revealed that a 30% decline in the WV mixing ratio resulted in a −3.8 W m^{-2} atmospheric effect. At the surface and TOA, the effect amounted to +1 and −2.8 W m^{-2}, respectively. Hence, the atmospheric column ARE (+4.5 W m^{-2}) outweighed the WV radiative effect (−3.8 W m^{-2}), with maximum weekly WV fluctuations over Ny-Ålesund (2–9 April) being lower than 30% at 5–15 km.

5. Discussion

An unusual aspect of the investigated long-range aerosol transport is its high altitude, with the majority of reported episodes over Ny-Ålesund confined below 5 km [14,24,60,61]. Aerosol climate models predict such transport with an increased black carbon burden at atmospheric levels higher than 500 hPa [62–64], but this is usually attributed to an underestimation of wet removal. With our study, however, we found observational evidence of the rare but persistent presence of such elevated aerosol transport [14,24,60,61]. North European sources affect, in general, lower altitudes due to their relative proximity to the Arctic [16]. The intrusion of north Eurasian air masses in early spring is promoted either by blocking patterns or strong diabatic cooling along transport over snow-covered areas. Thereby, aerosol trapped below temperature inversions can be transported over long distances [16–18]. Asian origin transport at such high tropospheric altitudes is seldom observed [18]. Lifting processes, nonetheless, that facilitate the export of air masses from the boundary layer to these altitudes are related to the warm conveyor belts of mid-latitude cyclones. Air masses ascend along the sloping isentropes and can penetrate the polar dome from above through radiative cooling [16–18]. In the future, lofted aerosol layers may appear more frequently over the Arctic in the context of projected intensification in strong extra-tropical surface cyclones [65,66]. Moreover, in view of the increasing long-term trend in blockings over the west Pacific [67], more frequent elevated aerosol layers can play an important role in high cloud formation processes.

The intra-Arctic aerosol modification of this episode demonstrates not only the effect of altered aerosol source regions but also the role of removal processes. Mechanisms promoting the aerosol depletion within pollution plumes are mixing with ambient air, coagulation, nucleation scavenging (activation to cloud condensation nuclei and ice nucleating particles as well as subsequent scavenging of interstitial aerosol), dry deposition (impaction, gravitational settling, Brownian motion), and wet scavenging [68]. In this transport episode, a transition towards smaller accumulation mode particles along with the depletion of the coarse mode occurred (Figure 9). The bigger aerosol is deposited more effectively due to gravitational settling and impaction is more efficient due to their higher inertia [69] (pp. 362–395). However, solely dry deposition could not account for the coarse mode depletion within a period of three days. The presence of clouds over Fram Strait (Figure S1) and Ny-Ålesund (Figure 4a) in the interim of the episode indicates that the transported aerosol underwent nucleation scavenging. Nucleation rates are proportional to the aerosol surface area [70].

The Fram Strait plume and its Ny-Ålesund counterpart consisted of aerosol with a similar refractive index. Biomass burning aerosol usually exhibits lower absorption along its transport, which is associated with the degree of oxidative processing on the particle surface [71]. However, in the course of this episode, a part of the aerosol plume (lower Ny-Ålesund sub-layer) was more absorbing, possibly due to interstitial and additionally supplied black carbon particles from neighboring industrial pollution sources (north Europe). Overall, the refractive indices are in line with reported Arctic haze events over Ny-Ålesund, which comprised either polluted ([72], 1.56 + 0.01 i) or mixtures of industrial pollution and biomass-burning aerosol from Siberia ([24], 1.6 + 0.007 i). The aforementioned Arctic haze events demonstrated a higher total number concentration ([72], 343 ± 30% cm^{-3}; [24], 480–950 cm^{-3}), probably due to the more efficient aerosol sources or less effective removal processes along their shorter transport paths (central and east Siberia). The accumulation mode at Ny-Ålesund presented similarities with those events ([72], r_{eff} = 0.18 μm; [24], r_{eff} = 0.19 μm). Average Arctic aerosol size distributions (derived from in situ measurements with a 0.5-μm cut-off diameter) peak around radii of 0.15 μm [8,9,73]. However, the accumulation mode particles at Fram Strait were untypically large. Bi-modal size distributions are also typical for the Arctic haze season of west Greenland [74], but their accumulation mode is shifted to smaller radii than the present episode.

Both biomass-burning and industrial pollution aerosol constitute combustion particles. The LR spectral dependency (higher LR_{532} than LR_{355}) at Ny-Ålesund is more similar to that of aged biomass-burning aerosol [71,75,76]. Overall, biomass-burning aerosol demonstrates diverse LR (LR_{355} = 40–89 sr and LR_{532} = 46–100 sr; [71,75–79]) due to the interplay of several parameters, such as vegetation type, transport time, combustion efficiency (temperature and oxygen availability), local meteorological conditions (RH), as well as mixing state of the particles and content of elemental and organic carbon. Similar factors formulate the light scattering properties of industrial pollution particles. These particles display higher LR in the ultraviolet spectral region (LR_{355} = 41–73 sr and LR_{532} = 23–74 sr, [77,79,80] but generally exhibit lower r_{eff} than ours. The LR_{532} (15 sr) at Fram Strait was lower compared to the reported literature values for biomass-burning and industrial pollution particles. From a microphysical perspective, the particle size distribution exhibits less sensitivity to the visible light, since this is optically dominated by the coarse mode. As illustrated in Figure S3 the coarse mode particles dominate the extinction and backscatter cross-sections. By performing Mie calculations, we estimated that the highest LR is produced in the near-infrared (LR_{1064} = 36 sr), whereas in the visible considerably lower values were revealed (LR_{532} = 19 sr). During PAMARCMiP2018, an increased BC concentration, reaching 117 ng m^{-3}, was revealed over Fram Strait by aircraft in situ measurements [81]. In general, black carbon has a longer lifetime due to its low deposition ability [82]. Thus, interstitial black carbon aerosol that survived the nucleation scavenging and dry deposition processes in combination with a further supply of pollution particles from north Europe could have resulted in the increased LR over Ny-Ålesund (64–82 sr).

The similarity of biomass-burning and industrial pollution particles in their optical and microphysical properties makes their distinction difficult [83]. Our study shows that under specific solar geometry and surface albedo conditions, these similarities are also reflected upon the ARE. The aerosol-related parameters, which are incorporated into radiative transfer simulations, are the SSA, the asymmetry factor, and the aerosol extinction coefficient. We will attempt to illuminate the role of each on the resulting ARE. From the SSA viewpoint, we cannot effectively tell any difference in the aerosol properties between Fram Strait and Ny-Ålesund. Our sensitivity analysis revealed that a 10% SSA perturbation can modulate the atmospheric ARE by 2.1 W m^{-2}, with the SSA modification being lower between the two locations. The asymmetry parameter did not reflect the pronounced coarse mode modification, since it has different sensitivity compared to other size relevant parameters, such as the Angström exponent [10]. Finally, despite the diminished extinction in the near-infrared (from 20 to 4–7 Mm^{-1}), the impact on the spectrally integrated irradiance was not considerable, since this spectral region corresponds to low solar spectral irradiance. At the same time, the visible spectrum extinction, which is sensitive to the accumulation mode and the dominating visible solar irradiance, was almost constant. Therefore, the optically and microphysically similar accumulation mode led to an indistinguishable SW radiative footprint over the two locations.

Aerosol modification was reflected more intensely on the aerosol backscatter coefficient compared to the extinction. On the one hand, extinction, comprising scattering and absorption, is more representative for radiative transfer processes compared to solely scattering that mainly leads to the re-distribution of radiation. On the other hand, Lidar-derived extinction is characterized by high uncertainties due to the weak Raman cross-sections. Backscatter is accompanied by lower uncertainties and can also be derived from simple elastic Lidar systems [84], such as ceilometers, but it cannot be directly incorporated into aerosol climate models. Therefore, the aerosol retrieval uncertainties and limitations need to be thoroughly explored so as to better constrain the vertically resolved ARE from Lidar-inferred aerosol properties.

The derived aerosol properties, concentrations, and its change in time could, in principle, be compared to outcomes of chemical transport models. This has already been done successfully in the Arctic by Warneke et al. [12] and Zielinski et al. [85]. Despite the differences in the aerosol size distribution at the two sites, similar and stable results for the ARE have been found. Apparently, the forcing of the aerosol is more dependent on the solar zenith angle, surface albedo, and the geometrical height of the aerosol layer, as these parameters were similar for our case. Aerosol indirect effects could not be constrained from this case study.

6. Summary and Conclusions

In this study, we exploited synergistic remote sensing observations (from Lidar and sun-photometer) of a long-range aerosol transport episode over two parts of the European Arctic in April 2018. Our aims were oriented towards investigating the intra-Arctic aerosol modification and assessing its influence on the local radiative budget. The main findings and conclusions can be summarized as follows:

- This transport episode stood out due to its considerable altitude, with such elevated aerosol layers predicted by aerosol climate models but rarely observed [14,60,61]. Generalizing the findings from our radiative transfer simulations, geometrically higher aerosol layers will produce a warming effect at higher tropospheric levels. Since these aerosol mixtures cool the surface, and hence near-surface air, but warm the surrounding air, the tropospheric stratification will be stabilized. Hence, the vertical propagation of heat and radiation fluxes will be suppressed. The latter applies for similar aerosol mixtures aloft bright surfaces during early spring.
- The aerosol size distribution was clearly modified between the two Arctic locations. The effective radius of the accumulation mode decreased, while the coarse mode was depleted (Figure 9). Industrial pollution and biomass-burning aerosol were funneled to Fram Strait and Ny-Ålesund, Svalbard. In the beginning, aged air masses originating from Asia dominated, while in the course of the episode they mixed with less mature air from northern Europe (Figure 6). Solely, the modified aerosol sources and dry deposition could not account for the elimination of the coarse mode. However, high-level clouds in the interim of the two observations indicate the presence of nucleation scavenging (Figure S1).
- Along the intra-Arctic transport, particles presented a significantly higher Lidar ratio at visible wavelengths (LR_{532} increased from 15 sr to 64–82 sr). A possible increased absorption can be attributed to black carbon, which could be transported at longer distances as interstitial aerosol since it has low deposition ability [82] and, thus, a longer lifetime. Moreover, black carbon could have been advected through neighboring north European intrusions.
- Despite the coarse mode depletion and increased Lidar ratio, the microphysical similarities of the accumulation mode led to an indistinguishable short-wave radiative footprint in the two episode stages. The accumulation mode drives visible solar irradiance and, thus, dominates the short-wave radiative budget. Thus, the atmospheric column aerosol radiative effect bore similarities over the two locations, amounting to +4.4–4.9 W m^{-2} over the ice-covered Fram Strait and +4.5 W m^{-2} over the snow-covered Ny-Ålesund. This episode caused top-of-atmosphere warming accompanied by surface cooling, with implications for atmospheric stratification.
- The aerosol radiative effect was more significant compared to the ambient humidity effect on a weekly basis (30% WV mixing ratio perturbation).
- Our study suggests that within this Arctic haze episode, which was dominated by accumulation mode particles, the intra-Arctic aerosol modification (and hence, the precise aerosol microphysics) had little effect on the local radiative budget. However, in the context of retreating Arctic sea ice and declining surface albedo [86], the local aerosol radiative effect may change along individual transport episodes.

Supplementary Materials: The following are available online at http://www.mdpi.com/2072-4292/12/13/2112/s1, Figure S1: Track of Polar 5 with color indicated flight altitude (**a**) and corresponding AMALi range corrected signal on 3 April (**b**). Clouds entangled with aerosol features were observed over Fram Strait, implying the presence of nucleation scavenging processes, Figure S2: MODIS Fire and Thermal Anomalies product (red spots) overlaid on corrected reflectance from MODIS Terra, MODIS Acqua and VIIRS/Suomi NPP for 22 March 2018. https://worldview.earthdata.nasa.gov, Last Access: 19 March 2020). Limited fire hotspots together with a thin smoke plume can be seen over the Jilin Province, China, Figure S3: Particle volume size distribution over Fram Strait. Backscatter and extinction cross-sections for the Lidar relevant wavelengths were calculated with Mie theory. The extinction cross-sections and backscatter at near-infrared (1064 nm) of coarse mode particles outweighed the fine mode ones, indicating the dominant optical role of the coarse mode.

Author Contributions: Data curation, K.N., D.K., O.E., E.J., A.H., M.M. and R.N.; Formal analysis, K.N., C.B., D.K., O.E., V.R., L.M. and V.P.; Funding acquisition, R.N.; Investigation, K.N., C.R., C.B. and V.P.; Methodology, K.N., C.R., V.R., L.M., V.P. and E.J.; Resources, A.H., M.M. and R.N.; Software, O.E., V.R. and L.M.; Supervision, C.R., M.M. and R.N.; Validation, K.N.; Visualization, K.N.; Writing–original draft, K.N.; Writing–review & editing, K.N., C.R., C.B., D.K., O.E., L.M., V.P., E.J., M.M. and R.N. All authors have read and agreed to the published version of the manuscript.

Funding: This work was funded by the Deutsche Forschungsgemeinschaft (DFG, German Research Foundation)–Project-ID 268020496–TRR 172.

Acknowledgments: We would like to thank the operators of AMALi during PAMARCMiP 2018. Wilfried Ruhe and Ingo Beninga from Impres GmbH for providing long-term technical support at the AWIPEV research base and maintaining the KARL system. Siegrid Debatin for the quality assurance of the Schulz and Partner photometer data. Moreover, we thank Wilfried Ruhe for providing training in the operation of KARL and discussing several of its technical aspects. We gratefully acknowledge the funding by the Deutsche Forschungsgemeinschaft (DFG, German Research Foundation)–Project Number 268020496–TRR 172, within the Transregional Collaborative Research Center "ArctiC Amplification: Climate Relevant Atmospheric and SurfaCe Processes, and Feedback Mechanisms $(AC)^3$". Last but not least, we would like to thank the two anonymous reviewers for their constructive comments that improved this work.

Conflicts of Interest: The authors declare no conflict of interest.

References

1. Breider, T.J.; Mickley, L.J.; Jacob, D.J.; Ge, C.; Wang, J.; Payer Sulprizio, M.; Croft, B.; Ridley, D.A.; McConnell, J.R.; Sharma, S. Multidecadal trends in aerosol radiative forcing over the Arctic: Contribution of changes in anthropogenic aerosol to Arctic warming since 1980. *J. Geophys. Res. Atmos.* **2017**, *122*, 3573–3594. [CrossRef]
2. Bellouin, N.; Quaas, J.; Gryspeerdt, E.; Kinne, S.; Stier, P.; Watson-Parris, D.; Boucher, O.; Carslaw, K.S.; Christensen, M.; Daniau, A.-L. Bounding global aerosol radiative forcing of climate change. *Rev. Geophys.* **2019**, *58*, e2019RG000660. [CrossRef]
3. Sand, M.; Samset, B.H.; Balkanski, Y.; Bauer, S.; Bellouin, N.; Berntsen, T.K.; Bian, H.; Chin, M.; Diehl, T.; Easter, R. Aerosols at the poles: An AeroCom Phase II multi-model evaluation. *Atmos. Chem. Phys.* **2017**, *17*, 12197–12218. [CrossRef]
4. Najafi, M.R.; Zwiers, F.W.; Gillett, N.P. Attribution of Arctic temperature change to greenhouse-gas and aerosol influences. *Nat. Clim. Chang.* **2015**, *5*, 246–249. [CrossRef]
5. Serreze, M.C.; Barry, R.G. Processes and impacts of Arctic amplification: A research synthesis. *Glob. Planet. Chang.* **2011**, *77*, 85–96. [CrossRef]
6. Wendisch, M.; Brückner, M.; Burrows, J.P.; Crewell, S.; Dethloff, K.; Ebell, K.; Lüpkes, C.; Macke, A.; Notholt, J.; Quaas, J. Understanding causes and effects of rapid warming in the Arctic. *Eos* **2017**, *98*. [CrossRef]
7. Korhonen, H.; Carslaw, K.S.; Spracklen, D.V.; Ridley, D.A.; Ström, J. A global model study of processes controlling aerosol size distributions in the Arctic spring and summer. *J. Geophys. Res. Atmos.* **2008**, *113*. [CrossRef]
8. Croft, B.; Martin, R.V.; Leaitch, W.R.; Tunved, P.; Breider, T.J.; D'Andrea, S.D.; Pierce, J.R. Processes controlling the annual cycle of Arctic aerosol number and size distributions. *Atmos. Chem. Phys.* **2016**, *16*, 3665–3682. [CrossRef]
9. Freud, E.; Krejci, R.; Tunved, P.; Leaitch, R.; Nguyen, Q.T.; Massling, A.; Skov, H.; Barrie, L. Pan-Arctic aerosol number size distributions: Seasonality and transport patterns. *Atmos. Chem. Phys.* **2017**, *17*, 8101–8128. [CrossRef]
10. Schmeisser, L.; Backman, J.; Ogren, J.A.; Andrews, E.; Asmi, E.; Starkweather, S.; Uttal, T.; Fiebig, M.; Sharma, S.; Eleftheriadis, K. Seasonality of aerosol optical properties in the Arctic. *Atmos. Chem. Phys.* **2018**, *18*, 11599–11622. [CrossRef]
11. Quinn, P.K.; Shaw, G.; Andrews, E.; Dutton, E.G.; Ruoho-Airola, T.; Gong, S.L. Arctic haze: Current trends and knowledge gaps. *Tellus B Chem. Phys. Meteorol.* **2007**, *59*, 99–114. [CrossRef]
12. Warneke, C.; Bahreini, R.; Brioude, J.; Brock, C.A.; De Gouw, J.A.; Fahey, D.W.; Froyd, K.D.; Holloway, J.S.; Middlebrook, A.; Miller, L. Biomass burning in Siberia and Kazakhstan as an important source for haze over the Alaskan Arctic in April 2008. *Geophys. Res. Lett.* **2009**, *36*. [CrossRef]
13. Di Pierro, M.; Jaeglé, L.; Eloranta, E.W.; Sharma, S. Spatial and seasonal distribution of Arctic aerosols observed by CALIOP (2006–2012). *Atmos. Chem. Phys. Discuss.* **2013**, *13*, 7075–7095. [CrossRef]
14. Ritter, C.; Neuber, R.; Schulz, A.; Markowicz, K.M.; Stachlewska, I.S.; Lisok, J.; Makuch, P.; Pakszys, P.; Markuszewski, P.; Rozwadowska, A. 2014 iAREA campaign on aerosol in Spitsbergen–Part 2: Optical properties from Raman-lidar and in-situ observations at Ny-\AAlesund. *Atmos. Environ.* **2016**, *141*, 1–19. [CrossRef]

15. Shibata, T.; Shiraishi, K.; Shiobara, M.; Iwasaki, S.; Takano, T. Seasonal Variations in High Arctic Free Tropospheric Aerosols Over Ny-\AAlesund, Svalbard, Observed by Ground-Based Lidar. *J. Geophys. Res. Atmos.* **2018**, *123*, 12–353. [CrossRef]
16. Klonecki, A.; Hess, P.; Emmons, L.; Smith, L.; Orlando, J. Seasonal changes in the transport of pollutants into the Arctic troposphere-model study: Tropospheric Ozone Production about the Spring Equinox (TOPSE). *J. Geophys. Res.* **2003**, *108*, TOP15-1. [CrossRef]
17. Stohl, A. Characteristics of atmospheric transport into the Arctic troposphere. *J. Geophys. Res. Atmos.* **2006**, *111*, 0148–0227. [CrossRef]
18. Di Pierro, M.; Jaeglé, L. Satellite observations of aerosol transport from East Asia to the Arctic: Three case studies. *Atmos. Chem. Phys.* **2011**, *11*, 2225. [CrossRef]
19. Eleftheriadis, K.; Vratolis, S.; Nyeki, S. Aerosol black carbon in the European Arctic: Measurements at Zeppelin station, Ny-\AAlesund, Svalbard from 1998–2007. *Geophys. Res. Lett.* **2009**, *36*. [CrossRef]
20. Lisok, J.; Markowicz, K.M.; Ritter, C.; Makuch, P.; Petelski, T.; Chilinski, M.; Kaminski, J.W.; Becagli, S.; Traversi, R.; Udisti, R. 2014 iAREA campaign on aerosol in Spitsbergen–Part 1: Study of physical and chemical properties. *Atmos. Environ.* **2016**, *140*, 150–166. [CrossRef]
21. Moroni, B.; Arnalds, O.; Dagsson-Waldhauserová, P.; Crocchianti, S.; Vivani, R.; Cappelletti, D. Mineralogical and chemical records of Icelandic dust sources upon Ny-\AAlesund (Svalbard Islands). *Front. Earth Sci.* **2018**, *6*, 187. [CrossRef]
22. Ferrero, L.; Ritter, C.; Cappelletti, D.; Moroni, B.; Močnik, G.; Mazzola, M.; Lupi, A.; Becagli, S.; Traversi, R.; Cataldi, M. Aerosol optical properties in the Arctic: The role of aerosol chemistry and dust composition in a closure experiment between Lidar and tethered balloon vertical profiles. *Sci. Total Environ.* **2019**, *686*, 452–467. [CrossRef] [PubMed]
23. Law, K.S.; Stohl, A. Arctic air pollution: Origins and impacts. *Science* **2007**, *315*, 1537–1540. [CrossRef] [PubMed]
24. Stock, M.; Ritter, C.; Herber, A.; von Hoyningen-Huene, W.; Baibakov, K.; Graeser, J.; Orgis, T.; Treffeisen, R.; Zinoviev, N.; Makshtas, A. Springtime Arctic aerosol: Smoke versus haze, a case study for March 2008. *Atmos. Environ.* **2012**, *52*, 48–55. [CrossRef]
25. Quennehen, B.; Schwarzenboeck, A.; Matsuki, A.; Burkhart, J.F.; Stohl, A.; Ancellet, G.; Law, K.S. Anthropogenic and forest fire pollution aerosol transported to the Arctic: Observations from the POLARCAT-France spring campaign. *Atmos. Chem. Phys.* **2012**. [CrossRef]
26. Monks, S.A.; Arnold, S.R.; Emmons, L.K.; Law, K.S.; Turquety, S.; Duncan, B.N.; Flemming, J.; Huijnen, V.; Tilmes, S.; Langner, J. Multi-model study of chemical and physical controls on transport of anthropogenic and biomass burning pollution to the Arctic. *Atmos. Chem. Phys.* **2015**, *15*, 3575–3603. [CrossRef]
27. Evangeliou, N.; Balkanski, Y.; Hao, W.; Petkov, A.; Silverstein, R.P.; Corley, R.; Nordgren, B.L.; Urbanski, S.; Eckhardt, S.; Stohl, A. Wildfires in northern Eurasia affect the budget of black carbon in the Arctic-a 12-year retrospective synopsis (2002–2013). *Atmos. Chem. Phys.* **2016**, *16*, 7587–7604. [CrossRef]
28. Gjelten, H.M.; Nordli, Ø.; Isaksen, K.; Førland, E.J.; Sviashchennikov, P.N.; Wyszynski, P.; Prokhorova, U.V.; Przybylak, R.; Ivanov, B.V.; Urazgildeeva, A.V. Air temperature variations and gradients along the coast and fjords of western Spitsbergen. *Polar Res.* **2016**, *35*, 29878. [CrossRef]
29. Dahlke, S.; Hughes, N.E.; Wagner, P.M.; Gerland, S.; Wawrzyniak, T.; Ivanov, B.; Maturilli, M. The observed recent surface air temperature development across Svalbard and concurring footprints in local sea ice cover. *Int. J. Climatol.* **2020**. [CrossRef]
30. Dahlke, S.; Maturilli, M. Contribution of atmospheric advection to the amplified winter warming in the Arctic North Atlantic region. *Adv. Meteorol.* **2017**, 1687–9309. [CrossRef]
31. Woods, C.; Caballero, R.; Svensson, G. Large-scale circulation associated with moisture intrusions into the Arctic during winter. *Geophys. Res. Lett.* **2013**, *40*, 4717–4721. [CrossRef]
32. Graßl, S.; Ritter, C. Properties of Arctic Aerosol Based on Sun Photometer Long-Term Measurements in Ny-\AAlesund, Svalbard. *Remote Sens.* **2019**, *11*, 1362. [CrossRef]
33. Binietoglou, I.; D'Amico, G.; Baars, H.; Belegante, L.; Marinou, E. A methodology for cloud masking uncalibrated lidar signals. In Proceedings of the EPJ Web of Conferences; EDP Sciences: Les Ulis, France, 2018; Volume 176, p. 05048.
34. Böckmann, C. Hybrid regularization method for the ill-posed inversion of multiwavelength lidar data in the retrieval of aerosol size distributions. *Appl. Opt.* **2001**, *40*, 1329–1342. [CrossRef] [PubMed]

35. Böckmann, C.; Kirsche, A. Iterative regularization method for lidar remote sensing. *Comput. Phys. Commun.* **2006**, *174*, 607–615. [CrossRef]
36. Osterloh, L.; Böckmann, C.; Mamouri, R.-E.; Papayannis, A. An adaptive base point algorithm for the retrieval of aerosol microphysical properties. *Open Atmos. Sci. J.* **2011**, *5*, 61–73. [CrossRef]
37. Samaras, S.; Nicolae, D.; Böckmann, C.; Vasilescu, J.; Binietoglou, I.; Labzovskii, L.; Toanca, F.; Papayannis, A. Using Raman-lidar-based regularized microphysical retrievals and Aerosol Mass Spectrometer measurements for the characterization of biomass burning aerosols. *J. Comput. Phys.* **2015**, *299*, 156–174. [CrossRef]
38. Stachlewska, I.S.; Neuber, R.; Lampert, A.; Ritter, C.; Wehrle, G. AMALi the Airborne Mobile Aerosol Lidar for Arctic research. *Atmos. Chem. Phys.* **2010**, *10*, 2947–2963. [CrossRef]
39. Hoffmann, A. Comparative Aerosol Studies Based on Multi-Wavelength Raman LIDAR at Ny-\AAlesund, Spitsbergen. Ph.D. Thesis, University Potsdam, Potsdam, Germany, 2011.
40. Iarlori, M.; Madonna, F.; Rizi, V.; Trickl, T.; Amodeo, A. Effective resolution concepts for lidar observations. *Atmos. Meas. Tech. Discuss.* **2015**, *8*, 5157–5176. [CrossRef]
41. Fernald, F.G. Analysis of atmospheric lidar observations: Some comments. *Appl. Opt.* **1984**, *23*, 652–653. [CrossRef] [PubMed]
42. Klett, J.D. Stable analytical inversion solution for processing lidar returns. *Appl. Opt.* **1981**, *20*, 211–220. [CrossRef]
43. Ansmann, A.; Riebesell, M.; Weitkamp, C. Measurement of atmospheric aerosol extinction profiles with a Raman lidar. *Opt. Lett.* **1990**, *15*, 746–748. [CrossRef] [PubMed]
44. Mei, L.; Vandenbussche, S.; Rozanov, V.; Proestakis, E.; Amiridis, V.; Callewaert, S.; Vountas, M.; Burrows, J.P. On the retrieval of aerosol optical depth over cryosphere using passive remote sensing. *Remote Sens. Environ.* **2020**, *241*, 111731. [CrossRef]
45. Alexandrov, M.D.; Marshak, A.; Cairns, B.; Lacis, A.A.; Carlson, B.E. Automated cloud screening algorithm for MFRSR data. *Geophys. Res. Lett.* **2004**, *31*. [CrossRef]
46. Müller, D.; Böckmann, C.; Kolgotin, A.; Schneidenbach, L.; Chemyakin, E.; Rosemann, J.; Znak, P.; Romanov, A. Microphysical particle properties derived from inversion algorithms developed in the framework of EARLINET. *Atmos. Meas. Tech.* **2016**, *9*, 5007–5035. [CrossRef]
47. Sprenger, M.; Wernli, H. The LAGRANTO Lagrangian analysis tool—Version 2.0. *Geosci. Model Dev.* **2015**, *8*, 2569–2586. [CrossRef]
48. Wernli, B.H.; Davies, H.C. A Lagrangian-based analysis of extratropical cyclones. I: The method and some applications. *Q. J. R. Meteorol. Soc.* **1997**, *123*, 467–489. [CrossRef]
49. Ehrlich, A.; Wendisch, M. Reconstruction of high-resolution time series from slow-response broadband terrestrial irradiance measurements by deconvolution. *Atmos. Meas. Tech.* **2015**, *8*, 3671–3684. [CrossRef]
50. Herber, A. *Meteorological Observations during POLAR 5 Campaign PAMARCMIP 2018*; Alfred Wegener Institute, Helmholtz Centre for Polar and Marine Research: Bremerhaven, Germany, 2019.
51. Burba, G.G.; McDermitt, D.K.; Anderson, D.J.; Furtaw, M.D.; Eckles, R.D. Novel design of an enclosed CO_2/H_2O gas analyser for eddy covariance flux measurements. *Tellus B: Chem. Phys. Meteorol.* **2010**, *62*, 743–748. [CrossRef]
52. Maturilli, M.; Herber, A.; König-Langlo, G. Surface radiation climatology for Ny-\AAlesund, Svalbard (78.9 N), basic observations for trend detection. *Theor. Appl. Climatol.* **2015**, *120*, 331–339. [CrossRef]
53. Maturilli, M. *Basic and Other Measurements of Radiation at Station Ny-Ålesund (2018-04)*; Alfred Wegener Institute—Research Unit Potsdam: Potsdam, Germany, 2018.
54. Maturilli, M. *High Resolution Radiosonde Measurements from Station Ny-Ålesund (2018-04)*; Alfred Wegener Institute—Research Unit Potsdam: Potsdam, Germany, 2018.
55. Rozanov, V.V.; Rozanov, A.V.; Kokhanovsky, A.A.; Burrows, J.P. Radiative transfer through terrestrial atmosphere and ocean: Software package SCIATRAN. *J. Quant. Spectrosc. Radiat. Transf.* **2014**, *133*, 13–71. [CrossRef]
56. Rothman, L.S.; Barbe, A.; Benner, D.C.; Brown, L.R.; Camy-Peyret, C.; Carleer, M.R.; Chance, K.; Clerbaux, C.; Dana, V.; Devi, V.M. The HITRAN molecular spectroscopic database: Edition of 2000 including updates through 2001. *J. Quant. Spectrosc. Radiat. Transf.* **2003**, *82*, 5–44. [CrossRef]
57. Sinnhuber, B.-M.; Sheode, N.; Sinnhuber, M.; Chipperfield, M.P.; Feng, W. The contribution of anthropogenic bromine emissions to past stratospheric ozone trends: A modelling study. *Atmos. Chem. Phys.* **2009**, *9*, 2863–2871. [CrossRef]

58. Yamanouchi, T.; Treffeisen, R.; Herber, A.; Shiobara, M.; Yamagata, S.; Hara, K.; Sato, K.; Yabuki, M.; Tomikawa, Y.; Rinke, A. Arctic study of tropospheric aerosol and radiation (ASTAR) 2000: Arctic haze case study. *Tellus B: Chem. Phys. Meteorol.* **2005**, *57*, 141–152. [CrossRef]
59. Lazaridis, M.; Hov, Ø.; Eleftheriadis, K. Heterogeneous nucleation on rough surfaces: Implications to atmospheric aerosols. *Atmos. Res.* **2000**, *55*, 103–113. [CrossRef]
60. Hoffmann, A.; Osterloh, L.; Stone, R.; Lampert, A.; Ritter, C.; Stock, M.; Tunved, P.; Hennig, T.; Böckmann, C.; Li, S.-M. Remote sensing and in-situ measurements of tropospheric aerosol, a PAMARCMiP case study. *Atmos. Environ.* **2012**, *52*, 56–66. [CrossRef]
61. Ritter, C.; Angeles Burgos, M.; Böckmann, C.; Mateos, D.; Lisok, J.; Markowicz, K.M.; Moroni, B.; Cappelletti, D.; Udisti, R.; Maturilli, M. Microphysical properties and radiative impact of an intense biomass burning aerosol event measured over Ny-\AAlesund, Spitsbergen in July 2015. *Tellus B Chem. Phys. Meteorol.* **2018**, *70*, 1–23. [CrossRef]
62. Schacht, J.; Heinold, B.; Quaas, J.; Backman, J.; Cherian, R.; Ehrlich, A.; Herber, A.; Huang, W.T.; Kondo, Y.; Massling, A. The importance of the representation of air pollution emissions for the modeled distribution and radiative effects of black carbon in the Arctic. *Atmos. Chem. Phys.* **2019**, *19*, 11159–11183. [CrossRef]
63. Schwarz, J.P.; Weinzierl, B.; Samset, B.H.; Dollner, M.; Heimerl, K.; Markovic, M.Z.; Perring, A.E.; Ziemba, L. Aircraft measurements of black carbon vertical profiles show upper tropospheric variability and stability. *Geophys. Res. Lett.* **2017**, *44*, 1132–1140. [CrossRef]
64. Schwarz, J.P.; Samset, B.H.; Perring, A.E.; Spackman, J.R.; Gao, R.S.; Stier, P.; Schulz, M.; Moore, F.L.; Ray, E.A.; Fahey, D.W. Global-scale seasonally resolved black carbon vertical profiles over the Pacific. *Geophys. Res. Lett.* **2013**, *40*, 5542–5547. [CrossRef]
65. Mizuta, R. Intensification of extratropical cyclones associated with the polar jet change in the CMIP5 global warming projections. *Geophys. Res. Lett.* **2012**, *39*. [CrossRef]
66. Lambert, S.J.; Fyfe, J.C. Changes in winter cyclone frequencies and strengths simulated in enhanced greenhouse warming experiments: Results from the models participating in the IPCC diagnostic exercise. *Clim. Dyn.* **2006**, *26*, 713–728. [CrossRef]
67. Barriopedro, D.; García-Herrera, R.; Lupo, A.R.; Hernández, E. A Climatology of Northern Hemisphere Blocking. *J. Clim.* **2006**, *19*, 1042–1063. [CrossRef]
68. Garrett, T.; Zhao, C.; Novelli, P. Assessing the relative contributions of transport efficiency and scavenging to seasonal variability in Arctic aerosol. *Tellus B Chem. Phys. Meteorol.* **2010**, *62*, 190–196. [CrossRef]
69. Seinfeld, J.H.; Pandis, S.N. *Atmospheric Chemistry and Physics: From Air Pollution to Climate Change*; John Wiley & Sons: Hoboken, NJ, USA, 2016.
70. Fletcher, N.H.J. Size effect in heterogeneous nucleation. *J. Chem. Phys.* **1958**, *29*, 572–576. [CrossRef]
71. Nicolae, D.; Nemuc, A.; Müller, D.; Talianu, C.; Vasilescu, J.; Belegante, L.; Kolgotin, A. Characterization of fresh and aged biomass burning events using multiwavelength Raman lidar and mass spectrometry. *J. Geophys. Res. Atmos.* **2013**, *118*, 2956–2965. [CrossRef]
72. Hoffmann, A.; Ritter, C.; Stock, M.; Shiobara, M.; Lampert, A.; Maturilli, M.; Orgis, T.; Neuber, R.; Herber, A. Ground-based lidar measurements from Ny-\AAlesund during ASTAR 2007: A statistical overview. *Atmos. Chem. Phys. Discuss* **2009**, *9*, 15453–15510. [CrossRef]
73. Tunved, P.; Ström, J.; Krejci, R. Arctic aerosol life cycle: Linking aerosol size distributions observed between 2000 and 2010 with air mass transport and precipitation at Zeppelin station, Ny-\AA lesund, Svalbard. *Atmos. Chem. Phys.* **2013**, *13*, 3643–3660. [CrossRef]
74. Mei, L.; Rozanov, V.; Ritter, C.; Heinold, B.; Jiao, Z.; Vountas, M.; Burrows, J.P. Retrieval of aerosol optical thickness in the Arctic snow-covered regions using passive remote sensing: Impact of aerosol typing and surface reflection model. *IEEE Trans. Geosci. Remote Sens.* **2020**, *58*, 5117–5131. [CrossRef]
75. Janicka, L.; Stachlewska, I.S.; Veselovskii, I.; Baars, H. Temporal variations in optical and microphysical properties of mineral dust and biomass burning aerosol derived from daytime Raman lidar observations over Warsaw, Poland. *Atmos. Environ.* **2017**, *169*, 162–174. [CrossRef]
76. Ortiz-Amezcua, P.; Guerrero-Rascado, J.L.; Granados-Muñoz, M.J.; Benavent-Oltra, J.A.; Böckmann, C.; Samaras, S.; Stachlewska, I.S.; Janicka, L.; Baars, H.; Bohlmann, S. Microphysical characterization of long-range transported biomass burning particles from North America at three EARLINET stations. *Atmos. Chem. Phys.* **2019**, *17*, 5931–5946. [CrossRef]

77. Giannakaki, E.; Van Zyl, P.G.; Müller, D.; Balis, D.; Komppula, M. Optical and microphysical characterization of aerosol layers over South Africa by means of multi-wavelength depolarization and Raman lidar measurements. *Atmos. Chem. Phys.* **2016**, *16*, 8109–8123. [CrossRef]
78. Haarig, M.; Ansmann, A.; Baars, H.; Jimenez, C.; Veselovskii, I.; Engelmann, R.; Althausen, D. Depolarization and lidar ratios at 355, 532, and 1064 nm and microphysical properties of aged tropospheric and stratospheric Canadian wildfire smoke. *Atmos. Chem. Phys.* **2018**, *18*, 11847–11861. [CrossRef]
79. Janicka, L.; Stachlewska, I.S. Properties of biomass burning aerosol mixtures derived at fine temporal and spatial scales from Raman lidar measurements: Part I optical properties. *Atmos. Chem. Phys. Discuss* **2019**, 1–46. [CrossRef]
80. Groß, S.; Esselborn, M.; Weinzierl, B.; Wirth, M.; Fix, A.; Petzold, A. Aerosol classification by airborne high spectral resolution lidar observations. *Atmos. Chem. Phys.* **2013**, *13*, 2487. [CrossRef]
81. Donth, T.; Jäkel, E.; Ehrlich, A.; Heinold, B.; Schacht, J.; Herber, A.; Zanatta, M.; Wendisch, M. Combining atmospheric and snow layer radiative transfer models to assess the solar radiative effects of black carbon in the Arctic. *Atmos. Chem. Phys. Discuss.* **2020**, 1–26. [CrossRef]
82. Macdonald, K.M.; Sharma, S.; Toom, D.; Chivulescu, A.; Hanna, S.; Bertram, A.K.; Platt, A.; Elsasser, M.; Huang, L.; Tarasick, D. Observations of atmospheric chemical deposition to high Arctic snow. *Atmos. Chem. Phys.* **2017**, *17*, 5775–5788. [CrossRef]
83. Papagiannopoulos, N.; Alados Arboledas, L.; Guerrero-Rascado, J.L. An automatic observation-based aerosol typing method for EARLINET. *Atmos. Chem. Phys.* **2018**, *18*, 15879–15901. [CrossRef]
84. Giannakaki, E.; Kokkalis, P.; Marinou, E.; Bartsotas, N.S.; Amiridis, V.; Ansmann, A.; Komppula, M. The potential of elastic and polarization lidars to retrieve extinction profiles. *Atmos. Meas. Tech.* **2020**, *13*, 893–905. [CrossRef]
85. Zielinski, T.; Bolzacchini, E.; Cataldi, M.; Ferrero, L.; Graßl, S.; Hansen, G.; Mateos, D.; Mazzola, M.; Neuber, R.; Pakszys, P.; et al. Study of Chemical and Optical Properties of Biomass Burning Aerosols during Long-Range Transport Events toward the Arctic in Summer 2017. *Atmosphere* **2020**, *11*, 84. [CrossRef]
86. Döscher, R.; Vihma, T.; Maksimovich, E. Recent advances in understanding the Arctic climate system state and change from a sea ice perspective: A review. *Atmos. Chem. Phys.* **2014**, *14*, 13571–13600. [CrossRef]

remote sensing

MDPI

Article

Properties of Cirrus Clouds over the European Arctic (Ny-Ålesund, Svalbard)

Konstantina Nakoud [1,2], Christoph Ritter [1,*] and Iwona S. Stachlewska [3]

[1] Alfred Wegener Institute, Helmholtz Centre for Polar and Marine Research, Telegrafenberg A45, 14473 Potsdam, Germany; konstantina.nakoudi@awi.de

[2] Institute of Physics and Astronomy, University of Potsdam, Karl-Liebknecht 24/25, 14476 Potsdam, Germany

[3] Faculty of Physics, University of Warsaw (FUW), 00-927 Warsaw, Poland; Iwona.Stachlewska@fuw.edu.pl

* Correspondence: christoph.ritter@awi.de

Abstract: Cirrus is the only cloud type capable of inducing daytime cooling or heating at the top of the atmosphere (TOA) and the sign of its radiative effect highly depends on its optical depth. However, the investigation of its geometrical and optical properties over the Arctic is limited. In this work the long-term properties of cirrus clouds are explored for the first time over an Arctic site (Ny-Ålesund, Svalbard) using lidar and radiosonde measurements from 2011 to 2020. The optical properties were quality assured, taking into account the effects of specular reflections and multiple-scattering. Cirrus clouds were generally associated with colder and calmer wind conditions compared to the 2011–2020 climatology. However, the dependence of cirrus properties on temperature and wind speed was not strong. Even though the seasonal cycle was not pronounced, the winter-time cirrus appeared under lower temperatures and stronger wind conditions. Moreover, in winter, geometrically- and optically-thicker cirrus were found and their ice particles tended to be more spherical. The majority of cirrus was associated with westerly flow and westerly cirrus tended to be geometrically-thicker. Overall, optically-thinner layers tended to comprise smaller and less spherical ice crystals, most likely due to reduced water vapor deposition on the particle surface. Compared to lower latitudes, the cirrus layers over Ny-Ålesund were more absorbing in the visible spectral region and they consisted of more spherical ice particles.

Keywords: Arctic clouds; cirrus clouds; ice clouds; lidar

Citation: Nakoudi, K.; Ritter, C.; Stachlewska, I.S. Properties of Cirrus Clouds over the European Arctic (Ny-Ålesund, Svalbard). *Remote Sens.* **2021**, *13*, 4555. https://doi.org/10.3390/rs13224555

Academic Editor: Simone Lolli

Received: 20 September 2021
Accepted: 9 November 2021
Published: 12 November 2021

1. Introduction

Over the last decades, the rate of near-surface warming in the Arctic has been at least double than elsewhere on our planet (*Arctic amplification*) [1]. However, the relative contribution of different feedback processes to *Arctic amplification* is a topic of ongoing research. Clouds are a major modulator of the Arctic energy flow and surface energy budget [2]. Huang et al. [3] suggested that clouds are a driving force for sea ice melt from April to June, following the active coupling between the atmosphere and sea ice in early spring. Still clouds are a challenging puzzle, both within the Arctic climate system and through their role in *Arctic amplification* [2].

Generally, ice clouds cause a warming effect within the lower atmosphere, while a cooling effect occurs in higher levels[4]. At the top of the atmosphere (TOA) the global effect of ice clouds was approximated at 5.1 ± 3.8 Wm^{-2}, while the globally averaged surface effect was negative [4]. Cirrus is the only cloud type capable of inducing daytime cooling or heating at TOA [5]. In contrast, the rest of the clouds produce solely daytime cooling at TOA. The relative magnitude of short-wave (SW) cooling and infrared warming depends on the cloud properties, solar geometry, thermal contrast to the surface and surface albedo [6]. Moreover, cirrus clouds affect the redistribution of water vapor in the upper troposphere and its exchange with the lower stratosphere [7]. However, cirrus clouds are among the most uncertain components of general circulation models due to their wide range of cloud optical depth (COD) and occurrence altitudes [8]. There are quite a few studies on cirrus properties derived from aircraft campaigns (e.g., Gultepe and Starr [9]). Following the air masses by plane allows the detailed investigation of cirrus formation and evolution. Contrary, by ground-based remote sensing, the advection and temporal evolution of air masses are generally hard to study.

Over the European Arctic site of Ny-Ålesund, a negative annual average cloud radiative effect (CRE) (-16 Wm^{-2}) is estimated at TOA [10]. This region is characterized by a strong winter warming of up to 3 °C per decade [11]. While liquid-containing clouds dominate the surface CRE in summer, the effect of ice clouds is dominant during polar nights. However, the contribution of Arctic cirrus clouds to the local radiation budget has not been quantified yet. To this end, the first necessary step is a better understanding of the cirrus geometrical and optical properties under different ambient conditions. Passive radiometric sensors often fail to detect optically-thinner cirrus due to their low cloud mask sensitivity [12,13]. Using lidar observations over the sub-Arctic site of Fairbanks, Alaska, the daytime CRE at TOA was investigated for each cirrus regime [14]; opaque ($0.3 < COD < 3$), optically-thin ($0.03 < COD < 0.3$) and sub-visible ($COD < 0.03$) [15]. Opaque cirrus induced an exclusively negative CRE (from -3.4 to -0.3 Wm^{-2}) [14]. However, a fluctuating CRE between positive and negative values (from -1.1 to $+0.8$ Wm^{-2}) was derived when the optically-thin and sub-visible cirrus were considered. Thus, the sign of the cirrus cloud CRE is highly dependent on their COD. Therefore, the reliable detection of optically-thinner cirrus is crucial for the improvement of cirrus CRE estimates. The role of lidar observations is essential in this respect. Comprehensive statistics on the occurrence frequency of high-level ice clouds have been reported over different Arctic sites. Over Barrow, Eureka, and SHEBA ice clouds accounted for 30% of the total cloudiness above 6 km (Figure 3 from Shupe [16]), while over Ny-Ålesund the respective contribution was lower than 20% (Figure 8 from Nomokonova et al. [17]). However, the geometrical and optical properties of Arctic cirrus clouds are under-explored. Cloud-Aerosol lidar with Orthogonal Polarization (CALIOP) and Cloud Profiling Radar have partially filled the observational deficit in the last 15 years by providing observations up to 82°N [18]. Devasthale et al. [19] investigated the geometrical properties of optically-thin clouds over the Arctic using CALIOP observations. The authors found that the geometrical thickness (GT) of ice clouds fluctuated mainly between 0.4 and 1 km, with thicker clouds appearing in winter. Moreover, the cirrus cloud base (C_{base}) and top (C_{top}) occurrences peaked at 6–8 km and 7–9 km, respectively.

Placing emphasis on optically- and geometrically-thinner cirrus, Nakoudi et al. [20] developed an extended lidar-based retrieval scheme (more details in Section 2.2). In this work, this scheme is applied on a 10 year cirrus lidar data set over Ny-Ålesund, Svalbard. This work focuses on the long-term geometrical properties of cirrus clouds but also expands on their optical properties, which are investigated for the first time over an Arctic site. The aim of this work is to give an overview of cirrus geometrical and optical properties, as obtained by lidar, in different seasons over the European Arctic.

2. Methods and Measurement Site

2.1. Lidar and Radiosonde Measurements

The geometrical and optical properties of cirrus layers were retrieved from Koldewey Aerosol Raman lidar (KARL) observations between 2011 and 2020. KARL is installed at the Alfred Wegener Institute–Institute Paul Emile Victor (AWIPEV) research base, Ny-Ålesund (78.9° N, 11.9° E), Svalbard. KARL is well-suited for cloud investigations, as it can provide observations on high vertical (7.5 m) and temporal resolution (1.5 min—4096 pulses). KARL emits light at 1064, 532, and 355 nm (after frequency doubling and tripling) with 200 mJ of power per pulse. The light at 355 and 532 nm is analyzed in two orthogonal polarization planes (perpendicular and parallel with respect to the polarization of the emitted light) and the resulting light components are detected and recorded separately. Additionally, the rotational vibrational Raman lines of nitrogen can be detected (at 387 and 607 nm), allowing for the independent retrieval of the particle backscatter and extinction coefficient. It should be noted that KARL is not in 24-7 operation and its laser beam is not tilted due to its large receiving telescope (diameter of 70 cm). The field of view (FOV) of the telescope is 2.28 mrad. A detailed technical description of KARL can be found in Hoffmann [21].

In order to distinguish layers of ice crystals from those of supercooled liquid-water, a C_{top} temperature threshold (<-37 °C) was used following Campbell et al. [22] and Shupe [16]. The temperature information was obtained from collocated radiosonde ascents. Radiosondes (type Vaisala RS92 before 2018 and RS41 afterwards) are launched daily from AWIPEV or four times per day during dedicated measurement campaigns. The radiosonde manufacturer provides estimates of the combined uncertainties, these being 0.2–0.4 °C for temperature. An additional source of uncertainty, which affects the investigation of cirrus properties' meteorological dependencies, is the horizontal drift of the radiosondes and the temporal discrepancy to the lidar observations. The radiosonde measurements are available via the Pangea data repository (e.g., Maturilli [23]). Furthermore, information on the horizontal wind speed and wind direction were obtained from the radiosondes in order to investigate the dependence of cirrus properties on the meteorological conditions. By choosing this low temperature limit of -37 °C for ice clouds, possible temperature changes due to drift of the radiosondes or time difference between the nearest radiosonde and the lidar observations should not compromise the cloud phase. The relative humidity over ice was calculated from radiosonde measurements using the Goff-Gratch equation. It should be noted that we carefully distinguished cirrus clouds from polar stratospheric clouds, which occur at lower temperatures (below -78 °C) and high altitudes (mainly between 20 and 24 km) [24].

2.2. Retrieval of Cirrus Properties

The retrieval of cirrus geometrical and optical properties was based on a newly developed scheme [20]. The detection scheme extended the Wavelet Covariance Transform (WCT) method [25] by dynamic thresholds (*dynamic WCT*) so as to provide the cirrus C_{base} and C_{top}. The WCT profile is sensitive to signal gradients, which may designate the geometrical boundaries of aerosol [26] or cloud layers [27]. Additionally, the *dynamic WCT* is adaptable to the cloud strength, ambient conditions and lidar specifications. Therefore, the *dynamic WCT* proved sensitive to thin and faint cirrus layers that would have been partly or completely omitted by static thresholds.

The optical characterization scheme (*constrained Klett*) extended the Klett–Fernald retrieval [28,29] by an iterative lidar ratio (LR) selection. The iterative process was constrained by a reference value, which indicated the aerosol concentration beneath the cirrus layers (Section 2.3 and Figure 4 of Nakoudi et al. [20]). Contrary to existing approaches (e.g., double-ended Klett [30]), the aerosol-free assumption was not adopted, but the aerosol conditions were approximated by an initial guess. Thanks to the more realistic reference value, the accuracy of the optical properties was increased. The scheme is described in detail in Nakoudi et al. [20] and it is publicly available [31].

The *constrained Klett* yielded the vertically-constant LR of cirrus layers and subsequently their COD (Equation (4) Nakoudi et al. [20]). Additionally, the linear particle depolarization ratio (LPDR) and color ratio (CR) were derived for each cirrus layer. The LPDR was estimated from the ratio of the particle backscatter coefficients in the perpendicular over the parallel polarization plane (with respect to the polarization of the emitted light). The CR was obtained from the ratio of the total particle backscatter coefficient at two wavelengths (355 over 532 nm). The LPDR (proportional to the particle asphericity) provides qualitative information on the particle shape, while the CR (inversely proportional to the particle radius) offers qualitative information on the particle size.

Overall, the uncertainties in geometrical and optical properties are higher for thin cirrus layers. More specifically, for optically thin cirrus, the inherent uncertainties of the optical retrieval algorithm reached 50% (60–74%) in terms of COD (LR). However, for opaque cirrus COD (LR) uncertainties were lower than 10% (15%). Errors are also introduced due to statistical signal noise of the lidar signals and the procedure of temporal averaging. The aforementioned uncertainties affect the derived particle backscatter coefficient, which is used for the calculation of the LPDR and the CR.

2.3. *Quality Assurance of Optical Properties*

In cirrus layers with COD lower than 0.02, the *constrained Klett* could not perform a robust LR adjustment as the light attenuation was not strong enough to scale the solution [20]. Therefore, in this work only the optical properties of layers with COD higher than the *constrained Klett* sensitivity limit are further analyzed. However, it should be noted that the geometrical properties were reliably derived in both COD regimes.

In lidar systems, where the laser beam is not tilted by some degrees off-nadir, the potential effect of specular reflections should be considered. Specular reflections appear in the presence of horizontally-oriented hexagonal ice crystals and they can be identified by local maxima in the backscatter coefficient together with local minima in the LPDR profiles [32]. According to simulations, the LPDR of horizontally-oriented crystals is lower than 10% (Figures 8b and 9b from Okamoto et al. [33]). Accordingly, we identified LPDR profiles with local minima lower than 10% coincident with local maxima in the backscatter coefficient. The areas suspected of horizontal crystal orientation appeared mostly as thin filaments. Nevertheless, the whole cirrus layers were discarded since a distorted backscatter coefficient profile would impact on the LR and, by extension, on the COD. In total, we screened out 38% of the layers (319 out of 832), indicating a significant presence of hexagonal plate and column ice crystals over Ny-Ålesund.

Additionally, cirrus layers with unexpectedly high LR were screened out. Based on the simulations of Okamoto et al. [33], apparent LR values (before accounting for the multiple-scattering effect) higher than 50 sr are not expected. The exploration of such high values revealed a higher occurrence in spring and for vertically-inhomogeneous layers. The solar elevation is lower in spring compared to summer but the surface albedo is higher as the ground around Ny-Ålesund is still partly snow- or ice-covered [34]. Thus, the background illumination conditions can be relatively high in spring, resulting in lower lidar signal quality. In total, 301 out of 832 cirrus layers passed the quality assurance procedure.

As a final step, the multiple-scattering correction (MSC) was applied using the analytical model of Eloranta [35]. Big particles, such as ice crystals, exhibit a strong forward-scattering peak due to light diffraction. Thereby, the forward-scattered photons remain within the receiving telescope FOV and travel with the laser pulse. As a result, some of the backscattered photons can experience further forward-scattering (one or multiple times) before reaching the lidar detector. The multiple-scattering effect needs to be corrected as the lidar equation assumes that each detected photon originates from a single-scattering event in the atmosphere. The Eloranta model [35] simulates the ratio of photon power from higher order scattering events over single-scattering ones and provides a MSC factor. In this work, the vertically-averaged MSC factor was applied to the LR and COD of each quality assured cirrus layer. The multiple-scattering effect was found equally important for all cirrus regimes in terms of LR. The effect on the COD was higher for the opaque regime. More details are given in Section 2.4 of Nakoudi et al. [20].

2.4. *Measurement Site*

The analysis of this work is focused on Ny-Ålesund, a site in the west coast of Svalbard, between the Greenland and Barents Sea (Figure 1). Ny-Ålesund is embedded in the complex orography of fjord, glaciers, and mountain ranges of up to 800 m altitude. Micrometeorology plays an important role, especially in the lower troposphere [36,37]. The surrounding mountains create a wind channeling effect along the Kongsfjord with SE flow throughout the year and less frequently flow of the opposite NW direction. Above the orography the free troposphere is characterized by a general westerly flow [38]. Moreover, the warm West Spitsbergen current affects the local conditions through diabatic heating [39]. The Svalbard region is the focal point of winter warming with up to 3 °C decadal near-surface temperature increase in the past 20 years [11,40]. The positive temperature and humidity trends extend in the winter-time free troposphere, accompanied by an increasing occurrence of southerly flow [38]. The current conditions over Ny-Ålesund might be representative of a future warming Arctic. However, the Arctic cannot be treated as a uniform environment. The properties of short-lived climate forcing agents depend on several parameters such as the special characteristics of each site in terms of local and synoptic meteorological conditions, aerosol concentration, and proximity to the marginal ice zone.

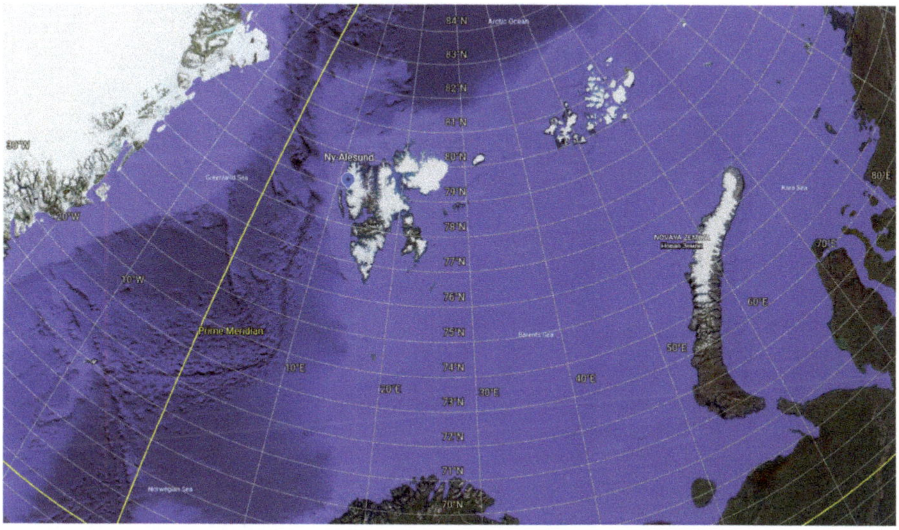

Figure 1. Map of the European Arctic. The site of Ny- Ålesund, which is located on the west coast of the Spitsbergen island, Svalbard archipelago, is indicated on the map. The map was produced with Google Earth.

3. Results

3.1. Cirrus Occurrence and Geometrical Properties

The vertically-resolved occurrence of cirrus (cirrus observations over lidar observation time) is presented on a monthly basis in Figure 2. The cirrus occurrence frequency was higher in winter (3.6 ± 0.7%) and spring (4.1 ± 4.3%) relative to summer (1.6 ± 0.3%) and autumn (1.2 ± 0.8%). This seasonality is in accordance with Nomokonova et al. [17], who reported higher occurrence of ice clouds over Ny-Ålesund in winter and spring using continuous remote sensing observations. The RH over ice (RHice) from the radiosondes revealed fairly constant conditions over seasons in the altitude range 5.5 km to 9.5 km, with mean RHice about 62%. From summer to autumn (July to November), a maximum of 65% was measured. However, the probability of finding moist conditions (defined by RHice > 75%) was highest in October (23%), followed by March (18%). Hence, the RHice from one radiosonde per day is not sufficient to explain the measured cirrus occurrence frequency in the lidar data. It should be noted that the highest occurrence frequency in March is possibly associated with the intensive measurement campaigns that take place at AWIPEV each year. In the rest of the year, the measurements are only performed in the absence of low-level clouds, while in March under a breadth of conditions.

The annual mean occurrence frequency amounted to 2.7 ± 1.8%. The occurrence frequencies reported here are negatively biased as they only include reliably detected layers. More specifically, neither weak or highly opaque cirrus (no C_{top} detection) nor cirrus with directly underlying mid-level clouds or multi-layered cirrus were analyzed. The cirrus occurrence frequencies reported here should not be confused with the total cloud cover, which is approximately 80% over Ny-Ålesund as reported by Yeo et al. [41] andNomokonova et al. [17].

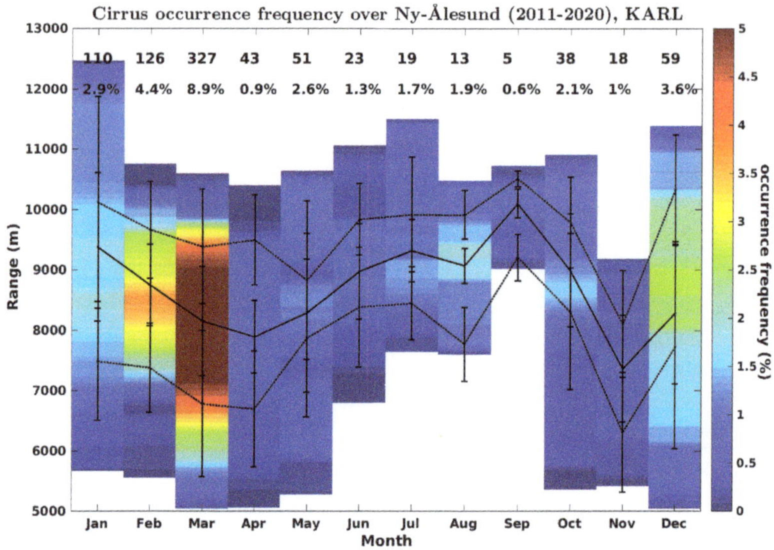

Figure 2. Cirrus monthly occurrence frequency (9 min–7.5 m) cirrus observations normalized to KARL observation hours). The numbers on top denote the detected cirrus layers and the cirrus occurrence frequency. The C_{base}, effective C_{mid} and C_{top} are overlaid (median and standard deviation).

An overview of the cirrus geometrical properties for each season is presented in Figure 3 as well as in Table A1. Thicker layers were observed in winter and spring (Figure 3a). The total GT of cirrus layers amounted to 2 ± 1.2 km and their majority was thinner than 4.5 km. Cirrus layers down to a thickness of 158 m were detected (in July), while the thickest layer extended over 6.3 km (in January). The distribution of GT was broader compared to mid-latitudes. Broader distributions have also been demonstrated from satellite observations during the local winter of the Polar regions [19,42].

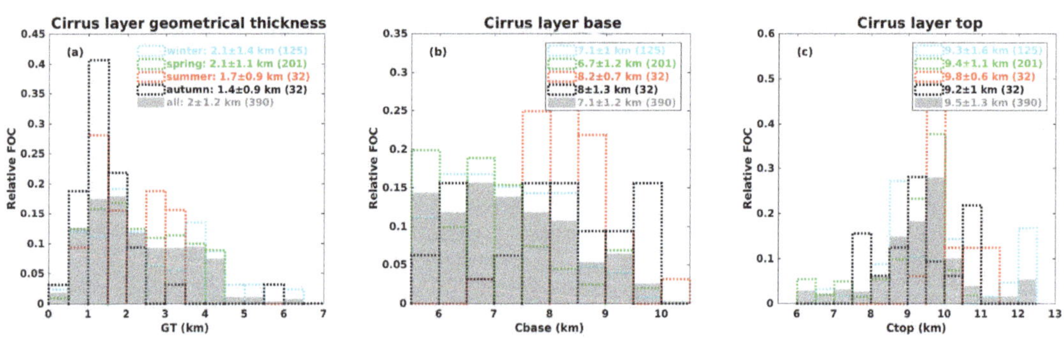

Figure 3. Relative frequency of occurrence (FOC) distribution of geometrical thickness (GT, (**a**)), C_{base} (**b**) and C_{top} (**c**). The median ± std (number of observations) is given in the legend.

The cirrus C_{base} (C_{top}) varied from 5.5 to 8.5 km (from 8.5 to 10.5 km). The C_{base} followed the seasonal cycle of the cirrus temperature regime (below −37 °C, Section 2.1) and appeared lower in winter and spring. The C_{top} was higher during summer, which is in accordance with the seasonality of the tropopause height. However, in December and

January, the C_{top} appeared in comparable or even higher altitudes than summer (Figure 2). This at first sight paradox is related to the frequently poorly-defined polar night tropopause. Maturilli and Kayser [38], using 22 years of radiosonde measurements over Ny-Ålesund, demonstrated that the winter-time thermal tropopause is often indistinguishable due to very low static stability. This regime is connected to radiative cooling in the absence of solar radiation and the governing polar vortex dynamics, while during the sun-lit period a well-defined tropopause regime appears [38]. On a more detailed analysis (not shown), we found that for a well-defined tropopause the relative humidity (RH) declined rapidly to nearly-zero levels due to positive temperature gradients, whereas for a poorly-defined tropopause the RH decrease was smoother in connection with nearly-neutral temperature gradients. A poor separation between the troposphere and the stratosphere is more likely to favor the intrusion of water vapor into the lowermost stratosphere, which, in turn, fosters cloud formation and persistence. Concerning the relation between cirrus occurrence and a poorly-defined tropopause, the following can be stated: the poorly-defined regime is necessary but not sufficient for cirrus to extend within the tropopause as the availability of water vapor is also a crucial factor for cloud occurrence. Cirrus clouds up to 12 km over the Arctic have also been observed by CALIOP (Figure 7 from Devasthale et al. [19]).

3.2. Meteorological Conditions

The meteorological parameters were derived from radiosonde ascents (Section 2.1). For days with multiple launches, each cirrus layer was matched with the temporally-closest radiosonde. Thereby, an overview of the meteorological conditions in the presence of the detected cirrus can be obtained. Moreover, the meteorological conditions throughout the study period (2011–2020) were examined. To this end, all available radiosonde ascents (4691 in total) were used and the measurements within the monthly cirrus envelope heights (as given in Figure 2) were extracted. It should be noted that the detected cirrus as well as those appearing when KARL was switched off are part of the climatological dataset. However, this approach can still assess the conditions in the presence of detected cirrus with respect to the overall meteorological conditions (Figure 4). Cirrus clouds were generally associated with colder and calmer wind conditions compared to the 2011–2020 climatology. However, under SE and NW flow strong winds also seemed conducive for cirrus occurrence.

In order to increase our confidence in the representativeness of radiosonde measurements for the meteorological conditions in the presence of cirrus, a temporal criterion was applied. Limiting the temporal difference between radiosonde and lidar measurements to 2 h, 148 layers remained for analysis and an overview of the meteorological conditions within their altitude range is presented in Figure 5. The median temperature ranged between −65 and −50 °C, while the $T_{C_{base}}$ ($T_{C_{top}}$) varied from −55 to −38 °C (from −70 to −55 °C). A broad range of wind conditions was found (mostly between 1 and 15 ms^{-1}), with an increasing wind speed tendency from the C_{base} towards the C_{top}. A similar tendency was also reported over mid-latitude (Utah [43]) and sub-tropical sites (Wuhan [44]). Concerning wind direction, the majority of cirrus was connected to SW flow, followed by NW and NE contributions. Apparently, the warmer and moister Atlantic air masses can provide the reservoir of humidity for cirrus formation when advected northwards. Hence, the cirrus occurrence frequency may increase over Svalbard if the advection from SW continues to rise as reported by Maturilli and Kayser [38]. Yeo et al. [41] showed that warm and moist air masses from the North Atlantic were also conducive for cloud occurrence below 2 km over Ny-Ålesund. Gierens et al. [45] reported that low-level mixed-phase clouds were most common under westerly winds over Ny-Ålesund. It should be noted, however, that for low-level clouds the role of orography and surface coupling is important. This is the first work that tries to shed light on the relation of cirrus occurrence with temperature and wind conditions over Ny-Ålesund. A more extended and statistically robust analysis is desirable in the future, for example using satellite lidar observations. Finally, a seasonal overview of the meteorological conditions is presented in Table A2. Colder cirrus tended to appear in winter and spring and winter-time cirrus were accompanied by markedly stronger wind conditions. Winter- and spring-time cirrus were associated with W–SW flow, whereas a E–NE flow appeared in summer. The most variable wind conditions appeared in autumn.

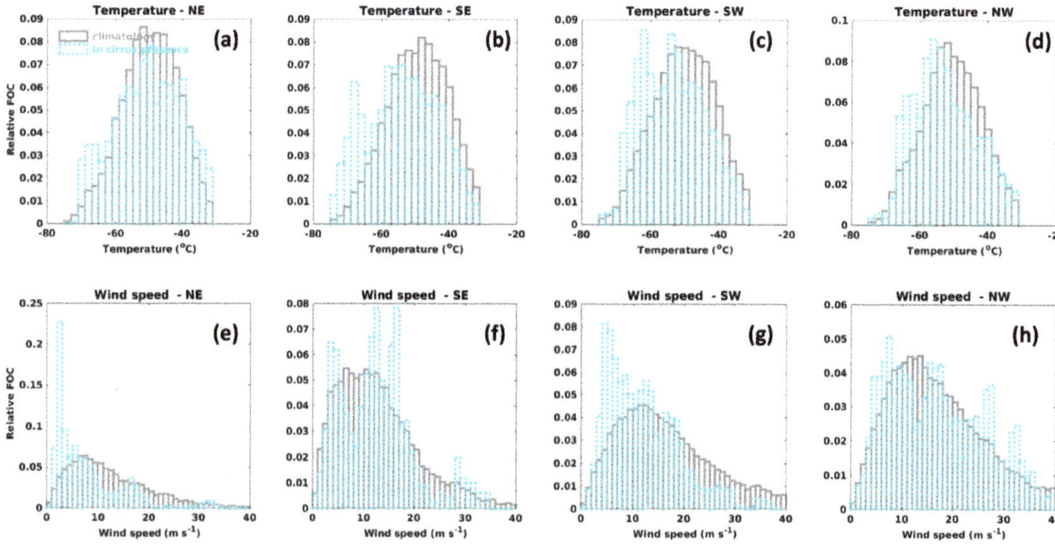

Figure 4. Relative FOC distribution of climatological meteorological conditions (gray) and conditions in the presence of cirrus (cyan). Temperature (**a–d**) and wind speed (**e–h**) within the monthly cirrus envelope (as given in Figure 2) are presented per wind sector, without any temporal restriction between the radiosonde and lidar measurements.

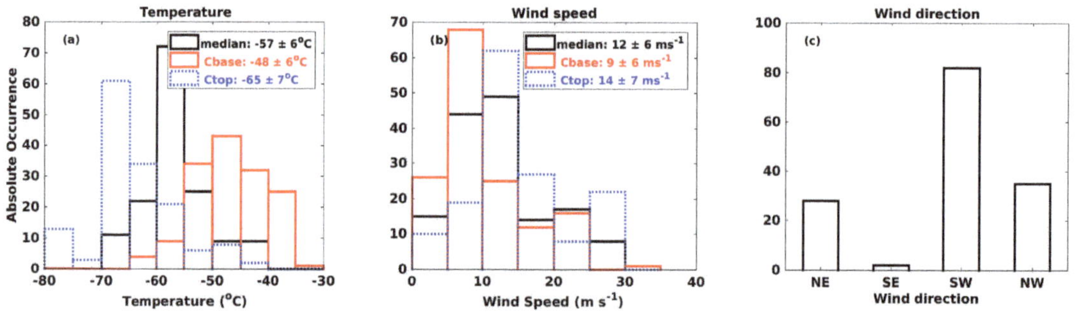

Figure 5. Absolute occurrence distribution of temperature (**a**), wind speed (**b**) and wind direction (**c**) within the cirrus layer altitude range, presented as layer median, at the C_{base} and at the C_{top}. The median ± standard deviation is given in the legend. Only observations satisfying the 2 h temporal criterion are considered.

3.3. Cirrus Optical Properties

An overview of the cirrus layer optical properties is presented in Figure 6 and Table A3. The MSC LR is given along with the layer median LPDR and CR. Data points that passed the quality assurance (Section 2.3) were only taken into account. Therefore, only a subset of the layers shown in Figure 2 are presented here. The LR_{532} was shifted to higher values in comparison to the LR_{355} (Figure 6a,b), indicating that the cirrus layers were more absorbing in the visible spectral region. The LR_{355} (33 ± 9 sr) over Ny-Ålesund was in great agreement with the average LR_{355} (33 ± 7 sr) over the sub-Arctic site of Kuopio [27].

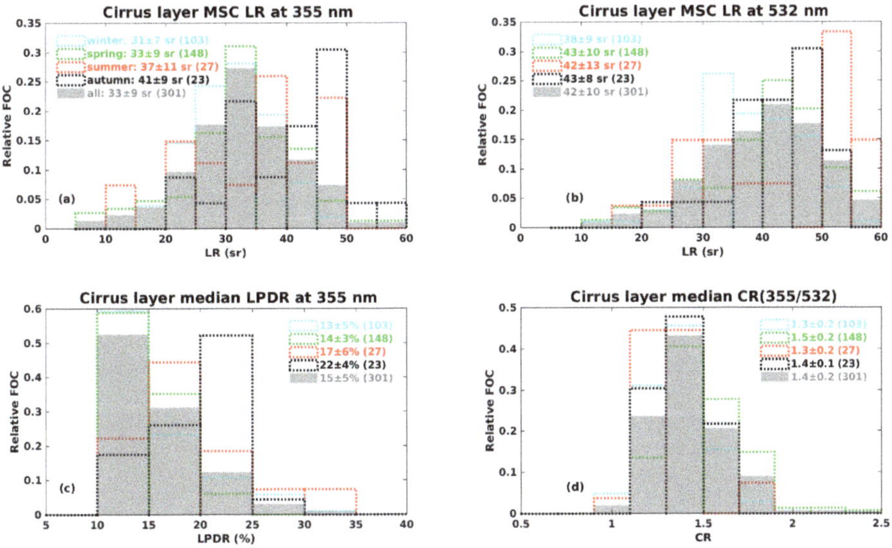

Figure 6. FOC distribution of MSC *LR* (**a**,**b**), LPDR (**c**) and CR (**d**). The median ± std (number of observations) is given in the legend. Only quality assured observations are considered.

The $LPDR_{355}$ mainly varied between 10 and 20% (Figure 6c). The LPDR was higher in summer and autumn. However, compared to other cirrus studies, the presented LPDR values are lower. For instance, over Kuopio an annual mean $LPDR_{355}$ of 38 ± 7% was reported but the analysis was limited to clouds with LPDR higher than 25% [27]. Wang et al. [44] found an annual mean LPDR of 30 ± 9% over Wuhan, China. For cirrus clouds, which originated from heavy air traffic sectors over the Atlantic Ocean and Europe, higher LPDR (47 ± 7%) was reported compared to cirrus from clearer regions (38 ± 6% [46]). The decreasing LPDR tendency for clearer origin cirrus is in the same direction with the lower LPDR over Ny-Ålesund.

Regarding the CR, it mainly varied between 1.2 and 1.6 (Figure 6d). The corresponding Ångström exponent, which is an alternative size indicating property, amounted to 0.8 ± 0.4. Unfortunately, it is hard to find comparable literature for the cirrus CR as various definitions and wavelength pairs are used. Voudouri et al. [27] used the same CR definition and the same wavelength pair (355–532 nm) as here. They reported an average CR of 1.1 ± 0.8 over Kuopio, indicating the presence of slightly bigger ice crystals compared to Ny-Ålesund. Finally, according to the simulations of Okamoto et al. [33] the CR ranged mainly between 0.7 and 2 (values from Figure 9a of Okamoto et al. [33] were inverted to match our definition).

The majority of cirrus observed by lidar over Ny-Ålesund were optically-thin (73%), followed by the sub-visible regime (20%). The domination of optically-thin cirrus was also observed over Kuopio [27] and Fairbanks [14]. However, it should be noted that the derived distribution of this study is likely to be positively biased towards optically-thin cirrus. Concerning sub-visible cirrus, their detection and optical evaluation was the most challenging [20]. Regarding the opaque regime, its underestimation is related to the instrumental limitations of lidar. More specifically, the lidar laser beam can be partly or completely attenuated within highly opaque layers. Sassen and Cho [15] suggested a COD of 3 as an upper lidar attenuation limit for cirrus layers. In this study the maximum COD, which was reliably derived, reached 1.8. Additionally, KARL is usually switched off in the presence of highly opaque clouds because the strong backscattered signal can damage its detectors. Keeping the aforementioned possible bias sources in mind, the mean ± standard deviation (median) MSC COD over Ny-Ålesund was equal to 0.17 ± 0.26 (0.1) at 355 nm.

Similar values were found at 532 nm. In winter the COD over Ny-Ålesund was higher compared to the rest of the year (Table A3). Optically-thicker cirrus layers were observed over Kuopio (COD_{355} = 0.25 ± 0.2 [27]) due to the lower contribution of sub-visible cirrus (only 3%) in that dataset.

3.4. Inter-Relations of Cirrus Properties

In this section ,the cirrus properties are investigated in relation to one another. The relation of LPDR with CR and LR is examined (Figure 7a,b). Individual measurements are presented together with the median ± standard deviation of different CR and LPDR classes. In order to facilitate visual guidance, linear regression lines are overlaid on the median values and only in cases where the coefficient of determination (R^2) exceeded 0.5. However, looking at individual observation points it is evident that the relations of cirrus properties are more complex than linear. The LPDR and CR exhibited an overall proportional relation (Figure 7a), which translates into bigger ice particles (lower CR) being more spherical (lower LPDR). Increased deposition of water vapor is more likely to have occurred on the surface of bigger particles, rendering their shape more spherical. For LPDR below 25%, the LR and LPDR were proportional (Figure 7b), while for higher LPDR an inversely proportional relation was displayed.

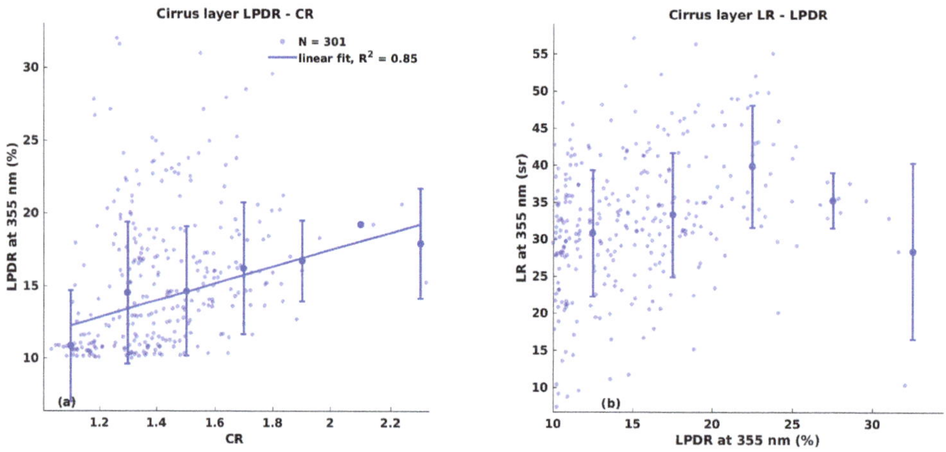

Figure 7. LPDR versus CR (**a**) and LR versus LPDR (**b**). The median ± standard deviation of LPDR (LR) in 0.2 CR (5% LPDR) classes is also given. The linear regression curve on the median values is overlaid only if the coefficient of determination (R^2) exceeded 0.5.

Subsequently, the dependence of COD on LPDR and CR (Figure 8a,b) was investigated. The dependence on LR and GT is not presented as these parameters affect the COD by definition. An inversely proportional relation was indicated between COD and LPDR as well as CR. This means that the COD tended to decrease for smaller (increasing CR) and less spherical ice particles (increasing LPDR), most likely due to reduced water vapor deposition on the ice particle surface. Similar results were obtained for COD_{532} (not shown).

The aforementioned relations were also investigated on a seasonal basis, but no clear difference was found (Figure A1). It should be noted that the distribution of observation points is not uniform across the created LPDR and CR classes, which also limits the statistical robustness of the presented relations. Finally, the relations discussed above may also depend on the ice particle shape, aerosol load, air mass origin, cirrus formation mechanism, relative humidity over ice, vertical wind speed and other parameters. However, such dependencies need to be investigated in the context of case studies, which is out of the scope of this work.

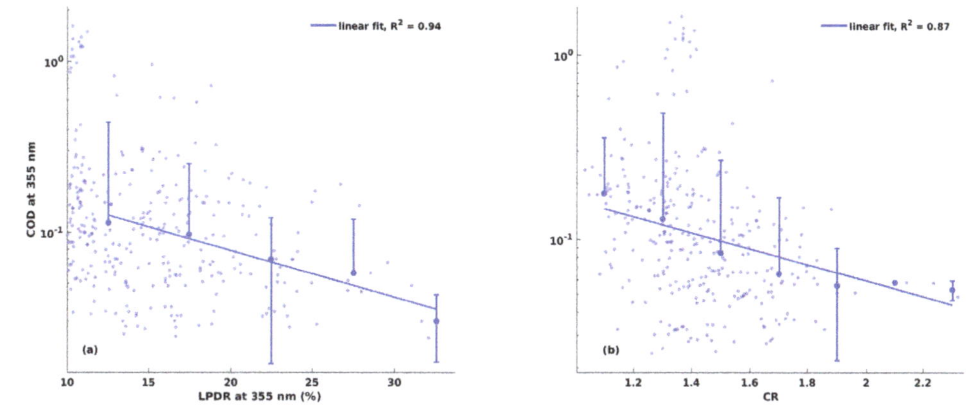

Figure 8. COD versus LPDR (**a**) and CR (**b**). The median ± standard deviation of COD in 5% LPDR and 0.2 CR classes is also given. The linear regression curve on the median values is overlaid only if the coefficient of determination (R^2) exceeded 0.5.

3.5. Meteorological Dependence of Cirrus Properties

The dependence of cirrus geometrical and optical properties on the meteorological conditions is examined here. The dependency of GT on the meteorological conditions is explored in Figure 9. The observations correspond to the high and low COD regimes (above and below the *constrained Klett* sensitivity limit, respectively), with the GT reliably derived in both regimes. In the *high COD regime* (blue), the GT decreased with temperature, which is surprising since layers forming at higher temperatures (lower altitudes) are generally expected to be geometrically-thicker due to their greater distance from the tropopause [44,47]. However, the high GT branch comprised mainly winter and spring cirrus, with those being thicker than 3 km and colder than −50 °C related to the poorly-defined tropopause regime discussed in Section 3.1. A decreasing GT with temperature has also been observed over Kuopio below −50 °C [27]. Only the GT of layers thinner than 2.5 km and colder than −55 °C increased with temperature.

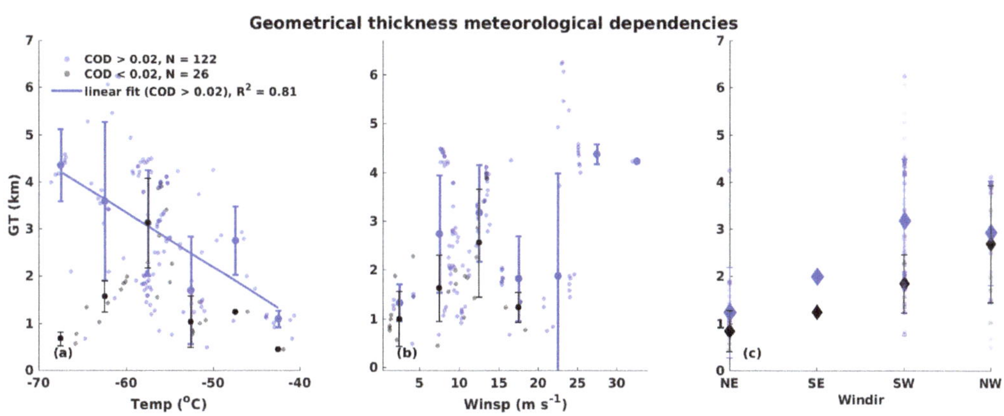

Figure 9. Cirrus layer GT versus layer median temperature (**a**), wind speed (**b**) and wind direction (**c**). Overlaid is the median ± standard deviation of GT per temperature (5 °C), wind speed (5 ms^{-1}) and wind direction classes. Observations corresponding to both COD regimes are shown. Only observations satisfying the 2 h temporal criterion are considered.

Concerning wind speed, a proportionality to GT was observed, especially for wind lower than 15 ms^{-1} (Figure 9b). This can be explained considering that wind speed is a proxy for advection of air masses. Assuming air masses of equal RH advected towards a region with similar ice nucleating particle type, number concentration, and size distribution, a stronger wind regime will result in intensified ice nucleation. During intensified condensation and deposition of water vapor, higher amounts of latent heat are released, which promote ascending air motion within the cirrus and, thus, lead to higher vertical extent. Regarding wind direction, geometrically-thicker cirrus appeared under westerly flow but the different GT classes fell within each other's variability (Figure 9c).

In the following, the dependence of cirrus optical properties on temperature (Figure 10) and wind speed (Figure 11) is investigated. The dependency of COD and LPDR was not clear (Figure 10a,c), similar to Kuopio [27]. However, over lower latitudes the COD–temperature relation was closer to linearly proportional [44,47,48] and the LPDR decreased with temperature [44,49]. Concerning the *LR*, a clear dependence on temperature has not been observed at any site yet [27,44,50], similarly to Ny-Ålesund (Figure 10b). Below −55 °C the CR increased slightly (smaller ice particles) with temperature, while the opposite was observed between −55 and −45 °C (Figure 10d).

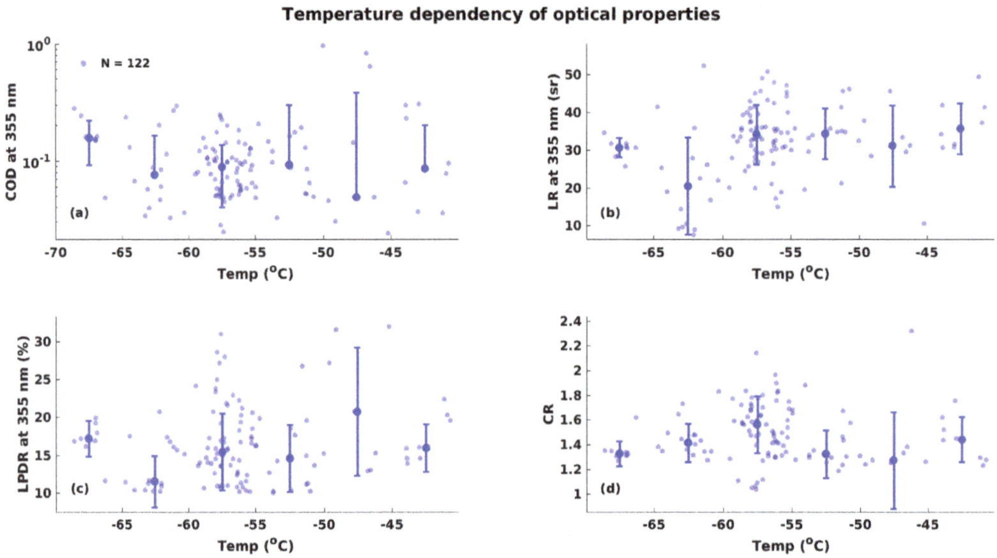

Figure 10. Cirrus layer optical properties versus median temperature. Overlaid are the median ± standard deviation optical properties per temperature (5 °C classes). Only observations satisfying the 2 h temporal criterion are considered.

The relation of COD with wind speed was weakly proportional (Figure 11a), while the *LR* decreased with wind speed (Figure 11b). The respective LPDR and CR relations were not clear. However, more spherical ice particles tended to appear for stronger winds. More specifically, the LPDR decreased (more spherical particles) in the wind regimes of 1–15 ms^{-1} and 15–35 ms^{-1} (Figure 11c,d). Stronger wind could be a proxy of intensified water vapor deposition on the ice particles, which increases their sphericity and the cirrus layer COD. Finally, the optical properties were similar for different wind sectors (not shown). It should be noted that the number of cloud observations is still small in order to draw statistically robust conclusions. However, the presented results can be a useful benchmark for relevant future studies.

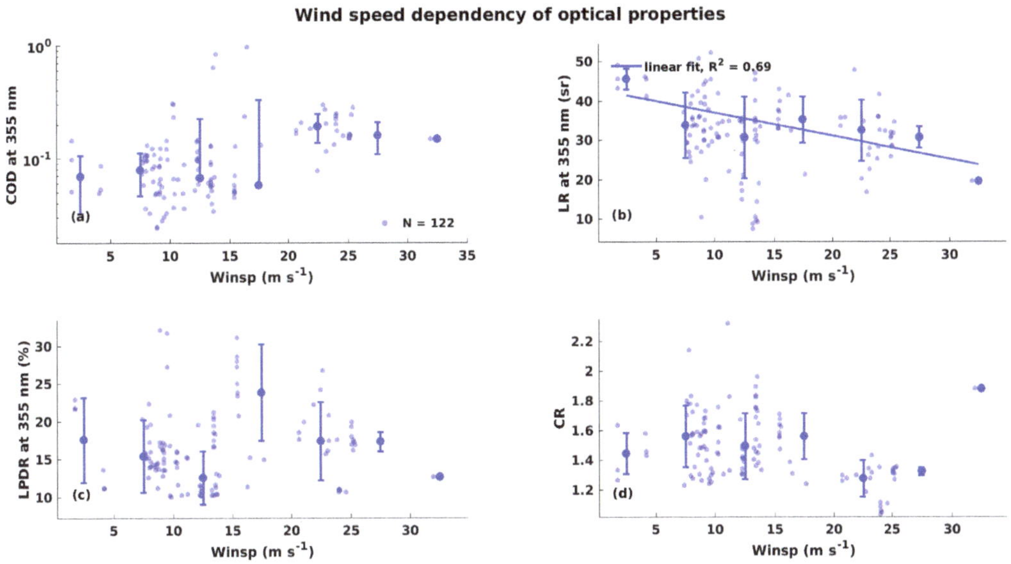

Figure 11. Same as Figure 10, except for dependence on wind speed (5 ms^{-1} classes). Only observations satisfying the 2 h temporal criterion are considered.

4. Discussion

The occurrence of winter-time cirrus clouds at high altitudes might have implications for the CRE. In the era of *Arctic Amplification*, an increasing thermal contrast between the warming surface and the higher and, thus, colder cirrus may drive a positive feedback process [14]. Cirrus clouds will receive more long-wave irradiance from the warmer surface but they will re-emit less irradiance towards the TOA, due to their lower temperature, and, thereby, intensify the positive CRE. By performing simplified calculations of the TOA CRE using the Corti–Peter model [51], we found a positive effect throughout the year. The CRE was maximum in winter (+9.4 Wm^{-2}) and minimum in summer (+2.5 Wm^{-2}) due to the higher contribution of the SW cooling term in the latter case. The overall TOA CRE over Ny-Ålesund is also positive in spring, autumn and winter (approximately +10 Wm^{-2} per month) but gets negative (varying from −180 to −10 Wm^{-2}) in the summer [10]. Thus, cirrus are expected to contribute to the warming CRE signal in spring, autumn, and winter but to counterbalance the summer-time cooling signal. A further CRE sensitivity analysis was performed with the COD, surface albedo, T_{Ctop} and surface temperature perturbed within representative ranges for the site of Ny-Ålesund. A predominantly positive effect was revealed and the effect turned negative only for highly opaque cirrus (COD higher than 10) and over tundra. Thus, over the high European Arctic, the COD along with the surface albedo seem to be the most critical parameters in determining the cirrus CRE sign at TOA. It should be noted that the discussed CRE are first order estimates. Comprehensive radiative transfer simulations are necessary to accurately estimate the cirrus contribution to the local radiative budget. However, this is a task for a dedicated future study.

The cirrus clouds observed over Ny-Ålesund seem colder compared to those over sub-Arctic sites. Over Fairbanks the $T_{C_{top}}$ rarely dropped below −60 °C. Over Kuopio the temperature was −50 ± 10 °C (for C_{mid} of 9.2 km), while over Ny-Ålesund the median layer temperature amounted to −57 ± 6 °C (for C_{mid} of 8.3 km). It should be noted, however, that the analysis for Fairbanks and Kuopio a priori considered warmer cirrus, as the $T_{C_{base}}$ threshold was set to −27 °C (for Ny-Ålesund this was set at −40 °C). Compared to tropical sites, the Ny-Ålesund cirrus tended to be slightly warmer due to their lower occurrence altitudes, constrained by the meridionally sloping tropopause. For instance,

$T_{C_{mid}}$ of -65 ± 12 °C was reported at 14.1 ± 2 km over tropical India [52]. During aircraft campaigns over different regions, more than half of the cirrus clouds were observed above −48 °C, except for the Tropics [53]. Therefore, the presented cirrus properties can be considered as more representative for cold cirrus. In the present study, the dependence of COD and LPDR on temperature seems to be weaker compared to tropical [49] and sub-tropical studies [44].

The LR_{532} over Ny-Ålesund seems to follow the latitudinally increasing LR tendency, which is reported by Voudouri et al. [27] (their Figure 7c). Nevertheless, an asymptotic behavior was found for LR_{355}. A poleward increase of the LR has been reported over the Northern Hemisphere for opaque ice clouds (Young et al. [54], using CALIOP observations). More specifically, the LR_{532} (following an increasing gradient of centroid temperature) started from approximately 25 sr over the Tropics and reached about 40 sr over the Arctic, with higher values in its Canadian and Russian parts (Young et al. [54], their Figure 7). However, a stable LR_{532} of 33 ± 5 sr has been derived over the 50° S–50° N ocean zone for semi-transparent cirrus (Josset et al. [55], using CALIOP–CloudSat observations).

As cirrus layers suspected of horizontal crystal orientation were screened out (Section 2.3), an attempt will be made to interpret the lower LPDR values compared to lower latitudes. Low LPDR values similar to Ny-Ålesund have been observed over Antarctica, Dumont d'Urville [56] and Sondakyla, north Finland [57]. The $LPDR_{532}$ was quite constant (13–23%, Figure 7 from Del Guasta [57]) as attributed to small and possibly columnar ice crystals, after excluding the presence of horizontal crystals and super-cooled droplets. Sassen and Zhu [58] investigated the geographical distribution of linear volume depolarization ratio at 532 nm in ice clouds (using off-nadir CALIOP observations). Although the present findings cannot be quantitatively compared to Sassen and Zhu [58], the latitudinal depolarization changes are worth mentioning. Sassen and Zhu [58] highlighted a depolarization decrease towards the high latitudes of both hemispheres, attributing the differences to possible changes in the ice nucleation mechanisms. Over the high latitudes, ice clouds occur at lower heights and, thus, the availability and properties of aerosol that serve as ice nucleating particles may differ [58].

Cirrus layers were carefully distinguished from aerosol layers in the upper troposphere—lower stratosphere region. Noteworthy are the Siberian fires' smoke layers of summer 2019. These layers were observed from 7–8 km up to 17–18 km between August 2019 and May 2020 over Ny-Ålesund as well as the central Arctic [59]. Thus, they partly overlapped with the cirrus occurrence altitude range. However, the optical properties of the smoke layers were different to those expected for cirrus clouds. The smoke layers consisted of nearly-spherical particles (LPDR lower than 4% both at 355 and 532 nm) that were in general more absorbing (LR355 = 55 sr and LR532 = 85 sr) than ice particles [59].

5. Conclusions

In this work, the long-term properties of cirrus clouds were analyzed for the first time over an Arctic site (Ny-Ålesund, Svalbard), using lidar and radiosonde observations from 2011 to 2020. The inter-relations of properties were examined together with their dependence on meteorological conditions. The main findings can be summarized as follows:

- In winter and spring cirrus layers were geometrically-thicker, they occurred more frequently and their C_{base} appeared lower compared to the rest of the year (Figure 2). The tropopause thermodynamic structure seemed to control the seasonality of cirrus C_{top} altitude.
- Cirrus clouds were generally associated with colder and calmer wind conditions compared to the 2011–2020 climatology. Their majority was associated with westerly flow and westerly cirrus tended to be geometrically-thicker.
- The lidar ratio (LR) in the ultraviolet spectral region (33 ± 9 sr) was in good agreement with a sub-Arctic study [27], while the LR in the visible region (42 ± 10 sr) was higher compared to lower latitudes. In winter, optically-thicker cirrus of lower LR were

observed compared to the rest of the year (Figure 6 and Table A3). In summer and autumn the linear particle depolarization ratio (LPDR) was higher, indicating less spherical ice particles (Figure 6c). Overall, more spherical particles were inferred compared to lower latitudes.

- Overall, the LPDR and color ratio (CR) were found to be proportional (Figure 7a), reflecting a tendency of bigger ice particles being more spherical. Increased deposition of water vapor had most likely occurred on the surface of bigger particles, rendering their shape more spherical. Conversely, the cloud optical depth (COD) tended to be lower for less spherical and smaller ice particles (Figure 8) most likely due to reduced deposition of water vapor on the particle surface. The aforementioned relations were also investigated on a seasonal basis but no clear difference was found (Figure A1). These relations need to be investigated with respect to ice particle shape, aerosol load, air mass origin, cirrus formation mechanism, relative humidity over ice, vertical wind speed, and other parameters in order to gain a better understanding of the underlying physical processes. Such processes need to be investigated in the context of future case studies.
- The overall dependency of cirrus optical properties on temperature and wind speed was not pronounced (Figures 10 and 11). However, the properties of winter-time cirrus were distinctive, accompanied by colder conditions and stronger winds (Table A2).

Over the 10 years of the analysis, the cirrus properties have not exhibited any clear temporal trend. Micro-pulse lidar measurements might be more suitable for detecting trends of cirrus occurrence and properties as they are conducted on a continuous basis. The lidar-derived cirrus properties can be compared to those from the cloud radar, which is installed at AWIPEV. More specifically, the geometrical boundary discrepancies between lidar and radar can be examined as well as discrepancies in the extinction coefficient.

This work aimed at providing a first-time overview of cirrus geometrical and optical properties over a site in the European part of the Arctic. However, there is still a need for more Arctic cirrus investigations, which can be based on satellite observations from CALIOP aboard CALIPSO or the upcoming ATLID aboard EarthCare [60]. Furthermore, similar studies to ours from other sites, in the American and Russian part of the Arctic, are necessary in order to assess possible intra-Arctic differences in cirrus properties.

Author Contributions: Conceptualization, C.R. and I.S.S.; data curation, K.N.; formal analysis, K.N.; funding acquisition, C.R. and I.S.S.; investigation, K.N.; methodology, K.N., C.R. and I.S.S.; resources, C.R.; software, K.N.; supervision, C.R. and I.S.S.; validation, I.S.S.; visualization, K.N.; writing—original draft, K.N.; writing—review & editing, K.N., C.R. and I.S.S. All authors have read and agreed to the published version of the manuscript.

Funding: This work was funded by the Deutsche Forschungsgemeinschaft (DFG, German Research Foundation)–Project-ID 268020496–TRR 172.

Data Availability Statement: The dataset related to this publication has been submitted to the PANGEA data repository and is expected to appear online soon.

Acknowledgments: This work was partly performed during the research stay of K. Nakoudi in the Faculty of Physics, University of Warsaw, funded by the Helmholtz Graduate School for Polar and Marine Research (POLMAR). We acknowledge the support of KARL by Wilfried Ruhe and Ingo Beninga at Impres GmbH. We gratefully acknowledge the funding by the Deutsche Forschungsgemeinschaft (DFG, German Research Foundation)—Project Number 268020496—TRR 172, within the Transregional Collaborative Research Center—ArctiC Amplification: Climate Relevant Atmospheric and SurfaCe Processes, and Feedback Mechanisms $(AC)^3$.

Conflicts of Interest: The authors declare no conflict of interest.

Appendix A. Seasonal Cirrus Properties

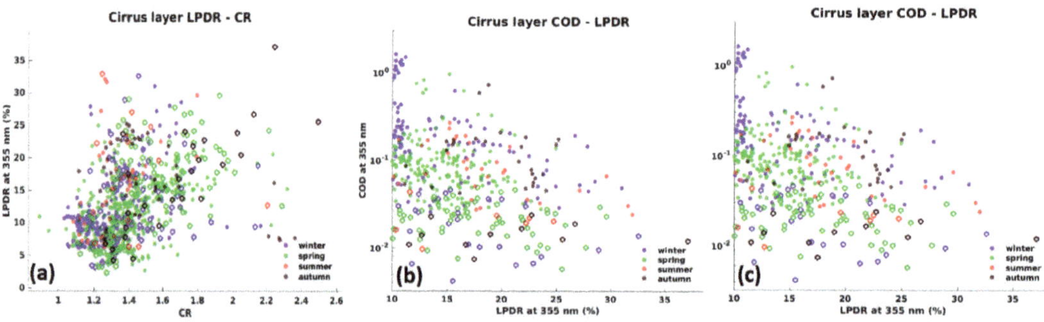

Figure A1. Inter-relation of cirrus optical properties on a seasonal basis. Cirrus layer LPDR-CR (**a**), cirrus layer COD-LPDR (**b**) and cirrus layer COD-LPDR (**c**).

The geometrical, optical, and meteorological cirrus properties over Ny-Ålesund are summarized in the following tables. The total along with the seasonal properties are given as median ± standard deviation unless stated otherwise.

Table A1. Total and seasonal descriptive statistics of cirrus layer geometrical properties, including both *COD regimes*.

	Total	Winter	Spring	Summer	Autumn
GT (km)	2 ± 1.2	2.1 ± 1.4	2.1 ± 1.1	1.7 ± 0.9	1.4 ± 0.9
C_{base} (km)	7.1 ± 1.2	7.1 ± 1	6.7 ± 1.2	8.2 ± 0.7	8 ± 1.3
C_{top} (km)	9.5 ± 1.3	9.3 ± 1.6	9.4 ± 1.1	9.8 ± 0.6	9.2 ± 1

Table A2. Total and seasonal descriptive statistics of cirrus layer meteorological properties, including both *COD regimes*.

	Total	Winter	Spring	Summer	Autumn
Median temperature (°C)	-57 ± 6	-61 ± 5	-56 ± 5	-44 ± 4	-53 ± 3
Temperature at C_{base} (°C)	-48 ± 6	-51 ± 7	-47 ± 5	-38 ± 3	-48 ± 4
Temperature at C_{top} (°C)	-65 ± 7	-66 ± 6	-66 ± 6	-49 ± 5	-57 ± 3
Median wind speed (ms^{-1})	12 ± 6	23 ± 5	10 ± 4	9 ± 1	4 ± 9
Wind speed at C_{base} (ms^{-1})	9 ± 6	18 ± 6	8 ± 4	9 ± 1	5 ± 8
Wind speed at C_{top} (ms^{-1})	14 ± 7	28 ± 6	13 ± 4	9 ± 1	10 ± 9
Median wind direction (°)	238 ± 85	243 ± 87	242 ± 77	57 ± 88	108 ± 216
Median wind sector	W/SW	W/SW	W/SW	E/NE	E/SE

Table A3. Total and seasonal descriptive statistics of cirrus layer optical properties for the *high COD regime*.

	Total	Winter	Spring	Summer	Autumn
COD_{355}	0.1 ± 0.26	0.16 ± 0.39	0.09 ± 0.13	0.08 ± 0.07	0.1 ± 0.17
Mean COD_{355}	0.17	0.3	0.11	0.11	0.16
LR_{355} (sr)	33 ± 9	31 ± 7	33 ± 9	37 ± 11	41 ± 9
LR_{532} (sr)	42 ± 10	38 ± 9	43 ± 10	42 ± 13	43 ± 8
$LPDR_{355}$ (%)	15 ± 5	13 ± 5	14 ± 3	17 ± 6	22 ± 4
CR	1.4 ± 0.2	1.3 ± 0.2	1.5 ± 0.2	1.3 ± 0.2	1.4 ± 0.1

References

1. Serreze, M.C.; Barry, R.G. Processes and impacts of Arctic amplification: A research synthesis. *Glob. Planet. Chang.* **2011**, *77*, 85–96. [CrossRef]
2. Wendisch, M.; Macke, A.; Ehrlich, A.; Lüpkes, C.; Mech, M.; Chechin, D.; Dethloff, K.; Velasco, C.B.; Bozem, H.; Brückner, M.; et al. The Arctic cloud puzzle: Using ACLOUD/PASCAL multiplatform observations to unravel the role of clouds and aerosol particles in Arctic amplification. *Bull. Am. Meteorol. Soc.* **2019**, *100*, 841–871. [CrossRef]
3. Huang, Y.; Dong, X.; Bailey, D.A.; Holland, M.M.; Xi, B.; DuVivier, A.K.; Kay, J.E.; Landrum, L.L.; Deng, Y. Thicker clouds and accelerated Arctic sea ice decline: The atmosphere-sea ice interactions in spring. *Geophys. Res. Lett.* **2019**, *46*, 6980–6989. [CrossRef]
4. Hong, Y.; Liu, G.; Li, J.L. Assessing the radiative effects of global ice clouds based on CloudSat and CALIPSO measurements. *J. Clim.* **2016**, *29*, 7651–7674. [CrossRef]
5. Lolli, S.; Campbell, J.R.; Lewis, J.R.; Gu, Y.; Marquis, J.W.; Chew, B.N.; Liew, S.C.; Salinas, S.V.; Welton, E.J. Daytime top-of-the-atmosphere cirrus cloud radiative forcing properties at Singapore. *J. Appl. Meteorol. Climatol.* **2017**, *56*, 1249–1257. [CrossRef]
6. Kienast-Sjögren, E.; Rolf, C.; Seifert, P.; Krieger, U.K.; Luo, B.P.; Krämer, M.; Peter, T. Climatological and radiative properties of midlatitude cirrus clouds derived by automatic evaluation of lidar measurements. *Atmos. Chem. Phys.* **2016**, *16*, 7605–7621. [CrossRef]
7. Kraemer, M.; Rolf, C.; Luebke, A.; Afchine, A.; Spelten, N.; Costa, A.; Meyer, J.; Zoeger, M.; Smith, J.; Herman, R.L.; et al. A microphysics guide to cirrus clouds—Part 1: Cirrus types. *Atmos. Chem. Phys.* **2016**, *16*, 3463–3483. [CrossRef]
8. Heymsfield, A.J.; Krämer, M.; Luebke, A.; Brown, P.; Cziczo, D.J.; Franklin, C.; Lawson, P.; Lohmann, U.; McFarquhar, G.; Ulanowski, Z.; et al. Cirrus clouds. *Meteorol. Monogr.* **2017**, *58*, 2.1–2.26. [CrossRef]
9. Gultepe, I.; Starr, D.O. Dynamical structure and turbulence in cirrus clouds: Aircraft observations during FIRE. *J. Atmos. Sci.* **1995**, *52*, 4159–4182. [CrossRef]
10. Ebell, K.; Nomokonova, T.; Maturilli, M.; Ritter, C. Radiative effect of clouds at Ny-Ålesund, Svalbard, as inferred from ground-based remote sensing observations. *J. Appl. Meteorol. Climatol.* **2020**, *59*, 3–22. [CrossRef]
11. Dahlke, S.; Maturilli, M. Contribution of atmospheric advection to the amplified winter warming in the Arctic North Atlantic region. *Adv. Meteorol.* **2017**, *2017*, 4928620. [CrossRef]
12. Holz, R.E.; Platnick, S.; Meyer, K.; Vaughan, M.; Heidinger, A.; Yang, P.; Wind, G.; Dutcher, S.; Ackerman, S.; Amarasinghe, N.; et al. Resolving ice cloud optical thickness biases between CALIOP and MODIS using infrared retrievals. *Atmos. Chem. Phys.* **2016**, *16*, 5075–5090. [CrossRef]
13. Marquis, J.W.; Bogdanoff, A.S.; Campbell, J.R.; Cummings, J.A.; Westphal, D.L.; Smith, N.J.; Zhang, J. Estimating infrared radiometric satellite sea surface temperature retrieval cold biases in the tropics due to unscreened optically thin cirrus clouds. *J. Atmos. Ocean. Technol.* **2017**, *34*, 355–373. [CrossRef]
14. Campbell, J.R.; Dolinar, E.K.; Lolli, S.; Fochesatto, G.J.; Gu, Y.; Lewis, J.R.; Marquis, J.W.; McHardy, T.M.; Ryglicki, D.R.; Welton, E.J. Cirrus cloud top-of-the-atmosphere net daytime forcing in the Alaskan subarctic from ground-based MPLNET monitoring. *J. Appl. Meteorol. Climatol.* **2021**, *60*, 51–63. [CrossRef]
15. Sassen, K.; Cho, B.S. Subvisual-thin cirrus lidar dataset for satellite verification and climatological research. *J. Appl. Meteorol.* **1992**, *31*, 1275–1285. [CrossRef]
16. Shupe, M.D. Clouds at Arctic atmospheric observatories. Part II: Thermodynamic phase characteristics. *J. Appl. Meteorol. Climatol.* **2011**, *50*, 645–661. [CrossRef]
17. Nomokonova, T.; Ebell, K.; Löhnert, U.; Maturilli, M.; Ritter, C.; O'Connor, E. Statistics on clouds and their relation to thermodynamic conditions at Ny-Ålesund using ground-based sensor synergy. *Atmo. Chem. Phys.* **2019**, *19*, 4105–4126. [CrossRef]
18. Winker, D.M.; Vaughan, M.A.; Omar, A.; Hu, Y.; Powell, K.A.; Liu, Z.; Hunt, W.H.; Young, S.A. Overview of the CALIPSO Mission and CALIOP Data Processing Algorithms. *J. Atmos. Ocean. Technol.* **2009**, *26*, 2310–2323. [CrossRef]
19. Devasthale, A.; Tjernstrã-M, M.; Karlsson, K.G.R.; Thomas, M.A.; Jones, C.; Sedlar, J.; Omar, A.H. The vertical distribution of thin features over the Arctic analysed from CALIPSO observations. *Tellus Ser. B Chem. Phys. Meteorol. B* **2011**, *63*, 77–85. [CrossRef]
20. Nakoudi, K.; Stachlewska, I.S.; Ritter, C. An extended lidar-based cirrus cloud retrieval scheme: first application over an Arctic site. *Opt. Express* **2021**, *29*, 8553–8580. [CrossRef]
21. Hoffmann, A. Comparative Aerosol Studies Based on Multi-Wavelength Raman LIDAR at Ny-Ålesund, Spitsbergen. Ph.D Thesis, Universität Potsdam, Potsdam, Germany, 2011.
22. Campbell, J.; Vaughan, M.; Oo, M.; Holz, R.; Lewis, J.; Welton, E. Distinguishing cirrus cloud presence in autonomous lidar measurements. *Atmos. Meas. Tech.* **2015**, *8*, 435–449. [CrossRef]
23. Maturilli, M. High resolution radiosonde measurements from station Ny-Ålesund (2017-04 et seq). *PANGAEA* **2020**. [CrossRef]
24. Massoli, P.; Maturilli, M.; Neuber, R. Climatology of Arctic polar stratospheric clouds as measured by lidar in Ny-Ålesund, Spitsbergen (79° N, 12° E). *J. Geophys. Res. Atmos.* **2006**, *111*. [CrossRef]
25. Gamage, N.; Hagelberg, C. Detection and analysis of microfronts and associated coherent events using localized transforms. *J. Atmos. Sci.* **1993**, *50*, 750–756. [CrossRef]
26. Nakoudi, K.; Giannakaki, E.; Dandou, A.; Tombrou, M.; Komppula, M. Planetary boundary layer height by means of lidar and numerical simulations over New Delhi, India. *Atmos. Meas. Tech.* **2019**, *12*, 2595–2610. [CrossRef]

27. Voudouri, K.A.; Giannakaki, E.; Komppula, M.; Balis, D. Variability in cirrus cloud properties using a Polly^{XT} Raman lidar over high and tropical latitudes. *Atmos. Chem. Phys.* **2020**, *20*, 4427–4444. [CrossRef]
28. Klett, J.D. Stable analytical inversion solution for processing lidar returns. *Appl. Opt.* **1981**, *20*, 211–220. [CrossRef]
29. Fernald, F.G. Analysis of atmospheric lidar observations: some comments. *Appl. Opt.* **1984**, *23*, 652–653. [CrossRef]
30. Ansmann, A.; Wandinger, U.; Riebesell, M.; Weitkamp, C.; Michaelis, W. Independent measurement of extinction and backscatter profiles in cirrus clouds by using a combined Raman elastic-backscatter lidar. *Appl. Opt.* **1992**, *31*, 7113–7131. [CrossRef]
31. Nakoudi, K.; Ritter, C. *AWI Cirrus Cloud Retrieval Scheme (v1.0.0)*; Zenodo: Geneve, Switzerland, 2020. [CrossRef]
32. Borovoi, A.; Balin, Y.; Kokhanenko, G.; Penner, I.; Konoshonkin, A.; Kustova, N. Layers of quasi-horizontally oriented ice crystals in cirrus clouds observed by a two-wavelength polarization lidar. *Opt. Express* **2014**, *22*, 24566–24573. [CrossRef]
33. Okamoto, H.; Sato, K.; Borovoi, A.; Ishimoto, H.; Masuda, K.; Konoshonkin, A.; Kustova, N. Wavelength dependence of ice cloud backscatter properties for space-borne polarization lidar applications. *Opt. Express* **2020**, *28*, 29178–29191. [CrossRef]
34. Maturilli, M.; Herber, A.; König-Langlo, G. Surface radiation climatology for Ny-Ålesund, Svalbard (78.9 N), basic observations for trend detection. *Theor. Appl. Climatol.* **2015**, *120*, 331–339. [CrossRef]
35. Eloranta, E. Practical model for the calculation of multiply scattered lidar returns. *Appl. Opt.* **1998**, *37*, 2464–2472. [CrossRef]
36. Jocher, G.; Schulz, A.; Ritter, C.; Neuber, R.; Dethloff, K.; Foken, T. The sensible heat flux in the course of the year at Ny-Ålesund, svalbard: Characteristics of eddy covariance data and corresponding model results. *Adv. Meteorol.* **2015**, *2015*, 852108. [CrossRef]
37. Rader, F.; Traversi, R.; Severi, M.; Becagli, S.; Müller, K.J.; Nakoudi, K.; Ritter, C. Overview of Aerosol Properties in the European Arctic in Spring 2019 Based on In Situ Measurements and Lidar Data. *Atmosphere* **2021**, *12*, 271. [CrossRef]
38. Maturilli, M.; Kayser, M. Arctic warming, moisture increase and circulation changes observed in the Ny-Ålesund homogenized radiosonde record. *Theor. Appl. Climatol.* **2017**, *130*, 1–17. [CrossRef]
39. Walczowski, W.; Piechura, J. Influence of the West Spitsbergen Current on the local climate. *Int. J. Climatol.* **2011**, *31*, 1088–1093. [CrossRef]
40. Gjelten, H.M.; Nordli, Ø.; Isaksen, K.; Førland, E.J.; Sviashchennikov, P.N.; Wyszynski, P.; Prokhorova, U.V.; Przybylak, R.; Ivanov, B.V.; Urazgildeeva, A.V. Air temperature variations and gradients along the coast and fjords of western Spitsbergen. *Polar Res.* **2016**, *35*, 29878. [CrossRef]
41. Yeo, H.; Park, S.J.; Kim, B.M.; Shiobara, M.; Kim, S.W.; Kwon, H.; Kim, J.H.; Jeong, J.H.; Park, S.S.; Choi, T. The observed relationship of cloud to surface longwave radiation and air temperature at Ny-Ålesund, Svalbard. *Tellus B Chem. Phys. Meteorol.* **2018**, *70*, 1–10. [CrossRef]
42. Nazaryan, H.; McCormick, M.P.; Menzel, W.P. Global characterization of cirrus clouds using CALIPSO data. *J. Geophys. Res. Atmos.* **2008**, *113*. [CrossRef]
43. Sassen, K.; Campbell, J.R. A midlatitude cirrus cloud climatology from the Facility for Atmospheric Remote Sensing. Part I: Macrophysical and synoptic properties. *J. Atmos. Sci.* **2001**, *58*, 481–496. [CrossRef]
44. Wang, W.; Yi, F.; Liu, F.; Zhang, Y.; Yu, C.; Yin, Z. Characteristics and Seasonal Variations of Cirrus Clouds from Polarization Lidar Observations at a 30° N Plain Site. *Remote Sens.* **2020**, *12*, 3998. [CrossRef]
45. Gierens, R.; Kneifel, S.; Shupe, M.D.; Ebell, K.; Maturilli, M.; Löhnert, U. Low-level mixed-phase clouds in a complex Arctic environment. *Atmos. Chem. Phys.* **2020**, *20*, 3459–3481. [CrossRef]
46. Urbanek, B.; Groß, S.; Wirth, M.; Rolf, C.; Krämer, M.; Voigt, C. High depolarization ratios of naturally occurring cirrus clouds near air traffic regions over Europe. *Geophys. Res. Lett.* **2018**, *45*, 13–166. [CrossRef]
47. Sassen, K.; Comstock, J.M. A Midlatitude Cirrus Cloud Climatology from the Facility for Atmospheric Remote Sensing. Part III: Radiative Properties. *J. Atmos. Sci.* **2001**, *58*, 2113–2127. [CrossRef]
48. Wang, Z.; Sassen, K. Cirrus Cloud Microphysical Property Retrieval Using Lidar and Radar Measurements. Part II: Midlatitude Cirrus Microphysical and Radiative Properties. *J. Atmos. Sci.* **2002**, *59*, 2291–2302. [CrossRef]
49. Sunilkumar, S.V.; Parameswaran, K. Temperature dependence of tropical cirrus properties and radiative effects. *J. Geophys. Res. Atmos.* **2005**, *110*, doi;10.1029/2004JD005426. [CrossRef]
50. Gouveia, D.A.; Barja, B.; Barbosa, H.M.; Seifert, P.; Baars, H.; Pauliquevis, T.; Artaxo, P. Optical and geometrical properties of cirrus clouds in Amazonia derived from 1 year of ground-based lidar measurements. *Atmos. Chem. Phys.* **2017**, *17*, 3619–3636. [CrossRef]
51. Corti, T.; Peter, T. A simple model for cloud radiative forcing. *Atmos. Chem. Phys.* **2009**, *9*, 5751–5758. [CrossRef]
52. Pandit, A.K.; Gadhavi, H.; Venkat Ratnam, M.; Raghunath, K.; Rao, S.; Jayaraman, A. Long-term trend analysis and climatology of tropical cirrus clouds using 16 years of lidar data set over Southern India. *Atmos. Chem. Phys.* **2015**, *15*, 13833–13848. [CrossRef]
53. Krämer, M.; Rolf, C.; Spelten, N.; Afchine, A.; Fahey, D.; Jensen, E.; Khaykin, S.; Kuhn, T.; Lawson, P.; Lykov, A.; et al. A microphysics guide to cirrus—Part 2: Climatologies of clouds and humidity from observations. *Atmos. Chem. Phys.* **2020**, *20*, 12569–12608. [CrossRef]
54. Young, S.A.; Vaughan, M.A.; Garnier, A.; Tackett, J.L.; Lambeth, J.D.; Powell, K.A. Extinction and optical depth retrievals for CALIPSO's Version 4 data release. *Atmos. Meas. Tech.* **2018**, *11*, 5701–5727. [CrossRef]
55. Josset, D.; Pelon, J.; Garnier, A.; Hu, Y.; Vaughan, M.; Zhai, P.W.; Kuehn, R.; Lucker, P. Cirrus optical depth and lidar ratio retrieval from combined CALIPSO-CloudSat observations using ocean surface echo. *J. Geophys. Res. Atmos.* **2012**, *117*. [CrossRef]
56. Del Guasta, M.; Morandi, M.; Stefanutti, L.; Brechet, J.; Piquad, J. One year of cloud lidar data from Dumont d'Urville (Antarctica): 1. General overview of geometrical and optical properties. *J. Geophys. Res. Atmos.* **1993**, *98*, 18575–18587. [CrossRef]

57. Del Guasta, M. Simulation of LIDAR returns from pristine and deformed hexagonal ice prisms in cold cirrus by means of "face tracing". *J. Geophys. Res. Atmos.* **2001**, *106*, 12589–12602. [CrossRef]
58. Sassen, K.; Zhu, J. A global survey of CALIPSO linear depolarization ratios in ice clouds: Initial findings. *J. Geophys. Res. Atmos.* **2009**, *114*. [CrossRef]
59. Ohneiser, K.; Ansmann, A.; Chudnovsky, A.; Engelmann, R.; Ritter, C.; Veselovskii, I.; Baars, H.; Gebauer, H.; Griesche, H.; Radenz, M.; et al. The unexpected smoke layer in the High Arctic winter stratosphere during MOSAiC 2019–2020. *Atmos. Chem. Phys.* **2021**, *21*, 15783–15808. [CrossRef]
60. Illingworth, A.J.; Barker, H.; Beljaars, A.; Ceccaldi, M.; Chepfer, H.; Clerbaux, N.; Cole, J.; Delanoë, J.; Domenech, C.; Donovan, D.P.; et al. The EarthCARE satellite: The next step forward in global measurements of clouds, aerosols, precipitation, and radiation. *Bull. Am. Meteorol. Soc.* **2015**, *96*, 1311–1332. [CrossRef]

MDPI
St. Alban-Anlage 66
4052 Basel
Switzerland
Tel. +41 61 683 77 34
Fax +41 61 302 89 18
www.mdpi.com

Remote Sensing Editorial Office
E-mail: remotesensing@mdpi.com
www.mdpi.com/journal/remotesensing